Ecological Connectivity of Forest Ecosystems

Katharina Lapin • Janine Oettel
Martin Braun • Heino Konrad
Editors

Ecological Connectivity of Forest Ecosystems

Waldfonds
Republik Österreich

Eine Initiative des Bundesministeriums
für Land- und Forstwirtschaft, Regionen
und Wasserwirtschaft

Editors
Katharina Lapin
Forest Biodiversity & Nature Conservation
Austrian Research Centre
for Forests (BFW)
Vienna, Austria

Martin Braun
Forest Biodiversity & Nature Conservation
Austrian Research Centre for Forests
Vienna, Austria

Janine Oettel
Forest Biodiversity & Nature Conservation
Austrian Research Centre for Forests
Vienna, Austria

Heino Konrad
Forest Biodiversity & Nature Conservation
Austrian Research Centre
for Forests (BFW)
Vienna, Austria

ISBN 978-3-031-82205-6   ISBN 978-3-031-82206-3 (eBook)
https://doi.org/10.1007/978-3-031-82206-3

This work was supported by
Waldfonds Republik Österreich; Project ConnectPLUS (BMLRT/III-2021-M10/5)

© The Editor(s) (if applicable) and The Author(s) 2025. This book is an open access publication.

Open Access This book is licensed under the terms of the Creative Commons Attribution 4.0 International License (http://creativecommons.org/licenses/by/4.0/), which permits use, sharing, adaptation, distribution and reproduction in any medium or format, as long as you give appropriate credit to the original author(s) and the source, provide a link to the Creative Commons license and indicate if changes were made.
The images or other third party material in this book are included in the book's Creative Commons license, unless indicated otherwise in a credit line to the material. If material is not included in the book's Creative Commons license and your intended use is not permitted by statutory regulation or exceeds the permitted use, you will need to obtain permission directly from the copyright holder.
The use of general descriptive names, registered names, trademarks, service marks, etc. in this publication does not imply, even in the absence of a specific statement, that such names are exempt from the relevant protective laws and regulations and therefore free for general use.
The publisher, the authors and the editors are safe to assume that the advice and information in this book are believed to be true and accurate at the date of publication. Neither the publisher nor the authors or the editors give a warranty, expressed or implied, with respect to the material contained herein or for any errors or omissions that may have been made. The publisher remains neutral with regard to jurisdictional claims in published maps and institutional affiliations.

This Springer imprint is published by the registered company Springer Nature Switzerland AG
The registered company address is: Gewerbestrasse 11, 6330 Cham, Switzerland

If disposing of this product, please recycle the paper.

Preface

It all began with a coffee break on an incredibly hot summer afternoon at the Austrian Research Centre for Forests' tiny forest in the heart of Vienna. We, a diverse group of researchers, had just come together as a newly formed team within the Department for Forest Biodiversity and Nature Conservation. Our backgrounds spanned a spectrum of expertise in biodiversity research, each of us looking back on more than a decade of experience in distinct fields.

Heino Konrad, an expert of population genetics, had crafted genetic monitoring programs for both endangered and common tree species. Janine Oettel brought her expertise on species-driven communities and intricate habitat assessments to the table. Martin Braun, immersed in forest ecosystem analysis and skilled in economic development predictions and big data management, enriched our group. Then there was me, Katharina Lapin, with my focus on forest conservation management, biodiversity indicators, and invasive biology. During these cheerful brainstorming sessions, an idea took root: ecological connectivity. Our collective passion for this concept swiftly united us across disciplines. And just as naturally, the concept evolved into a vision for a book—this book.

As our thoughts flowed that day, the realization dawned upon us that exploring ecological connectivity within forest ecosystems would require a global collective effort. It was to be an endeavor that would harness local experiences and insights from experts worldwide. So we reached out and found contributors—scientists, practitioners, and enthusiasts who shared our fascination with the interplay of ecological connectivity and forest ecosystems. Their engagement formed the bedrock of this book.

However, it should be clearly stated that this book—like the subject it delves into—is far from complete. But then again, a work of this nature is never truly finished. Instead, we simply hope it stands as a sturdy stepping stone in the global dialogue concerning the future management of ecological connectivity and its profound value to all life on Earth.

We express our sincere gratitude to every individual who has explored the realms of ecological connectivity—those who have ventured into its depths as well as those who are yet to do so. It is through the continuous discovery of new research findings and the sharing of knowledge spanning local, national, and global contexts across a variety of sectors from conservation and biology to forest science, landscape management, social science, and economics that we can effectively protect the movement of species and the flow of natural processes vital to our planet's well-being.

Of course, a project of this magnitude would never have reached fruition without the meticulous attention of our publishers and the unwavering support and financial backing from the Waldfonds of the Republic of Austria (Project ConnectPLUS, BMLRT/III-2021-M10/5), an initiative of the Austrian Federal Ministry for Agriculture, Forestry, Regions, and Water Management. And last but certainly not least, our heartfelt appreciation goes out to our families, friends, and colleagues who have been an endless source of encouragement and support throughout our journey to champion ecological connectivity and forest biodiversity through our research endeavors.

In closing, we offer a humble suggestion: Take a moment to share a cup of tea or coffee with your cherished colleagues. Allow your thoughts to meander and your ideas to flourish—preferably in a serene outdoor setting. You might be surprised where such moments can lead you. Thank you sincerely for embarking on this journey with us by reading this book.

The Editors of the book (from left to right): Katharina Lapin, Martin Braun, Janine Oettel, and Heino Konrad (Department of Forest Biodiversity & Nature Conservation, Austrian Research Centre for Forests, Seckendorff-Gudent-Weg 8, 1131 Vienna, Austria)

Warm regards,
Vienna, Austria   Katharina Lapin

Introduction to the Book

Aerial view of forest patches in the south of Austria (Photo: BFW/Florian Winter)

In today's world, climate change has emerged as a critical global concern posing a substantial risk to biodiversity at the planetary scale (Bonebrake et al., 2019; Deutsch et al., 2008; Pörtner et al., 2021). A key consequence of climate change is the migration of species (Krosby et al., 2010; Thompson & Gonzalez, 2017; Uroy et al., 2021), which are compelled to shift their distribution ranges due to the warming climate (Platts et al., 2019; Wilson, 2022). Remarkably, these shifts manifest with notable disparities between species, influenced by their respective ability to move (Honnay et al., 2002) as well as by external factors such as the availability of habitat in sufficient quantity and quality within the landscape and by temporal limitations related to climate change. Additionally, genetic diversity within species populations and their ability to navigate through fragmented landscapes play a crucial role.

The ongoing decline in biodiversity is frequently attributed to the prolonged effects of habitat loss and fragmentation stemming from human activities (Haddad et al., 2015). This fragmentation, in turn, impedes the movement of species in response to the anticipated impacts of climate change on their habitats (Taylor & Lindenmayer, 2020). Furthermore, populations face increased pressure from human appropriation (Doherty et al., 2021; Le Provost et al., 2021; Tucker et al., 2018), climate change, nitrogen deposition, and biotic exchange (Sala et al., 2000). As we deal with the simultaneous challenges of biodiversity loss and climate change, the importance of habitat connectivity as a vital asset in our efforts becomes increasingly clear.

It is important to recognize, however, that the benefits of enhancing ecological connectivity are not evenly distributed among all species (Mony et al., 2022) and hinge on demographic variables (Drake et al., 2022). To unlock advantages for species with limited dispersal abilities and small populations frequently unable to undertake migration journeys, specific planning tools, active monitoring approaches, and tailored management strategies are crucial. This involves embracing methods such as assisted migration and implementation of conservation measures both in their natural habitats and in controlled environments.

Forests, which cover 31% of Earth's land area (UNEP & FAO, 2020), occupy a central position in the discourse on ecological connectivity. Brimming with diverse plant, fungal, vertebrate, and invertebrate life forms, these ecosystems carry significant importance. Even though forests are incredibly important for biodiversity (Liang et al., 2016), climate regulation and ecosystem services, deforestation, and forest degradation remain ongoing issues that have yet to be resolved. Common and ongoing of deforestation and forest degradation lead to a significant decline in biodiversity, especially among specialist species (Sverdrup-Thygeson et al., 2017), exacerbated by the expansion of human land use as the primary cause of these problems (Andronache et al., 2019; Collins et al., 2009; Fahrig, 2003). In this context, ecological connectivity emerges as a vital lifeline for forest ecosystems, playing a crucial role in biodiversity restoration and facilitating adaptation to the rapidly changing climate conditions.

What Is Ecological Connectivity?

Ecological connectivity as defined on a global scale refers to the unimpeded movement of species and the flow of natural processes that sustain life on Earth (CBD, 2021). Therefore, it also indicates the importance of ecosystems remaining connected through ecological corridors without interruption. The concept encompasses two forms of connectivity: structural connectivity, which refers to the seamless transition between ecosystems, and functional connectivity, which refers to the movement of species or the flow of processes (Tischendorf & Fahrig, 2000). Over the last two decades, numerous scientific disciplines have elaborated these two principal perspectives with regard to habitat connectivity (Fletcher et al., 2016; Keeley et al., 2021):

The structural connectivity approach assesses the interconnectedness of landscape components, evaluating the extent to which habitat patches are physically intertwined. It quantifies habitat permeability contingent on the physical attributes of habitat patches, disturbances, and related elements (Saura et al., 2011; Taylor et al., 2006; Tischendorf & Fahrig, 2000). Models employing this perspective aim to determine areas conducive to the movement of diverse species, with a focus on ecologically minimally altered corridors presumed to accommodate species sensitive to human interference.

The functional connectivity perspective focuses on the actual dispersal capabilities of species along with habitat patch dimensions and distribution and land-use characteristics within the intervening matrix. A landscape might be functionally connected for one species, but not for another (Wang et al., 2018). Identifying present or future areas with functional connectivity based on known species movement (capabilities) delineates movement corridors (Adriaensen et al., 2003; Crooks & Sanjayan, 2006; Rudnick et al., 2012). In some instances, indicator or umbrella species assist in prioritizing areas of high ecological connectivity (Wang et al., 2018). Genetic methods are often used to monitor the functionality of landscape patterns (Balkenhol et al., 2015).

The preservation of connectivity includes a spectrum of ecological strategies aimed at connecting suitable habitat patches, thereby facilitating the interconnectivity of ecological processes across multiple scales. It also involves supporting evolutionary process connectivity, such as the exchange of genetic material (gene flow) between populations (Fung et al., 2017; Gaitán-Espitia & Hobday, 2021). Among these strategies, wildlife corridors emerge as a widely endorsed approach, serving as protective pathways for species migrations. While corridors represent linear elements, stepping stones are separate habitat patches that support the movement of species (Formann, 1995). Due to a growing awareness of the profound impacts of climate change, emphasis is being placed on the creation of climate corridors, which are particularly interesting along elevational gradients, enabling species to migrate in response to shifting temperature patterns (Beier, 2012; Krosby et al., 2018). In addition, they can function as linking elements between future climate refugia.

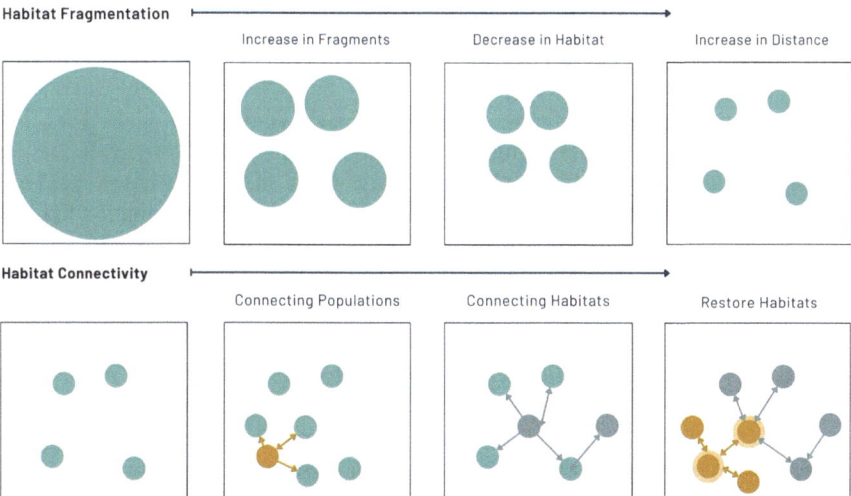

Fig. 1 Schemes of habitat fragmentation and habitat connectivity. Conservation efforts targeting restoration of habitats focus on connecting populations by aggregating patches around dispersal sources and/or on connecting habitats via evenly distributed patches within the landscape matrix

Why Should We Care About Ecological Connectivity?

"Everything in a healthy ecosystem is connected," as stated by the International Union for the Conservation of Nature (IUCN), a global network of conservation expects. This simple notion summarizes the overwhelming scientific evidence demonstrating the pivotal role of ecological connectivity in preserving biodiversity and sustaining life on Earth.

However, strategies to preserve and enhance ecological connectivity have been scattered and inconsistent. Countries around the globe, as well as regional and local policymakers and practitioners, are concurrently exploring strategies (Zeller et al., 2020) and management actions for ecological connectivity. These efforts not only address biodiversity loss but also position ecological connectivity as a strategic element for adaptation to climate change.

The significance of connectivity extends far beyond ensuring the survival of migratory species (Chap. 2); it directly relates to achieving all three objectives of the Convention on Biological Diversity (CBD). A recent assessment by IPBES (2019) underscored the importance of connectivity for the post-2020 framework, and its relevance extends to fulfilling the aims of other international agreements such as the Ramsar Convention on Wetlands, the UN Convention to Combat Desertification, and the World Heritage Convention. Connectivity positively influences ecosystem functions and services, ensuring species preservation by enabling movement, facilitating adaptation to environmental shifts, mitigating human–wildlife conflicts, and countering threats arising from barriers (Fletcher et al., 2016). The synergy between forest ecosystems and nature-based solutions is apparent (Cohen-Shacham et al., 2016, 2019; O'Brien et al., 2023). Connectivity serves as a foundation for numerous nature-based solutions, supporting the sustainable development of forest ecosystems.

Why Study Ecological Connectivity in Forest Ecosystems?

Forests host an impressive 80% of Earth's terrestrial plants and animals (UNEP & FAO, 2020). However, pinpointing an exact figure remains challenging due to the evolving understanding of global biodiversity. Over the last century, land-use changes have significantly reshaped forest landscapes. This transformation has had profound impacts on the structure of forests and their structural and functional connectivity, ultimately resulting in a substantial loss of biodiversity.

Understanding the scale, reasons, and consequences of forest connectivity is essential for conserving both forest biodiversity and the overall functionality of ecosystems. This includes investigating the extent of forest fragmentation, recognizing the drivers behind this fragmentation, and deciphering the cascading effects it has on species communities in forest ecosystems. A deep understanding of these dynamics enables us to formulate effective strategies for preserving and restoring diversity within forests and ensuring the sustainable use of resources dependent on forests. A key challenge is addressing threats as well as ecological and socioeconomic barriers to ecological connectivity in forest ecosystems (Aslan et al., 2021). The implementation of the corresponding findings is crucial for the success and effectiveness of connectivity conservation actions.

Moreover, the management of land beyond forest boundaries significantly impacts connectivity as well. Practices like agroforestry can serve as bridges between agriculture and sustainably managed forest landscapes, fostering connectivity and preserving habitat remnants. The reverse impact of well-connected and biodiversity-rich forests on the landscape and ecosystem processes outside of them, as well as their socioeconomic impacts on human health and well-being, remains largely unexplored.

What Are the Challenges to Forest Ecosystem Connectivity?

Forest ecosystem connectivity faces challenges on multiple fronts, including climate and land-use changes which disrupt migration routes and hinder species movement, ultimately resulting in fragmented habitats. Extreme events such as wildfires, windstorms, and droughts further exacerbate this development by destroying habitat and disrupting connectivity corridors. Invasive species pose a significant threat by altering habitat conditions and food webs, often outcompeting native species. Pollution from various sources degrades habitat quality and affects soil health, while unsustainable forest management practices fragment habitats through logging activities. In addition, the expansion of infrastructure creates barriers to species movement and increases mortality rates. Addressing the intricate interdisciplinary dynamics and barriers that affect species movement and ecological functions within and between ecosystems demands global strategies, collaborative efforts, and innovative solutions to promote forest conservation and sustainable land use.

To effectively conserve biodiversity in protected areas, it is essential to enhance ecological connectivity both within and among these areas. With the challenges posed by climate change, the significance of ecological connectivity becomes even more pronounced. This transition demands a fundamental shift in conservation practices, with objectives and actions needing to be redefined to adopt a more interconnected approach.

The concept of connectivity spans various fields and reflects the numerous factors influencing forest ecosystems. Physical barriers along with declining habitat quality and quantity contribute to shrinking forest areas, fragmented landscapes, and habitat loss (Fahrig, 2013). However, non-physical challenges such as diseases, invasive species, pollution, and climate change also significantly impact connectivity, potentially impeding species' reproduction and survival.

In environments dominated by humans, barriers act as filters allowing some species to move while blocking others. A significant challenge thus lies in accurately assessing the mobility and habitat needs of different species and understanding natural processes, which makes achieving forest ecosystem connectivity a complex endeavor.

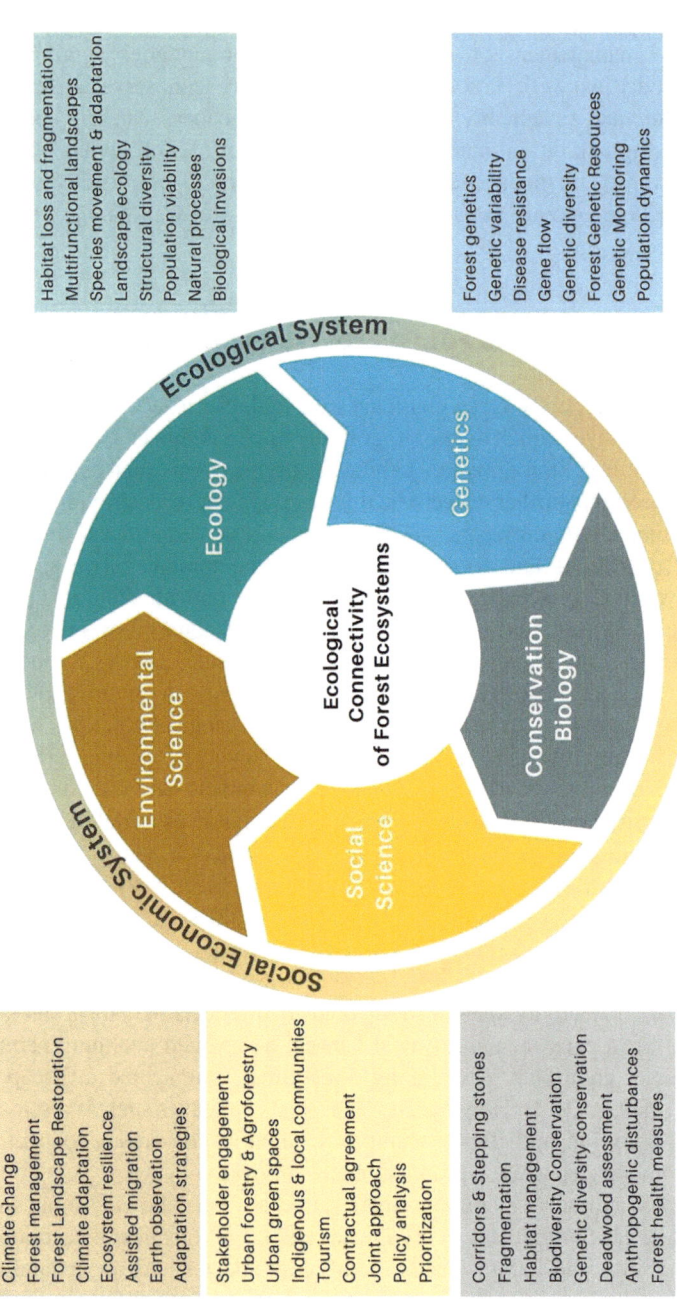

Fig. 2 Work on ecological connectivity in forest ecosystems requires an interdisciplinary approach, encompassing the scientific disciplines outlined in this framework. The summary of a keyword analysis discussed in the book is presented in the boxes

Organization of This Book

This volume aims to highlight the critical role of ecological connectivity in forest ecosystems for biodiversity conservation in the era of climate change. It is written for a diverse audience including students, teachers, conservation practitioners, forest managers, NGOs, researchers, policymakers, and interested citizens aiming to understand the complexities in the conservation of forest biodiversity.

Comprising four distinct sections and a total of 37 chapters—each authored and reviewed by a global consortium of experts specializing in ecological connectivity, forest biodiversity, and forest ecosystem management—this book provides a comprehensive and multifaceted exploration of its subject matter. With contributions by 125 authors, it stands as a collaborative compendium at the intersection of scientific inquiry and practical conservation action.

On its pages, readers will find a blend of theoretical concepts, real-world case studies, and pragmatic guidance. The intention behind this comprehensive structure is to provide an overview of each level of ecological connectivity, equipping readers with the necessary information for effective management implementations and offering general guidance for navigating the intricate realms of forest ecosystem connectivity.

Part I, "Understanding Ecological Connectivity," explores different concepts, measures, and models for assessing connectivity at different levels of biodiversity. We discuss species migration, range shifts, and dispersal as well as emphasize the significance of connectivity for saproxylic species. Furthermore, we examine the state of forest genetic diversity and conservation efforts, highlighting genetic connectivity and local adaptation in the face of climate change. Lastly, we address the role of soil in maintaining forest ecosystem connectivity, providing a comprehensive foundation for understanding this crucial aspect of ecological dynamics.

Part II, "Monitoring and Assessment Techniques," focuses on methods for monitoring and evaluating connectivity in forest ecosystems. It includes discussions on monitoring habitat fragmentation and biodiversity, as well as on assessing habitat quality and quantity using specific features and metrics. This section also explores both in situ and ex situ conservation measures and offers practical guidance for conducting rapid biodiversity assessment, providing valuable tools for effective monitoring and conservation efforts.

Part III, "Restoration, Social Dynamics, and Policy Frameworks," examines the restoration of forest landscape connectivity, addressing the reasons, locations, and methods involved. It discusses assisted migration as a strategy for adapting to climate change and the management of forest genetic resources under changing climate conditions. It also covers forest health management in connected landscapes and the control of invasive alien species in forest corridors. Furthermore, it explores ecological connectivity in urban and semi-urban forests along with its social-ecological implications and contributions to people. This section also presents conservation initiatives aimed at connecting landscapes with involvement of indigenous and local communities. Lastly, it discusses ecological connectivity perspectives for policy and practice, providing insights into effective conservation strategies.

Part IV, "Case Studies in Ecological Connectivity," presents 16 case studies from 17 countries across four continents, offering insights into ecological connectivity in forest ecosystems. The chapters discuss initiatives like Austria's national stepping stone network and forest reserves in Argentina. Challenges in Botswana's Kazuma Forest Reserve are addressed, as are projects in Brazil, Chile, and China. The studies also cover landscape connectivity in Ethiopia, best practices in transnational initiatives in Austria and Hungary as well as along the Sava River in Serbia and Croatia, and research hubs in central India. In addition, there are studies from Mongolia, Paraguay, and Tanzania as well as insights from the Republic of Korea, Tunisia, and Scotland.

Finally, *Ecological Connectivity of Forest Ecosystems* is more than just a book; it is a comprehensive exploration and a call to action. On its pages, readers will embark on a journey through the intricate pathways of ecological connectivity, hopefully allowing them to recognize and appreciate the pivotal role of ecological connectivity in shaping the future of our forests as they face the severe challenges of a changing climate. Uniting the knowledge of global experts, this volume invites all who engage with its contents to become stewards of ecological connectivity, ensuring the resilience of forest ecosystems and safeguarding their biodiversity for future generations to come.

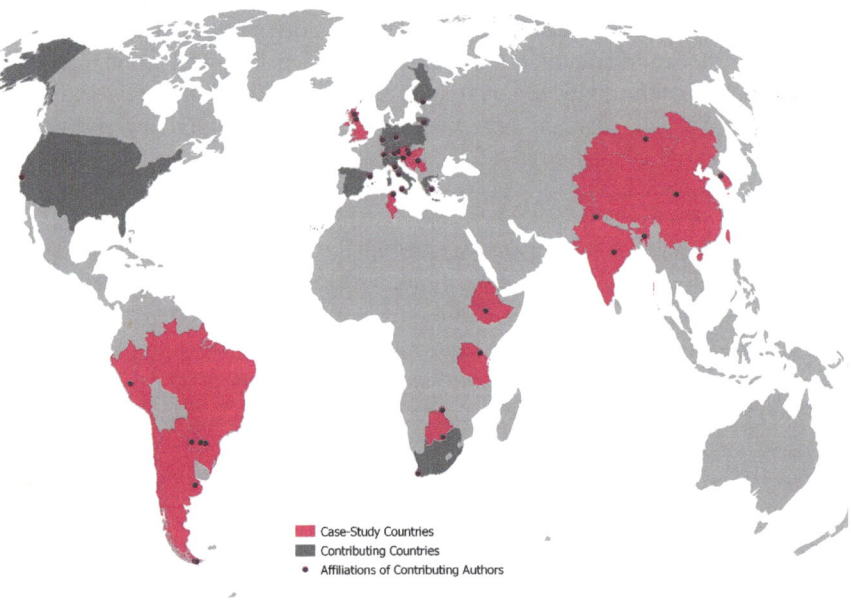

Fig. 3 Map of included case studies (pink) as well as countries (dark gray) and institutional affiliations (dots) of contributing authors

Department of Forest Biodiversity &
Nature Conservation, Austrian
Research Centre for Forests
Vienna, Austria

Katharina Lapin
Janine Oettel
Martin Braun
Heino Konrad

References

Adriaensen, F., Chardon, J. P., De Blust, G., Swinnen, E., Villalba, S., Gulinck, H., & Matthysen, E. (2003). The application of "least-cost" modelling as a functional landscape model. *Landscape and Urban Planning*, *64*(4), 233–247. https://doi.org/10.1016/S0169-2046(02)00242-6

Andronache, I., Marin, M., Fischer, R., Ahammer, H., Radulovic, M., Ciobotaru, A. M., Jelinek, H. F., Di Ieva, A., Pintilii, R. D., Drăghici, C. C., Herman, G. V., Nicula, A. S., Simion, A. G., Loghin, I. V., Diaconu, D. C., & Peptenatu, D. (2019). Dynamics of Forest Fragmentation and Connectivity Using Particle and Fractal Analysis. *Scientific Reports*, *9*(1), 1–9. https://doi.org/10.1038/s41598-019-48277-z

Aslan, C. E., Brunson, M. W., Sikes, B. A., Epanchin-Niell, R. S., Veloz, S., Theobald, D. M., & Dickson, B. G. (2021). Coupled ecological and management connectivity across administrative boundaries in undeveloped landscapes. *Ecosphere*, *12*(1). https://doi.org/10.1002/ecs2.3329

Balkenhol, N., Cushman, S. A., Storfer, A. T., & Waits, L. P. (2015). *Landscape genetics: Concepts, methods, applications*. https://doi.org/10.1002/9781118525258

Beier, P. (2012). Conceptualizing and designing corridors for climate change. *Ecological Restoration*, *30*(4), 312–319. https://doi.org/10.3368/er.30.4.312

Bonebrake, T. C., Guo, F., Dingle, C., Baker, D. M., Kitching, R. L., & Ashton, L. A. (2019). Integrating proximal and horizon threats to biodiversity for conservation. *Trends in Ecology and Evolution*, *34*(9), 781–788. https://doi.org/10.1016/j.tree.2019.04.001

CBD. (2021). *First draft of the post-2020 global biodiversity framework*.

Cohen-Shacham, E., Andrade, A., Dalton, J., Dudley, N., Jones, M., Kumar, C., Maginnis, S., Maynard, S., Nelson, C. R., Renaud, F. G., Welling, R., & Walters, G. (2019). Core principles for successfully implementing and upscaling nature-based solutions. *Environmental Science and Policy*, *98*, 20–29. https://doi.org/10.1016/j.envsci.2019.04.014

Cohen-Shacham, E., Walters, G., Janzen, C., & Maginnis, S. (2016). *Nature-based solutions to address global societal challenges*. https://doi.org/10.2305/iucn.ch.2016.13.en

Collins, C. D., Holt, R. D., & Foster, B. L. (2009). Patch size effects on plant species decline in an experimentally fragmented landscape. *Ecology*, *90*(9), 2577–2588. https://doi.org/10.1890/08-1405.1

Crooks, K. R., & Sanjayan, M. (2006). *Connectivity conservation* (Kd 14). Cambridge University Press.

Deutsch, C. A., Tewksbury, J. J., Huey, R. B., Sheldon, K. S., Ghalambor, C. K., Haak, D. C., & Martin, P. R. (2008). Impacts of climate warming on terrestrial ectotherms across latitude. *Proceedings of the National Academy of Sciences of the United States of America*, *105*(18), 6668–6672. https://doi.org/10.1073/pnas.0709472105

Doherty, T. S., Hays, G. C., & Driscoll, D. A. (2021). Human disturbance causes widespread disruption of animal movement. *Nature Ecology and Evolution*, *5*(4), 513–519. https://doi.org/10.1038/s41559-020-01380-1

Drake, J., Lambin, X., & Sutherland, C. (2022). The value of considering demographic contributions to connectivity: A review. *Ecography*, *2022*(6). https://doi.org/10.1111/ecog.05552

Fahrig, L. (2003). Effects of habitat fragmentation on biodiversity. *Annual Review of Ecology, Evolution, and Systematics*, *34*, 487–515. https://doi.org/10.1146/annurev.ecolsys.34.011802.132419

Fahrig, L. (2013). Rethinking patch size and isolation effects: The habitat amount hypothesis. *Journal of Biogeography*, *40*(9), 1649–1663. https://doi.org/10.1111/jbi.12130

Fletcher, R. J., Burrell, N. S., Reichert, B. E., Vasudev, D., & Austin, J. D. (2016). Divergent perspectives on landscape connectivity reveal consistent effects from genes to communities. *Current Landscape Ecology Reports*, *1*(2), 67–79. https://doi.org/10.1007/s40823-016-0009-6

Formann, R. T. T. (1995). Some general principles of landscape and regional ecology. *Landscape Ecology*, *10*(June), 133–142. https://doi.org/https://doi-org.uaccess.univie.ac.at/10.1007/BF00133027

Fung, E., Imbach, P., Corrales, L., Vilchez, S., Zamora, N., Argotty, F., Hannah, L., & Ramos, Z. (2017). Mapping conservation priorities and connectivity pathways under climate change for tropical ecosystems. *Climatic Change*, *141*(1), 77–92. https://doi.org/10.1007/s10584-016-1789-8

Gaitán-Espitia, J. D., & Hobday, A. J. (2021). Evolutionary principles and genetic considerations for guiding conservation interventions under climate change. *Global Change Biology*, *27*(3), 475–488. https://doi.org/10.1111/gcb.15359

Haddad, N. M., Brudvig, L. A., Clobert, J., Davies, K. F., Gonzalez, A., Holt, R. D., Lovejoy, T. E., Sexton, J. O., Austin, M. P., Collins, C. D., Cook, W. M., Damschen, E. I., Ewers, R. M., Foster, B. L., Jenkins, C. N., King, A. J., Laurance, W. F., Levey, D. J., Margules, C. R., … Townshend, J. R. (2015). Habitat fragmentation and its lasting impact on Earth's ecosystems. *Science Advances*, *1*(2). https://doi.org/10.1126/sciadv.1500052

Honnay, O., Verheyen, K., Butaye, J., Jacquemyn, H., Bossuyt, B., & Hermy, M. (2002). Possible effects of habitat fragmentation and climate change on the range of forest plant species. *Ecology Letters*, *5*(4), 525–530. https://doi.org/10.1046/j.1461-0248.2002.00346.x

IPBES. (2019). *Global assessment report on biodiversity and ecosystem services of the Intergovernmental Science-Policy Platform on Biodiversity and Ecosystem Services* (E. S. Brondizio, J. Settele, S. Díaz, & H. T. Ngo (toim)). https://doi.org/10.5281/zenodo.3831673

Keeley, A. T. H., Beier, P., & Jenness, J. S. (2021). Connectivity metrics for conservation planning and monitoring. *Biological Conservation*, *255*, 109008. https://doi.org/10.1016/j.biocon.2021.109008

Krosby, M., Tewksbury, J., Haddad, N. M., & Hoekstra, J. (2010). Ecological connectivity for a changing climate. *Conservation Biology*, *24*(6), 1686–1689. https://doi.org/10.1111/j.1523-1739.2010.01585.x

Krosby, M., Theobald, D. M., Norheim, R., & McRae, B. H. (2018). Identifying riparian climate corridors to inform climate adaptation planning. *PLoS ONE*, *13*(11), 1–18. https://doi.org/10.1371/journal.pone.0205156

Le Provost, G., Thiele, J., Westphal, C., Penone, C., Allan, E., Neyret, M., van der Plas, F., Ayasse, M., Bardgett, R. D., Birkhofer, K., Boch, S., Bonkowski, M., Buscot, F., Feldhaar, H., Gaulton, R., Goldmann, K., Gossner, M. M., Klaus, V. H., Kleinebecker, T., … Manning, P. (2021). Contrasting responses of above- and belowground diversity to multiple components of land-use intensity. *Nature Communications*, *12*(1). https://doi.org/10.1038/s41467-021-23931-1

Liang, J., Crowther, T. W., Picard, N., Wiser, S., Zhou, M., Alberti, G., Schulze, E. D., McGuire, A. D., Bozzato, F., Pretzsch, H., De-Miguel, S., Paquette, A., Hérault, B., Scherer-Lorenzen, M., Barrett, C. B., Glick, H. B., Hengeveld, G. M., Nabuurs, G. J., Pfautsch, S., … Reich, P. B. (2016). Positive biodiversity-productivity relationship predominant in global forests. *Science*, *354*(6309). https://doi.org/10.1126/science.aaf8957

Mony, C., Uroy, L., Khalfallah, F., Haddad, N., & Vandenkoornhuyse, P. (2022). Landscape connectivity for the invisibles. *Ecography*, *2022*(8). https://doi.org/10.1111/ecog.06041

O'Brien, P., Gunn, J. S., Clark, A., Gleeson, J., Pither, R., & Bowman, J. (2023). Integrating carbon stocks and landscape connectivity for nature-based climate solutions. *Ecology and Evolution*, *13*(1). https://doi.org/10.1002/ece3.9725

Platts, P. J., Mason, S. C., Palmer, G., Hill, J. K., Oliver, T. H., Powney, G. D., Fox, R., & Thomas, C. D. (2019). Habitat availability explains variation in climate-driven range shifts across multiple taxonomic groups. *Scientific Reports, 9*(1), 1–10. https://doi.org/10.1038/s41598-019-51582-2

Pörtner, H. O., Scholes, R. J., Agard, J., Archer, E., Arneth, A., Bai, X., Barnes, D., Burrows, M., Chan, L., Cheung, W. L., Diamond, S., Donatti, C., Duarte, C., Eisenhauer, N., Foden, W., Gasalla, M. A., Handa, C., Hickler, T., Hoegh-Guldberg, O., … Ngo, H. T. (2021). *IPBES-IPCC co-sponsored workshop report on biodiversity and climate change.* https://doi.org/10.5281/zenodo.4782538

Rudnick, D. A., Ryan, S. J., Beier, P., Cushman, S. A., Dieffenbach, F., Epps, C. W., Gerber, L. R., Hartter, J., Jenness, J. S., Kintsch, J., Merenlender, A. M., Perkl, R. M., Preziosi, D. V, & Trombulak, S. C. (2012). The role of landscape connectivity in planning and implementing conservation and restoration priorities. *Issues in Ecology* (Number 16).

Sala, O. E., Chapin, F. S., Armesto, J. J., Berlow, E., Bloomfield, J., Dirzo, R., Huber-Sanwald, E., Huenneke, L. F., Jackson, R. B., Kinzig, A., Leemans, R., Lodge, D. M., Mooney, H. A., Oesterheld, M., Poff, N. L. R., Sykes, M. T., Walker, B. H., Walker, M., & Wall, D. H. (2000). Global biodiversity scenarios for the year 2100. *Science, 287*(5459), 1770–1774. https://doi.org/10.1126/science.287.5459.1770

Saura, S., Estreguil, C., Mouton, C., & Rodríguez-Freire, M. (2011). Network analysis to assess landscape connectivity trends: Application to European forests (1990-2000). *Ecological Indicators, 11*(2), 407–416. https://doi.org/10.1016/j.ecolind.2010.06.011

Sverdrup-Thygeson, A., Skarpaas, O., Blumentrath, S., Birkemoe, T., & Evju, M. (2017). Habitat connectivity affects specialist species richness more than generalists in veteran trees. *Forest Ecology and Management, 403*(1432), 96–102. https://doi.org/10.1016/j.foreco.2017.08.003

Taylor, C., & Lindenmayer, D. B. (2020). Temporal fragmentation of a critically endangered forest ecosystem. *Austral Ecology, 45*(3), 340–354. https://doi.org/10.1111/aec.12863

Taylor, P. D., Fahrig, L., & With, K. A. (2006). 2 - Landscape connectivity: a return to the basics. K. R. Crooks & M. Sanjayan (Toim), *Connectivity conservation* (lk 29–43). Cambridge University Press. https://doi.org/10.1017/CBO9780511754821

Thompson, P. L., & Gonzalez, A. (2017). Dispersal governs the reorganization of ecological networks under environmental change. *Nature Ecology and Evolution, 1*(6). https://doi.org/10.1038/s41559-017-0162

Tischendorf, L., & Fahrig, L. (2000). On the usage and measurement of landscape connectivity. *Oikos, 90*(1), 7–19. https://doi.org/10.1034/j.1600-0706.2000.900102.x

Tucker, M. A., Böhning-Gaese, K., Fagan, W. F., Fryxell, J. M., Van Moorter, B., Alberts, S. C., Ali, A. H., Allen, A. M., Attias, N., Avgar, T., Bartlam-Brooks, H., Bayarbaatar, B., Belant, J. L., Bertassoni, A., Beyer, D., Bidner, L., Van Beest, F. M., Blake, S., Blaum, N., … Mueller, T. (2018). Moving in the Anthropocene: Global reductions in terrestrial mammalian movements. *Science, 359*(6374), 466–469. https://doi.org/10.1126/science.aam9712

UNEP & FAO. (2020). *The state of the World's forests 2020. Forests, biodiversity and people.* Rome.

Uroy, L., Alignier, A., Mony, C., Foltête, J. C., & Ernoult, A. (2021). How to assess the temporal dynamics of landscape connectivity in ever-changing landscapes: A literature review. *Landscape Ecology, 36*(9), 2487–2504. https://doi.org/10.1007/s10980-021-01277-9

Wang, F., McShea, W. J., Li, S., & Wang, D. (2018). Does one size fit all? A multispecies approach to regional landscape corridor planning. *Diversity and Distributions, 24*(3), 415–425. https://doi.org/10.1111/ddi.12692

Wilson, R. J. (2022). Northern wildlife feels the heat. *Nature Climate Change, 12*(6), 506–507. https://doi.org/10.1038/s41558-022-01378-6

Zeller, K. A., Lewison, R., Fletcher, R. J., Tulbure, M. G., & Jennings, M. K. (2020). Understanding the importance of dynamic landscape connectivity. *Land, 9*(9). https://doi.org/10.3390/LAND9090303

Contents

Part I Understanding Ecological Connectivity

1 **Concepts, Measures, and Models for Assessing Connectivity**......... 3
Janine Oettel, Frederik Sachser, Ana Isabel Martinez-Richart, and Manoj Kumar

2 **Species on the Move: Migration, Range Shifts, and Dispersal of Species**... 23
Katharina Lapin, Heino Konrad, Christoph Leeb, and Janine Oettel

3 **Do Saproxylic Species Need Habitats, Connectivity, or Connected Habitats?**... 39
Thibault Lachat, Janine Oettel, and Felix Meyer

4 **The State of Forest Genetic Diversity: Anthropogenic Impacts and Conservation Initiatives**................................. 55
Heino Konrad, Jan-Peter George, and Aglaia Szukala

5 **Genetic Connectivity and Local Adaptation of Forest Trees in the Face of Climate Change**............................. 91
Oliver Gailing, Katharina Birgit Budde, Ludger Leinemann, Markus Müller, and Selina Wilhelmi

6 **Forest Ecosystems Under Climate Change**..................... 115
Florian Kraxner, Dmitry Schepaschenko, Georg Kindermann, and Andrey Krasovsky

7 **Soil: The Foundation for Ecological Connectivity of Forest Ecosystems**.. 123
Owen Bradley, David Keßler, Josef Gadermaier, Mathias Mayer, and Ernst Leitgeb

Part II Monitoring and Assessment Techniques

8 **Monitoring Methods for the Protection of Connectivity in Forest Ecosystems**... 143
Janine Oettel, Bettina Thalinger, Aglaia Szukala, Linus Munishi, and Katharina Lapin

9	**Monitoring Habitat Fragmentation and Biodiversity in Forest Ecosystems**..	171
	Adriano Mazziotta, Saverio Francini, and Francesco Parisi	
10	**Habitat Quality and Quantity: Features and Metrics**..............	187
	Marcin K. Dyderski, Shubhadeep Roychoudhury, Katharina Lapin, Janine Oettel, and Martin Braun	
11	**In Situ and Ex Situ Conservation Measures**......................	213
	Barbara Fussi, Muhidin Šeho, and Darius Kavaliauskas	
12	**Practical Guidance for Rapid Biodiversity Assessment in Central European Forests**	241
	Janine Oettel, Cornelia Amon, Martin Steinkellner, Owen Bradley, Christoph Leeb, Frederik Sachser, and Katharina Lapin	

Part III Restoration, Social Dynamics, and Policy Frameworks

13	**Restoring Forest Landscape Connectivity: Why, Where, and How?**..	265
	Johanna A. Hoffmann, Demel Teketay, Mesele Negash, and Hafte Mebrahten Tesfay	
14	**Assisted Migration as a Climate Change Adaptation Strategy**	297
	Erik Szamosvári, Debojyoti Chakraborty, Silvio Schüler, and Marcela van Loo	
15	**Forest Genetic Resources Under Climate Change and International Framework: Conservation Measures of Serbia and Greece**...	311
	Branislav Trudić, Srđan Stojnić, Evangelia V. Avramidou, and Ermioni Malliarou	
16	**Managing Forest Health in Connected Landscapes**	331
	Gernot Hoch, Katharina Lapin, and Maarten de Groot	
17	**Managing Invasive Alien Species in Forest Corridors and Stepping Stones**	347
	Giuseppe Brundu, Maarten de Groot, Sabrina Kumschick, Jan Pergl, and Katharina Lapin	
18	**Ecological Connectivity in Urban and Semi-Urban Forests**.........	365
	Andrea Kodym, Katharina Lapin, and Debashis Sanyal	
19	**Connectivity in the Social-Ecological Context and Nature's Contribution to People**	383
	Rosina Soler, Verónica Chillo, Paula Rodríguez, Gimena Bustamante, and Matthew Ruggirello	

20 Conservation Initiatives to Connect the Landscape Across
    Indigenous and Local Communities: Perspectives from Chilean
    and Peruvian Biosphere Reserves 405
    Gabriela Albarracín-Llúncor, Eduardo Jackson Filomeno,
    Cesar Ipenza Peralta, Andrés Moreira-Muñoz,
    and Andrea A. Pino Piderit

21 Ecological Connectivity Perspectives for Policy and Practice 425
    Katharina Lapin, Janine Oettel, and Magda Bou Dagher Kharrat

Part IV Case Studies in Ecological Connectivity

22 Austria: The Austrian Stepping-Stone Program—A Bottom-Up
    Approach 439
    Janine Oettel and Katharina Lapin

23 Argentina: Balancing Connectivity and Production in Forest
    Reserves 451
    Rosina Soler, Dardo Paredes, Martin Parodi, Sebastián Farina,
    and Carolina Hernández

24 Botswana: Stand Structure and Hampered Regeneration
    of Woody Species in Kazuma Forest Reserve, the Busiest Elephant
    Corridor in Northern Botswana 467
    Demel Teketay, Witness Mojeremane, Lawrence Akanyang,
    Kamogelo Makgobota, Rampart Melusi, Ronnie Mmolotsi,
    David Monekwe, Ismael Kopong, Gosiame Neo-Mahupeleng,
    Topoyame Makoi, Kakambi Obusitswe, and Ednah Kgosiesele

25 Brazil: Applied Nucleation Through Key Microsites............... 487
    Bruna Elisa Trentin and Katharina Lapin

26 Chile: Increasing Connectivity for Nature and People in Highly
    Anthropogenic Landscapes 499
    Aníbal Pauchard, Eduardo Fuentes-Lillo, Darío Moreira-Arce,
    J. Cristóbal Pizarro, and Mónica Ortiz

27 China: Ecological Restoration Projects for Connected
    Landscapes... 507
    Qiuxiao Duan and Guangzhe Liu

28 Ethiopia: Enhancing Landscape Connectivity Through
    Agroforests... 521
    Hafte Mebrahten Tesfay and Mesele Negash

29 Hungary and Austria: Best Practice for Habitat and Species
    Connectivity: European Beech and Sessile Oak.................. 535
    Marcela van Loo, Erik Szamosvári, Anita Bálint,
    Anikó Neuvirthné Bilics, Heino Konrad, and László Nagy

30	**India: Hotspot of Connectivity Research and Conservation in Central India**... 545
	Trishna Dutta and Sandeep Sharma

31	**Republic of Korea: Predicting Shifts in Forest Biodiversity**......... 561
	Yuyoung Choi, Chul-Hee Lim, Hye In Chung, Yoonji Kim, Hyo Jin Cho, Jinhoo Hwang, Florian Kraxner, Gregory S. Biging, Woo-Kyun Lee, Jin Hyung Chon, and Seong Woo Jeon

32	**Mongolia: Connectivity Conservation Actions in the Khan Khentii Region**.. 573
	Jargalan Gerelsaikhan, Martin Braun, Tamir Mandakh, and Ochirvaani Soronzonbold

33	**Paraguay: Toward a Landscape Restoration of the Paraguayan Atlantic Forest**... 591
	Maria Laura Quevedo-Fernández, Haroldo Nicolás Silva-Imas, Lidia Florencia Pérez-de-Molas, Lila Mabel Gamarra-Ruiz-Díaz, Alba Liz González, Stella Mary Amarilla-Rodríguez, and Lucia Janet Villalba-Marin

34	**Serbia: Transnational Ecological Corridor Connectivity and Invasive Plant Species (Sava River Basin)**................... 609
	Alen Kiš and Klara Szabados

35	**Tanzania: The Eastern Arc Mountains Forests as World Natural Heritage—Status and Future Prospects**....................... 623
	Linus Munishi

36	**Tunisia: Genetic Diversity Assessment of Cork Oak Provenance Trials in the Context of Climate Change**....................... 635
	Boutheina Stiti, Issam Touhami, Awatef Slama, Amel Ennajah, Lamia Hamrouni, Mohamed Larbi Khouja, and Abdelhamid Khaldi

37	**United Kingdom/Scotland: Assisted Regeneration to Restore Lost Forests**... 647
	Philippa Gullett, Mark Hancock, Lucy Mason, and Andrew Weatherall

About the Editors

Katharina Lapin is the head of the Department for Forest Biodiversity and Nature Conservation at the Austrian Research Centre for Forests (BFW). She holds a PhD in Landscape Planning and Landscape Architecture from the University of Natural Resources and Life Sciences, Vienna. She has made significant contributions to forest biodiversity management, vegetation science, invasion biology, and biodiversity conservation through over 30 publications and participation in more than 20 research projects. Dr. Lapin is dedicated to translating scientific results into practical applications, particularly in conservation and restoration programs. She is actively involved in international collaboration and multidisciplinary networks, including the European Network for Environmental Citizenship, the IUCN SSC Invasive Species Specialist Group, and as the Austrian Focal Point for the Forest Invasive Species Network for Europe and Central Asia (REUFIS). She volunteers as an environmental consultant and educator and regularly engages in lectures and workshops to facilitate knowledge transfer. Her primary scientific motivations include assessing and promoting forest biodiversity, particularly in the context of climate change. Dr. Lapin is passionate about restoration efforts for forest-related biodiversity and ecosystem services and enjoys the challenges of international, multilingual collaboration. Her extensive expertise and experience make her an ideal editor for a book on the *Ecological Connectivity of Forest Ecosystems*. Her vision for the book focuses on enhancing understanding and communication, fostering transnational cooperation, and facilitating knowledge transfer to increase connectivity, aligning with her broader goals of improving conservation efforts and ensuring the sustainability of forest ecosystems globally.

Janine Oettel is the head of the Forest Biodiversity Unit at the Department of Forest Biodiversity and Nature Conservation at the Austrian Research Centre for Forests (BFW). She holds a PhD in Biology from the University of Vienna and has a strong background in forestry and nature conservation. Her research primarily focuses on forest-dwelling species and their habitats in the context of climate change, including deadwood, tree-related microhabitats, and other forest structures. She emphasizes the critical importance of these structures for biodiversity and conservation. Janine Oettel's expertise includes modeling the abiotic, biotic, and anthropogenic influences on forest-dwelling species and forest structures. Her doctoral thesis on the characteristics and dynamics of deadwood under forest management and climate change earned her the Stefan M. Gergely Award for Outstanding Dissertations in Environmental, Nature and Species Conservation. Since 2021, Dr. Oettel has led the Austrian stepping stone program, dedicated to the nationwide establishment and investigation of stepping stones in forests. This program aims to enhance habitat connectivity and promote biodiversity conservation across the country. Her practical experience and innovative research make her an ideal editor for this book.

Martin Braun is an environmental data analyst at the Austrian Research Centre for Forests (BFW) with substantiated experience in forest sector modeling, natural hazards modeling, forest economics, and forest ecology. For eight years Martin Braun worked on analyses relating to market and market dynamics in the forest-based sector. His research is focused on carbon dynamics and the analysis of economic interactions between different sectors of the wood products market. During his work, he made important contributions to the national greenhouse gas inventory and to the development of indicators and stylized facts on the forest-based sector. He has been working as an environmental data analyst at BFW's Institute for Forest Biodiversity and Nature Conservation for three years. There, he has contributed to research on deadwood dynamics, the role of deadwood-dwelling/decomposing insects, and the role of non-native trees in the Alpine space. His main areas of research are forest sector modeling with a focus on forest sector carbon budgets and flows as well as modeling market dynamics regarding harvested wood products, deadwood dynamics and forest biodiversity, rapid biodiversity assessment, and biodiversity indicators. Martin complements the editorial team through his interdisciplinary qualifications and years of experience working in Northern and Central Asia.

Heino Konrad, PhD, is the head of the Unit for Ecological Genetics at the Department of Forest Biodiversity and Nature Conservation at the Austrian Research Centre for Forests (BFW). With nearly two decades of extensive research experience, he has established himself as a leading expert in in situ and ex situ conservation strategies for forest tree species. His scholarly pursuits have led him to investigate the population history, genetic diversity, and gene flow patterns of tree species across four continents. He has made significant contributions to the field through more than 30 major peer-reviewed publications. In addition to his academic achievements, Heino Konrad is instrumental in the practical implementation of conservation measures in Austria. He oversees the establishment and management of gene conservation forests (in situ) and gene conservation seed orchards (ex situ), ensuring the preservation of genetic diversity within forest ecosystems. Heino Konrad is the national coordinator for Austria of the EUFORGEN network and FAO working group on forest genetic resources. A significant focus of his current research is the provision of forest reproductive material to align with the European Green Deal's objectives for climate change mitigation and biodiversity conservation and restoration. Heino Konrad's research emphasizes the critical importance of maintaining connectivity among populations of forest-dwelling organisms. He conducts genetic studies to monitor the impacts of habitat fragmentation, providing invaluable insights into the long-term survival and adaptability of these species. His comprehensive understanding of ecological genetics and practical conservation applications makes him an ideal candidate to co-edit this volume, bringing both academic rigor and practical expertise to the project.

Part I

Understanding Ecological Connectivity

Concepts, Measures, and Models for Assessing Connectivity

Janine Oettel, Frederik Sachser,
Ana Isabel Martinez-Richart, and Manoj Kumar

Aerial view of a Bavarian Forest near Grafenau, Germany (Photo: stgrafix/Adobe Stock)

J. Oettel (✉) · F. Sachser · A. I. Martinez-Richart
Department of Forest Biodiversity & Nature Conservation, Austrian Research Centre for Forests, Vienna, Austria
e-mail: janine.oettel@bfw.gv.at; frederik.sachser@bfw.gv.at; AnaIsabel.Martinez-Richart@bfw.gv.at

M. Kumar
Biodiversity & Climate Change Division, Indian Council of Forestry, Research & Education, Dehradun, India
e-mail: kumarmanoj@icfre.org

© The Author(s) 2025
K. Lapin et al. (eds.), *Ecological Connectivity of Forest Ecosystems*,
https://doi.org/10.1007/978-3-031-82206-3_1

Abstract

The concept of landscape connectivity involves species movement between habitat patches influenced by landscape features. It encompasses structural and functional connectivity as well as species-specific considerations. Structural connectivity analyzes spatial patterns of landscapes, while functional connectivity considers the response of organisms to the landscape. Evaluating habitats for connectivity requires accounting for their spatial and temporal variations. Temporal connectivity—often overlooked—is particularly essential for long-term population viability. Conservation planning should therefore integrate monitoring and assessment measures to achieve connectivity objectives in dynamic landscapes affected by land use and climate change. Measuring landscape connectivity considers landscape composition, structure, and heterogeneity as well as the presence of barriers, each varying among species and scales. Assessing connectivity across scales requires considering biological levels of organization from genetic flow to community processes. Modeling connectivity is complex and incorporates patch- and landscape-based approaches. Patch-based models focus on attributes of habitat patches, while landscape-based models consider movement behavior and resistance surfaces. Landscape connectivity research has expanded rapidly in recent decades, and its conceptual foundations are evolving. Recent advances integrate metapopulation dynamics with habitat configuration and movement behavior. Traditional static models are being replaced with dynamic models considering temporal variations in landscape attributes. Further technological advancements such as remote sensing and climate simulators allow more accurate representations of dynamic landscapes, promoting ecosystem understanding and supporting conservation planning.

Keywords

Corridors · Fragmentation · Habitat · Landscape matrix · Metapopulation · Patch concept · Resistance surface · Stepping stones

The Concept of Connectivity

The concept of landscape connectivity dates back to the 1970s and 1980s and was developed based on three key components (Fahrig et al. 2021). First, populations of many species are distributed across patches of habitat that are not connected (Den Boer 1968). Second, the persistence of populations depends on the movement of individuals, which enables gene exchange between different patches of habitat (Levins 1969). Third, the landscape features between these patches can either facilitate or hinder movement, which is crucial to the concept of connectivity (Merriam 1984). Accordingly, Merriam (1984) defined connectivity as the interaction between movement attributes and landscape structure that influences movement between patches, and thus population persistence. He described landscape connectivity as the degree to which absolute isolation is prevented by landscape elements, allowing organisms to move between different habitat patches. Later, Taylor et al. (1993) defined landscape connectivity as "the degree to which a landscape facilitates or impedes movement among resource patches," encompassing the spatial distribution of patches as well as the movement success of species in response to it. In much of the literature on landscape connectivity, movement success is assumed to be closely linked to the spatial distribution of habitats across landscapes, and movement is assumed to be strongly constrained by habitat (Fahrig et al. 2021). This has led to a focus on linear structures (habitat corridors), small patches of temporary habitat (stepping stones), and the distances between habitats (Forman 1995). Corridors are expected to be advantageous for species that specialize in certain habitats, rely on undisturbed habitats, and have limited mobility. On the other hand, stepping stones may not offer the same physical habitat continuity as corridors, but they can still be beneficial for more mobile species and those more resilient to habitat disturbance, as well as for species with wider ranges compared to those that benefit from corridors (With 2019). In conservation planning, small areas are often overlooked due to the assumption that their ecological value is limited. However, a global synthesis by Wintle et al. (2019) found that neglecting these smaller areas would lead to the irrevocable loss of numerous species that inhabit them exclusively.

Following Tischendorf and Fahrig (2000), the concept of landscape connectivity includes two basic aspects: *structural* and *functional* connectivity. Structural connectivity is based entirely on the spatial relationships of structural elements of a landscape, with no direct link to the behavioral characteristics of organisms (Saura et al. 2011; Taylor et al. 2006; Tischendorf and Fahrig 2000). Functional connectivity, on the other hand, explicitly relates these spatial arrangements of structural landscape elements to the ability of organisms to move or disperse through the landscape (Adriaensen et al. 2003; Crooks and Sanjayan 2006; Rudnick et al. 2012). In fact, connectivity is species-specific, and a suitable dispersal habitat or corridor for one species may not be favorable for others (Wang et al. 2018). Since each species has unique requirements and dispersal behaviors, the likelihood of different species reaching the same patch varies. Therefore, Salgueiro et al. (2021) recommended a multispecies approach to address communities and draw inferences for as many species as possible. Assessing the effectiveness of connectivity initiatives is also challenging since newly created habitat corridors or stepping stones connecting existing habitat fragments need time to develop before providing functional connectivity (Brouwers et al. 2010). However, evaluation periods are often insufficient and lack habitat information concerning the species of interest (e.g., specialized forest-dwelling species). Information about habitats encompasses several aspects such as habitat quality, use, and change in order to determine suitable habitats (Morris 2003; Morris et al. 2009) (see Chap. 10). Habitat suitability refers to the ability of a habitat to sustain a viable population over an ecological time scale and is considered part of functional connectivity (Hall et al. 1997; Kellner et al. 1992; Wang et al. 2008). Conventional approaches that assess habitats by linking changes in habitat to changes in species density, richness, and diversity may not be sufficient to ensure species persistence. Therefore, the evaluation of habitats should also consider the heterogeneity of the entire landscape and its spatiotemporal variability (Baudry et al. 2003).

Habitats vary both spatially and temporally, which poses a major challenge for conservation planning when it comes to measuring or modeling connectivity at relevant scales, leading in turn to difficulties in selecting appropriate protected areas or measures. Habitat connectivity can manifest over timescales ranging from hours to centuries, and understanding different species' dispersal and movement characteristics is crucial for integrating connectivity into spatial planning (Beger et al. 2022). To achieve conservation and connectivity objectives, it is essential to accompany them with corresponding monitoring and assessment measures (Pressey et al. 2021). In this context, the following sections will present a brief overview of the most commonly used measures and models for assessing connectivity, along with information about conservation planning in a dynamic landscape affected by changes in land use and climate.

Measuring Connectivity

The consensus in the literature is that connectivity is a species- and landscape-specific concept (e.g. Schumaker 1996; Tischendorf and Fahrig 2000; Wiens 1997). The duality of this definition—that is, the dynamic interaction between the characteristics of an organism and the landscape—adds complexity to measuring connectivity. On one side, numerous characteristics of the landscape such as composition, structure, heterogeneity, quality, possible barriers, and scale exert their influence (Fig. 1.1). On the other side, different species exhibit unique biological traits, dispersal abilities, or survival rates in non-habitat areas (Tischendorf and Fahrig 2000), and there are behavioral differences between life stages and even among individuals of the same species (Bélisle 2005). Due to these complexities, it is impossible to permanently classify areas as connected or disconnected; rather, classification depends on the specific species or process being considered, the landscape, and the scale at which connectivity is assessed (With 2019). In other words, the degree of connectivity for a given species can vary significantly across different landscapes (Kindlmann and Burel 2008), while conversely, the same landscape may exhibit different levels of connectivity for different organisms (Tischendorf and Fahrig 2000). To simplify the measurement of connectivity, two main approaches are commonly considered: structural and functional.

Structural connectivity, also known as physical connectivity, is determined by analyzing the spatial pattern of the landscape, including factors like the size, shape, and location of habitat patches. It refers to the adjacency of patches (spatial contagion) or the presence of physical linkages like corridors (With 2019). However, it does not encompass actual or functional habitat connectivity for species living in the landscape (Fagan and Calabrese 2006). While structural connectivity may not provide a comprehensive measure of connectivity, it offers a relatively straightforward and practical method for assessing connectivity, particularly as an initial evaluation or in cases where other approaches are not feasible. As a result, structural

Fig. 1.1 Aerial view of a rural intersection where a road crosses over a railway track amidst lush green fields in South Korea. The road curves around the railway, with clear markings and a small green structure nearby (Photo: Jacky Woo/Adobe Stock)

connectivity assessments are frequently used in restoration projects or to provide an approximation of the potential functional connectivity when species data is not available (Hilty et al. 2020).

Functional connectivity is a dimension that considers the response of organisms to landscape properties and can be split into two categories, namely potential connectivity and actual connectivity, depending on the level of detail of the data required and obtained from each measure (Calabrese and Fagan 2004). Potential connectivity combines metrics incorporating limited or indirect knowledge about an organism's dispersal ability and the spatial relationships among the landscape's physical attributes. For instance, indirect measures may involve estimating mobility based on body size or energy budgets, while limited data could include measurements with little spatial detail such as mean or maximum recapture distances from tagging or banding studies. Actual connectivity provides a more concrete estimate of the real linkages among landscape elements or habitat patches through direct observation. For example, actual connectivity metrics may involve observing how organisms respond to habitat edges or quantifying the movement of individuals through corridors, either into or out of focal patches or across a landscape (Fagan and Calabrese 2006). Understanding all three categories of connectivity—structural, potential, and actual—is crucial when determining the amount and type of information that a chosen measuring method will provide about spatial dynamics in ecological systems.

For a connectivity assessment, several decisions regarding the chosen approaches must be made. These include determining whether connectivity serves as a dependent or independent variable, opting for a structural or functional approach, and understanding how these choices interrelate with other factors. The considerations should also include deciding whether to conduct a single- or multispecies assessment, taking into account single or multiple habitat types, employing either a patch- or landscape-based approach, and considering spatial and temporal scales as well as biological levels of organization of the subjects under study. For a summary of the characteristics of patch and landscape approaches and their links to other aspects in a connectivity assessment, see Fig. 1.2.

Following Goodwin (2003), structural measurements treat connectivity as an independent variable. This approach involves directly measuring the physical characteristics of patches and empirically assessing their impact on biological variables such as species presence, abundance, and richness as responses. Another approach treats connectivity as a dependent variable that is modeled rather than obtained empirically from the data gathered about the landscape and the species (Goodwin 2003). This approach is often combined with the use of functional methods, which incorporate species movement parameters.

A connectivity assessment can employ either a single-species or a multispecies strategy. The single-species approach is chosen based on the understanding that a connectivity measure for one species might not be suitable for others (Wang et al. 2018). Typically, single-species methods rely on functional approaches and require precise data on the selected species' movement responses to landscape elements. If the selected species acts as an umbrella species, such connectivity assessments can potentially lead to the protection of several different species (Breckheimer et al. 2014). On the other hand, the multispecies approach has commonly been adopted when defining connectivity as the presence of corridors or other structural connecting elements. When using structural methods, no assumption is made about a particular species, and only structural data is considered the measure of connectivity (Kindlmann and Burel 2008). As a result, these findings can be interpreted as valid for multiple species. In addition, functional methods for measuring connectivity for multiple species exist, such as potential connectivity methods that use standardized values like the average dispersal distances (Santini et al. 2016). These approaches assess several species with similar characteristics simultaneously. However, care needs to be taken when generalizing connectivity results using standardized values or extrapolating multispecies connectivity from single-species studies. This practice may introduce bias by assuming that all species perceive landscape and barriers similarly, as highlighted by Salgueiro et al. (2021) after testing single- and multi-species models. In conclusion, while multispecies assessments using structural or potential approaches may be less precise compared to single-species assessments conducted with actual connectivity methods, they offer more practical advantages for large-scale studies and for identifying areas with potential conservation value for multiple species. Another way to approach a connectivity assessment is by considering the patch or landscape scales. Patch-based approaches are often measured using structural methods, while landscape-based approaches are more commonly associated with functional methods. One of the key assumptions of the structural approach is that species movement is restricted to the preferred habitat (Tischendorf

Approach	Patch-based	Landscape-based
Theories on which the approach is based	Metapopulation theory and island biogeography	Landscape ecology
General assumptions	Species movement is restricted to preferred habitat; patch-matrix edge has a boundary effect that species do not traverse	Species move across different land cover types that have different effects on species movement (travelling costs or matrix resistance)
Recommended for landscapes	Defined patch-habitat scenarios with homogeneous matrix effect (if not, matrix effect should be compensated in analysis)	Heterogeneous landscapes; better accounts for matrix effect through resistance surfaces
Recommended for species	Strict habitat specialists with highly restricted movement	Habitat generalists with wide habitat range (i.e., large predators)
Common measuring methods	Connectivity assessed using structural methods, presence of corridors or potential connectivity methods	Connectivity assessed using functional-potential methods
Key features in measurements	Euclidean distance, dispersal distance, habitat amount, occupancy	Matrix effect, travelling cost, habitat suitability, functional distance, movement behavior
Measurement scale	Produces single measures of patch connectivity at landscape scale	Studies with larger scale and replicability
Connectivity studied as	Independent variable (using structural methods)	Dependent variable (using functional methods)

Fig. 1.2 A comparison table outlines the characteristics of patch- and landscape-based approaches in ecological studies, correlated to the different perspectives possible within connectivity assessments

and Fahrig 2000). Accordingly, the landscape is understood as islands of habitat (patches) connected via dispersal within a matrix of non-habitat. The patch approach is built on the island biogeography and metapopulation theories (Moilanen and Nieminen 2002). It acknowledges the critical importance of taxon movement among patches for recolonizing habitats after local extinctions as well as for the colonization of new habitats. The underlying assumption is that small patches may be more susceptible to demographic, genetic, or environmental stochastic events leading to local extinctions. In this context, the role of dispersal becomes key for ecological equilibrium. Structural patch measures focus on determining connectivity based on dispersion, either through continuity or through dispersal distance. By contrast, the landscape approach based on landscape ecology (Howell et al. 2018) views the landscape as a heterogeneous mix of physical attributes, attempting to relate the effects of this heterogeneity to ecological processes and interactions such as connectivity for species. Since its fundamental idea is that given the heterogeneity of the landscape, its different parts will have different effects on species and energy flows that can be better determined using functional approaches rather than with structural methods. Despite their limitations, landscape-based measures offer an advantage over patch-based measures by enabling the assessment of different movement responses in heterogeneous landscapes. Another significant benefit of landscape-based methods is their capability to cover larger spatial scales. First, certain processes or species operate at the landscape level, thus rendering patch-based analysis incomplete (With 2019). Second, connectivity measurements of a patch cluster cannot always be easily extrapolated to the entire landscape, particularly when landscapes exhibit a hierarchical patch structure—patches embedded in other patches at different scales (Wu and David 2002). The spatial scales of connectivity studies span a wide range, from attempts to develop global connectivity coefficients (Larrey-Lassalle et al. 2018) to examining the smallest distances between veteran trees or deadwood logs for certain insect species (e.g., Ranius et al. 2011; Ruiz-Carbayo et al. 2017). Typically, connectivity assessments focus on a single type of habitat; however, with regard to species utilizing different habitats during their life cycle, multi-habitat connectivity can often provide a better understanding of species presence, abundance, and richness than single-habitat connectivity (Clauzel et al. 2024). Assessing connectivity in heterogeneous landscapes comprising different types of habitats has been challenging in the past, but the development of spatial models that interpret the landscape as a resistance surface has enabled researchers to better analyze connectivity in such landscapes.

In addition to spatial characteristics, the temporal continuity of a habitat plays a crucial role in determining the biodiversity it can support. As highlighted by Kindlmann and Burel (2008), a threshold for metapopulation extinction exists not only with regard to the amount of suitable habitat but also with regard to patch turnover. Despite its significance, temporal connectivity is often overlooked in connectivity assessments, as noted by Fahrig (1992). Most studies that explicitly assess temporal connectivity quantify the temporal changes in spatial connectivity across two or more varying time periods (Uroy et al. 2021). This approach may involve examining different seasons or years, such as when landscapes undergo rapid transformation due to intensification or before and after the removal of a specific element. More recently developed methods have begun to quantify spatiotemporal connectivity, thus accounting for both spatial and temporal dispersal, considering current and future climate scenarios (Huang et al. 2020).

When assessing connectivity, it is crucial to take into account the biological levels of the organization, including genetic flow, propagule dispersal (such as spores, pollen, or seeds), individual movements (ontogenetic movements and ecological interactions), populations, species, and communities. These aspects are intricately linked to abiotic cycles such as nutrient and water cycles (Beger et al. 2022). Considering these different levels is important because the temporal scales of responses are hierarchically nested. For instance, a temporal response may be shorter at an individual level than at a community level (Hylander and Ehrlén 2013). Adaptation of species to changes in their environment can occur rapidly, sometimes even within a single generation; it is a population—as well as an individual-level process (plasticity and dispersal). On the other hand, population growth and evolution involve longer time frames and consider populations as a whole (O'Connor et al. 2012). Acknowledging these hierarchical levels and temporal scales helps achieve a comprehensive understanding of how species and ecosystems respond to changes in connectivity.

Modeling Connectivity

Connectivity modeling aims to describe the spatiotemporal dynamics of diverse ecological processes and is implemented in fundamentally different fields of research. Most disciplines have developed specific modeling frameworks, resulting in a large number of varying approaches. This subchapter focuses on models addressing the movement of animals through landscapes and its implications for population dynamics, highlighting several well-known concepts that reflect the progression of connectivity modeling.

The analysis of animal movement encompasses simple measures as well as complex modeling techniques that often focus on specific characteristics of connectivity. Yet almost all of these models correspond to one of two basic concepts of connectivity originating from distinct ecological disciplines: (1) Population ecology typically focuses on features of distinct habitat patches and their effect on population dynamics through colonization and extinction and (2) landscape ecology generally aims to quantify the effects of landscape attributes on animal movement or connectivity per se (Howell et al. 2018). All ecological models simplify the real world by definition: The patch-based approach oversimplifies the effect of the landscape matrix between subpopulations, while the landscape approach neglects the links to population-level aspects beyond individual movement (Howell et al. 2018). The differences between the two concepts are reflected in their models, which differ in their data requirements as well as in terms of the research questions that can be addressed.

In the late 1960s, Levins (1969) introduced the theory of metapopulation dynamics and laid the foundation for a patch-based view of connectivity. The patch concept focuses on discrete habitat areas and how their spatial configuration as well as certain patch attributes affect colonization and extinction dynamics. Patch-based models typically ignore local population dynamics and reduce this information by using the occupancy data of patches. The theory behind this concept shares some similarities with the island biogeography theory described by MacArthur and Wilson (1969) (Hanski 2014; Moilanen and Nieminen 2002). Related models often assume that extinction is affected by the size of a habitat patch and that connectivity between patches determines the colonization probability (Moilanen and Nieminen 2002). Following Moilanen and Nieminen (2002), connectivity is described by combining properties of the focal patch (such as its size, shape, and habitat quality), the population of the source patch, and the intervening habitat matrix. However, very simple connectivity measures such as the nearest neighbor approach only include the distances to the nearest neighboring patches. Obviously, such measures ignore potential source patches within a reasonable migration range beyond an arbitrary number of neighboring patches closest to the focal patch. Buffer measures address this issue by including all occupied patches within a certain radius (Moilanen and Nieminen 2002), but since such buffers are specified via fixed radii instead of a probabilistic formulation using dispersal kernels, they cannot adequately incorporate rare long-distance dispersal events.

A subsequent approach known as incidence function models (IFMs) (Hanski 1994a, b) addresses this shortcoming by including various parameters such as patch area, interpatch distance, species dispersal ability, and other environmental variables as well as life history traits of the respective species to estimate extinction and colonization (Hanski 1994a; Prugh 2009). This model assumes constant colonization and extinction rates and estimates the probabilities with which occupied patches become extinct and unoccupied patches become colonized between discrete time steps (Hanski 1994a). The occupancy state of a single patch is described by a linear first-order Markov chain with two states (Hanski 1994a), and the long-term probability of a patch being occupied is called incidence. The IFM framework allows flexible implementation of covariates including landscape structure (Moilanen and Hanski 2001) and represents the most common spatially realistic metapopulation model (Risk et al. 2011). Moilanen (2002) identified three types of errors that typically occur in metapopulation datasets: (1) a biased estimation of the patch area, (2) incomplete identification of patches, and (3) misclassification of occupied patches as being unoccupied (i.e., imperfect detection resulting in false absences). Risk et al. (2011) extended the IFM to address these types of errors via a hierarchical formulation in a Bayesian framework. Sutherland et al. (2014) used an occupancy modeling framework to implement a spatially realistic metapopulation model of dispersal and connectivity while accounting for imperfect detection and additionally including demographic parameters to account for age class–specific contributions to both extinction and colonization processes. In summary, patch-based modeling approaches can be seen as a framework that incorporates movements of individuals between (but not within) local subpopulations and therefore connectivity as it affects the persistence and stability of metapopulations.

The assumed effect of the intervening landscape matrix on the movement behavior of organisms is fundamental to landscape ecology models and critical for our understanding of connectivity (Kindlmann and Burel 2008; Tischendorf and Fahrig 2000). A common representation of heterogeneous landscapes associated with an estimated cost of movement is termed the *resistance surface*. Resistance surfaces are raster maps in which each pixel features a specific value representing either a survival risk or the willingness of (respectively the physiological effort for) an individual to move through that pixel (Zeller et al. 2012). The creation of such maps is a two-step process including (1) the preparation of appropriate data considering species-specific environmental covariates as well as the temporal resolution and (2) the actual construction and optimization of the resistance surface, which may be based on expert opinion, literature review, or empirical data on the focal species (Dutta et al. 2022). It is important to consider that the spatial resolution as well as the level of detail of the underlying layers can seriously affect the results of such studies (Cushman and Landguth 2010), making critical reflection and justification of each step of the process highly recommendable (Zeller et al. 2012).

One of the first and most widely applied methods using resistance surfaces for functional connectivity analyses is least-cost modeling (Adriaensen et al. 2003; Correa Ayram et al. 2016). Least-cost models were developed in transport geography to determine optimal routes between pairs of locations (Etherington 2016). The underlying idea originates in graph theory, and the approach has been adapted many times to improve the framework for the analysis of animal movement (Diniz et al. 2020). In simple terms, the resulting least-cost path is a path between two predefined locations with minimal accumulated costs according to the resistance surface. Several variations exist, including a factorial implementation to compute the least-cost path for every possible pair of points simultaneously (e.g., Cushman et al. 2013) or the inclusion of suboptimal routes to account for imperfect knowledge of individual animals regarding landscape resistance (Pinto and Keitt 2009). Least-cost analysis has also been used to create undirected connectivity maps of entire landscapes. Such resistance kernels are constructed by calculating the least-cost path with a species-specific dispersal threshold from each source cell to every adjacent cell before summing up all values to estimate potential movement rates (Diniz et al. 2020). A further common approach utilizing resistance surfaces is circuit theory, which is based on electrical circuit theory (McRae et al. 2008). In circuit theory–based models, each raster cell that does not act as a complete barrier will be assigned an electrical node, and all adjacent cells are connected via resistors that represent dispersal. High current values of cells represent a high probability of individuals passing through when moving randomly from source to destination patches. Visualization of the results allows straightforward identification of pinch points in the landscape (McRae et al. 2008; Pelletier et al. 2014).

Another relevant modeling approach known as individual-based dispersal models (IBDMs) employs simulation and allows the incorporation of animal behavior affecting movement decisions (Diniz et al. 2020). In such models, simulated individuals are released on predefined cells and move with each time step of the simulation (Allen et al. 2016; Diniz et al. 2020). The behavioral component can be specified for any combination of landscape characteristics and state variables of the individual (Diniz et al. 2020). This process can be implemented via if-else statements or probability functions (Allen et al. 2016) and should be based on field data or reliable expert knowledge (Hauenstein et al. 2019). Further analysis is then based on the overlapping of a large number of simulated individuals and movement paths (Allen et al. 2016). Fletcher et al. (2019) developed a framework that includes the concept of resistance surfaces along with an additional mortality risk map to analyze the complementary effects of movement behavior and mortality risk. Based on spatial absorbing Markov chains and random walk theory, this framework allows movement steps to be specified via transient states and mortality via absorbing states of a Markov chain; it therefore does not require individual-based simulations (Marx et al. 2020). In summary, landscape-based connectivity models substantially contribute to our understanding of the linkages between animal movement behavior and landscape configuration and add valuable input for conservation strategies and research.

Recent advances in the field of connectivity modeling unify concepts of metapopulation dynamics with habitat configuration (structural connectivity component) and movement behavior (functional connectivity component; Drake et al. 2022). Building on a strong theoretical background, advanced modeling techniques such as spatially explicit hierarchical models allow metapopulation dynamics in heterogeneous landscapes to be addressed (e.g., Howell et al. 2018; Royle et al. 2018) and tackle long-standing challenges in the field of connectivity modeling (Drake et al. 2022).

Application in a Dynamic Landscape

The conceptual foundations of landscape connectivity models have undergone significant evolution in recent decades, as highlighted by Bishop-Taylor et al. (2018). Traditionally, these models lacked dynamic quantification of variations at seasonal, yearly, and decadal scales, often treating landscapes as static units when modeling species movement. However, there has been a shift toward incorporating the dynamic behavior of landscapes in recent years, resulting in more accurate representations. Three major characteristics of habitats in a landscape, namely (i) size, (ii) arrangement, and (iii) quality are often not static throughout an assessment time frame. Landscape structure and quality can change over time (Kindlmann and Burel 2008). These temporal variations may be intra- or interannual; for example, a landscape may alternate between dry and wet periods within a year, resulting in distinct patch qualities with varying spatial arrangements. In addition, natural events and human activities such as tree felling, wildfires, ecological disturbances (e.g., invasion), climate change, and other context-dependent effects contribute to the dynamism of landscapes. As a consequence, a landscape may undergo contrasting variations leading to spatiotemporal heterogeneity and changing landscape connectivity (Puckridge et al. 1998). Cushman et al. (2005) demonstrated that the temporal dynamics of landscapes significantly influence animal movement pathways. Despite the inherent dynamism of landscapes, many studies have evaluated landscapes based on a limited timeframe or a single point in time (Kaszta et al. 2021; Lorimer 2015; Unnithan Kumar et al. 2022).

The availability of fine-scale GIS data has paved the way to incorporating spatial heterogeneity with detailed parameterization when studying landscape connectivity (Cushman and Huettmann 2010; Kumar et al. 2019). With the increased availability of time series of remote sensing data, a new generation of landscape models can now also consider temporal dynamics, thereby further enhancing our understanding of species movement patterns. Increasing computational capabilities and empirical evidence supporting the influence of dynamic landscapes on species movement have inspired modelers to consider the dynamic parameters of landscapes in landscape connectivity models (Zeller et al. 2020). Basic parameterization in a dynamic model usually considers variables such as the patch quantity, quality, and arrangement as well as the dispersal success of the population at specific spatial and temporal scales aligned with the ecological processes of interest.

A growing number of imaging satellite programs are providing large amounts of data at different spatial resolutions for the entire planet. Remote sensing data can be processed to obtain land use, land cover, and habitat suitability maps at fine spatial and temporal scales, providing opportunities to detect landscape dynamics (see Chap. 9). In addition, land change models can utilize historical trends to predict future land use patterns (Baig et al. 2022; Weng 2002). Recent climatic data and its future projections using climate simulators have further increased data availability, enabling a more comprehensive definition of habitat dynamics. The increasing use of radio telemetry to track animal movements has also contributed to the wealth of data available for the systematic evaluation of landscape dynamics. However, data intensification raises the important question of how far the dynamics of a landscape should be taken into account in landscape connectivity models when aiming for the implementation of conservation plans (Zeller et al. 2020).

The dynamics of landscape connectivity are closely related to the processes influenced by changing spatial and temporal scales (Gurarie and Ovaskainen 2011). Consequently, the initial step in landscape connectivity planning is to decide on the spatial and temporal scales that align with the ecological questions under investigation. The second stage of planning entails identifying other variables that contribute to the dynamic structural and functional connectedness of the landscape. In a real scenario, predicting future conditions and disturbance dynamics is challenging, making conservation planning a complex task that often requires some level of approximation and flexibility in implementation (Zeller et al. 2020). To address such uncertainties, planners may wish to evaluate worst-case scenarios or establish disturbance thresholds for different conservation options (Van Teeffelen et al. 2012).

The modeling approach for landscape connectivity dynamics can be classified into distinct temporal scales (short, moderate, and long) as well as future projections. At smaller scales, the influence of species' responses is modeled for changes in daily temperature, precipitation, cloud cover, humidity, and other species-specific variables that may have an influence over shorter time periods. For example, Jarvis et al. (2019) studied amphibian connectivity between two sites connected via a road underpass in Yorkshire, England. In moderate-scale studies, the incorporation of seasonality is the most common approach: Seasonal changes in land cover may lead to variations in species movement patterns. Chibeya et al. (2021) identified elevation, land cover, and vegetation index as the most prominent predictors of elephant movement during the wet season in Sioma Ngwezi National Park, Zambia. By acknowledging landscape as a dynamic unit, researchers have begun to question whether static approaches identify the same important areas for connectivity as multi-seasonal and multiyear analyses. Changes occurring at longer time scales, such as climate change or land use change, alter the functional connectivity of a landscape. Enhancing connectivity at these scales provides climate change adaptation opportunities (Krosby et al. 2010). Species-specific dynamic landscape connectivity evaluations under projected scenarios of climate change are usually performed by employing niche modeling and species distribution modeling (see Chap. 6).

Understanding a landscape's dynamic nature may be simple, but effectively simulating the dynamic data in a modeling environment is challenging. In recent decades, significant progress has been made, thanks to technological advancements, allowing for the incorporation of dynamic entities in landscape connectivity models and thus making them more realistic for conservation planning. However, caution is advised when representing dynamics in a landscape connectivity model, as any inappropriate representation of dynamics can deliver misleading connectivity results. It is therefore essential to develop models capable of analyzing dynamic connectivity with minimal bias in order to improve conservation planning efforts with as little misdirected conservation investment as possible. Such models are invaluable tools for understanding ecosystem dynamics and informing management strategies for reserve planning, policymaking, and species conservation planning. The diverse perspectives in this field of research have led to the development of a plethora of techniques, emphasizing the need to explicitly define study objectives to select the optimal modeling approach and gather the required data (Diniz et al. 2020). Interdisciplinary collaboration is crucial to making informed decisions, prioritizing conservation actions, and allocating resources effectively. Integrating domain-specific ecological knowledge with advanced modeling techniques, as well as harnessing the potential of big data and remote sensing, allows us to gain profound insights into landscape dynamics. This collaborative approach enables us to develop more robust connectivity models, improving ecosystem understanding and aiding conservation planning.

References

Adriaensen F, Chardon JP, De Blust G, Swinnen E, Villalba S, Gulinck H, Matthysen E (2003) The application of "least-cost" modelling as a functional landscape model. Landsc Urban Plan 64(4):233–247. https://doi.org/10.1016/S0169-2046(02)00242-6

Allen CH, Parrott L, Kyle C (2016) An individual-based modelling approach to estimate landscape connectivity for bighorn sheep (Ovis canadensis). PeerJ 2016(5). https://doi.org/10.7717/peerj.2001

Baig MF, Mustafa MRU, Baig I, Takaijudin HB, Zeshan MT (2022) Assessment of land use land cover changes and future predictions using CA-ANN simulation for Selangor, Malaysia. Water (Switzerland) 14(3):402. https://doi.org/10.3390/w14030402

Baudry J, Burel F, Aviron S, Martin M, Ouin A, Pain G, Thenail C (2003) Temporal variability of connectivity in agricultural landscapes: do farming activities help? Landsc Ecol 18(3):303–314. https://doi.org/10.1023/A:1024465200284

Beger M, Metaxas A, Balbar AC, McGowan JA, Daigle R, Kuempel CD, Treml EA, Possingham HP (2022) Demystifying ecological connectivity for actionable spatial conservation planning. Trends Ecol Evol 37(12):1079–1091. https://doi.org/10.1016/j.tree.2022.09.002

Bélisle M (2005) Measuring landscape connectivity: the challenge of behavioral landscape ecology. Ecology 86(8):1988–1995. https://doi.org/10.1890/04-0923

Bishop-Taylor R, Tulbure MG, Broich M (2018) Evaluating static and dynamic landscape connectivity modelling using a 25-year remote sensing time series. Landsc Ecol 33(4):625–640. https://doi.org/10.1007/s10980-018-0624-1

Breckheimer I, Haddad NM, Morris WF, Trainor AM, Fields WR, Jobe RT, Hudgens BR, Moody A, Walters JR (2014) Defining and evaluating the umbrella species concept for conserving and restoring landscape connectivity. Conserv Biol 28(6):1584–1593. https://doi.org/10.1111/cobi.12362

Brouwers NC, Newton AC, Watts K, Bailey S (2010) Evaluation of buffer-radius modelling approaches used in forest conservation and planning. Forestry 83(4):409–421. https://doi.org/10.1093/forestry/cpq023

Calabrese JM, Fagan WF (2004) A comparison-shopper's guide to connectivity metrics. Front Ecol Environ 2(10):529–536. https://doi.org/10.1890/1540-9295(2004)002[0529:ACGTCM]2.0.CO;2

Chibeya D, Wood H, Cousins S, Carter K, Nyirenda MA, Maseka H (2021) How do African elephants utilize the landscape during wet season? A habitat connectivity analysis for Sioma Ngwezi landscape in Zambia. Ecol Evol 11(21):14916–14931. https://doi.org/10.1002/ece3.8177

Clauzel C, Godet C, Tarabon S, Eggert C, Vuidel G, Bailleul M, Miaud C (2024) From single to multiple habitat connectivity: the key role of composite ecological networks for amphibian conservation and habitat restoration. Biol Conserv 289:110418. https://doi.org/10.1016/j.biocon.2023.110418

Correa Ayram CA, Mendoza ME, Etter A, Salicrup DRP (2016) Habitat connectivity in biodiversity conservation: a review of recent studies and applications. Prog Phys Geogr 40(1):7–37. https://doi.org/10.1177/0309133315598713

Crooks KR, Sanjayan M (2006) Connectivity conservation (Kd 14). Cambridge University Press

Cushman SA, Huettmann F (2010) Spatial complexity, informatics, and wildlife conservation, 9784431877, 1–458. https://doi.org/10.1007/978-4-431-87771-4

Cushman SA, Landguth EL (2010) Scale dependent inference in landscape genetics. Landsc Ecol 25(6):967–979. https://doi.org/10.1007/s10980-010-9467-0

Cushman SA, Chase M, Griffin C (2005) Elephants in space and time. Oikos 109(2):331–341. https://doi.org/10.1111/j.0030-1299.2005.13538.x

Cushman SA, Lewis JS, Landguth EL (2013) Evaluating the intersection of a regional wildlife connectivity network with highways. Movem Ecol 1(1):12. https://doi.org/10.1186/2051-3933-1-12

Den Boer PJ (1968) Spreading of risk and stabilization of animal numbers. Acta Biotheor 18:165–194. https://doi.org/10.1007/BF01556726

Diniz MF, Cushman SA, Machado RB, De Marco Júnior P (2020) Landscape connectivity modeling from the perspective of animal dispersal. Landsc Ecol 35(1):41–58. https://doi.org/10.1007/s10980-019-00935-3

Drake J, Lambin X, Sutherland C (2022) The value of considering demographic contributions to connectivity: a review. Ecography 2022(6):e05552. https://doi.org/10.1111/ecog.05552

Dutta T, Sharma S, Meyer NFV, Larroque J, Balkenhol N (2022) An overview of computational tools for preparing, constructing and using resistance surfaces in connectivity research. Landsc Ecol 37(9):2195–2224. https://doi.org/10.1007/s10980-022-01469-x

Etherington TR (2016) Least-cost modelling and landscape ecology: concepts, applications, and opportunities. Curr Landsc Ecol Rep 1(1):40–53. https://doi.org/10.1007/s40823-016-0006-9

Fagan W, Calabrese J (2006) Quantifying connectivity: balancing metric performance with data requirements. In: Crooks KR, Sanjayan Toim M (eds) Connectivity conservation. Cambridge University Press, pp 297–317

Fahrig L (1992) Relative importance of spatial and temporal scales in a patchy environment. Theor Popul Biol 41(3):300–314. https://doi.org/10.1016/0040-5809(92)90031-N

Fahrig L, Arroyo-Rodríguez V, Cazetta E, Ford A, Lancaster J, Ranius T (2021) Landscape connectivity. In: The Routledge handbook of landscape ecology. Routledge, pp 67–88. https://doi.org/10.4324/9780429399480-5

Fletcher RJ, Sefair JA, Wang C, Poli CL, Smith TAH, Bruna EM, Holt RD, Barfield M, Marx AJ, Acevedo MA (2019) Towards a unified framework for connectivity that disentangles movement and mortality in space and time. Ecol Lett 22(10):1680–1689. https://doi.org/10.1111/ele.13333

Forman RTT (1995) Some general principles of landscape and regional ecology. Landsc Ecol 10(June):133–142. https://doi-org.uaccess.univie.ac.at/10.1007/BF00133027

Goodwin BJ (2003) Is landscape connectivity a dependent or independent variable? Landsc Ecol 18(7):687–699. https://doi.org/10.1023/B:LAND.0000004184.03500.a8

Gurarie E, Ovaskainen O (2011) Characteristic spatial and temporal scales unify models of animal movement. Am Nat 178(1):113–123. https://doi.org/10.1086/660285

Hall LS, Krausman PR, Morrison ML, Hall LS, Krausman PR, Morrison ML (1997) The Habitat Concept and a Plea for Standard Terminology Linked references are available on JSTOR for this article: the habitat concept and a plea for standard terminology. Wildl Soc Bull (1973–2006) 25(1):173–182

Hanski I (1994a) A practical model of metapopulation dynamics. J Anim Ecol 63(1):151. https://doi.org/10.2307/5591

Hanski I (1994b) Patch-occupancy dynamics in fragmented landscapes. Trends Ecol Evol 9(4):131–135. https://doi.org/10.1016/0169-5347(94)90177-5

Hanski I (2014) The theories of island biogeography and metapopulation dynamics. In: The theory of island biogeography revisited. Princeton University Press, pp 186–213. https://doi.org/10.1515/9781400831920.186

Hauenstein S, Fattebert J, Grüebler MU, Naef-Daenzer B, Pe'er G, Hartig F (2019) Calibrating an individual-based movement model to predict functional connectivity for little owls. Ecol Appl 29(4):1–16. https://doi.org/10.1002/eap.1873

Hilty J, Worboys G, Keeley A, Woodley S, Lausche B, Locke H, Carr M, Pulsford I, Pittock J, White W, Theobald D, Levine J, Reuling M, Watson J, Ament R, Tabor G (2020) Guidance for conserving connectivity through ecological networks and corridors 30

Howell PE, Muths E, Hossack BR, Sigafus BH, Chandler RB (2018) Increasing connectivity between metapopulation ecology and landscape ecology. Ecology 99(5):1119–1128. https://doi.org/10.1002/ecy.2189

Huang JL, Andrello M, Martensen AC, Saura S, Liu DF, He JH, Fortin MJ (2020) Importance of spatio-temporal connectivity to maintain species experiencing range shifts. Ecography 43(4):591–603. https://doi.org/10.1111/ecog.04716

Hylander K, Ehrlén J (2013) The mechanisms causing extinction debts. Trends Ecol Evol 28(6):341–346. https://doi.org/10.1016/j.tree.2013.01.010

Jarvis LE, Hartup M, Petrovan SO (2019) Road mitigation using tunnels and fences promotes site connectivity and population expansion for a protected amphibian. Eur J Wildl Res 65(2):27. https://doi.org/10.1007/s10344-019-1263-9

Kaszta Ż, Cushman SA, Slotow R (2021) Temporal non-stationarity of path-selection movement models and connectivity: an example of African Elephants in Kruger National Park. Front Ecol Evol 9:553263. https://doi.org/10.3389/fevo.2021.553263

Kellner CJ, Brawn JD, Karr JR (1992) What is habitat suitability and how should it be measured? In: McCullough DR, Toim RHB (eds) Wildlife 2001: populations. Springer, pp 476–477. https://doi.org/10.1007/978-94-011-2868-1_36

Kindlmann P, Burel F (2008) Connectivity measures: a review. Landsc Ecol 23(8):879–890

Krosby M, Tewksbury J, Haddad NM, Hoekstra J (2010) Ecological connectivity for a changing climate. Conserv Biol 24(6):1686–1689. https://doi.org/10.1111/j.1523-1739.2010.01585.x

Kumar M, Padalia H, Nandy S, Singh H, Khaiter P, Kalra N (2019) Does spatial heterogeneity of landscape explain the process of plant invasion? A case study of Hyptis suaveolens from Indian Western Himalaya. Environ Monit Assess 191:794. https://doi.org/10.1007/s10661-019-7682-y

Larrey-Lassalle P, Esnouf A, Roux P, Lopez-Ferber M, Rosenbaum RK, Loiseau E (2018) A methodology to assess habitat fragmentation effects through regional indexes: illustration with forest biodiversity hotspots. Ecol Indic 89(February):543–551. https://doi.org/10.1016/j.ecolind.2018.01.068

Levins R (1969) Some demographic and genetic consequences of environmental heterogeneity for biological control. Bull Entomol Soc Am 15:237–240

Lorimer J (2015) Wildlife in the anthropocene: Conservation after nature. University of Minnesota Press. https://doi.org/10.1080/14649365.2017.1288696

MacArthur RH, Wilson EO (1969) The theory of island biogeography

Marx AJ, Wang C, Sefair JA, Acevedo MA, Fletcher RJ (2020) samc: an R package for connectivity modeling with spatial absorbing Markov chains. Ecography 43(4):518–527. https://doi.org/10.1111/ecog.04891

McRae BH, Dickson BG, Keitt TH, Shah VB (2008) Using circuit theory to model connectivity in ecology, evolution, and conservation. Ecology 89(10):2712–2724. https://doi.org/10.1890/07-1861.1

Merriam G (1984) Connectivity: a fundamental ecological characteristic of landscape pattern. In: Brandt J, Toim PAA (eds) Proceedings of the first international seminar on methodology in landscape ecological research and planning. Rosskilde University Centre

Moilanen A (2002) Implications of empirical data quality to metapopulation model parameter estimation and application. Oikos 96(3):516–530. https://doi.org/10.1034/j.1600-0706.2002.960313.x

Moilanen A, Hanski I (2001) On the use of connectivity measures in spatial ecology. Oikos 95(1):147–151. https://doi.org/10.1034/J.1600-0706.2001.950116.X

Moilanen A, Nieminen M (2002) Simple connectivity measures in spatial ecology. Ecology 83(4):1131–1145. https://doi.org/10.1890/0012-9658(2002)083[1131:SCMISE]2.0.CO;2

Morris DW (2003) How can we apply theories of habitat selection to wildlife conservation and management? Wildl Res 30(4):303–319. https://doi.org/10.1071/WR02028

Morris DW, Kotler BP, Brown JS, Sundararaj V, Ale SB (2009) Behavioral indicators for conserving mammal diversity. Ann N Y Acad Sci 1162:334–356. https://doi.org/10.1111/j.1749-6632.2009.04494.x

O'Connor MI, Selig ER, Pinsky ML, Altermatt F (2012) Toward a conceptual synthesis for climate change responses. Glob Ecol Biogeogr 21(7):693–703. https://doi.org/10.1111/j.1466-8238.2011.00713.x

Pelletier D, Clark M, Anderson MG, Rayfield B, Wulder MA, Cardille JA (2014) Applying circuit theory for corridor expansion and management at regional scales: tiling, pinch points, and omnidirectional connectivity. PLoS One 9(1):e84135. https://doi.org/10.1371/journal.pone.0084135

Pinto N, Keitt TH (2009) Beyond the least-cost path: evaluating corridor redundancy using a graph-theoretic approach. Landsc Ecol 24(2):253–266. https://doi.org/10.1007/s10980-008-9303-y

Pressey RL, Visconti P, McKinnon MC, Gurney GG, Barnes MD, Glew L, Maron M (2021) The mismeasure of conservation. Trends Ecol Evol 36(9):808–821. https://doi.org/10.1016/j.tree.2021.06.008

Prugh LR (2009) An evaluation of patch connectivity measures. Ecol Appl 19(5):1300–1310

Puckridge JT, Sheldon F, Walker KF, Boulton AJ (1998) Flow variability and the ecology of large rivers. Mar Freshw Res 49(1):55–72. https://doi.org/10.1071/MF94161

Ranius T, Martikainen P, Kouki J (2011) Colonisation of ephemeral forest habitats by specialised species: beetles and bugs associated with recently dead aspen wood. Biodivers Conserv 20(13):2903–2915. https://doi.org/10.1007/s10531-011-0124-y

Risk BB, De Valpine P, Beissinger SR (2011) A robust-design formulation of the incidence function model of metapopulation dynamics applied to two species of rails. Ecology 92(2):462–474. https://doi.org/10.1890/09-2402.1

Royle JA, Fuller AK, Sutherland C (2018) Unifying population and landscape ecology with spatial capture–recapture. Ecography 41(3):444–456. https://doi.org/10.1111/ecog.03170

Rudnick DA, Ryan SJ, Beier P, Cushman SA, Dieffenbach F, Epps CW, Gerber LR, Hartter J, Jenness JS, Kintsch J, Merenlender AM, Perkl RM, Preziosi DV, Trombulak SC (2012) The role of landscape connectivity in planning and implementing conservation and rostoration priorities. Issues in ecology (Number 16)

Ruiz-Carbayo H, Bonal R, Espelta JM, Hernández M, Pino J (2017) Community assembly in time and space: the case of Lepidoptera in a Quercus ilex L. savannah-like landscape. Insect Conserv Divers 10(1):21–31. https://doi.org/10.1111/icad.12184

Salgueiro PA, Valerio F, Silva C, Mira A, Rabaça JE, Santos SM (2021) Multispecies landscape functional connectivity enhances local bird species' diversity in a highly fragmented landscape. J Environ Manag 284(January):112066. https://doi.org/10.1016/j.jenvman.2021.112066

Santini L, Saura S, Rondinini C (2016) A composite network approach for assessing multi-species connectivity: an application to road defragmentation prioritisation. PLoS One 11(10):e0164794. https://doi.org/10.1371/journal.pone.0164794

Saura S, Estreguil C, Mouton C, Rodríguez-Freire M (2011) Network analysis to assess landscape connectivity trends: application to European forests (1990–2000). Ecol Indic 11(2):407–416. https://doi.org/10.1016/j.ecolind.2010.06.011

Schumaker NH (1996) Using landscape indices to predict habitat connectivity. Ecology 77(4):1210–1225. https://doi.org/10.2307/2265590

Sutherland CS, Elston DA, Lambin X (2014) A demographic, spatially explicit patch occupancy model of metapopulation dynamics and persistence. Ecology 95(11):3149–3160. https://doi.org/10.1890/14-0384.1

Taylor PD, Fahrig L, Henein K, Merriam G (1993) Connectivity is a vital element of landscape structure. Oikos 68(December):571–573

Taylor PD, Fahrig L, With KA (2006) 2-Landscape connectivity: a return to the basics. In: Crooks KR, Toim MS (eds) Connectivity conservation. Cambridge University Press, pp 29–43. https://doi.org/10.1017/CBO9780511754821

Tischendorf L, Fahrig L (2000) On the usage and measurement of landscape connectivity. Oikos 90(1):7–19. https://doi.org/10.1034/j.1600-0706.2000.900102.x

Unnithan Kumar S, Turnbull J, Hartman Davies O, Hodgetts T, Cushman SA (2022) Moving beyond landscape resistance: considerations for the future of connectivity modelling and conservation science. Landsc Ecol 37(10):2465–2480. https://doi.org/10.1007/s10980-022-01504-x

Uroy L, Alignier A, Mony C, Foltête JC, Ernoult A (2021) How to assess the temporal dynamics of landscape connectivity in ever-changing landscapes: a literature review. Landsc Ecol 36(9):2487–2504. https://doi.org/10.1007/s10980-021-01277-9

Van Teeffelen AJA, Vos CC, Opdam P (2012) Species in a dynamic world: consequences of habitat network dynamics on conservation planning. Biol Conserv 153:239–253. https://doi.org/10.1016/j.biocon.2012.05.001

Wang YH, Yang KC, Bridgman CL, Lin LK (2008) Habitat suitability modelling to correlate gene flow with landscape connectivity. Landsc Ecol 23(8):989–1000. https://doi.org/10.1007/s10980-008-9262-3

Wang F, McShea WJ, Li S, Wang D (2018) Does one size fit all? A multispecies approach to regional landscape corridor planning. Divers Distrib 24(3):415–425. https://doi.org/10.1111/ddi.12692

Weng Q (2002) Land use change analysis in the Zhujiang Delta of China using satellite remote sensing, GIS and stochastic modelling. J Environ Manag 64(3):273–284. https://doi.org/10.1006/jema.2001.0509

Wiens JA (1997) The emerging role of patchiness in conservation biology. In: The ecological basis of conservation. Springer, Boston, MA, pp 93–107. https://doi.org/10.1007/978-1-4615-6003-6_10

Wintle BA, Kujala H, Whitehead A, Cameron A, Veloz S, Kukkala A, Moilanen A, Gordon A, Lentini PE, Cadenhead NCR (2019) Global synthesis of conservation studies reveals the importance of small habitat patches for biodiversity. Proc Natl Acad Sci 116(3):909–914

With KA (2019) Landscape connectivity. In: Essentials of landscape ecology (lk 206–2038). Oxford University Press

Wu J, David JL (2002) A spatially explicit hierarchical approach to modeling complex ecological systems: theory and applications. Ecol Model 153(1–2):7–26. https://doi.org/10.1016/S0304-3800(01)00499-9

Zeller KA, McGarigal K, Whiteley AR (2012) Estimating landscape resistance to movement: a review. Landsc Ecol 27(6):777–797. https://doi.org/10.1007/s10980-012-9737-0

Zeller KA, Lewsion R, Fletcher RJ, Tulbure MG, Jennings MK (2020) Understanding the importance of dynamic landscape connectivity. Land 9(9):303. https://doi.org/10.3390/LAND9090303

Open Access This chapter is licensed under the terms of the Creative Commons Attribution 4.0 International License (http://creativecommons.org/licenses/by/4.0/), which permits use, sharing, adaptation, distribution and reproduction in any medium or format, as long as you give appropriate credit to the original author(s) and the source, provide a link to the Creative Commons license and indicate if changes were made.

The images or other third party material in this chapter are included in the chapter's Creative Commons license, unless indicated otherwise in a credit line to the material. If material is not included in the chapter's Creative Commons license and your intended use is not permitted by statutory regulation or exceeds the permitted use, you will need to obtain permission directly from the copyright holder.

Species on the Move: Migration, Range Shifts, and Dispersal of Species

2

Katharina Lapin, Heino Konrad, Christoph Leeb, and Janine Oettel

The Wood Warbler (*Phylloscopus sibilatrix*) is a forest-dependent and migratory bird species that inhabits open deciduous and mixed forests, as well as beech forests; its wintering grounds are located in tropical Africa. (Photo: Samuel Schnierer).

K. Lapin (✉) · H. Konrad · J. Oettel
Department of Forest Biodiversity and Nature Conservation, Federal Research and Training Centre for Forests, Natural Hazards and Landscape, Vienna, Austria
e-mail: katharina.lapin@bfw.gv.at; heino.konrad@bfw.gv.at; janine.oettel@bfw.gv.at

C. Leeb
First Zoological Department, Herpetological Collection, Natural History Museum Vienna, Vienna, Austria
e-mail: Christoph.leeb@nhm.at

© The Author(s) 2025
K. Lapin et al. (eds.), *Ecological Connectivity of Forest Ecosystems*,
https://doi.org/10.1007/978-3-031-82206-3_2

Abstract

Migration is a fundamental biological phenomenon with significant impacts on the survival of populations and ecosystems. Its precise definition depends on the ecological discipline in focus, but it essentially entails the movement of an individual or population between habitats to ensure survival and enhance the likelihood of successful reproduction. Evolution has led to different types and mechanisms of migration ranging from passive to active movements, and from a few meter to migrations across biomes. Climate change has a strong effect on migration behaviour, leading to shifts in the ranges of populations and species as the quality and availability of habitats are altered. This chapter provides an overview of migratory behaviour and points out examples of variations caused by climate change. In addition, it discusses the context of migratory and ecological connectivity, both of which are essential for preserving biodiversity and facilitating adaptation. Understanding migration, range shifts, and dispersal is key to effective conservation and management efforts and should therefore be a primary focus of international collaborations.

Keywords

Active migration · Passive migration · Migratory connectivity · Long-distance migration · Migratory phenology shift · Vertical shift

Introduction

Migration is a natural phenomenon observed in almost any taxon worldwide; it involves individual movements that contribute to shaping population dynamics and influencing ecosystems (Dingle and Drake 2007). It is typically defined as locomotion by walking, swimming, or flying that results in a directional movement which is usually repeated periodically, often within a season. In plants and other non-mobile organisms, migration is mediated by the dispersal of seeds and other propagules. The reasons why species migrate vary, but at its core, migration is driven by the need to move between habitats that differ in resources, environmental conditions, competition, predation, or parasites (Joly et al. 2019) to ensure or increase the probability of survival and reproduction. The spectrum of migration behaviour across species is extensive as can be seen in bird species migration (see opening figure in Chap. 2). Certain species exhibit obligatory migration, requiring them to migrate each year regardless of local conditions. Examples of obligatory migration can be found in many bird species, especially in birds with long migration distances, where individuals usually spend the summer in the northern and the winter in the southern hemisphere. By contrast, the migration of species with facultative migration behaviour is triggered by local environmental conditions. Such facultative behaviour is typical for species with partial migration (Newton 2012), where parts of a population migrate while others do not (Chapman et al. 2011)—as found, for example, in blue tits (Nilsson et al. 2008). A special variation is multi-generation migration, where periodical migration occurs not within a single generation but across successive generations, with each undertaking different stages of the migratory route. A classic example of this is observed in the Monarch butterfly (*Danaus plexippus*). Certain North American populations perform migrations over thousands of kilometers to Central Mexico. Individuals usually do not survive the entire migration; instead, females lay eggs on the way back north and the next generation completes the journey (Reppert and de Roode 2018).

While most animals exhibit some kind of migration behaviour, alternative life histories can also be found. Nomadic behaviour characterized by irregular movements is observed in some species (Teitelbaum and Mueller 2019). Conversely, some organisms adopt a strictly stationary lifestyle, such as sponges or corals inhabiting aquatic environments. Similar stationary life stages are of course also characteristic for plants. When studying plant ecology, migration describes the movement of a population rather than the movement of individuals and is typically linked to the unidirectional spread of seeds and/or pollen (Ennos 1994). In both animal and plant ecology, such movements are referred to as dispersal. Consequently, the definition of migration depends on the ecological discipline under consideration.

Migration pressures vary across regions, areas, and local conditions. Global warming has emerged as one of the most urgent ecological challenges of our time, with tremendous effects on biodiversity. While certain species may be able to adapt to changing conditions under climate change, others may be forced to migrate to areas with a more favourable environment. Regions characterized by significant topographic and climatic heterogeneity have the potential to mitigate these pressures on local species (Barber et al. 2016; Loarie et al. 2009), potentially leading to only minor migrations within the same region. In other regions, climate change may force species into significant range shifts. In species capable of long-distance migrations like caribou, wolves, or some birds, these range shifts can traverse entire biomes within a short time. Conversely, for the large share of species such as the fire salamander (*Salamandra salamandra*) or the western capercaillie (*Tetrao urogallus*)—both forest-dependent species found in European temperate forests—which can only migrate over small distances, such range shifts can only occur gradually over multiple generations and are linked to the process of dispersal.

Understanding the complexities of species migration and the differing responses to climate change is essential for predicting future ecological scenarios, particularly in regard to the connectivity of ecosystems. By examining these patterns, researchers can develop effective strategies to mitigate the impacts of climate change on biodiversity and preserve heavily affected species.

Range and Migratory Phenology Shifts Under Climate Change

Climate change has a profound effect on species, influencing their coexistence within local environments as well as on a wider ecological level. The quality and availability of habitats are directly impacted by climate, thereby affecting species' survival rates and future occurrences (Davis and Shaw 2001). As the climate changes, many species face challenges in surviving in their current habitats. Consequently, they may undergo a shift in their geographical distribution—either within their existing range or by migrating to entirely new regions. These migratory patterns vary among species, with some traversing thousands of kilometers in search of suitable conditions while others remain restricted to their original habitat.

The migration of tree populations following postglacial warming is frequently cited as an illustrative example. Through the analysis of fossil pollen data, scientists can enhance their understanding of the postglacial re-colonization process originating from isolated refugial populations (Brewer et al. 2017; Giesecke et al. 2019). The last ice age started approximately 115,000 years ago and concluded with the onset of the Holocene around 11,700 years ago. Paleoecological and genetic evidence gathered in studies of postglacial re-colonization all across the Northern Hemisphere indicates that numerous tree species expanded their range (again) towards higher latitudes (Hao and Hao 2018; Kreft and Jetz 2007). Recent studies increasingly recognize migration complexity, emphasizing that postglacial migration is shaped not only by climatic factors but also by geomorphology, the availability of northerly microrefugia, and various limitations regarding species dispersal ranges: in North America, the southern distribution boundaries of many tree species shifted northward due to postglacial warming, whereas the European Alps formed an impassable barrier for species migration in many cases (Giesecke et al. 2019). Research involving 1016 European plant species revealed that climate played a crucial role for all of them, with over 50% being restricted in their ranges due to barriers in postglacial re-colonization. In particular, dispersal-limited species, e.g. those with glacial ranges in the south, seed-bearing plants as opposed to ferns, and species with small ranges in Southern Europe were greatly handicapped in their range expansion (Normand et al. 2011). One noteworthy conclusion drawn from a large-scale analysis of a wide range of available data sources is that northerly microrefugia were present during glaciation in some species and constituted a significant factor to explain differing re-migration rates among species; such outlier populations could become important for predicting potential future range shifts (Feurdean et al. 2013).

A commonly used unit to measure the velocity of range shifts (dispersal) is the migration rate, which estimates the distance a species covers over time, respectively the frequency of migration events, and the subsequent reduction in genetic differences between populations. Migration rates depend on species traits, competition, habitat availability, and climatic conditions (Meier et al. 2012). The postglacial migration rates for late-successional trees like the American beech (*Fagus grandifolia*) and red maple (*Acer rubrum*; see also opening figure in Chap. 1) typically fall within a range of approximately 100 meters per year (McLachlan et al. 2005), but for early colonizing species higher rates have been reported, e.g. at least 1000 meters year^{-1} in *Ulmus* spp. (Giesecke and Brewer 2018). In addition, a study utilizing the extensive collection of pollen diagrams available from the European Pollen Database indicates that high apparent rates of postglacial expansion in species like the alders (*Alnus* spp.) can be attributed to initial spread at low population density followed by later expansion (Giesecke and Brewer 2018).

However, the circumstances we are currently witnessing due to recent global warming are significantly different from the postglacial era (IPCC 2020). Based on the available evidence (see above), the migration rate of trees in North America to areas with suitable climate conditions is expected to be considerably faster, potentially reaching values exceeding 1000 meters per year (Malcolm et al. 2002). However, also additional factors such as population size, life history traits, or adaptive potential play a key role in the prediction of range shifts under the ongoing climate change scenario. Trees have been recognized as good adapters in the face of climate change. It has been demonstrated that gene expression—the process of translating the genetic code into the phenotype—is influenced not only by the genetic code itself but also by environmental and climatic factors. Such adaptations have been subsumed under the term "phenotypic plasticity" (Pigliucci et al. 2006; see also Chaps. 4 and 5 and one of the most important underlying mechanisms is epigenetic regulation (García-García et al. 2022). These regulatory mechanisms can be inherited, thus allowing for relatively rapid adaptation from one generation to the next. Nevertheless, critical views stress that the capacity of species to adapt to rapidly changing evolving environmental conditions has been overestimated and is more limited than initially anticipated (Pearson 2006; Zhu et al. 2012).

When considering external obstacles that impede species' migration and establishment in new regions, it appears urgent to develop conservation strategies that improve connectivity, enhance the overall quality of habitats (including breeding and feeding areas outside of existing protected areas), and encompass proactive management measures designed to ensure the maintenance of species in the face of climate change. Notably, the migration rates for the majority of species remain unknown even though they constitute crucial pieces of information for making accurate predictions regarding climate-induced range shifts and effectively managing and planning species conservation efforts (Estrada et al. 2016; Hovick et al. 2016).

Range shifts induced by climate change can also be observed in numerous animal species as rising temperatures degrade the quality of existing habitats and/or make previously unsuitable areas viable. An example is the hooded warbler (*Wilsonia citrina*), a bird species inhabiting North American forests. Its distribution range is generally heavily dependent on climatic conditions, and warmer temperatures have allowed its range to expand towards the north over the last decades (Melles et al. 2011). Also, other factors besides climate change can likewise lead to range shifts in a species. For the great spotted woodpecker (*Dendrocopos major*) and the black woodpecker (*Dryocopus martius*), changes in forest management practices that allow forest maturation have resulted in range expansions in the region of Catalonia (Spain). Besides the newly available habitat itself, its connectivity to the existing habitats plays an important role in realizing range shifts (Gil-Tena et al. 2013). In general, landscape connectivity is crucial for enabling range shifts and its improvement is therefore one of the most urgently recommended mitigation measures in the face of climate change (Littlefield et al. 2019).

However, not all species and populations exhibit range shifts to the north to maintain preferred environmental conditions. For some, such as amphibians from the Iberian Peninsula, it is easier to shift to habitats at higher altitudes (Enriquez-Urzelai et al. 2019). In general, cold temperatures and snow often restrict the occurrence of species at higher altitudes in mountainous or alpine ecosystems. Climate change has a significant influence on these conditions, increasingly creating more favourable environments for warm-adapted species in higher altitudes. The most prominent example of a vertical shift in alpine areas is the shift of the tree line (the upper boundary of the altitudinal belt where trees can grow). While global observations have noted shifts in tree lines within the last decades (e.g. in about half of 166 analysed sites around the world in a meta-analysis by Harsch et al. (2009), they seem to be frequently driven by land use changes, human activities, and other disturbances (Cudlín et al. 2017; Trant et al. 2020). Only a small (yet likely gradually increasing) proportion of these shifts can be directly attributed to climate change (Gehrig-Fasel et al. 2007). Nevertheless, the Global Observation Research Initiative in Alpine Environments (GLORIA), a standardized monitoring programme for vegetation in alpine study sites around the world, has documented a decline of cold-adapted species and an increase of warm-adapted species at high altitudes (Gottfried et al. 2012). This dynamic reflects a vertical range shift for alpine plant species. Expanding on this, an analysis of literature and observational data of 2133 taxa in the European Alps by Vitasse et al. (2021) found altitudinal shifts not only in plants but also in fungi and animals. These shifts were attributed to changing climate conditions. Among all investigated species groups, terrestrial insects exhibited the most pronounced shift, with an average upward shift of about 36 m per decade; however, vertebrates like reptiles also increasingly populate higher altitudes.

Climate change can induce shifts in the phenology of periodically repeated migration (e.g. Koleček et al. 2020; Lenzi et al. 2023; Van Buskirk et al. 2009). These changes are particularly observable in bird species, which exhibit the most prominent migration behaviour. For instance, Koleček et al. (2020) analysed the spring arrival time of 52 migrating birds in the Czech Republic, revealing that increasing temperature led to an earlier arrival. This shift was more pronounced in species with shorter migration distances, as these species can respond faster to increasing temperatures. The authors also found evidence that the shift in migration phenology had a direct positive effect on breeding success and thus on population trends. In line with this, failure to adapt to the timing of migration can lead to population declines (Møller et al. 2008). An earlier arrival of migrating birds was also observed when analysing the migrations of 78 bird species in Pennsylvania (USA) over 46 years (Van Buskirk et al. 2009). Lenzi et al. (2023) found that common toads (*Bufo bufo*) from a population in the Swiss Alps began breeding around 30 days earlier than they did four decades ago. Migration shifts are often contingent on the species and even on individual populations (Van Buskirk et al. 2009). Furthermore, additional factors besides temperature (Dalpasso et al. 2023; While and Uller 2014) also contribute to species migration; e.g., Seebacher and Post (2015) showed that the migration phenology of birds often correlates with shifts in the phenology of their prey's abundance, adding another layer to the complexities involved in these ecological processes.

Dispersal Distances and Mechanisms

Whether a species succeeds in shifting to a new range or not primarily depends on species-specific dispersal characteristics, as well as on the availability of suitable habitats being connected through a network of corridors and stepping stones (Hodgson et al. 2012). Dispersal is an ecological process that plays a pivotal role in shaping the distribution and population sizes of species as well as influencing the exchange of genetic variants among populations. In simple terms, dispersal can be described as the movement or spread of reproductive units, such as seeds or offspring, between suitable habitat patches (Clobert et al. 2004). This movement is crucial for species to establish and maintain populations in multiple locations. The specific dispersal traits of a species are a key factor in determining its capacity to migrate and adapt in response to the challenges posed by climate change (Clobert 2012; Johnson and Gaines 1990).

Dispersal mechanisms often determine the distance a species can move, and two main types of dispersal exist: active dispersal, which is mostly observed in animals, and passive dispersal, prevalent among plants and animals constrained in their self-mobility. Active dispersal involves self-propulsion, including activities like walking, flying, swimming, or other forms of self-driven locomotion. By contrast, passive dispersal relies on external factors such as wind (anemochory), water (hydrochory), animals (zoochory), humans (anthropochory), or environmental forces to transport organisms or their propagules to new locations. Direct observational data underscores the significant role of size in the dispersal strategies of different organisms. Actively dispersing organisms tend to be larger in size and can engage in self-propulsion over longer distances, while passively dispersing propagules are typically smaller and have limited overall dispersal capabilities (Jenkins et al. 2007).

Traditionally, it was believed that plants rarely achieve targeted dispersal because they lack specialized adaptations controlling the final destination of dispersed seeds. However, several studies have found that animal vectors often direct resources to disturbed areas or create favourable conditions for plants, implying that diffuse mutualisms can frequently lead to directed dispersal (Clobert et al. 2004; Jenkins et al. 2007; Mason et al. 2022). Besides the interaction of different dispersal mechanisms, environmental conditions strongly influence dispersal distance. Habitat specialists encounter more barriers during dispersal compared to generalists (Hansson 1991). Exploring dispersal strategies and their interactions is immensely valuable, especially when considering additional data on factors like dispersal distances and lifetime dispersal capabilities (Table 2.1).

Taking a closer look at dispersal mechanisms with a species-specific evolutionary history reveals distinct patterns: Passive dispersal of plant seeds by *anthropochory*—meaning dispersal through human activities such as agriculture, trade, or recreational activities—has resulted in the global spread of alien plant species such as the invasive ragweed (*Ambrosia artemisiifolia*) or the tree of heaven (*Ailanthus altissima*). By contrast, myrmecochorous plants such as the common hepatica (*Hepatica nobilis*) form symbiotic partnerships with ants. These plants produce

Table 2.1 Examples of species that can be found in forest ecosystems with brief descriptions of their migration behaviour and capabilities

Species	Description of migration behaviour and capabilities
	Some snail species have only very limited dispersal abilities. Edworthy et al. (2012) studied a population of the Oregon forestsnail (*Allogona townsendiana*) and documented movements of up to only 32.2 m over three years.
	The European stag beetle (*Lucanus cervus*) can fly to reach new habitats or find a mating partner. Telemetry data has documented flight distances of up to 1720 m for males and about 760 m for females, while the modelled dispersal abilities are 3 km and 1 km, respectively (Rink and Sinsch 2007).
	Female fire salamanders (*Salamandra salamandra*) migrate to water bodies, preferably small creeks in deciduous forests, to deposit their larvae. In some cases, migrations from the summer habitat to hibernation sites can also be observed in both sexes. Migration distances of up to 1900 m have been documented (Hendrix et al. 2017). Unfragmented forests are essential for the migration of this species.
	Wolves (*Canis lupus*) are known to migrate over hundreds of kilometers. For example, Mancinelli and Ciucci (2018) used GPS telemetry to document a 422.2 km migration within 41 days by a male individual in Italy. By analysing the genetic profiles of wolves, Andersen et al. (2015) revealed movements of over 800 km in north-western Europe. This dispersal ability allows wolves to recolonize areas from which the species had previously been extirpated, like France and Switzerland (Valière et al. 2003).
	Many bird species are capable of long-distance migration. Populations of the black stork (*Ciconia nigra*) that breed in Eastern Europe migrate to East Africa via the Bosporus and the Sinai, while most Central European populations migrate to West Africa via Gibraltar. For populations from the migratory divide, it is known that siblings from the same nest can take different routes to Africa (Literák et al. 2017). Migration can cover thousands of kilometers (Cano and Tellería 2013).

(continued)

Table 2.1 (continued)

Species	Description of migration behaviour and capabilities
	Species of the genus *Acer* (maple) have winged seeds (samara) that are transported by the wind, a passive dispersal mechanism called anemochory. Using wind dispersal models, simulations for the red maple (*Acer rubrum*) suggest median dispersal distances of seeds between 2.8 m under calm and 83.3 m under stormy wind conditions, with maximum distances of 11,371 m (Higgins et al. 2003).
	Like in most mistletoes, dispersal of the seeds of the European mistletoe (*Viscum album*) occurs by ingestion and defecation by birds, a passive dispersal mechanism that is called endozoochory. Since the time a seed spends in the digestive tract is usually quite short, the estimated dispersal distances are up to 20 km. Dispersal over longer distances is possible via epizoochory when seeds stick to the feathers of birds (Zuber and Widmer 2009)
	The seeds of some plants, like species of the genus *Arctium* (burdock), adhere to animals, a passive dispersal mechanism called epizoochory. For example, Picard and Baltzinger (2012) investigated the fur and hooves of 17 wild boars (*Sus scrofa*) in France and found seeds of 35 plant species. Actual dispersal distances via epizoochory are rarely reported but can reach hundreds of kilometers. On Macquarie Island (Australia), only plant species with epizoochorous dispersal have been found, suggesting dispersal by birds over at least 650 km (Taylor 1954).

seeds with elaiosomes, fleshy appendages that attract ants. The ants carry the seeds to their nests, consume the elaiosome, and discard the intact seed, facilitating its local dispersal within a confined habitat.

While trait databases for specific species groups such as mammalian herbivores (Teitelbaum et al. 2015) or European reptiles (Grimm et al. 2014) have enhanced our knowledge of species migration distances, the migration capacities of most species remain unknown. In some cases, estimations obtained through traditional methods like capture-recapture or telemetry have underestimated the respective species' actual capabilities. For example, telemetry studies revealed maximum annual movements of natterjack toads (*Epidalea calamita*) spanning up to 4.4 km, while modelling approaches estimate the dispersal capacity of the species at up to 12.2 km (Sinsch et al. 2012). In the case of the European stag beetle (*Lucanus cervus*), telemetry data showed maximum flight distances of 1720 m for males and about 760 m for females, whereas corresponding models estimated the dispersal abilities at 3 km and 1 km, respectively (Rink and Sinsch 2007). These differences underline the importance of further studies focusing on elucidating the true migration capacity of species.

From Dispersal to Ecological Connectivity

As shown, the ecological functioning of biodiversity is closely tied to effective dispersal, which allows population growth, as well as preserves populations by reducing intraspecific conflicts and preventing inbreeding (Lowe and Allendorf 2010; see also Chap. 4). It also serves as the foundation for adaptation to changing environmental conditions. Dispersal therefore acts as an indicator of a population's fitness and capability for evolutionary responses. Furthermore, ecological connectivity throughout a landscape is essential for enabling ecological processes to thrive by offering the necessary quantity and quality of suitable habitats for species.

Landscape connectivity measures how a landscape either facilitates or impedes movement across resource patches. It includes both structural connectivity, which involves physical distances, and functional connectivity, which encompasses the behavioural responses of organisms to the landscape matrix (Fahrig 2003; Suter and Schneller 1986; Taylor et al. 1993, 2006) (see Chap. 1). This concept plays a pivotal role in shaping species' temporal and frequency variations influenced by individual dispersal strategies, thereby affecting potential shifts in their ranges. Theoretical frameworks have striven to map the interaction between the distribution of spatial and temporal variations in disturbances within the landscape matrix that exert selective pressures on the evolution of dispersal (Baguette et al. 2013).

In forest ecosystems, for example succession plays a pivotal role in shaping the evolution of community structure over time. This process is driven by the interplay of nutrient competition in early succession and light competition in late succession, both triggered by disturbances (Clements 1936; Odum 1966; Tilman 1985). One such trend in forest ecosystem succession is the development of species diversity as ecosystems progress through successional stages (Huston and Smith 1987). These predictable variations between early and late-successional stages offer valuable insights into the dynamics of succession. Early successional species exhibit rapid responses to climate change, swiftly tracking its shifts, whereas mid- to late-successional species are anticipated to migrate at a considerably slower pace (Meier et al. 2012). The remarkable ability of early successional species to promptly adapt to climate change is evident in their nearly instantaneous migration patterns. This discrepancy can be attributed to specific traits associated with each group. Early successional tree species often possess attributes such as large-scale and long-distance seed dispersal mechanisms that facilitate rapid colonization of new areas. Common dispersal mechanisms of these species include wind- and bird-mediated dispersal. In addition, early successional species tend to have relatively short life spans, enabling quick adaptation to changing environmental conditions through the selection of recombined genotypes (Corlett 2011). On the contrary, mid- to late-successional species typically exhibit longer life spans and slower rates of reproduction. Their seeds are fewer in number but larger in size, reflecting more specialized dispersal mechanisms (Meier et al. 2010). Mid- to late-successional species predominantly migrate into established forest habitats where they encounter more inter-specific competition. As a result, the migration of these species occurs more slowly, requiring more time to establish in new habitats (Meier et al. 2012).

Migratory Connectivity in Long-Distance Migratory Animal Species

The presence, quality, and connectivity of suitable habitats for populations are vital for both long-distance migratory animal species and non-migratory resident species. Climate change, degradation, fragmentation, and loss of habitat pose substantial threats to biodiversity, impacting species regardless of their migration or dispersal strategies (Liu et al. 2015). For long-distance migratory species traversing continents and covering tens of thousands of kilometers annually, migratory connectivity emerges as a critical factor influencing their survival.

Migratory connectivity refers to the degree of connection between different regions within an animal's migratory range, based on the movement patterns of individual animals (Cohen et al. 2018; Gao et al. 2020). According to this concept, migratory species depend on crucial habitats during their seasonal journeys, including breeding and foraging locations. Ideally, continuous measurement of individual's locations within populations would enable the identification of crucial habitats, unveiling the strength of migratory connectivity throughout a year and facilitating the evaluation of consequences related to disturbances or disruptions. It will be essential to differentiate between sex and age groups to identify specific places and times when particular demographic groups are most vulnerable to the disruption of this connectivity (Briedis and Bauer 2018; Trierweiler et al. 2014).

In the context of long-distance migratory species, it becomes evident that ecological connectivity is a matter of global significance and should be a top priority on the biodiversity conservation agenda. Climate change is already causing significant shifts in population movements regardless of species' capacity to travel long or short distances. In particular, climate-induced changes necessitate international collaboration to facilitate, monitor, and safeguard range shifts within a connected landscape.

Brief Outlook for Land Managers

In our human-dominated world with prevalent habitat fragmentation, it is crucial to establish networks of interconnected habitats and populations to conserve and restore biodiversity. Understanding the dispersal capabilities of different species is important for establishing effective conservation measures such as the establishment of stepping stones or corridors designed to connect suitable habitats. While the complex interplay between dispersal strategies, species-specific characteristics, and the various components of the landscape at different forest successional stages can be daunting for conservation and forest managers, there is good news: The rapid development of user-friendly modelling and biodiversity assessment approaches as well as the growing availability of open biodiversity data (see Chaps. 19, 21, and 33), including data collected through citizen science platforms (see Chap. 18), is facilitating modelling and aiding decision-making processes.

These tools assist in identifying areas that improve migration and aid in range shifts, prioritizing management considerations, and predicting future scenarios for landscape connectivity.

References

Andersen LW, Harms V, Caniglia R, Czarnomska SD, Fabbri E, Jędrzejewska B, Kluth G, Madsen AB, Nowak C, Pertoldi C, Randi E, Reinhardt I, Stronen AV (2015) Long-distance dispersal of a wolf, Canis lupus, in northwestern Europe. Mamm Res 60(2):163–168. https://doi.org/10.1007/s13364-015-0220-6

Baguette M, Blanchet S, Legrand D, Stevens VM, Turlure C (2013) Individual dispersal, landscape connectivity and ecological networks. Biol Rev 88(2):310–326. https://doi.org/10.1111/brv.12000

Barber QE, Nielsen SE, Hamann A (2016) Assessing the vulnerability of rare plants using climate change velocity, habitat connectivity, and dispersal ability: a case study in Alberta, Canada. Reg Environ Chang 16:1433–1441. https://doi.org/10.1007/s10113-015-0870-6

Brewer S, Giesecke T, Davis BAS, Finsinger W, Wolters S, Binney H, de Beaulieu J-L, Fyfe R, Gil-Romera G, Kühl N (2017) Late-glacial and Holocene European pollen data. J Maps 13(2):921–928. https://doi.org/10.1080/17445647.2016.1197613

Briedis M, Bauer S (2018) Migratory connectivity in the context of differential migration. Biol Lett 14(12):20180679

Cano LS, Tellería JL (2013) Migration and winter distribution of Iberian and central European black storks Ciconia nigra moving to Africa across the Strait of Gibraltar: a comparative study. J Avian Biol 44(2):189–197. https://doi.org/10.1111/j.1600-048X.2012.05824.x

Chapman BB, Brönmark C, Nilsson J, Hansson L (2011) The ecology and evolution of partial migration. Oikos 120(12):1764–1775. https://doi.org/10.1111/j.1600-0706.2011.20131.x

Clements FE (1936) Nature and structure of the climax. J Ecol 24(1):252–284. https://doi.org/10.2307/2256278

Clobert J (2012) Dispersal ecology and evolution. Oxford University Press

Clobert J, Ims RA, Rousset F (2004) Causes, mechanisms and consequences of dispersal. In: Ecology, genetics and evolution of metapopulations. Elsevier, pp 307–335

Cohen EB, Hostetler JA, Hallworth MT, Rushing CS, Sillett TS, Marra PP (2018) Quantifying the strength of migratory connectivity. Methods Ecol Evol 9(3):513–524

Corlett RT (2011) Seed dispersal in Hong Kong, China: past, present and possible futures. Integr Zool 6(2):97–109

Cudlín P, Cudlín P, Cudlín P, Tognetti R, Malis F, Alados C, Bebi P, Grunewald K, Zhiyanski M, Andonowski V, La Porta N, Bratanova-Doncheva S, Kachaunova E, Edwards-Jonášová M, Ninot J, Rigling A, Hofgaard A, Hlásny T, Skalák P, Wielgolaski F (2017) Drivers of treeline shift in different European mountains. Clim Res 73(1–2):135–150. https://doi.org/10.3354/cr01465

Dalpasso A, Seglie D, Eusebio Bergò P, Ciracì A, Compostella M, Laddaga L, Manica M, Marino G, Pandolfo I, Soldato G, Falaschi M (2023) Effects of temperature and precipitation changes on shifts in breeding phenology of an endangered toad. Sci Rep 13(1):14573. https://doi.org/10.1038/s41598-023-40568-w

Davis MB, Shaw RG (2001) Range shifts and adaptive responses to Quaternary climate change. Science 292(5517):673–679

Dingle H, Drake VA (2007) What is migration? Bioscience 57(2):113–121

Edworthy AB, Steensma KMM, Zandberg HM, Lilley PL (2012) Dispersal, home-range size, and habitat use of an endangered land snail, the Oregon forestsnail (*Allogona townsendiana*). Can J Zool 90(7):875–884. https://doi.org/10.1139/z2012-056

Ennos R (1994) Estimating the relative rates of pollen and seed migration among plant populations. Heredity 72(3):250–259

Enriquez-Urzelai U, Bernardo N, Moreno-Rueda G, Montori A, Llorente G (2019) Are amphibians tracking their climatic niches in response to climate warming? A test with Iberian amphibians. Clim Chang 154(1–2):289–301. https://doi.org/10.1007/s10584-019-02422-9

Estrada A, Morales-Castilla I, Caplat P, Early R (2016) Usefulness of species traits in predicting range shifts. Trends Ecol Evol 31(3):190–203

Fahrig L (2003) Effects of habitat fragmentation on biodiversity. Annu Rev Ecol Evol Syst 34:487–515. https://doi.org/10.1146/annurev.ecolsys.34.011802.132419

Feurdean A, Bhagwat SA, Willis KJ, Birks HJB, Lischke H, Hickler T (2013) Tree migration-rates: narrowing the gap between inferred post-glacial rates and projected rates. PLoS One*** 8(8):e71797

Gao B, Hedlund J, Reynolds DR, Zhai B, Hu G, Chapman JW (2020) The 'migratory connectivity' concept, and its applicability to insect migrants. Movem Ecol 8:1–13

García-García I, Méndez-Cea B, Martín-Gálvez D, Seco JI, Gallego FJ, Linares JC (2022) Challenges and perspectives in the epigenetics of climate change-induced forests decline. Front Plant Sci 12:797958. https://doi.org/10.3389/fpls.2021.797958

Gehrig-Fasel J, Guisan A, Zimmermann NE (2007) Tree line shifts in the Swiss Alps: climate change or land abandonment? J Veg Sci 18(4):571–582. https://doi.org/10.1111/j.1654-1103.2007.tb02571.x

Giesecke T, Brewer S (2018) Notes on the postglacial spread of abundant European tree taxa. Veg Hist Archaeobotany 27(2):337–349

Giesecke T, Wolters S, van Leeuwen JFN, van Der Knaap PWO, Leydet M, Brewer S (2019) Postglacial change of the floristic diversity gradient in Europe. Nat Commun 10(1):5422

Gil-Tena A, Brotons L, Fortin M-J, Burel F, Saura S (2013) Assessing the role of landscape connectivity in recent woodpecker range expansion in Mediterranean Europe: forest management implications. Eur J For Res 132(1):181–194. https://doi.org/10.1007/s10342-012-0666-x

Gottfried M, Pauli H, Futschik A, Akhalkatsi M, Barančok P, Benito Alonso JL, Coldea G, Dick J, Erschbamer B, Fernández Calzado MR, Kazakis G, Krajči J, Larsson P, Mallaun M, Michelsen O, Moiseev D, Moiseev P, Molau U, Merzouki A et al (2012) Continent-wide response of mountain vegetation to climate change. Nat Clim Chang 2(2):111–115. https://doi.org/10.1038/nclimate1329

Grimm A, Prieto Ramírez AM, Moulherat S, Reynaud J, Henle K (2014) Life-history trait database of European reptile species. Nat Conserv 9:45–67. https://doi.org/10.3897/natureconservation.9.8908

Hansson L (1991) Dispersal and connectivity in metapopulations. Biol J Linn Soc 42(1–2):89–103

Hao Q, Hao Q (2018) Glacial refugia and the postglacial migration of dominant tree species in Northern China. In: The LGM distribution of dominant tree genera in Northern China's forest-steppe ecotone and their postglacial migration. Springer, pp 31–56

Harsch MA, Hulme PE, McGlone MS, Duncan RP (2009) Are treelines advancing? A global meta-analysis of treeline response to climate warming. Ecol Lett 12(10):1040–1049. https://doi.org/10.1111/j.1461-0248.2009.01355.x

Hendrix R, Schmidt BR, Schaub M, Krause ET, Steinfartz S (2017) Differentiation of movement behaviour in an adaptively diverging salamander population. Mol Ecol 26(22):6400–6413. https://doi.org/10.1111/mec.14345

Higgins SI, Nathan R, Cain ML (2003) Are long-distance dispersal events in plants usually caused by nonstandard means of dispersal? Ecology 84(8):1945–1956. https://doi.org/10.1890/01-0616

Hodgson JA, Thomas CD, Dytham C, Travis JMJ, Cornell SJ (2012) The speed of range shifts in fragmented landscapes. PLoS One 7(10):e47141

Hovick TJ, Allred BW, McGranahan DA, Palmer MW, Dwayne Elmore R, Fuhlendorf SD (2016) Informing conservation by identifying range shift patterns across breeding habitats and migration strategies. Biodivers Conserv 25:345–356

Huston M, Smith T (1987) Plant succession: life history and competition. Am Nat 130(2):168–198

IPCC. (o. J.). No Title. 2020

Jenkins DG, Brescacin CR, Duxbury CV, Elliott JA, Evans JA, Grablow KR, Hillegass M, Lyon BN, Metzger GA, Olandese ML (2007) Does size matter for dispersal distance? Glob Ecol Biogeogr 16(4):415–425

Johnson ML, Gaines MS (1990) Evolution of dispersal: theoretical models and empirical tests using birds and mammals. Annu Rev Ecol Syst 21(1):449–480

Joly K, Gurarie E, Sorum MS, Kaczensky P, Cameron MD, Jakes AF, Borg BL, Nandintsetseg D, Hopcraft JGC, Buuveibaatar B (2019) Longest terrestrial migrations and movements around the world. Sci Rep 9(1):1–10

Koleček J, Adamík P, Reif J (2020) Shifts in migration phenology under climate change: temperature vs. abundance effects in birds. Clim Chang 159(2):177–194. https://doi.org/10.1007/s10584-020-02668-8

Kreft H, Jetz W (2007) Global patterns and determinants of vascular plant diversity. Proc Natl Acad Sci 104(14):5925–5930

Lenzi O, Grossenbacher K, Zumbach S, Lüscher B, Althaus S, Schmocker D, Recher H, Thoma M, Ozgul A, Schmidt BR (2023) Four decades of phenology in an alpine amphibian: trends, stasis, and climatic drivers. Peer Commun J 3:e15. https://doi.org/10.24072/pcjournal.240

Literák I, Kafka P, Vrána J, Pojer F (2017) Migration of Black Storks *Ciconia nigra* at a migratory divide: two different routes used by siblings from one nest and two different routes used by one individual. Ringing Migr 32(1):19–24. https://doi.org/10.1080/03078698.2017.1332260

Littlefield CE, Krosby M, Michalak JL, Lawler JJ (2019) Connectivity for species on the move: supporting climate-driven range shifts. Front Ecol Environ 17(5):270–278. https://doi.org/10.1002/fee.2043

Liu J, Mooney H, Hull V, Davis SJ, Gaskell J, Hertel T, Lubchenco J, Seto KC, Gleick P, Kremen C (2015) Systems integration for global sustainability. Science 347(6225):1258832

Loarie SR, Duffy PB, Hamilton H, Asner GP, Field CB, Ackerly DD (2009) The velocity of climate change. Nature 462(7276):1052–1055

Lowe WH, Allendorf FW (2010) What can genetics tell us about population connectivity? Mol Ecol 19:3038–3051. https://doi.org/10.1111/j.1365-294X.2010.04688.x

Malcolm JR, Markham A, Neilson RP, Garaci M (2002) Estimated migration rates under scenarios of global climate change. J Biogeogr 29(7):835–849

Mancinelli S, Ciucci P (2018) Beyond home: preliminary data on wolf extraterritorial forays and dispersal in Central Italy. Mamm Biol 93:51–55. https://doi.org/10.1016/j.mambio.2018.08.003

Mason DS, Baruzzi C, Lashley MA (2022) Passive directed dispersal of plants by animals. Biol Rev 97(5):1908–1929

McLachlan JS, Clark JS, Manos PS (2005) Molecular indicators of tree migration capacity under rapid climate change. Ecology 86(8):2088–2098

Meier ES, Lischke H, Schmatz DR, Zimmermann NE (2012) Climate, competition and connectivity affect future migration and ranges of European trees. Glob Ecol Biogeogr 21(2):164–178

Meier ES, Kienast F, Pearman PB, Svenning JC, Thuiller W, Araújo MB et al (2010) Biotic and abiotic variables show little redundancy in explaining tree species distributions. Ecography 33(6):1038–1048. https://doi.org/10.1111/j.1600-0587.2010.06229.x

Melles SJ, Fortin M-J, Lindsay K, Badzinski D (2011) Expanding northward: influence of climate change, forest connectivity, and population processes on a threatened species' range shift. Glob Chang Biol 17(1):17–31. https://doi.org/10.1111/j.1365-2486.2010.02214.x

Møller AP, Rubolini D, Lehikoinen E (2008) Populations of migratory bird species that did not show a phenological response to climate change are declining. Proc Natl Acad Sci 105(42):16195–16200. https://doi.org/10.1073/pnas.0803825105

Newton I (2012) Obligate and facultative migration in birds: ecological aspects. J Ornithol 153(S1):171–180. https://doi.org/10.1007/s10336-011-0765-3

Nilsson ALK, Alerstam T, Nilsson J-Å (2008) Diffuse, short and slow migration among Blue Tits. J Ornithol 149(3):365–373. https://doi.org/10.1007/s10336-008-0280-3

Normand S, Ricklefs RE, Skov F, Bladt J, Tackenberg O, Svenning J-C (2011) Postglacial migration supplements climate in determining plant species ranges in Europe. Proc R Soc B Biol Sci 278(1725):3644–3653

Odum E (1966) The strategy of ecosystem development. Science 164(262.270)
Pearson RG (2006) Climate change and the migration capacity of species. Trends Ecol Evol 21(3):111–113
Picard M, Baltzinger C (2012) Hitch-hiking in the wild: should seeds rely on ungulates? Plant Ecol Evol 145(1):24–30. https://doi.org/10.5091/plecevo.2012.689
Pigliucci M, Murren CJ, Schlichting CD (2006) Phenotypic plasticity and evolution by genetic assimilation. J Exp Biol 209(12):2362–2367. https://doi.org/10.1242/jeb.02070
Reppert SM, de Roode JC (2018) Demystifying monarch butterfly migration. Curr Biol 28(17):R1009–R1022. https://doi.org/10.1016/j.cub.2018.02.067
Rink M, Sinsch U (2007) Radio-telemetric monitoring of dispersing stag beetles: implications for conservation. J Zool 272(3):235–243. https://doi.org/10.1111/j.1469-7998.2006.00282.x
Seebacher F, Post E (2015) Climate change impacts on animal migration. Clim Chang Responses 2(1):5. https://doi.org/10.1186/s40665-015-0013-9
Sinsch U, Oromi N, Miaud C, Denton J, Sanuy D (2012) Connectivity of local amphibian populations: modelling the migratory capacity of radio-tracked natterjack toads. Anim Conserv 15(4):388–396. https://doi.org/10.1111/j.1469-1795.2012.00527.x
Suter B, Schneller JJ (1986) Autokologische Untersuchungen an der Mauerraute (Asplenium ruta-muraria L.). Farnblatter 14:1–14
Taylor BW (1954) An example of long distance dispersal. Ecology 35(4):569–572. https://doi.org/10.2307/1931046
Taylor PD, Fahrig L, Henein K, Merriam G (1993) Connectivity is a vital element of landscape structure. Oikos:571–573
Taylor PD, Fahrig L, With KA (2006) Landscape connectivity: a return to the basics. Conservation Biology Series, Cambridge, 14, 29
Teitelbaum CS, Fagan WF, Fleming CH, Dressler G, Calabrese JM, Leimgruber P, Mueller T (2015) How far to go? Determinants of migration distance in land mammals. Ecol Lett 18(6):545–552. https://doi.org/10.1111/ele.12435
Teitelbaum CS, Mueller T (2019) Beyond migration: causes and consequences of nomadic animal movements. Trends Ecol Evol 34(6):569–581. https://doi.org/10.1016/j.tree.2019.02.005
Tilman D (1985) The resource-ratio hypothesis of plant succession. Am Nat 125(6):827–852
Trant A, Higgs E, Starzomski BM (2020) A century of high elevation ecosystem change in the Canadian Rocky Mountains. Sci Rep 10(1):9698. https://doi.org/10.1038/s41598-020-66277-2
Trierweiler C, Klaassen RH, Drent RH, Exo KM, Komdeur J, Bairlein F, Koks BJ (2014) Migratory connectivity and population-specific migration routes in a long-distance migratory bird. Proc R Soc B Biol Sci 281(1778):20132897
Valière N, Fumagalli L, Gielly L, Miquel C, Lequette B, Poulle M, Weber J, Arlettaz R, Taberlet P (2003) Long-distance wolf recolonization of France and Switzerland inferred from non-invasive genetic sampling over a period of 10 years. Anim Conserv 6(1):83–92. https://doi.org/10.1017/S1367943003003111
Van Buskirk J, Mulvihill RS, Leberman RC (2009) Variable shifts in spring and autumn migration phenology in North American songbirds associated with climate change. Glob Chang Biol 15(3):760–771. https://doi.org/10.1111/j.1365-2486.2008.01751.x
Vitasse Y, Ursenbacher S, Klein G, Bohnenstengel T, Chittaro Y, Delestrade A, Monnerat C, Rebetez M, Rixen C, Strebel N, Schmidt BR, Wipf S, Wohlgemuth T, Yoccoz NG, Lenoir J (2021) Phenological and elevational shifts of plants, animals and fungi under climate change in the E uropean A lps. Biol Rev 96(5):1816–1835. https://doi.org/10.1111/brv.12727
While GM, Uller T (2014) Quo vadis amphibia? Global warming and breeding phenology in frogs, toads and salamanders. Ecography 37(10):921–929. https://doi.org/10.1111/ecog.00521
Zhu K, Woodall CW, Clark JS (2012) Failure to migrate: lack of tree range expansion in response to climate change. Glob Chang Biol 18(3):1042–1052
Zuber D, Widmer A (2009) Phylogeography and host race differentiation in the European mistletoe (*Viscum album* L.). Mol Ecol 18(9):1946–1962. https://doi.org/10.1111/j.1365-294X.2009.04168.x

Open Access This chapter is licensed under the terms of the Creative Commons Attribution 4.0 International License (http://creativecommons.org/licenses/by/4.0/), which permits use, sharing, adaptation, distribution and reproduction in any medium or format, as long as you give appropriate credit to the original author(s) and the source, provide a link to the Creative Commons license and indicate if changes were made.

The images or other third party material in this chapter are included in the chapter's Creative Commons license, unless indicated otherwise in a credit line to the material. If material is not included in the chapter's Creative Commons license and your intended use is not permitted by statutory regulation or exceeds the permitted use, you will need to obtain permission directly from the copyright holder.

Do Saproxylic Species Need Habitats, Connectivity, or Connected Habitats?

3

Thibault Lachat, Janine Oettel, and Felix Meyer

Rosalia alpina on lying beech deadwood in an Austrian Natural Forest Reserve (Photo: BFW/Janine Oettel)

T. Lachat (✉)
Division of Forest Sciences, Bern University of Applied Sciences, School of Agricultural, Forest and Food Sciences, Zollikofen, Switzerland

Swiss Federal Institute for Forest, Snow and Landscape Research WSL, Birmensdorf, Switzerland
e-mail: thibault.lachat@bfh.ch

J. Oettel · F. Meyer
Department of Forest Biodiversity & Nature Conservation, Austrian Research Centre for Forests, Vienna, Austria
e-mail: janine.oettel@bfw.gv.at; felix.meyer@bfw.gv.at

© The Author(s) 2025
K. Lapin et al. (eds.), *Ecological Connectivity of Forest Ecosystems*,
https://doi.org/10.1007/978-3-031-82206-3_3

Abstract

The importance of saproxylic species within forest ecosystems cannot be overstated, as they span a wide range of taxa contributing to the recycling of dying and dead woody material. Originally defined as invertebrates reliant on decaying wood, wood-inhabiting fungi, or other saproxylics, the group has been expanded to include species involved in or dependent on moribund trees and wood decay processes. Since centuries, their habitat has faced loss and fragmentation from intensive forest management practices and land use changes, underscoring the urgency of conservation efforts. While habitat connectivity is crucial for species dispersal and colonization, evidence supporting its significance for saproxylic species conservation remains unclear. Dispersal abilities vary considerably across taxa, highlighting the importance of understanding these differences for effective forest management aiming at saproxylics conservation. Specialized species such as fungus-dwelling beetle *Bolitophagus reticulatus* demonstrate limited dispersal but robust recolonization capabilities facilitated by the availability of habitat, in this case, *Fomes fomentarius*. Similarly, saproxylic fungi with a broad dispersal ability such as *Fomitopsis rosea* rely on habitat amount for successful colonization. Efforts to increase the amount of deadwood at the landscape scale thus benefit species (re)colonization efforts. Prioritizing the preservation of large populations and distributing habitat patches are key strategies for supporting saproxylic biodiversity in forest ecosystems. Aggregating patches around dispersal sources can attract species of conservation concern, although identifying these sources remains challenging. Conversely, evenly distributed habitat patches throughout the forest landscape promote higher species diversity. A balanced approach combining both aggregation and distribution of habitats seems therefore essential for effective conservation efforts. However, scientific evidence tends to prioritize habitat quantity over habitat connectivity for the conservation of saproxylic species.

Keywords

Dispersal · Colonization · Habitat amount · Deadwood · Biodiversity conservation · Forest management

Saproxylic Species and Their Role in Forests

Saproxylics are a functional group of species that are perhaps more than any other reliant on trees and forest ecosystems, as they inhabit and thrive in dead and decaying trees. As a result, the significance of deadwood for biodiversity conservation cannot be overstated, as it provides essential habitats for thousands of species. Within forest ecosystems, the decay of wood is one of the major ecological processes alongside primary production. This process not only facilitates the recycling of deadwood but also ensures the long-term availability of essential nutrients.

Originally, saproxylic organisms have been defined as invertebrates that, at least during some part of their life cycle, rely on the deceased or decaying wood of dying or dead trees—whether standing or fallen—or on wood-inhabiting fungi, or on the presence of other saproxylic species (Speight 1989). This definition underwent revision by Alexander (2008), who shifted the focus toward the ecological functions of saproxylic organisms and defined them as species intricately involved in or dependent on the process of fungal decay of wood, or on the byproducts of decay. Moreover, saproxylics are associated with both living and dead woody material, including not only wood but also bark and sap at any stages of decay.

Saproxylic organisms represent a considerable share of forest biodiversity, encompassing a large range of species groups ranging from arthropods (such as insects, especially beetles) to birds (such as woodpeckers) and fungi (such as basidiomycetes). The estimated global number of saproxylic species ranges from 0.4 to 1 million. In well-explored Northern Europe, they represent up to 25% of all forest species, predominantly comprising mainly fungi and invertebrates (Stokland et al. 2012). Among these, insects and fungi are the two largest taxonomic groups, contributing between 30% and over 50% of the total saproxylic diversity in forests. This represents an important reason to address biodiversity in deadwood. Another reason is the alarming threat posed to this diversity by habitat loss in managed forests because of harvesting activities, as well as by loss and fragmentation of forested areas at the landscape level for centuries. Consequently, saproxylic species are of considerable conservation concern, as they exhibit high sensitivity to forest management practices that can alter their habitats by reducing the quantity and quality of deadwood habitats (Gossner et al. 2013a; Grove 2002; Müller et al. 2015; Seibold et al. 2017).

Gossner et al. (2013a) demonstrated a positive effect of deadwood enrichment initiatives in managed forests, where the presence and variety of saproxylic beetles are typically limited. The authors determined an immediate increase in species richness and a shift in guild composition upon increasing the deadwood amount, with the effect being even more pronounced in the tree canopy compared to the forest floor. Müller et al. (2015) showed that regional tree species composition influences the habitat preferences of early colonizing saproxylic beetle communities, showcasing local variations in their choice of host trees. With an increase in deadwood amount in European beech forests, Gossner et al. (2013b) observed a noticeable shift in assemblage composition, with larger species and those favoring deadwood of greater diameter and advanced decay stages becoming dominant. Plots with a mean deadwood amount ranging from 20 to 60 m^3ha^{-1} accommodated most species. Consequently, the authors recommend increasing the deadwood amount to more than 20 m^3ha^{-1}, refraining from the removal of large-diameter deadwood (>50 cm), facilitating the development of more deadwood in advanced decay stages, and establishing strict forest reserves characterized by exceptionally high deadwood amounts. These findings support the ecological thresholds reviewed by Müller and Bütler (2010). The authors highlighted deadwood amounts necessary for the conservation of saproxylic species with values ranging between 10 and 150 m^3ha^{-1} and peak values between 20 and 50 m³ha⁻¹ for the survival of most species, depending on the forest type.

Most saproxylic species are associated with specific habitat quality in terms of, e.g., decay stage (fresh to soft powdery dead wood), dimension (small branches to large logs), position (lying or standing deadwood), microclimate, and tree species or genus (Lachat et al. 2013; Lettenmaier et al. 2022; Vogel et al. 2020). As deadwood habitats are steadily evolving, saproxylic organisms face the challenge of compensating for local extinctions due to the loss of their habitat through wood decay. To maintain populations, they must colonize new deadwood structures that are adapted to their ecological requirements both local and across landscapes (Jonsson and Siitonen 2012). Their dispersal strategies are affected by numerous driving forces, including the spatial and temporal variability of habitat within the landscape, feeding strategy, resource competition, and avoidance of inbreeding (Feldhaar and Schauer 2018). The significance of each factor in shaping dispersal strategy varies among species, contingent upon their unique life histories, and interactions with the environment, such as longevity of deadwood habitat. This habitat turnover has to be considered in conservation measures for this functional group as well as in integrative and segregative approaches (Bollmann and Braunisch 2013; Doerfler et al. 2018).

Habitat Amount vs. Habitat Connectivity

In the course of reduction of natural habitats through human activities, local habitats get more and more fragmented. Such a fragmented landscape affects local populations in two ways. First, the distance between patches increases (pure isolation effect), second the total amount of habitat in a landscape decreases and thereby the population size of a species in the landscape (Fahrig 2013). While in most landscapes nowadays, both habitat amount and spatial connectivity of resource patches are correlated, it is important to disentangle both mechanisms, particularly for restoration.

In response to the habitat deficit for saproxylic species resulting from forest management practices, the conservation of this functional group has become a priority in landscapes dominated by managed forests. In this context, different measures to improve habitat availability and amount at the landscape scale can be implemented, including setting aside natural forest reserves, establishment of (smaller) stepping stones of old-growth forest patches, or the retention of habitat trees and deadwood within managed forests (Bütler and Lachat 2009; Komonen and Müller 2018). The latter involves preserving trees in managed forests until they decay entirely. The underlying concept of these conservation measures is to create a network of habitats capable of supporting saproxylic species that would otherwise struggle to persist in managed forests due to insufficient deadwood resources (see Fig. 3.1).

This concept draws on the theory of island biogeography proposed by MacArthur and Wilson (1963, 1967), which posits that smaller and more isolated habitats tend to harbor fewer species compared to larger and less isolated ones. This theory has prompted conservationists to prioritize the establishment of singular large areas as

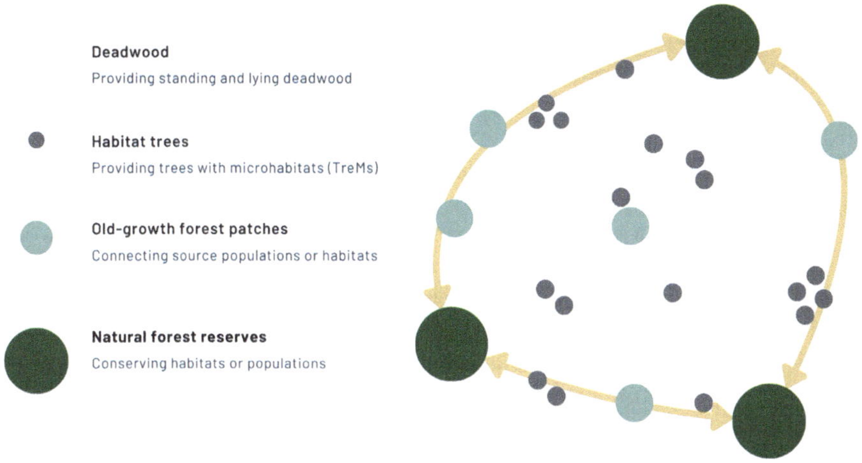

Fig. 3.1 Concept of habitat connectivity for the conservation of saproxylic species at the landscape level based on four different conservation measures: Natural forest reserves, old-growth forest patches, habitat trees, and deadwood, the last two distributed in a matrix of managed forests (adapted from Lachat and Bütler 2009)

the optimal conservation strategy for supporting biodiversity rather than implementing multiple smaller areas (Diamond 1975). In 1980, the IUCN developed a global conservation strategy emphasizing the importance of large contiguous areas in real-world conservation efforts (World Conservation Strategy).

Testing the validity of the "single large over several small" (SLOSS) theory, Simberloff and Abele (1982) conducted a comparative analysis of empirical data on species richness. They found that "not a single case" demonstrated superior species richness in single large habitats compared to several smaller ones covering the same total area. Conversely, most studies demonstrate that the same amount of habitat in a landscape with multiple small areas harbored more species than a single large one. Similar studies focusing on various species, such as Fahrig (2017), have echoed these findings, consistently indicating that species richness is higher in multiple small areas than in a single large one. The mechanisms behind this pattern range from increased habitat heterogeneity provided by more patches to ecological drift (Fahrig et al. 2022; Hovestadt et al. 2024).

Although scientific evidence consistently supports the superiority of several small protected habitats over a single large one, discussions among ecologists persist in part due to sampling biases in many SLOSS studies, where sampling intensity is not proportional to habitat size. Fahrig's (2020) research on species extinction challenges the assumption that larger habitats inherently support greater species richness because species depending on larger habitats for survival may become extinct in smaller habitats over time (process of selective extinction). However, even in studies with unbiased sampling efforts, several small areas exhibited higher species richness. This trend persisted when examining extinction rates in small areas, even within low-quality or hostile matrices, and held true for specialist species as well. These findings suggest that species richness may be driven more by individual species' minimum habitat requirements than by minimum area size requirements, as emphasized by the Habitat Amount Hypothesis (Fahrig 2013). This hypothesis advocates for considering the overall amount of available habitat in an area rather than the size of the area to estimate species richness. For example, a large managed forest may provide a lower amount of deadwood than a smaller area of primeval forest.

The fragmentation of forest landscapes inevitably leads to reduced deadwood habitat availability and increased distances between dead trees, which can be termed isolated or connected depending on perspective (Lachat and Müller 2018). However, for forest management and conservation, the spatial distribution of restoration efforts such as retention of habitat trees or establishment of old-growth-forest patches or natural forest reserves significantly influence timber production and ecosystem services depending on their location. Conservation measures often prioritize either the creation of new suitable habitats within a forest area or the enhancement of connectivity to reduce habitat isolation. These differing concepts entail significant planning and financial implications, which can vary considerably depending on the initial conditions of the forest area. A comprehensive understanding of these factors is therefore crucial for successful restoration efforts.

Spatial Arrangement of Conservation Measures

Although concepts for habitat connectivity are widely accepted and integrated into many regional or national conservation strategies (see Chap. 22), there exists limited scientific evidence supporting the need for connectivity regarding the conservation of saproxylic species. Dispersal abilities vary greatly across taxa, with only a few species' abilities well understood. Nevertheless, understanding these abilities is crucial and can greatly improve forest management and conservation efforts, particularly in terms of the required spatial and temporal distribution of suitable habitats to enhance species persistence (Oettel et al. 2023).

The distribution of habitat within a landscape, whether aggregated or evenly dispersed, can influence the ability to maintain biodiversity across the landscape. While direct comparisons between areas with aggregated and distributed habitats are lacking, studies focusing on habitat patches within different landscape contexts have shed light on this issue. These studies assess whether aggregated landscapes host greater species richness compared to dispersed ones. Their findings indicate that distributing conservation efforts evenly across smaller patches covers greater landscape heterogeneity, potentially resulting in higher species richness (Fahrig et al. 2022; Haeler et al. 2024; Müller and Goßner 2010; Rubene et al. 2015).

Examples of Saproxylic Beetles

Saproxylic beetles are excellent ecological indicators for the quantity and quality of their habitat (e.g. Lachat et al. 2013) and therefore well adapted for testing hypotheses on habitat amount and connectivity in forest ecosystems. Each piece of deadwood—including logs, snags, and other saproxylic habitats like tree hollows or dead branches—undergoes constant evolution. Throughout the decay processes, the biotic and abiotic conditions of deadwood habitats constantly alter, thus influencing species assemblages. Schmidl and Bussler (2004) have categorized them into different guilds, spanning from initial colonizers of fresh deadwood to inhabitants of aged, decayed wood, including species developing in the mold of rot-holes. Saproxylic species have evolved as a group capable of continuously seeking out new habitats adapted to their specific requirements, as individuals must relocate when deadwood decays further. This raises the fundamental question of how far saproxylic beetles are able to disperse once they have to relocate following changes in their habitats that render them unsuitable for their development.

The empirical evidence regarding the dispersal ability of saproxylic species and the importance of connectivity to them is mainly indirect. It relies on comparisons of species presence/absence or richness in forest stands with varying degrees of spatiotemporal isolation or in landscapes exhibiting different levels of fragmentation (Sverdrup-Thygeson et al. 2014). Although such occupancy studies can offer valuable insights into the importance of dispersal and connectivity, few have been able to differentiate between habitat connectivity and habitat amount (or between dispersal limitation and habitat limitation) at scales relevant to management (Fahrig 2013). An experiment by Seibold et al. (2017) with different amounts of local deadwood habitat in landscapes with different amounts of dead trees, found no evidence for isolation effects but for habitat amount effects on saproxylic beetle diversity. Taking together the current evidence on the importance of isolation and resource amount, the conservation efforts for saproxylic beetles should prioritize increasing the amount of deadwood wherever possible.

Peltis grossa, a specialist beetle of primeval forests, breeds in standing stems with brown-rotted wood of different coniferous tree species (Palm 1951; Saalas 1917). This species gets locally extinct because of the lack of suitable breeding substrate in managed forests (Weslien et al. 2011). Initially considered regionally extinct in Bavaria National Park, its return was triggered by forest dieback caused by windstorms and bark beetle invasion (Busse et al. 2022). In Sweden, initial findings on *P. grossa* reveal an increase in colonization rate and a decrease in extinction risk with higher connectivity (100–200 m) (Djupström et al. 2012, 2024). However, this association was only evident when using connectivity-based information on species abundance. Structural measures of connectivity, such as the number of snags, failed to show any significant relationship. Despite generally optimal habitat conditions in Bavaria National Park, including a substantial increase in deadwood until the 1990s, *P. grossa* was absent. This may be attributable to irregularly distributed amounts of deadwood, with a less pronounced increase in deadwood in the refuge area of a smaller source population in the south. Bark beetle invasion in subsequent years led to a population explosion of *P. grossa*, resulting in rare long-distance dispersal events and eventual colonization of the national park. By reconstructing the recolonization in Bavaria National Park, Busse et al. (2022) determined dispersal distances of up to 10–40 km, above which colonization rates significantly decreased. This indicates that the source population in the south before the bark beetle invasion was insufficient to fully recolonize the new habitats within the national park.

The rare hermit beetle *Osmoderma* spp. is associated with hollow trees, where its larvae develop in mold. For this rare species, different methods yielded similar dispersal distances (see Fig. 3.2), even though dispersal rates and distances seemed smaller for Swedish populations compared to those in more southern European regions. Populations of *O. barnabita* exhibited positive kinship up to 10 km, indicating a limit to their dispersal at this scale. The estimated average dispersal distance is much larger than the results for *Osmoderma* spp. using other methods (Oleksa et al. 2013). A study on the colonization-extinction dynamics of *O. eremita* over 25 years (Lindman et al. 2020) showed that colonization rates increased with connectivity at a 60-m scale and with tree characteristics indicative of early successional stages, leading to higher occurrence frequencies per tree. Conversely, extinction rates increased with larger tree diameter, indicating late succession stages. Most *O. eremita* individuals stay in the same hollow tree throughout their lifetime, with dispersal typically occurring between trees at distances of less than 250 m from one another (Dodelin et al. 2017). Although long-distance dispersals are rare, this does not mean that the species cannot fly further. These findings underscore the importance of connectivity, as *O. eremita* is more likely to colonize or recolonize habitats within about 200 m. However, these initial observations regarding its flight capabilities have been revised with subsequent studies, indicating that *O. eremita* is indeed capable of flights spanning several kilometers (Fig. 3.1).

Tenebrio opacus is known to inhabit hollow oak trees in pastures (Ranius and Fahrig 2006). A notable threshold regarding the frequency of presence per tree emerges in small areas with fewer than 10 hollow trees, where less than 5%

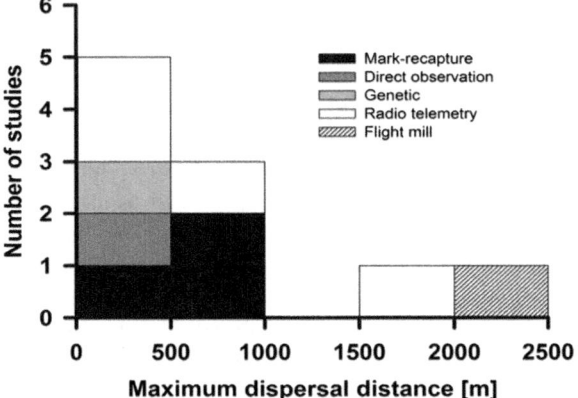

Fig. 3.2 Documented maximum flight distances in studies of the hermit beetle *Osmoderma eremita* and *O. barnabita* with different methods (*n* = 10 studies). Direct observation refers to dispersing individuals followed by foot. In the genetic study, the distance is average, not maximum. (Source: Komonen and Müller 2018)

occupancy is observed. However, this threshold significantly increases to over 40% occupancy per tree in areas with 11 or more hollow trees. This phenomenon is closely linked to the persistence of habitat over time. Highly specialized species like *T. opacus* may have developed strategies to thrive in long-lasting habitats such as hollow trees over generations. However, the spatial scale for measuring habitat amount involved oak trees located less than 250 m apart, indicating a sensitivity to habitat fragmentation (Ranius 2002).

Examples of Saproxylic Fungi

Wood-inhabiting fungi produce billions of minuscule spores daily with deposition rates ranging from 10 to 1000 spores per m^2 per day (Edman et al. 2004a). Notably, some old-growth forest species, like Fomitopsis rosea, can deposit even higher amounts, exceeding 5000 spores per m2 daily. F. rosea also exhibits a considerable dispersal ability. Even in landscapes with low proportions of old-growth forests, this species has been found to have spores present in sufficient quantities, suggesting potential for long-distance dispersal (Edman et al. 2004b). Despite this figure appears substantial, the precise threshold required for successful colonization is unknown (Edman et al. 2004a). Moreover, colonization of deadwood appeared to rely on neighboring occurrences of the species (Edman et al. 2004b; Jönsson et al. 2008).

Edman et al. (2004b) discovered that sites rich in deadwood generally harbor greater species richness, with several species being abundant due to high spore deposition from the local species pool. For instance, species like *Asterodon ferrugi-nosus, Phellinus ferrugineofuscus, P. viticola,* and *Phlebia centrifuga* exhibited a preference for colonizing deadwood near previously occupied pieces. Jönsson et al. (2008) demonstrated the significant influence of both local deadwood characteristics within patches and connectivity between patches in an old-growth boreal Norway spruce forest on fungi species dynamics. According to the authors, substrate decay and resource disappearance are the main causes of local extinctions. This is in line with findings by Norros et al. (2012), showing a higher colonization probability on logs within a distance of up to 60 m from sporocarps. However, fungi still distribute spores beyond this distance, suggesting low or no dispersal limitations extending over several kilometers.

The duration of persistence on one resource depends on the specific habitat requirements of fungal species. Early successional species like *P. ferrugineofuscus* and *Stereum sanguinolentum* display high annual extinction rates, whereas late successional species like *Phellinus nigrolimitatus* exhibit lower rates. Jonsson (2012) summarizes time windows of up to 20 years for occurrences of deadwood-inhabiting fungi, with typical durations of less than 8 years. However, persistence time analyses for wood-inhabiting fungi are still scarce.

Biotic Interactions of Fungi and Beetles

Successful colonization requires both dispersal and establishment. Establishment relies on the availability of suitable habitat (substrate). The suitability of a substrate is determined by its characteristics, biotic competitive interactions, and abiotic conditions (Jonsson 2012). Characteristics encompass aspects like tree species, decay stage, genesis history, moisture content, chemical composition of deadwood, and temperature. Furthermore, the establishment of some species of fungi relies on biotic interactions. Indeed, some fungi species utilize specific vectors for establishment; for instance, *Amylostereum areolatum* depends on *Sirex* wood wasps for the transfer to suitable substrates. Other examples include blue-stain fungi and bark beetles inducing establishment through vector interactions. Sequencing fungi from saproxylic beetles revealed a high diversity of fungi transported by beetles (Seibold et al. 2019). Experiments have shown that the presence of saproxylic insects can affect the community of fungi (Jacobsen et al. 2018; Zou et al. 2023). Nonetheless, the relative importance of insects under natural conditions remains unclear (Sverdrup-Thygeson et al. 2014).

Zytynska et al. (2018) examined the fungus-dwelling beetle *Bolitophagus reticulatus* breeding on *Fomes fomentarius* in a broadleaf forest of southern Germany. The study revealed that the population genetic structuring is largely influenced by forest management history. Nevertheless, low isolation-by-distance and limited relatedness among beetles collected from the same trees or fungus occurrences suggest robust dispersal enabling recolonization across considerable distances. It is expected that genetic structuring will continue to decrease in the future, emphasizing that the increase of deadwood amount—regardless of its spatial arrangement—can foster species recolonization. As long as relic populations persist, the increase of deadwood amount and diversity can support to increase population sizes sufficient for the dispersal and recolonization of habitats.

Komonen and Müller (2018) gathered evidence demonstrating species-specific dispersal limitations for saproxylic insects and fungi at different spatial scales ranging from local (<50 m) to continental (>500 km) scale. Adult beetles disperse primarily through active flight, while fungi rely mainly on wind dispersal over longer distances (anemochory), making them more effective dispersers. Based on this review, insects show a majority of dispersal limitations at larger scales such as regional and continental scales, whereas even rare fungal species hardly exhibited any dispersal constraints. The authors concluded that while systematic and species-specific differences exist, most saproxylic species face colonization and establishment limitations in terms of finding suitable habitats rather than true dispersal limitations at management-relevant scales (see Fig. 3.3).

Implications for the Conservation of Saproxylic Species

Saproxylic species exhibit significant variations in their dispersal abilities and habitat requirements. For instance, species associated with ephemeral habitats like fresh deadwood, such as early successional bark beetles (Scolytinae), can disperse over

Fig. 3.3 Documented maximum flight distances of saproxylic (**a**) insects and (**b**) fungi in experimental and genetic studies (*n* = 25). 'Counts' indicate the number of studies. (Source: Komonen and Müller 2018)

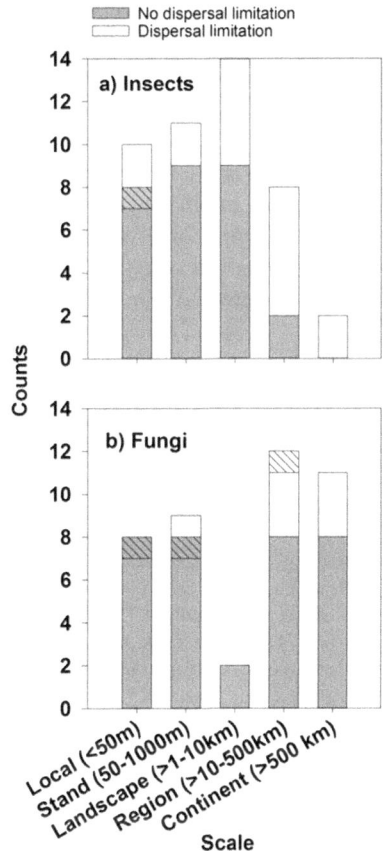

large distances and swiftly locate and colonize new habitats. Additionally, these species are characterized by high reproductive rates (r-strategies) and can produce up to four generations per year (Perny et al. 2008; Steyrer et al. 2020). In contrast, species linked with long-lived habitats, such as hollow trees with mold, are less mobile and have lower reproduction rates (k-strategies). Such long-lasting habitats can harbor populations over multiple generations spanning decades, resulting in reduced dispersal necessity and limited individual mobility. However, dispersal is often underestimated. New research frequently reveals higher dispersal rates and distances than previously anticipated. Nevertheless, dispersal limitations exist among saproxylic beetles beyond 10 km. Other insect species in deadwood as some syrphids show similar dispersal abilities as beetles up to many kilometers. For fungi, which together with beetles represent the most diverse saproxylic group, the situation is even less critical. The production of huge quantities of spores highlights the importance of suitable habitat rather than the necessity of connectivity for the conservation of this group.

Accumulated knowledge for both species groups—beetles and fungi—suggests that the availability of habitat amount in the landscape is by far more critical than habitat connectivity. This holds true for the spatial scale at which forest managers operate. It is important to note that an increase in habitat amount can also enhance connectivity by enhancing population sizes, thereby increasing the number of dispersing individuals or serving as the most important component for dispersal. Moreover, it contributes to a reduction in spatial distances between deadwood objects. As more effective and better grounded by ecological studies, current knowledge suggests a focus on habitat amount management rather than on spatial distances in deadwood management.

References

Alexander KNA (2008) Tree biology and saproxylic Coleoptera: issues of definitions and conservation language. Rev Ecol (Terre Vie) 63(Suppl. 10):9–13

Bollmann K, Braunisch V (2013) To integrate or to segregate: balancing commodity production and biodiversity conservation in European forests. In: Kraus D, Krumm (Toim) F (eds) Integrative approaches as an opportunity for conservation of forest biodiversity (pp 18–31)

Busse A, Cizek L, Čížková P, Drag L, Dvorak V, Foit J, Heurich M, Hubený P, Kašák J, Kittler F, Kozel P, Lettenmaier L, Nigl L, Procházka J, Rothacher J, Straubinger C, Thorn S, Müller J (2022) Forest dieback in a protected area triggers the return of the primeval forest specialist Peltis grossa (Coleoptera, Trogossitidae). Conserv Sci Pract 4(2):e612. https://doi.org/10.1111/csp2.612

Bütler R, Lachat T (2009) Wälder ohne Bewirtschaftung: eine Chance für die saproxylische Biodiversität | Forests without harvesting: an opportunity for the saproxylic biodiversity. Schweizerische Zeitschrift fur Forstwesen 160(2009):324–333. https://doi.org/10.3188/szf.2009.0324

Diamond JM (1975) The island dilemma: lessons of modern biogeographic studies for the design of natural reserves. Biol Conserv 7(2):129–146. https://doi.org/10.1016/0006-3207(75)90052-X

Djupström LB, Weslien J, Hoopen JT, Schroeder LM (2012) Restoration of habitats for a threatened saproxylic beetle species in a boreal landscape by retaining dead wood on clear-cuts. Biol Conserv 155:44–49. https://doi.org/10.1890/03-5103

Djupström LB, Johansson V, Lindman L, Schroeder M, Weslien J, Ranius T (2024) Density of dispersal sources affects to what extent restored habitat is used: a case study on a red-listed wood-dependent beetle. For Ecol Manag 555:121716. https://doi.org/10.1016/j.foreco.2024.121716

Dodelin B, Gaudet S, Fantino G (2017) Spatial analysis of the habitat and distribution of Osmoderma eremita (Scop.) in trees outside of woodlands. Nat Conserv 19:149–170. https://doi.org/10.3897/natureconservation.19.12417

Doerfler I, Gossner MM, Müller J, Seibold S, Weisser WW (2018) Deadwood enrichment combining integrative and segregative conservation elements enhances biodiversity of multiple taxa in managed forests. Biol Conserv 228(November 2017):70–78. https://doi.org/10.1016/j.biocon.2018.10.013

Edman M, Gustafsson M, Stenlid J, Jonsson BG, Ericson L (2004a) Spore deposition of wood-decaying fungi: Importance of landscape composition. Ecography 27(1):103–111. https://doi.org/10.1111/j.0906-7590.2004.03671.x

Edman M, Kruys N, Jonsson BG (2004b) Local dispersal sources strongly affect colonization patterns of wood-decaying fungi on spruce logs. Ecol Appl 14(3):893–901. https://doi.org/10.1890/03-5103

Fahrig L (2013) Rethinking patch size and isolation effects: the habitat amount hypothesis. J Biogeogr 40(9):1649–1663. https://doi.org/10.1111/jbi.12130

Fahrig L (2017) Ecological responses to habitat fragmentation per Se. Annu Rev Ecol Evol Syst 48:1–23. https://doi.org/10.1146/annurev-ecolsys-110316-022612

Fahrig L (2020) Why do several small patches hold more species than few large patches? Glob Ecol Biogeogr 29(4):615–628. https://doi.org/10.1111/geb.13059

Fahrig L, Watling JI, Arnillas CA, Arroyo-Rodríguez V, Jörger-Hickfang T, Müller J, Pereira HM, Riva F, Rösch V, Seibold S, Tscharntke T, May F (2022) Resolving the SLOSS dilemma for biodiversity conservation: a research agenda. Biol Rev 97(1):99–114. https://doi.org/10.1111/brv.12792

Feldhaar H, Schauer B (2018) Dispersal of saproxylic insects. In: Toim MDU (ed) Saproxylic insects. Springer, pp 515–546

Gossner MM, Floren A, Weisser WW, Linsenmair KE (2013a) Effect of dead wood enrichment in the canopy and on the forest floor on beetle guild composition. For Ecol Manag 302:404–413. https://doi.org/10.1016/j.foreco.2013.03.039

Gossner MM, Lachat T, Brunet J, Isacsson G, Bouget C, Brustel H, Brandl R, Weisser WW, Müller J (2013b) Current near-to-nature forest management effects on functional trait composition of saproxylic beetles in beech forests. Conserv Biol 27(3):605–614. https://doi.org/10.1111/cobi.12023

Grove SJ (2002) Saproxylic insect ecology and the sustainable management of forests. Annu Rev Ecol Syst 33:1–23. https://doi.org/10.1146/annurev.ecolsys.33.010802.150507

Haeler E, Stillhard J, Hindenlang Clerc K, Pellissier L, Lachat T (2024) Dead wood distributed in different-sized habitat patches enhances diversity of saproxylic beetles in a landscape experiment. J Appl Ecol 61(2):316–327. https://doi.org/10.1111/1365-2664.14554

Hovestadt T, Poethke H-J, Müller J, Mitesser O (2024) Species diversity and habitat fragmentation per se: the influence of local extinctions and species clustering. Am Nat 203:655–667. https://doi.org/10.1086/729620

IUCN (1980) World conservation strategy (Number September)

Jacobsen RM, Sverdrup-Thygeson A, Kauserud H, Mundra S, Birkemoe T (2018) Exclusion of invertebrates influences saprotrophic fungal community and wood decay rate in an experimental field study. Funct Ecol 32(11):2571–2582. https://doi.org/10.1111/1365-2435.13196

Jonsson BG (2012) Population dynamics and evolutionary strategies. In: Stokland JN, Siitonen J, Toim BGJ (eds) Biodiversity in deadwood. Cambridge University Press, pp 338–355

Jonsson BG, Siitonen J (2012) Deadwood and sustainable forest management. In: Biodiversity in Deadwood (pp 302–337)

Jönsson MT, Edman M, Jonsson BG (2008) Colonization and extinction patterns of wood-decaying fungi in a boreal old-growth Picea abies forest. J Ecol 96(5):1065–1075. https://doi.org/10.1111/j.1365-2745.2008.01411.x

Komonen A, Müller J (2018) Dispersal ecology of deadwood organisms and connectivity conservation. Conserv Biol 32(3):535–545. https://doi.org/10.1111/cobi.13087

Lachat T, Bütler R (2009) Identifying conservation and restoration priorities for saproxylic and old-growth forest species: a case study in Switzerland. Environ Manag 44(1):105–118. https://doi.org/10.1007/s00267-009-9281-0

Lachat T, Müller J (2018) Importance of primary forests for the conservation of saproxylic insects (pp 581–605). https://doi.org/10.1007/978-3-319-75937-1_17

Lachat T, Bouget C, Bütler R, Müller J (2013) Totholz: quantitative und qualitative Voraussetzungen für die Erhaltung der biologischen Vielfalt von Xylobionten. In: Kraus D, Krumm (Toim) F (eds), Integrative Ansätze als Chance für die Erhaltung der Artenvielfalt in Wäldern (pp 96–106)

Lettenmaier L, Seibold S, Bässler C, Brandl R, Gruppe A, Müller J, Hagge J (2022) Beetle diversity is higher in sunny forests due to higher microclimatic heterogeneity in deadwood. Oecologia 198(3):825–834. https://doi.org/10.1007/s00442-022-05141-8

Lindman L, Larsson MC, Mellbrand K, Svensson GP, Hedin J, Tranberg O, Ranius T (2020) Metapopulation dynamics over 25 years of a beetle, Osmoderma eremita, inhabiting hollow oaks. Oecologia 194(4):771–780. https://doi.org/10.1007/s00442-020-04794-7

MacArthur RH, Wilson EO (1963) An equilibrium theory of insular zoogeography. Evolution 17(4):373. https://doi.org/10.2307/2407089

MacArthur RH, Wilson EO (1967) The theory of island biogeography. Princeton University Press. http://www.jstor.org/stable/j.ctt19cc1t2

Müller J, Bütler R (2010) A review of habitat thresholds for dead wood: a baseline for management recommendations in European forests. Eur J For Res 129(6):981–992. https://doi.org/10.1007/s10342-010-0400-5

Müller J, Goßner MM (2010) Three-dimensional partitioning of diversity informs state-wide strategies for the conservation of saproxylic beetles. Biol Conserv 143(3):625–633. https://doi.org/10.1016/j.biocon.2009.11.027

Müller J, Wende B, Strobl C, Eugster M, Gallenberger I, Floren A, Steffan-Dewenter I, Linsenmair KE, Weisser WW, Gossner MM (2015) Forest management and regional tree composition drive the host preference of saproxylic beetle communities. J Appl Ecol 52(3):753–762. https://doi.org/10.1111/1365-2664.12421

Norros V, Penttilä R, Suominen M, Ovaskainen O (2012) Dispersal may limit the occurrence of specialist wood decay fungi already at small spatial scales. Oikos 121(6):961–974. https://doi.org/10.1111/j.1600-0706.2012.20052.x

Oettel J, Zolles A, Gschwantner T, Lapin K, Kindermann G, Schweinzer KM, Gossner MM, Essl F (2023) Dynamics of standing deadwood in Austrian forests under varying forest management and climatic conditions. J Appl Ecol 60:696–713. https://doi.org/10.1111/1365-2664.14359

Oleksa A, Chybicki IJ, Gawroński R, Svensson GP, Burczyk J (2013) Isolation by distance in saproxylic beetles may increase with niche specialization. J Insect Conserv 17(2):219–233. https://doi.org/10.1007/s10841-012-9499-7

Palm T (1951) Die Holz- und Rindenkäfer der nordschwedischen Laubbäume. Meddelanden fran Statens Skogsforsknings- institut, 40

Perny B, Krehan H, Steyrer G (2008) Borkenkäferarten. BFW Praxisinfo 17:30–31

Ranius T (2002) Influence of stand size and quality of tree hollows on saproxylic beetles in Sweden. Biol Conserv 103(1):85–91. https://doi.org/10.1016/S0006-3207(01)00124-0

Ranius T, Fahrig L (2006) Targets for maintenance of dead wood for biodiversity conservation based on extinction thresholds. Scand J For Res 21(3):201–208. https://doi.org/10.1080/02827580600688269

Rubene D, Schroeder M, Ranius T (2015) Diversity patterns of wild bees and wasps in managed boreal forests: effects of spatial structure, local habitat and surrounding landscape. Biol Conserv 184:201–208. https://doi.org/10.1016/j.biocon.2015.01.029

Saalas U (1917) Die fichtenkäfer Finnlands: Studien uber die Entwichlungsstadien, Lebensweise und geographische Verbreitung der an Picea excelsa. Die fichtenkäfer Finnlands: Studien uber die Entwichlungsstadien, Lebensweise und geographische Verbreitung der an Picea excelsa. https://doi.org/10.5962/bhl.title.38683

Schmidl J, Bussler H (2004) Ökologische Gilden xylobionter Käfer Deutschlands. Naturschutz und Landschaftsplanung 36(January 2004):202–218

Seibold S, Bässler C, Brandl R, Fahrig L, Förster B, Heurich M, Hothorn T, Scheipl F, Thorn S, Müller J (2017) An experimental test of the habitat-amount hypothesis for saproxylic beetles in a forested region. Ecology 98(6):1613–1622. https://doi.org/10.1002/ecy.1819

Seibold S, Müller J, Baldrian P, Cadotte MW, Štursová M, Biedermann PHW, Krah FS, Bässler C (2019) Fungi associated with beetles dispersing from dead wood – Let's take the beetle bus! Fungal Ecol 39:100–108. https://doi.org/10.1016/j.funeco.2018.11.016

Simberloff D, Abele LG (1982) Refuge design and island biogeographic theory: effects of fragmentation. Am Nat 120(1):41–50. https://doi.org/10.1086/283968

Speight MCD (1989) Saproxylic invertebrates and their conservation. Council of Europe

Steyrer G, Cech TL, Fürst A, Hoch G, Perny B (2020) Waldschutzsituation 2019 in Österreich: Schäden durch Borkenkäfer weiter extrem hoch. Forstschutz Aktuell 64:33–44

Stokland JN, Siitonen J, Jonsson BG (2012) Biodiversity in deadwood. https://doi.org/10.1017/CBO9781139025843

Sverdrup-Thygeson A, Gustafsson L, Kouki J (2014) Spatial and temporal scales relevant for conservation of dead-wood associated species: current status and perspectives. Biodivers Conserv 23(3):513–535. https://doi.org/10.1007/s10531-014-0628-3

Vogel S, Gossner MM, Mergner U, Müller J, Thorn S (2020) Optimizing enrichment of deadwood for biodiversity by varying sun exposure and tree species: an experimental approach. J Appl Ecol 57(10):2075–2085. https://doi.org/10.1111/1365-2664.13648

Weslien J, Djupström LB, Schroeder M, Widenfalk O (2011) Long-term priority effects among insects and fungi colonizing decaying wood. J Anim Ecol 80(6):1155–1162. https://doi.org/10.1111/j.1365-2656.2011.01860.x

Zou JY, Cadotte MW, Bässler C, Brandl R, Baldrian P, Borken W, Stengel E, Luo YH, Müller J, Seibold S (2023) Wood decomposition is increased by insect diversity, selection effects, and interactions between insects and microbes. Ecology 104(12):e4184. https://doi.org/10.1002/ecy.4184

Zytynska SE, Doerfler I, Gossner MM, Sturm S, Weisser WW, Müller J (2018) Minimal effects on genetic structuring of a fungus-dwelling saproxylic beetle after recolonisation of a restored forest. J Appl Ecol 55(6):2933–2943. https://doi.org/10.1111/1365-2664.13160

Open Access This chapter is licensed under the terms of the Creative Commons Attribution 4.0 International License (http://creativecommons.org/licenses/by/4.0/), which permits use, sharing, adaptation, distribution and reproduction in any medium or format, as long as you give appropriate credit to the original author(s) and the source, provide a link to the Creative Commons license and indicate if changes were made.

The images or other third party material in this chapter are included in the chapter's Creative Commons license, unless indicated otherwise in a credit line to the material. If material is not included in the chapter's Creative Commons license and your intended use is not permitted by statutory regulation or exceeds the permitted use, you will need to obtain permission directly from the copyright holder.

The State of Forest Genetic Diversity: Anthropogenic Impacts and Conservation Initiatives

4

Heino Konrad, Jan-Peter George, and Aglaia Szukala

Fragmented forest landscape in Upper Austria, Amstetten area (Photo: BFW/Florian Winter)

H. Konrad (✉) · A. Szukala
Department of Forest Biodiversity & Nature Conservation, Austrian Research Centre for Forests (BFW), Vienna, Austria
e-mail: heino.konrad@bfw.gv.at; aglaia.szukala@bfw.gv.at

J.-P. George
Natural Resources Institute Finland, Production Systems, Helsinki, Finland
e-mail: jan-peter.george@luke.fi

Abstract

Genetic diversity is a key pillar of biodiversity, underlying the adaptive potential of any species. At the same time, it is difficult to quantify this component of biodiversity, and we know very little about the genetic diversity status of most species. A wealth of studies points toward a substantial decline, which is also apparent in numerous forest species. Connectivity is fundamental to maintaining high levels of genetic diversity and adaptability via gene flow between populations. In this chapter, we attempt to elucidate the importance of genetic diversity for the entire forest ecosystem with a focus on its main components: tree species. We elaborate on the anthropogenic factors impacting forest diversity, like exploitation and artificial regeneration, climate change, and introduced pests. An overview of molecular methods for studying genetic diversity and connectivity is presented. Regular genetic monitoring is imperative for optimizing conservation strategies such as the creation of stepping stones to counteract population fragmentation. We highlight the importance of taking genetic diversity into account when sourcing plant material for forest and landscape restoration projects. Finally, international efforts to conserve genetic diversity are presented along with recommendations on suitable indicators to monitor it. Scientists working on genetic diversity are encouraged to actively participate in national and international processes to incorporate genetic principles into policy development as well as conservation and restoration efforts.

Keywords

Climate change · Forest genetic resources · Forest reproductive material · Genetic rescue · Human impact

The Importance of Genetic Diversity for Forest Ecosystems

Processes Affecting Genetic Diversity in a Changing Climate

Genetic diversity is an intrinsic and essential element of biodiversity, encompassing the variation of genetic traits within a population, a species, or an entire ecosystem. All forms of biological diversity are based on the genetic variation found within populations. The variability of genetic traits can be seen as the "raw material" of evolution since the change, further development, and *adaptation* (see Box 4.1) of a species in response to changing environmental conditions depend on it. Without a sufficient amount of genetic diversity, the long-term survival of any species is at risk. It is decisive for species' ability to cope with and adapt to new stresses such as changing site and climate conditions or novel pests (Fisher 1930; Pitelka 1988; Pease et al. 1989; Burger and Lynch 1995; Burdon and Thrall 2001; Etterson 2004; Reusch et al. 2005; Schaberg et al. 2008). Importantly, genetic diversity is a direct function of population and ecosystem connectivity, which determines the extent of genetic variants exchanged between patches in a fragmented habitat (Lowe and Allendorf 2010).

When it comes to the analysis and conservation of genetic diversity, the *population* emerges as the most significant unit. Population genetic diversity is affected by a range of factors, among which *genetic drift* and *inbreeding* can be considered the most important for biodiversity conservation (Allendorf et al. 2022) since they play a major role in small populations (Fig. 4.1). Drift affects all members of a population equally and has the greatest influence when the *effective population size* (N_e) is small (see also Chap. 5). In the majority of species, some populations have undergone a *bottleneck* during their more or less recent evolutionary history, for example, due to range contraction as a consequence of periods of glaciation. These fluctuations in population size lead to recurrent reductions in the genetic variation of populations. The random recruitment of genetic variants by drift causes changes in the genetic composition of populations that may have a negative effect on their fitness and adaptive potential. Therefore, a large N_e is most important for population and species survival. As discussed in more detail in the following, N_e is considered a suitable indicator for evaluating whether a population needs conservative intervention (prioritization of conservation efforts), but it is also very difficult to determine in practice (Waples 2002; Santos-del-Blanco et al. 2022).

Through the exchange of genetic variants, *gene flow* shapes the genetic composition of species, maintaining shared diversity across different populations. On the other hand, gene flow also homogenizes genetic variation across populations, thus counteracting *local adaptation*; however, it is generally regarded as beneficial in terms of preserving genetic diversity, enhancing the fitness of fragmented populations, and minimizing the effects of drift and inbreeding (Ralls et al. 2018). Most importantly, gene flow is an immediate and fundamental measure of population connectivity. By reducing the ability of a species to migrate and exchange genes among populations, reduced connectivity exacerbates the effects of climate change. In the geological past, forest species have modified their distribution ranges through migration to more favorable environments multiple times (Bernabo and Webb 1977; Webb III 1981; Davis 1983; Huntley and Birks 1983; and review by Geber and Dawson 1993; Huntley and Webb III 1988). However, the current climate change and habitat fragmentation processes are likely too fast to be compensated by natural migration rates (Huntley 1991; Davis and Shaw 2001; Jump Peñuelas 2005).

Natural selection acts against individuals with low fitness, purging deleterious genetic variants while favoring beneficial ones. It can thus change the allelic composition of a population rapidly, for example, in the case of the appearance of a novel pathogen causing a *selective sweep*. Such adaptive shifts leave recognizable signatures at the genomic level (e.g., Pritchard et al. 2010) and can describe certain adaptive patterns in natural populations. Nevertheless, phenotypic traits are often *polygenic* with a complex *genetic architecture* and do not respond to selection via pronounced frequency changes of few genetic variants (Höllinger et al. 2019). Due to *redundancy* in polygenic traits, most genetic loci likely contribute only transiently to a change in the phenotype (Yeaman 2015), meaning that the adaptive trait architecture (Barghi et al. 2019; Pritchard et al. 2010) varies strongly over time and

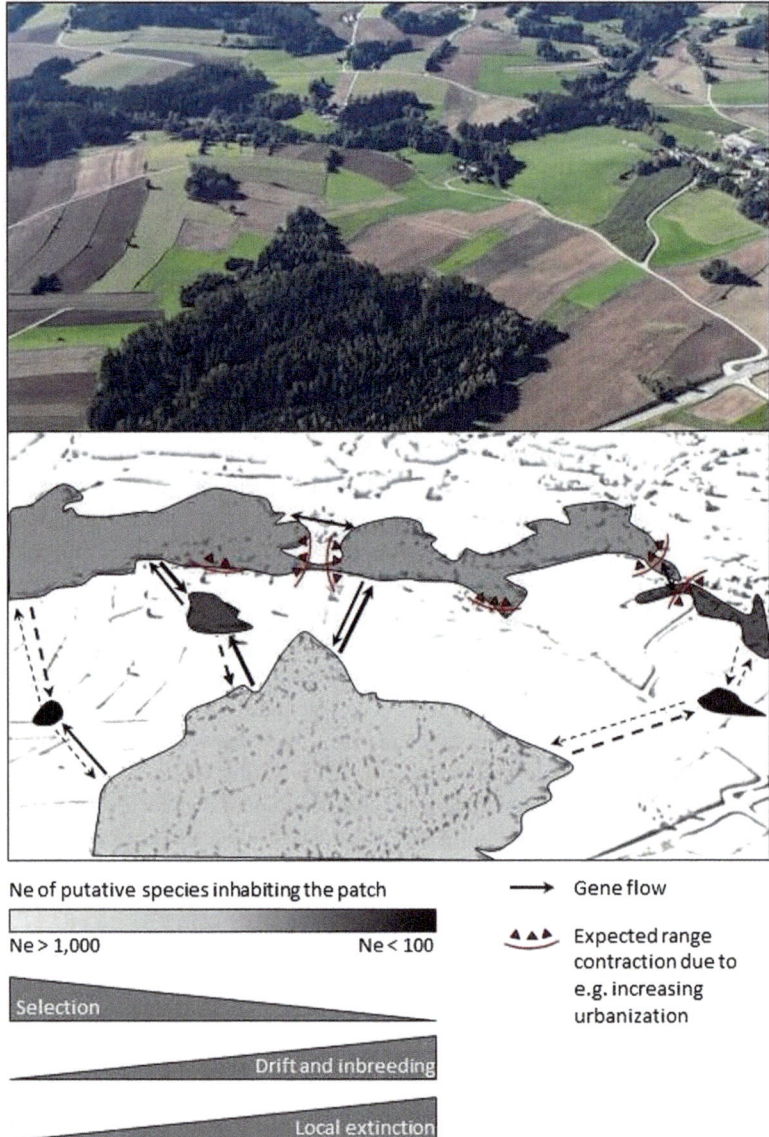

Fig. 4.1 A forest landscape in Upper Austria (Schönau im Mühlkreis), typically fragmented by agricultural and urban land (top), along with an idealized and simplistic representation of the degree of gene flow between the forest patches (bottom). Note that the depicted processes are species-specific and cannot be realistically represented for multiple species in a single illustration. Thicker, thinner, and dashed arrows indicate more frequent, less frequent, and rare gene flow, respectively. Note that gene flow between two specific patches is often asymmetrical. Red curves and arrows symbolize the direction of plausible future range contraction in already urbanized areas, leading to further reduction and fragmentation of existing patches. The gray color gradient indicates larger (lighter) to smaller (darker) Ne, which is accompanied by a parallel decrease in the strength of selection along with increases in the effects of drift and inbreeding as well as the probability of local extinction. (Drone photography by Florian Winter)

multiple molecular mechanisms are available to organisms during adaptation (Leinonen et al. 2013; Luo et al. 2014; Szukala et al. 2022; Yeaman 2022). Several studies have found theoretical and experimental evidence that a sufficient amount of genetic variation must be present in populations for adaptation to take place via small adaptive shifts, as expected under the framework of polygenic adaptation and redundancy (Bakker et al. 2010; Fagny and Austerlitz 2021; Sinclair-Waters et al. 2020; Thornton 2019). Moreover, Bakker et al. (2010) demonstrated that prolonged habitat fragmentation with limited gene flow severely limits the adaptive potential of species, given that adaptive genetic variation is scattered across patches, reducing the adaptive potential of individual subpopulations. Thus, given the complexity of the genetic basis of adaptation, the amount of genetic variation available must be "large enough" (see section "Connectivity Conservation Strategies and Actions" below for further insights on this concept) and/or subpopulations sufficiently connected for adaptation to occur.

Species can also react to differing environmental conditions by way of *phenotypic plasticity*, changing a phenotypic parameter like tree height or behavior in reaction to a change in the environment (Nussey et al. 2005). These phenotypic adjustments are reversible and occur if the underlying genetic architecture permits it and the energetic cost is not too high (DeWitt et al. 1998). Under certain circumstances, plasticity can be adaptive, bringing the phenotype closer to an optimum (e.g., Nicotra et al. 2015), but neutral and even maladaptive effects of plasticity have been documented as well (e.g., Arnold et al. 2019). How and when phenotypic plasticity supplies the variation targeted by selection and contributes to adaptation is a matter of debate (Wund 2012; Levis and Pfennig 2016; Fox et al. 2019; Szukala et al. 2023). In the context of biodiversity conservation, it is important to note that plasticity has intrinsic physiological limits (DeWitt et al. 1998) that are likely exceeded by the demands of climate change for most species, as suggested by several studies (e.g., Forcada et al. 2008; Reed et al. 2011).

Three key means by which species can react to climate change can be determined: (1) Dispersion of seed or vegetative propagules into a more favorable environment (i.e., similar to the native environment prior to climatic change); (2) changes in allelic composition in response to natural selection, resulting in better adaptation to the new environment (Burdon and Thrall 2001; Reusch et al. 2005); or, when the genetic architecture permits it, (3) phenotypic plasticity changes to cope with the new conditions. As mentioned above, both the capacity to migrate—at least, without human intervention—and the ability to respond plastically to changing conditions are limited: Migration rates and plastic reaction norms are mostly exceeded by climate change and human-induced habitat modification. Maintaining sufficient genetic variation through population connectivity is therefore essential for enabling adaptation—or, in a human-driven context, for delivering the material for artificial selection, marker-assisted breeding, assisted migration, and even gene editing. It is important to note that evidence of the positive effects of selection on fitness in the wild has been reported (Bonnet et al. 2022), suggesting that many species do indeed have the potential to adapt to ongoing climate change. Nevertheless, we must also remember that every species has its own reproductive and genetic systems that interact with the described evolutionary processes in specific ways to shape that species' genetic diversity. Therefore, a profound understanding of species-specific population genetic dynamics and demographic and evolutionary histories is required to carefully evaluate the state of individual species and optimize conservation efforts.

Box 4.1 Glossary of classical population genetic terms used in the current chapter

Adaptation	An evolutionary process that increases an individual's probability of survival and reproduction in a given environment.
Bottleneck	A sharp reduction in population size due to stochastic events.
Demographic connectivity	The effects of dispersal on population growth and mortality (Lowe and Allendorf 2010).
Effective population size, N_e	The size of a theoretical population in which the genetic composition is affected by the same degree of random change (drift) and inbreeding as the observed population.
Gene flow	Effective transfer of genetic material by pollen containing the haploid male gametophyte resulting in fertilization and development of a diploid embryo.
Genetic architecture	The sum of all genetic loci contributing to a trait of interest, including their effect size on the phenotype, their position in the genome, and their interaction (i.e., linkage disequilibrium, epistatic interaction, and pleiotropy).
Genetic connectivity	The effects of gene flow on population evolution (Lowe and Allendorf 2010).
Genetic drift	Changes in allele frequency due to random effects (e.g., natural catastrophes, such as wildfires and storms).
Inbreeding	The mating of individuals that are genetically closely related, including self-fertilization.
Local adaptation	The process by which a population evolves to be more suited and better adapted to its local environment than other populations within the same species.
Natural selection	The process by which individuals with higher fitness are more likely to survive and produce progeny.
Phenotypic plasticity	The same genotype can modify the expression of the phenotype in different environments (Schlichting and Pigliucci 1998).
Population	A group of interbreeding individuals of the same species that live in the same place at the same time.
Polygenic	Governed by large amounts of genetic variants.
Redundancy	Different combinations of genetic variants can produce the optimal phenotype (Goldstein and Holsinger 1992).
Selective sweep	The process by which positive selection increases the frequency of a beneficial mutation in the population, leading to a frequency increase of other mutations linked to the beneficial one (so-called *hitchhikers*) and, in consequence, to a decrease of genetic diversity in the genomic region surrounding the beneficial mutation.

Genetic Diversity of Forest Foundation Species

Foundation species, such as trees in forest ecosystems, have been defined as species that structure a species community by creating locally stable conditions for other species as well as by modulating and stabilizing fundamental ecosystem processes (e.g., Dayton 1972; Whitham et al. 2003). The field of community genetics has shown that the genetic diversity of the main tree species in a forest can affect the community of dependent species (e.g., microbes, plants, arthropods, birds, and mammals) (Whitham et al. 2003). To cite an example, Tovar-Sánchez et al. (2013) found evidence of higher arthropod diversity in the crown of oaks positively associated with the within-population genetic variation of the host plant. In addition, accumulated evidence shows direct impacts of foundation species' intraspecific genetic diversity on several ecosystem processes such as primary productivity, population recovery from disturbance, interspecific competition, community structure, and energy and nutrient flows (Crutsinger et al. 2007; Hughes et al. 2008). Studying the interactions between the genetic diversity of foundation species and several ecosystem components should allow us to develop better strategies for preserving biodiversity and ecosystem function in the face of forest fragmentation, climate change, and introduced pests (Whitham et al. 2006). Therefore, the conservation of genetic diversity and population connectivity of trees merits special attention. Indeed, the genetic diversity of foundation tree species has been the focus of multiple studies pursuing a better understanding of forest genetic resources (FGR) and how to improve their conservation (e.g., Geburek and Konrad 2008).

Trees are among the organisms with the highest genetic diversity (Hamrick and Godt 1990). This diversity within and among natural populations provides the foundation for forest ecosystem stability in variable and changing environments (Gregorius 1996; Petit and Hampe 2006), as well as for relatively rapid adaptive responses to environmental challenges (Alberto et al. 2013; DeHayes et al. 2000; Davis and Shaw 2001). For example, forest trees? reacted to dramatic changes in climate and other stresses several times during the quaternary period, both through adaptation and migration (Davis and Shaw 2001; Petit et al. 2004). Nevertheless, within-species tree genetic diversity must be complemented by diversity at the species level of another organismal group to ensure tree population stability; for example, trees rely on other organisms for their survival, needing pollinators (e.g., insects, bats, and birds) and seed dispersers (e.g., birds and mammals). The survival and genetic diversity of these dispersers are therefore likewise important for the long-term resilience of forest ecosystems. The number of studies on the genetic diversity of organisms associated with forests is large, yet these species are still understudied and the results are scattered among the body of conservation genetics literature. Nevertheless, a large body of literature points toward a strong decline in genetic diversity of pollinator and seed disperser species (Exposito-Alonso et al. 2022; Hoban et al. 2023a, b).

Populations experiencing a rate of environmental change beyond the rate at which they can adapt or disperse are threatened by decline (Lynch and Lande 1993; Burger and Lynch 1995; Visser 2008). At present, anthropogenic climate change is quickly altering the natural environment: The current biodiversity crisis is manmade and requires active interference to be halted and reversed. We have reached a state in which fragmentation is reducing effective population sizes and impeding gene flow among populations (Fig. 4.1). As a result, the genetic diversity of forest species is under considerable threat due to multiple anthropogenic factors, with climate change, the introduction of novel pests and diseases, and forest fragmentation representing the main threats (which will be discussed in more detail in the following section). Connectivity between forest populations is pivotal to the conservation of genetic diversity and species survival, but it is also one of the aspects most heavily affected by human impact. Every effort needs to be made to maintain forest ecosystems. In particular, since most current threats to biodiversity and the genetic diversity underlying it are manmade, it is our obligation to mitigate anthropogenic impacts and take action to increase or at least preserve the remaining levels of genetic diversity to prevent further loss and, eventually, the collapse of forest ecosystems.

In this chapter, we attempt to review the current state of forest genetic diversity and connectivity, with a special focus on (i) the major anthropogenic threats impacting them, (ii) the available tools used for their assessment, including the respective advantages and limitations, and (iii) the guidelines currently in use for genetic monitoring and global conservation initiatives. We emphasize that genetic diversity and connectivity are difficult to assess, especially on a large scale and for many species, but nevertheless necessary to inform conservation practices. Evaluation is demanding in terms of time, funding, and other resources, especially when molecular methods are used (which is necessary for most species). At the same time, genetic monitoring and easy-to-apply indicators are urgently needed to inform policy makers and define conservation priorities.

Anthropogenic Impacts on Genetic Diversity in Forest Ecosystems

Climate Change

Human alteration and management of forest ecosystems have heavily impacted the genetic diversity of forests; in some regions, this impact has been ongoing for millennia. Only a brief overview of this topic, which has been more broadly assessed, e.g., by Ledig (1992), Savolainen and Kärkkäinen (1992), and Lefèvre (2004), can be presented here. At present, the primary threat is climate change. It is expected that population dieback and/or increased selection pressure—e.g., due to prolonged periods of drought—may erase genetic variation in populations that have already suffered a loss of genetic diversity due to human activity in the past (Alsos et al. 2012; Armbruster and Reed 2005; Pauls et al. 2012). Due to its crucial importance, the multiple effects of climate change on forest ecosystems are discussed in detail in Chaps. 5 and 6. Here, we will focus on a specific aspect that also impacts reproductive biology, genetic diversity, and the survival of forest species, namely the influence of climate change on the masting behavior of forest trees.

Mast seeding—the large-scale, synchronized interannual variation in seed production among populations of forest trees—is a phenomenon exhibited by many of the foundation tree species in temperate forests, including oak, beech, spruce, and pine. The enhancement of pollination efficiency and the decrease of seed predation in mast years have been suggested as the main benefits of masting for tree reproduction (Kelly et al. 2001). There is now increasing evidence that climate change disrupts masting dynamics, likely leading to a decrease in the long-term regeneration of masting plants (Foest et al. 2024). In turn, decreased regeneration potential is expected to impede the rapid geographic range shifts required to maintain species within suitable habitats under the scenario of climate change (Walther et al. 2002; Chen et al. 2011). Alarming evidence was provided by Bogdziewicz et al. (2020), who showed that climate warming decreased the interannual variability of seed production and the reproductive synchrony among individuals in European beech. These effects led to lower pollination success as well as greater loss of seeds to predators, offsetting the benefits of masting dynamics even though climate warming has increased the overall seed production of European beech over the last few years. Moreover, the authors showed that seed viability is decreasing, especially in old trees. In connection with the effects of forest fragmentation like increased inbreeding and lower fitness among offspring (see below), these findings offer a pessimistic outlook for the future of many forest species if no countermeasures are taken.

Deforestation

The most profound and direct impact on all levels of forest biodiversity is caused by deforestation, meaning the permanent removal of forest growth, which is equivalent to habitat loss for all forest species. Deforestation has accompanied cultural development since prehistoric times and has resulted in a steady decrease in forest cover on a global scale. Deforestation for agricultural and human development has been less frequent at higher latitudes and high elevations while being most severe in tropical countries (Balboni et al. 2023)—for example in Ethiopia, where 60% of the forest area recorded at the end of the nineteenth century was lost during the twentieth century (Oljirra 2019). Unfortunately, this process is still ongoing. Globally, we are currently still losing more forest than is being restored (ca. ten million ha per year; FAO 2022). The majority of forest area is destroyed for agricultural purposes (including pastures) in the southern hemisphere. By contrast, deforestation has been halted in most countries of the northern hemisphere, where the total forest area is presently increasing again (FAO 2022). Forest restoration is one of the key approaches to mitigating the effects of climate change, with the additional benefit of restoring forest biodiversity, at least in the long term (e.g., European Green Deal; EC 2019). The effects of deforestation on genetic diversity and species survival are most severe when whole populations are lost, or even entire species go extinct. The latter case has primarily been observed for endemic species with restricted distribution, especially on islands (e.g., Madagascar; Allnutt et al. 2008), and should not be underestimated since such events are impossible to reverse.

Fragmentation

A common result of deforestation is forest fragmentation, which is currently viewed as the greatest issue concerning long-term species survival, particularly in the face of climate change. Forest fragmentation subdivides populations into smaller sections and creates barriers to migration. Eventually, this process leads to inbreeding in the forest fragments, which has long-term negative effects on population fitness and adaptability to environmental change. Moreover, demographic stochasticity or drift effects (the decline of populations due to random effects like browsing or fire) can have an earlier and more immediate effect in small forest fragments, thereby quickly leading to local extinction (Lande 1988). The size of forest fragments is therefore very important: Larger fragments can hold more individuals of more species as well as attract more migrants, which in turn reduces the effects of population isolation. Most tree species have developed mechanisms for long-distance dispersal of their propagules. Nevertheless, a review of the effects of fragmentation on the genetic diversity of plant species shows that tree species are affected by disruptions of their gene flow in the same ways as herbaceous species (Vranckx et al. 2012; Aguilar et al. 2019). The only likely exception to this general rule are species pollinated by vertebrates (birds and bats; mostly in tropical regions), which can fly long distances and are thus able to counteract fragmentation by maintaining pollen flow among population fragments (Hadley and Betts 2009).

The effects of fragmentation extend further than the mere loss of alleles and increased rates of inbreeding in adult individuals: It also causes changes in reproductive output. Due to pollen limitation, both the reproductive output and genetic diversity of the offspring are reduced (Aguilar et al. 2006; Leimu et al. 2006; Honnay and Jacquemyn 2007; Vranckx et al. 2012), and inbreeding strongly affects the performance of the progeny—in other words, fewer offspring with a lower chance of survival are produced (Aguilar et al. 2008; Vranckx et al. 2012). These factors work together to increase seedling mortality, which impedes natural regeneration and leads to local extinction (Charlesworth and Willis 2009; González-Varo et al. 2010; Ashworth and Martí 2011; Aguilar et al. 2012). The quality of planting material is also decisive for the success of forest and landscape restoration (FLR) (Maginnis and Jackson 2007), but this is often neglected in practice (Jalonen et al. 2018). Implications for forest restoration efforts and possible ways to improve the situation are further elaborated and discussed in Box 4.2. A variety of management strategies have been developed to counteract fragmentation, including the creation of habitat corridors and stepping stones, under- and overpasses for animals, and supplementation of populations by adding individuals from other populations to increase genetic diversity and population size, a strategy termed "genetic rescue" (Allendorf et al. 2022).

Overexploitation

A further anthropogenic impact is overexploitation. With regard to forest trees in particular, excessive harvesting can severely impact a species' genetic diversity, primarily due to a significant reduction in population sizes. Overexploitation will not result in species extinction but can profoundly impact the size of the gene pool, equivalent to a population bottleneck (Allendorf et al. 2022). For instance, selective logging of a large part of a tree population while leaving a small number of undesirably shaped individuals as seed trees can have a lasting negative effect on the growth of future tree generations (Ledig 1992). Research has also shown that while thinning of stands as a silvicultural management practice may not affect the genetic diversity of the main target species, it can, however, have a detrimental effect on the associated secondary tree species when their population size is strongly reduced (El-Kassaby and Benowicz 2000).

Another effect of overexploitation can be an increase in hybridization rates among species that would not have occurred in undisturbed habitats. For example, extensive hybridization between *Picea rubens* and *P. mariana* has been observed as a consequence of extensive logging and wildfires in the maritime provinces of Canada (Ledig 1992). In a situation where *P. rubens* was heavily harvested and simultaneously affected by fires, *P. mariana* was not overharvested and less affected by forest fires due to its more humid habitat; as a consequence, *P. mariana* outnumbered and was able to fertilize *P. rubens*, resulting in large-scale establishment of hybrid offspring in the clearcut areas. This hybridization does not occur to the same extent in undisturbed habitats, and hybrids do not establish as easily, since they tend to grow slower and be more susceptible to pests. A similar example has been reported with regard to *Pinus palustris* and *P. taeda* (Namkoong 1966). Climate change may have a similar effect in certain situations (e.g., Lind-Riehl and Gailing 2017). Hybridization can threaten the genetic integrity of a species, but on the other hand can also provide the genetic variation necessary to adapt to new climate conditions (Brauer et al. 2023; see also Chap. 5).

Translocation

The anthropogenic translocation of forest plants, animals, and microorganisms has heavily affected the appearance and composition of current forests. Unfortunately, the unintentional introduction of novel pests and diseases can significantly impact the genetic diversity and even survival of many species, and of forest tree species in particular. Among the most severe such cases is the introduction of the chestnut blight pathogen (*Cryphonectria parasitica*) to eastern North America, which has basically eradicated *Castanea dentata* as one of the main tree species from the forests in this geographic region (Anagnostakis 1988). Other examples are the spread of Dutch elm disease (*Ophiostoma novo-ulmi*) to Europe and North America (Brasier 2000), the appearance of the gypsy moth in eastern North America (Liebhold et al. 1992), and more recently the introduction of the ash dieback fungus

(*Hymenoscyphus fraxineus*) to Europe (McMullan et al. 2018) and the emerald ash borer (*Agrilus planipennis*) to North America (Herms and McCullough 2014). These introductions have decimated the population sizes of the affected tree species, causing massive reductions in extant genetic diversity. In combination with fragmentation and climate change, these factors pose further threats to long-term species survival. Efforts should be undertaken to prevent such intercontinental spreading of pests and pathogens by applying strict phytosanitary measures.

The introduction of non-native tree species as an alternative to autochthonous trees for wood production in the face of climate change is a hotly debated topic. Introduced species might become invasive and outcompete native ones with large-scale and long-lasting negative effects on forest ecosystems. Therefore, careful evaluation of species performance is necessary before such introductions are recommended (Brundu et al. 2020). For example, *Robinia pseudoacacia* was introduced to Europe from North America in the seventeenth century and has since invaded many Central European ecosystems, becoming a typical element of the landscape. Though its effects on native forests have mostly been described as negative, this species currently represents an important component of forests of Southeastern Europe, where its eradication has become impossible (and no longer even desirable) for forest managers. The European populations spread out from a small number of founding trees possessing a restricted share of the genetic diversity present in *R. pseudoacacia*'s native range. Interestingly, the seeds of these most successful trees were shown to have significantly higher germination rates than the average native American populations, as well as low pre-adaptation germination requirements (Bouteiller et al. 2021); this provides evidence that anthropogenic selection has favored these provenances without their invasive potential being taken into account.

The translocation of native forest plants as part of reforestation and forest management efforts likewise impacts genetic diversity. Planting can change local patterns of variation, modify the mating system, and impact natural populations in the vicinity through pollen flow and seed migration by reducing local adaptation (outbreeding depression; Frankham et al. 2011). Since often no records are kept on which planting material has been used in which locations, it is difficult to assess associated negative and positive impacts on local performance and diversity patterns. In general, however, the available examples show that effects on local populations seem to be limited or even beneficial (Lefèvre 2004). Complementary planting in small populations of endangered species can actually be seen as a measure of assisted gene flow or genetic rescue, since it provides demographic support (census size) and the introduced migrants reduce the effects of inbreeding depression in the respective population fragment (Allendorf et al. 2022). However, it is essential for this practice to be carried out with proper forest reproductive material (FRM) containing sufficient genetic diversity to avoid negative effects on extant populations: Excessive planting of a rare species to increase its distribution under the neglect of the native gene pool and the genetic diversity of the reproductive material used can result in further erosion of genetic diversity and eventual loss of the entire population. An example is provided by Lefèvre (2004) for *Sorbus domestica* when plants with low genetic diversity are planted on a large scale, outnumbering the native gene pool (genetic swamping; García-Ramos and Kirkpatrick 1997; Lenormand 2002). Awareness for this issue is increasing and it is becoming common practice to check the genetic diversity of rare species as well as potential source populations before large-scale planting is conducted (e.g., for *Sorbus torminalis*, Kavaliauskas et al. 2021; or *Acer platanoides*, Lazic et al. 2022). In situations where an endangered population needs to be conserved, special attention should be paid to the influx of unwanted immigrant pollen or seeds (Unger et al. 2016).

Widespread plantation failures, poor performance, bole shape of ill-adapted forest tree provenances, and loss of productivity resulting from the use of low-quality planting material led to higher standards being applied by forest managers in the sourcing and employment of FRM. In the European Union, a directive on baselines for the collection of and trade in FRM is in place that is aligned with the respective OECD standards (OECD 2023). Extant guidelines on FRM translocation are mostly based on the assumption that "local is best." All FRM needs to be labeled properly from seed to plant, and a database of all registered FRM sources has been created (FOREMATIS). The "local is best" paradigm is presently being challenged by the advent of climate change. One generally accepted way to help forests adapt to climate change is the development of assisted migration (or assisted gene flow; Aitken and Whitlock 2013) by planting putatively better-adapted FRM (i.e., provenances from portions of the species' range that are closer to future climatic predictions for the target site) in addition to natural regeneration. In this way, forests should have the possibility to adapt through natural selection in the long term. Modeling approaches regarding optimal provenance selection are already available and steadily being improved to include various site conditions (Poupon et al. 2021). A more in-depth review of the concept of assisted migration is provided in Chap. 14.

Artificial Regeneration

Artificial regeneration is considered beneficial to local genetic diversity if the employed FRM is of proper origin and derived from a sufficiently high number of mother trees. In contrast to natural regeneration, planted trees are often unrelated. Plantations can thus feature high genetic diversity, especially when seeds from a large number of unrelated clones are used (Lefèvre 2004). On the other hand, natural regeneration can potentially capture new genetic variation from a larger cohort of pollen donors (Raja et al. 1998). Despite artificial regeneration being challenging for several reasons (see the review on oak artificial regeneration by Dey et al. 2008), it becomes necessary when natural regeneration is too slow or does not meet the required objectives; for instance, after forest decline in large areas. Therefore, both management systems should be practiced and complement each other with the aim of enhancing genetic diversity and fitness (see also preceding section). Importantly, Lefèvre (2004) showed that the majority of economically important tree species harbor moderate to high levels of genetic diversity so that human selection and breeding do not significantly reduce their genetic diversity, with corresponding effects mostly detectable only in a few genomic regions.

In addition to the selection of FRM sources, cultivation in nurseries can have different effects on the genetic diversity of the resulting plant lots. For instance, common practices such as seed sorting for efficient container plant production and thinning out surplus (i.e., smaller) seedlings after seeding numerous seeds per container have effects equivalent to directional selection for large-sized seedlings, a trait that does not correlate with improved performance in later life stages (Edwards and El-Kassaby 1996). Moreover, seed lots have differing requirements for stratification and substrate conditions, which are mostly not taken into account in nurseries. This becomes particularly problematic when seedling selection is based on performance in the nursery environment rather than at the planting site. Finally, additional challenges are posed by the methods of fertilization, mycorrhization, growth sorting, cold storage, and undercutting of nursery stock and their effects on later performance of trees and their progenies (see the review by Himanen et al. 2021). Direct sowing should be considered an option for artificial regeneration, since it appears to maintain within-seed-source diversity better than planting and because the higher selection intensity during seed germination and seedling establishment results in the establishment of the seedlings best adapted to the respective site (Lefèvre 2004).

How Can Genetic Diversity and Population Connectivity Be Measured Using Molecular Tools?

The field of conservation genetics was established to detect and monitor anthropogenic influences on natural populations and develop conservation schemes encompassing genetic principles (Holderegger et al. 2019). As discussed in the previous paragraphs, connectivity is an essential feature affecting the genetic makeup and dynamics of populations in fragmented habitats. The metapopulation concept introduced by Levins (1970) has become a valuable model for conservation genetics (e.g., Wade 2016). A metapopulation is a group of populations of different sizes occupying similar habitat patches within a specific region that are connected via the exchange of migrants. The different patches vary in population size and migrant exchange and are thus affected to differing degrees by drift and inbreeding, with the possibility of local extinction and recurrent natural recolonization of patches (Allendorf et al. 2022; Fig. 4.1). Bigger fragments harboring larger populations, as well as subpopulations located closer to each other, will exchange more migrants. Migration between habitat patches is particularly important because it counteracts population decline and helps to recolonize patches in which individual species have gone extinct. In conservation genetics, the estimation of gene dispersal distances and the rate of exchange among patches is therefore of particular interest, as these factors are decisive for overall (meta-)population survival.

Plant research places a key focus on current rates of pollen and seed exchange to obtain a reliable measure of population connectivity. To achieve this goal, data on gene dispersal processes at a local, regional, and species-range scale is required (Kremer et al. 2012). Every species has different dispersal capabilities, and a broad range of analytical methods to assess gene flow has been developed (Manel et al. 2005; Smouse and Sork 2004; Burczyk et al. 2006; Robledo-Arnuncio 2012). In general, the available methods can be classified as direct or indirect. Indirect methods offer information on historic, intrinsic dispersal capabilities of a species, both at the intra- and interpopulation level. Indirect approaches to assessing historical gene flow within continuous populations employ the principle that the spatial genetic structure (SGS) displayed by neutral genetic markers is essentially caused by local drift, the effect of which is balanced out by gene dispersal. Under this isolation by distance model, the decay of genetic relatedness with distance has been shown to be inversely proportional to the effective population density (i.e., a measure of population density based on N_e as defined in Vekemans and Hardy 2004). When a reliable estimate of the latter is available, methods to estimate gene dispersal distance from patterns of SGS exist (Vekemans and Hardy 2004; Rousset 2000).

Despite the importance of non-time-specific estimation methods, most present-day research aims to assess the contemporary gene flow among populations. Estimates of current levels of pollen and seed exchange between populations deliver valuable information for ecological monitoring and conservation management in a wide range of scenarios (Lenormand 2002). These include *genetic and demographic connectivity* after landscape fragmentation, containment of allochthonous (or even genetically modified) populations, or potentially adaptive long-distance gene flow across heterogeneous habitats under climate change (Robledo-Arnuncio 2012). A plant-specific maximum-likelihood approach to jointly estimating contemporary pollen and seed exchange rates was established by Robledo-Arnuncio (2012) and developed further by Unger et al. (2016). This method employs genetic markers that are biparentally inherited (nuclear microsatellite markers) and sampled from the putative target and source populations sequentially, i.e., before and after a reference dispersal event. A major limitation is that the methodology cannot be applied to species with a continuous distribution, including those with long-range dispersal—which includes a majority of economically and ecologically important tree species.

Alternatively, patterns of current gene flow can be estimated "directly" using genetic fingerprint and parentage analyses to assign seedlings to their parent trees and thus infer species-specific pollen and seed dispersal curves (Oddou-Muratorio and Klein 2008; Oddou-Muratorio et al. 2010). Although such results are locally derived, very detailed information about the reproductive system and reproductive biology of a species can be obtained. In plant populations, parentage analysis consists of genotyping a sample of dispersed seeds or established seedlings as well as all reproductive individuals within a circumscribed area using a set of shared polymorphic markers (see Chap. 5 for a more detailed description) to determine the parents of each seedling (Meagher 1986). To distinguish between male and female parentage of seeds and seedlings, maternally inherited tissues collected on dispersed seeds can be genotyped (Jones et al. 2005; Jordano et al. 2007). When already established seedlings are studied, the average effective pollen/seed dispersal distance can be directly estimated from parent–offspring genotype data by model fitting, such as the neighborhood model (Burczyk et al. 2006). These approaches also enable the detection of ecological factors that are likely to influence patterns of gene flow and relatedness, e.g., parental phenotypic traits (González-Martínez et al. 2006), seed disperser behavior (Jordano et al. 2007), or spatial environmental heterogeneity (Jones et al. 2005). The most advanced approaches additionally allow estimation of selection gradients and detection of fitness-related traits that enhance the reproductive and dispersal success of parental individuals (Chybicki 2018, 2023).

Clustering and assignment methods have also been used to estimate dispersal between population fragments (Berry et al. 2004; Gagnaire et al. 2015); these have mostly been applied to animal populations but are likewise useful in plant studies (e.g., Bizoux et al. 2009; Kassa et al. 2017). In these approaches, a set of reference populations is defined *a priori* and individuals are assigned to their respective population of origin based on multilocus genotypes. In this way, immigrants can be identified when the sampling location and genetic group of origin do not match. These methods also allow the offspring of immigrants and later-generation descendants to be identified (Wilson and Rannala 2003). Rates and direction of dispersal among the studied populations can thus be estimated. Moreover, it is possible to detect natural or anthropogenic barriers as well as other factors with an impact on gene flow (e.g., wind direction and migration corridors). The results of these investigations can inform conservation strategies to restore connectivity (Balkenhol et al. 2015). The research field focused on these aspects by combining population genetics and landscape ecology has been termed "landscape genetics" see also Chap. 5.

There is a wide variety of measures of genetic diversity (see also Chaps. 5 and 8) the most useful for genetic monitoring are allelic diversity (or allelic richness when comparing different sample sizes) and expected heterozygosity. The latter is most sensitive to reductions in N_e and thus the most suitable for detecting population decline (Allendorf et al. 2022). The estimation of N_e is particularly important to glean information on the genetic status of a population; for example, a population's census size (N_c, the total number of individuals in a population) can be high even though N_e is low (e.g., when few parents have sired a large part of the population). The effective population size N_e was introduced by Sewall Wright (1931, 1933) and can be defined as *"the size of an ideal population that would experience the same magnitude of genetic drift and inbreeding as the studied population"*; in other words, it informs us about the "true" size of a population in terms of genetic diversity (Allendorf et al. 2022). An in-depth review of existing methods to infer N_e based on demographic parameters and genetic data was performed by Wang et al. (2016). For practical applications, thresholds have been suggested (see below) for how large N_e and N_c should be to enable a population to survive in the long term. Nevertheless, N_e estimation is particularly difficult and generally approximate for large populations of forest trees with a continuous distribution (Santos-del-Blanco et al. 2022). Additional limitations of this measure include the impossibility to know whether an estimated N_e refers to a single population from which samples are taken or to the metapopulation it is a part of, as well as the time point reflected by the measurement (Wang et al. 2016).

Most existing studies on genetic diversity and gene flow have been performed using neutral nuclear microsatellite markers since they are straightforward to score and the obtained results are repeatable and usually show a high number of alleles per studied locus. Technological progress in the development of molecular methods has been rapid thanks to next-generation sequencing, which allows the study of thousands to millions of single nucleotide polymorphisms (SNPs) for any given species (Ellegren 2014). At present, the use of microsatellite markers for parentage analyses is rapidly declining in favor of next-generation sequencing markers—mostly SNPs (Flanagan and Jones 2018). Major challenges arising in the context of next-generation markers include (i) whether the existing software products can handle much greater numbers of markers, (ii) whether the methods can deal with the higher degree of uncertainty of genotype estimation at single loci (e.g., genotype likelihoods), and (iii) whether additional sources of error should be considered (Flanagan and Jones 2018). Moreover, with the steadily increasing number of available reference genomes, whole-genome resequencing of large numbers of individuals is becoming feasible (and affordable) for an increasing number of laboratories. Most of the methods listed above have already been extended to allow the use of SNP data to study connectivity and gene flow as well as considering genotyping errors and rare allele frequencies (e.g., Korneliussen et al. 2014; Chybicki 2018, 2023; Heena et al. 2023).

Although genotyping and sequencing technologies are advancing quickly, monitoring the genetic diversity of one or multiple species remains very resource-intensive. New technologies that allow biological information to be collected for a large number of individuals simultaneously are emerging as a possible solution to this limitation. Remote sensing via imaging spectrometry—i.e., detection of the physical features of an area from its reflected and emitted radiation—provides extensive biochemical information on natural ecosystems including forests, and is increasingly publicly available (see e.g. www.geo.uzh.ch/en/units/rs.htmland; www.genesinspace.org). This data can support an indirect assessment of within-species genetic diversity for many tree species (Cavender-Bares et al. 2022; Jung et al. 2021; Wang and Gamon 2019). Indicators of genetic diversity based on estimated census sizes of reproductively mature individuals in a population (see below) do not require genetic data for estimation. It is therefore potentially feasible to quantify species abundance and characterize populations using remote sensing data from satellite or aircraft images, and to integrate this information with existing field-based knowledge on habitat type and extent. The superposition of spectral information and environmental data can thus provide proxies of genetic variation and identify areas of high or low genetic diversity, as shown by an increasing number of studies (Kivinen et al. 2020; Madritch et al. 2014; D'Odorico et al. 2023; Yamasaki et al. 2017). Despite several technical limitations that still need to be overcome (e.g., Jung et al. 2021; Khanal et al. 2020; Tran et al. 2022), remote sensing data represents a promising new approach for the future of genetic monitoring.

How Much Genetic Diversity Is Needed?

Genetic monitoring can detect long-term changes in genetic diversity. Both *in situ* and *ex situ* approaches to conserving genetic diversity have been suggested and implemented, at least for some of the main forest tree species (e.g., Hoban and Schlarbaum 2014; Lindenmayer and Laurance 2017; Mounce et al. 2017). These concepts are presented and discussed in Chap. 11, and will not be elaborated on any further here. However, the number of populations conserved in either fashion is a good indicator of the conservation status of a species of interest (Hoban et al. 2020).

Most data accumulated so far indicates only rare losses of genetic diversity among the primary tree species in temperate forest ecosystems, mostly due to their large population sizes, frequent wind-pollination, and predominately outcrossing mating systems. Generally, tropical tree species exhibit more evident population structure and enhanced genetic differentiation between populations of the same species than temperate tree species (Dick et al. 2008). This difference is likely caused by low but significant rates of self-fertilization and biparental inbreeding in tropical species, probably due to lower population densities and predominant pollination by insects (Dick et al. 2008). Comparatively little is known in this regard about secondary (i.e., economically less important) tree species, and even less about the vast number of other forest-dwelling organisms. While relatively high levels of genetic diversity are still found even in rare tree species within fragmented habitats, these results are probably strongly biased: Most studies to date have analyzed adult trees, which do not realistically reflect the effects of fragmentation and anthropogenic impact over the last 100 years but rather preceding and historic patterns of genetic diversity. This phenomenon has been termed the "extinction debt" (Aguilar et al. 2008; Vranckx et al. 2012) and describes the time lag between population decline and subsequent measurable changes in genetic diversity; there is increasing evidence showing that offspring generations generally have lower genetic diversity (Aguilar et al. 2018; see also above under fragmentation impacts).

But the question remains how much genetic diversity is needed for a population to survive, or for conservation efforts to be necessary or meaningful. Are small populations doomed because all individuals will be the result of inbreeding after a few generations? These questions cannot be easily answered. There exist examples of populations with very low diversity but no apparent effects of inbreeding depression (Allendorf et al. 2022). The importance of inbreeding depression for the fate of populations has been debated (e.g., Nonaka et al. 2019), but the accumulated evidence clearly indicates that inbreeding depression needs to be considered in the context of the persistence of populations (Frankham et al. 2011; Spurgin and Gage 2019; Allendorf et al. 2022). It is therefore safe to say that small populations will generally have a lower ability to adapt to changing environments or novel pests and be more strongly affected by stochastic effects (drift). Genetic data can provide precise information about the state of a given population: Is its diversity lower than that of other populations of the same species? Is the population connected to other populations or is it relatively isolated?

As a general guideline for management decisions, the so-called 50/500 rule (Franklin 1980) postulates that N_e should not fall below 50 individuals in the short term and 500 individuals in the long term to prevent erosion of genetic diversity. Later studies have suggested changing these thresholds to 100 and 1000 individuals, respectively (Frankham et al. 2014). In theory, the long-term adaptability of a population remains stable if less than 0.1 of its heterozygosity is lost per generation (Frankham et al. 2010). This preservation of diversity per generation is estimated to be achieved with approximately 1000 randomly mating individuals with a balanced sex ratio (i.e., N_e of 1000). In nature, N_e is around 10% of the N_c of a population on average; this means that approximately 10,000 individuals would be needed in the long term for a population to remain stable under the influence of factors that erode genetic diversity (Frankham et al. 2014). Such general rules have been widely criticized as being too approximative and far from reality (e.g., Fady and Bozzano 2021; Franklin et al. 2014), and because they do not take species-specific assumptions and constraints into consideration (Flather et al. 2011). Most likely, a more meaningful approach involves avoiding such thresholds as targets but instead considering them as indicators of potential risk of decline (Allendorf et al. 2022). In this way, N_e approximated on the basis of N_c can be applied as a suitable pragmatic indicator of "genetic health state" for any species, including forest trees. Such indicators are necessary for decision-making and the prioritization of conservation efforts, since the alternative is to make entirely unscientifically based decisions at the political and bureaucratic levels (Brook et al. 2011).

Practical indicators for monitoring the status and trend of genetic diversity within species have been developed under the Post-2020 Global Biodiversity Framework (GBF; e.g., Hoban et al. 2023a, b) of the Convention on Biological Diversity's (CBD, i.e., the main legally binding agreement with respect to biodiversity conservation; see also next section). These indicators are partly based on the abovementioned relationships between N_e and N_c and were ultimately recommended by the CBD at the Kunming-Montreal Summit in 2022. The purpose of Indicator 1 (fraction of populations with $N_e > 500$ or $N_c > 5000$) is to provide a baseline for conserving sufficient within-population diversity in case of rapidly changing environmental conditions. By contrast, Indicator 2 (fraction of populations still existing) measures the temporal trends in among-population diversity with a view to providing diverse options for the future adaptability of the species. Indicator 3 is a binary value describing whether the species has been monitored by means of molecular markers or any DNA data that could guide future conservation actions. These three indicators are illustrated for a hypothetical species in Fig. 4.2. While Indicators 1 and 3 are based on a present state (i.e., only current data is required), Indicator 2 is more challenging because it requires historic population data and monitoring efforts. Although such data is presumably available for some of the commercially important tree species (e.g., from national forest inventories), the indicator will likely be difficult to assess for rare species with scattered distributions.

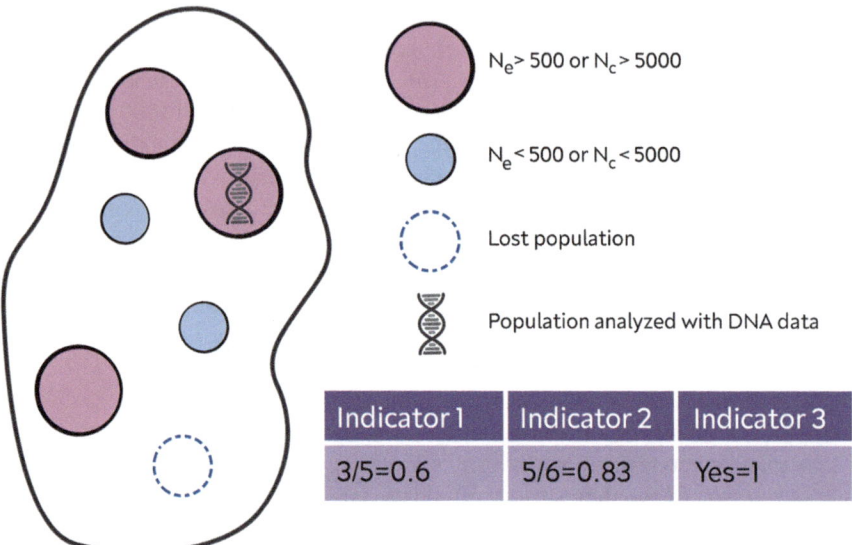

Fig. 4.2 Illustrative and simplified example of the application of three indicators suggested for monitoring genetic diversity within a hypothetical species. The indicators relate to goals and targets defined by the CBD under the Post-2020 Global Biodiversity Framework: the fraction of populations with Ne > 500 or Nc > 5000 (Indicator 1), the fraction of populations still existing (Indicator 2), and the presence of populations for which genetic data exists (Indicator 3). The black outline represents the species distribution, and the circles inside represent individual populations

The State of Genetic Diversity at a Global Scale, and Initiatives to Conserve It

The available data on anthropogenic damage to forests and other ecosystems offers a desolating perspective. Hoban et al. (2023a, b) provided an overview of the loss of genetic diversity in the recent past by reporting DNA-based studies documenting high genetic diversity losses over the past 50–100 years, particularly in island species (28% loss), and harvested fish species (14% loss) (Pinsky and Palumbi 2014; Leigh et al. 2019). Over the past few decades, the genetic diversity of International Union for Conservation of Nature (IUCN) Threatened species has declined by 9–33% on average (estimates are based on a mathematical relationship between population loss and genetic diversity loss in several plant and animal species; Exposito-Alonso et al. 2022). Hoban et al. (2021) predicted that, based on population genetic theory and the Living Planet Index (www.livingplanetindex.org), populations may ultimately lose an average of 19 to 66% of their genetic (allelic) diversity within the next few decades without interventions to stop and reverse species' population declines. More specific numbers for forest ecosystems are hitherto not available. The most striking example of forest habitat loss is South America. Most forest ecosystems in this region are rapidly declining, and the predictions mentioned above are therefore presumably valid for forest species as well. Projections by Frankham (2022) have shown that a loss of "only" 10% of genetic diversity within a given species in the long term (more than 100 years) will already give rise to increased levels of inbreeding that can severely debilitate that species' evolutionary potential to adapt to a changing environment. Policymakers are urged to take measures accordingly.

The CBD marked a historic milestone as the world's first international treaty uniting nearly all nations in a common mission to preserve biodiversity and promote sustainable utilization and equitable distribution of the benefits it generates. Since entering into force in 1993, the CBD has developed multiple frameworks; preparations are currently underway for the Post-2020 GBF (CBD 2022). The Post-2020 GBF is expected to include four high-level goals for 2050 related to the state of nature resulting from conservation, nature's contribution to people and its sustainable use, shared benefits arising from biodiversity, and means of implementation and resource mobilization, along with 22 action targets for changes in human society and activities required by 2030 to achieve those goals (Hoban et al. 2023a, b). The GBF is currently still being negotiated and must be agreed upon by all parties; it therefore reflects scientific input, political negotiation, perceived feasibility, and compromise. The CBD encourages countries to develop strategies and action plans to conserve and sustainably manage their forest biodiversity.

The signatory states have committed to monitoring and reporting on biodiversity development in their countries. In the original CBD, genetic resources were mentioned, but not explicitly with regard to the conservation of genetic diversity of wild animals and plants. In the new Post-2020 GBF development, this topic is to be extended to include wild animals, plants, and fungi (Laikre et al., 2020). Defining implementable indicators is a prerequisite for countries to report on their respective status, but discussions on this matter are still in progress (Hoban et al. 2023a, b). Frankham (2022) describes the process and recommends that goals, milestones, and targets in the GBF should mention as core elements: (i) the maintenance of sufficiently large populations (rather than permitting an "acceptable loss of genetic diversity"), (ii) sufficient and appropriate genetic exchange among populations (connectivity) and (iii) active monitoring and management of genetic diversity, as well as (iv) no loss of populations. Respective indicators have been described in the previous section. Furthermore, such indicators have been recommended and are also required for reporting under other biodiversity conservation schemes. A more extensive treatment of the topic of biodiversity indicators is presented in Chap. 8.

In connection to the CBD, the European Commission has launched the EU Forest and Biodiversity Strategy 2030 (European Commission 2020), a comprehensive, ambitious, and long-term plan to protect nature and reverse the degradation of ecosystems. The strategy aims to place Europe's biodiversity on a path to recovery by 2030 and encompasses specific actions and commitments. It is the proposal for the EU's contribution to the upcoming international negotiations on the Post-2020 GBF and a core element of the European Green Deal (EC 2019). In particular, it calls for the establishment of ecological corridors to prevent genetic isolation, allow for species migration, and maintain and enhance healthy ecosystems.

Another important initiative primarily concerned with the genetic diversity of forest trees and other woody species that are of realized or potential economic, environmental, scientific, or societal value is the FAO scheme on forest genetic resources. Work on forest genetic resources at FAO was initiated in the 1950s, and since then FAO has supported countries in their efforts to improve the management of forest genetic resources and promoted regional and international cooperation. Within FAO, the Commission on Genetic Resources for Food and Agriculture requested countries to provide input to and guide the preparation of a report on "The State of the World's Forest Genetic Resources" (FAO 2014; a second updated report is currently in preparation). Furthermore, FAO agreed on strategic priorities which the FAO Conference adopted in June 2013 as the Global Plan of Action for the Conservation, Sustainable Use, and Development of Forest Genetic Resources. The results of the 2014 report show that studies have thus far described genetic parameters for less than 1 percent of tree species and that no data is available for many countries. Although the Global Plan of Action recognizes that both the number of (molecular genetic) studies and the number of species studied have increased significantly over the past 20 years, it regrets that little of the accumulating knowledge has direct application in management, improvement, or conservation. The report shows that most research on forest genetic resources has been concentrated on temperate conifers, eucalypts, several acacia species, teak, and a few other broadly adapted, widely planted, and rapidly growing species—mostly with the aim of describing genetic resources for breeding rather than for conservation. Genomic or marker-assisted selection is close to being realized, but major gaps still exist in phenotyping and data management. The report also states that many of the species identified as priorities, especially for local use, have received little or no research attention, indicating a need to associate funding with priority-setting practices. An Intergovernmental Technical Working Group on Forest Genetic Resources (ITWG-FGR) was also established within FAO to address issues relevant to the conservation and sustainable use of forest genetic resources as well as advising and making recommendations concerning the report preparation process (see also Chap. 15).

In Europe, a specific network on FGR conservation and use is in place. EUFORGEN—the European Forest Genetic Resources Programme—is an international cooperation program that promotes the conservation and sustainable use of forest genetic resources as an integral part of sustainable forest management. It was established in 1994 as a result of a resolution adopted in 1990 by the first Ministerial Conference of the Forest Europe process. Experts from member countries come together within EUFORGEN to exchange information and experience, analyze policies and practice, and develop science-based strategies, tools, and methods to improve the management of FGR.

The International Union for Conservation of Nature (IUCN) is the main global organization providing expertise, assessments, and guidelines for conservation efforts. It publishes the IUCN Red List of Threatened Species, which includes information on the conservation status of various groups of organisms (mainly plant and animal species, including many forest species). The Red List is a key data source that also informs CBD reporting, and the respective data could be used as a baseline for genetic monitoring. The Red List also shows that we still lack data on biodiversity: Only 6% and 15% of all known plant and animal species, respectively, have been assessed for their conservation status. The available data show, for example, that an alarming 34% of all conifer species are threatened by extinction and listed in the IUCN Red List (IUCN 2019).

The Living Planet Index and Living Planet Report are published by the World Wildlife Fund (WWF) biannually (Almond et al. 2022), reporting trends in biodiversity for animal species on a global scale. The 2022 edition shows a 69% global decline in the relative abundance of monitored wildlife populations between 1970 and 2018. Latin America exhibits the greatest regional decline in average population abundance (94%), while freshwater species populations have seen the greatest overall global decline (83%). Data for the report are partially provided by IUCN and the Intergovernmental Panel on Climate Change (IPCC).

The United Nations Collaborative Program on Reducing Emissions from Deforestation and Forest Degradation (UN-REDD) supports countries in their efforts to reduce emissions from deforestation and forest degradation while promoting sustainable forest management. It is the UN reference knowledge and advisory platform on forest-related solutions to the climate crisis. UN-REDD promotes approaches that ensure the environmental integrity of carbon emissions reductions while supporting non-carbon benefits—from safeguarding biodiversity to supporting local livelihoods and promoting the rights of indigenous peoples. Genetic diversity preservation is a part of the broader conservation objectives of this program (UN-REDD Program, www.un-redd.org).

The initiatives mentioned above, along with many others, contribute to the global effort to protect and conserve the genetic diversity of forest ecosystems, acknowledging the importance of biodiversity for ecosystem resilience, sustainability, and human well-being.

Active Participation of Conservation Geneticists in Policy Development Is Needed

In this chapter, we presented an overview of the knowledge on and status of forest genetic diversity, highlighting the anthropogenic influences on it and the associated policy initiatives to conserve it. Policy developments, implementations, and conservation decisions need to be based on scientific research. To conserve and improve biological and genetic diversity, scientists should actively work to suggest and improve related policy development in close cooperation with policymakers. This can be a difficult process for both sides (Hoban et al. 2023a, b). Typically, foresters and scientists are not involved in political processes, while policymakers are usually not familiar with concepts specific to the field of research. The integration of these two spheres of action is very time-consuming and demanding, yet still frequently does not deliver the output needed by scientists (e.g., peer-reviewed publications).

Nevertheless, steady interaction with decision-makers is the only path toward sustainable and long-term stabilization and conservation of genetic diversity and ecosystem functioning. Research questions often do not directly address the needs of forest or conservation managers (Taylor and Dizon 1999; Geburek and Konrad 2008; Holderegger and Segelbacher 2016). This is currently changing as genetic diversity and connectivity are becoming increasingly recognized as important parts of successful conservation and restoration strategies (Jalonen et al. 2018; Aguilar et al. 2019). Nevertheless, interactions and dialogue need to be intensified to work toward the common goal of sustainable conservation of forest biodiversity. In addition, the general public constantly needs to be informed on initiatives and principles to generate the required attention and (financial) support.

Box 4.2 Providing reproductive material for sustainable forest and landscape restoration (FLR)

The purpose of forest and landscape restoration (FLR) is to restore ecological processes at the landscape scale to maintain biodiversity and ecosystem functions and enhance resilience to environmental change. FLR has become the aim of a range of multi-million-hectare commitments in many parts of the world to mitigate climate change effects and halt the loss of biodiversity (e.g., European Green Deal; EC 2019). To achieve these ambitious goals, billions of seedlings are needed, yet the provision of seeds has often received little attention in the planning of restoration projects, and Jalonen et al. (2018) reported widespread use of unsuitable reproductive material for FLR. As explained in this chapter, reproductive material to be used in FLR (i.e. seed, seedlings, or vegetatively produced propagules) needs to be locally adapted and provide sufficient genetic diversity to build stable, adaptable, disease-resistant, and self-reliant forests. However, due to a lack of awareness in restoration practices, FLR projects often use seeds that are ill-adapted to the local conditions or offer insufficient genetic variation, for example, when they are collected from a small pool of mother trees (Broadhurst 2013; Liu et al. 2008; Navascués and Emerson 2007; Thomas et al. 2014). On the other hand, habitat loss, fragmentation of source populations, and climate change have a joint negative influence on the genetic diversity of seed lots and their actual availability (Aguilar et al. 2019). To counteract this development, Jalonen et al. (2018) recommend the following policy interventions:
1. *Assembling a national assessment of seed supply and demand for meeting FLR targets.* Identification of gaps in seed supply and development of strategies for sustainable sourcing should occur beyond specific project demand. Seed supply assessments should consider quantity, genetic

(continued)

quality (diversity), and geographic origin. This strategy will be most efficient if applied across national borders.

2. *Adjusting FLR targets and funding cycles.* Building up long-term seed supplies goes beyond the average FLR project duration. FLR projects should include investments in seed availability and access to quality seeds to avoid unfit selection and deployment of low-quality plants. In this context, funding schemes and projects should allow long enough durations to avoid unsuccessful FLR efforts.
3. *Exchanging of knowledge and experiences regarding seed selection and supply options.* The unprecedented amounts of seed and plant material currently needed to meet FLR targets require knowledge sharing among actors to identify functioning approaches for different species and socioeconomic contexts, as well as which actors are most efficient at which stage of the process. Multi-stakeholder platforms can efficiently bring together these different actors (e.g., seed suppliers, restoration practitioners, and policymakers). Such platforms already exist in some countries and could be expanded and developed further (Melo et al. 2013).
4. *Facilitating seed exchange across landscapes.* In many cases, seeds for FLR are collected from origins as close as possible to the deployment site, often at the cost of genetic diversity and quality of the seeds. This approach needs to be shifted toward genetically more viable seed sources. Documentation of employed seed sources should also become common practice to allow the performance of different seed origins to be compared. Use of multiple different seed sources and stimulation of natural gene flow by restoring landscape connectivity is also recommended (e.g., Sgrò et al. 2011).
5. *Establishing regulations on seed quality and strengthening capacities for compliance.* Other than in regular forestry, where regulations on the collection and marketing of forest reproductive material exist (e.g., the OECD schemes on forest reproductive material or the EU Directive 105/1999/EU on Forest Reproductive Material in the European Union), this is not the case in many countries where FLR is implemented. Accreditation of seed sources and nurseries is an important step toward ensuring the availability of high-quality plant material from known sources. It should become common practice to only use seeds from accredited sources for subsidized FLR projects.

References

Aguilar R, Ashworth L, Galetto L, Aizen MA (2006) Plant reproductive susceptibility to habitat fragmentation: review and synthesis through a meta-analysis. Ecol Lett 9(8):968–980. https://doi.org/10.1111/j.1461-0248.2006.00927.x

Aguilar R, Quesada M, Ashworth L, Herrerias-Diego YVONNE, Lobo J (2008) Genetic consequences of habitat fragmentation in plant populations: susceptible signals in plant traits and methodological approaches. Mol Ecol 17(24):5177–5188. https://doi.org/10.1111/j.1365-294X.2008.03971.x

Aguilar R, Ashworth L, Calviño A, Quesada M (2012) What is left after sex in fragmented habitats? Assessing the quantity and quality of progeny in the endemic tree *Prosopis caldenia* (Fabaceae). Biol Conserv 152:81–89. https://doi.org/10.1016/j.biocon.2012.03.021

Aguilar R, Calviño A, Ashworth L, Aguirre-Acosta N, Carbone LM, Albrieu-Llinas G et al (2018) Unprecedented plant species loss after a decade in fragmented subtropical Chaco Serrano forests. PLoS One 13(11):e0206738. https://doi.org/10.1371/journal.pone.0206738

Aguilar R, Cristóbal-Pérez EJ, Balvino-Olvera FJ, de Jesús Aguilar-Aguilar M, Aguirre-Acosta N, Ashworth L et al (2019) Habitat fragmentation reduces plant progeny quality: a global synthesis. Ecol Lett 22(7):1163–1173. https://doi.org/10.1111/ele.13272

Aitken SN, Whitlock MC (2013) Assisted gene flow to facilitate local adaptation to climate change. Annu Rev Ecol Evol Syst 44(1):367–388. https://doi.org/10.1146/annurev-ecolsys-110512-135747

Alberto FJ, Aitken SN, Alía R, González-Martínez SC, Hänninen H, Kremer A et al (2013) Potential for evolutionary responses to climate change – evidence from tree populations. Glob Chang Biol 19:1645–1661. https://doi.org/10.1111/gcb.12181

Allendorf FW, Funk WC, Aitken SN, Byrne M, Luikart G (2022) Conservation and the genomics of populations, 3rd edn. Oxford University Press, p 746. https://doi.org/10.1093/oso/9780198856566.001.0001

Allnutt TF, Ferrier S, Manion G, Powell GV, Ricketts TH, Fisher BL et al (2008) A method for quantifying biodiversity loss and its application to a 50-year record of deforestation across Madagascar. Conserv Lett 1(4):173–181. https://doi.org/10.1111/j.1755-263X.2008.00027.x

Almond REA, Grooten M, Juffe Bignoli D, Petersen T (2022) Living Planet Report 2022 – Building a nature-positive society. WWF, Gland

Alsos IG, Ehrich D, Thuiller W, Eidesen PB, Tribsch A, Schönswetter P et al (2012) Genetic consequences of climate change for northern plants. Proc R Soc Lond B Biol Sci 279(1735):2042–2051. https://doi.org/10.1098/rspb.2011.2363

Armbruster P, Reed DH (2005) Inbreeding depression in benign and stressful environments. Heredity 95:235–242. https://doi.org/10.1038/sj.hdy.6800721

Arnold PA, Kruuk LEB, Nicotra AB (2019) How to analyse plant phenotypic plasticity in response to a changing climate. New Phytol 222:1235–1241. https://doi.org/10.1111/nph.15656

Ashworth L, Martí ML (2011) Forest fragmentation and seed germination of native species from the Chaco Serrano Forest. Biotropica 43(4):496–503. https://doi.org/10.1111/j.1744-7429.2010.00721.x

Anagnostakis SL (1988) *Cryphonectria parasitica*, cause of chestnut blight. In: Sidhu GS (ed) Advances in plant pathology. Academic Press, London, pp 123–136. https://doi.org/10.1016/B978-0-12-033706-4.50011-6

Bakker J, van Rijswijk MEC, Weissing FJ, Bijlsma R (2010) Consequences of fragmentation for the ability to adapt to novel environments in experimental *Drosophila* metapopulations. Conserv Genet 11:435–448. https://doi.org/10.1007/s10592-010-0052-5

Balboni C, Berman A, Burgess R, Olken BA (2023) The economics of tropical deforestation. Ann Rev Econ 15:723–754. https://doi.org/10.1146/annurev-economics-090622-024705

Balkenhol N, Cushman SA, Storfer A, Waits LP (2015) Introduction to landscape genetics – concepts, methods, applications. In: Balkenhol N, Cushman SA, Storfer AT, Waits LP (eds) Landscape genetics (pp 1–8). https://doi.org/10.1002/9781118525258.ch01

Barghi N, Tobler R, Nolte V, Jakšić AM, Mallard F, Otte KA et al (2019) Genetic redundancy fuels polygenic adaptation in *Drosophila*. PLoS Biol 17(2):1–31. https://doi.org/10.1371/journal.pbio.3000128

Berry O, Tocher MD, Sarre SD (2004) Can assignment tests measure dispersal? Mol Ecol 13(3):551–561. https://doi.org/10.1046/j.1365-294X.2004.2081.x

Bernabo JC, Webb T III (1977) Changing patterns in the Holocene pollen record of northeastern North America: a mapped summary. Quat Res 8(1):64–96. https://doi.org/10.1016/0033-5894(77)90057-6

Bouteiller XP, Moret F, Ségura R, Klisz M, Martinik A, Monty A, Pino J, van Loo M, Wojda T, Porté AJ, Mariette S (2021) The seeds of invasion: enhanced germination in invasive European populations of black locust (Robinia pseudoacacia L.) compared to native American populations. Plant Biol 23(6):1006–1017. https://doi.org/10.1111/plb.13332

Burczyk J, Adams WT, Birkes DS, Chybicki IJ (2006) Using genetic markers to directly estimate gene flow and reproductive success parameters in plants on the basis of naturally regenerated seedlings. Genetics 173(1):363–372. https://doi.org/10.1534/genetics.105.046805

Bizoux JP, Daïnou K, Bourland N, Hardy OJ, Heuertz M, Mahy G, Doucet JL (2009) Spatial genetic structure in *Milicia excelsa* (Moraceae) indicates extensive gene dispersal in a low-density wind-pollinated tropical tree. Mol Ecol 18(21):4398–4408. https://doi.org/10.1111/j.1365-294X.2009.04365.x

Bogdziewicz M, Kelly D, Thomas PA, Lageard JG, Hacket-Pain A (2020) Climate warming disrupts mast seeding and its fitness benefits in European beech. Nat Plants 6(2):88–94. https://doi.org/10.1038/s41477-020-0592-8

Bonnet T, Morrissey MB, De Villemereuil P, Alberts SC, Arcese P, Bailey LD et al (2022) Genetic variance in fitness indicates rapid contemporary adaptive evolution in wild animals. Science 376(6596):1012–1016. https://doi.org/10.1126/science.abk0853

Brasier CM (2000) Intercontinental spread and continuing evolution of the Dutch Elm disease pathogens. In: Dunn CP (ed) The elms: breeding, conservation, and disease management. Springer, Boston, MA, pp 61–72. https://doi.org/10.1007/978-1-4615-4507-1_4

Brauer CJ, Sandoval-Castillo J, Gates K, Hammer MP, Unmack PJ, Bernatchez L, Beheregaray LB (2023) Natural hybridization reduces vulnerability to climate change. Nat Clim Chang 13(3):282–289. https://doi.org/10.1038/s41558-022-01585-1

Broadhurst LM (2013) A genetic analysis of scattered Yellow Box trees (*Eucalyptus melliodora* A. Cunn. ex Schauer, Myrtaceae) and their restored cohorts. Biol Conserv 161:48–57. https://doi.org/10.1016/j.biocon.2013.02.016

Brook BW, Bradshaw CJ, Traill LW, Frankham R (2011) Minimum viable population size: not magic, but necessary. Trends Ecol Evol 26(12):619–620. https://doi.org/10.1016/j.tree.2011.09.006

Brundu G, Pauchard A, Pyšek P, Pergl J, Bindewald AM, Brunori A et al (2020) Global guidelines for the sustainable use of non-native trees to prevent tree invasions and mitigate their negative impacts. NeoBiota 61:65–116. https://doi.org/10.3897/neobiota.65.58380

Burdon JJ, Thrall PH (2001) The demography and genetics of host-pathogen interactions. In: Silvertown J, Antonovics J (eds) Integrating ecology and evolution in a spatial context. Blackwell Science, London, pp 197–217

Burger R, Lynch M (1995) Evolution and extinction in a changing environment: a quantitative-genetic analysis. Evolution 49(1):151–163. https://doi.org/10.1111/j.1558-5646.1995.tb05967.x

Cavender-Bares J, Schneider FD, Santos MJ, Armstrong A, Carnaval A, Dahlin KM et al (2022) Integrating remote sensing with ecology and evolution to advance biodiversity conservation. Nat Ecol Evol 6:506–519. https://doi.org/10.1038/s41559-022-01702-5

CBD (2022) Report of the open-ended working group on the post-2020 global biodiversity framework on its fourth meeting. CBD/WG2020/4/4. UNEP. https://www.cbd.int/doc/c/3303/d892/4fd11c27963bd3f826a961e1/wg2020-04-04-en.pdf

Chen I-C, Hill JK, Ohlemüller R, Roy DB, Thomas CD (2011) Rapid range shifts of species associated with high levels of climate warming. Science 333:1024–1026. https://doi.org/10.1126/science.1206432

Charlesworth D, Willis JH (2009) The genetics of inbreeding depression. Nat Rev Genet 10(11):783–796. https://doi.org/10.1038/nrg2664

Chybicki IJ (2018) NMπ—improved re-implementation of NM+, a software for estimating gene dispersal and mating patterns. Mol Ecol Resour 18(1):159–168. https://doi.org/10.1111/1755-0998.12710

Chybicki IJ (2023) NMπ 2.0: software update to minimize the risk of false positives among determinants of reproductive success. Mol Ecol Resour 23(5):1168–1181. https://doi.org/10.1111/1755-0998.13767

Crutsinger GM, Collins MD, Fordyce JA, Sanders NJ (2007) Temporal dynamics in non-additive responses of arthropods to host-plant genotypic diversity. Oikos 117:255–264. https://doi.org/10.1111/j.2007.0030-1299.16276.x

D'Odorico P, Schuman MC, Kurz M, Csilléry K (2023) Discerning oriental from European beech by leaf spectroscopy: operational and physiological implications. Forest Ecol Manag 541:121056. https://doi.org/10.1016/j.foreco.2023.121056

Davis MB (1983) Quaternary history of deciduous forests of eastern North America and Europe. Ann Mo Bot Gard 70(3):550–563. https://doi.org/10.2307/2992086

Davis MB, Shaw RG (2001) Range shifts and adaptive responses to Quaternary climate change. Science 292(5517):673–679. https://doi.org/10.1126/science.292.5517.673

Dayton PK (1972) Toward an understanding of community resilience and the potential effects of enrichments to the benthos at McMurdo Sound, Antarctica. In: Parker BC (ed) Proceedings of the colloquium on conservation problems in Antarctica. Allen Press, Lawrence, pp 81–89

DeHayes DH, Jacobson GL Jr, Schaberg PG, Bongarten B, Iverson L, Dieffenbacher-Krall AC (2000) Forest responses to changing climate: lessons from the past and uncertainty for the future. In: Mickler RA, Birdsey RA, Hom JL (eds) Responses of northern U.S. forests to environmental change. Springer, New York, pp 495–540. https://doi.org/10.1007/978-1-4612-1256-0_14

DeWitt TJ, Sih A, Wilson DS (1998) Costs and limits of phenotypic plasticity. Trends Ecol Evol 13(2):77–81. https://doi.org/10.1016/S0169-5347(97)01274-3

Dey DC, Jacobs D, McNabb K, Miller G, Baldwin V, Foster G (2008) Artificial regeneration of major oak (Quercus) species in the eastern United States—a review of the literature. For Sci 54(1):77–106. https://doi.org/10.1093/forestscience/54.1.77

Dick CW, Hardy OJ, Jones FA, Petit RJ (2008) Spatial scales of pollen and seed-mediated gene flow in tropical rain forest trees. Trop Plant Biol 1:20–33. https://doi.org/10.1007/s12042-007-9006-6

Edwards DG, El-Kassaby YA (1996) The biology and management of coniferous forest seeds: genetic aspects. For Chron 72(5):481–484. https://doi.org/10.5558/tfc72481-5

European Commission (2020) EU biodiversity strategy for 2030: bringing nature back into our lives. COM/2020/380. https://eur-lex.europa.eu/legal-content/EN/TXT/?uri=celex:52020DC0380

Exposito-Alonso M, Booker TR, Czech L et al (2022) Genetic diversity loss in the Anthropocene. Science 377:1431–1435. https://doi.org/10.1126/science.abn5642

El-Kassaby YA, Benowicz A (2000) Effects of commercial thinning on genetic, plant species and structural diversity in second growth Douglas-fir *(Pseudotsuga menziesii* (Mirb.) Franco) stands. For Genet 7(3):193–203

Ellegren H (2014) Genome sequencing and population genomics in non-model organisms. Trends Ecol Evol 29(1):51–63. https://doi.org/10.1016/j.tree.2013.09.008

Etterson JR (2004) Evolutionary potential of *Chamaecrista fasciculata* in relation to climate change. I. Clinal patterns of selection along an environmental gradient in the Great Plains. Evolution 58(7):1446–1458. https://doi.org/10.1111/j.0014-3820.2004.tb01726.x

European Commission (2019) Communication from the commission to The European Parliament, The European Council, The Council, The European Economic and Social Committee and the Committee of the Regions. The European Green Deal. Brussel. European Comission, Brussels, 11.12.2019

Fady B, Bozzano M (2021) Effective population size does not make a practical indicator of genetic diversity in forest trees. Biol Conserv 253:108904. https://doi.org/10.1016/j.biocon.2020.108904

Fagny M, Austerlitz F (2021) Polygenic adaptation: integrating population genetics and gene regulatory networks. Trends Genet 37(7):631–638. https://doi.org/10.1016/j.tig.2021.03.005

FAO (2014) The state of the world's forest genetic resources. United Nations Food and Agriculture Organization

FAO (2022) The State of the World's Forests 2022. Forest pathways for green recovery and building inclusive, resilient and sustainable economies. Rome, FAO

Fisher RA (1930) The genetical theory of natural selection. Oxford University Press, Oxford

Flanagan SP, Jones AG (2018) The future of parentage analysis: from microsatellites to SNPs and beyond. Mol Ecol 28(3):544–567. https://doi.org/10.1111/mec.14988

Flather CH, Hayward GD, Beissinger SR, Stephens PA (2011) Minimum viable populations: is there a 'magic number' for conservation practitioners? Trends Ecol Evol 26:307–316. https://doi.org/10.1016/j.tree.2011.03.001

Franklin IR (1980) Evolutionary change in small populations. In: Soulé ME, Wilcox BA (eds) Conservation biology: an evolutionary-ecological perspective. Sinauer, Sunderland, MA, pp 135–149

Foest JJ, Bogdziewicz M, Pesendorfer MB, Ascoli D, Cutini A, Nussbaumer A, Verstraeten A, Beudert B, Chianucci F, Mezzavilla F, Gratzer G, Kunstler G, Meesenburg H, Wagner M, Mund M, Cools N, Vacek S, SchmidtW, Vacek Z, Hacket-Pain A (2024) Widespread breakdown in masting in European beech due to rising summer temperatures. Global Change Biology 30(5):e17307. https://doi.org/10.1111/gcb.17307

Forcada J, Trathan PN, Murphy EJ (2008) Life history buffering in Antarctic mammals and birds against changing patterns of climate and environmental variation. Glob Chang Biol 14:2473–2488. https://doi.org/10.1111/j.1365-2486.2008.01678.x

Fox RJ, Donelson JM, Schunter C, Ravasi T, Gaitan-Espitia JD (2019) Beyond buying time: the role of plasticity in phenotypic adaptation to rapid environmental change. Philos Trans R Soc B Biol Sci 374:20180174. https://doi.org/10.1098/rstb.2018.0174

Frankham R (2022) Evaluation of proposed genetic goals and targets for the convention on biological diversity. Conserv Genet 23(5):865–870. https://doi.org/10.1007/s10592-022-01459-1

Frankham R, Ballou JD, Briscoe DA (2010) Introduction to conservation genetics. Cambridge University Press, Cambridge

Frankham R, Ballou JD, Eldridge MD, Lacy RC, Ralls K, Dudash MR, Fenster CB (2011) Predicting the probability of outbreeding depression. Conserv Biol 25(3):465–475. https://doi.org/10.1111/j.1523-1739.2011.01662.x

Frankham R, Bradshaw CJA, Brook BW (2014) Genetics in conservation management: revised recommendations for the 50/500 rules, Red List criteria and population viability analyses. Biol Conserv 170:56–63. https://doi.org/10.1016/j.biocon.2013.12.036

Franklin IR, Allendorf FW, Jamieson IG (2014) The 50/500 rule is still valid—reply to Frankham et al. Biol Conserv 176:284–285. https://doi.org/10.1016/j.biocon.2014.05.004

Gagnaire PA, Broquet T, Aurelle D, Viard F, Souissi A, Bonhomme F, Arnoud-Haond S, Bierne N (2015) Using neutral, selected, and hitchhiker loci to assess connectivity of marine populations in the genomic era. Evol Appl 8(8):769–786. https://doi.org/10.1111/eva.12288

García-Ramos G, Kirkpatrick M (1997) Genetic models of adaptation and gene flow in peripheral populations. Evolution 51(1):21–28. https://doi.org/10.1111/j.1558-5646.1997.tb02384.x

Geburek T, Konrad H (2008) Why the conservation of forest genetic resources has not worked. Conserv Biol 22(2):267–274. https://doi.org/10.1111/j.1523-1739.2008.00900.x

Goldstein DB, Holsinger KE (1992) Maintenance of polygenic variation in spatially structured populations: roles for local mating and genetic redundancy. Evolution 46(2):412–429. https://doi.org/10.2307/2409861

Geber MA, Dawson TE (1993) Evolutionary responses of plants to global change. In: Kareiva PM, Kingsolver JG, Huey RB (eds) Biotic interactions and global change. Sinauer, Sunderland, MA, pp 179–197

González-Martínez SC, Burczyk J, Nathan RAN, Nanos N, Gil L, Alia R (2006) Effective gene dispersal and female reproductive success in Mediterranean maritime pine (Pinus pinaster Aiton). Mol Ecol 15(14):4577–4588. https://doi.org/10.1111/j.1365-294X.2006.03118.x

González-Varo JP, Albaladejo RG, Aparicio A, Arroyo J (2010) Linking genetic diversity, mating patterns and progeny performance in fragmented populations of a Mediterranean shrub. J Appl Ecol 47(6):1242–1252. https://doi.org/10.1111/j.1365-2664.2010.01879.x

Gregorius HR (1996) The contribution of the genetics of populations to ecosystem stability. Silvae Genetica 45(5):267–271

Hamrick JL, Godt MJ (1990) Allozyme diversity in plant species. In: Brown AHD, Clegg MT, Kahler AL, Weir BS (eds) Plant population genetics, breeding and genetic resources. Sinauer Associates, Sunderland, MA, pp 43–63

Hadley AS, Betts MG (2009) Tropical deforestation alters hummingbird movement patterns. Biol Lett 5(2):207–210. https://doi.org/10.1098/rsbl.2008.0691

Holderegger R, Segelbacher G (2016) Naturschutzgenetik: Ein Handbuch für die Praxis. Haupt, Bern. 247pp

Honnay O, Jacquemyn H (2007) Susceptibility of common and rare plant species to the genetic consequences of habitat fragmentation. Conserv Biol 21(3):823–831. https://doi.org/10.1111/j.1523-1739.2006.00646.x

Huntley B (1991) How plants respond to climate change: migration rates, individualism and the consequences for plant communities. Ann Bot 67:15–22

Huntley B, Birks HJB (1983) An atlas of past and present pollen maps for Europe: 0–13000 BP. Cambridge University Press

Huntley B, Webb T III (eds) (1988) Vegetation history, Handbook of vegetation science, vol 7. 828 pp. Springer

Heena S, Kandarp J, Upasana M, Sanober W, Gulnaz Z, Shravani R et al (2023) Next-generation sequencing technology: current trends and advancements. Biology 12(7):997. https://doi.org/10.3390/biology12070997

Herms DA, McCullough DG (2014) Emerald ash borer invasion of North America: history, biology, ecology, impacts, and management. Annu Rev Entomol 59:13–30. https://doi.org/10.1146/annurev-ento-011613-162051

Himanen K, Kennedy S, Bordács S, Yüksel T, Kraigher H, Gömöry D (2021) Nursery practices. In: Gömöry D, Himanen K, Tollefsrud MM, Uggla C, Kraigher H, Bordács S, Alizoti P, A'Hara S, Frank A, Proschowsky GF, Frýdl J, Geburek T, Guibert M, Ivanković M, Jurše A, Kennedy S, Kowalczyk J, Liesebach H, Maaten T, Pilipović A, Proietti R, Schneck V, Servais A, Skúlason B, Sperisen C, Wolter F, Yüksel T, Bozzano M (eds) Genetic aspects in production and use of forest reproductive material: collecting scientific evidence to support the development of guidelines and decision support tools. European Forest Genetic Resources Programme (EUFORGEN), European Forest Institute, pp 92–104

Hoban S, Schlarbaum S (2014) Optimal sampling of seeds from plant populations for *ex-situ* conservation of genetic biodiversity, considering realistic population structure. Biol Conserv 177:90–99. https://doi.org/10.1016/j.biocon.2014.06.014

Hoban S, Bruford M, D'Urban Jackson J, Lopes-Fernandes M, Heuertz M, Hohenlohe PA et al (2020) Genetic diversity targets and indicators in the CBD post-2020 Global Biodiversity Framework must be improved. Biol Conserv 248:108654. https://doi.org/10.1016/j.biocon.2020.108654

Hoban S, Bruford M, Funk WC, Galbusera P, Griffith MP, Grueber CE et al (2021) Global commitments to conserving and monitoring genetic diversity are now necessary and feasible. Bioscience 71(9):964–976. https://doi.org/10.1093/biosci/biab054

Hoban S, Bruford M, da Silva JM, Funk WC, Frankham R, Gill MJ et al (2023a) Genetic diversity goals and targets have improved, but remain insufficient for clear implementation of the post-2020 global biodiversity framework. Conserv Genet 24:181–191. https://doi.org/10.1007/s10592-022-01492-0

Hoban S, da Silva JM, Mastretta-Yanes A, Grueber CE, Heuertz M, Hunter ME et al (2023b) Monitoring status and trends in genetic diversity for the convention on biological diversity: an ongoing assessment of genetic indicators in nine countries. Conserv Lett 16(3):e12953

Höllinger I, Pennings PS, Hermisson J (2019) Polygenic adaptation: from sweeps to subtle frequency shifts. PLoS Genet 15(3):e1008035. https://doi.org/10.1371/journal.pgen.1008035

Holderegger R, Balkenhol N, Bolliger J, Engler JO, Gugerli F, Hochkirch A et al (2019) Conservation genetics: linking science with practice. Mol Ecol 28:3848–3856. https://doi.org/10.1111/mec.15202

Hughes AR, Inouye BD, Johnson TJ, Underwood N, Vellend M (2008) Ecological consequences of genetic diversity. Ecol Lett 11(6):1–15. https://doi.org/10.1111/j.1461-0248.2008.01179.x

IUCN (2019) The IUCN red list of threatened species. Version 2022–2. https://www.iucnredlist.org (accessed 1stSeptember 2023)

Jalonen R, Valette M, Boshier D, Duminil J, Thomas E (2018) Forest and landscape restoration severely constrained by a lack of attention to the quantity and quality of tree seed: Insights from a global survey. Conserv Lett 11(4):e12424. https://doi.org/10.1111/conl.12424

Jung J, Maeda M, Chang A, Bhandari M, Ashapure A, Landivar-Bowles J (2021) The potential of remote sensing and artificial intelligence as tools to improve the resilience of agriculture production systems. Curr Opin Biotechnol 70:15–22. https://doi.org/10.1016/j.copbio.2020.09.003

Jones FA, Chen J, Weng GJ, Hubbell SP (2005) A genetic evaluation of seed dispersal in the neotropical tree Jacaranda copaia (Bignoniaceae). Am Nat 166(5):543–555. https://doi.org/10.1086/491661

Jordano P, García C, Godoy JA, Garcia-Castaño J (2007) Differential contribution of frugivores to complex seed dispersal patterns. Proc Natl Acad Sci 104(9):3278–3282. https://doi.org/10.1073/pnas.0606793104

Jump AS, Peñuelas J (2005) Running to stand still: adaptation and the response of plants to rapid climate change. Ecol Lett 8(9):1010–1020. https://doi.org/10.1111/j.1461-0248.2005.00796.x

Kassa A, Konrad H, Geburek T (2017) Landscape genetic structure of *Olea europaea* subsp. *cuspidata* in Ethiopian highland forest fragments. Conserv Genet 18:1463–1474. https://doi.org/10.1007/s10592-017-0993-z

Kavaliauskas D, Šeho M, Baier R, Fussi B (2021) Genetic variability to assist in the delineation of provenance regions and selection of seed stands and gene conservation units of wild service tree (*Sorbus torminalis* (L.) Crantz) in southern Germany. Eur J For Res 140(3):551–565. https://doi.org/10.1007/s10342-020-01352-x

Kelly D, Hart DE, Allen RB (2001) Evaluating the wind pollination benefits of mast seeding. Ecology 82(1):117–126. https://doi.org/10.1890/0012-9658(2001)082[0117:ETWPBO]2.0.CO;2

Kremer A, Ronce O, Robledo-Arnuncio JJ, Guillaume F, Bohrer G, Nathan R, Bridle JR, Gomulkiewicz R, Klein EK, Ritland K, Kuparinen A, Gerber S, Schueler S (2012) Long-distance gene flow and adaptation of forest trees to rapid climate change. Ecol Lett 15(4):378–392. https://doi.org/10.1111/j.1461-0248.2012.01746.x

Khanal S, Kushal KC, Fulton JP, Shearer S, Ozkan E (2020) Remote sensing in agriculture—accomplishments, limitations, and opportunities. Remote Sens 12(22):3783. https://doi.org/10.3390/rs12223783

Kivinen S, Koivisto E, Keski-Saari S, Poikolainen L, Tanhuanpää T, Kuzmin A et al (2020) A keystone species, European aspen (*Populus tremula L.*), in boreal forests: ecological role, knowledge needs and mapping using remote sensing. Forest Ecol Manag 462:118008. https://doi.org/10.1016/j.foreco.2020.118008

Korneliussen TS, Albrechtsen A, Nielsen R (2014) ANGSD: analysis of next generation sequencing data. BMC Bioinf 15:356. https://doi.org/10.1186/s12859-014-0356-4

Lazic D, George JP, Rusanen M, Ballian D, Pfattner S, Konrad H (2022) Population differentiation in *Acer platanoides* L. at the regional Scale — Laying the basis for effective conservation of Its genetic resources in Austria. Forests 13(4):552. https://doi.org/10.3390/f13040552

Lande R (1988) Genetics and demography in biological conservation. Science 241(4872):1455–1460. https://doi.org/10.1126/science.3420403

Leinonen T, McCairns S, O'Hara B, Merilä J (2013) QST-FST comparisons: evolutionary and ecological insights from genomic heterogeneity. Nat Rev Genet 14:179–190. https://doi.org/10.1038/nrg3395

Lenormand T (2002) Gene flow and the limits to natural selection. Trends Ecol Evol 17(4):183–189. https://doi.org/10.1016/S0169-5347(02)02497-7

Levis NA, Pfennig DW (2016) Evaluating 'plasticity-first' evolution in nature: key criteria and empirical approaches. Trends Ecol Evol 31:563–574. https://doi.org/10.1016/j.tree.2016.03.012

Ledig FT (1992) Human impacts on genetic diversity in forest ecosystems. Oikos 63:87–108. https://doi.org/10.2307/3545518

Lefèvre F (2004) Human impacts on forest genetic resources in the temperate zone: an updated review. For Ecol Manag 197(1–3):257–271. https://doi.org/10.1016/j.foreco.2004.05.017

Leigh DM, Hendry AP, Vázquez-Domínguez E, Friesen VL (2019) Estimated six per cent loss of genetic variation in wild populations since the industrial revolution. Evol Appl 12(8):1505–1512. https://doi.org/10.1111/eva.12810

Leimu R, Mutikainen PIA, Koricheva J, Fischer M (2006) How general are positive relationships between plant population size, fitness and genetic variation? J Ecol 94(5):942–952. https://doi.org/10.1111/j.1365-2745.2006.01150.x

Levins R (1970) Some mathematical problems in biology: extinction. The American Mathematical Society, Providence, pp 77–107

Liebhold AM, Halverson JA, Elmes GA (1992) Gypsy moth invasion in North America: a quantitative analysis. J Biogeogr 19(5):513–520. https://doi.org/10.2307/2845770

Lind-Riehl J, Gailing O (2017) Adaptive variation and introgression of a CONSTANS-like gene in North American red oaks. Forests 8(1):3. https://doi.org/10.3390/f8010003

Lindenmayer DB, Laurance WF (2017) The ecology, distribution, conservation and management of large old trees. Biol Rev 92(3):1434–1458. https://doi.org/10.1111/brv.12290

Liu MH, Chen XY, Zhang X, Shen DW (2008) A population genetic evaluation of ecological restoration with the case study on *Cyclobalanopsis myrsinaefolia* (Fagaceae). Plant Ecol 197(1):31–41. https://doi.org/10.1007/s11258-007-9357-y

Lowe WH, Allendorf FW (2010) What can genetics tell us about population connectivity? Mol Ecol 19:3038–3051. https://doi.org/10.1111/j.1365-294X.2010.04688.x

Luo Y, Widmer A, Karrenberg S (2014) The roles of genetic drift and natural selection in quantitative trait divergence along an altitudinal gradient in *Arabidopsis thaliana*. Heredity 114(2):220–228. https://doi.org/10.1038/hdy.2014.89

Lynch M, Lande R (1993) Evolution and extinction in response to environmental change. In: Kareiva PM, Kingsolver JG, Huey RB (eds) Biotic interactions and global change. Sinauer Associates, Sunderland, MA, pp 234–250

Madritch MD, Kingdon CC, Singh A, Mock KE, Lindroth RL, Townsend PA (2014) Imaging spectroscopy links aspen genotype with below-ground processes at landscape scales. Philos Trans R Soc Lond B Biol Sci 369(1643):20130194. https://doi.org/10.1098/rstb.2013.0194

Manel S, Gaggiotti OE, Waples RS (2005) Assignment methods: matching biological questions with appropriate techniques. Trends Ecol Evol 20(3):136–142. https://doi.org/10.1016/j.tree.2004.12.004

Meagher TR (1986) Analysis of paternity within a natural population of Chamaelirium luteum. 1. Identification of most-likely male parents. Am Nat 128(2):199–215. https://doi.org/10.1086/284554

Maginnis S, Jackson W (2007) What is FLR and how does it differ from current approaches? In: Rietbergen-McCracken J, Maginnis S, Sarre A (eds) The forest landscape restoration handbook. Sterling, VA, London, pp 5–20

McMullan M, Rafiqi M, Kaithakottil G, Clavijo BJ, Bilham L, Orton E et al (2018) The ash dieback invasion of Europe was founded by two genetically divergent individuals. Nat Ecol Evol 2(6):1000–1008. https://doi.org/10.1038/s41559-018-0548-9

Melo FPL, Pinto SRR, Brancalion PHS, Castro PS, Rodrigues RR, Aronson J, Tabarelli M (2013) Priority setting for scaling-up tropical forest restoration projects: early lessons from the Atlantic Forest Restoration Pact. Environ Sci Pol 33:395–404. https://doi.org/10.1016/j.envsci.2013.07.013

Mounce R, Smith P, Brockington S (2017) Ex situ conservation of plant diversity in the world's botanic gardens. Nat Plants 3:795–802. https://doi.org/10.1038/s41477-017-0019-3

Namkoong G (1966) Statistical analysis of introgression. Biometrics 22(3):488–502. https://doi.org/10.2307/2528184

Nussey DH, Postma E, Gienapp P, Visser ME (2005) Selection on heritable phenotypic plasticity in a wild bird population. Science 310(5746):304–306. https://doi.org/10.1126/science.1117004

Navascués M, Emerson BC (2007) Natural recovery of genetic diversity by gene flow in reforested areas of the endemic Canary Island pine, Pinus canariensis. Forest Ecol Manag 244(1–3):122–128. https://doi.org/10.1016/j.foreco.2007.04.009

Nonaka E, Sirén J, Somervuo P, Ruokolainen L, Ovaskainen O, Hanski I (2019) Scaling up the effects of inbreeding depression from individuals to metapopulations. J Anim Ecol 88:1202–1214. https://doi.org/10.1111/1365-2656.13011

Nicotra AB, Segal DL, Hoyle GL, Schrey AW, Verhoeven KJF, Richards CL (2015) Adaptive plasticity and epigenetic variation in response to warming in an Alpine plant. Ecol Evol 5:634–647. https://doi.org/10.1002/ece3.1329

Oddou-Muratorio S, Bontemps A, Klein EK, Chybicki I, Vendramin GG, Suyama Y (2010) Comparison of direct and indirect genetic methods for estimating seed and pollen dispersal in Fagus sylvatica and Fagus crenata. For Ecol Manag 259(11):2151–2159. https://doi.org/10.1016/j.foreco.2010.03.001

Oddou-Muratorio S, Klein EK (2008) Comparing direct vs. indirect estimates of gene flow within a population of a scattered tree species. Mol Ecol 17(11):2743–2754. https://doi.org/10.1111/j.1365-294X.2008.03783.x

OECD (2023) OECD forest seed and plant scheme – OECD scheme for the certification of forest reproductive material moving in international trade. OECD 2023, Paris. https://www.oecd.org/agriculture/forest/documents/FINAL%20FOREST%20RULES%20ENGLISH%202023.pdf

Oljirra A (2019) The causes, consequences and remedies of deforestation in Ethiopia. J Degrad Mining Lands Manag 6(3):1747. https://doi.org/10.15243/jdmlm.2019.063.1747

Pauls SU, Nowak C, Bálint M, Pfenninger M (2012) The impact of global climate change on genetic diversity within populations and species. Mol Ecol 22(4):925–946. https://doi.org/10.1111/mec.12152

Pease CM, Lande R, Bull JJ (1989) A model of population growth, dispersal and evolution in a changing environment. Ecology 70:1657–1664. https://doi.org/10.2307/1938100

Petit RJ, Bialozyt R, Garnier-Gere P, Hampe A (2004) Ecology and genetics of tree invasions: from recent introductions to Quaternary migrations. Forest Ecol Manag 197(1–3):117–137. https://doi.org/10.1016/j.foreco.2004.05.009

Petit RJ, Hampe A (2006) Some evolutionary consequences of being a tree. Ann Rev Ecol Evol Syst 37:187–214. https://doi.org/10.1146/annurev.ecolsys.37.091305.110215

Pinsky ML, Palumbi SR (2014) Meta-analysis reveals lower genetic diversity in overfished populations. Mol Ecol 23(1):29–39. https://doi.org/10.1111/mec.12509

Pitelka LF (1988) Evolutionary responses of plants to anthropogenic pollutants. Trends Ecol Evol 3(9):233–236. https://doi.org/10.1016/0169-5347(88)90165-6

Poupon V, Chakraborty D, Stejskal J, Konrad H, Schueler S, Lstibůrek M (2021) Accelerating adaptation of forest trees to climate change using individual tree response functions. Front Plant Sci 12:758221. https://doi.org/10.3389/fpls.2021.758221

Pritchard JK, Pickrell JK, Coop G (2010) The genetics of human adaptation: hard sweeps, soft sweeps, and polygenic adaptation. Curr Biol 20(4):R208–R215. https://doi.org/10.1016/j.cub.2009.11.055

Raja RG, Tauer CG, Wittwer RF, Huang Y (1998) Regeneration methods affect genetic variation and structure in shortleaf pine (*Pinus echinata* Mill.). For Genet 5(3):171–178

Ralls K, Ballou JD, Dudash MR, Eldridge MD, Fenster CB, Lacy RC et al (2018) Call for a paradigm shift in the genetic management of fragmented populations. Conserv Lett 11(2):e12412. https://doi.org/10.1111/conl.12412

Reed TE, Schindler DE, Waples RS (2011) Interacting effects of phenotypic plasticity and evolution on population persistence in a changing climate. Conserv Biol 25(1):56–63. https://doi.org/10.1111/j.1523-1739.2010.01552.x

Reusch TB, Ehlers A, Hämmerli A, Worm B (2005) Ecosystem recovery after climatic extremes enhanced by genotypic diversity. Proc Natl Acad Sci 102(8):2826–2831. https://doi.org/10.1073/pnas.0500008102

Robledo-Arnuncio JJ (2012) Joint estimation of contemporary seed and pollen dispersal rates among plant populations. Mol Ecol Resour 12(2):299–311. https://doi.org/10.1111/j.1755-0998.2011.03092.x

Rousset F (2000) Genetic differentiation between individuals. J Evol Biol 13:58–62. https://doi.org/10.1046/j.1420-9101.2000.00137.x

Santos-del-Blanco L, Olsson S, Budde KB, Grivet D, González-Martínez SC, Alía R, Robledo-Arnuncio JJ (2022) On the feasibility of estimating contemporary effective population size (Ne) for genetic conservation and monitoring of forest trees. Biol Conserv 273:109704. https://doi.org/10.1016/j.biocon.2022.109704

Savolainen O, Kärkkäinen K (1992) Effect of forest management on gene pools. In: Adams WT, Strauss SH, Copes DL, Griffin AR (eds) Population genetics of Forest trees, Forestry sciences, vol 42. Springer, Dordrecht, pp 329–345. https://doi.org/10.1007/978-94-011-2815-5_17

Smouse PE, Sork VL (2004) Measuring pollen flow in forest trees: an exposition of alternative approaches. For Ecol Manag 197(1–3):21–38. https://doi.org/10.1016/j.foreco.2004.05.049

Spurgin LG, Gage MJ (2019) Conservation: the costs of inbreeding and of being inbred. Curr Biol 29(16):R796–R798. https://doi.org/10.1016/j.cub.2019.07.023

Schaberg PG, DeHayes DH, Hawley GJ, Nijensohn SE (2008) Anthropogenic alterations of genetic diversity within tree populations: Implications for forest ecosystem resilience. For Ecol Manag 256:855–862. https://doi.org/10.1016/j.foreco.2008.06.038

Schlichting C, Pigliucci M (1998) Phenotypic evolution: a reaction norm perspective. Sinauer Associates, Sunderland, Massachusetts

Sgrò CM, Lowe AJ, Hoffmann AA (2011) Building evolutionary resilience for conserving biodiversity under climate change. Evol Appl 4(2):326–337. https://doi.org/10.1111/j.1752-4571.2010.00157.x

Sinclair-Waters M, Ødegård J, Korsvoll SA, Moen T, Lien S, Primmer CR, Barson NJ (2020) Beyond large-effect loci: large-scale GWAS reveals a mixed large-effect and polygenic architecture for age at maturity of Atlantic salmon. Genet Sel Evol 52:9. https://doi.org/10.1186/s12711-020-0529-8

Szukala A, Lovegrove-Walsh J, Luqman H, Fior S, Wolfe T, Frajman B, Schönswetter P, Paun O (2022) Polygenic routes lead to parallel altitudinal adaptation in *Heliosperma pusillum* (Caryophyllaceae). Mol Ecol 32(8):1832–1847. https://doi.org/10.1111/mec.16393

Szukala A, Bertel C, Frajman B, Schönswetter P, Paun O (2023) Parallel adaptation to lower altitudes is associated with enhanced plasticity in *Heliosperma pusillum* (Caryophyllaceae). Plant J 115:1619–1632. https://doi.org/10.1111/tpj.16342

Taylor BL, Dizon AE (1999) First policy then science: why a management unit based solely on genetic criteria cannot work. Mol Ecol 8:S11–S16. https://doi.org/10.1046/j.1365-294X.1999.00797.x

Thomas BL, Jalonen R, Loo J, Boshier D, Gallo L, Cavers S et al (2014) Genetic considerations in ecosystem restoration using native tree species. For Ecol Manag 333:66–75. https://doi.org/10.1016/j.foreco.2014.07.015

Thornton KR (2019) Polygenic adaptation to an environmental shift: temporal dynamics of variation under gaussian stabilizing selection and additive effects on a single trait. Genetics 213(4):1513–1530. https://doi.org/10.1534/genetics.119.302662

Tovar-Sánchez E, Valencia-Cuevas L, Castillo-Mendoza E, Mussali-Galante P, Pérez-Ruiz RV, Mendoza A (2013) Association between individual genetic diversity of two oak host species and canopy arthropod community structure. Eur J For Res 132(1):165–179. https://doi.org/10.1007/s10342-012-0665-y

Tran TV, Reef R, Zhu X (2022) A review of spectral indices for mangrove remote sensing. Remote Sens 14(19):4868. https://doi.org/10.3390/rs14194868

Unger GM, Heuertz M, Vendramin GG, Robledo-Arnuncio JJ (2016) Assessing early fitness consequences of exotic gene flow in the wild: a field study with Iberian pine relicts. Evol Appl 9(2):367–380. https://doi.org/10.1111/eva.12333

Vekemans X, Hardy OJ (2004) New insights from fine-scale spatial genetic structure analyses in plant populations. Mol Ecol 13(4):921–935. https://doi.org/10.1046/j.1365-294X.2004.02076.x

Visser ME (2008) Keeping up with a warming world; assessing the rate of adaptation to climate change. Proc R Soc Lond B Biol Sci 275(1635):649–659. https://doi.org/10.1098/rspb.2007.0997

Vranckx GUY, Jacquemyn H, Muys B, Honnay O (2012) Meta-analysis of susceptibility of woody plants to loss of genetic diversity through habitat fragmentation. Conserv Biol 26(2):228–237. https://doi.org/10.1111/j.1523-1739.2011.01778.x

Walther G-R, Post E, Convey P, Menzel A, Parmesan C, Beebee TJC et al (2002) Ecological responses to recent climate change. Nature 416:389–395. https://doi.org/10.1038/416389a

Wang J, Santiago E, Caballero A (2016) Prediction and estimation of effective population size. Heredity 117:193–206. https://doi.org/10.1038/hdy.2016.43

Wang R, Gamon JA (2019) Remote sensing of terrestrial plant biodiversity. Remote Sens Environ 231:111218. https://doi.org/10.1016/j.rse.2019.111218

Whitham TG, Young WP, Martinsen GD, Gebring CA, Schweitzer JA, Shuster SM et al (2003) Community and ecosystem genetics: consequence of the extended phenotype. Ecology 84(3):559–573. https://doi.org/10.1890/0012-9658(2003)084[0559:CAEGAC]2.0.CO;2

Webb T III (1981) The past 11,000 years of vegetational change in eastern North America. Bioscience 31(7):501–506. https://doi.org/10.2307/1308492

Wright S (1931) Evolution in Mendelian populations. Genetics 16(2):97–159. https://doi.org/10.1093/genetics/16.2.97

Wright S (1933) Inbreeding and homozygosis. Proc Natl Acad Sci 19(4):411–420. https://doi.org/10.1073/pnas.19.4.411

Wund MA (2012) Assessing the impacts of phenotypic plasticity on evolution. Integr Comp Biol 52:5–15. https://doi.org/10.1093/icb/ics050

Wade MJ (2016) Adaptation in Metapopulations: how interaction changes evolution. University of Chicago Press, Chicago, IL. 240 pp.

Waples RS (2002) Definition and estimation of effective population size in the conservation of endangered species. In: Beissinger SR, McCullough DR (eds) Population Viability Analysis. University of Chicago Press, Chicago, IL, pp 147–168

Whitham TG, Bailey JK, Schweitzer JA, Shuster SM, Bangert RK, LeRoy CJ, Lonsdorf EV, Gery JA, DiFazio SP, Potts BM, Fischer DG, Gehring CA, Lindroth RL, Marks JC, Hart SC, Wimp GM, Wooley SC (2006) A framework for community and ecosystem genetics: from genes to ecosystems. Nat Rev Genet 7(7):510–523. https://doi.org/10.1038/nrg1877

Yamasaki E, Altermatt F, Cavender-Bares J, Schuman MC, Zuppinger-Dingley D, Garonna I et al (2017) Genomics meets remote sensing in global change studies: monitoring and predicting phenology, evolution and biodiversity. Curr Opin Environ Sustain 29:177–186. https://doi.org/10.1016/j.cosust.2018.03.005

Wilson GA, Rannala B (2003) Bayesian inference of recent migration rates using multilocus genotypes. Genetics 163(3):1177–1191. https://doi.org/10.1093/genetics/163.3.1177

Yeaman S (2015) Local adaptation by alleles of small effect. Am Nat 186:S74–S89. https://doi.org/10.1086/682405

Yeaman S (2022) Evolution of polygenic traits under global *vs* local adaptation. Genetics 220(1):iyab134. https://doi.org/10.1093/genetics/iyab134

Open Access This chapter is licensed under the terms of the Creative Commons Attribution 4.0 International License (http://creativecommons.org/licenses/by/4.0/), which permits use, sharing, adaptation, distribution and reproduction in any medium or format, as long as you give appropriate credit to the original author(s) and the source, provide a link to the Creative Commons license and indicate if changes were made.

The images or other third party material in this chapter are included in the chapter's Creative Commons license, unless indicated otherwise in a credit line to the material. If material is not included in the chapter's Creative Commons license and your intended use is not permitted by statutory regulation or exceeds the permitted use, you will need to obtain permission directly from the copyright holder.

Genetic Connectivity and Local Adaptation of Forest Trees in the Face of Climate Change

5

Oliver Gailing, Katharina Birgit Budde, Ludger Leinemann, Markus Müller, and Selina Wilhelmi

Pollen of Norway Spruce tree in springtime (Photo: Jaroslav Moravcik/Adobe Stock)

O. Gailing (✉) · L. Leinemann · M. Müller · S. Wilhelmi
Forest Genetics and Forest Tree Breeding, University of Göttingen, Göttingen, Germany
e-mail: ogailin@gwdg.de; lleinem@gwdg.de; markus.mueller@forst.uni-goettingen.de; selina.wilhelmi@uni-goettingen.de

K. B. Budde
Forest Genetics and Forest Tree Breeding, University of Göttingen, Göttingen, Germany

Northwest German Forest Research Institute, Hann. Münden, Germany
e-mail: Katharina.Budde@nw-fva.de

© The Author(s) 2025
K. Lapin et al. (eds.), *Ecological Connectivity of Forest Ecosystems*,
https://doi.org/10.1007/978-3-031-82206-3_5

Abstract

The long-term survival of populations depends on genetic variation in traits related to survival and reproductive fitness. The polygenic architecture of traits is thought to facilitate adaptive shifts, but whether tree species will be able to adapt to the currently rapidly changing climatic conditions remains a subject of debate. On the other hand, trees are characterized by considerable phenotypic plasticity that allows them to grow under different or variable environmental conditions caused by global climate change. Phenotypic plasticity may thus help populations survive by "buying time" until genetic adaptation to the new environmental conditions occurs. One of the most important mechanisms underlying phenotypic plasticity is epigenetic regulation—stable altered gene expression without changes to the DNA sequence. Efficient dispersal mechanisms and the high fecundity of forest trees can promote genetic connectivity and facilitate the spread of adaptive genes and the colonization of new habitats. However, the colonization of new areas in response to a shift in suitable habitats, for example by northward migration, requires the dispersal of diploid sporophytes by seeds or fruits. Natural dispersal of tree species is therefore likely largely lagging behind the expansion of potentially suitable habitats dependent on the genetic system of species.

Keywords

Adaptation strategies · Gene flow · Environmental adaptation · Extinction risks · Conservation · Management

Population Genetics and Evolutionary Factors

Biological evolution is estimated to have given rise to approximately 73,000 tree species on Earth (Gatti et al. 2022). The vast diversity of tree species that are morphologically distinct and often adapted to different environmental conditions reflects adaptive evolution. Evolution causes changes in inheritable characteristics such as the morphology or habitat preferences of organisms and can lead to speciation in the long term (Coyne and Orr 2004; Nosil 2012). However, evolutionary processes like local adaptation to a specific habitat occur within species and even within populations and are often difficult to detect. The field of population genetics uses genetic markers to study these evolutionary processes by investigating changes in the genetic constitution of populations across space and time.

Our current understanding of evolutionary theory is mostly based on the synthesis of Darwin's description of evolution through natural selection (Darwin 1859) and Mendel's inheritance principles (Mendel 1866). A crucial prerequisite for evolution is the presence of genetic variation. New genetic variants arise randomly through *mutations*. Depending on the location within the genome and the type of mutation, new genetic variants can have no fitness effect at all on the organism carrying them (neutral genetic variants), or they can be beneficial or detrimental under certain conditions. Most mutations accumulating in the genomes of a species are selectively neutral (e.g., synonymous mutations or mutations in intergenic regions), while beneficial mutations are rare and deleterious mutations are purged by natural selection. One could argue that neutral genetic variants are not important for evolution, but this is not true. Recent studies have shown that "neutral" mutations affect evolutionary potential by facilitating phenotypic change through subsequent mutations that would otherwise not have occurred, or by increasing the mutability of flanking DNA regions (Tenaillon and Matic 2020). The fate of a new mutation within a population is determined by a complex interplay of population dynamics and environmental pressures mediated by evolutionary factors (Travis 1990). These evolutionary factors—namely mutation, genetic drift, gene flow/migration, and selection—are processes that change the genetic composition of a population (Fig. 5.1).

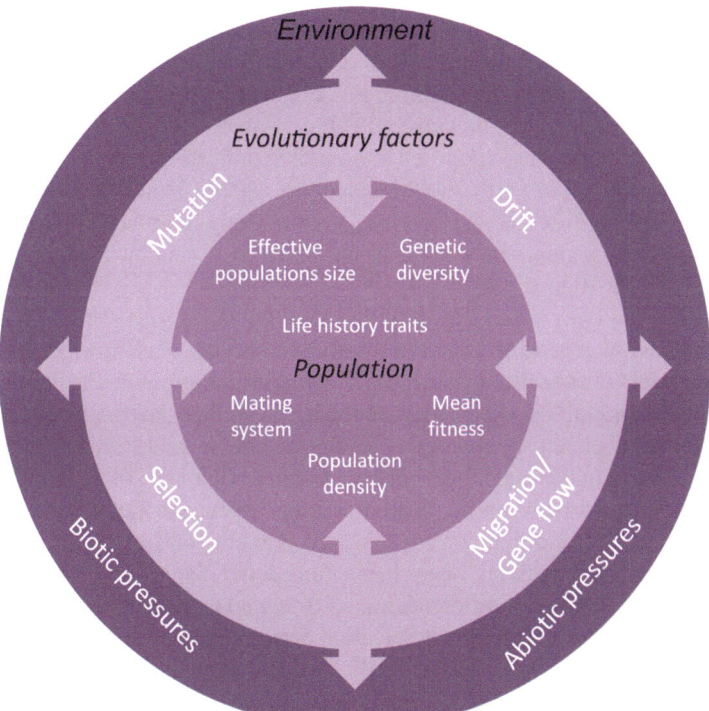

Fig. 5.1 Evolutionary factors shape the genetic constitution of populations. The complex interplay of the factors of mutation, migration/gene flow, selection, and genetic drift in combination with environmental pressures and the specific characteristics of a population and the life history traits of the species determine the population's genetic setup. The arrows indicate interactions between the population and the local environment. On the one hand, the environment shapes the population's genetic constitution through evolutionary factors, but on the other hand, the population also affects the environment, for example, by creating habitats for associated species and influencing microenvironmental conditions

Independent of their fitness effect, all genetic variants are equally affected by *genetic drift*, which describes a random change in allele frequencies over time (Crow 2010; Fisher 1930; Wright 1931). These random fluctuations in allele frequencies are undirected and independent of environmental conditions but strongly influenced by population size and allele frequency. Especially in small populations, rare alleles are likely to be lost from one generation to the next simply by chance (Ellstrand and Elam 1993). Genetic drift will therefore lead to fixation at certain loci, thus reducing genetic diversity. Populations evolving independently without gene flow will become more differentiated from each other over time, meaning that the same loci will show different allele frequencies and different alleles will be fixed at some loci in each of the populations. As mentioned above, genetic drift is especially noticeable in small and isolated populations with only a few reproductive individuals. Another risk for small populations that further reduces their viability is mating between related individuals, also known as inbreeding. Inbreeding causes an increase in homozygosity, which can lead to inbreeding depression, a reduction in fitness in inbred individuals, particularly in predominantly outcrossing species (Charlesworth and Charlesworth 1987). Although genetic drift and inbreeding do not cause directed change, the resulting loss of genetic variation and reduced fitness will cause a decrease in the adaptive potential of populations. Effective population size (N_e) has therefore been proposed as a suitable indicator for the genetic diversity of populations (Charlesworth 2009; Hoban et al. 2021). However, this measure is difficult to estimate—especially in large, continuously distributed species such as many forest tree species—and should be interpreted with caution (Fady and Bozzano 2021; Santos-del-Blanco et al. 2022).

Gene flow and *migration* can introduce new genetic variants into a population and thus increase genetic diversity. In sessile tree species, gene flow occurs via pollen dispersal between distant individuals while migration happens through long-range seed dispersal. The efficiency and range of pollen flow and seed dispersal depend on factors like the conspecific density within the landscape and specific life history traits (e.g., dispersal modes) of each species (Ghazoul 2005). However, gene flow and migration between populations are rare events by definition, since high levels of gene flow and migration would have a homogenizing effect, resulting in spatially separated groups no longer being considered distinct populations.

Selection is the result of differences in survival and reproductive success (fitness) of genetically distinct individuals. Selection acts on phenotypes and can cause the death of maladapted individuals (viability selection) under certain conditions or lead to higher (or lower) reproductive success of certain individuals (fertility selection). In contrast to random genetic drift, selection causes a directed change in allele frequencies enabling local adaptation of populations to specific environmental conditions, pests, or pathogens. Tree species typically experience highly heterogeneous biotic and abiotic conditions over time and across their distribution range due to fluctuating environmental conditions or the episodic occurrence of certain pests and pathogens. This heterogeneity of selection pressures is aggravated by the longevity of trees and the vast diversity of organisms like mycorrhizal fungi, endophytes, pests, and pathogens with which they interact (Boege and Marquis 2005; Linhart and Grant 1996). This causes a complex interplay of selection pressures that may include frequency-dependent, balancing, and/or episodic selection as potential contributors to the maintenance of genetic diversity (Petit and Hampe 2006).

Due to their long lifespan, one might think that evolution in tree species would be slow—but studies have shown that tree species can exhibit quick adaptive responses (Petit et al. 2004). Certain characteristics of trees may explain this paradox (reviewed in Petit and Hampe 2006): Tree populations are typically characterized by high levels of standing genetic variation (Hamrick and Godt 1996; Nybom 2004) even though they experience lower rates of molecular evolution and speciation compared to herbaceous species (Smith and Donoghue 2008). This may be due to the often very large effective population sizes, which also enabled the maintenance of high levels of genetic variation throughout adverse climatic conditions in the past (Milesi et al. 2024). Furthermore, tree species are predominantly outcrossing and have a very high lifetime reproductive output, and selection pressures are particularly strong during their early life stages (Petit and Hampe 2006). Differentiation between populations at neutral genetic markers is usually low due to wide-ranging gene flow, while differentiation at quantitative traits is often pronounced, reflecting local adaptation (Alberto et al. 2013a, b; McKay and Latta 2002). Furthermore, quantitative traits are typically polygenic, encoded by many loci with moderate to small effect sizes (Yeaman 2022). This polygenic architecture is thought to facilitate fast adaptive shifts, though there is general concern about whether tree species will be able to adapt to the currently rapidly changing climatic conditions (Lind et al. 2018).

Phenotypic Plasticity and Epigenetic Effects

Phenotypic plasticity is the ability of a given genotype to produce different phenotypes under different environmental conditions (Pigliucci et al. 2006). These different phenotypes may be expressed along environmental gradients or between years with different environmental conditions (Gailing et al. 2021). A prominent example for the investigation of phenotypic plasticity in tree species is the establishment of response functions for a given trait along environmental gradients. Response functions test the performance of provenances or genotypes between sites with different environmental conditions (Poupon et al. 2021). For instance, these functions could test how the height of trees differs along a temperature gradient. Tree species can usually grow within a wide range of temperatures but tend to exhibit their best performance at a specific temperature. This temperature optimum does not necessarily overlap with the temperature at the population's area of origin. For example, Rehfeldt et al. (2002) developed response functions for 110 *Pinus sylvestris* populations growing at 47 planting sites in Eurasia and North America. The authors inferred different growth potentials of the populations, which also exhibited different climatic optima. The populations tended to inhabit climates colder than their respective optimum (Rehfeldt et al. 2002). Phenotypic plasticity thus allows tree species to grow under different or variable environmental conditions, which may be of great significance in the light of global climate change. However, rapidly changing environmental conditions may not allow tree species that usually have long generation times to genetically adapt fast enough. Therefore, phenotypic plasticity may help populations to survive by "buying time" to allow genetic adaptation to the new environmental conditions to occur (Diamond and Martin 2021). Challis et al. (2022) found intraspecific differences in drought tolerance due to adaptive phenotypic plasticity in marri (*Corymbia calophylla*; a south-west Australian foundation tree species) saplings. The authors detected significant plasticity in a population originating from warm, dry climatic conditions in response to water deficit and therefore enhanced drought tolerance compared to a population originating from cool and wet climatic conditions (Challis et al. 2022). Nevertheless, it should also be mentioned that phenotypic plasticity can also prevent adaptation (Ghalambor et al. 2007). This can be the case when the new phenotype is close to the one that would be favored by selection under the new environmental conditions, thus leading to the persistence of the population rather than adaptation by directional selection (Ghalambor et al. 2007). Trees often exhibit high phenotypic plasticity, which can be more impactful than genetic differences between provenances. For instance, Gárate-Escamilla et al. (2019) investigated trait variation among European beech (*Fagus sylvatica*) trees in common gardens throughout the distribution range of the species, invariably finding a higher contribution of phenotypic plasticity to trait variation than that of local adaptation. A higher phenotypic plasticity (for height increment) compared to genetic differentiation was also detected in a large translocation experiment with European beech in Germany (Müller et al. 2020).

One of the most important mechanisms underlying phenotypic plasticity is epigenetic regulation (García-García et al. 2022). Epigenetics describes processes that alter gene expression without changes to the DNA sequence. There are three main groups of epigenetic mechanisms that affect gene expression: DNA methylation, histone modification, and processes mediated by noncoding RNAs (Kurpisz and Pawłowski 2022). DNA methylation describes the covalent addition of a methyl group to cytosine (and sometimes adenosine) and has been shown to be involved in transposon silencing and gene regulation (García-García et al. 2022; Kurpisz and Pawłowski 2022; Sow et al. 2018). Histone modifications (e.g., acetylation, methylation, phosphorylation, and ubiquitination) influence the compaction of chromatin and can thus affect transcription (e.g., open chromatin is accessible to the transcription machinery while condensed chromatin may not be) (García-García et al. 2022). Noncoding RNAs can be involved in different processes including DNA methylation (RNA-directed DNA methylation, RdDM) and gene silencing (Kurpisz and Pawłowski 2022). An increasing number of studies has been investigating the role and mechanisms of epigenetic modifications in tree species. Most studies focus on methylation since it has been shown to be involved in important biological processes and several different methods for studying DNA methylation are available (García-García et al. 2022; Müller et al. 2023). For instance, differences in methylation patterns and correlations with environmental conditions have been detected among populations of European beech and valley oak (*Quercus lobata*) (Gugger et al. 2016; Hrivnák et al. 2017). Guevara et al. (2022) analyzed genetic and epigenetic variation among European beech populations from Germany, Spain, and Sweden using amplified fragment length polymorphisms (AFLPs) and methylation-sensitive amplified polymorphisms (MSAPs). They found lower genetic and epigenetic diversity in Spanish provenances compared to provenances from Germany and Sweden, with 15% and 16% of the variance among populations associated with genetic and epigenetic variation, respectively (Guevara et al. 2022). Besides the identification of epigenetic variation involved in environmental adaptation, there is also an interest in finding epigenetic variants associated with (quantitative) traits that are important in breeding programs. Lu et al. (2020) used quantitative trait locus (QTL) mapping based on epigenetic markers (MSAPs) to identify epigenetic quantitative trait loci (epiQTLs) underlying growth and wood property traits in *Populus*. The authors identified 163 epiQTLs that explained between 1.7% and 44.5% of phenotypic variation. Other studies have investigated small RNAs (sRNA) in forest tree species that may be involved in epigenetic mechanisms (e.g., Liu and El-Kassaby 2017; Yakovlev et al. 2016; Yakovlev and Fossdal 2017). Yakovlev and Fossdal (2017) analyzed sRNA in embryogenic tissues of Norway spruce (*Picea abies*) that was produced under different epitype-inducing temperatures. They identified 654 micro RNAs (miRNAs) that were differentially expressed in the different tissues. Modesto et al. (2022) identified 105 miRNAs that were responsive to pine wilt disease caused by the pinewood nematode *Bursaphelenchus xylophilus*.

Adaptation to Climate Change

Whereas the preceding section described adaptation at the individual level, focusing on phenotypic plasticity and epigenetic effects, this section will examine long-term genetic and evolutionary adaptation. Genetic adaptation refers to the process by which the genetic composition of a population changes over time in response to environmental pressures. It is through genetic adaptation that forest trees are able to thrive in a variety of ecological niches ranging from dense tropical rainforests to dry, cold-to-temperate regions. In particular, because of their nature as sessile organisms and their long lifespan, the adaptation of trees to changing environmental conditions is highly relevant (Aitken and Bemmels 2016). Furthermore, as a result of ongoing climate change, trees are exposed to steadily changing and sometimes extreme environmental conditions during their lifetimes.

The basis for genetic adaptation is genetic diversity, especially in genes expressing and regulating important adaptive traits. As forest trees usually exhibit high genetic diversity within populations, their adaptive potential is estimated to be relatively high (Aitken et al. 2008; Kremer and Hipp 2020; Savolainen et al. 2007). Genetic diversity can be measured based on the comparison of individual genomes. Defined as the totality of an organism's genetic material including all its genes, the genome is characterized by a unique sequence of deoxyribonucleic acid (DNA) containing a unique combination of genomic variations. The most common types of genetic variations are single nucleotide polymorphisms (SNPs)—nucleotide changes at a single position. Originally arising as random mutations, most genomic variations are functionally neutral and thus not subject to selection (Gutschick and BassiriRad 2003). However, when environmental conditions change, the presence of certain genetic variants may become important as a prerequisite for selection and thus possible adaptation.

This can be shown in the context of diebacks caused by newly emerging pathogens or insect pests, which is particularly important because pathogens also adapt to changing abiotic conditions in the course of climate change and globalization, and their distribution areas may change as a result. A prime example is the dieback of European ash caused by the invasive pathogenic fungus *Hymenoscyphus fraxineus*, which was most likely introduced to Eastern Europe in the mid-1990s through the import of *Fraxinus mandshurica* plants from Eastern Asia (Budde et al. 2016; McKinney et al. 2014). However, it was observed that some trees carried a partial resistance in their genome and thus survived. McKinney et al. (2014) estimated the frequency of resistant individuals in natural populations to be between 1% and 5%. Although this number seems relatively low, it means that resistant individuals can be expected in native populations, illustrating the adaptive potential of ash trees.

To identify genetic variants potentially causing or regulating an adaptive trait such as bud burst, drought stress, or parasite resistance, the method of choice is usually to determine statistical associations of genomic markers like SNPs with phenotype measurements. These genome-wide association studies (GWAS) show that adaptive traits are usually controlled by a very large number of genes, each of which has only a minor effect on the phenotype (Alberto et al. 2013a, b; Kremer and Hipp 2020; Neale and Kremer 2011). Besides the high genetic diversity in forest trees, the abundance of these so-called complex traits with their complementary contribution of many genes to the trait variation is an indicator of high adaptive potential of forest trees to variable environmental conditions (Kremer and Hipp 2020). Various studies have investigated the genomic sequences and identified candidate genes for adaptive traits. In European ash trees, for example, more than 50 candidate genes have been identified for the resistance to the ash dieback fungus—and numerous homologs of them have been determined to be related to pathogen response in other plant species (Stocks et al. 2019).

In a review article on abiotic genetic adaptation in the Fagaceae family, candidate genes were identified that are found across species and could potentially affect multiple adaptive traits simultaneously (Müller and Gailing 2019). For example, the *CONSTANS*-like (*COL*) gene was identified as a candidate for bud burst timing in *Q. petraea* (Alberto et al. 2013a, b) and *F. sylvatica* (Müller et al. 2015, 2017), as well as appearing as a drought-related candidate gene in *Q. rubra* and *Q. ellipsoidalis* (Lind-Riehl et al. 2014).

Furthermore, investigation of data collected along environmental gradients such as temperature or precipitation can provide important insights into climatic adaptation processes. In their study on *Q. petraea*, Alberto et al. (2013a, b) identified clinal patterns along a latitudinal and altitudinal gradient, determining one SNP located in the 59-adenylylsulfate reductase (*APS*) gene that was significantly correlated with temperature in both gradients. The enzyme APS plays a key role in the sulfate reduction pathway, which is involved in biotic and abiotic stress defense (Alberto et al. 2013a, b; Rennenberg et al. 2007). The same study also identified the circadian clock gene *GIGANTEA*, which has been associated with precipitation in *Q. petraea*, *Q. robur*, and *Q. pubescens* (Rellstab et al. 2016) and determined to be a main candidate gene for local adaptation in Norway spruce (Caré et al. 2020).

To assess the behavior of populations under future environmental conditions, predictive approaches have recently been aiming to measure the difference between a current genomic composition and that required to cope with a changing environment, for example, due to climate change, at a set of putative adaptive loci ("genomic offset") (Dauphin et al. 2023). In general, conservation of genetic variance is of utmost importance to maintain the high genetic and evolutionary adaptation potential in forest trees. However, particularly in southern and warmer marginal populations, genetic variation and adaptive potential may not be sufficient to adapt rapidly to drought and higher temperatures, making these populations particularly vulnerable to climate change (Fréjaville et al. 2020; Müller and Gailing 2019; Tegel et al. 2014). Therefore, identification of genes regulating the expression of adaptive traits is important not only for targeted gene conservation measures but also for the future of forests, especially at the southern margins of their distribution areas that are under increasing environmental stress.

Gene Flow and Migration

In the face of climate change, forest trees can also cope with changing environmental conditions through the dispersal of seeds (migration) or pollen (gene flow). Even though trees and their female gametes are immobile, forest trees possess efficient mechanisms to disseminate genetic information via seeds (diploid sporophytes after fertilization) and pollen (haploid male gametophyte before fertilization) within and between populations (Finkeldey and Hattemer 2007). Efficient dispersal mechanisms and the high fecundity of forest trees can promote genetic connectivity and facilitate the spread of adaptive genes and the colonization of new habitats (Kremer et al. 2012). Accordingly, genetic variation in wind-pollinated tree species can be comparatively high even in marginal populations at the species' northern distribution edges (Götz et al. 2022; Hampe et al. 2013). While effective pollen dispersal, especially in wind-pollinated species, can occur over great distances—for example, more than 80 km in pedunculate oaks (Buschbom et al. 2011; Kremer et al. 2012)— long-distance seed dispersal is a rare event, e.g., in dominant tree species in temperate forests with heavy seeds such as oaks, beech, and conifer species (Hampe 2011; Hampe et al. 2013). However, the colonization of new habitats (e.g., by northward migration) in response to the shifting of suitable conditions requires the dispersal of the diploid sporophyte via seeds or fruits (Hampe 2011). Natural dispersion of tree species thus likely generally lags behind the expansion of potentially suitable habitats owing to limited seed dispersal, competing vegetation, and topographical features as well as the associated microclimates (Moracho et al. 2016).

Efficient transmission of genetic information among populations increases their adaptive potential and is the prerequisite for adaptation to rapidly changing environmental conditions through natural selection. On the other hand, gene influx of maladapted alleles along steep environmental gradients or from non-native planted populations introduced from a different environment (allochthonous populations) (Caré et al. 2020) can slow down adaptation processes. For example, 70% to 92% of immigrant pollen from neighboring plantations was observed based on paternity analyses in individual tree progeny in a locally adapted autochthonous high-elevation seed stand of *Picea abies* (Caré et al. 2020).

While most forest tree species possess reproductive characteristics like high fecundity and efficient means of gene dispersal that make them resilient to a certain level of fragmentation, severe fragmentation and low conspecific tree density can result in diminished gene exchange between forest fragments, reduction in the number of reproducing trees, mating between related individuals (inbreeding), and genetic differentiation among fragments as a result of selection and inbreeding (Ellstrand 1992; Ledig et al. 2001). It is therefore well demonstrated that gene flow patterns and mating systems (selfing *versus* outcrossing rates) in trees are affected by population density and fragmentation (Bodare et al. 2017; Ismail et al. 2014a, b, 2017). Lower population density can cause higher genetic differentiation of effective pollen contributions to individual seed trees, as well as higher selfing rates (Goncalves et al. 2022; Murawski and Hamrick 1991; Tarazi et al. 2013). Increased selfing rates and mating among related individuals can in turn affect the survival of the progeny (inbreeding depression), especially in mainly outcrossing species such as forest trees (Duminil et al. 2016; Eriksson et al. 2020). Especially in insect-pollinated rare tropical tree species or lianas with low conspecific density, very high inbreeding coefficients and levels of selfing were observed in isolated populations (e.g., *Ancistrocladus korupensis* (Foster and Sork 1997), *Pananga* spp., (Shapcott 1999; cit. in Finkeldey and Hattemer 2007)), while in other more common tree species, comparatively low levels of inbreeding (e.g., in *Swietenia macrophylla* (Lemes et al. 2003; cit. in Finkeldey and Hattemer 2007)) and efficient long-distance pollen dispersal (e.g., in the bee-pollinated tropical tree *Dinizia excelsa*, with a maximum distance of 3.2 km (Dick 2001)) was observed. As an extreme example, among isolated population fragments of the desert tree *Ficus sycomorus*, pollen flow distances of up to 160 km mediated by small wind-borne, host-specific wasps (mean distance of 88.6 km) were observed in the Namib Desert, Namibia (Ahmed et al. 2009), suggesting that pollen flow may be effective over very large distances in highly fragmented landscapes. Likewise, while preferential mating between neighboring trees within sampling sites and comparatively large numbers of full-sibs in individual tree progeny were evident for the insect-pollinated temperate tree species *Gleditsia triacanthos* L. in a highly fragmented agricultural landscape, pollen flow occurred mostly from outside the plots and over very long distances (>12 km) (Owusu et al. 2016). On the other hand, converse effects of forest fragmentation on genetic variation were observed in two West African tree species with different successional status. Severe effects of human impact and forest fragmentation were observed in the late-successional species *Mansonia altissima*, while no adverse effects were detected in the co-occurring pioneer tree species *Triplochiton scleroxylon* that preferentially grows in open forests (Akinnagbe et al. 2019).

Biodiversity conservation as a supporting ecosystem service is threatened by increased land use change and fragmentation (Foley et al. 2005). Human impacts affect the abundance and composition of species as well as evolutionary factors, genetic variation, and the potential of populations to adapt to new and constantly changing environments (Finkeldey et al. 2020). However, a diverse mosaic of forest fragments might also maintain species diversity in general as well as gene flow between conspecific trees in a highly fragmented landscape, and thus the evolutionary potential of these tree populations. Furthermore, connectivity among pockets of tree populations depends on the landscape matrix, in which urban forest patches can function as stepping stones, thereby facilitating gene exchange for some species even in a highly fragmented landscape (Van Rossum and Triest 2012). Finally, depending on their reproductive strategies, gene dispersal mechanisms, adaptive potential, abundance, and distribution, tree species are affected differently by land use change and habitat fragmentation (e.g., Akinnagbe et al. 2019).

Conservation and silvicultural measures should contribute to promoting rare tree species and maintaining high genetic variation in natural regeneration for natural selection to act upon. For example, measures to support the natural regeneration of species at high risk of browsing and maintaining connectivity between stands through migration and gene flow can promote natural adaptation processes (Gailing et al. 2021). Conservation measures need to be adapted based on the reproductive strategies and dispersal mechanisms of individual tree species, as well as on patterns of genetic variation indicative of these mechanisms and past impacts of evolutionary processes. For example, widely distributed and predominantly wind-pollinated species such as spruce, beech, and oak produce large amounts of pollen that can be dispersed over long distances (e.g., Nascimento de Sousa et al. 2010), so comparatively high genetic variation is observed in these species even in peripheral populations, e.g., at the northern edge of their range (Götz et al. 2022). However, the transfer distance of pollen and seeds from more southern populations to northern populations may not be sufficient to enable adaptation to extreme conditions in the face of climate change. Therefore, for species with comparatively low heat and drought resistance in temperate regions, the admixture of nonlocal origins (assisted migration, Aitken and Whitlock 2013; Aitken and Bemmels 2016) based on climate projections and results from provenance trials should be considered and is generally recommended (Mauri et al. 2023).

The shift of species distribution ranges caused by climate change will potentially generate new contact zones between closely related interfertile tree species. In Germany, for example, suitable habitats for *Q. pubescens*, whose range is currently restricted to southwestern Germany, will likely increase, as will the level of gene flow from the drought-adapted *Q. pubescens* into *Q. petraea* and *Q. robur*. Hybridization and introgression may generate positive effects through the transfer of adaptive alleles (Arnold 2004), thus reducing the vulnerability of species with narrow environmental ranges (Brauer et al. 2023), but they may also have negative effects (outbreeding depression, Whitlock et al. 2013) on the adaptive potential of tree populations, especially for rare taxa (e.g., high mountain species/subspecies such as *Pinus sylvestris* subsp. *nevadensis*, Gómez et al. 2015).

When the ranges of closely related species overlap (as in the case of interfertile oak species), interspecific gene flow (hybridization) can occur; however, the frequency of effective gene flow is dependent on the environment, likely due to postzygotic selection (Khodwekar and Gailing 2017; Lind-Riehl and Gailing 2017; Lind-Riehl et al. 2014). As a result of continuing interspecific gene flow, genome-wide genetic differentiation between hybridizing species is low, except for genes that have a possible function in different species adaptations, for example to drought stress (outlier genes) (Leroy et al. 2020b). The exchange of genes between hybridizing oak species may thus be environment-dependent and favor adaptation to changing environmental conditions (Leroy et al. 2020a). For example, hybridization frequently occurs among closely related oak species in contact zones within intermediate environments (Lepais and Gerber 2011; Lind and Gailing 2013). Likewise, introgression of outlier adaptive genes between closely related North American red oaks with different drought tolerance was found to correlate with water availability in the transition zone between species (Khodwekar and Gailing 2017). Gene exchange between species or ecotypes may thus be another often-understudied mechanism for relatively rapid adaptation of forest tree populations to new environments (Hamilton and Miller 2016; Chan et al. 2019). On the other hand, rare tree species may be threatened by hybridization with more common native species or introduced species through genetic assimilation or outcrossing depression (Carney et al. 2000).

Decline and Extinction Risk

Decline and extinction risk refers to the potential of a species or population to decrease in numbers or face the threat of extinction. Various factors including habitat loss, climate change, pollution, overexploitation, and disease can contribute to decline and extinction risk (Contreras-Hermosilla 2000; Lindenmayer 2023; Soulé 1983; Woo 2010).

Decline—Forest Decline

Forest decline describes a gradual reduction or deterioration of forest conditions. In the context of forest decline or forest dieback, the term goes back to the large-scale dieback of spruce forests in Germany in the 1970s and 1980s. It is defined, for example, as a marked loss of vitality of many trees in an ecosystem, or as a widespread decline in growth in soft- and hardwood ecosystems. These and similar definitions are not based on a universal standard, however, nor can the term "forest decline" be explained monocausally (Innes 1992; Manion 1985; Mueller-Dombois 1992; Woo 2010).

The Process of Extinction, the Risk of Extinction

"*Extinction can be a natural process, and one we might not regret if it occurred at a rate balanced by an equivalent rate of origin of new species*" (Schonewald-Cox et al. 1983). The process of extinction can most likely be attributed to a gradual loss of fitness and is difficult to analyze. Among other things, it is important to note that population extinction and species extinction are not the same. Because the process of extinction is difficult to describe and the state of being extinct can rarely be accurately determined, the term "extinction probability" or "extinction risk" is often used in studies (Balmford et al. 2003; Kéry et al. 2006; Soulé 1983). Extinction risk refers to the likelihood that a species will become extinct soon. It is assessed based on various factors, including population size, distribution, reproductive rates, genetic diversity, and environmental threats. The International Union for Conservation of Nature (IUCN) has developed the Red List of Threatened Species, which categorizes species into different levels of extinction risk ranging from "Least Concern" to "Extinct" (Mace et al. 2008).

Both "decline" and "extinction risk" generally refer to the problem of dynamic degradation of forest conditions that eventually leads to extinction (Collen et al. 2011). In terms of solving this problem, further questions need to be addressed, such as whether the ecosystem is affected, whether individual species are more likely to be affected than others, whether the dynamics are localized, and how fast they are progressing. The criteria for assessing extinction risk are provided in the IUCN Red List (Mace et al. 2008).

With respect to global climatic changes, the only processes by which tree species and their defined ecosystems can avoid possible extinction are adaptation and migration. The extent to which the genetic system of tree species is able to compensate for a dramatic loss of genetic variation must be analyzed individually for each species. In this context, the identification of thresholds—e.g., with regard to genetic variation—is fundamental (Aitken et al. 2008; Hamrick 2004; Trumbore et al. 2015). Particularly important for the development of action strategies is to determine when the process of decline actually enters the phase of extinction, at which point the process may no longer be reversible. Since not all species and ecosystems are protected in the same way, measuring these dynamics and defining priorities is essential (Kéry et al. 2006; Myers et al. 2000; Trumbore et al. 2015).

How to Measure?

A variety of methodological approaches ranging from global remote sensing via satellites to local vegetation surveys of individual species exist. The data thus collected also form the basis for meta-analyses, for example, to identify biodiversity hotspots or model species' extinction scenarios. Furthermore, extinction probabilities and rates can be measured and estimated indirectly, for example, by considering the effects of habitat loss. Artificial intelligence is also being used to estimate the conservation status of species worldwide. In addition to direct observation in field studies, a further method for analyzing the genetic basis of biodiversity is the use of gene markers (Balmford et al. 2003; Bredemeier et al. 2007; Fussi et al. 2016; Pimm and Raven 2000; Silva et al. 2022).

In general, time series with repeated measurements are used to measure the dynamics of ecosystem and species decline or extinction. This can lead to biases and errors related to the probability of observations for individual species. To minimize such errors, many authors point out that the most accurate measurement of species decline or extinction can only be based on permanent monitoring using well-defined areas and methods (Balmford et al. 2003; Fussi et al. 2016; Lindenmayer 2023; Trumbore et al. 2015).

What to Measure? Indicators and Verifiers

For assessing forest management, Stork et al. (1997) proposed a concept based on indicators and verifiers. An indicator is a variable or component of the forest ecosystem or the relevant management systems used to infer attributes relating to the sustainability of the resource and its utilization. Verifiers are data or information enhancing the specificity or ease of assessment of an indicator. They may define the limits of a hypothetical zone from which recovery can still safely take place (performance threshold/target).

Later, this concept was used by forest geneticists and conservation biologists (Boyle 2000; Namkoong et al. 2002) to describe the dynamics of genetic structures in populations based on four indicators.

Indicator 1: Levels of variation
Indicator 2: Directional change in allele or genotype frequencies
Indicator 3: Migration/gene flow among populations
Indicator 4: Reproductive processes/mating system

To determine critical thresholds, a combination of comparatively easy-to-observe demographic and genetic characteristics is used. For example, verifiers concerning indicator 1 are the number of sexually mature individuals (demographic) and genetic diversity (genetic). In combination with a monitoring system, these indicators can be used to detect changes in genetic structures with potentially deleterious effects on the adaptive potential of forest trees early on (Fussi et al. 2016). However, the debate about the optimal indicators or indicator combinations is still ongoing (Graudal et al. 2014).

Finally, it is crucial to address decline and extinction risk on a global scale through collaborative efforts among governments, conservation organizations, scientists, and the public. Implementing conservation strategies and promoting sustainable practices could reduce the decline and extinction risk of many species.

Box 5.1 Terminology
Gene flow. Effective transfer of genetic information by pollen containing the haploid male gametophyte resulting in fertilization and development of a diploid embryo.
Migration. Transfer of genetic information via seeds containing the diploid embryo.
Hybridization. Effective gene flow between taxa resulting in interspecific first-generation (F_1) hybrids.
Introgression. Backcrossing of an F_1 or later generation hybrid with one of the parental taxa.
Adaptation. An evolutionary process that increases an individual's probability of survival and reproduction in a given environment.
Evolutionary/adaptive potential. The ability of a population or species to adapt to new environments as a result of natural selection. The genetic variation in adaptive genes is directly related to a population's adaptive potential.
Adaptive trait. A trait that increases an individual's probability of survival and reproduction in a given environment.
Phenotypic plasticity. The ability of an organism to physiologically adapt to changing environmental conditions.
Epigenetic regulation. Stable altered gene expression without changes in the DNA sequence through DNA methylation, histone modification, and processes mediated by noncoding RNAs.
Genetic system. The way of transmission of genetic material from parent to filial generations.

References

Ahmed S, Compton SG, Butlin RK, Gilmartin PM (2009) Wind-borne insects mediate directional pollen transfer between desert fig trees 160 kilometers apart [Article]. Proc Natl Acad Sci USA 106(48):20342–20347. https://doi.org/10.1073/pnas.0902213106

Aitken SN, Bemmels JB (2016) Time to get moving: assisted gene flow of forest trees [Review]. Evol Appl 9(1):271–290. https://doi.org/10.1111/eva.12293

Aitken SN, Whitlock MC (2013) Assisted gene flow to facilitate local adaptation to climate change. Annu Rev Ecol Evol Syst 44(1):367–388. https://doi.org/10.1146/annurev-ecolsys-110512-135747

Aitken SN, Yeaman S, Holliday JA, Wang T, Curtis-McLane S (2008) Adaptation, migration or extirpation: climate change outcomes for tree populations. Evol Appl 1(1):95–111. https://doi.org/10.1111/j.1752-4571.2007.00013.x

Akinnagbe A, Gailing O, Finkeldey R, Lawal A (2019) Towards conservation of genetic variation of tropical tree species with differing successional status: the case of *Mansonia altissima* A. Chev and *Triplochiton scleroxylon* K. Schum [Article]. Trop Conserv Sci 12:9. https://doi.org/10.1177/1940082919864267

Alberto FJ, Aitken SN, Alía R, González-Martínez SC, Hänninen H, Kremer A, Lefèvre F, Lenormand T, Yeaman S, Whetten R, Savolainen O (2013a) Potential for evolutionary responses to climate change – evidence from tree populations. Glob Chang Biol 19(6):1645–1661. https://doi.org/10.1111/gcb.12181

Alberto FJ, Derory J, Boury C, Frigerio J-M, Zimmermann NE, Kremer A (2013b) Imprints of natural selection along environmental gradients in phenology-related genes of *Quercus petraea*. Genetics 195(2):495–512. https://doi.org/10.1534/genetics.113.153783

Arnold ML (2004) Transfer and origin of adaptations through natural hybridization: were Anderson and Stebbins right? Plant Cell 16(3):562–570. https://doi.org/10.1105/tpc.160370

Balmford A, Green RE, Jenkins M (2003) Measuring the changing state of nature. Trends Ecol Evol 18(7):326–330. https://doi.org/10.1016/S0169-5347(03)00067-3

Bodare S, Ravikanth G, Ismail SA, Patel MK, Spanu I, Vasudeva R et al (2017) Fine- and local-scale genetic structure of *Dysoxylum malabaricum*, a late-successional canopy tree species in disturbed forest patches in the Western Ghats, India. Conserv Genet 18(1):1–15. https://doi.org/10.1007/s10592-016-0877-7

Boege K, Marquis RJ (2005) Facing herbivory as you grow up: the ontogeny of resistance in plants. Trends Ecol Evol 20(8):441–448. https://doi.org/10.1016/j.tree.2005.05.001

Boyle TJ (2000) Criteria and indicators for the conservation of genetic diversity. In: Young A, Boshier D, Boyle T (eds) Forest conservation genetics, principles and practice. CSIRO Publishing, pp 239–251

Brauer CJ, Sandoval-Castillo J, Gates K, Hammer MP, Unmack PJ, Bernatchez L, Beheregaray LB (2023) Natural hybridization reduces vulnerability to climate change. Nat Clim Chang 13(3):282–289. https://doi.org/10.1038/s41558-022-01585-1

Bredemeier M, Dennis P, Sauberer N, Petriccione B, Torok K, Cocciufa C et al (2007) Biodiversity assessment and change – the challenge of appropriate methods. Issu Environ Sci Technol 25:217–251. https://doi.org/10.1039/9781847557650-00217

Budde KB, Nielsen LR, Ravn HP, Kjær ED (2016) The natural evolutionary potential of tree populations to cope with newly introduced pests and pathogens-lessons learned from forest health catastrophes in recent decades. Curr For Rep 2(1):18–29. https://doi.org/10.1007/s40725-016-0029-9

Buschbom J, Yanbaev Y, Degen B (2011) Efficient long-distance gene flow into an isolated relict oak stand. J Hered 102(4):464–472. https://doi.org/10.1093/jhered/esr023

Caré O, Gailing O, Müller M, Krutovsky KV, Leinemann L (2020) Crown morphology in Norway spruce (*Picea abies* [Karst.] L.) as adaptation to mountainous environments is associated with single nucleotide polymorphisms (SNPs) in genes regulating seasonal growth rhythm. Tree Genet Genomes 16(1):4. https://doi.org/10.1007/s11295-019-1394-x

Carney SE, Wolf DE, Rieseberg LH (2000) Hybridization and forest conservation. In: Young A, Boshier D, Boyle T (eds) Forest conservation genetics, principles and practice. CSIRO Publishing, pp 167–182

Challis A, Blackman C, Ahrens C, Medlyn B, Rymer P, Tissue D (2022) Adaptive plasticity in plant traits increases time to hydraulic failure under drought in a foundation tree. Tree Physiol 42(4):708–721. https://doi.org/10.1093/treephys/tpab096

Chan WY, Hoffmann AA, van Oppen MJH (2019) Hybridization as a conservation management tool. Conserv Lett 12(5):e12652. https://doi.org/10.1111/conl.12652

Charlesworth B (2009) Effective population size and patterns of molecular evolution and variation. Nat Rev Genet 10(3):195–205. https://doi.org/10.1038/nrg2526

Charlesworth D, Charlesworth B (1987) Inbreeding depression and its evolutionary consequences. Annu Rev Ecol Syst 18:237–268. https://doi.org/10.1146/annurev.es.18.110187.001321

Collen B, McRae L, Deinet S, De Palma A, Carranza T, Cooper N et al (2011) Predicting how populations decline to extinction. Phil Trans R Soc B Biol Sci 366(1577):2577–2586. https://doi.org/10.1098/rstb.2011.0015

Contreras-Hermosilla A (2000) The underlying causes of forest decline. CIFOR, Occasional paper NO. 30. Bogor, Indonesia. http://www.jstor.org/stable/resrep02260.4

Coyne JA, Orr HA (2004) Speciation. Sinauer Associates, Sunderland

Crow JF (2010) Wright and Fisher on inbreeding and random drift. Genetics 184(3):609–611. https://doi.org/10.1534/genetics.109.110023

Darwin C (1859) On the origin of species by means of natural selection, or preservation of favoured races in the struggle for life. John Murray, London. https://search.library.wisc.edu/catalog/9934839413602122

Dauphin B, Rellstab C, Wüest RO, Karger DN, Holderegger R, Gugerli F, Manel S (2023) Re-thinking the environment in landscape genomics. Trends Ecol Evol 38(3):261–274. https://doi.org/10.1016/j.tree.2022.10.010

Diamond SE, Martin RA (2021) Buying time: plasticity and population persistence. In: Phenotypic plasticity & evolution. CRC Press, pp 185–209

Dick CW (2001) Genetic rescue of remnant tropical trees by an alien pollinator [Article]. Proc R Soc B Biol Sci 268(1483):2391–2396. https://doi.org/10.1098/rspb.2001.1781

Duminil J, Daïnou K, Kaviriri DK, Gillet P, Loo J, Doucet JL, Hardy OJ (2016) Relationships between population density, fine-scale genetic structure, mating system and pollen dispersal in a timber tree from African rainforests. Heredity 116(3):295–303. https://doi.org/10.1038/hdy.2015.101

Ellstrand NC (1992) Gene flow by pollen: implications for plant conservation genetics. Oikos 63:77–86

Ellstrand NC, Elam DR (1993) Population genetic consequences of small population size: implications for plant conservation. Annu Rev Ecol Syst 24(1):217–242. https://doi.org/10.1146/annurev.es.24.110193.001245

Eriksson G, Ekberg I, Clapham D (2020) Genetics applied to forestry – an introduction. Department of Plant Biology SLU

Fady B, Bozzano M (2021) Effective population size does not make a practical indicator of genetic diversity in forest trees. Biol Conserv 253:108904. https://doi.org/10.1016/j.biocon.2020.108904

Finkeldey R, Hattemer HH (2007) Tropical forest genetics. Springer, Berlin

Finkeldey R, Müller M, de Melo Moura CC, Gailing O (2020) Understanding and exploiting genetics of tropical tree species for restoration of tropical forests. In: Blaser J, Hardcastle P (eds) Achieving sustainable management of tropical forests. Part 4 Monitoring and management techniques in sustainable forest management (SFM). Burleigh Dodds Science Publishing, pp 447–468. https://doi.org/10.19103/AS.2020.0074.32

Fisher RA (1930) The genetical theory of natural selection. Clarendon Press. https://www.biodiversitylibrary.org/item/69976

Foley JA, DeFries R, Asner GP, Barford C, Bonan G, Carpenter SR et al (2005) Global consequences of land use. Science 309(5734):570–574. https://doi.org/10.1126/science.1111772

Foster PF, Sork VL (1997) Population and genetic structure of the West African rain forest liana *Ancistrocladus korupensis* (Ancistrocladaceae). Am J Bot 84(8):1078–1091. https://doi.org/10.2307/2446151

Fréjaville T, Vizcaíno-Palomar N, Fady B, Kremer A, Benito Garzón M (2020) Range margin populations show high climate adaptation lags in European trees. Glob Chang Biol 26(2):484–495. https://doi.org/10.1111/gcb.14881

Fussi B, Westergren M, Aravanopoulos F, Baier R, Kavaliauskas D, Finzgar D et al (2016) Forest genetic monitoring: an overview of concepts and definitions. Environ Monit Assess 188(8):493. https://doi.org/10.1007/s10661-016-5489-7

Gailing O, Budde KB, Müller M (2021) Veränderung genetischer Variationsmuster von Waldbäumen unter Gesichtspunkten des Klimawandels. Allgemeine Forst- und Jagdzeitung 192(5/6):93–105. https://doi.org/10.23765/afjz0002079

Gárate-Escamilla H, Hampe A, Vizcaíno-Palomar N, Robson TM, Benito Garzón M (2019) Range-wide variation in local adaptation and phenotypic plasticity of fitness-related traits in *Fagus sylvatica* and their implications under climate change. Glob Ecol Biogeogr 28(9):1336–1350. https://doi.org/10.1111/geb.12936

García-García I, Méndez-Cea B, Martín-Gálvez D, Seco JI, Gallego FJ, Linares JC (2022) Challenges and perspectives in the epigenetics of climate change-induced forests decline. Front Plant Sci 12:797958. https://doi.org/10.3389/fpls.2021.797958

Gatti RC, Reich PB, Gamarra JGP, Crowther T, Hui C, Morera A, Bastin JF, de-Miguel S, Nabuurs GJ, Svenning JC, Serra-Diaz JM, Merow C, Enquist B, Kamenetsky M, Lee J, Zhu J, Fang J, Jacobs DF, Pijanowski B, Liang J (2022) The number of tree species on Earth. Proc Natl Acad Sci USA 119(6):e2115329119. https://doi.org/10.1073/pnas.2115329119

Ghalambor CK, McKay JK, Carroll SP, Reznick DN (2007) Adaptive versus non-adaptive phenotypic plasticity and the potential for contemporary adaptation in new environments. Funct Ecol 21(3):394–407. https://doi.org/10.1111/j.1365-2435.2007.01283.x

Ghazoul J (2005) Pollen and seed dispersal among dispersed plants. Biol Rev Camb Philos Soc 80(3):413–443. https://doi.org/10.1017/S1464793105006731

Gómez JM, González-Megías A, Lorite J, Abdelaziz M, Perfectti F (2015) The silent extinction: climate change and the potential hybridization-mediated extinction of endemic high-mountain plants. Biodivers Conserv 24(8):1843–1857. https://doi.org/10.1007/s10531-015-0909-5

Goncalves AL, García MV, Barrandeguy ME, González-Martínez SC, Heuertz M (2022) Spatial genetic structure and mating system in forest tree populations from seasonally dry tropical forests: a review. Tree Genet Genomes 18(3):18. https://doi.org/10.1007/s11295-022-01550-1

Götz J, Rajora OP, Gailing O (2022) Genetic structure of natural northern range-margin mainland, peninsular, and island populations of northern red oak (*Quercus rubra* L.) [Original Research]. Front Ecol Evol 10:907414. https://doi.org/10.3389/fevo.2022.907414

Graudal L, Aravanopoulos F, Bennadji Z, Changtragoon S, Fady B, Kjaer ED et al (2014) Global to local genetic diversity indicators of evolutionary potential in tree species within and outside forests. For Ecol Manag 333:35–51. https://doi.org/10.1016/j.foreco.2014.05.002

Guevara MÁ, Sánchez-Gómez D, Vélez MD, De María N, Díaz LM, Ramírez-Valiente JA, Mancha JA, Aranda I, Cervera MT (2022) Epigenetic and genetic variability in contrasting latitudinal *Fagus sylvatica* L. provenances. Forests 13(12):1971. https://doi.org/10.3390/f13121971

Gugger PF, Fitz-Gibbon S, Pellegrini M, Sork VL (2016) Species-wide patterns of DNA methylation variation in *Quercus lobata* and their association with climate gradients. Mol Ecol 25(8):1665–1680. https://doi.org/10.1111/mec.13563

Gutschick VP, BassiriRad H (2003) Extreme events as shaping physiology, ecology, and evolution of plants: toward a unified definition and evaluation of their consequences. New Phytol 160(1):21–42. https://doi.org/10.1046/j.1469-8137.2003.00866.x

Hamilton JA, Miller JM (2016) Adaptive introgression as a resource for management and genetic conservation in a changing climate. Conserv Biol 30(1):33–41. https://doi.org/10.1111/cobi.12574

Hampe A (2011) Plants on the move: the role of seed dispersal and initial population establishment for climate-driven range expansions. Acta Oecol 37(6):666–673. https://doi.org/10.1016/j.actao.2011.05.001

Hampe A, Pemonge MH, Petit RJ (2013) Efficient mitigation of founder effects during the establishment of a leading-edge oak population [Article]. Proc R Soc B Biol Sci 280(1764):20131070. https://doi.org/10.1098/rspb.2013.1070

Hamrick JL (2004) Response of forest trees to global environmental changes. For Ecol Manag 197(1):323–335. https://doi.org/10.1016/j.foreco.2004.05.023

Hamrick JL, Godt MJW (1996) Effects of life history traits on genetic diversity in plant species. Phil Trans R Soc B Biol Sci 351(1345):1291–1298. https://doi.org/10.1098/rstb.1996.0112

Hoban S, Paz-Vinas I, Aitken S, Bertola LD, Breed MF, Bruford MW, Funk WC, Grueber CE, Heuertz M, Hohenlohe P, Hunter ME, Jaffé R, Fernandes ML, Mergeay J, Moharrek F, O'Brien D, Segelbacher G, Vernesi C, Waits L, Laikre L (2021) Effective population size remains a suitable, pragmatic indicator of genetic diversity for all species, including forest trees. Biol Conserv 253:108906. https://doi.org/10.1016/j.biocon.2020.108906

Hrivnák M, Krajmerová D, Frýdl J, Gömöry D (2017) Variation of cytosine methylation patterns in European beech (*Fagus sylvatica* L.). Tree Genet Genomes 13(6):117. https://doi.org/10.1007/s11295-017-1203-3

Innes JL (1992) Forest decline. Progr Phys Geogr Earth Environ 16(1):1–64. https://doi.org/10.1177/030913339201600101

Ismail SA, Ghazoul J, Ravikanth G, Kushalappa CG, Uma Shaanker R, Kettle CJ (2014a) Forest trees in human modified landscapes: ecological and genetic drivers of recruitment failure in *Dysoxylum malabaricum* (Meliaceae). PLoS One 9(2):e89437. https://doi.org/10.1371/journal.pone.0089437

Ismail SA, Ghazoul J, Ravikanth G, Kushalappa CG, Uma Shaanker R, Kettle CJ (2014b) Fragmentation genetics of *Vateria indica*: implications for management of forest genetic resources of an endemic dipterocarp. Conserv Genet 15(3):533–545. https://doi.org/10.1007/s10592-013-0559-7

Ismail SA, Ghazoul J, Ravikanth G, Kushalappa CG, Shaanker RU, Kettle CJ (2017) Evaluating realized seed dispersal across fragmented tropical landscapes: a two-fold approach using parentage analysis and the neighbourhood model [Article]. New Phytol 214(3):1307–1316. https://doi.org/10.1111/nph.14427

Kéry M, Spillmann JH, Truong C, Holderegger R (2006) How biased are estimates of extinction probability in revisitation studies? J Ecol 94(5):980–986. https://doi.org/10.1111/j.1365-2745.2006.01151.x

Khodwekar S, Gailing O (2017) Evidence for environment-dependent introgression of adaptive genes between two red oak species with different drought adaptations [Article]. Am J Bot 104(7):1088–1098. https://doi.org/10.3732/ajb.1700060

Kremer A, Hipp AL (2020) Oaks: an evolutionary success story. New Phytol 226(4):987–1011. https://doi.org/10.1111/nph.16274

Kremer A, Ronce O, Robledo-Arnuncio JJ, Guillaume F, Bohrer G, Nathan R et al (2012) Long-distance gene flow and adaptation of forest trees to rapid climate change. Ecol Lett 15(4):378–392. https://doi.org/10.1111/j.1461-0248.2012.01746.x

Kurpisz B, Pawłowski TA (2022) Epigenetic mechanisms of tree responses to climatic changes. Int J Mol Sci 23(21):13412. https://doi.org/10.3390/ijms232113412

Ledig FT, Capó-Arteaga MA, Hodgskiss PD, Sbay H, Flores-López C, Thompson Conkle M, Bermejo-Velázquez B (2001) Genetic diversity and the mating system of a rare Mexican piñon, *Pinus pinceana*, and a comparison with *Pinus maximartinezii* (Pinaceae). Am J Bot 88(11):1977–1987. https://doi.org/10.2307/3558425

Lemes MR, Gribel R, Proctor J, Grattapaglia D (2003) Population genetic structure of mahogany (*Swietenia macrophylla* King, Meliaceae) across the Brazilian Amazon, based on variation at microsatellite loci: implications for conservation. Mol Ecol 12(11):2875–2883. https://doi.org/10.1046/j.1365-294X.2003.01950.x

Lepais O, Gerber S (2011) Reproductive patterns shape introgression dynamics and species succession within the European white oak complex [Article]. Evolution 65(1):156–170. https://doi.org/10.1111/j.1558-5646.2010.01101.x

Leroy T, Louvet J-M, Lalanne C, Le Provost G, Labadie K, Aury J-M et al (2020a) Adaptive introgression as a driver of local adaptation to climate in European white oaks. New Phytol 226:1171–1182. https://doi.org/10.1111/nph.16095

Leroy T, Rougemont Q, Dupouey J-L, Bodénès C, Lalanne C, Belser C et al (2020b) Massive postglacial gene flow between European white oaks uncovered genes underlying species barriers. New Phytol 226(4):1183–1197. https://doi.org/10.1111/nph.16039

Lind J, Gailing O (2013) Genetic structure of *Quercus rubra* L. and *Q. ellipsoidalis* E. J. Hill populations at gene-based EST-SSR and nuclear SSR markers. Tree Genet Genomes 9:707–722. https://doi.org/10.1007/s00606-012-0656-y

Lind BM, Menon M, Bolte CE, Faske TM, Eckert AJ (2018) The genomics of local adaptation in trees: are we out of the woods yet? Tree Genet Genomes 14:29. https://doi.org/10.1007/s11295-017-1224-y

Lindenmayer DB (2023) Forest biodiversity declines and extinctions linked with forest degradation: a case study from Australian tall, wet forests. Land 12(3):528. https://doi.org/10.3390/land12030528

Lind-Riehl J, Gailing O (2017) Adaptive variation and introgression of a CONSTANS-like gene in North American red oaks. Forests 8(1):3. https://doi.org/10.3390/f8010003

Lind-Riehl JF, Sullivan AR, Gailing O (2014) Evidence for selection on a CONSTANS-like gene between two red oak species. Ann Bot 113(6):967–975. https://doi.org/10.1093/aob/mcu019

Linhart YB, Grant MC (1996) Evolutionary significance of local genetic differentiation in plants. Annu Rev Ecol Syst 27(1):237–277. https://doi.org/10.1146/annurev.ecolsys.27.1.237

Liu Y, El-Kassaby YA (2017) Global analysis of small RNA dynamics during seed development of *Picea glauca* and *Arabidopsis thaliana* populations reveals insights on their evolutionary trajectories. Front Plant Sci 8:1719. https://doi.org/10.3389/fpls.2017.01719

Lu W, Xiao L, Quan M, Wang Q, El-Kassaby YA, Du Q, Zhang D (2020) Linkage-linkage disequilibrium dissection of the epigenetic quantitative trait loci (epiQTLs) underlying growth and wood properties in *Populus*. New Phytol 225(3):1218–1233. https://doi.org/10.1111/nph.16220

Mace GM, Collar NJ, Gaston KJ, Hilton-Taylor C, AkÇAkaya HR, Leader-Williams N et al (2008) Quantification of extinction risk: IUCN's system for classifying threatened species. Conserv Biol 22(6):1424–1442. https://doi.org/10.1111/j.1523-1739.2008.01044.x

Manion PD (1985) Factors contributing to the decline of forests. In: Stubbs HS (ed) Air pollutants effects on forest ecosystems. Acid Rain Foundation, pp 63–73

Mauri A, Girardello M, Forzieri G, Manca F, Beck PSA, Cescatti A, Strona G (2023) Assisted tree migration can reduce but not avert the decline of forest ecosystem services in Europe. Glob Environ Chang 80:102676. https://doi.org/10.1016/j.gloenvcha.2023.102676

McKay JK, Latta RG (2002) Adaptive population divergence: markers, QTL and traits. Trends Ecol Evol 17(6):285–291. https://doi.org/10.1016/S0169-5347(02)02478-3

McKinney LV, Nielsen LR, Collinge DB, Thomsen IM, Hansen JK, Kjaer ED (2014) The ash dieback crisis: genetic variation in resistance can prove a long-term solution. Plant Pathol 63(3):485–499. https://doi.org/10.1111/ppa.12196

Mendel G (1866) Experiments in plant hybridazation (Versuche über Pflanzenhybriden). J R Hortic Soc, 26

Milesi P, Kastally C, Dauphin B, Cervantes S, Bagnoli F, Budde KB, Cavers S, Ojeda DI, Fady B, Faivre-Rampant P, González-Martínez SC, Grivet D, Gugerli F, Jorge V, Lesur-Kupin I, Olsson S, Opgenoorth L, Pinosio S, Plomion C, ... Pyhäjärvi T (2024) Resilience of genetic diversity in forest trees over the Quaternary. Nature Communications 15:8538. https://doi.org/10.1038/s41467-024-52612-y

Modesto I, Inácio V, Van de Peer Y, Miguel CM (2022) MicroRNA-mediated post-transcriptional regulation of *Pinus pinaster* response and resistance to pinewood nematode. Sci Rep 12:5160

Moracho E, Moreno G, Jordano P, Hampe A (2016) Unusually limited pollen dispersal and connectivity of peduncolate oak (*Quercus robur*) refugial populations at the species' southern range margin. Mol Ecol 25(14):3319–3331. https://doi.org/10.1111/mec.13692

Mueller-Dombois D (1992) A global perspective on forest decline. Environ Toxicol Chem 11(8):1069–1076. https://doi.org/10.1002/etc.5620110804

Müller M, Gailing O (2019) Abiotic genetic adaptation in the Fagaceae. Plant Biol 21(5):783–795. https://doi.org/10.1111/plb.13008

Müller M, Seifert S, Finkeldey R (2015) A candidate gene-based association study reveals SNPs significantly associated with bud burst in European beech (Fagus sylvatica L.). Tree Genet Genomes 11(6):116. https://doi.org/10.1007/s11295-015-0943-1

Müller M, Seifert S, Finkeldey R (2017) Comparison and confirmation of SNP-bud burst associations in European beech populations in Germany. Tree Genet Genomes 13(3):59. https://doi.org/10.1007/s11295-017-1145-9

Müller M, Kempen T, Finkeldey R, Gailing O (2020) Low population differentiation but high phenotypic plasticity of European beech in Germany. Forests 11(12):1354. https://doi.org/10.3390/f11121354

Müller M, Kües U, Budde KB, Gailing O (2023) Applying molecular and genetic methods to trees and their fungal communities. Appl Microbiol Biotechnol 107(9):2783–2830. https://doi.org/10.1007/s00253-023-12480-w

Murawski DA, Hamrick JL (1991) The effect of the density of flowering individuals on the mating systems of nine tropical tree species. Heredity 67:167–174

Myers N, Mittermeier RA, Mittermeier CG, da Fonseca GAB, Kent J (2000) Biodiversity hotspots for conservation priorities [Article]. Nature 403(6772):853–858. https://doi.org/10.1038/35002501

Namkoong G, Boyle T, El-Kassaby YA, Palmberg-Lerche C, Eriksson G, Gregorius HR, ... Prabhu R (2002) Criteria and indicators for sustainable forest management: assessment and monitoring of genetic variation (pp FGR 37). FAO working paper: Forest Genetic Resources

Nascimento de Sousa S, Gailing O, Finkeldey R, Ziehe M, Hattemer HH (2010) Reproduction system of Norway spruce (*Picea abies* L. Karst.) in pure and mixed stands with different density. Forstarchiv 81:218–227

Neale DB, Kremer A (2011) Forest tree genomics: growing resources and applications. Nat Rev Genet 12(2):111–122. https://doi.org/10.1038/nrg2931

Nosil P (2012) Ecological speciation. Oxford University Press, Oxford

Nybom H (2004) Comparison of different nuclear DNA markers for estimating intraspecific genetic diversity in plants. Mol Ecol 13(5):1143–1155. https://doi.org/10.1111/j.1365-294X.2004.02141.x

Owusu SA, Schlarbaum S, Carlson JE, Gailing O (2016) Gene flow analyses and identification of full-sib families in isolated populations of *Gleditsia triacanthos* L. Botany-Botanique 94:523–532. https://doi.org/10.1139/cjb-2015-0244

Petit RJ, Hampe A (2006) Some evolutionary consequences of being a tree. Annu Rev Ecol Evol Syst 37(1):187–214. https://doi.org/10.1146/annurev.ecolsys.37.091305.110215

Petit RJ, Bialozyt R, Garnier-Géré P, Hampe A (2004) Ecology and genetics of tree invasions: from recent introductions to Quaternary migrations. For Ecol Manag 197(1–3):117–137. https://doi.org/10.1016/j.foreco.2004.05.009

Pigliucci M, Murren CJ, Schlichting CD (2006) Phenotypic plasticity and evolution by genetic assimilation. J Exp Biol 209(12):2362–2367. https://doi.org/10.1242/jeb.02070

Pimm SL, Raven P (2000) Extinction by numbers. Nature 403(6772):843–845. https://doi.org/10.1038/35002708

Poupon V, Chakraborty D, Stejskal J, Konrad H, Schueler S, Lstibůrek M (2021) Accelerating adaptation of forest trees to climate change using individual tree response functions. Front Plant Sci 12:758221. https://doi.org/10.3389/fpls.2021.758221

Rehfeldt GE, Tchebakova NM, Parfenova YI, Wykoff WR, Kuzmina NA, Milyutin LI (2002) Intraspecific responses to climate in *Pinus sylvestris*. Glob Chang Biol 8(9):912–929. https://doi.org/10.1046/j.1365-2486.2002.00516.x

Rellstab C, Zoller S, Walthert L, Lesur I, Pluess AR, Graf R, Bodénès C, Sperisen C, Kremer A, Gugerli F (2016) Signatures of local adaptation in candidate genes of oaks (*Quercus* spp.) with respect to present and future climatic conditions. Mol Ecol 25(23):5907–5924. https://doi.org/10.1111/mec.13889

Rennenberg H, Herschbach C, Haberer K, Kopriva S (2007) Sulfur metabolism in plants: are trees different? Plant Biol 9(5):620–637. https://doi.org/10.1055/s-2007-965248

Santos-del-Blanco L, Olsson S, Budde KB, Grivet D, González-Martínez SC, Alía R, Robledo-Arnuncio JJ (2022) On the feasibility of estimating contemporary effective population size (N_e) for genetic conservation and monitoring of forest trees. Biol Conserv 273:109704. https://doi.org/10.1016/j.biocon.2022.109704

Savolainen O, Pyhäjärvi T, Knürr T (2007) Gene flow and local adaptation in trees. Annu Rev Ecol Evol Syst 38(1):595–619. https://doi.org/10.1146/annurev.ecolsys.38.091206.095646

Schonewald-Cox C, Chambers S, MacBryde B, Thomas L (1983) Genetics and conservation: a reference for managing wild animal and plant populations. Benjamin Cummings. http://pubs.er.usgs.gov/publication/81492

Silva SV, Andermann T, Zizka A, Kozlowski G, Silvestro D (2022) Global estimation and mapping of the conservation status of tree species using artificial intelligence [Original Research]. Front Plant Sci 13:839792. https://doi.org/10.3389/fpls.2022.839792

Smith SA, Donoghue MJ (2008) Rates of molecular evolution are linked to life history in flowering plants. Science 322(5898). https://doi.org/10.1126/science.1163197

Soulé ME (1983) What do we really know about extinction? In: Schonewald-Cox CM, Chambers SM, MacBryde B, Thomas WL (eds) Genetics and conservation. A reference for managing wild animal and plant populations. Benjamin/Cummings, pp 111–124

Sow MD, Allona I, Ambroise C, Conde D, Fichot R, Gribkova S, Jorge V, Le-Provost G, Pâques L, Plomion C, Salse J, Sanchez-Rodriguez L, Segura V, Tost J, Maury S (2018) Chapter twelve—Epigenetics in forest trees: state of the art and potential implications for breeding and management in a context of climate change. In: Mirouze M, Bucher E, Gallusci P (eds) Advances in botanical research, vol 88. Academic Press, pp 387–453

Stocks JJ, Metheringham CL, Plumb WJ, Lee SJ, Kelly LJ, Nichols RA, Buggs RJA (2019) Genomic basis of European ash tree resistance to ash dieback fungus. Nat Ecol Evol 3(12):1686–1696. https://doi.org/10.1038/s41559-019-1036-6

Stork NE, Boyle TJB, Dale V, Eeley H, Finegan B, Lawes M, … Soberon J (1997) Criteria and indicators for assessing the sustainability of forest management: conservation of biodiversity. In CIFOR Working Paper (Vol. 17, pp 1–29)

Tarazi R, Sebbenn AM, Kageyama PY, Vencovsky R (2013) Edge effects enhance selfing and seed harvesting efforts in the insect-pollinated Neotropical tree *Copaifera langsdorffii* (Fabaceae). Heredity 110(6):578–585. https://doi.org/10.1038/hdy.2013.8

Tegel W, Seim A, Hakelberg D, Hoffmann S, Panev M, Westphal T, Büntgen U (2014) A recent growth increase of European beech (*Fagus sylvatica* L.) at its Mediterranean distribution limit contradicts drought stress. Eur J For Res 133(1):61–71. https://doi.org/10.1007/s10342-013-0737-7

Tenaillon O, Matic I (2020) The impact of neutral mutations on genome evolvability. Curr Biol 30(10):R527–R534. https://doi.org/10.1016/j.cub.2020.03.056

Travis J (1990) The interplay of population dynamics and the evolutionary process. Phil Trans R Soc Lond B 330:253–259. https://doi.org/10.1098/rstb.1990.0196

Trumbore S, Brando P, Hartmann H (2015) Forest health and global change. Science 349(6250):814–818. https://doi.org/10.1126/science.aac6759

Van Rossum F, Triest L (2012) Stepping-stone populations in linear landscape elements increase pollen dispersal between urban forest fragments. Plant Ecol Evol 145(3):332–340. https://doi.org/10.5091/plecevo.2012.737

Whitlock R, Stewart GB, Goodman SJ, Piertney SB, Butlin RK, Pullin AS, Burke T (2013) A systematic review of phenotypic responses to between-population outbreeding. Environ Evid 2(1):13. https://doi.org/10.1186/2047-2382-2-13

Woo S-Y (2010) Forest decline of the world: a linkage with air pollution and global warming. Afr J Biotechnol 825:7409–7414. https://doi.org/10.5897/AJB2009.000-9576

Wright S (1931) Evolution in Mendelian populations. Genetics 16(97):97–159. https://doi.org/10.1016/S0092-8240(05)80011-4

Yakovlev IA, Fossdal CG (2017) *In silico* analysis of small RNAs suggest roles for novel and conserved miRNAs in the formation of epigenetic memory in somatic embryos of Norway spruce. Front Physiol 8:674. https://doi.org/10.3389/fphys.2017.00674

Yakovlev IA, Carneros E, Lee Y, Olsen JE, Fossdal CG (2016) Transcriptional profiling of epigenetic regulators in somatic embryos during temperature induced formation of an epigenetic memory in Norway spruce. Planta 243(5):1237–1249. https://doi.org/10.1007/s00425-016-2484-8

Yeaman S (2022) Evolution of polygenic traits under global vs local adaptation. Genetics 220(1):iyab134. https://doi.org/10.1093/genetics/n134

Open Access This chapter is licensed under the terms of the Creative Commons Attribution 4.0 International License (http://creativecommons.org/licenses/by/4.0/), which permits use, sharing, adaptation, distribution and reproduction in any medium or format, as long as you give appropriate credit to the original author(s) and the source, provide a link to the Creative Commons license and indicate if changes were made.

The images or other third party material in this chapter are included in the chapter's Creative Commons license, unless indicated otherwise in a credit line to the material. If material is not included in the chapter's Creative Commons license and your intended use is not permitted by statutory regulation or exceeds the permitted use, you will need to obtain permission directly from the copyright holder.

Forest Ecosystems Under Climate Change

Florian Kraxner, Dmitry Schepaschenko, Georg Kindermann, and Andrey Krasovsky

Mountains, trails, and forest on hills in French Alps. Pointe d'Anterne Top (Photo: mzabarovsky/Adobe Stock)

F. Kraxner (✉) · D. Schepaschenko · G. Kindermann · A. Krasovsky
Agriculture, Forestry, and Ecosystem Services (AFE), International Institute for Applied Systems Analysis (IIASA), Laxenburg, Austria
e-mail: kraxner@iiasa.ac.at; schepd@iiasa.ac.at; kinder@iiasa.ac.at; krasov@iiasa.ac.at

Abstract

Forest ecosystems are significantly impacted by climate change, particularly through drought and increased weather variability. Forests are characterized by their long-lived vegetation, making it essential to consider climate projections when planning forest management actions—especially those involving the selection of tree species for reforestation and afforestation. In this context, the following pages present two examples: (1) A global estimation of trends in forest biomass change from 2020 to 2100 utilizing the Global Forest Model (G4M, Kindermann et al., Carbon Balance Manag 8: 2, 2013) and (2) an assessment of tree species suitability within the European Alps.

Keywords

Climate change · Alps · Global forest · Black locust · Douglas fir · Norway spruce · Forest biomass

Projection of Forest Biomass Change Using the Global Forest Model

The IIASA Global Forest Model (G4M, https://iiasa.ac.at/g4m) estimates the productivity of five forest types (evergreen needleleaf, evergreen broadleaf, deciduous needleleaf, deciduous broadleaf, and woody savannas) across four ecoregions: Tropical, subtropical, temperate, and boreal. The estimation is based on dynamic site characteristics such as monthly temperature, precipitation, radiation, and CO_2 concentration; semi-dynamic factors including water holding capacity and soil depth as well as nitrogen, phosphorus, salinity, and pH values; and static attributes such as air pressure.

By combining forest productivity with different management regimes (e.g., maintaining current stock, maximizing harvests, maximizing stock, or avoiding harvests), which can be enabled or disabled to change the current species to better adapt to potentially altered site characteristics, the model shows the development of increment (carbon sequestration), stock (stored carbon), and harvests (potential substitutes for nonrenewable products, also storing carbon).

The projection of forest biomass change (tC/ha) under the current management (without deforestation) and assuming the RCP2.6 climate scenario (a mild emissions pathway with projected temperature increases limited to below 2 °C compared to pre-industrial levels) is presented in Fig. 6.1. A significant increase in biomass is possible through sustainable forest use in the tropics. Biomass loss is mainly projected in dry and semi-dry areas such as central China, eastern Siberia, sub-Saharan Africa, Australia, eastern Brazil, and the central USA. This loss occurs even under the relatively mild climate change scenario RCP2.6, highlighting the increasing risk of droughts in these regions. The changes in biomass are increasing gradually, and the main trends are already visible in the short-term forecast by 2030, although losses naturally occur faster than accumulation (Fig. 6.1).

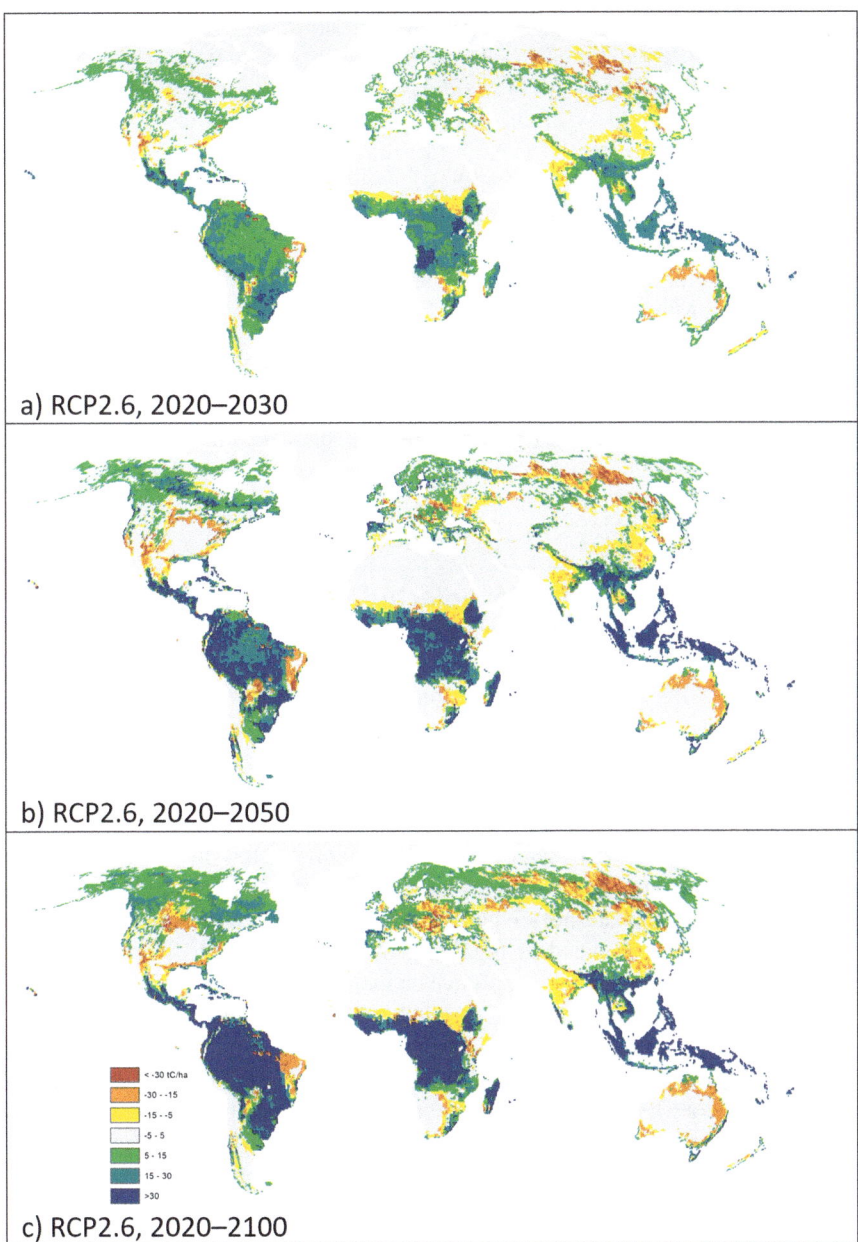

Fig. 6.1 Projected forest biomass change (tC/ha) assuming continuation of current management (without deforestation) and RCP2.6 climate scenario for (**a**) short-term, (**b**) medium-term, and (**c**) long-term development

A comparison of projected biomass changes until the end of the twenty-first century under different climate scenarios RCP2.6 (a), RCP6.0 (b), and RCP8.5 (c) is presented in Fig. 6.2. Even in the mildest climate change scenario (RCP2.6), 28% of the forest area would experience a decrease in biomass, with 8% suffering substantial losses of more than 15 tC/ha. Under the harshest considered climate change scenario (RCP8.5), the forest area with decreased biomass reaches 65%, with 38% experiencing significant losses. Notably, only a few environments remain favorable for forests by the end of the twenty-first century; this includes regions with a high elevation and/or at high latitudes as well as certain moist tropical areas with secure water supplies. Major threats to tropical forests associated with climate change are also confirmed by Doughty et al. (2023)

The results provided by the G4M model highlight areas of risk where conventional forest management may not be able to sustain forest biomass and productivity under changing climate conditions. Addressing these challenges requires more comprehensive strategies such as climate-smart management and assisted migration of suitable tree species (see Chap. 14).

Mapping Area Suitability for Selected Tree Species Under Various Climate Change Scenarios in the European Alps

Climate projections predict substantial changes in temperature and precipitation for the European Alps over the coming years (Fig. 6.3). Most areas will experience reduced amounts of precipitation. Even slight precipitation increases in some areas cannot compensate for the concomitant rise in temperature.

For mapping habitat suitability of selected tree species, data from two sources was used: (1) A systematic sample from the national forest inventories (NFI) of Alpine countries that provides insights into the presence and absence of species and (2) crowdsourced global species occurrence data from iNaturalist (https://www.inaturalist.org/). While the data regarding tree species in NFI may be limited, crowdsourcing offers a global perspective on their distribution (see Fig. 6.4). Global data is particularly valuable since it provides information on species distribution and suitability across diverse climatic conditions. Using these data sources, we work with presence-only data (Engler et al. 2004), meaning that we only possess information about locations in which species were observed, but no data on their absence in other areas. Several methods developed for utilizing occurrence data in modeling are discussed in the literature, including the Random Forest Classifier (Valavi et al. 2021) and MaxEnt (Della Pietra et al. 1997; Phillips et al. 2006). We then compare these data with maps of ecological parameters relating to climate, soil, and geomorphology. This step aids in delineating ecological boundaries for species distribution and identifying the preferable range of parameters. In the final stage, we employ Alpine maps of ecological conditions to determine species suitability under current and future climates. Figure 6.3 outlines the analysis workflow.

Examples of native (Norway spruce) and introduced (Douglas fir and black locust) species are presented in Fig. 6.5 for the current climate period (2001–2010)

6 Forest Ecosystems Under Climate Change

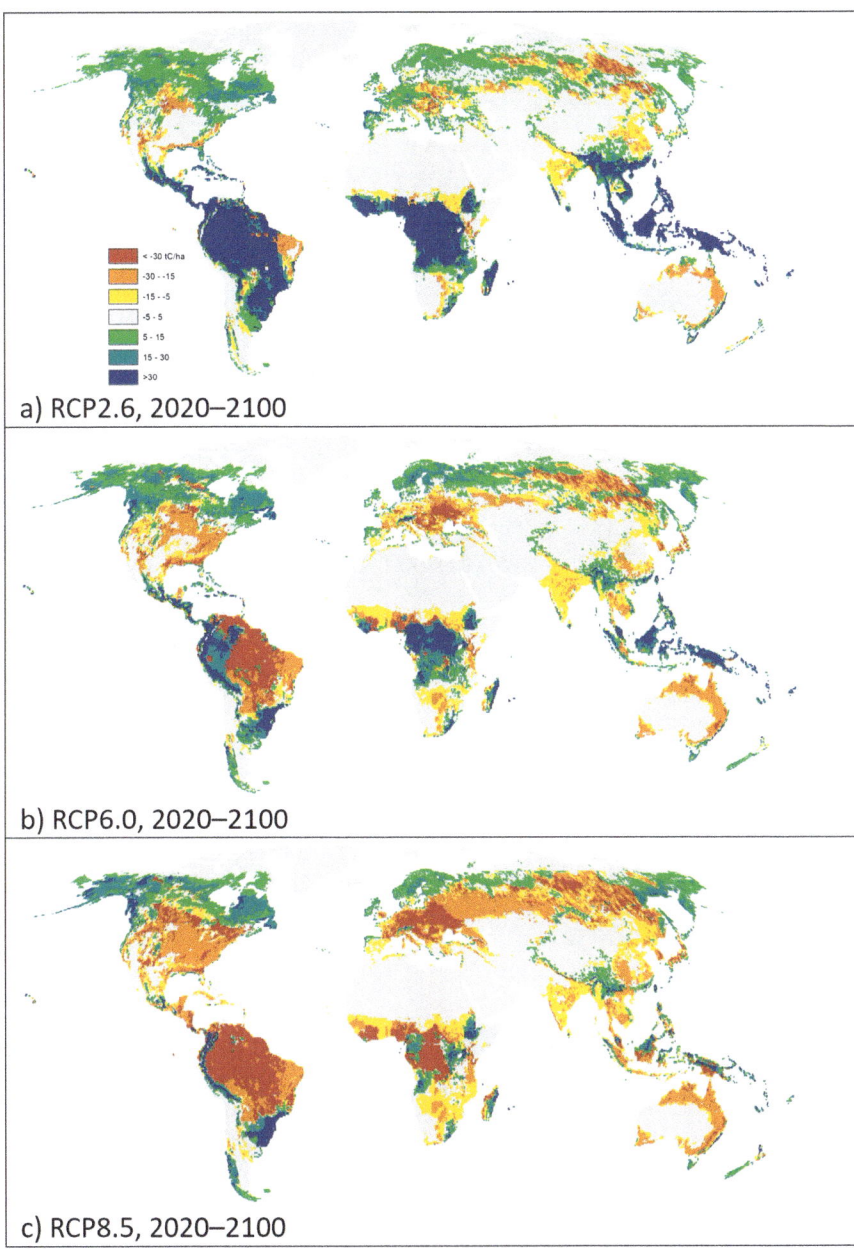

Fig. 6.2 Projected forest biomass change (tC/ha) by the end of the century under different climate scenarios, assuming continuation of current management. (**a**) RCP2.6—a mild emissions pathway with projected temperature increases by 2100 limited to below 2 °C compared to pre-industrial levels, (**b**) RCP6.0—projected temperature increases of around 3 °C, and (**c**) RCP8.5—temperature increase exceeding 4 °C

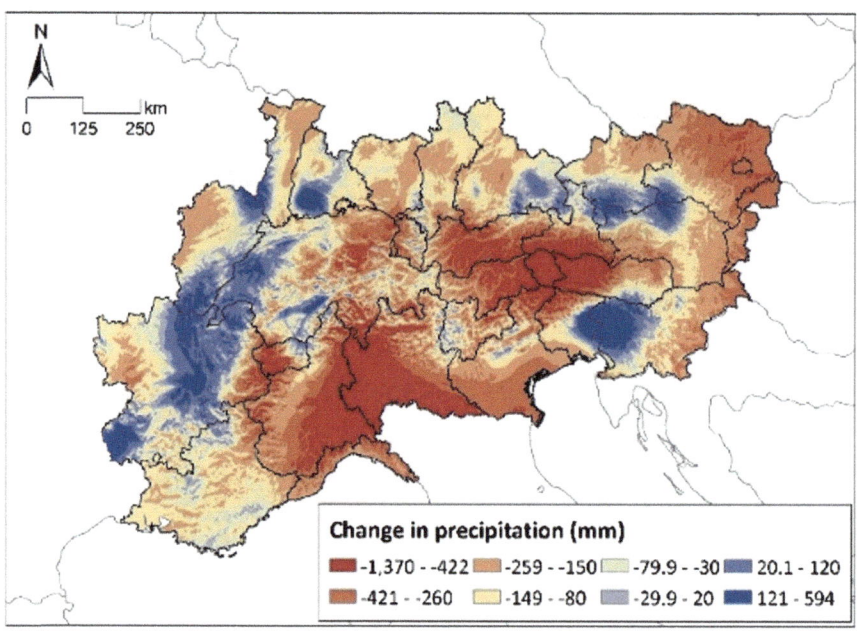

Fig. 6.3 Projected mean annual precipitation changes in the Alpine region by 2050 compared to the base year 2010 assuming an RCP 8.5 scenario (Compilation for the Alpine Space based on CHELSA climate data)

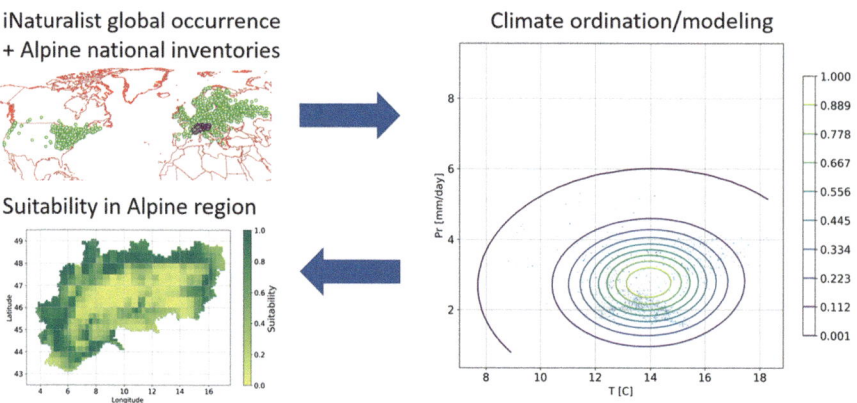

Fig. 6.4 Integrated risk modeling (Norway spruce)

and the RCP8.5 climate projection for the end of the twenty-first century (2081–2090). Although the Douglas fir appears most promising under current climate conditions, it faces the severest challenges among the three analyzed tree species by the end of the century, particularly under the extreme RCP8.5 climate change scenario (see Fig. 6.5).

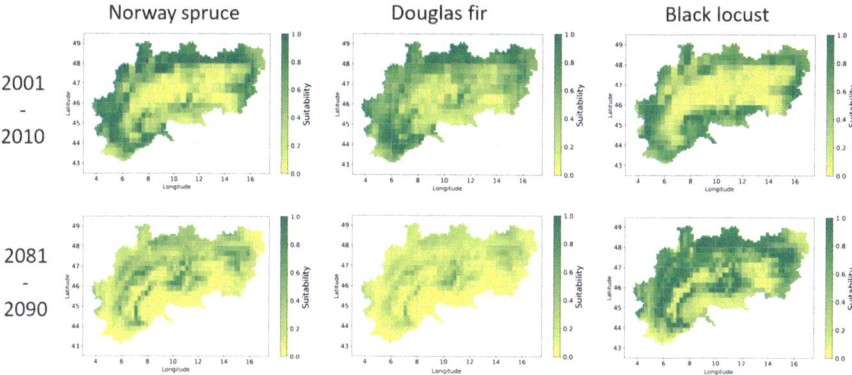

Fig. 6.5 Climate risks for three different tree species under assumption of RCP8.5 in the Alpine region

To identify the climate risks for selected tree species, we model the changes in their suitability under various climatic conditions. This provides essential information for policymakers to manage the future distribution of tree species in the Alpine region. While future climate conditions may pose a risk to some native tree species such as the Norway spruce, non-native tree species may become more suitable in certain areas.

References

Della Pietra S, Della Pietra V, Lafferty J (1997) Inducing features of random fields. IEEE Trans Pattern Anal Mach Intell 19(4):1–13

Doughty CE, Keany JM, Wiebe BC et al (2023) Tropical forests are approaching critical temperature thresholds. Nature. https://doi.org/10.1038/s41586-023-06391-z

Engler R, Guisan A, Rechsteiner L (2004) An improved approach for predicting the distribution of rare and endangered species from occurrence and pseudo-absence data. J Appl Ecol 41:263–274

Kindermann GE, Schörghuber S, Linkosalo T et al (2013) Potential stocks and increments of woody biomass in the European Union under different management and climate scenarios. Carbon Balance Manag 8:2. https://doi.org/10.1186/1750-0680-8-2

Phillips SJ, Anderson RP, Schapire RE (2006) Maximum entropy modeling of species geographic distribution. Ecol Model 190:231–259. https://doi.org/10.1016/j.ecolmodel.2005.03.026

Valavi R, Elith J, Lahoz-Monfort JJ, Guillera-Arroita G (2021) Modelling species presence-only data with random forests. Ecography 44(12):1731–1742. https://doi.org/10.1111/ecog.05615

Open Access This chapter is licensed under the terms of the Creative Commons Attribution 4.0 International License (http://creativecommons.org/licenses/by/4.0/), which permits use, sharing, adaptation, distribution and reproduction in any medium or format, as long as you give appropriate credit to the original author(s) and the source, provide a link to the Creative Commons license and indicate if changes were made.

The images or other third party material in this chapter are included in the chapter's Creative Commons license, unless indicated otherwise in a credit line to the material. If material is not included in the chapter's Creative Commons license and your intended use is not permitted by statutory regulation or exceeds the permitted use, you will need to obtain permission directly from the copyright holder.

Soil: The Foundation for Ecological Connectivity of Forest Ecosystems

7

Owen Bradley, David Keßler, Josef Gadermaier, Mathias Mayer, and Ernst Leitgeb

Soil sampling site in a mixed deciduous forest in the southeast of Austria (Photo: BFW/Owen Bradley)

O. Bradley (✉)
Department of Forest Biodiversity & Nature Conservation, Austrian Research Centre for Forests, Vienna, Austria
e-mail: owen.bradley@bfw.gv.at

D. Keßler · E. Leitgeb
Department of Forest Ecology & Soil, Austrian Research Centre for Forests, Vienna, Austria
e-mail: david.kessler@bfw.gv.at; ernst.leitgeb@bfw.gv.at

J. Gadermaier
Department of Ecosystem Management, Climate and Biodiversity, Institute of Forest Ecology, BOKU University, Vienna, Austria
e-mail: josef.gadermaier@boku.ac.at

© The Author(s) 2025
K. Lapin et al. (eds.), *Ecological Connectivity of Forest Ecosystems*,
https://doi.org/10.1007/978-3-031-82206-3_7

Abstract

Soil, ranking third in importance after air and water for supporting life on land, provides habitat, nutrients, water, and a physical foundation for plants, animals, fungi, and microorganisms. Thus, forest ecosystems, like all land-based ecosystems, are entirely dependent on soil for their existence. Consequently, soil health is critical to ecosystem connectivity, since without healthy soils, there are no healthy ecosystems or species to connect with one another. Therefore, a foundational knowledge of soil properties, its formation, and its role in shaping forest ecosystems is essential to comprehending the concept of forest ecosystem connectivity.

Soil is formed by the weathering of Earth's rocky surface and reflects past climates, geology, and vegetation. By studying soil, we can learn about previous environmental conditions and predict which plants may grow on it now and in the future. Soil not only supports plants but is teeming with complex assemblages of diverse and abundant life. A handful of healthy forest soil can contain as many individual organisms as all the people on Earth.

Plants play three pivotal roles in impacting soil properties, by shaping its physical structure, chemical composition, and the habitats it provides for other organisms. However, not only plants but all terrestrial organisms, from tiny microbes to larger animals, influence soil structure and function. Through burrowing, bioturbation, and microbial activity, soil organisms help shape the complex structure of soil, optimising it for their own needs. This chapter aims to describe the most important properties, services, and interactions of soil within forest ecosystems, underscoring its importance in maintaining forest health and connectivity.

Keywords

Forest soil characteristics · Soil ecosystem services · Soil biodiversity · Humus · Nutrient cycling · Hydrology · Soil organic carbon

Characteristics of Forest Soils

Soil is a complex and dynamic system that results from various interactions over time. Soil properties are the characteristics of soil that describe its chemical, physical, and biological attributes. Specifically, forest soils are characterised by a natural profile structure resulting from moderate intensity of silvicultural management

---

M. Mayer
Department of Ecosystem Management, Climate and Biodiversity, Institute of Forest Ecology, BOKU University, Vienna, Austria

Forest Soils and Biogeochemistry, Swiss Federal Institute for Forest, Snow and Landscape Research (WSL), Birmensdorf, Switzerland
e-mail: mathias.mayer@boku.ac.at

(Stolte et al. 2015). Due to the long rotation period typical in forestry, the nutrient cycles tend to be closed, with the soil acting as a nutrient store and playing a central role in the recycling of organic matter. The forests of the world (boreal, temperate, and tropical forests) range across a variety of climatic conditions and soil properties, with a high variation across steep spatial gradients (Binkley and Fisher 2013).

Parent Material

Rock is the most common source of parent material for soil formation. It is a solid aggregate of minerals with distinctive characteristics and origins. Rock can be classified into three main types: Igneous, metamorphic, and sedimentary (Brady and Weil 2016).

Igneous rock is formed by the cooling and solidification of magma or lava. Examples of igneous rocks are andesite, basalt, diorite, gabbro, and granite.

Metamorphic rock is formed by the transformation of existing rock when subjected to high pressure and temperature. Examples of metamorphic rock are amphibolite, gneiss, serpentinite, and schist.

Sedimentary rock is formed by the accumulation and compaction of sediments, which are fragments of rocks, minerals, or organic material that have been eroded, transported, and deposited by water, wind, ice, or gravity. Sedimentary rock has a variable mineral content depending on the source and type of sediments. Examples of sedimentary rock are breccias, conglomerates, limestone, and sandstone.

The parent material determines the initial chemical composition of a soil, including the type and amount of minerals and various physical properties such as soil texture. As some materials are more resistant or susceptible to weathering than others, the parent material also influences the rate and direction of soil formation.

Soil Formation

Soil formation or pedogenesis is the process of soil development from parent material in combination with climate, topography, organisms, and time. It involves the transformation of parent material into soil horizons, which are layers of soil with distinct properties such as colour, texture, and nutrient availability (Brady and Weil 2016; Scheffer and Schachtschabel 2018). The translocation and redistribution of soil constituents such as water, nutrients, and organic matter within and between horizons likewise play a crucial role in the formation of soil and affect its chemical, physical, and biological attributes. In general, soil formation is the result of dynamic and continuous processes that can be affected by natural or human-induced changes in the environment (Jenny 1994). Among these processes, the following are crucial:

Weathering is the process of physical and chemical breakdown of parent material into smaller particles and soluble substances caused by diverse factors, such as hydrology, temperature, carbon dioxide, acids, and organisms.

Decomposition is the process of biological breakdown of organic matter into its constituent components and elements. Organic matter includes all dead plant and animal residues as well as synthetic substances such as organic waste or pesticides. Decomposition is differentiated into humification and mineralisation.

Humification is the process of transforming organic matter into humic substances. These are stable, dark substances that contribute to the formation of a humus layer (see Box 7.1) and play a significant role in the nutrient and water balance in the top- and subsoil.

Mineralisation is the process by which microorganisms break down organic matter into dissolved inorganic compounds, making them available to the nutrient cycle and allowing them to be absorbed by plant roots.

Translocation is the process of vertical and horizontal movement of soil constituents within and between horizons. Translocation can be caused by various forces, exerted by water, gravity, wind, and organisms.

Gleying is the process of reduction of iron and manganese compounds in waterlogged soils. Anoxic conditions change the oxidation state and solubility of iron and manganese ions, resulting in greenish-blue-grey soil horizons. The transition zone between anoxic and oxic conditions exhibits mottles of reddish, yellow, and orange colours along with the colours of the anoxic horizon.

Podsolisation is the process of formation of acidic soils in cold to temperate humid climates with high precipitation and under plant species with low nutrient requirements. This results in the accumulation of organic matter, translocation (leaching) and destruction of clay minerals, and the accumulation of iron and aluminium.

Soil Physical Properties

Soil physical properties are the characteristics of soil that describe its physical structure and behaviour. They are primarily determined by the size, shape, and arrangement of soil particles, which affect water retention and movement, aeration and gas exchange, temperature and heat transfer, and erosion and compaction processes. The features of soil particles are influenced by the nature and origin of the parent material, as well as by the processes of weathering, decomposition, translocation, gleying, and podsolisation. The size and shape of soil particles determine the soil texture, which is the relative proportion of sand, silt, and clay particles in a soil sample (Fig. 7.1). Soil texture in turn affects the surface area, porosity, permeability, and water holding capacity of soil (Hillel 2003). The arrangement of soil particles is influenced by the processes of translocation and aggregation that modify and transform the soil structure. The soil structure is the spatial organisation of soil particles and pores into aggregates or peds, which are units of soil featuring distinct shapes and sizes.

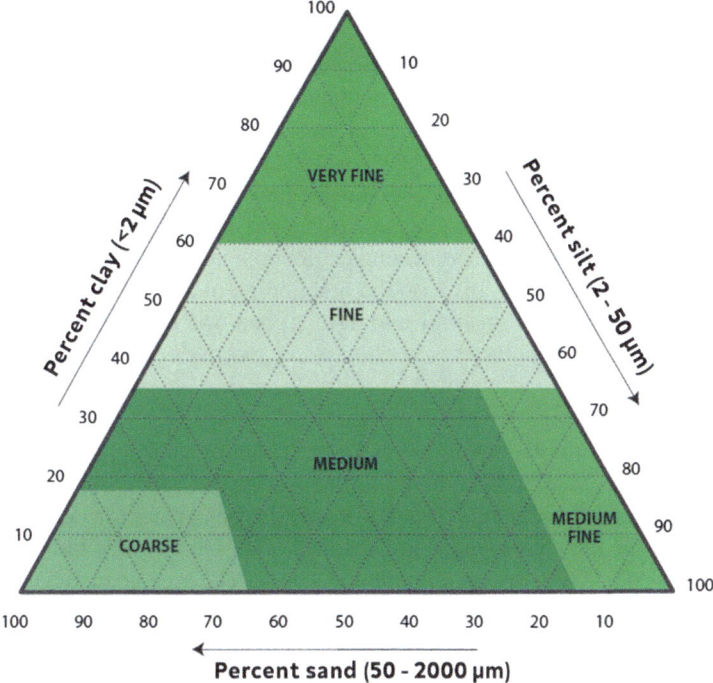

Fig. 7.1 Soil texture triangle used to define soils by texture (After Montanarella et al. 2010)

Soil Chemical Properties

Soil chemical properties are the characteristics of soil that describe its chemical composition and reactions. The chemical properties of soil are primarily determined by its mineralogical composition, the origin and nature of the parent material, and the processes of soil formation that modify and transform it. Weathering alters the mineral composition of the parent material and the soil matrix and releases nutrients and ions into the soil solution. The release of organic matter and nutrients through decomposition and the redistribution of minerals, organic matter, and nutrients between and within soils by translocation also affects chemical properties such as pH, cation exchange capacity, base saturation, and organic matter content. The pH is a measure of the acidity or alkalinity of the soil solution, which affects the nutrient status of plants and microorganisms. The cation exchange capacity describes the ability of the soil to retain positively charged ions (cations) such as calcium, magnesium, and potassium, which are essential plant nutrients. Base saturation is a measure of the proportion of cations that are bases (alkaline) rather than acids on the surface of soil particles. Organic matter content is a measure of the amount of carbon-containing compounds derived from living organisms in the soil, which affects the nutrient cycling, water retention, and biological activity of soil (Brady and Weil 2016).

> **Box 7.1 Humus**
>
> Humus is composed of organic material at various stages of decomposition, which is transformed into complex compounds such as humins, fulvic acids, and humic acids through the process of humification. It is generally found at the surface of the soil but also plays a significant role in the mineral subsoil. This first layer of soil generally contains the greatest abundance and diversity of organisms that decompose organic matter. It is also a major driver and indicator of processes, such as decomposition rate and carbon storage, which shape entire ecosystems. To grasp the functioning of a forest ecosystem at a fundamental level, it is imperative to have a good understanding of different humus forms and what they reveal about the respective ecosystem.

Generally, the different humus forms are the result of a changing rate of decomposition. The rate of decomposition is influenced by all soil formation processes (see **Soil formation**), with the effect of climatic gradient, parent material, and vegetation being the most profound. **Mull** consists of an **OL** (organic litter) layer and sometimes also an **OF** (organic fermented) horizon. The organic material falling to the ground as litter is quickly broken down and integrated into the upper (**A**) mineral soil horizon by macrofauna such as earthworms. **Amphi** and **Moder** humus types, found on calcareous and siliceous parent material respectively, are typically thicker forms of humus with three distinct horizons: **OL, OF,** and **OH** (organic humus), which indicate slower rates of decomposition. **Amphi**, which is generally less acidic, tends to have high zoogenic activity across the spectrum of mega-, meso-, and microfauna along with slower rates of decomposition resulting solely from a cooler climate. **Moder**, on the other hand, tends to be rather acidic (pH < 5) owing to the combination of noncarbonate parent material and ligneous-rich litter input from conifers, with fungi dominating the primary decomposers in terms of biomass. **Tangel** and **Mor** humus forms exhibit the slowest transformation of litter into mineral soil. **Tangel** is a thick (>15 cm), woven-together, and slowly decomposing humus found on high alpine hard calcareous rock where decomposer activity is frozen for most of the year. **Mor** is usually very acidic due to a combination of siliceous parent material and ligneous litter input from conifers, with decomposition further slowed by cold climate. Due to the harshness of conditions, both forms tend to have a lower total and relative abundance of biogenic activity (Fig. 7.2).

Humus is extremely important for many of the ecosystem services provided by soil. It is the first horizon to accumulate and store soil organic carbon (SOC) after afforestation; the process of decomposition that occurs in the humus provides food for the bulk of all soil biodiversity; and it can hold 80%–90% of its dry weight in moisture, thus helping mitigate drought and flood events.

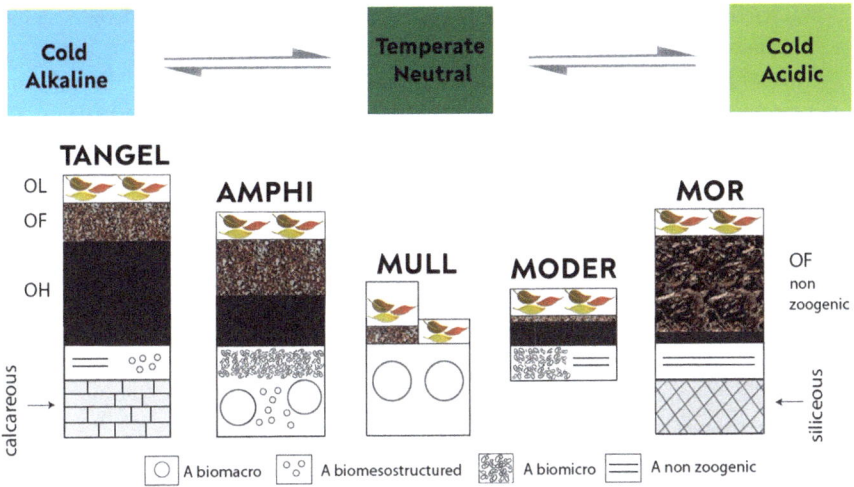

Fig. 7.2 Schematic of humus forms in the terrestrial environment (Adapted from Montanarella et al. 2010)

Soil Ecosystem Services

Soil ecosystem services are the various benefits humans gain from the ground. They can be separated into three broad categories: The principal category is the direct provisioning of materials such as food, water, fuel, raw building materials, and medicinal plants. The secondary category encompasses regulatory services such as nutrient cycling, water purification, habitat provision, and carbon storage. The third and less quantifiable group of services may be termed as "natural beauty". These services are important to humans by helping to satisfy our aesthetic, cultural, recreational, spiritual, and scientific needs (Geitner et al. 2019). Following air and water, soil is the third most important component for supporting life on Earth, and essential for the survival of all terrestrial organisms. There is a long list of ecosystem services and functions provided by soil, which include:

Water storage, runoff regulation, and purification: Water availability for plants and soil biota. Precipitation uptake, flood mitigation, and removal of pollutants to provide drinking water.

Nutrient cycling: Storage and exchange of macro- and micronutrients with plants and microorganisms.

Global climate regulation: The ability of soil to store and potentially sequester carbon from the atmosphere.

Habitat and biodiversity: Soil is home to an immense range of biotic diversity.

Agricultural biomass production: Food, fodder, technical fibre, medicinal plants, and energy biomass.

Forest biomass production: Construction timber, fuelwood, and non-timber forest products.

Microclimate regulation: Local air-cooling effect from plant evapotranspiration.

Cultural and natural archives: Soil can preserve objects from the past and offer us insights into past events and processes through the interpretation of its current form.

Recreational and spiritual services: Soil is the surface upon which many sporting, cultural, and spiritual events take place.

This section provides information on three of the more relevant services that soil provides in terms of ecosystem connectivity: water storage, runoff regulation and purification, nutrient cycling, and soil's role in global climate regulation. All fall into the broad category of regulatory ecosystem services.

Soil Hydrology

Hydrologic processes are fundamental to both soil habitat quality and the mitigation of extreme hydrological events, with a strong feedback loop between the two aspects. Due to the essential role of water in all biological processes, soil hydrology significantly influences soil biodiversity across various scales. It affects life in soils directly via water availability and indirectly via air availability and is among the most critical factors affecting the diversity, abundance, and composition of soil organisms. At the microscale level, soil moisture regulates the metabolic activity of microorganisms. Water is a vital component in soil biochemical reactions, facilitating enzymatic reactions and transporting nutrients required for microbial growth. Changes in soil moisture affect microbial diversity and community composition, leading to altered soil processes (Drenovsky et al. 2004). At a larger scale, water affects soil structure, porosity, and aeration, all of which influence the diversity and distribution of soil fauna. Soil moisture conditions have a direct impact on animal activity, burrowing, and reproduction. Conversely, however, the activity of soil micro- and macro-organisms also alters a soil's hydraulic properties. High microbial activity generally leads to increased stabilisation of soil organic matter and reduced bulk density. These two effects positively affect the water storage capacity of soil and facilitate its rewetting. The water drainage function created by earthworms or burrows of larger soil-dwelling organisms is likewise important for infiltration of water into the soil and, by extension, prevention of surface runoff (Védère et al. 2022). Active and diverse forest soils therefore provide the important function of water retention and water storage and are crucial elements in the hydrological cycle. Such forest ecosystems can effectively fulfil their mediating role in the hydrological cycle, serving as a buffer against flooding or periods of drought-induced stress.

Different soil organisms are adapted to different hydrological soil regimes such as waterlogged soils, seasonally flooded systems, or soils with access to groundwater or seawater, to name just a few. However, land-use changes and climate change are nowadays the main drivers of soil transformation (Berhe 2019), altering the hydrological regime of forest soils in many regions of the world. For example, the draining of wetlands and intensive forest management practices cause vast changes in forest hydrological cycles (Bredemeier et al. 2010) that can have a severe impact on the soil organism composition. It is estimated that 80% of previously extant wetlands in Europe have disappeared (Finlayson and Spiers 1999). Riparian habitats are rich in biodiversity at temperate latitudes, and from a biodiversity conservation and connectivity perspective, their preservation and restoration are extremely valuable (Muys et al. 2022).

Soil Nutrient Cycling

> "The same circulation exists on the surface of the earth as in the sea; there is unceasing change – a perpetual destruction and re-establishment of equilibrium."
> —Justus Liebig, 1849

Forest nutrient cycling is the exchange of elements between living and non-living components of ecosystems. Nutrients such as nitrogen, phosphorus, and carbon are essential for plant growth and survival, and their cycling in forest ecosystems is regulated by numerous factors including climate, plant species community structure, soil type, and topography (Foster and Bhatti 2006). The processes of forest nutrient cycling are nutrient uptake and storage in vegetation, litter production, decomposition, nutrient transformations by soil organisms, atmospheric inputs, mineral weathering, and nutrient export from the soil.

Climate arguably plays the most significant role in nutrient cycling, with latitude and altitude representing a rough proxy for temperature. Terrestrial primary production generally increases from colder boreal forests through mild temperate forests to warmer tropical forests, while forest soil (specifically topsoil and organic horizon) nutrient content and residence time increase in the opposite direction. The cold average annual temperatures in subarctic woodland soils and taiga forests result in slow nutrient cycling rates. On the contrary, warm tropical forests have high microbial activity and turnover leading to fast decomposition and nutrient cycling rates. Variation in nutrient cycles is also influenced by biotic factors such as tree species-specific regulation of resource use: Different tree species have varying nutrient requirements and strategies of nutrient use affecting their general patterns of nutrient accumulation, partitioning, and recycling (Coleman et al. 1983).

Abiotic factors such as soil character, topography, and parent material also influence nutrient cycling in forests. Soils developed from different parent materials vary greatly in terms of nutrient content and availability. The constraints of soil properties and microclimatic variation define the potential forest plant community structure, and thus productivity and nutrient cycling processes. For example, soils with a low pH tend to promote trees with low-quality, high-lignin-content litter that decomposes slowly.

Disturbances like fire, harvesting, and natural events have lasting impacts on nutrient cycling. Forest fire can redistribute ecosystem nutrients through ash deposition, mineralisation of nutrients bound in organic matter, and charcoal formation, while harvesting interrupts nutrient cycling by removing nutrients bound in wood biomass from the system. Forest management practices can affect nutrient cycling positively or negatively depending on their impact on soil and vegetation (Rahman et al. 2013). For example, extensive forest management practices, such as selective harvesting, can be employed to reduce large interruptions of the nutrient cycle, thereby enhancing the habitat quality of forest ecosystems.

Soil Organic Carbon

Soil organic carbon (SOC) is with approximately 1500 Pg carbon, the largest store of terrestrial carbon on the planet (Petrokofsky et al. 2012; Scharlemann et al. 2014), and plays a crucial role in maintaining the structure, fertility, and functioning of forest soils. Carbon constitutes approximately 58% of all soil organic matter (SOM), which is a derivative of decaying plant and animal material, and soil microorganisms and their exudates (Perie and Ouimet 2008). This organic material supplies energy and nutrients to soil organisms and binds soil particles together, thereby helping to reduce erosion and improve the water holding capacity of soil (Heimann and Reichstein 2008). In addition, SOC is usually positively correlated to soil pH, which increases the availability of essential plant nutrients such as nitrogen, phosphorus, and sulphur.

With approximately 20 Pg of total biomass in global soils, soil biota not only makes up a significant fraction of SOC, but is also responsible for processing and integrating organic matter into long-term stores of SOC through their necromass and secretions (Crowther et al. 2019). The more healthy and active forest soils are, the more potential they have to effectively process, humify, and mineralise this material into long-term stores (Chertov et al. 2017).

At a global level, forest SOC is vital to mitigating the effects of climate change (Heimann and Reichstein 2008). Forests grow and accumulate biomass by absorbing carbon dioxide (CO_2) from the atmosphere through photosynthesis. In death and decomposition, plant biomass carbon is cycled into SOC, reducing atmospheric CO_2 and associated greenhouse effects. Further research into the soil food web, and specifically carbon nutrient flow, is required to better understand how biomass carbon is cycled into long-term stores of SOC and how soil microbial community composition mediates this process.

Soil Biodiversity

Soil biodiversity is the diversity of living organisms within the soil. As previously established, soils are extremely complex systems that vary greatly in how they are formed, which influences their chemical and physical structure over space and time. The impact of climate and site conditions, plus the effects of plant communities that grow on them, both past and present, adds to their heterogeneity. Such an immense variety of ecological niches leads to a superabundance of species potentially present in a small amount of soil. This species richness can be hard to comprehend, with the

7 Soil: The Foundation for Ecological Connectivity of Forest Ecosystems

total number of species present in one handful of healthy forest soil approximating the estimated number of the world's plant, animal, and insect species combined—around ten million (Montanarella et al. 2010). Plants represent the largest proportion of biomass of any group of organisms living in the soil, and due to rooting patterns and litter input, they have the greatest influence on the physical and chemical structure of soil. Soil type and plant community together strongly influence the potential diversity and structure of additional soil biota (Berg and Smalla 2009).

Soil is home to and necessary for the life of all terrestrial plants, animals, fungi, and single-celled organisms. Soil-specific biota can generally be divided into three groups based on size (Montanarella et al. 2010): Macro- and megafauna larger than two millimetres (earthworms, ants, woodlice, centipedes, amphibians, reptiles, mammals, and birds); mesofauna from two millimetres down to one hundred micrometres (0.1 mm), comprising tardigrades (water bears), collembola (springtails), and mites; finally, microfauna and microflora between one and one hundred micrometres (0.001–0.1 mm), including nematodes, bacteria, fungi, protozoa, and other single-celled organisms (Fig. 7.3).

Soil Mega-, Macro-, and Mesofauna

Megafauna: Soil megafauna species are not big compared to other large animals, and rarely exceed a total mass of one kilogram per individual. They are typically vertebrates adapted to life underground, with slender bodies, efficient digging apparatus, and particularly sensitive noses that often have the capacity to detect

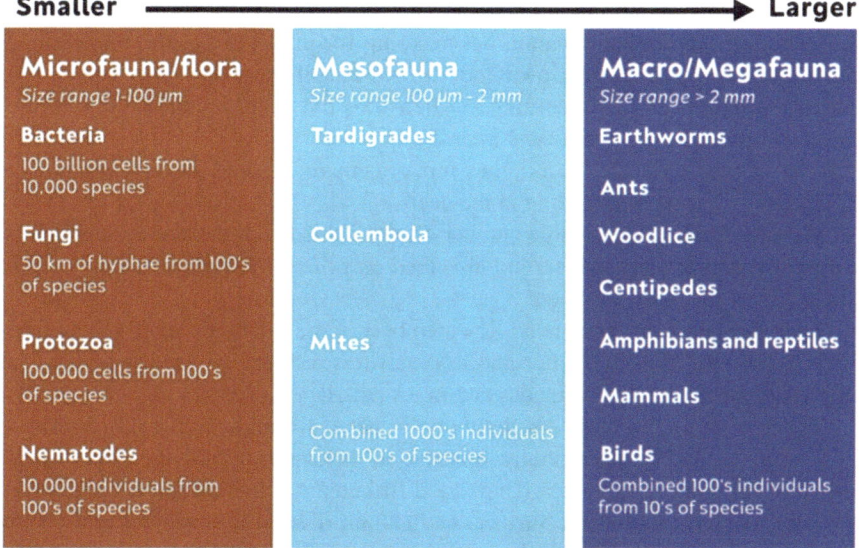

Fig. 7.3 Soil biota divided into three groups based on size (Adapted from Montanarella et al. 2010)

bioelectric signals—for example, moles, shrews, salamanders, and blind snakes. Some mammals and birds nest in the ground but are not considered truly euedaphic, meaning they are specifically adapted to subterranean life. Soil megafauna influences the invertebrate community structure by predation, carrion, and faeces and by modifying the soil structure through burrowing and bioturbation.

Macrofauna or "ecosystem engineers" are the principal litter transformers on the forest floor. Here are a few of the more important species and some of their key features:

Ants (formicidae): Form nests in trees, on the ground surface, or underground. They have often evolved remarkable symbiotic relationships with other organisms in their home environments, such as leaf-cutter ants, who build exceptionally large nests up to 300 m^2 where they store harvested leaves on which they grow a fungus as their main food source. Other ants "care" for certain aphid species in order to harvest nutrient-rich secretions from the fattened aphids.

Termites: Feed on dead plant material and are extremely accomplished nest builders, with some termitaria estimated to have been continuously occupied for more than 50,000 years.

Isopods or woodlice: Occupy all terrestrial landscapes, from seashores to high alpine environments. They are often found under stones, deadwood, and bark and represent important detritivores that digest leaf litter and mediate microbial communities and nutrient cycles.

Myriapods or centipedes and millipedes: Myriapods are arthropods with elongated bodies that have up to several tens of similarly shaped segments, each bearing one or two pairs of legs. They tend to be more commonly found in calcareous soils and are important contributors to the destruction of litter in the first phase of decomposition.

Earthworms: Found in soils all around the world. They are not particularly adept at digesting organic matter, but make up for this with prolific bioturbation, processing 10–30 times their own body weight in soil every day. They can be subdivided into three groups according to where they live in the solum:
Epigeic—In the upper humus and litter layers.
Anecic—Dwelling in the topsoil, they mix organic matter into the soil system.
Endogenic—Living and feeding in the deep soil.

Mesofauna: Includes a large number of organisms that break down plant debris, digest soil and organic matter, and also feed on primary decomposers. Key representatives of this group include:

Enchytraeidae or potworms: Key members of the soil biotic food web by feeding on decaying organic matter and microbivores, as well as serving as food for other soil fauna such as centipedes and mites (acari).

Acari: Part of the class Arachnida and one of the most numerous arthropod groups in the soil, with potentially hundreds of thousands of individuals per square metre at particularly rich sites. They are distributed throughout the solum, though typically concentrated at the surface, and exhibit diverse feeding strategies. They not only primarily decompose decaying organic material, but also engage in parasitic behaviour. Additionally, they are known to symbiotically carry bacteria and fungi into humus material to aid in its breakdown and consumption.

Collembola: Also known as "springtails" due to their method of propulsion by releasing tension in their curved abdomen to fling themselves away from potential predators. They are believed to be the most abundant hexapods on Earth and feed on organic detritus in all forms. Interestingly, collembola have been found to aid moss fertilisation by accidentally carrying moss sperm (which must otherwise swim) on their bodies to fertile moss archegonia (Cronberg et al. 2006).

Nematodes: The most abundant multicellular organism on Earth, with densities of up to ten million individuals per square metre. About 30,000 species are known to science, but this number is estimated to be only 3% of the total (Abebe et al. 2011). They take the form of cylindrical tubes and inhabit the water film around soil particles. Due to their distinct mouth morphology, they are typically classified into five groups based on their dietary preferences: Bacterivores, fungivores, omnivores, plant parasites, and predators.

Rotifers: These minute animals (0.2–0.4 mm in length) are superabundant in the soil surface, and like most soil mesofauna, require some amount of moisture to live and move around. They are important members of the soil food web who feed almost exclusively on bacteria, algae, and yeasts.

Tardigrades: Commonly known as water bears due to their bear-like appearance, they are renowned for their incredible resistance to extreme environments. They can endure temperatures of up to 100 °C and periods of desiccation for more than 20 years after which, when rehydrated, they can reanimate, lay eggs, and carry on with life (Møbjerg and Neves 2021). A community of tardigrades was even sent into low Earth orbit and exposed to the extreme conditions of the vacuum of space, after which several of them laid eggs and resumed life as usual (Fig. 7.4).

Soil Microorganisms

Soil microorganisms form highly diverse and complex communities and are commonly grouped into bacteria, fungi, archaea, protists, and viruses (Crowther et al. 2019; Fierer 2017). A single gram of soil can contain more than 50,000 different

Fig. 7.4 *Aporrectodea smaragdina*, an epigeic green rainworm observed in the Southern Limestone Alps of Austria (Photo: BFW/Nele Wolter)

species and more than 0.5 mg of microbial biomass carbon (Banerjee and Van Der Heijden 2023; Gao et al. 2022). Globally, soil microorganisms store about 27 Pg, 4 Pg, and 2 Pg of carbon, nitrogen, and phosphorus in their biomass, respectively (Gao et al. 2022). The most abundant microbial taxa in soils are bacteria and fungi, which contribute to a variety of important below-ground processes (Fierer 2017). For example, bacteria and fungi are the primary decomposers of dead organic matter, making them major drivers of carbon and nutrient cycling in terrestrial ecosystems (Van Der Heijden et al. 2008). Bacteria also interact with plant roots and mediate multiple critical steps in the nitrogen cycle, including nitrogen fixation from the atmosphere (Lladó et al. 2017). Many fungal taxa form mycorrhizal symbioses with plant roots, supplying their hosts with growth-limiting nutrients and water from the soil in exchange for photosynthetically fixed carbon (Read and Perez-Moreno 2003; Smith and Read 2010). Among mycorrhizal associations, ectomycorrhiza, arbuscular mycorrhiza, and ericoid mycorrhiza are the geographically most important types (Read and Perez-Moreno 2003; Soudzilovskaia et al. 2019; Ward et al. 2022). Another key functional group of soil microbiota are pathogens, which can have a strong influence on plant diversity and community composition (Van Der Heijden et al. 2008). Climatic changes including elevated CO_2 levels, higher temperatures, and increased drought incidence are expected to strongly affect the composition and functioning of soil microbial communities (Jansson and Hofmockel 2020).

Summary "Soil: The Foundation of Forest Ecosystems"

Soil plays a critical role in providing habitat for forest species and supporting ecosystem functionality. Soil properties strongly influence potential species community composition within forests. Research suggests that forest structural diversity is positively correlated with soil organism diversity (Lang et al. 2023). However, soil properties, particularly SOC content and pH, are the strongest predictors for variation in taxonomic richness and soil community composition (Crowther et al. 2019; Högberg et al. 2007; Lladó et al. 2017; Soudzilovskaia et al. 2019). Moreover, the type of humus and litter composition significantly impact microbial and insect communities residing in the soil (Asplund et al. 2019; Ponge and Chevalier 2006; Salmon et al. 2005).

So we see that the relationship between soil ecosystem functionality and soil community composition forms strong feedback loops. Changes in soil moisture affect microbial diversity and community composition, leading to alterations in soil processes (Drenovsky et al. 2004). Furthermore, soil biota and soil nutrient input are connected in another feedback loop mediating soil community composition (Aponte et al. 2013), with SOC, in particular, playing a profound role due to its wider implications for climate change mitigation.

With our world's rapidly warming climate and land-use intensification, habitat destruction threatens forest ecosystems and the species they support. Maintaining habitat connectivity is essential for species survival, as it facilitates the transfer of energy, resources, and genetic material throughout the ecosystem. Soil, along with its associated functions and biota, is a fundamental component of forest ecosystems. Assessing connectivity in forest habitats requires consideration of the landscape structure, species movement, and habitat quality.

A rarely studied concept is the connectivity of soil organisms themselves. There are very few studies which specifically focus on how habitat connectivity affects soil organism connectivity through movement or genetic connectedness. Below is a brief description of the three most relevant studies discovered in the literature search for this chapter. In Rantalainen et al. 2005, Rantalainen et al. showed how over a three-year time period, enhanced habitat connectivity of small circular patches (Ø 120 cm) on a 50 × 50 m plot of soil increases the colonisation efficacy of soil fungi. A later study loosely relevant to soils showed that forest fragment connective structures, in this case hedgerows, were shown to be important habitats for rare forest plant species (Wehling and Diekmann 2009). The most recent and relevant study showed that SOC was the most common driver of soil biota richness, while habitat connectivity had a positive impact on larger soil faunal organism richness (Lopezosa et al. 2023). There are obviously large knowledge gaps in this area of study with lots of potential for specific research questions relating to the effect that forest habitat connectivity has on soil community composition.

References

Abebe E, Mekete T, Thomas WK (2011) A critique of current methods in nematode taxonomy. Afr J Biotechnol 10(3):312–323
Aponte C, García LV, Marañón T (2013) Tree species effects on nutrient cycling and soil biota: a feedback mechanism favouring species coexistence. For Ecol Manag 309:36–46
Asplund J, Kauserud H, Ohlson M, Nybakken L (2019) Spruce and beech as local determinants of forest fungal community structure in litter, humus and mineral soil. FEMS Microbiol Ecol 95(2):fiy232
Banerjee S, Van Der Heijden MG (2023) Soil microbiomes and one health. Nat Rev Microbiol 21(1):6–20
Berg G, Smalla K (2009) Plant species and soil type cooperatively shape the structure and function of microbial communities in the rhizosphere. FEMS Microbiol Ecol 68(1):1–13
Berhe AA (2019) Drivers of soil change. In: Developments in soil science, vol 36. Elsevier, pp 27–42
Binkley D, Fisher RF (2013) Ecology and management of forest soils. Wiley
Brady NC, Weil RR (2016) The nature and properties of soils, 15th edn. Pearson
Bredemeier M, Cohen S, Godbold DL, Lode E, Pichler V, Schleppi P (eds) (2010) Forest management and the water cycle: an ecosystem-based approach, vol 212. Springer Science & Business Media
Chertov O, Komarov A, Shaw C, Bykhovets S, Frolov P, Shanin V et al (2017) Romul_Hum—a model of soil organic matter formation coupling with soil biota activity. II. Parameterisation of the soil food web biota activity. Ecol Model 345:125–139
Coleman DC, Reid CPP, Cole CV (1983) Biological strategies of nutrient cycling in soil systems. In: Advances in ecological research, vol 13. Academic Press, pp 1–55
Cronberg N, Natcheva R, Hedlund K (2006) Microarthropods mediate sperm transfer in mosses. Science 313(5791):1255–1255
Crowther TW, Van den Hoogen J, Wan J, Mayes MA, Keiser AD, Mo L et al (2019) The global soil community and its influence on biogeochemistry. Science 365(6455):eaav0550

Drenovsky RE, Vo D, Graham KJ, Scow KM (2004) Soil water content and organic carbon availability are major determinants of soil microbial community composition. Microb Ecol 48:424–430

Fierer N (2017) Embracing the unknown: disentangling the complexities of the soil microbiome. Nat Rev Microbiol 15(10):579–590

Finlayson CM & Spiers AG (eds) 1999. Global review of wetland resources and priorities for wetland inventory. Supervising Scientist Report 144 / Wetlands Internation Publication 53, Suvervising Scientist, Canberra, https://doi.org/10.1016/j.ecoleng.2013.03.006

Foster NW, Bhatti JS (2006) Forest ecosystems: nutrient cycling. Encyclopedia of soil science, 718721

Gao D, Bai E, Wang S, Zong S, Liu Z, Fan X et al (2022) Three-dimensional mapping of carbon, nitrogen, and phosphorus in soil microbial biomass and their stoichiometry at the global scale. Glob Chang Biol 28(22):6728–6740

Geitner, C., Freppaz, M., Lesjak, J., Schaber, E., Stanchi, S., d'Amico, M., & Vrscaj, B. (2019). Soil Ecosystem Services in the Alps-An introduction for decision-makers (pp. 1–78). Agricultural Institute of Slovenia

Heimann M, Reichstein M (2008) Terrestrial ecosystem carbon dynamics and climate feedbacks. Nature 451(7176):289–292

Hillel D (2003) Introduction to environmental soil physics. Elsevier

Högberg MN, Högberg P, Myrold DD (2007) Is microbial community composition in boreal forest soils determined by pH, C-to-N ratio, the trees, or all three? Oecologia 150:590–601

Jansson JK, Hofmockel KS (2020) Soil microbiomes and climate change. Nat Rev Microbiol 18(1):35–46

Jenny H (1994) Factors of soil formation: a system of quantitative pedology. Courier Corporation

Lang AK, LaRue EA, Kivlin SN, Edwards JD, Phillips RP, Gallion J et al (2023) Forest structural diversity is linked to soil microbial diversity. Ecosphere 14(11):e4702

Lladó S, López-Mondéjar R, Baldrian P (2017) Forest soil bacteria: diversity, involvement in ecosystem processes, and response to global change. Microbiol Mol Biol Rev 81(2):10–1128

Lopezosa P, Berdugo M, Morales-Márquez J, Pastor E, Delgado-Baquerizo M, Bonet A et al (2023) On the relative importance of resource availability and habitat connectivity as drivers of soil biodiversity in Mediterranean ecosystems. J Ecol 111(7):1455–1467

Møbjerg N, Neves RC (2021) New insights into survival strategies of tardigrades. Comp Biochem Physiol A Mol Integr Physiol 254:110890

Montanarella L, Marmo L, Miko L, Ritz K, Peres G, Römbke J, Van der Putten W (2010) European atlas of soil biodiversity. European Commission, Publications Office of the European Union, Luxembourg

Muys B, Angelstam P, Bauhus J, Bouriaud L, Jactel H, Kraigher H, … & Van Meerbeek K (2022) Forest biodiversity in Europe. From science to policy 13

Perie C, Ouimet R (2008) Organic carbon, organic matter and bulk density relationships in boreal forest soils. Can J Soil Sci 88(3):315–325

Petrokofsky G, Kanamaru H, Achard F, Goetz SJ, Joosten H, Holmgren P et al (2012) Comparison of methods for measuring and assessing carbon stocks and carbon stock changes in terrestrial carbon pools. How do the accuracy and precision of current methods compare? A systematic review protocol. Environ Evid 1:1–21

Ponge JF, Chevalier R (2006) Humus Index as an indicator of forest stand and soil properties. For Ecol Manag 233(1):165–175

Rahman MM, Tsukamoto J, Tokumoto Y, Shuvo MAR (2013) The role of quantitative traits of leaf litter on decomposition and nutrient cycling of the forest ecosystems. J Forest Environ Sci 29(1):38–48

Rantalainen ML, Fritze H, Haimi J, Pennanen T, Setälä H (2005) Colonisation of newly established habitats by soil decomposer organisms: the effect of habitat corridors in relation to colonisation distance and habitat size. Appl Soil Ecol 28(1):67–77

Read DJ, Perez-Moreno J (2003) Mycorrhizas and nutrient cycling in ecosystems—a journey towards relevance? New Phytol 157(3):475–492

Salmon S, Geoffroy JJ, Ponge JF (2005) Earthworms and collembola relationships: effects of predatory centipedes and humus forms. Soil Biol Biochem 37(3):487–495

Scharlemann JP, Tanner EV, Hiederer R, Kapos V (2014) Global soil carbon: understanding and managing the largest terrestrial carbon pool. Carbon Manag 5(1):81–91

Scheffer F, Schachtschabel P (2018) Lehrbuch der Bodenkunde. 17, überarbeitete und ergänzte Auflage

Smith SE, Read DJ (2010) Mycorrhizal symbiosis. Academic Press

Soudzilovskaia NA, van Bodegom PM, Terrer C, Zelfde MVT, McCallum I, Luke McCormack M et al (2019) Global mycorrhizal plant distribution linked to terrestrial carbon stocks. Nat Commun 10(1):5077

Stolte J, Tesfai M, Øygarden L, Kværnø S, Keizer J, Verheijen F et al (eds) (2015) Soil threats in Europe. Publications Office, Luxembourg

Van Der Heijden MG, Bardgett RD, Van Straalen NM (2008) The unseen majority: soil microbes as drivers of plant diversity and productivity in terrestrial ecosystems. Ecol Lett 11(3):296–310

Védère C, Lebrun M, Honvault N, Aubertin ML, Girardin C, Garnier P et al (2022) How does soil water status influence the fate of soil organic matter? A review of processes across scales. Earth Sci Rev 234:104214

Ward EB, Duguid MC, Kuebbing SE, Lendemer JC, Bradford MA (2022) The functional role of ericoid mycorrhizal plants and fungi on carbon and nitrogen dynamics in forests. New Phytol 235(5):1701–1718

Wehling S, Diekmann M (2009) Importance of hedgerows as habitat corridors for forest plants in agricultural landscapes. Biol Conserv 142(11):2522–2530

Open Access This chapter is licensed under the terms of the Creative Commons Attribution 4.0 International License (http://creativecommons.org/licenses/by/4.0/), which permits use, sharing, adaptation, distribution and reproduction in any medium or format, as long as you give appropriate credit to the original author(s) and the source, provide a link to the Creative Commons license and indicate if changes were made.

The images or other third party material in this chapter are included in the chapter's Creative Commons license, unless indicated otherwise in a credit line to the material. If material is not included in the chapter's Creative Commons license and your intended use is not permitted by statutory regulation or exceeds the permitted use, you will need to obtain permission directly from the copyright holder.

Part II
Monitoring and Assessment Techniques

Monitoring Methods for the Protection of Connectivity in Forest Ecosystems

8

Janine Oettel, Bettina Thalinger, Aglaia Szukala, Linus Munishi, and Katharina Lapin

Stag beetle (*Lucanus cervus*). (Photo: Michal/Adobe Stock)

J. Oettel (✉) · A. Szukala · K. Lapin
Department of Forest Biodiversity & Nature Conservation, Austrian Research Centre for Forests, Vienna, Austria
e-mail: janine.oettel@bfw.gv.at; aglaia.szukala@bfw.gv.at; katharina.lapin@bfw.gv.at

B. Thalinger
Applied Animal Ecology Research Unit, Department of Zoology, University of Innsbruck, Innsbruck, Austria
e-mail: Bettina.Thalinger@uibk.ac.at

L. Munishi
School of Life Sciences and Bioengineering, Nelson Mandela-African Institution of Science and Technology (NM-AIST), Arusha, Tanzania
e-mail: linus.munishi@nm-aist.ac.tz

© The Author(s) 2025
K. Lapin et al. (eds.), *Ecological Connectivity of Forest Ecosystems*,
https://doi.org/10.1007/978-3-031-82206-3_8

Abstract

Forest ecosystems face increasing threats from climate change, resource exploitation, and other anthropogenic disturbances causing biodiversity loss and habitat fragmentation. The conservation priority of connected, healthy forests necessitates robust monitoring that covers the landscape, ecosystem, species, and genetic levels and employs direct as well as indirect methods. Connectivity objectives encompass patch colonization, prioritization, and landscape assessment at multiple scales. Monitoring landscapes and forest ecosystems involves assessing their physical attributes and functional diversity to understand biodiversity, land-use changes, and threats like deforestation and climate impacts. Remote sensing offers large-scale data collection, while terrestrial surveys including laser scanning provide detailed insights into forest dynamics. Challenges include scale issues, standardization, and potential oversights in finer-scale variations. While species monitoring captures long-term shifts in abundance or distribution, it can be resource-intensive and challenging for elusive species. Alternatively, molecular methods such as the use of environmental DNA (eDNA) can be effective for community monitoring, with DNA analysis being particularly effective for detecting the presence of endangered or elusive organisms and providing spatial and temporal high-resolution data for effective conservation and management. Gene-based monitoring traces changes in individual species' genetic parameters over time. Genetic indicators, which have recently been included in biodiversity monitoring standards, provide essential insights into connectivity and adaptive capacity. Landscape genetics combines conservation genetics and ecology to understand gene flow barriers and facilitators: population synchrony signals functional connectivity. Although genetic monitoring demands great technical expertise, it is less time-consuming than conventional methods. For future forest connectivity monitoring, a combination of various approaches is conceivable. Existing connectivity indicators need rigorous evaluation in terms of their sensitivity to environmental impacts. Dynamic models and novel indicators along with data sharing and collaboration will be crucial for future efforts in connectivity monitoring.

Keywords

Molecular methods · Forest structure · Functional group · Genetic diversity · Trophic level

Introduction

Forest ecosystems are critical habitats that support a vast number of plant and animal species. They are essential for global environmental well-being owing to their provision of ecosystem services and regulating factors (Cardinale et al. 2012; Emmett Duffy 2009) but are undervalued in economic systems (FAO 2022). The land area covered by forests and trees is also an important indicator in the monitoring of forest ecosystems and the assessment of environmental conditions (Keenan et al. 2015). However, forests around the world are under increasing pressure from climate change (Gaston 2000), resource exploitation, and anthropological disturbances (Holzwarth et al. 2020).

Globally, both multilateral institutions and national governments have recognized the urgent need for immediate action to conserve and restore ecological connectivity so as to help combat the alarming decline of biodiversity (Keenan et al. 2015; Pither et al. 2023). Results of recent assessments indicate dramatic increases in deforestation and the loss of connected habitats—and thus ultimately of forest species and overall biodiversity (Barnosky et al. 2011; Ceballos et al. 2017; FAO 2022; Mittermeier et al. 2011). Recent data also confirm that agricultural expansion drives almost 90 percent of global deforestation, with an estimated 289 million ha of land facing deforestation between 2016 and 2050 in the tropics alone (FAO 2022). Forest loss rates are highest in low-income countries (Keenan et al. 2015).

Essential characteristics of biodiversity include the composition, structure, and function integrating different levels of organization of organisms (Noss 1990). The term therefore encompasses the diversity of living entities on different levels of organization—from molecular, genetic, individual, and species to populations, communities, biomes, ecosystems, and landscapes. The interrelation of biodiversity with species community composition, nutrient cycling, and ecosystem productivity highlights its importance in maintaining the integrity and resilience of ecosystems (Gaston 2000; Maclaurin and Sterelny 2008).

Conservation plans now increasingly focus on maintaining a connected network of healthy and resilient areas (Keeley et al. 2021). For instance, linkages between conserved areas and new target sites to establish well-connected protected area systems are becoming common goals. Ensuring connectivity within and among forests is essential for ecological balance, biodiversity, and ecosystem resilience in the face of environmental changes (Kacic and Kuenzer 2022; Pearson et al. 2021). To effectively achieve conservation goals, the implementation of robust forest ecological and biodiversity monitoring methods at different spatial scales (local, regional, and global) is crucial (e.g., Evans et al. 2018). While increasing the amount of forested area tends to take center stage in public discussions involving forest monitoring data and reports, the state of monitoring methods and their effectiveness is critical to understanding ecological connectivity and the social-ecological benefits of forests and forestry. A distinction can be made between direct monitoring methods, which include habitat or species mapping, and indirect methods involving the modeling of species distributions and spatial patterns, including functional diversity (Gillespie et al. 2008; Nagendra 2001). Targeted monitoring also allows the assessment of conservation objectives for connectivity, as highlighted by Keeley et al. (2021). These objectives include (i) evaluating the connectivity of a specific patch to predict patch colonization, (ii) prioritizing areas for conservation and restoration efforts, (iii) quantifying the contribution of a specific site to the overall connectivity of the landscape, and (iv) evaluating the connectivity of an existing network of sites or an entire landscape. In this chapter, we will focus on monitoring approaches at four different levels of organization—namely landscape, ecosystem, species, and genes—incorporating compositional, structural, and functional components.

Monitoring Landscapes and Forest Ecosystems

A landscape is characterized by its visible and physical features like landforms, vegetation, and land-use types (Urban et al. 1987). The structural elements of a landscape are essential for providing habitats for plant and animal species, and the arrangement and configuration of these elements significantly influence habitat connectivity (Ernst 2014). Key landscape parameters such as patch size, heterogeneity, perimeter–area ratio, and connectivity serve as significant determinants of species composition and abundance (Noss 1990). In addition, the composition of the landscape—including the type and proportions of specific habitats—is of critical importance. Noss (1990) emphasized that the "functional combination" of habitats within the landscape mosaic is crucial for animals that rely on multiple habitat types. This includes ecotones and species assemblages that transition gradually along environmental gradients.

The arrangement of trees within a forest, together with other vegetation, terrain, and water, determines its stand structure (Seidler 2023), which encompasses the physical geography of the forest considered at different spatial scales. The stand structure includes characteristics like canopy cover and understory diversity, species distribution patterns, soil characteristics, age structures, and species composition. It is also crucial for shaping the biodiversity and functionality of an ecosystem. The presence of different tree species of varying age classes and distribution patterns significantly influences the overall health and resilience of a forest (Franklin et al. 2002). A diverse stand structure creates a plethora of niches and habitats supporting a wide array of plant and animal species adapted to specific conditions within the forest environment—from animals dependent on dead trees for nesting to others that browse on plants in light gaps, from bark- and wood-boring insects to those that consume root fungi (Boyle et al. 2016). As highlighted by Seidler (2023), changes in a forest's stand structure often stem from shifts in species composition and the age structure of trees. This can lead to alterations in the overall ecosystem structure, affecting biodiversity and ecosystem functions alike (Seidler 2023; Valbuena et al. 2012).

Approaches to Monitoring and Evaluating Landscapes and Forest Ecosystems

Earth Observation

Remote sensing technologies such as satellite imagery and aerial surveys have revolutionized landscape and forest monitoring. These tools enable us to collect vast amounts of data over large areas, enabling visualization and analysis of land cover changes, deforestation rates, and forest health (see also Chap. 9). Integration of geographic information systems (GIS) facilitates informed decision-making for conservation and land-use planning. The Copernicus Land Monitoring Service offers a high-resolution forestry layer with three types of products available: tree cover density, dominant leaf type (deciduous, coniferous, etc.), and a forest-type product following the forest definitions of the Food and Agriculture Organization (FAO) (Copernicus Programme 2023). With the availability of higher spatial resolution and area-covering datasets, we can derive detailed information on nationwide forest cover distribution and dynamics. This includes assessments of forest loss; changes in species composition; disturbances due to droughts, fires, storms, and plagues; and forest recovery and regrowth (Holzwarth et al. 2020).

Thanks to the increasing availability of remote sensing data and user-friendly processing software, remote sensing imagery has become a highly relevant and important tool for monitoring land cover dynamics. The consistent and repeatable measurements of remote sensors offer cost-effective solutions for large-scale biodiversity monitoring. Moreover, satellite imagery allows us to assess vegetation conditions in inaccessible, remote areas (Gillespie et al. 2008; Nagendra et al. 2013). While in situ monitoring of physical parameters as well as their modeling across landscapes and forests remains crucial, there is a growing emphasis on monitoring functional diversity (Wang and Gamon 2019). Functional diversity encompasses a broad spectrum of attributes such as reproductive, developmental, life history, dietary, ecological, and other functions that distinct species exhibit within an ecosystem (Mason and Mouillot 2013). Understanding functional diversity helps reveal the interplay between species and their ecological functions such as pollination, seed dispersal, and predation. By monitoring functional interactions, we gain deeper insights into ecosystem functioning and the underlying mechanisms that sustain ecosystem health and resilience.

Terrestrial Surveys

Terrestrial surveys involve the systematic observation and assessment of the physical environment within forest ecosystems. This approach helps identify changes in land use, vegetation cover, and habitat quality that directly impact connectivity. Such scrutiny of forest structure and biodiversity has traditionally been conducted as field surveys organized in small plot units from which general conclusions about the overall environmental conditions are drawn (Hui et al. 2019; Palmer et al. 2002). Ecological studies have adopted variables such as stand structure degrees, tree species composition, and population distribution patterns, while forestry studies have used variables such as tree height and diameter distribution and canopy cover (Hui et al. 2019). These variables directly reflect the different aspects of forest structure. Over the past two decades, there has been a growing interest in the use of terrestrial laser scanning as a tool for forest plot measurements. It enables nondestructive quantification of forest development and provides valuable insights into the dynamics of biodiversity and ecosystem function mechanisms at high temporal resolution

(Guimarães-Steinicke et al. 2019). Efforts have been made to replace traditional plot-scale measurements (Lovell et al. 2003; Newnham et al. 2015). According to Newnham et al. (2015), terrestrial laser scanning enables the assessment of tree volume, growth, and foliage development. Furthermore, it facilitates the scaling-up of ground-based measurements with airborne laser data to create detailed 3D models. These models may lead to more precise and comprehensive assessments of vegetation and ecosystem dynamics.

Advantages and Disadvantages of Landscape and Forest Ecosystem Monitoring

Monitoring landscapes and forest ecosystems offers significant advantages, such as the determination of large-scale patterns and assessment of ecosystem dynamics, biodiversity, and land-use changes. In addition, it allows early detection of potential threats like deforestation, habitat loss, and climate change impacts, enabling timely intervention. The scale of monitoring can pose challenges, however (Keeley et al. 2021), requiring standardized methodologies and advanced data processing. What is more, broad-scale monitoring may overlook finer-scale variations and specific ecological interactions within an ecosystem. In situ field sampling can also be time-consuming and expensive when conducted on a large scale. To overcome these limitations, it is crucial to combine landscape-level monitoring with localized and fine-scale assessments to achieve a more comprehensive understanding of ecosystem dynamics and better inform conservation and management decisions. Additionally, variations in survey methods among experts and disciplines can lead to a scarcity of standardized data, impeding cross-disciplinary synthesis and management goals (Lõhmus et al. 2018). To enhance the effectiveness of ecosystem monitoring, efforts are undertaken to address these limitations through collaborative research and data standardization initiatives.

Monitoring Species and Communities

Species monitoring as defined by Moussy et al. (2022) involves the systematic and repeated collection of data to detect long-term changes in the abundance or distribution of one or more taxa or taxonomic groups. Such monitoring is crucial for conservation practice and policy since many species are interlinked, fulfilling essential ecosystem functions and offering valuable ecosystem services (Liu et al. 2018). Noss (1990) emphasized the importance of monitoring multiple species or groups of species deserving special conservation effort, such as (i) indicator species, which predict the impact of perturbations on other species with similar habitat requirements, (ii) keystone species shaping the diversity of their respective community whose decline can lead to cascading effects on the entire ecosystem, and (iii) umbrella species with extensive habitat requirements whose protection benefits numerous other species sharing the same habitat (e.g., Mills et al. 1993; Roberge and Angelstam 2004; Simberloff 1998). In addition to these well-established categories, we emphasize the inclusion of a further group for forest ecosystems: (iv) forest-related species. This category encompasses species that either form an integral part of forest ecosystems or depend on forests for their daily living or reproductive needs (CBD 2023). By prioritizing the monitoring of forest-related species, we can simultaneously protect the habitats and resources they rely on, promoting sustainable forest management and long-term ecological balance.

Recently, Banker et al. (2022) underlined the importance of understanding trophic positions when planning restoration activities. In some cases, human activities may have altered a community to the extent that restoring a specific species or its interactions may no longer be feasible. In such instances, it becomes crucial to explore alternative approaches to monitor and restore vital ecosystem functions that may have become compromised. One such alternative encompassing a broader range of taxa is the assessment of functional groups or species communities (Brunialti 2014). A functional group is a set of species coexisting within a given community that share similar functional characteristics, particularly with regard to providing specific ecosystem services. These functional groups are commonly known as "plant functional types" in vegetation science and "guilds" in animal science (Pla et al. 2012).

In the context of forest ecosystems, the most prevalent functional groups can be categorized based on their trophic levels, which include producers, consumers, and decomposers (Egerton 2007; Elton 1927). Figure 8.1 illustrates an example of linking taxa that have been identified as highly relevant for monitoring within the corresponding trophic level. These groups play pivotal roles in essential processes within forests, such as wood and litter decomposition, pollination, predation, phytophagy, or overall biomass production (Schuldt et al. 2018). The functional composition undergoes deterministic changes, meaning that functional groups tend to exhibit shifts over time. However, the abundances of individual species within these functional groups may also drift randomly (Rubio and Swenson 2022). Exploring the interplay between deterministic and neutral dynamics within functional groups may provide a more comprehensive perspective on how forests respond to environmental changes, species invasions, and other disturbances—and thus enhance our ability to predict and manage forest ecosystems in the face of ongoing global environmental challenges.

Approaches for Assessing Species and Communities

Key monitoring methods for species and species groups encompass a diverse range of techniques including human observations, camera trapping, passive acoustic monitoring, GPS tracking, and DNA-based tools.

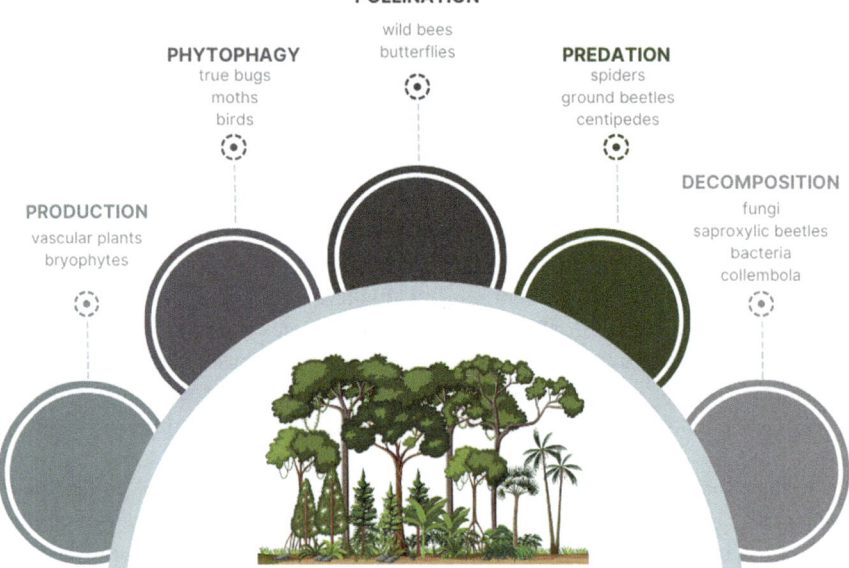

Fig. 8.1 Illustration of main trophic levels in forest ecosystems highlighting functional types and guilds identified as relevant for monitoring

Human Observations

Human observations are a fundamental approach to recording and documenting species data. Trained observers or citizen scientists look at species and their behaviors, distributions, and interactions, enabling real-time data collection. Observation by humans can be both direct and indirect: direct observation involves spotting a species in its natural habitat, while indirect observation focuses on recording signs such as nests, tracks, or feces (for further details, see Chap. 12) (Buckland et al. 2001; Thompson et al. 1994). As highlighted by Richard-Hansen et al. (2015), direct observations tend to be biased toward mammal and bird species that are easily detectable due to their vocalizations, size, and habits. By contrast, species that are rare, small, nocturnal, or cryptic are less likely to be observed. Direct observation requires highly skilled observers, and observer bias may arise as a result of differences in expertise and interests. These issues can be mitigated through careful training, limiting the length of monitoring sessions, and reducing the number of tasks assigned to each observer (Emlen and DeJong 1992). Consequently, direct field observations are best suited to highly detectable species and may not be ideal for community assessments that require broad taxonomic coverage (Roberts 2011). Indirect observation, on the other hand, offers several advantages: signs left by animals are more abundant than the animals themselves and generally remain visible for a certain time (Zwerts et al. 2021).

Camera Trapping

Camera trapping is a highly effective and noninvasive method used to capture images and videos of wild animals in their habitat. It is particularly valuable for studying elusive or nocturnal species, providing essential data on species presence, abundance, and behavior within forest habitats. The popularity of camera trapping has grown significantly over the past three decades (Glover-Kapfer et al. 2019). Equipped with passive infrared sensors (Welbourne et al. 2016), camera traps can record a wide range of wildlife spanning various sizes and taxonomic groups, including mammals (Tobler et al. 2008), birds (O'Brien and Kinnaird 2008), and reptiles (Ariefiandy et al. 2013; Hobbs and Brehme 2017). Time-lapse photography and specialized camera traps can even be used to survey arthropods (Collett and Fisher 2017; Hobbs and Brehme 2017). In general, camera trapping is a suitable technique, especially for monitoring active and less vocally communicative terrestrial animals of medium to large size. Its versatility extends beyond these taxa; however, validating the data from this is critical for accuracy assessments, thus allowing it to also be used to survey smaller, cryptic, and rare species in remote areas (e.g., Bessone et al. 2020; Khwaja et al. 2019).

Passive Acoustic Monitoring

Similarly to camera trapping, passive acoustic monitoring is a noninvasive and powerful technique used to detect and analyze sounds emitted by wildlife, including calls, vocalizations, and other acoustic signals. This method proves particularly effective for the study of nocturnal animals or species that primarily communicate through sound. By capturing and analyzing these sounds, passive acoustic monitoring facilitates the identification of species presence and activity levels, especially in dense forest environments. Passive acoustic monitoring is a rapidly developing and growing monitoring method for terrestrial wildlife (Darras et al. 2019), expanding its range of applications beyond the marine environments in which it is most commonly used. By deploying acoustic recording units (ARUs), researchers can continuously record the soundscape of a specific area over extended periods. The resulting data comprises a diverse range of sounds from biotic (animals), abiotic (water and wind), and anthropogenic (traffic) sources, as emphasized by Pijanowski et al. (2011). The versatility of passive acoustic monitoring extends its capacity to monitor all species that produce identifiable calls or sounds, thus offering an advantage over other monitoring techniques. Even elusive species like insects (Aide et al.

2017; Ganchev and Potamitis 2007) can be effectively monitored with this technique, enabling valuable insights into their presence and behavior.

GPS Tracking and Telemetry

The use of GPS tracking devices or telemetry tags has revolutionized the field of animal ecology by enabling the study of species across various landscapes and habitats. By tracking animals' locations in real time, researchers can gain a comprehensive understanding of their migration patterns, territory usage, and habitat preferences (Frair et al. 2010). This data is particularly valuable for identifying critical habitats as well as corridors and barriers affecting animal movements, thus facilitating connectivity planning (e.g., Bastille-Rousseau and Wittemyer 2021; Stewart et al. 2019). Several studies have showcased the effectiveness of GPS tracking for studying animal movements and behavior (Eriksen et al. 2011; Knopff et al. 2009). For instance, studies on large carnivores like wolves (Planella et al. 2016) and cougars (Maletzke et al. 2017) have demonstrated the significance of GPS technology in understanding their spatial ecology and interactions with human landscapes. GPS tracking has also been instrumental in studying migratory birds such as raptors (Katzner et al. 2012) and seabirds (Wakefield et al. 2013), providing critical data on their migration routes and stopover locations. By equipping animals with GPS-enabled collars or tags, researchers can gain remarkable insights into the lives of wildlife, and the continuous advancement of GPS technology has significantly enhanced our ability to track and study animals across diverse landscapes and habitats.

DNA-Based Methods

In recent years, the use of molecular methods for species identification and detection has seen a tremendous increase fueled by technological advances such as high-throughput sequencing (Bruce et al. 2021; Cristescu 2014). This section summarizes the main trajectories of these developments and discusses their application in the context of monitoring ecological connectivity in forests. Two decades ago, Hebert et al. (2003) formally proposed using molecular methods for the large-scale systematic identification of species. Molecular tools offer the potential to overcome the limitations of morphological identification, such as the difficulty of accounting for phenotypic plasticity and the inability to discern morphologically cryptic taxa (Cristescu 2014; DeSalle and Goldstein 2019; Fišer Pečnikar and Buzan 2014). DNA-based methods mitigate two practical shortcomings of conventional species identification as they can be applied to all life stages of an organism and offer an efficient alternative to the morphological identification of many taxa extending beyond family categorization (Fišer Pečnikar and Buzan 2014; Hebert et al. 2003).

Originally, the DNA of individual specimens was extracted, followed by amplification and sequencing of a short region of the mitochondrial genome, the so-called

DNA barcode (Hebert et al. 2003). After a DNA barcode is generated from a morphologically identified specimen, it can be used as a reference sequence for future monitoring efforts (DeSalle and Goldstein 2019). The generation of comprehensive DNA barcode databases from morphologically identified specimens is a prerequisite for the large-scale application of DNA barcoding. Considerable efforts have been made to generate such databases (e.g., the Barcode of Life Data System (BOLD); Ratnasingham and Hebert 2007), and recent endeavors have focused on barcoding the immense biodiversity of tropical regions and of previously underrepresented taxa (Hobern 2021; Janzen and Hallwachs 2019). However, DNA barcoding of a single gene is usually not sufficient to correctly determine the phylogenetic relationship between taxa. The technique is therefore primarily used to detect species diversity and its temporal and spatial dynamics (Cristescu 2014; DeSalle and Goldstein 2019), but frequently, reference specimens or DNA extracts are preserved to enable future investigations at the population and individual levels (Hendrich et al. 2015).

In recent years, advances in high-throughput sequencing techniques have enabled the processing of bulk samples (i.e., extraction, amplification, and sequencing from hundreds of individual specimens at once), giving rise to the term "metabarcoding" (Cristescu 2014; Ji et al. 2013; Yu et al. 2012). During high-throughput sequencing, millions of sequences are generated by amplification and parallel sequencing of the barcoding regions. These sequences are subsequently subjected to bioinformatic pipelines for the delimitation of individual samples and taxonomic assignment of the generated barcodes (Bik et al. 2012; Porter and Hajibabaei 2018). Metabarcoding drastically reduces the individual handling time of specimens (Cristescu 2014; Ji et al. 2013; Porter and Hajibabaei 2018) while delivering high-quality data on species assemblages and their structure (Bruce et al. 2021; Bush et al. 2019). However, metabarcoding is not devoid of shortcomings. They include amplification bias introduced by mismatches at the priming site (Clarke et al. 2014), underrepresentation of sequences originating from small specimens (Elbrecht et al. 2017), PCR and sequencing errors (Turon et al. 2020), and low-quality sequences hampering taxonomic assignment (Rivera et al. 2020). For this reason, many algorithms have been developed to optimize bioinformatic processing and taxonomic assignment for different taxa and sequencing platforms (e.g., Boyer et al. 2016; Buchner et al. 2022; Callahan et al. 2016), rendering metabarcoding a promising and efficient approach to assessing biodiversity and its changes in light of the global biodiversity crisis (Cristescu 2014). Substantial efforts are currently underway to test the implementation of metabarcoding data in routine monitoring programs and establish standards for their generation and use (Aylagas et al. 2018; Bruce et al. 2021; Geiger et al. 2016; Gueuning et al. 2019).

Whenever the monitoring of species is challenging, an identification or detection problem is likely the root cause (Bruce et al. 2021). DNA barcoding and metabarcoding have a clear focus on the identification aspect, while analysis of environmental DNA (eDNA) has the potential to improve the detection aspect. eDNA is defined as free DNA, cell components, secretions, tissue fragments, and the like released into the environment by organisms (Bohmann et al. 2014; Thomsen and Willerslev

2015). Filtered air (e.g., Lynggaard et al. 2022), water (e.g., Lamb et al. 2022), and soil (e.g., Vasar et al. 2023) are commonly used environmental sample types containing extraorganismal DNA (e.g., mucus, feces, or dander) as well as organismal DNA (i.e., individuals of small species contained in the environmental sample) (Deiner et al. 2017; Rodriguez-Ezpeleta et al. 2021).

Two techniques are used for the detection of eDNA: i) targeted amplification with species-specific assays and ii) the metabarcoding approach (Bruce et al. 2021). For both approaches, the molecular sondes (i.e., primers) need to be carefully designed and validated to avoid amplification of taxa outside the scope of the study from the environmental sample (Takahashi et al. 2023; Thalinger et al. 2021). Due to the sensitivity of the employed molecular methods, great care has to be taken to avoid sample contamination during the individual processing steps (Goldberg et al. 2016). Originally, detections via eDNA were only scored as the presence/absence of data, but quantification of eDNA signals has recently evolved into a routine practice for targeted approaches employing quantitative PCR or digital PCR assays (Butler et al. 1994; Goldberg et al. 2016; Thalinger et al. 2021). A positive relationship between read number and species abundance has also been confirmed in many case studies for eDNA metabarcoding, although the complex nature of the amplification, sequencing, and taxonomic identification processes still precludes the drawing of general quantitative conclusions (Deagle et al. 2019; Tsuji et al. 2022).

In the context of forest connectivity, DNA-based methods can represent powerful tools. They can provide a detailed inventory of species present at a location, ranging across the entire tree of life, and provide high-resolution data on spatial and temporal changes in species distribution. By establishing comprehensive reference databases (Weigand et al. 2019) and linking the detected taxa with their functional roles, molecular methods can be suitable for the calculation of biotic indices and ecosystem assessment (Brantschen et al. 2021; Dalongeville et al. 2022; Meyer et al. 2020). Sampling of distinct microhabitats (e.g., the canopy (Aucone et al. 2023; Macher et al. 2023) or soil (Allen et al. 2023) and specific processes (e.g., decomposition or pollination (Evans and Kitson 2020) as well as testing for the prevalence of pest species (Young et al. 2021) can further improve the resolution along with the chances of detecting indicator species and rare taxa, thereby providing even more precise data for management decisions. A taxonomic and ecological inventory of a location can thus be generated and used as a basis for tracking changes over time with extremely high sensitivity. In the context of forest connectivity, molecular methods provide a viable tool to monitor the use of individual habitats (e.g., stepping-stone biotopes) and migrations. Additionally, the success of renaturation and restoration measures can be efficiently monitored with DNA-based methods, as can changes in local biodiversity and ecosystem functioning induced by climate change and forest management.

Advantages and Disadvantages of Species and Community-Based Approaches

Species monitoring offers advantages in terms of understanding and conserving biodiversity by systematically and repeatedly collecting data. It allows researchers and conservationists to detect long-term changes in species abundance or distribution, which can be crucial for informing effective management strategies (Moussy et al. 2022). Moreover, monitoring facilitates the assessment of ecosystem health and the impact of human activities, assisting in the implementation of adaptive management approaches. On the other hand, species monitoring may be resource-intensive, requiring considerable time, effort, and financial investment, if monitoring methods are not chosen appropriately. For example, certain species may be challenging to monitor conventionally due to their elusive nature or remote habitats, potentially leading to incomplete data or limiting options to mono-temporal surveys (Lõhmus et al. 2018; Palmer et al. 2002). Here, the implementation of eDNA-based techniques with their superior sensitivity and cost-effectiveness in these situations is recommendable (Bohmann et al. 2014; Fediajevaite et al. 2021; Lampa et al. 2008). Furthermore, monitoring can cause disturbances to sensitive species or habitats, necessitating ethical considerations, careful study design, and standardized reporting guidelines (Pawlowski et al. 2018; Soulsbury et al. 2020). Despite these challenges, the large-scale application of species monitoring may finally provide researchers, policymakers, and managers with the much-needed high-resolution datasets required for the conservation of forest connectivity and the implementation of efficient management measures, thus far outweighing the drawbacks (Kéry and Schmidt 2008).

Monitoring Genetic Diversity

Forest genetic monitoring describes and tracks changes in the population genetic parameters of forest species over time by means of suitable indicators (Aravanopoulos 2011; Graudal et al. 2021). Despite the widespread use of population genetics and genomics in conservation studies (Allendorf et al. 2022; Barnes and Turner 2016; Taberlet et al. 2018), indicators of genetic diversity have only begun to be included in international biodiversity monitoring standards during the past decade (Graudal et al. 2014). This is even more surprising given that estimates of gene flow and population genetic structure (Slatkin 1987) represent a direct measure of present and past dispersal as well as functional connectivity between areas, especially when coupled with demographic information (Lowe and Allendorf 2010). Intraspecific patterns of genetic variation are the result of processes affecting population distribution and connectivity. Genetic indicators allow a long-term perspective on ecosystem and species establishment (Thomas et al. 2014), thereby facilitating the assessment of the future potential of species and ecosystem connectivity.

Efforts to improve landscape connectivity (e.g., via the creation of ecological corridors for dispersal) have a direct impact on the conservation of genetic diversity, counteracting processes such as population isolation and reduction in the effective population size including bottlenecks, inbreeding, random genetic drift, and local population extinction (Young et al. 2000). The genetic monitoring of tree populations adapted to specific environmental conditions and designated as genetic conservation units (Lefèvre et al. 2020; see also the EUFGIS portal, http://portal.eufgis.org) can thus provide information on connectivity by means of population genetic parameters. Importantly, the exchange of genetic material via gene flow (i.e., via intraspecific pollen exchange or seed dispersal) or introgressive hybridization (i.e., gene exchange between species), likely facilitated by enhanced landscape connectivity, is a means of transferring adaptive material (Leroy et al. 2020) and increasing the potential of populations and species to adapt to future climatic conditions. Assuming a stable rate of climate change, the potential of forests to adapt to future climate conditions depends on (1) the existence of (adaptive) genetic diversity in populations and (2) the possibility of sharing or exchanging (adaptive) genetic variants among populations via gene flow (Fady et al. 2016; Kremer et al. 2012). Nevertheless, it is important to note that genetic connectivity can be both beneficial and harmful in relation to the conservation of genetic units and tree breeding. Since gene flow modifies the allelic composition of populations, it can introduce but also remove adaptive variation from populations (Savolainen et al. 2007). For these reasons, it is important to include and carefully evaluate genetic indicators and verifiers in the methods and measures employed to improve ecological connectivity.

Approaches for Assessing Genetic Indicators

Genetic indicators that describe trends in species and population dispersal, the rate of gene exchange, and the genetic state of examined populations can be used to assess functional connectivity (Aravanopoulos 2016; Bajc et al. 2021; FAO 2014) (Table 8.1). Landscape genetics (Holderegger and Wagner 2006) is the field that combines population genetics and landscape ecology, explaining spatial genetic variation in connection with landscape features (Balkenhol et al. 2015; Manel et al. 2003). Storfer et al. (2010) reported that among all landscape genetics studies published by the time of their evaluation, almost 60% addressed research questions on connectivity. In the first place, landscape genetics aims to understand which landscape features facilitate or impede gene flow.

Gene flow is a function of the population size and the migration rate of individuals between populations and is strongly determined by the respective mating system (Bajc et al. 2021). It can be estimated using different statistics (Zheng and Janke 2018) and a variety of genetic markers ranging from microsatellites to next-generation sequencing (NGS) data (see also Chap. 4 for further details). One major limitation in the estimation of gene flow is the distinction between inferred and realized gene flow in genomic studies (Colosimo et al. 2014), with the latter representing a more realistic estimator of present connectivity between populations. Trends in population dispersal and connectivity can be further verified using estimates of multi-locus population outcrossing rate (i.e., the proportion of outcrossed progeny

Table 8.1 List of indicators and verifiers described in forest genetic monitoring and landscape genetics that can be used to assess genetic connectivity between demes. Note that the same descriptor (e.g., gene flow) is reported as an indicator or a verifier depending on the literature source consulted. The table describes two key indicators that can be potentially used to track the progress of a given target related to connectivity. Parameters that can be measured using genetic markers within each indicator are reported as verifiers.

Indicator of connectivity	Genetic verifier (i.e., the measure of the indicator)	Period of assessment	Geographic scale of assessment	Reference literature describing the indicator
Trends in gene exchange between populations in relation to landscape features	Gene flow Outcrossing rate Inbreeding Spatial genetic structure Population differentiation (e.g., by means of F_{st}) Genetic drift	Species-specific (e.g., every 10 years in most tree species)	Local/regional	Aravanopoulos (2011), (2016), Bajc et al. (2021), FAO (2014), Konnert et al. (2011), Manel et al. (2003)
Trends in population genetic conditions that affect synchrony between populations	Effective population size (N_e) Population genetic diversity (e.g., allelic richness, heterozygosity, etc.)	Species-specific (e.g., every 10 years in most tree species)	Local/regional	

produced by a population), inbreeding (i.e., the proportion of genetic variance of the population contained in a single individual), genetic drift, strength of selection, and measures of population genetic differentiation due to genetic structure (see Table 8.1).

Population genetic parameters can also be used to assess population synchrony (Table 8.1), which is defined as a positive correlation of the annual variation in population trends (e.g., abundance exemplified by the effective population size, N_e) between separate populations (Blomfield et al. 2023). Population synchrony has been shown to be a good proxy of functional connectivity (Powney et al. 2011), even though some limitations of this approach in regard to long-distance dispersal have been discussed (Blomfield et al. 2023). The effective population size is a key parameter of genetic monitoring defined as the number of crossbreeding individuals in a population that contribute genes to the next generation. Small values of N_e imply that stochastic processes (i.e., genetic drift) have a stronger effect on the genetic composition of a population than selective processes, resulting in a higher probability of inbreeding and, consequently, the two first processes lower a population's adaptability due to decreased amounts of genetic variation and low selection coefficients. Thus, despite N_e not being a direct estimator of connectivity, it is strongly influenced by it and is fundamental to the understanding of connectivity patterns.

A variety of genetic markers can be used for assessing forest connectivity. Popular markers used to monitor genetic variation in and between populations include microsatellites, allozymes, SNPs, AFLP loci, mitochondrial and chloroplast DNA, and the Y chromosome, with the latter three being particularly suitable for estimating dispersal and gene flow due to their uniparental inheritance (Cruzan and Hendrickson 2020; McCauley 1995).

Advantages and Disadvantages of Gene-Based Monitoring

The ever-decreasing cost of genotyping and steady improvement of bioinformatic and statistical tools, most of which are open-source, makes the use of genetic monitoring increasingly simple and affordable for most of the research and governmental institutions interested in monitoring forest connectivity. It is important to note that different markers deliver information on genetic parameters constrained by their own molecular features shaped, for instance, by their mode of inheritance and recombination rate. Similarly, any statistical method is constrained by underlying assumptions, which limits its usage to certain markers or study systems. For example, assuming random mating excludes partially or wholly selfing organisms. Given that natural populations mostly violate method assumptions, analytical approaches should be validated by the use of simulations (Manel et al. 2003). Genetic monitoring thus requires great technical expertise: the choice of which genetic markers and statistical pipelines to use must be carefully evaluated at the start of a project.

The fieldwork as well as laboratory and data analyses required for genetic studies in forest genetic monitoring is described in detail by Bajc et al. (2021) (Fig. 8.2). Compared to conventional methods for assessing dispersal, such as mark–release–recapture studies (Turlure et al. 2018; Zimmermann et al. 2011), genetic monitoring is less time-consuming, requires less field work, and is less invasive with regard to animal species. Several reviews propose a monitoring interval of around 10 years in tree species (e.g., Aravanopoulos et al. 2015; Bajc et al. 2021) for most population genetic verifiers (Table 8.1). This is likewise more feasible compared to other forest genetic verifiers tracking parameters like regeneration abundance and reproductive fitness, which require annual or biennial assessment. It is important to note that the appropriate frequency of monitoring is species-specific depending on the specific objectives and characteristics of the species being monitored like generation time, life history traits, and conservation status. Overall, despite several genetic descriptors, indicators, and verifiers being described or reported in individual studies and guidelines (Table 8.1), a well-established reference framework for best practices is still missing, as are unified databases collecting comparable data across countries, even at the European level.

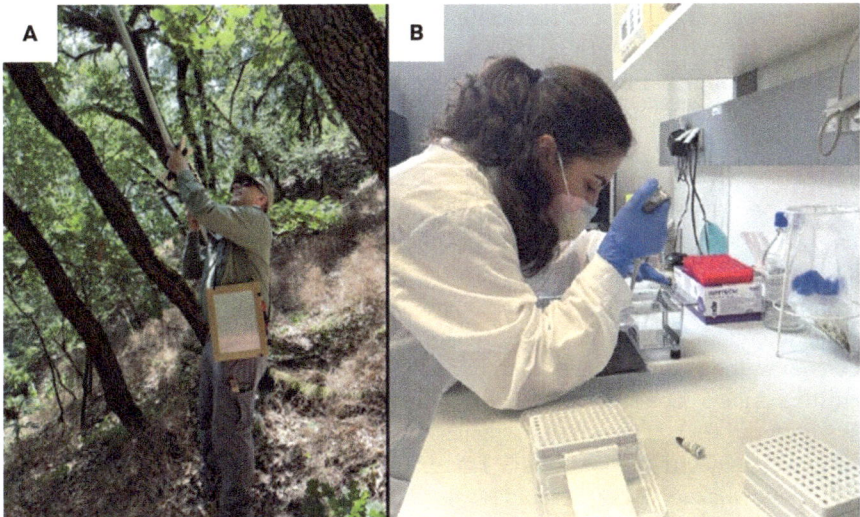

Fig. 8.2 Two-part image showing (**A**) fieldwork: leaf material sampling from a tree of *Quercus pubescens* for DNA extraction using a telescopic tree cutter and (**B**) loading of DNA extraction samples for quality checking on an electrophoresis gel. Both images show the scientific staff of the Austrian Research Centre for Forests. (Photos: BFW)

Outlook

The potential future of monitoring methods aimed at assessing connectivity in forest ecosystems is characterized by a combination of diverse approaches enabling a comprehensive understanding of ecosystem dynamics and their connectivity. Advancements in landscape monitoring technologies, including high-resolution satellite imagery and Light Detection and And Ranging (LiDAR), provide a detailed portrayal of forest structures and their changes over time. These data facilitate the identification of critical corridors and bottlenecks affecting species movement and gene flow. Moreover, linking landscape knowledge with species-specific data or population dispersal patterns offers insights into how individual species respond to landscape alterations. This helps reveal behavioral patterns and allows the assessment of species' ability to traverse fragmented landscapes.

The widespread implementation of eDNA techniques coupled with high-throughput sequencing enables rapid assessment of species presence and biodiversity, especially for elusive or rare species that may be challenging to observe morphologically. As DNA sequencing techniques become more accessible and affordable, the utilization of genetic data offers deeper insights into genetic diversity, population structure, and gene flow, ultimately allowing the evaluation of functional connectivity. To harness the full potential of these monitoring methods, it is essential to prioritize data sharing. The promotion of robust data-sharing platforms and open-access databases along with collaboration among researchers, practitioners, and policymakers on a global scale becomes pivotal. This collective effort will create a wealth of information for cross-disciplinary analysis, leading to more informed conservation strategies.

Existing indicators must be subjected to rigorous evaluation to ensure they encompass the dynamics of connectivity including genetic diversity, species interactions, and landscape structure while remaining adaptable to shifting environmental conditions. In the face of the ongoing impact of climate change on forest ecosystems, monitoring techniques will need to evolve. Dynamic models incorporating climate projections will help forecast connectivity in response to changing climatic conditions. Potentially novel indicators need to be defined and tested to determine the resilience and adaptive capacity of ecosystems in alignment with existing policies and agreements such as the CBD's Post-2020 Global Biodiversity Framework. The science–policy interface will play a crucial role in shaping the future of connectivity monitoring. Transparency and collaboration between scientists, policymakers, and stakeholders are required to ensure that monitoring methods for connectivity are aligned with conservation goals and integrate the latest scientific insights.

References

Aide TM, Hernández-Serna A, Campos-Cerqueira M, Acevedo-Charry O, Deichmann JL (2017) Species richness (of insects) drives the use of acoustic space in the tropics. Remote Sens (Basel) 9(11). https://doi.org/10.3390/rs9111096

Allen MC, Kwait R, Vastano A, Kisurin A, Zoccolo I, Jaffe BD, Angle JC, Maslo B, Lockwood JL (2023) Sampling environmental DNA from trees and soil to detect cryptic arboreal mammals. Sci Rep 13(1):1–13. https://doi.org/10.1038/s41598-023-27512-8

Allendorf FW, Funk WC, Aitken SN, Byrne M, Luikart G, Antunes A (2022) Conservation and the genomics of populations. https://doi.org/10.1093/oso/9780198856566.001.0001

Aravanopoulos FA (2011) Genetic monitoring in natural perennial plant populations. Botany 89(2):75–81. https://doi.org/10.1139/B10-087

Aravanopoulos FA (2016) Conservation and monitoring of tree genetic resources in temperate forests. Current Forestry Reports 2(2):119–129. https://doi.org/10.1007/s40725-016-0038-8

Aravanopoulos FA, Tollefsrud MM, Graudal L, Koskela J, Kätzel R, Soto A, Nagy L, Pilipovic A, Zhelev P, Božic G, Bozzano M (2015) Development of genetic monitoring methods for genetic conservation units of forest trees in Europe (European F). Biodivers Int

Ariefiandy A, Purwandana D, Seno A, Ciofi C, Jessop TS (2013) Can camera traps monitor Komodo Dragons a large ectothermic predator? PLoS One 8(3). https://doi.org/10.1371/journal.pone.0058800

Aucone E, Kirchgeorg S, Valentini A, Pellissier L, Deiner K, Mintchev S (2023) Drone-assisted collection of environmental DNA from tree branches for biodiversity monitoring. Sci Robot 8(74):eadd5762. https://doi.org/10.1126/scirobotics.add5762

Aylagas E, Borja Á, Muxika I, Rodríguez-Ezpeleta N (2018) Adapting metabarcoding-based benthic biomonitoring into routine marine ecological status assessment networks. Ecol Indic 95:194–202. https://doi.org/10.1016/j.ecolind.2018.07.044

Bajc M, Aravanopoulos F, Westergren M, Fussi B, Kavaliauskas D, Alizoti P, Kiourtsis F, Kraigher H (2021) Manual for forest genetic monitoring (Slovenian). Silva Slovenica Publishing Centre. https://doi.org/10.20315/sfs.167

Balkenhol N, Cushman SA, Storfer AT, Waits LP (2015) Landscape genetics: concepts, methods, applications. https://doi.org/10.1002/9781118525258

Banker RMW, Dineen AA, Sorman MG, Tyler CL, Roopnarine PD (2022) Beyond functional diversity: the importance of trophic position to understanding functional processes in community evolution. Front Ecol Evol 10. https://doi.org/10.3389/fevo.2022.983374

Barnes MA, Turner CR (2016) The ecology of environmental DNA and implications for conservation genetics. Conserv Genet 17(1):1–17. https://doi.org/10.1007/s10592-015-0775-4

Barnosky AD, Matzke N, Tomiya S, Wogan GOU, Swartz B, Quental TB, Marshall C, McGuire JL, Lindsey EL, Maguire KC, Mersey B, Ferrer EA (2011) Has the earth's sixth mass extinction already arrived? Nature 471(7336):51–57. https://doi.org/10.1038/nature09678

Bastille-Rousseau G, Wittemyer G (2021) Characterizing the landscape of movement to identify critical wildlife habitat and corridors. Conserv Biol 35(1):346–359. https://doi.org/10.1111/cobi.13519

Bessone M, Kühl HS, Hohmann G, Herbinger I, N'Goran KP, Asanzi P, Da Costa PB, Dérozier V, Fotsing EDB, Beka BI, Iyomi MD, Iyatshi IB, Kafando P, Kambere MA, Moundzoho DB, Wanzalire MLK, Fruth B (2020) Drawn out of the shadows: surveying secretive forest species with camera trap distance sampling. J Appl Ecol 57(5):963–974. https://doi.org/10.1111/1365-2664.13602

Bik HM, Porazinska DL, Creer S, Caporaso JG, Knight R, Thomas WK (2012) Sequencing our way towards understanding global eukaryotic biodiversity. Trends Ecol Evol 27(4):233–243. https://doi.org/10.1016/j.tree.2011.11.010

Blomfield A, Menéndez R, Wilby A (2023) Population synchrony indicates functional connectivity in a threatened sedentary butterfly. Oecologia 201(4):979–989. https://doi.org/10.1007/s00442-023-05357-2

Bohmann K, Evans A, Gilbert MTP, Carvalho GR, Creer S, Knapp M, Yu DW, de Bruyn M (2014) Environmental DNA for wildlife biology and biodiversity monitoring. Trends Ecol Evol 29(6):358–367. https://doi.org/10.1016/j.tree.2014.04.003

Boyer F, Mercier C, Bonin A, Le Bras Y, Taberlet P, Coissac E (2016) obitools: a unix-inspired software package for DNA metabarcoding. Mol Ecol Resour 16(1):176–182. https://doi.org/10.1111/1755-0998.12428

Boyle JR, Tappeiner JC, Waring RH, Tattersall Smith C (2016) Sustainable forestry: ecology and silviculture for resilient forests. Reference module in earth systems and environmental sciences. https://doi.org/10.1016/b978-0-12-409548-9.09761-x

Brantschen J, Blackman RC, Walser J-C, Altermatt F (2021) Environmental DNA gives comparable results to morphology-based indices of macroinvertebrates in a large-scale ecological assessment. PLoS One 16(9):e0257510. https://doi.org/10.1371/journal.pone.0257510

Bruce K, Blackman R, Bourlat SJ, Hellström AM, Bakker J, Bista I, Bohmann K, Bouchez A, Brys R, Clark K, Elbrecht V, Fazi S, Fonseca V, Hänfling B, Leese F, Mächler E, Mahon AR, Meissner K, Panksep K et al (2021) A practical guide to DNA-based methods for biodiversity assessment. In: A practical guide to DNA-based methods for biodiversity assessment. Pensoft Publishers. https://doi.org/10.3897/ab.e68634

Brunialti G (2014) Integrative approaches as an opportunity for the conservation of forest biodiversity. Int J Environ Stud 71(2):226–227. https://doi.org/10.1080/00207233.2014.889472

Buchner D, Macher T-H, Leese F (2022) APSCALE: advanced pipeline for simple yet comprehensive analyses of DNA metabarcoding data. Bioinformatics 38(20):4817–4819. https://doi.org/10.1093/bioinformatics/btac588

Buckland S, Anderson D, Burnham K, Laake J, Borchers D, Thomas L (2001) Introduction to distance sampling: estimating abundance of biological populations. Oxford University Press

Bush A, Compson ZG, Monk WA, Porter TM, Steeves R, Emilson E, Gagne N, Hajibabaei M, Roy M, Baird DJ (2019) Studying ecosystems with DNA metabarcoding: lessons from biomonitoring of aquatic macroinvertebrates. Front Ecol Evol 7:484625. https://doi.org/10.3389/fevo.2019.00434

Butler JM, McCord BR, Jung JM, Wilson MR, Budowle B, Allen RO (1994) Quantitation of polymerase chain reaction products by capillary electrophoresis using laser fluorescence. J Chromatogr B Biomed Sci Appl 658(2):271–280. https://doi.org/10.1016/0378-4347(94)00238-X

Callahan BJ, McMurdie PJ, Rosen MJ, Han AW, Johnson AJA, Holmes SP (2016) DADA2: High-resolution sample inference from Illumina amplicon data. Nat Methods 13(7):581–583. https://doi.org/10.1038/nmeth.3869

Cardinale BJ, Duffy JE, Gonzalez A, Hooper DU, Perrings C, Venail P, Narwani A, Mace GM, Tilman D, Wardle DA (2012) Biodiversity loss and its impact on humanity. Nature 486(7401):59–67

CBD (2023) Forest biodiversity/definitions. Indicative definitions taken from the report of the ad hoc technical expert group on forest biological diversity

Ceballos G, Ehrlich PR, Dirzo R (2017) Biological annihilation via the ongoing sixth mass extinction signaled by vertebrate population losses and declines. Proc Natl Acad Sci U S A 114(30):E6089–E6096. https://doi.org/10.1073/pnas.1704949114

Clarke LJ, Soubrier J, Weyrich LS, Cooper A (2014) Environmental metabarcodes for insects: in silico PCR reveals potential for taxonomic bias. Mol Ecol Resour 14(6):1160–1170. https://doi.org/10.1111/1755-0998.12265

Collett RA, Fisher DO (2017) Time-lapse camera trapping as an alternative to pitfall trapping for estimating activity of leaf litter arthropods. Ecol Evol 7(18):7527–7533. https://doi.org/10.1002/ece3.3275

Colosimo G, Knapp CR, Wallace LE, Welch ME (2014) Inferred vs realized patterns of gene flow: an analysis of population structure in the Andros Island Rock Iguana. PLoS One 9(9):1–8. https://doi.org/10.1371/journal.pone.0106963

Copernicus Porgramme (2023) Copernicus land monitoring service—high resolution layers—forests

Cristescu ME (2014) From barcoding single individuals to metabarcoding biological communities: towards an integrative approach to the study of global biodiversity. Trends Ecol Evol 29(10):566–571. https://doi.org/10.1016/j.tree.2014.08.001

Cruzan MB, Hendrickson EC (2020) Landscape genetics of plants: challenges and opportunities. Plant Commun 1(6). https://doi.org/10.1016/j.xplc.2020.100100

Dalongeville A, Boulanger E, Marques V, Charbonnel E, Hartmann V, Santoni MC, Deter J, Valentini A, Lenfant P, Boissery P, Dejean T, Velez L, Pichot F, Sanchez L, Arnal V, Bockel T, Delaruelle G, Holon F, Milhau T et al (2022) Benchmarking eleven biodiversity indicators based on environmental DNA surveys: more diverse functional traits and evolutionary lineages inside marine reserves. J Appl Ecol 59(11):2803–2813. https://doi.org/10.1111/1365-2664.14276

Darras K, Batáry P, Furnas BJ, Grass I, Mulyani YA, Tscharntke T (2019) Autonomous sound recording outperforms human observation for sampling birds: a systematic map and user guide. Ecol Appl 29(6). https://doi.org/10.1002/eap.1954

Deagle BE, Thomas AC, McInnes JC, Clarke LJ, Vesterinen EJ, Clare EL, Kartzinel TR, Eveson JP (2019) Counting with DNA in metabarcoding studies: how should we convert sequence reads to dietary data? Mol Ecol 28(2):391–406. https://doi.org/10.1111/mec.14734

Deiner K, Bik HM, Mächler E, Seymour M, Lacoursière-Roussel A, Altermatt F, Creer S, Bista I, Lodge DM, de Vere N, Pfrender ME, Bernatchez L (2017) Environmental DNA metabarcoding: transforming how we survey animal and plant communities. Mol Ecol 26(21):5872–5895. https://doi.org/10.1111/mec.14350

DeSalle R, Goldstein P (2019) Review and interpretation of trends in DNA barcoding. Front Ecol Evol 7:460426. https://doi.org/10.3389/fevo.2019.00302

Egerton FN (2007) Understanding food chains and food webs, 1700–1970. Bull Ecol Soc Am 88(1):50–69. https://doi.org/10.1890/0012-9623(2007)88[50:ufcafw]2.0.co;2

Elbrecht V, Peinert B, Leese F (2017) Sorting things out: assessing effects of unequal specimen biomass on DNA metabarcoding. Ecol Evol 7(17):6918–6926. https://doi.org/10.1002/ece3.3192

Elton CS (1927) *Animal ecology*. Sidgwick and Jackson

Emlen JT, DeJong MJ (1992) Counting birds: the problem of variable hearing abilities. J Field Ornithol 63(1):26–31

Emmett Duffy J (2009) Why biodiversity is important to the functioning of real-world ecosystems. Front Ecol Environ 7(8):437–444. https://doi.org/10.1890/070195

Eriksen A, Wabakken P, Zimmermann B, Andreassen HP, Arnemo JM, Gundersen H, Liberg O, Linnell J, Milner JM, Pedersen HC, Sand H, Solberg EJ, Storaas T (2011) Activity patterns of predator and prey: a simultaneous study of GPS-collared wolves and moose. Anim Behav 81(2):423–431. https://doi.org/10.1016/j.anbehav.2010.11.011

Ernst BW (2014) Quantifying landscape connectivity through the use of connectivity response curves. Landsc Ecol 29(6):963–978. https://doi.org/10.1007/s10980-014-0046-7

Evans DM, Kitson JJ (2020) Molecular ecology as a tool for understanding pollination and other plant–insect interactions. Curr Opin Insect Sci 38:26–33. https://doi.org/10.1016/j.cois.2020.01.005

Evans K, Guariguata MR, Brancalion PHS (2018) Participatory monitoring to connect local and global priorities for forest restoration. Conserv Biol 32(3):525–534. https://doi.org/10.1111/cobi.13110

Fady B, Cottrell J, Ackzell L, Alía R, Muys B, Prada A, González-Martínez SC (2016) Forests and global change: what can genetics contribute to the major forest management and policy challenges of the twenty-first century? Reg Environ Change 16(4):927–939. https://doi.org/10.1007/s10113-015-0843-9

FAO (2014) The state of the world's forest genetic resources. United Nations Food and Agriculture Organization

FAO (2022) The state of the world's forests 2022. https://doi.org/10.4060/cb9360en

Fediajevaite J, Priestley V, Arnold R, Savolainen V (2021) Meta-analysis shows that environmental DNA outperforms traditional surveys, but warrants better reporting standards. Ecol Evol 11(9):4803–4815. https://doi.org/10.1002/ece3.7382

Fišer Pečnikar Ž, Buzan EV (2014) 20 years since the introduction of DNA barcoding: from theory to application. J Appl Genet 55(1):43–52. https://doi.org/10.1007/s13353-013-0180-y

Frair JL, Fieberg J, Hebblewhite M, Cagnacci F, DeCesare NJ, Pedrotti L (2010) Resolving issues of imprecise and habitat-biased locations in ecological analyses using GPS telemetry data. Philos Trans R Soc B Biol Sci 365(1550):2187–2200. https://doi.org/10.1098/rstb.2010.0084

Franklin JF, Spies TA, Van Pelt R, Carey AB, Thornburgh DA, Berg DR, Lindenmayer DB, Harmon ME, Keeton WS, Shaw DC, Bible K, Chen J (2002) Disturbances and structural development of natural forest ecosystems with silvicultural implications, using Douglas-fir forests as an example. For Ecol Manage 155(1–3):399–423. https://doi.org/10.1016/S0378-1127(01)00575-8

Ganchev T, Potamitis I (2007) Automatic acoustic identification of singing insects. Bioacoustics 16(3):281–328. https://doi.org/10.1080/09524622.2007.9753582

Gaston KJ (2000) Global patterns in biodiversity. Nature 405(6783):220–227. https://doi.org/10.1038/35012228

Geiger MF, Moriniere J, Hausmann A, Haszprunar G, Wägele W, Hebert PDN, Rulik B (2016) Testing the Global Malaise Trap Program – how well does the current barcode reference library identify flying insects in Germany? Biodivers Data J 4(1). https://doi.org/10.3897/BDJ.4.e10671

Gillespie TW, Foody GM, Rocchini D, Giorgi AP, Saatchi S (2008) Measuring and modelling biodiversity from space. Prog Phys Geogr 32(2):203–221. https://doi.org/10.1177/0309133308093606

Glover-Kapfer P, Soto-Navarro CA, Wearn OR (2019) Camera-trapping version 3.0: current constraints and future priorities for development. Remote Sens Ecol Conserv 5(3):209–223. https://doi.org/10.1002/rse2.106

Goldberg CS, Turner CR, Deiner K, Klymus KE, Thomsen PF, Murphy MA, Spear SF, McKee A, Oyler-McCance SJ, Cornman RS, Laramie MB, Mahon AR, Lance RF, Pilliod DS, Strickler KM, Waits LP, Fremier AK, Takahara T, Herder JE, Taberlet P (2016) Critical considerations for the application of environmental DNA methods to detect aquatic species. Methods in Ecology and Evolution 7(11):1299–1307. https://doi.org/10.1111/2041-210X.12595

Graudal L, Aravanopoulos F, Bennadji Z, Changtragoon S, Fady B, Kjær ED, Loo J, Ramamonjisoa L, Vendramin GG (2014) Global to local genetic diversity indicators of evolutionary potential in tree species within and outside forests. For Ecol Manage 333:35–51. https://doi.org/10.1016/j.foreco.2014.05.002

Graudal L, Loo J, Fady B, Vendramin G, Aravanopoulos F (2021) Indicators of the genetic diversity of trees—state, pressure, benefit and response: the state of the world's forest genetic resources—thematic study. United Nations Food and Agriculture Organization

Gueuning M, Ganser D, Blaser S, Albrecht M, Knop E, Praz C, Frey JE (2019) Evaluating next-generation sequencing (NGS) methods for routine monitoring of wild bees: metabarcoding, mitogenomics or NGS barcoding. Mol Ecol Resour 19(4):847–862. https://doi.org/10.1111/1755-0998.13013

Guimarães-Steinicke C, Weigelt A, Ebeling A, Eisenhauer N, Duque-Lazo J, Reu B, Roscher C, Schumacher J, Wagg C, Wirth C (2019) Terrestrial laser scanning reveals temporal changes in biodiversity mechanisms driving grassland productivity. Adv Ecol Res 61:133–161. https://doi.org/10.1016/bs.aecr.2019.06.003

Hebert PDN, Cywinska A, Ball SL, DeWaard JR (2003) Biological identifications through DNA barcodes. Proc Roy Soc B Biol Sci 270(1512):313–321. https://doi.org/10.1098/rspb.2002.2218

Hendrich L, Morinière J, Haszprunar G, Hebert PDN, Hausmann A, Köhler F, Balke M (2015) A comprehensive DNA barcode database for Central European beetles with a focus on Germany: adding more than 3500 identified species to BOLD. Mol Ecol Resour 15(4):795–818. https://doi.org/10.1111/1755-0998.12354

Hobbs MT, Brehme CS (2017) An improved camera trap for amphibians, reptiles, small mammals, and large invertebrates. PLoS One 12(10):1–15. https://doi.org/10.1371/journal.pone.0185026

Hobern D (2021) Bioscan: DNA barcoding to accelerate taxonomy and biogeography for conservation and sustainability. Genome 64(3):161–164. https://doi.org/10.1139/gen-2020-0009

Holderegger R, Wagner HH (2006) A brief guide to landscape genetics. Landsc Ecol 21(6):793–796. https://doi.org/10.1007/s10980-005-6058-6

Holzwarth S, Thonfeld F, Abdullahi S, Asam S, Canova EDP, Gessner U, Huth J, Kraus T, Leutner B, Kuenzer C (2020) Earth observation based monitoring of forests in germany: a review. Remote Sens (Basel) 12(21):1–43. https://doi.org/10.3390/rs12213570

Hui G, Zhang G, Zhao Z, Yang A (2019) Methods of forest structure research: a review. Curr For Rep 5(3):142–154. https://doi.org/10.1007/s40725-019-00090-7

Janzen D, Hallwachs W (2019) How a tropical country can DNA barcode itself. iBOL Barcode Bull 9(1). https://doi.org/10.21083/ibol.v9i1.5526

Ji Y, Ashton L, Pedley SM, Edwards DP, Tang Y, Nakamura A, Kitching R, Dolman PM, Woodcock P, Edwards FA, Larsen TH, Hsu WW, Benedick S, Hamer KC, Wilcove DS, Bruce C, Wang X, Levi T, Lott M et al (2013) Reliable, verifiable and efficient monitoring of biodiversity via metabarcoding. Ecol Lett 16(10):1245–1257. https://doi.org/10.1111/ele.12162

Kacic P, Kuenzer C (2022) Forest biodiversity monitoring based on remotely sensed spectral diversity—a review. Remote Sens (Basel) 14(21):1–32. https://doi.org/10.3390/rs14215363

Katzner TE, Brandes D, Miller T, Lanzone M, Maisonneuve C, Tremblay JA, Mulvihill R, Merovich GT (2012) Topography drives migratory flight altitude of golden eagles: implications for on-shore wind energy development. J Appl Ecol 49(5):1178–1186. https://doi.org/10.1111/j.1365-2664.2012.02185.x

Keeley ATH, Beier P, Jenness JS (2021) Connectivity metrics for conservation planning and monitoring. Biol Conserv 255:109008. https://doi.org/10.1016/j.biocon.2021.109008

Keenan RJ, Reams GA, Achard F, de Freitas JV, Grainger A, Lindquist E (2015) Dynamics of global forest area: results from the FAO global forest resources assessment 2015. For Ecol Manage 352:9–20. https://doi.org/10.1016/j.foreco.2015.06.014

Kéry M, Schmidt BR (2008) Imperfect detection and its consequences for monitoring for conservation. Community Ecol 9(2):207–216. https://doi.org/10.1556/ComEc.9.2008.2.10

Khwaja H, Buchan C, Wearn OR, Bahaa-el-din L, Bantlin D, Bernard H, Bitariho R, Bohm T, Borah J, Brodie J, Chutipong W, du Preez B, Ebang-Mbele A, Edwards S, Fairet E, Frechette JL, Garside A, Gibson L, Giordano A et al (2019) Pangolins in global camera trap data: implications for ecological monitoring. Glob Ecol Conserv 20. https://doi.org/10.1016/j.gecco.2019.e00769

Knopff KH, Knopff AA, Warren MB, Boyce MS (2009) Evaluating global positioning system telemetry techniques for estimating cougar predation parameters. J Wildlife Manag 73(4):586–597. https://doi.org/10.2193/2008-294

Konnert M, Maurer W, Degen B, Kätzel R (2011) Genetic monitoring in forests – early warning and controlling system for ecosystemic changes. IForest 4:77–81. https://doi.org/10.3832/ifor0571-004

Kremer A, Ronce O, Robledo-Arnuncio JJ, Guillaume F, Bohrer G, Nathan R, Bridle JR, Gomulkiewicz R, Klein EK, Ritland K, Kuparinen A, Gerber S, Schueler S (2012) Long-distance gene flow and adaptation of forest trees to rapid climate change. Ecol Lett 15(4):378–392. https://doi.org/10.1111/j.1461-0248.2012.01746.x

Lamb PD, Fonseca VG, Maxwell DL, Nnanatu CC (2022) Systematic review and meta-analysis: water type and temperature affect environmental DNA decay. Mol Ecol Resour 22(7):2494–2505. https://doi.org/10.1111/1755-0998.13627

Lampa S, Gruber B, Henle K, Hoehn M (2008) An optimisation approach to increase DNA amplification success of otter faeces. Conservation Genetics 9(1):201–210. https://doi.org/10.1007/s10592-007-9328-9

Lefèvre F, Alia R, Fjellstad Bakkebø K, Graudal L, Oggioni S, Rusanen M, Vendramin M, Bozzano M (2020) Dynamic conservation and utilization of forest tree genetic resources: indicators for in situ and ex situ genetic conservation and forest reproductive materia (European F). European Forest Institute

Leroy T, Louvet JM, Lalanne C, Le Provost G, Labadie K, Aury JM, Delzon S, Plomion C, Kremer A (2020) Adaptive introgression as a driver of local adaptation to climate in European white oaks. New Phytol 226(4):1171–1182. https://doi.org/10.1111/nph.16095

Liu CLC, Kuchma O, Krutovsky KV (2018) Mixed-species versus monocultures in plantation forestry: development, benefits, ecosystem services and perspectives for the future. Global Ecology and Conservation 15:e00419. https://doi.org/10.1016/j.gecco.2018.e00419

Lõhmus A, Lõhmus P, Runnel K (2018) A simple survey protocol for assessing terrestrial biodiversity in a broad range of ecosystems. PLoS One 13(12):1–24. https://doi.org/10.1371/journal.pone.0208535

Lovell JL, Jupp DLB, Culvenor DS, Coops NC (2003) Using airborne and ground-based ranging lidar to measure canopy structure in Australian forests. Can J Remote Sens 29(5):607–622. https://doi.org/10.5589/m03-026

Lowe WH, Allendorf FW (2010) What can genetics tell us about population connectivity? Mol Ecol 19(15):3038–3051. https://doi.org/10.1111/j.1365-294X.2010.04688.x

Lynggaard C, Bertelsen MF, Jensen CV, Johnson MS, Frøslev TG, Olsen MT, Bohmann K (2022) Airborne environmental DNA for terrestrial vertebrate community monitoring. Curr Biol 32(3):701–707.e5. https://doi.org/10.1016/j.cub.2021.12.014

Macher T, Schütz R, Hörren T, Beermann AJ, Leese F (2023) It's raining species: rainwash eDNA metabarcoding as a minimally invasive method to assess tree canopy invertebrate diversity. Environ DNA 5(1):3–11. https://doi.org/10.1002/edn3.372

Maclaurin J, Sterelny K (2008) What is biodiversity. University of Chicago Press

Maletzke B, Kertson B, Swanson M, Koehler G, Beausoleil R, Wielgus R, Cooley H (2017) Cougar response to a gradient of human development. Ecosphere 8(7). https://doi.org/10.1002/ecs2.1828

Manel S, Schwartz MK, Luikart G, Taberlet P (2003) Landscape genetics: combining landscape ecology and population genetics. Trends Ecol Evol 18(4):189–197. https://doi.org/10.1016/S0169-5347(03)00008-9

Mason NWH, Mouillot D (2013) Functional diversity measures. In: Encyclopedia of biodiversity: second edition, pp 597–608. https://doi.org/10.1016/B978-0-12-384719-5.00356-7

McCauley DE (1995) The use of chloroplast DNA polymorphism in studies of gene flow in plants. Trends Ecol Evol 10(5):198–202. https://doi.org/10.1016/S0169-5347(00)89052-7

Meyer JM, Leempoel K, Losapio G, Hadly EA (2020) Molecular ecological network analyses: an effective conservation tool for the assessment of biodiversity, trophic interactions, and community structure. Front Ecol Evol 8:588430. https://doi.org/10.3389/fevo.2020.588430

Mills SL, Soulé ME, Doak DF (1993) The keystone-species consept in ecology and conservation. BioScience 43(4):219–224

Mittermeier RA, Turner WR, Larsen FW, Brooks TM, Gascon C (2011) Global biodiversity conservation: the critical role of hotspots. Biodivers Hotspots:3–22. https://doi.org/10.1007/978-3-642-20992-5_1

Moussy C, Burfield IJ, Stephenson PJ, Newton AFE, Butchart SHM, Sutherland WJ, Gregory RD, McRae L, Bubb P, Roesler I, Ursino C, Wu Y, Retief EF, Udin JS, Urazaliyev R, Sánchez-Clavijo LM, Lartey E, Donald PF (2022) A quantitative global review of species population monitoring. Conserv Biol 36(1). https://doi.org/10.1111/cobi.13721

Nagendra H (2001) Using remote sensing to assess biodiversity. Int J Remote Sens 22(12):2377–2400. https://doi.org/10.1080/01431160117096

Nagendra H, Lucas R, Pradinho J, Jongman RHG, Tarantino C, Adamo M, Mairota P (2013) Remote sensing for conservation monitoring: assessing protected areas, habitat extent, habitat condition, species diversity, and threats. Ecol Indic 33:45–59

Newnham GJ, Armston JD, Calders K, Disney MI, Lovell JL, Schaaf CB, Strahler AH, Mark Danson F (2015) Terrestrial laser scanning for plot-scale forest measurement. Current Forestry Reports 1(4):239–251. https://doi.org/10.1007/s40725-015-0025-5

Noss RF (1990) Indicators for monitoring biodiversity: a hierarchical approach. Conserv Biol 4(4):355–364. https://doi.org/10.1111/j.1523-1739.1990.tb00309.x

O'brien TG, Kinnaird MF (2008) A picture is worth a thousand words: the application of camera trapping to the study of birds. Bird Conserv Int 18:S144–S162. https://doi.org/10.1017/S0959270908000348

Palmer MW, Earls PG, Hoagland BW, White PS, Wohlgemuth T (2002) Quantitative tools for perfecting species lists. Environmetrics 13(2):121–137. https://doi.org/10.1002/env.516

Pawlowski J, Kelly-Quinn M, Altermatt F, Apothéloz-Perret-Gentil L, Beja P, Boggero A, Borja A, Bouchez A, Cordier T, Domaizon I, Feio MJ, Filipe AF, Fornaroli R, Graf W, Herder J, van der Hoorn B, Iwan Jones J, Sagova-Mareckova M, Moritz C et al (2018) The future of biotic indices in the ecogenomic era: Integrating (e)DNA metabarcoding in biological assessment of aquatic ecosystems. Sci Total Environ 637–638:1295–1310. https://doi.org/10.1016/j.scitotenv.2018.05.002

Pearson RM, Schlacher TA, Jinks KI, Olds AD, Brown CJ, Connolly RM (2021) Disturbance type determines how connectivity shapes ecosystem resilience. Sci Rep 11(1):1–8. https://doi.org/10.1038/s41598-021-80987-1

Pijanowski BC, Farina A, Gage SH, Dumyahn SL, Krause BL (2011) What is soundscape ecology? An introduction and overview of an emerging new science. Landsc Ecol 26(9):1213–1232. https://doi.org/10.1007/s10980-011-9600-8

Pither R, O'Brien P, Brennan A, Hirsh-Pearson K, Bowman J (2023) Predicting areas important for ecological connectivity throughout Canada. PLoS One 18. https://doi.org/10.1371/journal.pone.0281980

Pla L, Casanoves F, Di Rienzo J (2012) Functional groups, pp 9–25. https://doi.org/10.1007/978-94-007-2648-2_2

Planella A, Palacios V, García EJ, Llaneza L, García-Domínguez F, Muñoz-Igualada J, López-Bao JV (2016) Influence of different GPS schedules on the detection rate of wolf feeding sites in human-dominated landscapes. Eur J Wildlife Res 62(4):471–478. https://doi.org/10.1007/s10344-016-1020-2

Porter TM, Hajibabaei M (2018) Scaling up: a guide to high-throughput genomic approaches for biodiversity analysis. Mol Ecol 27(2):313–338. https://doi.org/10.1111/mec.14478

Powney GD, Roy DB, Chapman D, Brereton T, Oliver TH (2011) Measuring functional connectivity using long-term monitoring data. Methods Ecol Evol 2(5):527–533. https://doi.org/10.1111/j.2041-210X.2011.00098.x

Ratnasingham S, Hebert PDN (2007) BOLD: the barcode of life data system: barcoding. Mol Ecol Notes 7(3):355–364. https://doi.org/10.1111/j.1471-8286.2007.01678.x

Richard-Hansen C, Jaouen G, Denis T, Brunaux O, Marcon E, Guitet S (2015) Landscape patterns influence communities of medium-to large-bodied vertebrates in undisturbed terra firme forests of French Guiana. J Trop Ecol 31(5):423–436. https://doi.org/10.1017/S0266467415000255

Rivera SF, Vasselon V, Bouchez A, Rimet F (2020) Diatom metabarcoding applied to large scale monitoring networks: optimization of bioinformatics strategies using Mothur software. Ecol Indic 109:105775. https://doi.org/10.1016/j.ecolind.2019.105775

Roberge J-M, Angelstam P (2004) Usefulness of the umbrella species concept as a conservation tool. Conserv Biol 18(1):76–85

Roberts NJ (2011) Investigation into survey techniques of large mammals: surveyor competence and camera-trapping vs. transect-sampling. Biosci Horiz 4(1):40–49. https://doi.org/10.1093/biohorizons/hzr006

Rodriguez-Ezpeleta N, Morissette O, Bean CW, Manu S, Banerjee P, Lacoursière-Roussel A, Beng KC, Alter SE, Roger F, Holman LE, Stewart KA, Monaghan MT, Mauvisseau Q, Mirimin L, Wangensteen OS, Antognazza CM, Helyar SJ, Boer H, Monchamp M et al (2021) Trade-offs between reducing complex terminology and producing accurate interpretations from environmental DNA: Comment on "Environmental DNA: What's behind the term?" by Pawlowski et al. (2020). Mol Ecol 30(19):4601–4605. https://doi.org/10.1111/mec.15942

Rubio VE, Swenson NG (2022) Functional groups, determinism and the dynamics of a tropical forest. J Ecol 110(1):185–196. https://doi.org/10.1111/1365-2745.13795

Savolainen O, Pyhäjärvi T, Knürr T (2007) Gene flow and local adaptation in trees. Annu Rev Ecol Evol Syst 38:595–619. https://doi.org/10.1146/annurev.ecolsys.38.091206.095646

Schuldt A, Assmann T, Brezzi M, Buscot F, Eichenberg D, Gutknecht J, Härdtle W, He JS, Klein AM, Kühn P, Liu X, Ma K, Niklaus PA, Pietsch KA, Purahong W, Scherer-Lorenzen M, Schmid B, Scholten T, Staab M et al (2018) Biodiversity across trophic levels drives multifunctionality in highly diverse forests. Nat Commun 9(1). https://doi.org/10.1038/s41467-018-05421-z

Seidler R (2023) Biodiversity in anthropogenically altered forests. Ref Mod Life Sci. https://doi.org/10.1016/b978-0-12-822562-2.00084-0

Simberloff D (1998) Flagships, umbrellas, and keystones: Is single-species management passe in the landscape era? Biol Conserv 83(3):247–257. https://doi.org/10.1016/S0006-3207(97)00081-5

Slatkin M (1987) Gene flow and the geographic structure of natural populations. Science 236(4803):787–792. https://doi.org/10.1126/science.3576198

Soulsbury CD, Gray HE, Smith LM, Braithwaite V, Cotter SC, Elwood RW, Wilkinson A, Collins LM (2020) The welfare and ethics of research involving wild animals: a primer. Methods in Ecology and Evolution 11(10):1164–1181. https://doi.org/10.1111/2041-210X.13435

Stewart FEC, Darlington S, Volpe JP, McAdie M, Fisher JT (2019) Corridors best facilitate functional connectivity across a protected area network. Sci Rep 9(1):1–9. https://doi.org/10.1038/s41598-019-47067-x

Storfer A, Murphy MA, Spear SF, Holderegger R, Waits LP (2010) Landscape genetics: where are we now? Mol Ecol 19(17):3496–3514. https://doi.org/10.1111/j.1365-294X.2010.04691.x

Taberlet P, Bonin A, Zinger L, Coissac E (2018) Environmental DNA: for biodiversity research and monitoring. In: Environmental DNA: for biodiversity research and monitoring, pp 1–253. https://doi.org/10.1093/oso/9780198767220.001.0001

Takahashi M, Saccò M, Kestel JH, Nester G, Campbell MA, van der Heyde M, Heydenrych MJ, Juszkiewicz DJ, Nevill P, Dawkins KL, Bessey C, Fernandes K, Miller H, Power M, Mousavi-Derazmahalleh M, Newton JP, White NE, Richards ZT, Allentoft ME (2023) Aquatic environmental DNA: a review of the macro-organismal biomonitoring revolution. Sci Total Environ 873:162322. https://doi.org/10.1016/j.scitotenv.2023.162322

Thalinger B, Deiner K, Harper LR, Rees HC, Blackman RC, Sint D, Traugott M, Goldberg CS, Bruce K (2021) A validation scale to determine the readiness of environmental DNA assays for routine species monitoring. Environmental DNA:1–14. https://doi.org/10.1002/edn3.189

Thomas E, Jalonen R, Loo J, Boshier D, Gallo L, Cavers S, Bordács S, Smith P, Bozzano M (2014) Genetic considerations in ecosystem restoration using native tree species. For Ecol Manage 333:66–75. https://doi.org/10.1016/j.foreco.2014.07.015

Thompson SK, Buckland ST, Anderson DR, Burnham KP, Laake JL (1994) Distance sampling: estimating abundance of biological populations. Biometrics 50(3):891. https://doi.org/10.2307/2532812

Thomsen PF, Willerslev E (2015) Environmental DNA – an emerging tool in conservation for monitoring past and present biodiversity. Biol Conserv 183:4–18. https://doi.org/10.1016/j.biocon.2014.11.019

Tobler MW, Carrillo-Percastegui SE, Leite Pitman R, Mares R, Powell G (2008) An evaluation of camera traps for inventorying large- and medium-sized terrestrial rainforest mammals. Anim Conserv 11(3):169–178. https://doi.org/10.1111/j.1469-1795.2008.00169.x

Tsuji S, Inui R, Nakao R, Miyazono S, Saito M, Kono T, Akamatsu Y (2022) Quantitative environmental DNA metabarcoding shows high potential as a novel approach to quantitatively assess fish community. Sci Rep 12(1):1–11. https://doi.org/10.1038/s41598-022-25274-3

Turlure C, Pe'er G, Baguette M, Schtickzelle N (2018) A simplified mark–release–recapture protocol to improve the cost effectiveness of repeated population size quantification. Methods in Ecology and Evolution 9(3):645–656. https://doi.org/10.1111/2041-210X.12900

Turon X, Antich A, Palacín C, Præbel K, Wangensteen OS (2020) From metabarcoding to metaphylogeography: separating the wheat from the chaff. Ecol Appl 30(2):e02036. https://doi.org/10.1002/eap.2036

Urban DL, O'Neill RV, Shugart HH (1987) Landscape ecology. BioScience 37:119–127

Valbuena R, Packalén P, Martín-Fernández S, Maltamo M (2012) Diversity and equitability ordering profiles applied to study forest structure. For Ecol Manage 276:185–195. https://doi.org/10.1016/j.foreco.2012.03.036

Vasar M, Davison J, Moora M, Sepp S-K, Anslan S, Al-Quraishy S, Bahram M, Bueno CG, Cantero JJ, Fabiano EC, Decocq G, Drenkhan R, Fraser L, Oja J, Garibay-Orijel R, Hiiesalu I, Koorem K, Mucina L, Öpik M et al (2023) Metabarcoding of soil environmental DNA to estimate plant diversity globally. Front Plant Sci 14:1106617. https://doi.org/10.3389/fpls.2023.1106617

Wakefield ED, Bodey TW, Bearhop S, Blackburn J, Colhoun K, Davies R, Dwyer RG, Green JA, Grémillet D, Jackson AL, Jessopp MJ, Kane A, Langston RHW, Lescroël A, Murray S, Le Nuz M, Patrick SC, Péron C, Soanes LM et al (2013) Space partitioning without territoriality in gannets. Science 341(6141):68–70. https://doi.org/10.1126/science.1236077

Wang R, Gamon JA (2019) Remote sensing of terrestrial plant biodiversity. Remote Sens Environ 231. https://doi.org/10.1016/j.rse.2019.111218

Weigand H, Beermann AJ, Čiampor F, Costa FO, Csabai Z, Duarte S, Geiger MF, Grabowski M, Rimet F, Rulik B, Strand M, Szucsich N, Weigand AM, Willassen E, Wyler SA, Bouchez A, Borja A, Čiamporová-Zaťovičová Z, Ferreira S et al (2019) DNA barcode reference libraries for the monitoring of aquatic biota in Europe: gap-analysis and recommendations for future work. Sci Total Environ 678:499–524. https://doi.org/10.1016/j.scitotenv.2019.04.247

Welbourne DJ, Claridge AW, Paull DJ, Lambert A (2016) How do passive infrared triggered camera traps operate and why does it matter? Breaking down common misconceptions. Remote Sensing in Ecology and Conservation 2(2):77–83. https://doi.org/10.1002/rse2.20

Young A, Boshier D, Boyle T (2000) Forest conservation genetics: principles and practice. Csiro Publishing

Young RG, Milián-García Y, Yu J, Bullas-Appleton E, Hanner RH (2021) Biosurveillance for invasive insect pest species using an environmental DNA metabarcoding approach and a high salt trap collection fluid. Ecol Evol 11(4):1558–1569. https://doi.org/10.1002/ece3.7113

Yu DW, Ji Y, Emerson BC, Wang X, Ye C, Yang C, Ding Z (2012) Biodiversity soup: metabarcoding of arthropods for rapid biodiversity assessment and biomonitoring. Methods in Ecology and Evolution 3(4):613–623. https://doi.org/10.1111/j.2041-210X.2012.00198.x

Zheng Y, Janke A (2018) Gene flow analysis method, the D-statistic, is robust in a wide parameter space. BMC Bioinformatics 19(1). https://doi.org/10.1186/s12859-017-2002-4

Zimmermann K, Fric Z, Jiskra P, Kopeckova M, Vlasanek P, Zapletal M, Konvicka M (2011) Mark-recapture on large spatial scale reveals long distance dispersal in the marsh fritillary, Euphydryas aurinia. Ecol Entomol 36(4):499–510. https://doi.org/10.1111/j.1365-2311.2011.01293.x

Zwerts JA, Stephenson PJ, Maisels F, Rowcliffe M, Astaras C, Jansen PA, van der Waarde J, Sterck LEHM, Verweij PA, Bruce T, Brittain S, van Kuijk M (2021) Methods for wildlife monitoring in tropical forests: comparing human observations, camera traps, and passive acoustic sensors. Conserv Sci Pract 3(12). https://doi.org/10.1111/csp2.568

Open Access This chapter is licensed under the terms of the Creative Commons Attribution 4.0 International License (http://creativecommons.org/licenses/by/4.0/), which permits use, sharing, adaptation, distribution and reproduction in any medium or format, as long as you give appropriate credit to the original author(s) and the source, provide a link to the Creative Commons license and indicate if changes were made.

The images or other third party material in this chapter are included in the chapter's Creative Commons license, unless indicated otherwise in a credit line to the material. If material is not included in the chapter's Creative Commons license and your intended use is not permitted by statutory regulation or exceeds the permitted use, you will need to obtain permission directly from the copyright holder.

Monitoring Habitat Fragmentation and Biodiversity in Forest Ecosystems

9

Adriano Mazziotta, Saverio Francini, and Francesco Parisi

Aerial view of Bavarian Forest, Germany (Photo: Andrey Kuzmin/Adobe Stock)

A. Mazziotta (✉)
Natural Resources Institute Finland (Luke), Helsinki, Finland
e-mail: adriano.mazziotta@luke.fi

S. Francini
Department of Science and Technology of Agriculture and Environment (DISTAL),
University of Bologna, Bologna, Italy
e-mail: saverio.francini@unibo.it

F. Parisi
Department of Bioscience and Territory, Molise University, Pesche, Italy

NBFC, National Biodiversity Future Center, Palermo, Italy
e-mail: francesco.parisi@unimol.it

© The Author(s) 2025
K. Lapin et al. (eds.), *Ecological Connectivity of Forest Ecosystems*,
https://doi.org/10.1007/978-3-031-82206-3_9

Abstract

The current biodiversity crisis is primarily caused by habitat loss and fragmentation, which are exacerbated by global population expansion and land use intensification. The techniques applied to evaluate the impact of habitat loss and fragmentation in forest ecosystems tend to measure changes in landscape patterns induced by forest degradation. Earth observation techniques and remotely sensed imagery are crucial tools for the large-scale monitoring of forest habitat loss and fragmentation along with related changes in forest biodiversity characteristics. Recently, the relevance of remote sensing for monitoring forest fragmentation has been further amplified by new satellite missions providing up-to-date and high-resolution open-access data available on cloud computing platforms. However, while satellite programmes like Landsat that employ remote sensing techniques are suitable for large-scale monitoring of forest species distribution, they cannot capture micro-spatial variations, since their sensors cannot disentangle forest heterogeneity. Finally, remotely sensed canopy-level information alone cannot fully explain biodiversity patterns. Integration of remote sensing and ground survey activities may help to overcome the limitations of these techniques, providing solutions for designing and optimizing monitoring strategies to tackle forest fragmentation and biodiversity loss in forest ecosystems.

Keywords

Remote sensing · Google earth engine · Habitat loss and fragmentation · Land use change · Landsat

The Need to Monitor Forest Habitat Fragmentation

Human appropriation of the planet is restricting ecological connectivity for species and ecosystems and thus causing habitat loss and fragmentation, which are considered key drivers of the current biodiversity crisis together with pollution, overexploitation, and climate change (Bae et al. 2019; Muys et al. 2022). Habitat fragmentation (also known as habitat subdivision or patchiness) means the breaking apart of habitats into multiple patches (Fahrig 2003). It can compound habitat loss by reducing the size of the habitat area, increasing edge effects, and causing habitat isolation but it is also responsible for increasing habitat heterogeneity (but see the debate between Fletcher et al. 2018 and Fahrig et al. 2019). Smaller habitat patches can lead to population decline, as resources in smaller patches may be more limited. In addition, habitat fragmentation increases the isolation of remaining habitat areas, decreasing habitat connectivity which relates to the ability of species as well as ecological resources and processes to move through landscapes (Lindenmayer et al. 2008).

Habitat loss and fragmentation determine landscape degradation (Fahrig 2003), which Fischer and Lindenmayer (2007) define as the gradual deterioration of habitat quality. For example, logging is one of the main factors inducing degradation of intact forest habitats. Habitat degradation is a consequence of the impact of multiple anthropogenic stressors and transforms landscapes by reducing the size and connectivity of species' habitats. Its effects become visible at different scales, and its impact is not ubiquitous but rather species- and ecosystem-specific. The techniques and approaches commonly employed to evaluate the impact of habitat loss and fragmentation on biological systems tend to measure changes in landscape patterns induced by habitat degradation (Lindenmayer and Hobbs 2008).

Natural and anthropogenic disturbances affect habitat availability (amount) and configuration (connectivity) for biodiversity in forest landscapes (Thom et al. 2017). Timber harvesting as well as abiotic and biotic disturbances like windthrow, wildfires, insect outbreaks, diseases, pathogens, and drought may alter both the structural and functional connectivity of forest habitats, as well as the amount of habitat available for biodiversity. For example, windthrow events can quickly create large volumes of deadwood, while logging can isolate old-growth forest remnants. Forest management is an example of disturbance that can have positive as well as negative effects on the amount of connected habitat usable by forest species (Oettel and Lapin 2021). For example, intensive forest management conducted in the form of clear-cutting degrades the habitats of saproxylic species by reducing habitat amount through the removal of deadwood substrates on which they complete their life cycle as well as habitat connectivity through the creation of forest gaps impeding their dispersion (Mönkkönen et al. 2014; Mazziotta et al. 2023; Oettel et al. 2023). On the other hand, close-to-nature forest management (Bauhus et al. 2013)—for example, continuous cover forestry by means of selective logging (Peura et al. 2018)—improves habitat quality for species dwelling in semi-natural forests. The creation of forest gaps by selective logging increases habitat heterogeneity through the removal of large logs in the otherwise homogeneous mature forest, creating habitat for species developing in standing and lying deadwood associated with sunny microclimatic conditions. In doing so, selective logging also increases habitat connectivity by creating a heterogeneous forest matrix that facilitates the dispersion of saproxylic species associated with sunny microclimate. In both cases, forest management is changing the structure of the landscape by altering the amount of habitat available and the connectivity of suitable habitat patches for different species (Nordén et al. 2013; Undin et al. 2022). However, species that have declined due to forestry mostly require maintaining large living and dead trees, which cannot be preserved by continuous cover forestry alone. A mosaic of different management regimes may provide complementary ways to maintain valuable and connected habitats for forest species (Koivula et al. 2025; Rautio et al. 2025).

Since habitat loss and fragmentation induced by natural and anthropogenic disturbances take place on the level of the entire landscape rather than that of individual stands, tackling these changes requires earth observation techniques like remote sensing that can monitor variations in the characteristics of forests and their spatial patterns on a broad scale (Francini et al. 2022, 2023a) (Fig. 9.1).

Fig. 9.1 ESA Sentinel—2 (European Space Agency)

The Role of Remote Sensing in Habitat Fragmentation Monitoring

Habitat loss and fragmentation are historically monitored by means of ground surveys. Field analysis and detailed information acquisition are effective strategies for collecting exhaustive and comprehensive information about these two drivers of landscape degradation. On the other hand, ground surveys are subject to several shortcomings. First, acquiring data on the ground is time-consuming and consequently expensive. As a result, such data is acquired with long remeasurement intervals and only from small areas, limiting its effectiveness for estimating forest changes quickly and precisely (Zald et al. 2016). This is a crucial issue with regard to monitoring the rapid forest changes induced by global warming and frequent anthropogenic disturbances. Second, ground data can be aggregated to provide estimates, but it cannot be employed alone to produce detailed, spatially explicit maps useful for habitat loss and fragmentation assessment.

Remote sensing offers an effective alternative to ground surveys for mapping the processes of habitat loss and fragmentation. For example, active and passive remote sensing data can be used to obtain land cover and forest disturbance maps and to track changes in forest cover and health status (Hao et al. 2019; Francini et al. 2022). Landscape metrics such as patch size, shape, and connectivity are numerical indices quantifying landscape patterns (McGarigal 2015) and can be calculated from these maps to quantify habitat fragmentation (Liu et al. 2021). In the meantime, photosynthetic activity indices (e.g., the Normalized Difference Vegetation Index, NDVI) can be calculated from remotely sensed optical imagery to monitor changes in vegetation status and assess habitat condition and degradation over space and time (Guo et al. 2019).

In addition, remote sensing provides valuable information for guiding conservation and land management efforts (Tayyebi et al. 2020). For example, remote sensing data can be used to identify areas where conservation efforts are most needed, track the effectiveness of conservation interventions, and prioritize areas for habitat restoration (Cord et al. 2018; Schwieder et al. 2019).

The relevance of remote sensing for monitoring habitat loss and fragmentation has been further amplified by three recent innovations and advancements. First, new satellite missions such as Sentinel, PlanetScope, and Pléiades Neo play a crucial role in this context by providing new high-resolution data with shorter revisitation times compared to previous missions. This is a key advantage in the context of highly fragmented regions where the pixel sizes of medium-resolution imagery may not be small enough to reveal subtle habitat changes. Second, several satellite missions have begun to provide data under free open-access licences (e.g., Sentinel-2, Landsat). The third factor is the development of cloud computing platforms including Sentinel Hub, Open Data Cube, SEPAL, JEODPP, pipsCloud, OpenEO, and Google Earth Engine (Gomes et al. 2020). Combining the high-resolution open-access data available from new satellite missions with cloud computing platforms enables the application of complex algorithms detecting changes across very large areas (Woodcock et al. 2008). Among the mentioned cloud computing platforms, Google Earth Engine (GEE) is particularly suitable for monitoring habitat fragmentation at large scales. GEE combines a catalogue of satellite imagery and geospatial datasets with planetary-scale analysis capabilities (Gorelick et al. 2017) for aspects including forest change assessment (Hansen et al. 2013) and surface water extent and dynamics (Pekel et al. 2016). GEE has three key strengths compared to the other mentioned cloud computing platforms: The first is its flexibility allowing users to apply different algorithms to the data and use high-level programming languages and high-performance computing. The second is scientific reproducibility together with storage and process scalability. The final advantage is its processing performance, which can be scaled by adding more resources without users needing to alter their approach or code.

GEE has already implemented several algorithms relating to forest disturbance detection and exploiting the analysis-ready satellite data: (i) LandTrendr (Kennedy et al. 2012, 2018), (ii) Continuous Change Detection and Classification (CCDC; Zhu and Woodcock 2014), (iii) Exponentially Weighted Moving Average Change Detection (Brooks et al. 2014), (iv) Vegetation Change Tracker (VCT; Huang et al. 2010), and (v) the Verdet forest change detection algorithm (Hughes et al. 2017). Although some of these algorithms can use imagery from different satellite missions, they were all originally designed to work with Landsat data. Zhu (2017) and Francini et al. (2020, 2021) provide comprehensive reviews of these temporal segmentation algorithms, and a brief overview of the most commonly used remote sensing approaches for monitoring forest disturbances is provided in Box 9.1.

Box 9.1 Key Remote Sensing Approaches to Monitoring Changes in Forest Cover

During the past decade, the two most commonly used remote sensing techniques worldwide with the capability to monitor changes in forest cover have been LandTrendr (LT) (Kennedy et al. 2010) and the Global Forest Change (GFC) data set (Hansen et al. 2013).

LT consists of a temporal segmentation approach that predicts changes by identifying breakpoints in trajectories of a photosynthetic index (like NDVI) calculated over several consecutive years from a Landsat imagery time series. It requires calibration of input parameters for each ecosystem (Hudak et al. 2013; Fragal et al. 2016; Yang et al. 2018). Because LT is based on yearly time-series analyses, accuracy decreases for extremes of the time series and for near past detection applications.

By contrast, GFC data is constructed using more than 600,000 Landsat scenes and a hierarchic classifier based on recursive partitioning. The data consists of annual global maps of tree cover extent, loss, and gain. GFC was used together with aerial images to analyse harvested sites in mountainous boreal forests in Norway, but up to 30% omission errors were reported (Rossi et al. 2019). GFC has also been proven inaccurate in Mediterranean coppice forests in Italy, with an average precision of about 50% (Giannetti et al. 2020). Despite these shortcomings, GFC was recently used by Ceccherini et al. (2020) to assess the temporal trend of forest logging in Europe; however, several limitations were discovered by Palahí et al. (2021).

Further remote sensing algorithms include Continuous Change Detection and Classification (Zhu and Woodcock 2014), Breaks for Additive Season and Trend Monitor (Verbesselt et al. 2012), and Space-Time Extremes and Features (Hamunyela et al. 2017), most of which are likewise Landsat-based algorithms at 30-meter spatial resolution. Recently, new methods have been implemented for predicting forest disturbances at finer scales using Sentinel-2 and PlanetScope imagery (Francini et al. 2021, 2022).

Predictions of forest biomass loss due to disturbance have been possible through the combination of maps based on remote sensing with data from the Global Ecosystem Dynamics Investigation (GEDI) sensor (Francini et al. 2023a). Finally, remote sensing data has proven effective not just for forest disturbance monitoring but also for the detection and estimation of afforestation areas (Cavalli et al. 2022). Measuring afforestation rate is a key aspect considering that forest area is increasing in several world regions and that afforestation represents the main land cover change in Europe (Palmero-Iniesta et al. 2021).

Remote Sensing to Monitor the Impact of Habitat Fragmentation on Biodiversity

The rapid pace at which habitat loss and fragmentation occur worldwide is one of the main causes for the fast decline in species populations. In the 2022 IUCN global Red List, 28% of all assessed species were classified as threatened with extinction, belonging to the critically endangered, endangered, or vulnerable categories (www.iucnredlist.org). Within this context, biodiversity monitoring is a major concern in forest ecosystems, as they cover a third of the world's total land area and host a high species diversity amounting to three quarters of all terrestrial plant, fungus, and animal species (Forest Europe 2020).

In order to guide policies and management strategies for biodiversity conservation, regular, reliable, and standardized data on the state of biodiversity is required. Since the term 'biodiversity' encompasses the biological diversity of organisms in terms of composition, structure, and functionality all the way from genes to ecosystems, hundreds of variables can be measured to study it (Muys et al. 2022). The best indicator to measure biodiversity within an ecosystem, for example, would be the measurement of species diversity. However, since it is impossible to record all species present in an area, the use of readily observable, measurable, and quantifiable proxies and indicators is essential (McElhinny et al. 2005; Ozdemir et al. 2018).

Historically, the most commonly used indicators for assessing biodiversity fall into two main groups: habitat-based and taxon-based indicators (Paillet et al. 2024). The former represent environmental and structural variables considered to be proxies of the richness, composition, or diversity of species, while the latter are linked to the presence or abundance of indicator species (Lindenmayer et al. 2014). Although biodiversity monitoring using taxon-based approaches is more reliable for describing local species patterns, these monitoring methods still rely on traditional sampling methods and plot-level ground surveys. This means that they remain costly and time-consuming, especially when applied to large areas. Moreover, they require a lot of human resources and can easily be biased by human error, even when experts are involved in species identification (Wang and Gamon 2019). Among the habitat-based indicators, the monitoring of forest attributes related to forest structural complexity is certainly pivotal (Ćosović et al. 2020). These attributes include variability in canopy cover, tree diameter, tree height, and understory vegetation, which support the occurrence of diverse ecological niches for wildlife (Zellweger et al. 2013). Multiple studies have highlighted the existence of a link between forest structure and several groups of species (see Zeller et al. 2023 for a review), including vascular plants (Burrascano et al. 2008), bryophytes (Madžule et al. 2012), lichens (Moning et al. 2009), and wood-inhabiting fungi (Mazziotta et al. 2016) (Ruokolainen et al. 2018), birds (Herniman et al. 2020), insects (e.g., Parisi et al. 2023, 2024), and bats (Vogeler et al. 2022). For example, using a database of forest stands in southern Sweden, Hedwall et al. (2019) have found that the cover and species richness of understory vascular plants increased with an increasing proportion of birch and decreased with increasing forest density, while the cover of bryophytes decreased with an increasing proportion of birch and increasing forest density.

It is in this context that remote sensing can play a fundamental role in assessing habitat-based indicators of forest biodiversity at large scales (Fig. 9.2). Remote sensing represents a powerful and efficient instrument for monitoring forest characteristics and can efficiently support monitoring by providing open-access, up-to-date, and repeatable data that can be used to estimate and predict the abundance and diversity of different taxonomic groups at various scales of time and space (Parisi et al. 2022, 2024). One of the most important advantages of remotely sensed habitat-based indicators is that they are mapped 'wall-to-wall', meaning that they offer a continuous biodiversity assessment across the entire forest landscape (Ozdemir et al. 2018). The use of remote sensing techniques in biodiversity monitoring and mapping has only become more popular during the last three decades, with the first scientific studies regarding the topic emerging in the 1990s (Wang and Gamon 2019). Over this time, a variety of remote sensing sources ranging from passive—i.e. satellite imagery—to active methods like Light Detection And Ranging (LiDAR) have been developed and implemented.

Early uses of remote sensing in biodiversity assessment included landscape or habitat mapping through optical data (Wang and Gamon 2019). The sensors of the Sentinel-2 (ESA Copernicus programme) and Landsat (USGS/NASA) missions

Fig. 9.2 Remote sensing facilitates biodiversity assessment by monitoring forest characteristics (Francesco Parisi)

allow the calculation of spectral indices (Kacic and Kuenzer 2022). In particular, vegetation indices such as NDVI offer information about canopy cover and tree species diversity (Arekhi et al. 2017). This data can be used for diversity monitoring following the spectral variation hypothesis, according to which greater spectral heterogeneity in an image corresponds to greater tree species richness on the ground (Ozdemir et al. 2018). For example, Parisi et al. (2023) have analysed time series for Sentinel-2 harmonic metrics to relate changes in NDVI remotely sensed via Landsat images with biodiversity indices for the taxonomic groups of beetles, birds, and lichens (Fig. 9.3). Graf et al. (2005) made use of LiDaR remote sensing to evaluate the availability of habitat of a forest grouse species (capercaillie) at multiple spatial scales.

Moreover, the implementation of LiDAR systems (Fig. 9.2) has expanded the range of data that can be remotely sensed (Wang and Gamon 2019). Especially in forest ecosystems, laser technology is a powerful tool for biodiversity monitoring as it can collect information and metrics regarding vegetation structure (Moudrý et al. 2023). Recently, Airborne Laser Scanning (ALS) (Fig. 9.4) performed using LiDAR sensors aboard aircraft has enabled simultaneous detection of both vegetation

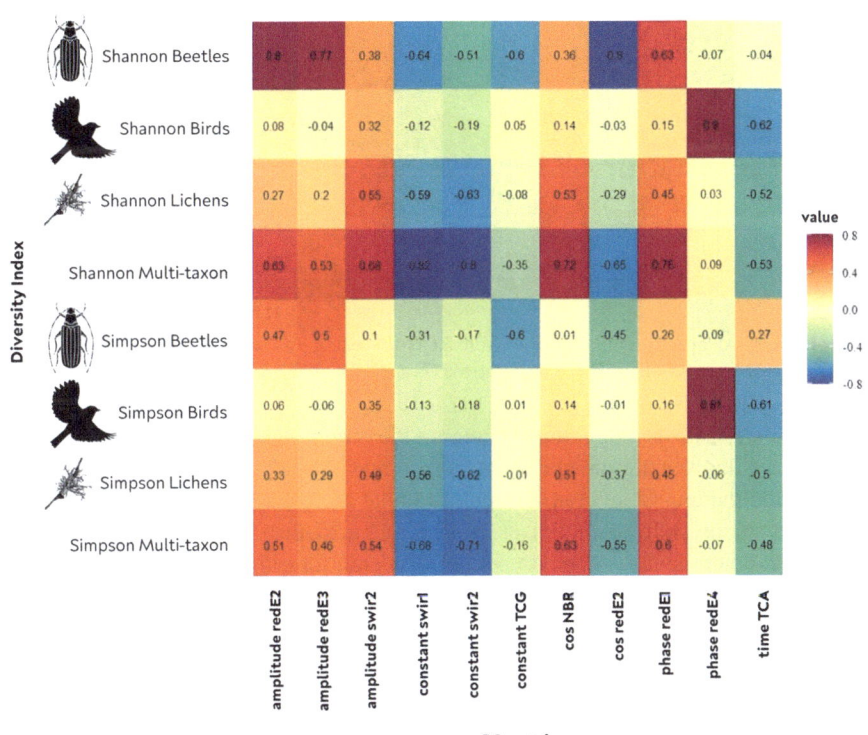

Fig. 9.3 Remote sensing supports detection of biodiversity patterns. Correlations between biodiversity indices of several taxa and the best Sentinel-2-derived temporal metrics (Parisi et al. 2023)

Fig. 9.4 The term LiDAR is an acronym for Light Detection And Ranging and it refers to sensors used to capture point clouds from both static and mobile methods (https://geolabforest.com/)

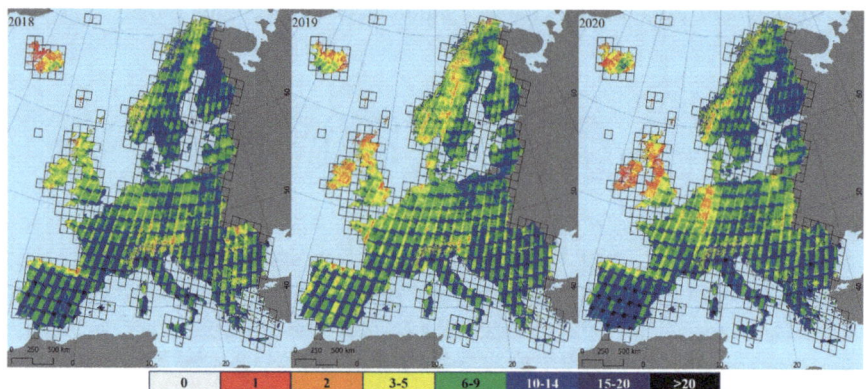

Fig. 9.5 An example of temporal series from Remote sensing via Landsat-7. Number of per-pixel valid observations per analysis tile are reported (Francini et al. 2023)

biochemistry and structure, thus becoming the primary method for collecting accurate terrain and vegetation data across large areas (Moudrý et al. 2023).

Despite the important contribution of remote sensing to the habitat-based monitoring of biodiversity, Sabatini et al. (2016) found that indicators such as stand structural heterogeneity alone do not perform well for estimating overall landscape biodiversity. This is because different taxa respond to a particular set of structural variables in different ways due to their habitat requirements (Burrascano et al. 2023). Complementary use of habitat-based and taxon-based approaches is therefore necessary to enable comprehensive assessment of the status of biodiversity (Blasi et al. 2010; Burrascano et al. 2018) (Fig. 9.5).

In forest ecosystems, few existing studies have focused on combining measurements of habitat-based indicators based on remotely sensed vegetation indices with multi-taxon biodiversity assessments (Vogeler et al. 2022); instead, most research has been limited to individual taxonomic groups, such as vascular plants (Moudrý et al. 2023), birds (Alaniz et al. 2021), and butterflies (de Vries et al. 2021).

Limitations of Monitoring via Remote Sensing and the Way Forward

To summarize, we have shown that remote sensing is an effective technique for detecting processes of landscape fragmentation as well as for evaluating changes in forest landscapes and consequent alterations to biodiversity patterns. It can be used at different spatial scales, is highly repeatable, and facilitates monitoring purposes as data can be easily compared over time. The recent advancements in terms of availability of high-resolution satellite images and global Landsat images (e.g., NASA Geocover dataset; Tucker et al. 2004) enable estimations of productivity using vegetation indices while simultaneously examining the relationships between these estimates and biodiversity indicators (Turner et al. 2015).

Despite the excellent opportunities offered by remote sensing and the long Landsat time-series data in particular, certain limitations should also be considered. First, the spatial resolution of Landsat is not adequate for capturing micro-spatial variations in the distribution of wood-dwelling species, which have poor dispersion capacity, making the Landsat data suitable only for monitoring biodiversity at large-scale resolution. Second, due to the well-known saturation effect of multispectral data, the Landsat sensor is not sensitive to multilayer canopy cover, dense forests, or complex topographic features (Chirici et al. 2020; Vangi et al. 2021; D'Amico et al. 2022) affecting NDVI values. Third, satellite data cannot fully explain biodiversity patterns since it only provides canopy-level information.

The integration of remote sensing approaches and ground monitoring activities within forest monitoring guidelines may overcome these limitations, helping to design and optimize monitoring strategies to tackle forest fragmentation and biodiversity loss in forest ecosystems. Although remote sensing data cannot replace fieldwork or identify individual species along with their rarity and composition, we assume that processing and analysing such data will become highly affordable in the future given the valuable insights provided by these images. In this regard, the availability of the GEE cloud platform allows an unprecedented view of forest areas worldwide.

In conclusion, despite the abovementioned limitations, the provided examples showcase that remote sensing data has great potential for supporting conservation planning and decision making in forest ecosystems. Remote sensing can help to identify hotspots for biodiversity and ecosystem services (de Araujo Barbosa et al. 2015) and even detect climate change refugia (Dubinin et al. 2018), thereby providing practical support for cost-effective biodiversity monitoring and nature-based forest management in complex silvicultural systems.

References

Alaniz AJ, Carvajal MA, Fierro A, Vergara-Rodríguez V, Toledo G, Ansaldo D et al (2021) Remote-sensing estimates of forest structure and dynamics as indicators of habitat quality for Magellanic woodpeckers. Ecol Indic 126:107634

Arekhi M, Yılmaz OY, Yılmaz H, Akyüz YF (2017) Can tree species diversity be assessed with Landsat data in a temperate forest? Environ Monit Assess 189:1–14

Bae S, Levick SR, Heidrich L, Magdon P, Leutner BF, Wöllauer S et al (2019) Radar vision in the mapping of forest biodiversity from space. Nat Commun 10(1):4757

Bauhus J, Puettmann KJ, Kühne C (2013) Close-to-nature forest management in Europe: does it support complexity and adaptability of forest ecosystems? In: Messier C, Puettmann KL, Coates KD (eds) Managing forests as complex adaptive systems: building resilience to the challenge of global change, pp 187–213

Blasi C, Marchetti M, Chiavetta U, Aleffi M, Audisio P, Azzella MM et al (2010) Multi-taxon and forest structure sampling for identification of indicators and monitoring of old-growth forest. Plant Biosyst 144(1):160–170

Brooks E, Wynne RH, Thomas VA, Blinn CE, Coulston J (2014) Exponentially weighted moving average change detection around the country (and the world). In: AGU fall meeting abstracts, vol 2014, p B51L-01

Burrascano S, Lombardi F, Marchetti M (2008) Old-growth forest structure and deadwood: Are they indicators of plant species composition? A case study from central Italy. Plant Biosyst 142(2):313–323

Burrascano S, De Andrade RB, Paillet Y, Ódor P, Antonini G, Bouget C et al (2018) Congruence across taxa and spatial scales: are we asking too much of species data? Glob Ecol Biogeogr 27(8):980–990

Burrascano S, Chianucci F, Trentanovi G, Kepfer-Rojas S, Sitzia T, Tinya F et al (2023) Where are we now with European forest multi-taxon biodiversity and where can we head to? Biol Conserv 284:110176

Cavalli A, Francini S, Cecili G, Cocozza C, Congedo L, Falanga V et al (2022) Afforestation monitoring through automatic analysis of 36-years Landsat best available composites. iForest-Biogeosci For 15(4):220

Ceccherini G, Duveiller G, Grassi G, Lemoine G, Avitabile V, Pilli R, Cescatti A (2020) Abrupt increase in harvested forest area over Europe after 2015. Nature 583(7814):72–77

Chirici G, Giannetti F, McRoberts RE, Travaglini D, Pecchi M, Maselli F et al (2020) Wall-to-wall spatial prediction of growing stock volume based on Italian National Forest Inventory plots and remotely sensed data. Int J Appl Earth Obs Geoinf 84:101959

Cord AF, Brauman KA, Chaplin-Kramer R et al (2018) Priorities to advance monitoring of ecosystem services using earth observation. Ecol Indic 105:577–588

Ćosović M, Bugalho MN, Thom D, Borges JG (2020) Stand structural characteristics are the most practical biodiversity indicators for forest management planning in Europe. Forests 11(3):343

D'Amico G, McRoberts RE, Giannetti F, Vangi E, Francini S, Chirici G (2022) Effects of lidar coverage and field plot data numerosity on forest growing stock volume estimation. Eur J Remote Sens 55(1):199–212

de Araujo Barbosa CC, Atkinson PM, Dearing JA (2015) Remote sensing of ecosystem services: a systematic review. Ecol Indic 52:430–443

de Vries JPR, Koma Z, WallisDeVries MF, Kissling WD (2021) Identifying fine-scale habitat preferences of threatened butterflies using airborne laser scanning. Divers Distrib 27(7):1251–1264

Dubinin V, Svoray T, Dorman M, Perevolotsky A (2018) Detecting biodiversity refugia using remotely sensed data. Landsc Ecol 33:1815–1830

Fahrig L (2003) Effects of habitat fragmentation on biodiversity. Annu Rev Ecol Evol Syst:487–515

Fahrig L, Arroyo-Rodríguez V, Bennett JR, Boucher-Lalonde V, Cazetta E, Currie, DJ et al (2019) Is habitat fragmentation bad for biodiversity?. Biol Conserv 230:179–186

Fischer J, Lindenmayer DB (2007) Landscape modification and habitat fragmentation: a synthesis. Glob Ecol Biogeogr 16(3):265–280

Fletcher Jr RJ, Didham RK, Banks-Leite C, Barlow J, Ewers RM, Rosindell J et al (2018) Is habitat fragmentation good for biodiversity?. Biol Conserv 226:9-15

Forest Europe (2020) State of Europe's forests 2020. Ministerial conference on the protection of forests in Europe. Liaison Unit Bratislava. https://foresteurope.org/wp533content/uploads/2016/08/SoEF_2020.pdf

Fragal EH, Silva TSF, Novo EMLDM (2016) Reconstructing historical forest cover change in the lower Amazon floodplains using the LandTrendr algorithm. Acta Amazon 46:13–24

Francini S, McRoberts RE, Giannetti F, Mencucci M, Marchetti M, Scarascia Mugnozza G, Chirici G (2020) Near-real time forest change detection using PlanetScope imagery. Eur J Remote Sens 53(1):233–244

Francini S, McRoberts RE, Giannetti F, Marchetti M, Scarascia Mugnozza G, Chirici G (2021) The three indices three dimensions (3I3D) algorithm: a new method for forest disturbance mapping and area estimation based on optical remotely sensed imagery. Int J Remote Sens 42(12):4693–4711

Francini S, D'Amico G, Vangi E, Borghi C, Chirici G (2022) Integrating GEDI and Landsat: spaceborne lidar and four decades of optical imagery for the analysis of forest disturbances and biomass changes in Italy. Sensors 22(5):2015

Francini S, Hermosilla T, Coops NC, Wulder MA, White JC, Chirici G (2023) An assessment approach for pixel-based image composites. ISPRS J Photogramm Remote Sens 202:1–12

Francini S, Cavalli A, D'Amico G, McRoberts RE, Maesano M, Munafò M et al (2023a) Reusing remote sensing-based validation data: comparing direct and indirect approaches for afforestation monitoring. Remote Sens (Basel) 15(6):1638

Giannetti F, Pegna R, Francini S, McRoberts RE, Travaglini D, Marchetti M et al (2020) A new method for automated clearcut disturbance detection in Mediterranean coppice forests using Landsat time series. Remote Sens (Basel) 12(22):3720

Gomes VC, Queiroz GR, Ferreira KR (2020) An overview of platforms for big earth observation data management and analysis. Remote Sens (Basel) 12(8):1253

Gorelick N, Hancher M, Dixon M, Ilyushchenko S, Thau D, Moore R (2017) Google earth engine: planetary-scale geospatial analysis for everyone. Remote Sens Environ 202:18–27

Graf RF, Bollmann K, Suter W, Bugmann H (2005) The importance of spatial scale in habitat models: capercaillie in the Swiss Alps. Landsc Ecol 20:703–717

Guo Y, Li J, Chen X et al (2019) The applicability of vegetation indices in assessing habitat quality: a review. Ecol Indic 105:440–449

Hamunyela E, Reiche J, Verbesselt J, Herold M (2017) Using space-time features to improve detection of forest disturbances from Landsat time series. Remote Sens (Basel) 9(6):515

Hansen MC, Potapov PV, Moore R et al (2013) High-resolution global maps of 21st-century forest cover change. Science 342(6160):850–853

Hao Y, Lei Y, Liu X et al (2019) Monitoring land use change and ecosystem service value in a typical karst area of Southwest China. Ecol Indic 103:88–98

Hedwall P-O, Holmström E, Lindbladh M, Felton A (2019) Concealed by darkness: how stand density can override the biodiversity benefits of mixed forests. Ecosphere 10(8):e02835

Herniman S, Coops NC, Martin K, Thomas P, Luther JE, van Lier OR (2020) Modelling avian habitat suitability in boreal forest using structural and spectral remote sensing data. Remote Sens Appl Soc Environ 19:100344

Huang C, Goward SN, Masek JG, Thomas N, Zhu Z, Vogelmann JE (2010) An automated approach for reconstructing recent forest disturbance history using dense Landsat time series stacks. Remote Sens Environ 114(1):183–198

Hudak AT, Bright BC, Kennedy RE (2013) Predicting live and dead basal area from LandTrendr variables in beetle-affected forests. In: MultiTemp 2013: 7th international workshop on the analysis of multi-temporal remote sensing images. IEEE, pp 1–4

Hughes MJ, Kaylor SD, Hayes DJ (2017) Patch-based forest change detection from Landsat time series. Forests 8(5):166

Kacic P, Kuenzer C (2022) Forest biodiversity monitoring based on remotely sensed spectral diversity—a review. Remote Sens (Basel) 14(21):5363

Kennedy RE, Yang Z, Cohen WB (2010) Detecting trends in forest disturbance and recovery using yearly Landsat time series: 1. LandTrendr—temporal segmentation algorithms. Remote Sens Environ 114(12):2897–2910

Kennedy RE, Yang Z, Cohen WB, Pfaff E, Braaten J, Nelson P (2012) Spatial and temporal patterns of forest disturbance and regrowth within the area of the northwest forest plan. Remote Sens Environ 122:117–133

Kennedy RE, Yang Z, Gorelick N, Braaten J, Cavalcante L, Cohen WB, Healey S (2018) Implementation of the LandTrendr algorithm on google earth engine. Remote Sens (Basel) 10(5):691

Knuff AK, Staab M, Frey J, Dormann CF, Asbeck T, Klein AM (2020) Insect abundance in managed forests benefits from multi-layered vegetation. Basic Appl Ecol 48:124–135

Koivula M, Felton A, Jönsson M, Löfroth T, Høistad Schei F, Siitonen J, Sjögren J (2025) Continuous cover forestry in boreal Nordic countries: biodiversity. Continuous Cover Forestry in Boreal Nordic Countries

Lindenmayer DB, Hobbs RJ (eds) (2008) Managing and designing landscapes for conservation: moving from perspectives to principles. Wiley

Lindenmayer D, Hobbs RJ, Montague-Drake R, Alexandra J, Bennett A, Burgman M et al (2008) A checklist for ecological management of landscapes for conservation. Ecol Lett 11(1):78–91

Lindenmayer DB, Barton PS, Lane PW, Westgate MJ, McBurney L, Blair D et al (2014) An empirical assessment and comparison of species-based and habitat-based surrogates: a case study of forest vertebrates and large old trees. PLoS One 9(2):e89807

Liu Y, Wu J, Duan Y et al (2021) Patch analysis in fragmented habitats: a review of methods and applications. Ecol Indic 122:107319

Madžule L, Brūmelis G, Tjarve D (2012) Structures determining bryophyte species richness in a managed forest landscape in boreo-nemoral Europe. Biodivers Conserv 21:437–450

Mazziotta A, Heilmann-Clausen J, Bruun HH, Fritz Ö, Aude E, Tøttrup AP (2016) Restoring hydrology and old-growth structures in a former production forest: modelling the long-term effects on biodiversity. For Ecol Manage 381:125–133

Mazziotta A, Borges P, Kangas A, Halme P, Eyvindson K (2023) Spatial trade-offs between ecological and economical sustainability in the boreal production forest. J Environ Manage 330:117144

McElhinny C, Gibbons P, Brack C, Bauhus J (2005) Forest and woodland stand structural complexity: its definition and measurement. For Ecol Manage 218(1–3):1–24

McGarigal K (2015) FRAGSTAT help. University of Massachusetts, Amherst, p 182

Mönkkönen M, Juutinen A, Mazziotta A, Miettinen K, Podkopaev D, Reunanen P et al (2014) Spatially dynamic forest management to sustain biodiversity and economic returns. J Environ Manage 134:80–89

Moning C, Werth S, Dziock F, Bässler C, Bradtka J, Hothorn T, Müller J (2009) Lichen diversity in temperate montane forests is influenced by forest structure more than climate. For Ecol Manage 258(5):745–751

Moudrý V, Cord AF, Gábor L, Laurin GV, Barták V, Gdulová K et al (2023) Vegetation structure derived from airborne laser scanning to assess species distribution and habitat suitability: the way forward. Divers Distrib 29(1):39–50

Muys B, Angelstam P, Bauhus J, Bouriaud L, Jactel H, Kraigher H et al (2022) Forest biodiversity in Europe. European Forest Institute, Joensuu, p 79

Nordén J, Penttilä R, Siitonen J, Tomppo E, Ovaskainen O (2013) Specialist species of wood-inhabiting fungi struggle while generalists thrive in fragmented boreal forests. J Ecol 101(3):701–712

Oettel J, Lapin K (2021) Linking forest management and biodiversity indicators to strengthen sustainable forest management in Europe. Ecol Indic 122:107275

Oettel J, Zolles A, Gschwantner T, Lapin K, Kindermann G, Schweinzer KM et al (2023) Dynamics of standing deadwood in Austrian forests under varying forest management and climatic conditions. J Appl Ecol 60(4):696–713

Ozdemir I, Mert A, Ozkan UY, Aksan S, Unal Y (2018) Predicting bird species richness and microhabitat diversity using satellite data. For Ecol Manage 424:483–493

Palahí M, Valbuena R, Senf C, Acil N, Pugh TA, Sadler J et al (2021) Concerns about reported harvests in European forests. Nature 592(7856):E15–E17

Palmero-Iniesta M, Pino J, Pesquer L, Espelta JM (2021) Recent forest area increase in Europe: expanding and regenerating forests differ in their regional patterns, drivers and productivity trends. Eur J For Res 140(4):793–805

Parisi F, Vangi E, Francini S, Chirici G, Travaglini D, Marchetti M, Tognetti R (2022) Monitoring the abundance of saproxylic red-listed species in a managed beech forest by Landsat temporal metrics. For Ecosyst 9:100050

Parisi F, Mazziotta A, Vangi E, Tognetti R, Travaglini D, Marchetti M, D´Amico G, Francini S, Borghi C, Chirici G (2023) Exposure elevation and forest structure predict the abundance of saproxylic beetles' communities in mountain managed beech forests. iFOREST 16(3):155–164

Parisi F, Mazziotta A, Chirici G, D'amico G, Vangi E, Francini S, Travaglini D (2024) Effects of forest management on beetle (Coleoptera) communities in beech forests (Fagus sylvatica) in the Apennines of Central Italy (Tuscany). Forests 15(7):1085

Paillet Y, Zapponi L, Schall P, Monnet JM, Ammer C, Balducci L et al (2024) One to rule them all? Assessing the performance of sustainable forest management indicators against multitaxonomic data for biodiversity conservation. Biol Conserv 300:110874

Pekel JF, Cottam A, Gorelick N, Belward AS (2016) High-resolution mapping of global surface water and its long-term changes. Nature 540(7633):418–422

Peura M, Burgas D, Eyvindson K, Repo A, Mönkkönen M (2018) Continuous cover forestry is a cost-efficient tool to increase multifunctionality of boreal production forests in Fennoscandia. Biol Conserv 217:104–112

Rautio P, Routa J, Huuskonen S, Holmström E, Cedergren J, Kuehne C (2025) Continuous cover forestry in boreal Nordic countries (p. 292). Springer Nature

Ruokolainen A, Shorohova E, Penttilä R, Kotkova V, Kushnevskaya H (2018) A continuum of dead wood with various habitat elements maintains the diversity of wood-inhabiting fungi in an old-growth boreal forest. Eur J For Res 137:707–718

Rossi F, Breidenbach J, Puliti S, Astrup R, Talbot B (2019) Assessing harvested sites in a forested boreal mountain catchment through global forest watch. Remote Sens (Basel) 11(5):543

Sabatini FM, Burrascano S, Azzella MM, Barbati A, De Paulis S, Di Santo D et al (2016) One taxon does not fit all: Herb-layer diversity and stand structural complexity are weak predictors of biodiversity in Fagus sylvatica forests. Ecol Indic 69:126–137

Schwieder M, Mueller M, Huttinger C et al (2019) Earth observation based ecosystem services assessments: opportunities and challenges in a transdisciplinary approach. Ecol Indic 98:76–87

Tayyebi A, Wilkinson B, Goodrich DC (2020) An overview of remote sensing for environmental monitoring and management. Environ Monit Assess 192(2):121

Thom D, Rammer W, Dirnböck T, Müller J, Kobler J, Katzensteiner K et al (2017) The impacts of climate change and disturbance on spatio-temporal trajectories of biodiversity in a temperate forest landscape. J Appl Ecol 54(1):28–38

Tucker CJ, Grant DM, Dykstra JD (2004) NASA's global orthorectified Landsat data set. Photogramm Eng Remote Sens 70(3):313–322

Turner W, Rondinini C, Pettorelli N, Mora B, Leidner AK, Szantoi Z et al (2015) Free and open-access satellite data are key to biodiversity conservation. Biol Conserv 182:173–176

Undin M, Atrena A, Carlsson F, Edman M, Jonsson BG, Sandström J (2022) To what extent does surrounding landscape explain stand-level occurrence of conservation-relevant species in fragmented boreal and hemi-boreal forest?–a systematic review protocol. Environ Evid 11(1):1–14

Vangi E, D'Amico G, Francini S, Giannetti F, Lasserre B, Marchetti M et al (2021) The effect of forest mask quality in the wall-to-wall estimation of growing stock volume. Remote Sens (Basel) 13(5):1038

Verbesselt J, Zeileis A, Herold M (2012) Near real-time disturbance detection using satellite image time series. Remote Sens Environ 123:98–108

Vogeler AVB, Otte I, Ferger S, Helbig-Bonitz M, Hemp A, Nauss T et al (2022) Associations of bird and bat species richness with temperature and remote sensing-based vegetation structure on a tropical mountain. Biotropica 54(1):135–145

Wang R, Gamon JA (2019) Remote sensing of terrestrial plant biodiversity. Remote Sens Environ 231:111218

Woodcock CE, Allen R, Anderson M, Belward A, Bindschadler R, Cohen W et al (2008) Free access to Landsat imagery. Science 320:1011

Yang Y, Erskine PD, Lechner AM, Mulligan D, Zhang S, Wang Z (2018) Detecting the dynamics of vegetation disturbance and recovery in surface mining area via Landsat imagery and LandTrendr algorithm. J Clean Prod 178:353–362

Zald HS, Wulder MA, White JC, Hilker T, Hermosilla T, Hobart GW, Coops NC (2016) Integrating Landsat pixel composites and change metrics with lidar plots to predictively map forest structure and aboveground biomass in Saskatchewan, Canada. Remote Sens Environ 176:188–201

Zeller L, Förster A, Keye C, Meyer P, Roschak C., Ammer C (2023) What does literature tell us about the relationship between forest structural attributes and species richness in temperate forests?–A review. Ecol Indic 153:110383

Zellweger F, Braunisch V, Baltensweiler A, Bollmann K (2013) Remotely sensed forest structural complexity predicts multi species occurrence at the landscape scale. For Ecol Manage 307:303–312

Zhu Z (2017) Change detection using Landsat time series: a review of frequencies, preprocessing, algorithms, and applications. ISPRS J Photogramm Remote Sens 130:370–384

Zhu Z, Woodcock CE (2014) Continuous change detection and classification of land cover using all available Landsat data. Remote Sens Environ 144:152–171

Open Access This chapter is licensed under the terms of the Creative Commons Attribution 4.0 International License (http://creativecommons.org/licenses/by/4.0/), which permits use, sharing, adaptation, distribution and reproduction in any medium or format, as long as you give appropriate credit to the original author(s) and the source, provide a link to the Creative Commons license and indicate if changes were made.

The images or other third party material in this chapter are included in the chapter's Creative Commons license, unless indicated otherwise in a credit line to the material. If material is not included in the chapter's Creative Commons license and your intended use is not permitted by statutory regulation or exceeds the permitted use, you will need to obtain permission directly from the copyright holder.

Habitat Quality and Quantity: Features and Metrics

10

Marcin K. Dyderski, Shubhadeep Roychoudhury,
Katharina Lapin, Janine Oettel, and Martin Braun

Case study Poland-Fallen tree, Bialowieza Forest, Bialowieza National Park, Poland (Photo: Maciej/Adobe Stock)

M. K. Dyderski (✉)
Department of Ecology, Institute of Dendrology, Polish Academy of Sciences, Kórnik, Poland
e-mail: mdyderski@man.poznan.pl

S. Roychoudhury
Department of Life Science and Bioinformatics, Assam University, Silchar, India

K. Lapin · J. Oettel · M. Braun
Department of Forest Biodiversity & Nature Conservation, Austrian Research Centre for Forests, Vienna, Austria
e-mail: katharina.lapin@bfw.gv.at; janine.oettel@bfw.gv.at; martin.braun@bfw.gv.at

© The Author(s) 2025
K. Lapin et al. (eds.), *Ecological Connectivity of Forest Ecosystems*,
https://doi.org/10.1007/978-3-031-82206-3_10

Abstract

Habitat quality and quantity play a vital role in maintaining ecosystems and populations of target species, and a deep understanding of features and metrics within the landscape is required to determine them. This chapter reviews the complexities involved in the assessment of these features and metrics to support evidence-based conservation strategies and long-term ecosystem sustainability. Evaluating habitat quality is related to structural diversity, management, natural disturbance legacy, and species richness and diversity as well as the presence of indicator or umbrella species. Generally, complex stand structures and the abundance of veteran trees and deadwood indicate ecosystems with higher levels of overall biodiversity and stability. The non-linear relationship between population viability and habitat size emphasizes the need for landscape-level management. Viability decreases significantly when the ecological capacity for a minimum viable population is exceeded, and recognizing this tipping point is therefore crucial for evidence-based conservation. This means that habitat size assessed as a single variable is insufficient to determine habitat quality, and a wider range of metrics like structural diversity and connectivity should be considered in population management at the landscape level.

Keywords

Island biogeography · Landscape ecology · Forest management · Connectivity · Community ecology · Structural diversity · Population viability

Introduction

A habitat is a specific area or environment representing a physical location wherein a particular organism or community of organisms naturally resides—for example, a forest area in which a community of living organisms thrives. It includes the surrounding abiotic (such as temperature, humidity, and soil type) and biotic factors (such as other species, food sources, and predators) that provide necessary resources and conditions for survival, growth, and reproduction. Habitats can vary widely in size and complexity, ranging from small tree cavities to vast ocean ecosystems or terrestrial forests. Each habitat has its distinct characteristics and supports a unique set of organisms that have adapted to its specific conditions. From the viewpoint of conservation biology, an entire forest area covered by various communities of living organisms may be referred to as a habitat. Smaller structures known as mesohabitats are physiographic or physiognomic features of habitats commonly comprising clusters of microhabitats, which are the smallest landscape unit making up the regional landscape mosaic (Vitt and Belland 1997). For example, in an oak forest habitat, streams, rocks, and trees are mesohabitats while logs, stumps, and tree cavities are microhabitats (Fig. 10.1). Smaller species and their communities are more affiliated with meso- and microhabitats, allowing characteristics of these smaller habitat structures to be used in their assessment.

10 Habitat Quality and Quantity: Features and Metrics

Fig. 10.1 Nested occurrence of meso- and microhabitats within a habitat. A habitat (*Fagus sylvatica* mountain forest) is composed of mesohabitats (red frames) like streams, boulders, logs, and trees; each individual mesohabitat (e.g., a tree) can include microhabitats (smaller images, blue arrows)—e.g., (**a**) cavity nests, (**b**) trunk-base rot holes and epiphytes, or (**c**) dendrotelmata (tree hollow accumulating rainwater)

Assessing forest habitats involves both quantitative and qualitative approaches. The quantitative approach rooted in landscape ecology focuses on evaluating habitat extent, shape, and spatial relationships within the broader landscape matrix. While this approach provides valuable information on the presence of habitats, it does not capture the qualitative characteristics necessary to determine their suitability for specific species or organism guilds.

This chapter delves into the features and metrics used to evaluate both habitat quality and quantity. By examining the evolving understanding of habitats and their conservation, it aims to explain the complexities involved in assessing and managing these vital ecosystems. Understanding the features and metrics related to habitat quality and quantity is crucial for effective conservation strategies and the long-term sustainability of natural environments. In order to inform decision-making with regard to the protection of valuable habitats and the enhancement of ecosystem connectivity, several descriptors of habitat quantity and quality will be introduced, and their respective relationships and conservation implications described (see Table 10.1).

Table 10.1 Quantitative indicators of habitat quality and quantity

Parameter	Indicator	Significance	References
Habitat size	Size of patch	Positively related to population size, viability, and species richness	(Banul et al. 2018; Gibb and Hochuli 2002; Godefroid and Koedam 2003; Hodgson et al. 2011; Weibull and Rydin 2005)
Habitat shape	Length of perimeter, shape indices (e.g., perimeter–area ratio, circularity, perimeter expansion index)	Modifies area effects on populations and communities, determines edge effect sizes	(Banul et al. 2018; Brosi et al. 2007; Cherkaoui et al. 2009; Ewers and Didham 2005)
Edge effects	Fraction of edge zone, related to habitat area, level of fragmentation (e.g., edge distance, edge density)	Ecotone area, where conditions typical of habitat transition into conditions typical of landscape matrix, decreased habitat quality and population size	(Babak and He 2009; Govaert et al. 2020; Jędrzejewski et al. 1994; Meeussen et al. 2020; Riitters et al. 2016)
Temporal dynamics	Microhabitat lifespan, decomposition rate, stand age, rotation age	Number of generations, ability to complete the life cycle of a generation, habitat stability	(Barkman 1958; Gdula et al. 2021; Snäll et al. 2004; Wesołowski 2011)
Neighborhood characteristics	Proportion of particular types of habitats in a buffer around the given habitat, habitat suitability (from species distribution models) within the matrix	Matrix permeability, colonization by species from other habitats, alteration of environmental conditions, connectivity	(Aslan et al. 2012; Czortek and Pielech 2019; Kopeć et al. 2011; Thiele et al. 2008)
Isolation	Shortest distance to next habitat within landscape matrix, shortest distance to source population, presence of biogeographic (scale-relevant) barriers	Dispersal limitation, colonization probability, population genetic diversity	(Ashrafzadeh et al. 2018; Dzwonko 1993; Kirschner et al. 2020; Ricketts 2001)
Connectivity	Number and quality of corridors or stepping stones, level of gene flow between patches	Overcoming negative effects of isolation	(Baldwin et al. 2010; Gilbert-Norton et al. 2010; Snäll et al. 2004; Thiele et al. 2008)
Spatial heterogeneity	Stand age, basal area, dead tree volume and diameters, stand structure index, diameter structure	Habitat and microhabitat availability, connectivity	(Lassauce et al. 2011; Oettel et al. 2023; Oettel and Lapin 2021; Sabatini et al. 2015; Wyka et al. 2023)

(continued)

Table 10.1 (continued)

Parameter	Indicator	Significance	References
Disturbance legacy	Presence of pit-and-mound structures, charcoal and fire scars, landslides, regeneration clumps	Niche availability for (re)colonization, habitat stability	(Clarke et al. 2015; Czortek et al. 2018; Godziek and Pawlik 2023; Johnstone et al. 2010; Marozas et al. 2007)
Management legacy	Number of stumps, stand structure (diameter distribution), clear-cut area, mean clear-cut area, proportion of stand removal	Niche availability, habitat quality, niche persistence	(Chabrerie et al. 2008; Oettel and Lapin 2021; Orczewska et al. 2019)

Quantitative Features of Habitat

Habitat size is the basic metric of habitat quantity within a landscape matrix (Table 10.1 and Fig. 10.2). For forest habitats, it refers to the area covered by a particular forest type and determines population sizes (Banul et al. 2018; Hanski 1997). This relationship between species abundance and habitat size has to do with the space required by each individual (Wesołowski 2007), respectively, by communities (Dengler et al. 2009; Godefroid and Koedam 2003). Species sensitivity to habitat size is greater in higher trophic levels, as its occurrence is determined by prey availability (Gibb and Hochuli 2002). Habitat size matters not only for forest habitats but also for meso- and microhabitats (Fig. 10.1). Microhabitat size determines the number of species (Weibull and Rydin 2005) as well as the probability of species occurrence, even in microhabitats as small as oak acorns (Myczko et al. 2018).

The size of a habitat and its spatial arrangement in the landscape matrix play a crucial role in determining its ability to sustain a specific number of individuals, thereby influencing whether it can serve as a population source or sink. The concept of source and sink populations stems from population ecology and refers to the dynamics of individuals moving between different habitat patches (Furrer and Pasinelli 2016). Habitat size directly impacts the availability of resources such as food, shelter, and breeding sites within a given area. Larger habitats generally have greater resource availability and can thus support larger populations of organisms. Consequently, such habitats are more likely to accommodate source populations producing a surplus of individuals that can disperse and contribute to other habitats within the landscape. On the other hand, smaller habitats with limited resources may only support a smaller number of individuals. These habitats may act as population sinks, meaning that they rely on immigration from source populations to maintain their population size. The capacity of a habitat to sustain a particular number of individuals is influenced not only by its size but also by factors such as habitat quality, fragmentation, and connectivity. This is related to matrix permeability, i.e. ability of the matrix to be penetrated by a target species. Although a matrix cannot

Fig. 10.2 Spatial arrangement of forest areas in various metrics: (**a**) linear woodland patches along streams in the Masai Mara savanna (Kenya) with linear shapes resulting in a large proportion of edge zone; (**b**) medium and large forest complexes in Beskid Niski (Poland); (**c**) forest island in a reservoir created in Khao Sok National Park (Thailand), an example of an impermeable matrix; (**d**) woodland patches in Babki Forest Inspectorate (Poland) consisting mostly of edge zone despite their round shape, due to their small size

serve as a habitat for a particular species, its properties determine whether it is a barrier or not to species movement. This can be related to either abiotic (e.g., insolation or flooding) or biotic conditions (e.g., shelter by vegetation or predator activity). High-quality habitats with abundant resources and suitable conditions can support larger populations even in smaller areas, while poor-quality habitats may offer limited capacity regardless of their size (MacArthur and Wilson 2001).

In addition, the population size of a habitat affects its genetic diversity (Prober and Brown 1994; Tsuzuki et al. 2022) and therefore determines the minimum viable population, i.e. the minimum number of individuals needed to sustain a population for a certain time (Shaffer 1981). Smaller populations are more exposed to three random fluctuations: i) demographic, related to the probability of population decline if a new generation is of a single sex; ii) genetic, connected to a higher probability of lethal alleles and homozygotic recessive genes occurring; and iii) environmental, relating to natural catastrophes and the variability of environmental conditions (e.g., cold winters or dry summers). The relationship between population viability and habitat size is non-linear, however, decreasing drastically beyond the threshold of ecological capacity for the minimum viable population.

Habitat area and shape determine the ratio between core and edge zones (Fig. 10.2) within a habitat (Banul et al. 2018). In forests, the core zone represents conditions typical of the forest interior, related to the presence of two biologically active surfaces (the soil with the understory and the canopy). The canopy layer intercepts solar radiation and precipitation (Breuer et al. 2003; Jagodziński et al. 2019; Niinemets 2010), buffering temperatures and humidity (von Arx et al. 2012; Zellweger et al. 2020) and thereby affecting all dependent organisms (De Lombaerde et al. 2020; Jagodziński et al. 2018; Mueller et al. 2016) as well as moderating ecosystem functioning (Govaert et al. 2021; Hobbie et al. 2010; Rawlik et al. 2019). Edge zones significantly differ from the interior in terms of greater light availability (Niinemets 2010), less stable microclimate (M. Schmidt et al. 2017, 2019), and a more diverse structure with lower tree heights, higher density, and higher shrub cover (Meeussen et al. 2020; Wyka et al. 2023). As a transitional zone between forest and non-forest vegetation, forest edges frequently feature an outer belt of shrubs and tall herbs (Ellenberg 1988; Govaert et al. 2020) hosting generalists and specialists, with an increasing alpha diversity (Wesołowski et al. 2022) and forest understory vegetation (Govaert et al. 2020).

Forest fragmentation leads to a decrease in total forest area accompanied by an increase in edge zones and a decrease in core zones required by forest specialists (Riitters et al. 2016). Forest edges can also affect the reproductive success and behavior of species (Jędrzejewski et al. 1994). For example, edges are suboptimal habitats for forest specialists (Babak and He 2009). The shape of a forest habitat directly influences the proportion and distribution of core areas as relatively undisturbed regions of the forest and edge areas (Banul et al. 2018; Ewers and Didham 2005). Various metrics are employed to quantify habitat shape, most of which are based on the level of perimeter complexity as well as directional evenness (Hesselbarth et al. 2019). Forest habitats with regular shapes and straight borders have a higher proportion of core zone and less edge zone. The latter usually comprise a zone deep by a one stand height. Increasing complexity or length of the habitat border increases the area subject to edge effects (Cherkaoui et al. 2009; Ewers and Didham 2005). This means that two habitats with the same area, with one having an elongated shape and the other a round shape, can be occupied by different communities (Brosi et al. 2007; Cherkaoui et al. 2009). The effect of habitat shape decreases with greater habitat area since the absolute area of the interior zone increases, making it large enough to allow the corresponding species to thrive even in the case of a high proportion of edge zone (Banul et al. 2018).

Temporal patterns of habitats are also crucial determinants of their biota (Fig. 10.3). Succession dynamics shape the replacement of generalists by specialists (Connell and Slatyer 1977; Walker and Chapin 1987), which is related to specific forest conditions including microclimate and light availability (Dzwonko 2001; Zellweger et al. 2020). This is especially important for dispersal-limited specialists, which struggle to quickly recolonize deforested sites (De Frenne et al. 2011; Orczewska 2009). Stand age, frequently used as a metric of habitat age, determines the diversity of numerous groups of organisms (del Moral and Wood 1993; Fritz et al. 2009; Majer et al. 2007; Prach et al. 2001). Furthermore, it modifies the

Fig. 10.3 Veteran trees and deadwood are important habitats for numerous organisms. (**a**) An enormously large *Fraxinus excelsior*, protected in Białowieża Forest Strict Reserve (Poland); (**b**) a large *Quercus robur*, remnant of a riparian forest, is conserved as a sacred tree in the orthodox monastery at Jabłeczna (Poland); two large dead logs are historical trees from Białowieża Forest: (**c**) Jagiełło's oak (*Quercus robur*), estimated to be 450 years old and knocked over by wind in 1974 and (**d**) Scots pine (*Pinus sylvestris*) used as a hive, with remnants of the hive entrance, a tree died before 1888 when bee hiving was abandoned

differentiation of spatial structure as well as being crucial for the development of specific meso- and microhabitats (Bütler et al. 2020; Snäll et al. 2004). Temporal patterns are also observed in the colonization of meso- and microhabitats—e.g., decomposing litter (Urbanowski et al. 2018) or perennial fungal fruiting bodies (Gdula et al. 2021). Lifespan is crucial for certain species colonizing these habitats. For example, wood decomposes at mass loss rates of 28.2%, 6.3%, and 3.3% per year in tropical, temperate, and boreal biomes, respectively (Seibold et al. 2021), which affects the attractiveness of the habitat for particular species. Similarly, non-excavated holes used as nests by birds have a median lifespan of 12 years, while those excavated by woodpeckers last only 7 to 10 years (Wesołowski 2011, 2012), thus affecting temporal availability for birds. Furthermore, tree species with loose and easily shedding bark host fewer epiphytes than those with more persistent bark (Barkman 1958; Jagodziński et al. 2018).

Qualitative Features of Habitats

Quality refers to the structural and functional features of habitats that support biodiversity and ecosystem processes, improving the persistence and growth of target species populations (Hodgson et al. 2011). These factors are species- or guild-specific, depending on individual ecological requirements (Austin 2013; Carroll et al. 2011; Oksanen and Minchin 2002). For this reason, it is crucial to define target groups of organisms and relate habitat quality to their requirements. Many global environmental changes are common processes that reduce habitat quality for various groups of organisms, like pollution, fragmentation, biological invasions, or overharvesting. Moreover, numerous groups of organisms are not sufficiently recognized or their identification requires considerable labor effort and expertise (Kiebacher et al. 2016). For this reason, general descriptors of habitat quality linked to greater diversity of specific taxa and umbrella species are utilized as metrics to measure the effectiveness of conservation efforts. The presence and viability of indicator species within a habitat are predictors for the ecological integrity and suitability of the environment for a range of other species. By focusing conservation efforts on protecting and managing habitats that support umbrella species (Roberge and Angelstam 2004)—that is, widely known species sharing ecological requirements with many other less recognized species—numerous other species with the same ecological requirements can indirectly be safeguarded. By linking information about the presence and abundance of these species with general descriptors of habitat quality, conservationists can obtain insights on the overall health and biodiversity of studied habitats, evaluate the success of conservation initiatives, and make informed decisions regarding the management and protection of habitats. Metrics for habitat quality enable targeted actions to preserve biodiversity and maintain ecological balance (Oettel and Lapin 2021). The former is based on niche availability and the assumption that the presence of conditions supporting particular taxa is crucial for maintaining their populations. This can be expressed by various indicators relating to nutrient availability, regeneration niches, biodiversity, structural diversity, or management legacy and intensity (Oettel and Lapin 2021). By contrast, the concept of umbrella species (Roberge and Angelstam 2004) assumes that monitoring and conserving easily identifiable and often popular species can be used as a proxy for wider groups of species with similar ecological requirements. This approach allows the conclusion that forest habitat quality determines management thresholds (Oettel and Lapin 2021) or predicts guild responses to climate change (Wierzcholska et al. 2020).

Trees are foundational elements of a forest ecosystem (Ellison et al. 2005) that determine its functioning. Owing to their biomass dominance and longevity, trees regulate the rate of matter cycling via species-specific chemical composition and traits of litterfall (Godoy et al. 2010; Hobbie et al. 2006; Horodecki et al. 2019), as well as via solar energy access to the soil and understory (Jagodziński et al. 2019; Niinemets 2010). Dominant tree species thus determine the composition of dependent organisms as well as the structure of microhabitats. For example, wood density and chemical composition affect the lifespan of deadwood and the succession of insects, fungi, and bryophytes (Chećko et al. 2015; Štursová et al. 2014) while bark water capacity, pH, and tree lifespan affect the composition of epiphytic bryophytes and lichens (Barkman 1958; Jagodziński et al. 2018; Łubek et al. 2020). As a result, habitat quality is largely determined by the dominant tree species, and the presence of geographically alien tree species can alter habitat suitability for native species (Dyderski and Jagodziński 2021; García et al. 2023; Wohlgemuth et al. 2022). Similarly, planting native tree species mismatched to local soil conditions can also affect biodiversity by decreasing habitat quality for particular species. An example is the planting of coniferous trees in habitats typical for broadleaved forests (Felton et al. 2010), which not only affects dependent biota (Pharo and Lindenmayer 2009; Zerbe and Wirth 2006) but also acidifies the soil and leaches nutrients from it (Augusto et al. 2002; Finzi et al. 1998). Therefore, assessments of habitat quality in terms of stand species composition should be not only quantitative (i.e., based on species richness and diversity) but also qualitative (i.e., based on tree species identity or functional traits).

Stand structure diversity is crucial for the existence of numerous guilds of organisms (Fig. 10.4). Former forest management practices led to the development of even-aged, single-layer monocultures (Brockerhoff et al. 2008; Felton et al. 2010)

Fig. 10.4 Structural diversity of forests—multi-age, multi-strata, and multi-species: (**a**) temperate lowland primeval forest in Białowieża Strict Reserve (Poland); (**b**) temperate mountain forest in Beskid Niski (Poland); (**c**) tropical coastal forest in Arabuko Sokoke National Park (Kenya); and (**d**) southern boreal mountain forest with *Nothofagus* spp. in Bernardo O'Higgins National Park (Chile)

providing few ecological niches for other organisms. Natural forests are usually characterized by higher species richness, the presence of multiple generations, and greater variability of tree dimensions (Lindenmayer et al. 2006; Pretzsch 2009; Sabatini et al. 2015). Thus, the presence of tree species providing similar substrates for dependent organisms increases the stability of an ecosystem. For example, in Białowieża Primeval Forest in Poland, 88.4% of epiphytic lichens associated with European ash (*Fraxinus excelsior*) also occur on alternative hosts, indicating that ash dieback will not cause cascading local extinction of these lichens (Łubek et al. 2019). Moreover, high trees species richness is usually related to greater understory diversity (Ampoorter et al. 2015).

High structural diversity maintains the presence of both old and young trees, a continuity of regeneration (Dyderski et al. 2023), and diversity in resource availability connected to gap dynamics (Dobrowolska et al. 2022; Doyle 1981). The presence of veteran trees or habitat trees (Fig. 10.3)—i.e., large trees often partially damaged by previous disturbances (Gutowski et al. 2022)—is crucial due to their substantial role in providing numerous microhabitats (Larrieu et al. 2018; Sever and Nagel 2019; Winter et al. 2015) and hosting many specialized species (Fritz et al. 2009; Király et al. 2013). Even solitary broadleaved trees can serve as significant hotspots for birds (Pustkowiak et al. 2021) or bryophytes (Wierzcholska et al. 2018). The diameter distribution of natural stands reveals a J-shaped pattern with a high density of smaller trees, while managed, even-aged stands are characterized by Gaussian distribution of diameter at breast height (DBH) (Johnson 1997; Szmyt et al. 2020; Szmyt and Tarasiuk 2018). Stand structure diversity increases with stand age (Sabatini et al. 2015) and differs between life stages (Li et al. 2023) (Fig. 10.4). Structural diversity can also be expressed by spatial patterns of aggregation (clumping), respectively, uniformity of tree distribution (Szmyt 2014; Szmyt and Tarasiuk 2018).

Another crucial metric for stand structural diversity is the quantity and quality of deadwood (Gutowski et al. 2022; Lassauce et al. 2011; Lindenmayer et al. 2000; Oettel and Lapin 2021). Deadwood quantity depends on tree mortality rate and tree size, but high quantity does not always equate to high quality (Gutowski et al. 2022; Humphrey et al. 2002; Oettel et al. 2023; Oettel and Lapin 2021), since larger logs and snags take longer to decompose compared to smaller ones (Holeksa et al. 2008) (Fig. 10.3). Deadwood serves as a habitat for saproxylic insects and fungi (Gutowski et al. 2022; Lassauce et al. 2011; Štursová et al. 2014) as well as for epixylic plants and lichens (Barkman 1958; Chećko et al. 2015; Humphrey et al. 2002; Király et al. 2013; Wierzcholska et al. 2018), with larger-diameter deadwood enabling the survival of species with long development cycles. Moreover, Kuijper et al. (2013) found that larger logs and snags in Białowieża Primeval Forest can simultaneously serve as a shelter for wolves and escape impediment for ungulates, increasing habitat quality for wolves and shaping regeneration niches for trees by reducing browsing-related mortality.

The presence of veteran trees and deadwood may be a legacy of preceding natural disturbances or past management (Gutowski et al. 2022; Johnstone et al. 2010; Lindenmayer et al. 2000). Disturbances cause damage to a substantial proportion of ecosystem biomass, pushing the ecosystem to regenerate (Fox 1979; Herben et al. 2016; Johnstone et al. 2016) (Fig. 10.5). Severe pollution affects tree growth and leads to mortality, which is reflected in future stand structure (Vacek et al. 2020). Windthrows lead to the development of pit-and-mound structures, increasing the diversity of meso- and microhabitats as well as providing regeneration niches (Czortek et al. 2018; Godziek and Pawlik 2023) (Fig. 10.5). Forest fires juvenilize the stand structure and open the canopy, enabling the regeneration of pioneer trees (Clarke et al. 2015; Johnstone et al. 2010).

The effects of former land use on ecosystems can be long-lasting and require a significant amount of time for recovery (Fig. 10.6). Even after the original land use has ceased, remnants of past activities can persist for centuries, leaving visible traces in the landscape. Examples of these remnants include the presence of road networks, small areas of trees of the same age, or traditional beehives carved in trees, e.g. pines and oaks (Jaroszewicz et al. 2019). More recent legacies can be related to post-thinning stumps and a low amount of deadwood as well as veteran and habitat trees (Baran et al. 2020; Lindenmayer et al. 2000; Oettel and Lapin 2021). The historical land use associated with industrial or agricultural activities can have significant consequences for habitat biodiversity as well. Recolonization

Fig. 10.5 Natural disturbance legacies in forests: (**a**) *Picea abies* forest after a bark-beetle outbreak in the Tatra Mountains (Poland); (**b**) *Betula pendula* forest after waterlogging due to beaver dam building in Poznań (Poland); (**c**) *Betula pubescens-Prunus tremula* pioneer forest emerging in a raised bog drained and burned 30 years ago near Czaplinek (Poland); (**d**) *Fagus sylvatica* cliff forest exposed to chronic winds in Orzechowo Morskie (Poland)

Fig. 10.6 Management legacies in forests: (**a**) post-agricultural *Pinus sylvestris* forest with low stand density and an understory dominated by grasses and shrubs in Ośno Lubuskie (Poland); (**b**) post-mining *Pinus sylvestris* forest with bare soil and low-decomposed litter in Bełchatów (Poland); (**c**) pits left by peat harvesting in raised bog *Betula pubescens* forest near Złocieniec (Poland); and (**d**) uniform age and dimension structure of *Pinus sylvestris* forest after clearcutting in Tleń (Poland)

of such sites from forest remnants is slow, as specialized species usually have slow migration rates (Hermy et al. 1999; Peterken 1974). Such legacies can lead to decreased abundance and numbers of specialized species (De Frenne et al. 2011; Hüttl and Weber 2001; Rawlik et al. 2018; Woźniak et al. 2022) or decreased viability of populations (Woziwoda et al. 2021).

Habitat Quality and Quantity in the Landscape Matrix

Most species are in movement, with individual populations joined within a metapopulation considered a "population" of instable local populations and inhabiting separated patches of habitat (Hanski 1998). Similar to population dynamics, which are shaped by mortality and recruitment, metapopulation is shaped by local extinctions and recolonization. The ability to migrate among subpopulations is therefore crucial for metapopulation persistence (Fig. 10.7). Even migrations of single individuals between subpopulations can minimize the loss of polymorphism and heterozygosity within them (Mills and Allendorf 1996). In contrast to the classical

Fig. 10.7 Examples of forest patch connectivity: (**a**) stepping stones of mature forest remnants in the matrix of *Pinus sylvestris* regeneration in Babki Forest Inspectorate (Poland); (**b**) strip of mature forest serving as a corridor across *Pinus sylvestris* regeneration in Tleń Forest Inspectorate (Poland); (**c**) strips of *Alnus glutinosa* connecting small riparian woodland patches in the matrix of meadows in Gorzkie Pole (Poland); (**d**) linear corridors of riparian forests connecting the city center with suburban forest along the Warta river in Poznań (Poland)

metapopulation model assuming random mortality in equally important subpopulations, metapopulations are now usually considered to comprise source and sink populations with positive and negative demographic balances, respectively. This difference results from various habitat qualities affecting demographic balance and lack of habitat selectiveness in high population densities (Allee 1929). In order to avoid negative density-dependent effects (Janzen 1970; Turczański et al. 2022), individuals colonize suboptimal patches of habitat, creating source subpopulations. Metapopulation can be extended by simplification into island biogeography theory, assuming that the mainland is a source population while more or less separated islands host sink populations (MacArthur and Wilson 2001). This theory assumes that species richness within each habitat island is determined by its area and distance from the mainland (source population). Such an approach can be useful for assessing fragmented forest habitats within a matrix of non-forested areas (Fig. 10.7). However, instead of a single source population, there are often multiple sources potentially connected by ecological corridors (Banul et al. 2018; Forman and Godron 1981).

Assessing habitat spatial structure is important in order to distinguish habitat loss from habitat fragmentation, which is usually a consequence of the former (Hanski 1998). While habitat loss and fragmentation often occur together, their impacts are distinctly different. Habitat loss directly decreases the size and viability of populations, reducing the ecological capacity of the affected area (Ewers and Didham 2005; Hanski 1998), whereas habitat fragmentation is characterized by a decrease in the number and size of core habitat patches, leading to an increase in the proportion of edge area (Fig. 10.7). On the other hand, fragmentation can also reduce the risk of total habitat destruction during catastrophic events. Ultimately, the specific effects of fragmentation thus depend on the spatial structure and connectivity of patches. The varying impacts of habitat quantity, quality, and connectivity on population size and viability pose a dilemma when designing networks for habitat conservation. This dilemma arises from the diverse needs and expectations of stakeholders and the trade-offs between different management approaches within limited land resources. These constraints may prevent the strict conservation of all potential habitats. Consequently, when faced with a limited proportion of habitat available for conservation, a decision must be made whether to conserve a single large patch or multiple smaller patches; this is commonly referred to as "single large or several small" (SLOSS) theory (Diamond 1975; Simberloff and Able 1976; Fahrig et al. 2022). The choice depends on factors such as landscape connectivity, species requirements, and the desired conservation outcomes. Both approaches have their advantages and drawbacks, and the optimal choice depends on the specific context and objectives of the conservation effort. Striking a balance between habitat size, connectivity, and overall conservation of the landscape is crucial for effectively preserving biodiversity and maintaining viable populations in the face of habitat limitations (Cherkaoui et al. 2009; Lomolino 1994; Ovaskainen 2002). Moreover, several small patches will have smaller core zones and more edge zones, while a single large patch will not disperse the risk of destruction during stochastic events. A recent meta-analysis of metacommunities by Riva and Fahrig (2023) found a twofold greater accumulation of species richness when smaller patches than larger ones were conserved, suggesting that biodiversity conservation may be most effective if habitats are composed of as many small patches as possible, plus a few large ones.

Moreover, the position of habitat patches in the landscape matrix determines species migration ability and resource availability (Banul et al. 2018; Forman and Godron 1981; Thiele et al. 2008). Population viability can depend on habitat quantity, habitat quality, and the spatial arrangement of habitats (Hodgson et al. 2011), while habitat connectivity is crucial for effective dispersal (Iverson et al. 2004) and gene flow (Guiller et al. 2023). The effective dispersal distance between subpopulations depends on organism size and dispersal rate. For example, the scale will be in centimeters for soil mesofauna, in meters for bryophytes, and in kilometers for large mammals. Species can migrate directly from one patch to another through the surrounding matrix over shorter distances, or through continuous linear structures (corridors) or discontinuous patches of habitat (stepping stones) that increase potential movement range between patches by 50% (Gilbert-Norton et al. 2010). Ecological corridors can increase the probability of recolonization after catastrophes as well as increasing gene flow and overall habitat quantity. On the other hand, they can also lead to a decline in local adaptation in previously isolated subpopulations and facilitate the spread of diseases and invasive species (see Chap. 17). This can be the case with forest remnants connected by rivers, which spread both forest specialists and ubiquitous invasive species (Burkart 2001; Dyderski et al. 2017; Johansson et al. 1996). Corridors are especially important for specialized species with narrow ecological requirements and limited dispersal rates (Dzwonko 1993; Hermy et al. 1999; Orczewska and Fernes 2011). Their importance also increases when habitats are fragmented by dangerous linear objects like motorways (Garrah et al. 2015; Moore et al. 2023). Matrix quality affects connectivity by increasing the energy required to overcome unsuitable habitats (Ricketts 2001). For this reason, modeling connectivity also requires an assessment of matrix permeability (Baldwin et al. 2010), e.g. using the framework of species distribution models (Ashrafzadeh et al. 2018; K. Schmidt et al. 2023). Moreover, matrix characteristics affect migration and recolonization in the case of isolated patches (Czortek and Pielech 2019; Dyderski et al. 2017).

> **Box 10.1 Epiphytic Bryophytes: Indicators of Habitat Quality and Connectivity**
> Epiphytic bryophytes (Fig. 10.8) are an example of forest specialists indicating high landscape quality and requiring habitat connectivity for persistence (Snäll et al. 2004; Wierzcholska et al. 2018). As poikilohydric organisms, they depend on air and substrate moisture and usually have a narrow ecological amplitude (Rydin 2008). Therefore, their presence depends on bark physical and chemical properties as well as on tree species–specific light interception (Barkman 1958; Jagodziński et al. 2018). Bark properties evolve through tree growth, and some species can thus appear only on large or old trees that offer habitats for them. Despite their high dispersal capacity (e.g., small spores are carried by wind over long distances), bryophytes' persistence in managed forests is limited by a small number of large trees forming habitat patches for them in the forest landscape matrix. As bark can be damaged by biotic and abiotic disturbances, connectivity between patches of suitable habitat determines the size of the source population and its viability. In old-growth forests with many large trees, populations are stable (Fritz et al. 2009; Király et al. 2013). However, given habitat connectivity, ancient woodland specialists can also colonize adjected new forests—even when they are comprised of alien tree species—and cover post-mining sites, forming sink populations (Jagodziński et al. 2018).
>
> (continued)

Fig. 10.8 Epiphytic bryophytes associated with ancient temperate forests: (**a**) *Porella platyphylla*; (**b**) *Dicranum viride*; (**c**) *Frullania dilatata* within a patch of *Homalothecium sericeum*; (**d**) *Homalia trichomanoides*. All images taken in Białowieża National Park (Poland)

Conclusions

This chapter reviewed the role of habitat quality and quantity in maintaining ecosystems and populations of target species, discussing the features and metrics of habitats and their landscape context in general. Habitat quantity depends not only on the size of an occupied habitat patch but also on its shape, proportion of core and edge area, and temporal dynamics. Habitat quality is related to structural diversity, management, natural disturbance legacies, and species richness as well as diversity. The non-linear relationship between population viability and habitat requires the consideration of landscape scale in conservation management. It is necessary to understand the complexities inherent in assessing these features and metrics in order to support evidence-based conservation strategies and long-term ecosystem sustainability.

References

Allee WC (1929) Studies in animal aggregations: natural aggregations of the isopod, Asellus Communis. Ecology 10(1):14–36. https://doi.org/10.2307/1940510

Ampoorter E, Baeten L, Vanhellemont M, Bruelheide H, Scherer-Lorenzen M, Baasch A, Erfmeier A, Hock M, Verheyen K (2015) Disentangling tree species identity and richness effects on the herb layer: first results from a German tree diversity experiment. J Veg Sci 26(4):742–755. https://doi.org/10.1111/jvs.12281

Ashrafzadeh M-R, Khosravi R, Ahmadi M, Kaboli M (2018) Landscape heterogeneity and ecological niche isolation shape the distribution of spatial genetic variation in Iranian brown bears, Ursus arctos (Carnivora: Ursidae). Mamm Biol 93:64–75. https://doi.org/10.1016/j.mambio.2018.08.007

Aslan CE, Rejmánek M, Klinger R (2012) Combining efficient methods to detect spread of woody invaders in urban–rural matrix landscapes: an exploration using two species of Oleaceae. J Appl Ecol 49(2):331–338. https://doi.org/10.1111/j.1365-2664.2011.02097.x

Augusto L, Ranger J, Binkley D, Rothe A (2002) Impact of several common tree species of European temperate forests on soil fertility. Annals of Forest Science 59(3):233–253. https://doi.org/10.1051/forest:2002020

Austin MP (2013) Vegetation and environment: discontinuities and continuities. In: van der Maarel WE, Franklin J (eds) Vegetation ecology. Wiley. http://onlinelibrary.wiley.com, pp 71–106. https://doi.org/10.1002/9781118452592.ch3/summary

Babak P, He F (2009) A neutral model of edge effects. Theor Popul Biol 75(1):76–83. https://doi.org/10.1016/j.tpb.2008.11.002

Baldwin RF, Perkl RM, Trombulak SC, Burwell WB (2010) Modeling ecoregional connectivity. In: Trombulak WSC, Baldwin RF (eds) Landscape-scale conservation planning. Springer, pp 349–367. https://doi.org/10.1007/978-90-481-9575-6_16

Banul R, Kosewska A, Borkowski J (2018) Animal occurrence in fragmented forest habitats—important factors at the patch and landscape scale. For Res Papers 79(1):89–100. https://doi.org/10.48538/FRP-2018-0010

Baran J, Pielech R, Kauzal P, Kukla W, Bodziarczyk J (2020) Influence of forest management on stand structure in ravine forests. For Ecol Manage 463:118018. https://doi.org/10.1016/j.foreco.2020.118018

Barkman JJ (1958) Phytosociology and ecology of cryptogamic epiphytes. Van Gorcum and Company

Breuer L, Eckhardt K, Frede HG (2003) Plant parameter values for models in temperate climates. Ecol Model 169(2):237–293. https://doi.org/10.1016/S0304-3800(03)00274-6

Brockerhoff EG, Jactel H, Parrotta JA, Quine CP, Sayer J (2008) Plantation forests and biodiversity: oxymoron or opportunity? Biodivers Conserv 17(5):925–951. https://doi.org/10.1007/s10531-008-9380-x

Brosi BJ, Daily GC, Shih TM, Oviedo F, Durán G (2007) The effects of forest fragmentation on bee communities in tropical countryside: Bee communities and tropical forest fragmentation. J Appl Ecol 45(3):773–783. https://doi.org/10.1111/j.1365-2664.2007.01412.x

Burkart M (2001) River corridor plants (Stromtalpflanzen) in Central European lowland: a review of a poorly understood plant distribution pattern. Glob Ecol Biogeogr 10(5):449–468. https://doi.org/10.1046/j.1466-822x.2001.00270.x

Bütler R, Lachat T, Krumm F, Kraus D, Larrieu L (2020) Field guide to tree-related microhabitats. Descriptions and size limits for their inventory. Swiss Federal Institute for Forest, Snow and Landscape Research WSL

Carroll IT, Cardinale BJ, Nisbet RM (2011) Niche and fitness differences relate the maintenance of diversity to ecosystem function. Ecology 92(5):1157–1165. https://doi.org/10.1890/10-0302.1

Chabrerie O, Verheyen K, Saguez R, Decocq G (2008) Disentangling relationships between habitat conditions, disturbance history, plant diversity, and American black cherry (*Prunus serotina*

Ehrh.) invasion in a European temperate forest. Divers Distrib 14(2):204–212. https://doi.org/10.1111/j.1472-4642.2007.00453.x

Chećko E, Jaroszewicz B, Olejniczak K, Kwiatkowska-Falińska AJ (2015) The importance of coarse woody debris for vascular plants in temperate mixed deciduous forests. Can J For Res 45(9):1154–1163. https://doi.org/10.1139/cjfr-2014-0473

Cherkaoui I, Selmi S, Boukhriss J, Hamid R-I, Mohammed D (2009) Factors affecting bird richness in a fragmented cork oak forest in Morocco. Acta Oecologica 35(2):197–205. https://doi.org/10.1016/j.actao.2008.10.002

Clarke PJ, Keith DA, Vincent BE, Letten AD (2015) Post-grazing and post-fire vegetation dynamics: long-term changes in mountain bogs reveal community resilience. J Veg Sci 26(2):278–290. https://doi.org/10.1111/jvs.12239

Connell JH, Slatyer RO (1977) Mechanisms of succession in natural communities and their role in community stability and organization. Am Nat:1119–1144

Czortek P, Pielech R (2019) Surrounding landscape influences functional diversity of plant species in urban parks. Urban Forestry and Urban Greening:126525. https://doi.org/10.1016/j.ufug.2019.126525

Czortek P, Ratyńska H, Dyderski MK, Jagodziński AM, Orczewska A, Jaroszewicz B (2018) Cessation of livestock grazing and windthrow drive a shift in plant species composition in the Western Tatra Mts. Tuexenia 37:177–196

De Frenne P, Baeten L, Graae BJ, Brunet J, Wulf M, Orczewska A, Kolb A, Jansen I, Jamoneau A, Jacquemyn H, Hermy M, Diekmann M, De Schrijver A, De Sanctis M, Decocq G, Cousins SAO, Verheyen K (2011) Interregional variation in the floristic recovery of post-agricultural forests. J Ecol 99(2):600–609. https://doi.org/10.1111/j.1365-2745.2010.01768.x

De Lombaerde E, Blondeel H, Baeten L, Landuyt D, Perring MP, Depauw L, Maes SL, Wang B, Verheyen K (2020) Light, temperature and understorey cover predominantly affect early life stages of tree seedlings in a multifactorial mesocosm experiment. For Ecol Manage 461:117907. https://doi.org/10.1016/j.foreco.2020.117907

del Moral R, Wood DM (1993) Early primary succession on the Volcano Mount St. Helens. J Veg Sci 4(2):223–234. https://doi.org/10.2307/3236108

Dengler J, Löbel S, Dolnik C (2009) Species constancy depends on plot size—a problem for vegetation classification and how it can be solved. J Veg Sci 20(4):754–766. https://doi.org/10.1111/j.1654-1103.2009.01073.x

Diamond JM (1975) The Island Dilemma: lessons of modern biogeographic studies for the design of natural reserves. Biol Conserv 7(2):129–146. https://doi.org/10.1016/0006-3207(75)90052-X

Dobrowolska D, Piasecka Ż, Kuberski Ł, Stereńczak K (2022) Canopy gap characteristics and regeneration patterns in the Białowieża forest based on remote sensing data and field measurements. For Ecol Manage 511:120123. https://doi.org/10.1016/j.foreco.2022.120123

Doyle TW (1981) The role of disturbance in the gap dynamics of a montane rain forest: an application of a tropical forest succession model. In: West WDC, Shugart HH, Botkin DB (eds) Forest succession. Springer, pp 56–73. https://doi.org/10.1007/978-1-4612-5950-3_6

Dyderski MK, Jagodziński AM (2021) Impacts of invasive trees on alpha and beta diversity of temperate forest understories. Biol Invasions 23:235–252. https://doi.org/10.1007/s10530-020-02367-6

Dyderski MK, Tyborski J, Jagodziński AM (2017) The utility of ancient forest indicator species in urban environments: a case study from Poznań, Poland. Urban For Urban Green 27:76–83. https://doi.org/10.1016/j.ufug.2017.06.016

Dyderski MK, Pawlik Ł, Chwistek K, Czarnota P (2023) Tree aboveground biomass increment and mortality in temperate mountain forests: tracing dynamic changes along 25-year monitoring period. For Ecol Manage 540:121054. https://doi.org/10.1016/j.foreco.2023.121054

Dzwonko Z (1993) Relations between the floristic composition of isolated young woods and their proximity to ancient woodland. J Veg Sci 4(5):693–698. https://doi.org/10.2307/3236135

Dzwonko Z (2001) Assessment of light and soil conditions in ancient and recent woodlands by Ellenberg indicator values. J Appl Ecol 38(5):942–951. https://doi.org/10.1046/j.1365-2664.2001.00649.x

Ellenberg H (1988) Vegetation ecology of central Europe. Cambridge University Press

Ellison AM, Bank MS, Clinton BD, Colburn EA, Elliott K, Ford CR, Foster DR, Kloeppel BD, Knoepp JD, Lovett GM, Mohan J, Orwig DA, Rodenhouse NL, Sobczak WV, Stinson KA, Stone JK, Swan CM, Thompson J, Von Holle B, Webster JR (2005) Loss of foundation species: consequences for the structure and dynamics of forested ecosystems. Front Ecol Environ 3(9):479–486. https://doi.org/10.1890/1540-9295(2005)003[0479:LOFSCF]2.0.CO;2

Ewers RM, Didham RK (2005) Confounding factors in the detection of species responses to habitat fragmentation. Biol Rev 81(1):117. https://doi.org/10.1017/S1464793105006949

Fahrig L, Watling JI, Arnillas CA, Arroyo-Rodríguez V, Jörger-Hickfang T, Müller J, Pereira HM, Riva F, Rösch V, Seibold S, Tscharntke T, May F (2022) Resolving the SLOSS dilemma for biodiversity conservation: a research agenda. Biol Rev 97(1):99–114. https://doi.org/10.1111/brv.12792

Felton A, Lindbladh M, Brunet J, Fritz Ö (2010) Replacing coniferous monocultures with mixed-species production stands: an assessment of the potential benefits for forest biodiversity in northern Europe. For Ecol Manage 260(6):939–947. https://doi.org/10.1016/j.foreco.2010.06.011

Finzi AC, Canham CD, Van Breemen N (1998) Canopy tree-soil interactions within temperate forests: Species effects on pH and cations. Ecol Appl 8(2):447–454. https://doi.org/10.1890/1051-0761(1998)008[0447:CTSIWT]2.0.CO;2

Forman RTT, Godron M (1981) Patches and structural components for a landscape ecology. BioScience 31(10):733–740. https://doi.org/10.2307/1308780

Fox JF (1979) Intermediate-disturbance hypothesis. Science 204(4399):1344–1345. https://doi.org/10.1126/science.204.4399.1344

Fritz Ö, Niklasson M, Churski M (2009) Tree age is a key factor for the conservation of epiphytic lichens and bryophytes in beech forests. Appl Veg Sci 12(1):93–106. https://doi.org/10.1111/j.1654-109X.2009.01007.x

Furrer RD, Pasinelli G (2016) Empirical evidence for source–sink populations: a review on occurrence, assessments and implications. Biol Rev 91(3):782–795. https://doi.org/10.1111/brv.12195

García RA, Fuentes-Lillo E, Cavieres L, Cóbar-Carranza AJ, Davis KT, Naour M, Núñez MA, Maxwell BD, Lembrechts JJ, Pauchard A (2023) *Pinus contorta* alters microenvironmental conditions and reduces plant diversity in patagonian ecosystems. Diversity 15(3) Article 3. https://doi.org/10.3390/d15030320

Garrah E, Danby RK, Eberhardt E, Cunnington GM, Mitchell S (2015) Hot spots and hot times: wildlife road mortality in a regional conservation corridor. Environ Manag 56(4):874–889. https://doi.org/10.1007/s00267-015-0566-1

Gdula, A. K., Skubała, P., Zawieja, B., and Gwiazdowicz, D. J. (2021). Mite communities (Acari: Mesostigmata, Oribatida) in the red belt conk, *Fomitopsis pinicola* (Polyporales), in Polish forests experimental and applied acarology, 84(3), 543–564. https://doi.org/10.1007/s10493-021-00635-1

Gibb H, Hochuli DF (2002) Habitat fragmentation in an urban environment: large and small fragments support different arthropod assemblages. Biol Conserv 106(1):91–100. https://doi.org/10.1016/S0006-3207(01)00232-4

Gilbert-Norton L, Wilson R, Stevens JR, Beard KH (2010) A meta-analytic review of corridor effectiveness. Conserv Biol 24(3):660–668. https://doi.org/10.1111/j.1523-1739.2010.01450.x

Godefroid S, Koedam N (2003) How important are large vs. Small forest remnants for the conservation of the woodland flora in an urban context? Glob Ecol Biogeogr 12(4):287–298. https://doi.org/10.1046/j.1466-822X.2003.00035.x

Godoy O, Castro-Díez P, Logtestijn RSPV, Cornelissen JHC, Valladares F (2010) Leaf litter traits of invasive species slow down decomposition compared to Spanish natives: a broad phylogenetic comparison. Oecologia 162(3):781–790. https://doi.org/10.1007/s00442-009-1512-9

Godziek J, Pawlik Ł (2023) Indicators of wind-driven forest disturbances—pit–mound topography, its automatic detection and significance. Catena 221:106757. https://doi.org/10.1016/j.catena.2022.106757

Govaert S, Meeussen C, Vanneste T, Bollmann K, Brunet J, Cousins SAO, Diekmann M, Graae BJ, Hedwall P, Heinken T, Iacopetti G, Lenoir J, Lindmo S, Orczewska A, Perring MP, Ponette Q, Plue J, Selvi F, Spicher F et al (2020) Edge influence on understorey plant communities depends on forest management. J Veg Sci 31(2):281–292. https://doi.org/10.1111/jvs.12844

Govaert S, Vangansbeke P, Blondeel H, De Lombaerde E, Verheyen K, De Frenne P (2021) Forest understorey plant responses to long-term experimental warming, light and nitrogen addition. Plant Biol 23(6):1051–1062. https://doi.org/10.1111/plb.13330

Guiller A, Decocq G, Kichey T, Poli P, Vandepitte K, Dubois F, Honnay O, Closset-Kopp D (2023) Spatial genetic structure of two forest plant metapopulations in dynamic agricultural landscapes. Landsc Urban Plan 231:104648. https://doi.org/10.1016/j.landurbplan.2022.104648

Gutowski JM, Bobiec A, Ciach M, Kujawa A, Zub K, Pawlaczyk P (2022) Drugie życie drzewa. Wydanie II. Fundacja WWF Polska

Hanski I (1997) Metapopulation dynamics: from concepts and observations to predictive models. In: Hanski WI, Gilpin ME (eds) Metapopulation biology. Academic Press, pp 69–91. https://doi.org/10.1016/B978-012323445-2/50007-9

Hanski I (1998) Metapopulation dynamics. Nature 396(6706) Article 6706. https://doi.org/10.1038/23876

Herben T, Chytrý M, Klimešová J (2016) A quest for species-level indicator values for disturbance. J Veg Sci 27(3):628–636. https://doi.org/10.1111/jvs.12384

Hermy M, Honnay O, Firbank L, Grashof-Bokdam C, Lawesson JE (1999) An ecological comparison between ancient and other forest plant species of Europe, and the implications for forest conservation. Biol Conserv 91(1):9–22. https://doi.org/10.1016/S0006-3207(99)00045-2

Hesselbarth MHK, Sciaini M, With KA, Wiegand K, Nowosad J (2019) landscapemetrics: an open-source R tool to calculate landscape metrics. Ecography 42(10):1648–1657. https://doi.org/10.1111/ecog.04617

Hobbie SE, Reich PB, Oleksyn J, Ogdahl M, Zytkowiak R, Hale C, Karolewski P (2006) Tree species effects on decomposition and forest floor dynamics in a common garden. Ecology 87(9):2288–2297. https://doi.org/10.1890/0012-9658(2006)87[2288:TSEODA]2.0.CO;2

Hobbie SE, Oleksyn J, Eissenstat DM, Reich PB (2010) Fine root decomposition rates do not mirror those of leaf litter among temperate tree species. Oecologia 162(2):505–513. https://doi.org/10.1007/s00442-009-1479-6

Hodgson JA, Moilanen A, Wintle BA, Thomas CD (2011) Habitat area, quality and connectivity: striking the balance for efficient conservation. J Appl Ecol 48(1):148–152. https://doi.org/10.1111/j.1365-2664.2010.01919.x

Holeksa J, Zielonka T, Żywiec M (2008) Modeling the decay of coarse woody debris in a subalpine Norway spruce forest of the West Carpathians, Poland. Can J For Res 38(3):415–428. https://doi.org/10.1139/X07-139

Horodecki P, Nowiński M, Jagodziński AM (2019) Advantage of mixed tree stands in restoration of upper soil layers on post-mining sites: a five-year leaf litter decomposition experiment. Land Degrad Dev 30(1):3–13. https://doi.org/10.1002/ldr.3194

Humphrey JW, Davey S, Peace AJ, Ferris R, Harding K (2002) Lichens and bryophyte communities of planted and semi-natural forests in Britain: the influence of site type, stand structure and deadwood. Biol Conserv 107(2):165–180. https://doi.org/10.1016/S0006-3207(02)00057-5

Hüttl RF, Weber E (2001) Forest ecosystem development in post-mining landscapes: a case study of the Lusatian lignite district. Naturwissenschaften 88(8):322–329. https://doi.org/10.1007/s001140100241

Iverson LR, Schwartz MW, Prasad AM (2004) How fast and far might tree species migrate in the eastern united states due to climate change? Glob Ecol Biogeogr 13(3):209–219

Jagodziński AM, Wierzcholska S, Dyderski MK, Horodecki P, Rusińska A, Gdula AK, Kasprowicz M (2018) Tree species effects on bryophyte guilds on a reclaimed post-mining site. Ecol Eng 110:117–127. https://doi.org/10.1016/j.ecoleng.2017.10.015

Jagodziński AM, Dyderski MK, Horodecki P, Knight KS, Rawlik K, Szmyt J (2019) Light and propagule pressure affect invasion intensity of *Prunus serotina* in a 14-tree species forest common garden experiment. NeoBiota 46:1–21. https://doi.org/10.3897/neobiota.46.30413

Janzen DH (1970) Herbivores and the number of tree species in tropical forests. Am Nat 104(940):501–528

Jaroszewicz B, Cholewińska O, Gutowski JM, Samojlik T, Zimny M, Latałowa M (2019) Białowieża forest—a relic of the high naturalness of european forests. Forests 10(10):849. https://doi.org/10.3390/f10100849

Jędrzejewski W, Szymura A, Jędrzejewska B (1994) Reproduction and food of the Buzzard *Buteo buteo* in relation to the abundance of rodents and birds in Białowieża National Park, Poland. Ethol Ecol Evol 6(2):179–190. https://doi.org/10.1080/08927014.1994.9522993

Johansson ME, Nilsson C, Nilsson E (1996) Do rivers function as corridors for plant dispersal? J Veg Sci 7(4):593–598. https://doi.org/10.2307/3236309

Johnson JB (1997) Stand structure and vegetation dynamics of a subalpine wooded fen in Rocky Mountain National Park, Colorado. J Veg Sci 8(3):337–342. https://doi.org/10.2307/3237322

Johnstone JF, Hollingsworth TN, Chapin FS, Mack MC (2010) Changes in fire regime break the legacy lock on successional trajectories in Alaskan boreal forest. Glob Chang Biol 16(4):1281–1295. https://doi.org/10.1111/j.1365-2486.2009.02051.x

Johnstone JF, Allen CD, Franklin JF, Frelich LE, Harvey BJ, Higuera PE, Mack MC, Meentemeyer RK, Metz MR, Perry GL, Schoennagel T, Turner MG (2016) Changing disturbance regimes, ecological memory, and forest resilience. Front Ecol Environ 14(7):369–378. https://doi.org/10.1002/fee.1311

Kiebacher T, Keller C, Scheidegger C, Bergamini A (2016) Hidden crown jewels: the role of tree crowns for bryophyte and lichen species richness in sycamore maple wooded pastures. Biodivers Conserv 25(9):1605–1624. https://doi.org/10.1007/s10531-016-1144-4

Király I, Nascimbene J, Tinya F, Ódor P (2013) Factors influencing epiphytic bryophyte and lichen species richness at different spatial scales in managed temperate forests. Biodivers Conserv 22(1):209–223. https://doi.org/10.1007/s10531-012-0415-y

Kirschner P, Záveská E, Gamisch A, Hilpold A, Trucchi E, Paun O, Sanmartín I, Schlick-Steiner BC, Frajman B, Arthofer W, The STEPPE Consortium, Steiner FM, Schönswetter P (2020) Long-term isolation of European steppe outposts boosts the biome's conservation value. Nat Commun 11(1):1968. https://doi.org/10.1038/s41467-020-15620-2

Kopeć D, Zając I, Halladin-Dąbrowska A (2011) The influence of surrounding vegetation on the flora of post-mining area. Biodiv Res Conserv 24:29–38

Kuijper DPJ, de Kleine C, Churski M, van Hooft P, Bubnicki J, Jędrzejewska B (2013) Landscape of fear in Europe: wolves affect spatial patterns of ungulate browsing in Białowieża primeval forest, Poland. Ecography 36(12):1263–1275. https://doi.org/10.1111/j.1600-0587.2013.00266.x

Larrieu L, Paillet Y, Winter S, Bütler R, Kraus D, Krumm F, Lachat T, Michel AK, Regnery B, Vandekerkhove K (2018) Tree related microhabitats in temperate and Mediterranean European forests: a hierarchical typology for inventory standardization. Ecol Indic 84:194–207. https://doi.org/10.1016/j.ecolind.2017.08.051

Lassauce A, Paillet Y, Jactel H, Bouget C (2011) Deadwood as a surrogate for forest biodiversity: meta-analysis of correlations between deadwood volume and species richness of saproxylic organisms. Ecol Indic 11(5):1027–1039. https://doi.org/10.1016/j.ecolind.2011.02.004

Li Y, Ye S, Bai W, Zhang G (2023) Species diversity patterns differ by life stages in a pine-oak mixed forest. Dendrobiology 88:138–149. https://doi.org/10.12657/denbio.088.010

Lindenmayer DB, Margules CR, Botkin DB (2000) Indicators of biodiversity for ecologically sustainable forest management. Conserv Biol 14(4):941–950. https://doi.org/10.1046/j.1523-1739.2000.98533.x

Lindenmayer DB, Franklin JF, Fischer J (2006) General management principles and a checklist of strategies to guide forest biodiversity conservation. Biol Conserv 131(3):433–445. https://doi.org/10.1016/j.biocon.2006.02.019

Lomolino MV (1994) An evaluation of alternative strategies for building networks of nature reserves. Biol Conserv 69(3):243–249. https://doi.org/10.1016/0006-3207(94)90423-5

Łubek A, Kukwa M, Czortek P, Jaroszewicz B (2019) Impact of *Fraxinus excelsior* dieback on biota of ash-associated lichen epiphytes at the landscape and community level. Biodivers Conserv. https://doi.org/10.1007/s10531-019-01890-w

Łubek A, Kukwa M, Jaroszewicz B, Czortek P (2020) Identifying mechanisms shaping lichen functional diversity in a primeval forest. For Ecol Manage 475:118434. https://doi.org/10.1016/j.foreco.2020.118434

MacArthur RH, Wilson EO (2001) The theory of island biogeography. Princeton University Press

Majer JD, Brennan KEC, Moir ML (2007) Invertebrates and the restoration of a forest ecosystem: 30 years of research following bauxite mining in Western Australia. Restor Ecol 15(s4):S104–S115. https://doi.org/10.1111/j.1526-100X.2007.00298.x

Marozas V, Racinskas J, Bartkevicius E (2007) Dynamics of ground vegetation after surface fires in hemiboreal *Pinus sylvestris* forests. For Ecol Manage 250(1–2):47–55. https://doi.org/10.1016/j.foreco.2007.03.008

Meeussen C, Govaert S, Vanneste T, Calders K, Bollmann K, Brunet J, Cousins SAO, Diekmann M, Graae BJ, Hedwall P-O, Krishna Moorthy SM, Iacopetti G, Lenoir J, Lindmo S, Orczewska A, Ponette Q, Plue J, Selvi F, Spicher F et al (2020) Structural variation of forest edges across Europe. For Ecol Manage 462:117929. https://doi.org/10.1016/j.foreco.2020.117929

Mills LS, Allendorf FW (1996) The one-migrant-per-generation rule in conservation and management. Conserv Biol 10(6):1509–1518. https://doi.org/10.1046/j.1523-1739.1996.10061509.x

Moore LJ, Petrovan SO, Bates AJ, Hicks HL, Baker PJ, Perkins SE, Yarnell RW (2023) Demographic effects of road mortality on mammalian populations: a systematic review. Biol Rev 98(4):1033–1050. https://doi.org/10.1111/brv.12942

Mueller KE, Eisenhauer N, Reich PB, Hobbie SE, Chadwick OA, Chorover J, Dobies T, Hale CM, Jagodziński AM, Kałucka I, Kasprowicz M, Kieliszewska-Rokicka B, Modrzyński J, Rożen A, Skorupski M, Sobczyk Ł, Stasińska M, Trocha LK, Weiner J et al (2016) Light, earthworms, and soil resources as predictors of diversity of 10 soil invertebrate groups across monocultures of 14 tree species. Soil Biol Biochem 92:184–198. https://doi.org/10.1016/j.soilbio.2015.10.010

Myczko Ł, Dylewski Ł, Mitrus S, Sparks TH (2018) Invasive Northern Red Oaks benefit *Temnothorax crassispinus* (Hymenoptera: Formicidae) ant colonies. Myrmecol News 27:25–31. https://doi.org/10.25849/myrmecol.news_027:025

Niinemets Ü (2010) A review of light interception in plant stands from leaf to canopy in different plant functional types and in species with varying shade tolerance. Ecol Res 25(4):693–714. https://doi.org/10.1007/s11284-010-0712-4

Oettel J, Lapin K (2021) Linking forest management and biodiversity indicators to strengthen sustainable forest management in Europe. Ecol Indic 122:107275. https://doi.org/10.1016/j.ecolind.2020.107275

Oettel J, Zolles A, Gschwantner T, Lapin K, Kindermann G, Schweinzer K-M, Gossner MM, Essl F (2023) Dynamics of standing deadwood in Austrian forests under varying forest management and climatic conditions. J Appl Ecol 60(4):696–713. https://doi.org/10.1111/1365-2664.14359

Oksanen J, Minchin PR (2002) Continuum theory revisited: What shape are species responses along ecological gradients? Ecol Model 157(2–3):119–129. https://doi.org/10.1016/S0304-3800(02)00190-4

Orczewska A (2009) The impact of former agriculture on habitat conditions and distribution patterns of ancient woodland plant species in recent black alder (*Alnus glutinosa* (L.) Gaertn.) woods in south-western Poland. For Ecol Manage 258(5):794–803. https://doi.org/10.1016/j.foreco.2009.05.021

Orczewska A, Fernes M (2011) Migration of herb layer species into the poorest post-agricultural pine woods adjacent to ancient pine forests. Pol J Ecol 59(1):75–85

Orczewska A, Czortek P, Jaroszewicz B (2019) The impact of salvage logging on herb layer species composition and plant community recovery in Białowieża forest. Biodivers Conserv 28:3407–3428. https://doi.org/10.1007/s10531-019-01795-8

Ovaskainen O (2002) Long-term persistence of species and the SLOSS problem. J Theor Biol 218(4):419–433. https://doi.org/10.1006/jtbi.2002.3089

Peterken GF (1974) A method for assessing woodland flora for conservation using indicator species. Biol Conserv 6(4):239–245. https://doi.org/10.1016/0006-3207(74)90001-9

Pharo EJ, Lindenmayer DB (2009) Biological legacies soften pine plantation effects for bryophytes. Biodivers Conserv 18(7):1751–1764. https://doi.org/10.1007/s10531-008-9556-4

Prach K, Pyšek P, Bastl M (2001) Spontaneous vegetation succession in human-disturbed habitats: a pattern across seres. Applied Vegetation Science 4(1):83–88. https://doi.org/10.1111/j.1654-109X.2001.tb00237.x

Pretzsch H (2009) Forest dynamics, growth, and yield. From measurement to model. Springer

Prober SM, Brown AHD (1994) Conservation of the grassy white box woodlands: population genetics and fragmentation of *Eucalyptus albens*. Conserv Biol 8(4):1003–1013

Pustkowiak S, Kwieciński Z, Lenda M, Żmihorski M, Rosin ZM, Tryjanowski P, Skórka P (2021) Small things are important: the value of singular point elements for birds in agricultural landscapes. Biol Rev 96(4):1386–1403. https://doi.org/10.1111/brv.12707

Rawlik M, Kasprowicz M, Jagodziński AM (2018) Differentiation of herb layer vascular flora in reclaimed areas depends on the species composition of forest stands. For Ecol Manage 409:541–551. https://doi.org/10.1016/j.foreco.2017.11.055

Rawlik M, Kasprowicz M, Jagodziński AM, Rawlik K, Kaźmierowski C (2019) Slope exposure and forest stand type as crucial factors determining the decomposition rate of herbaceous litter on a reclaimed spoil heap. Catena 175:219–227. https://doi.org/10.1016/j.catena.2018.12.008

Ricketts TH (2001) The matrix matters: effective isolation in fragmented landscapes. Am Nat 158(1):87–99. https://doi.org/10.1086/320863

Riitters K, Wickham J, Costanza JK, Vogt P (2016) A global evaluation of forest interior area dynamics using tree cover data from 2000 to 2012. Landsc Ecol 31(1):137–148. https://doi.org/10.1007/s10980-015-0270-9

Riva F, Fahrig L (2023) Obstruction of biodiversity conservation by minimum patch size criteria. Conserv Biol. https://doi.org/10.1111/cobi.14092

Roberge JM, Angelstam PER (2004) Usefulness of the umbrella species concept as a conservation tool. Conserv Biol 18(1):76–85

Rydin H (2008) Population and community ecology of bryophytes. In: Goffinet WB, Shaw AJ (eds) Bryophyte biology, 2nd edn. Cambridge University Press, pp 393–444

Sabatini FM, Burrascano S, Lombardi F, Chirici G, Blasi C (2015) An index of structural complexity for Apennine beech forests. IForest Biogeosci For 8(3):314. https://doi.org/10.3832/ifor1160-008

Schmidt M, Jochheim H, Kersebaum K-C, Lischeid G, Nendel C (2017) Gradients of microclimate, carbon and nitrogen in transition zones of fragmented landscapes—a review. Agric For Meteorol 232:659–671. https://doi.org/10.1016/j.agrformet.2016.10.022

Schmidt M, Lischeid G, Nendel C (2019) Microclimate and matter dynamics in transition zones of forest to arable land. Agric For Meteorol 268:1–10. https://doi.org/10.1016/j.agrformet.2019.01.001

Schmidt K, Górny M, Jędrzejewski W (2023) Effect of microhabitat characteristics for predicting habitat suitability for a stalking large carnivore—The Eurasian lynx in middle Europe. Anim Conserv 26(6):851–864. https://doi.org/10.1111/acv.12873

Seibold S, Rammer W, Hothorn T, Seidl R, Ulyshen MD, Lorz J, Cadotte MW, Lindenmayer DB, Adhikari YP, Aragón R, Bae S, Baldrian P, Barimani Varandi H, Barlow J, Bässler C, Beauchêne J, Berenguer E, Bergamin RS, Birkemoe T et al (2021) The contribution of insects to global forest deadwood decomposition. Nature 597(7874) Article 7874. https://doi.org/10.1038/s41586-021-03740-8

Sever K, Nagel TA (2019) Patterns of tree microhabitats across a gradient of managed to old-growth conditions. Acta Silvae et Ligni 118:29–40. https://doi.org/10.20315/ASetL.118.3

Shaffer ML (1981) Minimum population sizes for species conservation. BioScience 31(2):131–134. https://doi.org/10.2307/1308256

Simberloff DS, Able LG (1976) Island biogeography theory and conservation practice. Science 191:285–286. https://doi.org/10.1126/science.191.4224.285

Snäll T, Hagström A, Rudolphi J, Rydin H (2004) Distribution pattern of the epiphyte *Neckera pennata* on three spatial scales—importance of past landscape structure, connectivity and local conditions. Ecography 27(6):757–766. https://doi.org/10.1111/j.0906-7590.2004.04026.x

Štursová M, Šnajdr J, Cajthaml T, Bárta J, Šantrůčková H, Baldrian P (2014) When the forest dies: The response of forest soil fungi to a bark beetle-induced tree dieback. ISME J 8(9) Article 9. https://doi.org/10.1038/ismej.2014.37

Szmyt J (2014) Spatial statistics in ecological analysis: From indices to functions. Silva Fennica 48(1):#1008. https://doi.org/10.14214/sf.1008

Szmyt J, Tarasiuk S (2018) Species-specific spatial structure, species coexistence and mortality pattern in natural, uneven-aged Scots pine (*Pinus sylvestris* L.)-dominated forest. Eur J For Res 137(1):1–16. https://doi.org/10.1007/s10342-017-1084-x

Szmyt J, Barzdajn W, Kowalkowski W, Korzeniewicz R (2020) Moderate diversity in forest structure and its low dynamics are favored by uneven-aged silviculture—the lesson from medium-term experiment. Forests 11(1) Article 1. https://doi.org/10.3390/f11010057

Thiele J, Schuckert U, Otte A (2008) Cultural landscapes of Germany are patch-corridor-matrix mosaics for an invasive megaforb. Landsc Ecol 23(4):453–465. https://doi.org/10.1007/s10980-008-9202-2

Tsuzuki Y, Sato MP, Matsuo A, Suyama Y, Ohara M (2022) Genetic consequences of habitat fragmentation in a perennial plant *Trillium camschatcense* are subjected to its slow-paced life history. Popul Ecol 64(1):5–18. https://doi.org/10.1002/1438-390X.12093

Turczański K, Dyderski MK, Andrzejewska A (2022) Drivers of ash (*Fraxinus excelsior* L.) natural regeneration spread into suboptimal sites—Refugee or dead end? For Ecol Manage:119870. https://doi.org/10.1016/j.foreco.2021.119870

Urbanowski C, Horodecki P, Kamczyc J, Skorupski M, Jagodziński A (2018) Succession of mite assemblages (Acari, Mesostigmata) during decomposition of tree leaves in forest stands growing on reclaimed post-mining spoil heap and adjacent forest habitats. Forests 9(11):718. https://doi.org/10.3390/f9110718

Vacek Z, Vacek S, Prokůpková A, Bulušek D, Podrázský V, Hůnová I, Putalová T, Král J (2020) Long-term effect of climate and air pollution on health status and growth of *Picea abies* (L.) Karst. Peaty forests in the Black Triangle region. Dendrobiology 83:1–19. https://doi.org/10.12657/denbio.083.001

Vitt DH, Belland RJ (1997) Attributes of rarity among Alberta mosses: patterns and prediction of species diversity. Bryologist 100(1):1–12. https://doi.org/10.2307/3244382

von Arx G, Dobbertin M, Rebetez M (2012) Spatio-temporal effects of forest canopy on understory microclimate in a long-term experiment in Switzerland. Agric For Meteorol 166–167:144–155

Walker LR, Chapin FS (1987) Interactions among processes controlling successional change. Oikos 50(1):131–135

Weibull H, Rydin H (2005) Bryophyte species richness on boulders: relationship to area, habitat diversity and canopy tree species. Biol Conserv 122(1):71–79. https://doi.org/10.1016/j.biocon.2004.07.001

Wesołowski T (2007) Primeval conditions—what can we learn from them? Ibis 149(s2):64–77. https://doi.org/10.1111/j.1474-919X.2007.00721.x

Wesołowski T (2011) "Lifespan" of woodpecker-made holes in a primeval temperate forest: a thirty year study. For Ecol Manage 262(9):1846–1852. https://doi.org/10.1016/j.foreco.2011.08.001

Wesołowski T (2012) "Lifespan" of non-excavated holes in a primeval temperate forest: a 30year study. Biol Conserv 153:118–126. https://doi.org/10.1016/j.biocon.2012.04.017

Wesołowski T, Czeszczewik D, Hebda G, Maziarz M, Mitrus C, Rowiński P, Neubauer G (2022) Long-term changes in breeding bird community of a primeval temperate forest: 45 years of censuses in the Białowieża National Park (Poland). Acta Ornithol 57(1):71–100. https://doi.org/10.3161/00016454AO2022.57.1.005

Wierzcholska S, Dyderski MK, Pielech R, Gazda A, Smoczyk M, Malicki M, Horodecki P, Kamczyc J, Skorupski M, Hachułka M, Kałucka I, Jagodziński AM (2018) Natural forest remnants as refugia for bryophyte diversity in a transformed mountain river valley landscape. Sci Total Environ 640–641:954–964. https://doi.org/10.1016/j.scitotenv.2018.05.340

Wierzcholska S, Dyderski MK, Jagodziński AM (2020) Potential distribution of an epiphytic bryophyte depends on climate and forest continuity. Global Planet Change 193:103270. https://doi.org/10.1016/j.gloplacha.2020.103270

Winter S, Höfler J, Michel AK, Böck A, Ankerst DP (2015) Association of tree and plot characteristics with microhabitat formation in European beech and Douglas-fir forests. Eur J For Res 134(2):335–347. https://doi.org/10.1007/s10342-014-0855-x

Wohlgemuth T, Gossner MM, Campagnaro T, Marchante H, van Loo M, Vacchiano G, Castro-Díez P, Dobrowolska D, Gazda A, Keren S, Keserű Z, Koprowski M, Porta NL, Marozas V, Nygaard PH, Podrázský V, Puchałka R, Reisman-Berman O, Straigytė L et al (2022) Impact of non-native tree species in Europe on soil properties and biodiversity: a review. NeoBiota 78:45–69. https://doi.org/10.3897/neobiota.78.87022

Woziwoda B, Dyderski MK, Jagodziński AM (2021) Forest land use discontinuity and northern red oak *Quercus rubra* introduction change biomass allocation and life strategy of lingonberry *Vaccinium vitis-idaea*. For Ecosyst 8(1):9. https://doi.org/10.1186/s40663-021-00287-y

Woźniak G, Chmura D, Dyderski MK, Błońska A, Jagodziński AM (2022) How different is the forest on post-coal mine heap regarded as novel ecosystem? For Ecol Manage 515:120205. https://doi.org/10.1016/j.foreco.2022.120205

Wyka J, Piechnik Ł, Grzędzicka E, Lešo P, Dyderski MK, Kajtoch Ł (2023) The vertical form of the common ivy *Hedera helix* L. is associated with diverse and semi-natural forests in Central European highlands. For Ecol Manage 530:120750. https://doi.org/10.1016/j.foreco.2022.120750

Zellweger F, Frenne PD, Lenoir J, Vangansbeke P, Verheyen K, Bernhardt-Römermann M, Baeten L, Hédl R, Berki I, Brunet J, Calster HV, Chudomelová M, Decocq G, Dirnböck T, Durak T, Heinken T, Jaroszewicz B, Kopecký M, Máliš F et al (2020) Forest microclimate dynamics drive plant responses to warming. Science 368(6492):772–775. https://doi.org/10.1126/science.aba6880

Zerbe S, Wirth P (2006) Non-indigenous plant species and their ecological range in Central European pine (*Pinus sylvestris* L.) forests. Ann For Sci 63(2):189–203. https://doi.org/10.1051/forest:2005111

Open Access This chapter is licensed under the terms of the Creative Commons Attribution 4.0 International License (http://creativecommons.org/licenses/by/4.0/), which permits use, sharing, adaptation, distribution and reproduction in any medium or format, as long as you give appropriate credit to the original author(s) and the source, provide a link to the Creative Commons license and indicate if changes were made.

The images or other third party material in this chapter are included in the chapter's Creative Commons license, unless indicated otherwise in a credit line to the material. If material is not included in the chapter's Creative Commons license and your intended use is not permitted by statutory regulation or exceeds the permitted use, you will need to obtain permission directly from the copyright holder.

In Situ and *Ex Situ* Conservation Measures

11

Barbara Fussi, Muhidin Šeho, and Darius Kavaliauskas

Abies alba seed orchard managed by AWG. (Photo: T. Hase, StMELF)

B. Fussi (✉) · M. Šeho
Bavarian Office for Forest Genetics (AWG), Teisendorf, Germany
e-mail: barbara.fussi@awg.bayern.de; muhidin.seho@awg.bayern.de

D. Kavaliauskas
Vytautas Magnus University Agriculture Academy, Akademija, Lithuania
e-mail: darius.kavaliauskas@vdu.lt

© The Author(s) 2025
K. Lapin et al. (eds.), *Ecological Connectivity of Forest Ecosystems*,
https://doi.org/10.1007/978-3-031-82206-3_11

Abstract

The effects of climate change and biodiversity loss are global and not limited by national borders, with forest ecosystems, in particular, suffering under increasing pressure. To preserve and maintain species genetic diversity, well-considered *in situ* and *ex situ* measures are needed. We present a step-by-step guide outlining the key processes for structuring, performing, and selecting appropriate sustainable use and conservation measures for forest genetic resources (FGR). Two case studies focusing on European white elm (*Ulmus laevis* Pall.) and wild service tree (*Sorbus torminalis* (L.) Crantz), for which the guide was followed, demonstrate that differences in the regional genetic pattern should be followed when designing FGR sustainable use and conservation efforts. The fact that seed orchards can maintain high genetic diversity and provide high-quality, genetically diverse seed material makes them an optimal supplement to *in situ* genetic conservation units. Especially for scattered or threatened tree species, *ex situ* measures are of utmost importance. In light of a severe pest outbreak, i.e., ash dieback, preserving less susceptible ash trees (*Fraxinus excelsior* L.) is crucial to enable their reproduction and facilitate gene flow among them to prevent crucial loss of genetic diversity and eventually the species itself. Therefore, forest genetic monitoring should be used more intensively to observe, measure, and assess the long-term FGR conservation efforts as genetic variation is an integral part of biological diversity, which requires special attention.

Keywords

Forest genetic resources · Sustainable use of FGR · Genetic diversity · Conservation · GCU · Disease resistance · Gene flow · Forest genetic monitoring · FGM

Introduction

The effects of climate change and biodiversity loss are global and not limited by national borders, with forest ecosystems, in particular, suffering under increasing pressure (European Green Deal 2019; Lefevre et al. 2024). According to the New EU Forest Strategy for 2030, the adaptation of forests to climate change will require large quantities of appropriate forest reproductive material (FRM) from *in situ* and *ex situ* forest genetic resources/conservation units. Therefore, coordinated efforts following "best practice" examples will be needed to conserve and sustainably use forest genetic resources (FGR) on which more climate-adapted forestry depends. Furthermore, to increase the production and availability of such FRM, policymakers will need to raise support for research to develop principles and application methods for *in situ* and *ex situ* conservation as well as for assisted forest species migration where needed (New EU Forest Strategy for 2030).

> **Box 11.1 Definition of In Situ and *Ex Situ***
> By definition, *in situ* conservation means "the conservation of ecosystems and natural habitats and the maintenance and recovery of viable populations of species in their natural surroundings and, in the case of domesticated or cultivated species, in the surroundings where they have developed their distinctive properties," while *ex situ* conservation is defined as the conservation of components of biological diversity outside their natural habitats (Convention on Biological Diversity, Art. 2; https://www.cbd.int/convention/articles/?a=cbd-02).

Thus, conservation of FGR can be *in situ* dynamic, allowing evolutionary forces such as selection, mutation, recombination, and gene flow to act; *ex situ* dynamic when species or populations are moved and planted outside of their original habitats or even outside of their natural distribution range (*ex situ* conservation stand and seed orchards); or static (*ex situ*) when genetic material (seed, pollen, or plant tissues) is stored in gene banks (Forest Genetic Resources Strategy for Europe, 2021). Following the Forest Genetic Resources Strategy for Europe (2021), *in situ* conservation should be a priority and the primary conservation strategy for FGR since it enables evolution and local adaptation.

However, effective conservation of FGR requires a combination of both *in situ* and *ex situ* (dynamic and static) methods to be applied, as outlined in the Forest Genetic Resources Strategy for Europe (2021). This strategy is integrated, for example, into the concept for the conservation and sustainable use of FGR in Bavaria (Generhaltungskonzept 2015), which is based on the "Concept for the Conservation of Forest Genetic Resources in the Federal Republic of Germany" (Paul et al. 2000). Coordination of forest genetic resources selection, use, and conservation efforts in the Federal Republic of Germany has been led by a federal state working group (BLAG-FGR) since 1985. With the development of the "Recommendations for the Designation of Gene Conservation Units, Taking Minimum Criteria into Account," the procedure for recording and designating *in situ* gene conservation was defined (BLAG-FGR, status 01/2017). These recommendations align with the guidelines set forth by the European Forest Genetic Resources Conservation Programme (EUFORGEN).

Both *ex situ* and *in situ* conservation measures require decisions on which populations to select or sample (Ledig 1986). To optimize the conservation network and its management activities, it is important to accumulate relevant knowledge on species ecology, biology, distribution, and patterns of genetic variation (Ledig 1986 and references therein). According to Neel and Cummings (2003), if a selection of gene conservation units (GCU) is made without data on genetic variation, then larger sample sizes within each population and larger numbers of populations are needed to ensure the conservation of genetic diversity and representation of rare and common alleles. Thus, following Ledig (1986), a common method to select GCUs without genetic information would be to select populations representing different environments or habitats, which are most likely to contain most of the species' genetic variation. However, this type of conservation requires intensive selection of larger conservation units from different environments, and the consent of different forest owners is often necessary. In addition, the protection of individual trees as well as small and marginal populations of scattered tree species should also be considered in the process of selection for conservation since they can possess specific genetic variants and are needed to ensure connectivity among population fragments (Ledig, 1986; Bednorz 2007).

Since biotechnologies including various genotyping techniques have become cheaper and more efficient, DNA analysis after systematic sampling from different environments can be used to map the patterns of spatial genetic structure (gene pools) across a species' natural distribution range. Collected information and genetic diversity measures provide information on genetic variability within and among populations, allowing the optimized choice of populations for species conservation and the use of FGR to be optimized (Marshall and Brown 1975; Ledig, 1986; Hettemer 1995; Petit et al. 1998; Rajora and Mosseler 2001; Caballero et al. 2010; Koskela et al. 2013; Schueler et al. 2014; de Vries et al. 2015). Extensive sampling combined with DNA analysis can thus lead to a well-supported number and selection of populations needed for actual conservation measures (Ledig 1986 and references therein). There are numerous such examples for various tree species in Europe, like *Sorbus torminalis* (Demesure et al. 2000; Hoebee et al. 2006; Bednorz and Kosiński 2006; Bednorz 2007; Kučerova et al. 2010; Kavaliauskas et al. 2021; etc.), *Ulmus laevis* (Collin and Bozzano 2015; Collin et al. 2020; Kavaliauskas et al. 2022a; etc.), *Taxus baccata* (Šeho et al. 2022; Komárková et al. 2022; Linares 2013; Klumpp et al. 2011; Dubreuil et al. 2010), *Pinus* spp. (González-Martınez et al. 2004; Dzialuk et al. 2014; Pyhäjärvi et al. 2020; Kavaliauskas et al. 2022b; etc.), *Fagus sylvatica* (Lefevre et al. 2013; von Wuehlisch 2008; Vornam et al. 2004), and *Quercus* spp. (de Dato et al. 2018; Lefevre et al. 2013; Ducousso and Bordacs 2004).

IUCN has formulated different management categories for protected areas to comply with the purpose of protection (IUCN 1994; Dudley et al. 2010). Several conservation programs exist for forest trees (White et al. 2007). However, the selection of conservation strategies and GCUs for specific tree species depends on many factors such as species biology (e.g., pollination type), ecology (e.g., species rarity and scatteredness), genetic structure (e.g., genetic variation within and among populations), and external factors including forestry policies, ownership, economic importance, risk of extinction, etc. The Forest Genetic Resources Strategy for Europe (2021) underlines that GCUs (*in situ* and *ex situ* together) are the key elements in the pan-European network for the dynamic conservation of FGR and that the distribution and coverage of GCUs must be expanded to cover as many species and their genetic diversity as possible, including rare tree species and marginal populations at the boundaries of natural distributions. Furthermore, active GCU monitoring and maintenance through management, if necessary, must maintain evolutionary processes and increase the adaptability of selected tree populations. Regular genetic monitoring is needed to assess the effective population size and

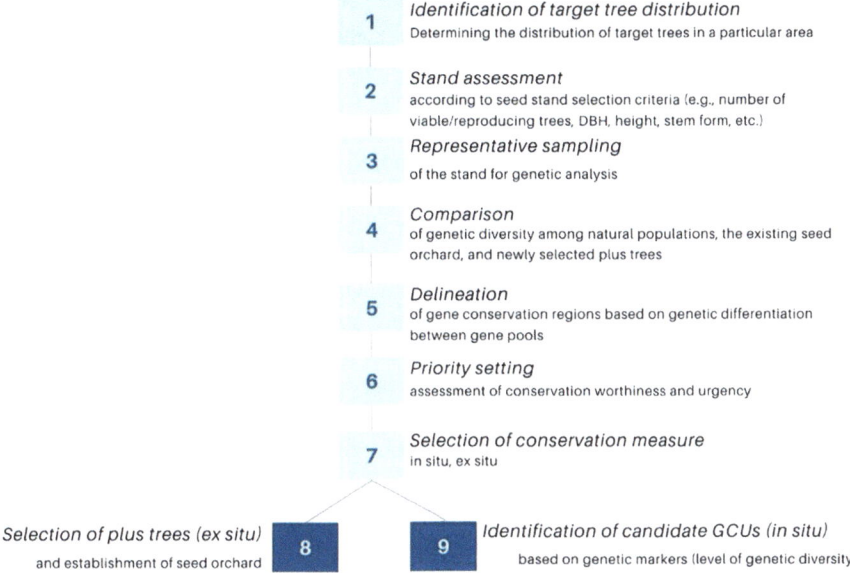

Fig. 11.1 A step-by-step guide to conservation of FGR including the choice of measure (*in situ* or *ex situ*)

reproduction capabilities of GCUs and their ability to adapt under changing environmental conditions (Fussi et al. 2016; Bajc et al. 2020, Liesebach et al. 2024, Forest Genetic Resources Strategy for Europe 2021). In the following, we will discuss *in situ* and *ex situ* conservation measures with examples for dominant and scattered tree species (Fig. 11.1).

Anthropogenic disturbances have negative consequences for genetic diversity and the adaptive potential of species (Aravanopoulos 2018; Kavaliauskas et al. 2018; Gautam et al. 2021, Lefevre et al. 2024). A high level of genetic diversity is the basis for the adaptation and adaptability of FGR and must therefore be maintained for the future. Conservation of FGR aims to improve the genetic diversity and adaptability of forest stands over time, and forest genetic monitoring (FGM) provides us with information on the conditions and changes in the forest genetic system (Fussi et al. 2016, Bajc et al. 2020). Modern forestry is based on the prudent use of FGR. The basis of silvicultural activity is the conservation and improvement of the various forest functions and the conservation of biological diversity at all levels. Assignment of a forest stand as a seed stand can enhance the usage and spread of a species, thereby increasing its distribution and preventing further fragmentation of its range.

Following the flowchart in Fig. 11.1, certain steps should be taken prior to making decisions on species conservation. First, information on the existing populations of the target species should be gathered. Species distribution maps, data from forest inventories, and surveys among forest owners can help collect all the relevant details on the species in a certain area. Based on the collected data, stands should be sorted according to minimum requirements and then visited and evaluated using a predesigned routine. Photographs of the potential stands facilitate comparison between them. Stand size and tree numbers depend on the tree species and the conservation objectives. Koskela et al. (2013) suggested minimum sizes of the GCUs to be set at 500, 50, or 15 reproducing individuals for stand-forming scattered and endangered tree species, respectively. Potential stands should be vital and possess good to very good quality (e.g., height growth, form, and vigor), and a minimum of 50 individuals of the target species should be present in each stand or within gene flow distance. These 50 trees should be of reproductive age. If flowering or fructification cannot be assessed directly, the DBH or crown size can serve as a proxy (DBH > 20 cm for most tree species). The more reproducing trees are present, the higher the potential for new genetic genotypes/combinations. Gene conservation for forestry purposes may also include attributes like the ability to harvest seeds from the stand or a minimum distance to poor phenotypes of 400 m (recommendations according to German FoVG, 2003). Information on the autochthony of the origin of the populations is also beneficial. Following this initial evaluation of the stands, populations to be sampled for genetic analysis should be preselected. In addition, species distribution should be considered for representative sampling. Sampling for genetic analysis should ideally be systematic to cover the entire population and include at least 50 reproducing trees with a minimum distance of 30 m between sampled individuals. The genetic analysis provides an overview of the species' genetic variation and spatial genetic structure and can help identify hotspots of genetic diversity. Several genetic diversity parameters can be used for comparison and prioritization of different populations. Following Marshall and Brown (1975), Petit et al. (1998), Rajora et al. (2000), Rajora and Mosseler (2001), and Caballero et al. (2010), we consider the allelic richness (Ar) value (number of different alleles segregating in the population) to be of key importance in conservation programs, especially for subdivided and scattered tree species. Therefore, we suggest Ar as the main parameter for GCU selection. In addition, the measure of effective population size (Ne) is a critical verifier in the selection and conservation of forest genetic resources (Lande and Barrowclough 1987, Santos-del-Blanco et al. 2022, Hoban et al. 2023, Liesebach et al. 2024, etc.). Estimates of contemporary effective population size (Ne) can provide valuable information for genetic conservation and monitoring, pinpointing populations at higher risk of genetic erosion, decreased fitness, maladaptation, and ultimately, demographic decline (Santos-del-Blanco et al. 2022 and references therein). In conservation genetics, Ne is important because it influences the rate at which genetic diversity is lost (e.g., due to genetic drift). Thus, small populations of scattered and threatened tree species are susceptible to genetic drift, leading to a faster loss of genetic diversity and an increased risk of inbreeding depression and the accumulation of deleterious mutations. Therefore, Ne should serve as a crucial measure for FGR selection and is essential for preserving genetic diversity, ensuring the long-term viability of populations, and formulating necessary action plans (Santos-del-Blanco et al. 2022, Pérez-Pereira et al. 2022 and references therein, Hoban et al. 2023).

The required number of populations to conserve as GCUs within a species depends on how diversity is divided within and among populations, which measure of diversity is chosen, and how much of the total diversity is considered sufficient (Neel and Cummings 2003). According to Neel and Cummings (2003), a greater number of populations is required to conserve genetic diversity if GCUs are selected without genetic information. The selection of GCUs should therefore be based on collective data of genetic diversity, environmental data, and tree/population distribution in certain regions to ensure gene flow among populations and sufficiently high genetic diversity. Moreover, genetic differentiation among populations may be used as one of the measures for GCU selection, even if some populations do not exhibit the required level of genetic diversity, since some of them may possess rare alleles and genotypes (Bednorz 2007; Jost et al. 2018). However, due to the high costs of genetic analysis, target areas and species might be prioritized and selected.

Species and population prioritization consider endangerment status and allow decisions regarding the urgency of conservation measures. From an ecological or economic point of view, valuable populations of a species that are adapted to their site conditions and harbor high genetic variability should be the focus for genetic conservation. For dominant tree species, in addition to vitality and the presence of natural regeneration, above-average stem quality and growth characteristics can also be considered important criteria for selection for conservation. An assessment of conservation necessity should be conducted for populations at high risk of extinction or decline. Within the group of rare and ecologically important tree species, the focus lies on the assessment of conservation urgency at the species level. For example, the guideline for the conservation and sustainable use of forest genetic resources in Bavaria (Germany) defines three priority levels:

1. Species of high priority—gene conservation measures are urgently needed.
2. Species of medium priority—gene conservation measures are needed but not urgent.
3. Species of low priority—gene conservation measures are currently not needed.

The endangerment of a population can be determined by estimating the following risk factors, some of which mutually influence each other or increase multiple risks when they coincide:

- Actual population size, minimum viable population size, and effective population size: For example, the minimum size for a viable population often has been used as a crucial measure in conservation practice to determine the extinction risks of populations and species. An effective population size of at least 50 unrelated and reproducing individuals ($N_e = 50$) is needed to minimize the risk of extinction due to inbreeding depression, and N_e of 500 individuals is required for a population's long-term survival (Pérez-Pereira et al. 2022 and references therein). The smaller the population, the higher the risk of extinction through drift effects (e.g., natural disaster and disease incidence).
- Deviation from the potential natural vegetation: The further the current population is from the optimal distribution (suitable ecological niche) for its species, the higher the risk of extinction (Aitken et al. 2008; Gougherty et al. 2021, etc.).

- Weakness in competition: Competition from precocious, more competitive species increases the risk of decline.
- Disposition to diseases/disasters: Species whose existence is threatened by diseases/disasters are at increased risk.
- Hybridization with cultivars: Plantations of cultivars in the immediate vicinity of gene conservation objects pose a threat to the gene resource through introgression (e.g., wild cherry, wild apple, wild pear, small-leaved lime, etc.).
- Loss of forest areas leads to a reduction in effective population size, as well as making gene flow more difficult due to habitat fragmentation.
- Browsing of natural regeneration limits the genetic preservation of GCU since establishing new tree generations becomes challenging.

Following the prioritization of populations, the appropriate conservation measures must be chosen. These can be *in situ* or *ex situ* measures or a combination of both depending on the circumstances; possible approaches are described in the following.

In Situ Conservation Measures

Species with an adequate number of viable populations within their natural distribution range can be considered for *in situ* conservation. The objective should be to conserve the genetic variation, facilitate the natural regeneration of the GCU, or conduct artificial regeneration with reproductive material taken from the same population. In situ measures offer the advantage that they can be integrated into regular forest management (forest management planning corresponding to GCU conservation aims).

By contrast, rare tree species should generally be completely protected from exploitation, with only minor exceptions granted if populations meet specific criteria, which ensures population continuity, such as the presence of natural regeneration, compliance with viable population size standards, etc. Different populations of a rare species often have unique genetic compositions, making the loss of any population an irreparable genetic loss (e.g., because of lost locally adapted genetic variation, interrupted connectivity, and gene flow). For most rare species, GCUs are an absolute necessity to protect the FGR, and additional *ex situ* preservation is also required to reduce the risk of sudden loss through drift effects (e.g., catastrophic events like fire or landslides). Common tree species are often genetically variable but are generally exploited more intensively, and population-specific (rare and private) alleles can therefore be lost if management methods do not consider the species' genetic peculiarities (Danusevičius et al. 2016; Kavaliauskas et al. 2018, Danusevičius et al. 2023, etc.). Certain forest areas with populations of target tree species should be set aside to allow natural evolutionary processes to act (natural regeneration is mandatory, harvest techniques enhancing genetic diversity, reproduction, selection, etc.). GCUs may be influenced by undesirable pollen flow from surrounding plantations regenerated with seedlings of off-site parentage. However, geneticists disagree strongly on the impact of pollen contamination (Nienstaedt, 1980). While some reports show that wide buffer zones are necessary (Yeatman, 1973), thus making GCUs impractically large, others report that pollen migration will be ineffectual if immigrants carry maladapted genes since proper harvesting can assure dense regeneration, and natural selection can be relied upon to eliminate unfit progeny (Adams and Burczyk 2000). On the other hand, if migration introduces selectively advantageous genes, then contamination poses no problem for the preservation of adapted populations (Fitzpatrick et al. 2015, Tigano and Friesen, 2016). The substitution of neutral alleles may be a problem if the management unit is so small that pollen influx swamps the local pollen contribution. However, many management alternatives can be devised to reduce the problem of pollen contamination; they include surrounding the management unit with plantations of different species, resorting to mass pollination with appropriate pollen, or controlled pollination, and artificial regeneration with progeny known to have originated from crossing among native trees (Ledig, 1986). Trees tend to have high levels of heterozygosity, but it is not clear whether they have the evolutionary potential to withstand climate change and whether their phenotypic plasticity enables them to respond to short- and medium-term climate shifts (Kelleher, 2018). Since only 1% of all tree species have been genetically analyzed, the use of indicators (e.g., allelic richness and effective population size) is needed to obtain a more comprehensive picture of global forest genetic resources. In any case, the best practice for the conservation of large, long-lived organisms like trees is a dynamic or near-natural in situ conservation approach facilitating evolution (Kelleher, 2018).

Ex Situ Conservation Measures

In some specific cases, established *in situ* GCUs may not survive, particularly in areas where environmental conditions are changing rapidly. Moreover, there are species exhibiting highly fragmented populations like *Sorbus torminalis, Sorbus domestica,* or *Pyrus pyraster*, where traditional *in situ* conservation approaches might be ineffective because of low target tree species densities in certain areas (Forest Genetic Resources Strategy for Europe 2021). Thus, for such species with a low number of viable populations and low tree density per population, *ex situ* measures should be favored. *Ex situ* measures include the establishment of populations outside a species' natural surroundings, conservation seed orchards, clone collections, *ex situ* conservation stands, storage of seeds and pollen, storage of plants and plant parts, and permanent vegetative and in vitro propagation.

Ex situ conservation is appropriate when a tree species is rare and grows only in small patches, when the habitat is threatened, or when natural regeneration is impeded (e.g., by game browsing or occurrence of clones). The principal method for *ex situ* conservation of forest tree species is the establishment of various tree collections (e.g., seedling or clonal seed orchards, clonal tree archives, *ex situ* conservation stands, and other tree collections). Individual trees are either propagated by grafting and the grafts moved to clone collections or by collecting seeds from the

Fig. 11.2 Long-term seed storage at AWG Teisendorf. (Photo: AWG/ Muhidin Seho)

original trees to establish so-called family collections with seedlings. In these family collections, several seedlings of one family are planted near each other, and thinning is carried out so that only one tree representing the family is ultimately left growing. The families representing the same forest are distributed within the collection so that they can also be used for seed production after thinning. One of the benefits of such tree collections is that a certain amount of selection takes place within them. They produce well-adapted and genetically variable seeds that can be used in forest regeneration or landscaping (Koskela et al. 2007).

In addition to *ex situ* conservation in living collections, mentioned in the paragraph above, static *ex situ* conservation in gene banks is an important method among gene conservation measures. Thus, long-term seed storage is an effective tool for ex situ conservation, especially for conifer tree species and for most small-fruited deciduous tree species. For example, in Bavaria (Germany) this form of gene conservation has been practiced in a forestry gene bank at the Bavarian Office for Forest Genetics (AWG) since 1989. AWG gene bank is structured into two storage facilities; in one of them, the temperature is −10 °C and in the other, −20 °C, with a total storage area of 55 m^2 (Fig. 11.2, Generhaltungskonzept 2015). Currently (data from 2024), AWG gene bank stores 158 seed lots of 24 tree species and 53 provenances, out of which 23.8 million plants can be produced, if necessary, in case of an emergency.

However, to reach the most effective gene conservation aims, both *in situ* and *ex situ* conservation methods must be combined and used depending on the threats to each population or species overall (Forest Genetic Resources Strategy for Europe 2021).

Box 11.2 Forest Genetic Monitoring
Forest genetic monitoring (FGM) is used as a tool to measure and observe the long-term conservation of FGR. Safeguarding the sustainability of forest ecosystems with their habitat variability and all their functions is of the utmost importance, and the long-term adaptability of forest ecosystems to a changing environment must therefore be secured, for example, through sustainable forest management. High adaptability is based on biological variation starting at the genetic level of each species. Monitoring of biological diversity over time allows changes that threaten biological resources to be detected. Genetic variation as an integral part of biological diversity requires special attention, and its monitoring can ensure effective conservation (Pearman et al. 2024). FGM is an important module of biodiversity monitoring and has been defined as the assessment of a forest population's capacity to survive, reproduce, and persist in the long term despite rapid environmental changes (Fussi et al. 2016).

Long-term genetic monitoring of specific tree species in selected areas (so far, priority has been given to silver fir, beech, ash, and spruce in Germany) allows changes in the formation, preservation, and transmission of genetic information to the next generation to be assessed (Kätzel et al. 2005; Konnert et al. 2011). A recent Germany-wide study reports on the transmission of genetic diversity to the offspring in 12 German beech and ten spruce stands (Liesebach et al. 2024). Effective population size values showed clear differences between the stands and generations studied. Natural regeneration was similar to adult trees, while seed samples revealed a clear bottleneck effect. Seeds were harvested from only 20 trees, which points to the fact that this number is too small to represent the genetic diversity of the stand. From these results, conclusions can be drawn for sustainable natural regeneration management in forest stands for the approval of seed stands and appropriate seed harvesting.

These observations fulfill the function of an "early warning system" for ecosystem changes at higher levels (e.g., stand composition, vitality, and regeneration success), which only occur much later but are determined by the genetic system. In Bavaria, multiple assessments are currently being conducted at regular intervals (bud break and flowering intensity every year, seed testing every 5 years, and genetic analysis every 10 years) on four FGM plots (two beech, one silver fir, and one Norway spruce) (Fig. 11.3).

Several genetic monitoring concepts have been proposed during the past decade, with a major step forward being achieved within the LIFEGENMON project, in which a comprehensive manual for forest genetic monitoring and detailed practical guidelines for seven species have been developed (Bajc et al. 2020). Within the project, several indicators and verifiers for forest genetic monitoring on three levels (basic level, standard level, and intensive level) have been defined, and the appropriate level can be applied depending on capacity and financial support.

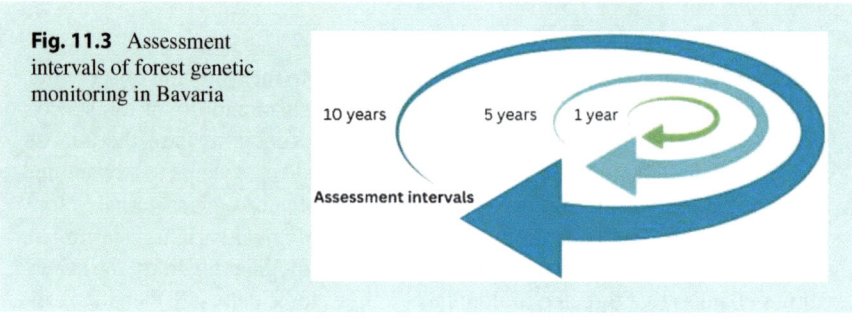

Fig. 11.3 Assessment intervals of forest genetic monitoring in Bavaria

Case Study of Species Genetic Conservation and Sustainable Use

In the following chapter, we are presenting two case studies focusing on the European white elm (*Ulmus laevis*) and wild service tree (*Sorbus torminalis*) in Bavaria (Germany), where a step-by-step guide outlining key processes for structuring, performing, and selecting appropriate forest genetic resources (FGR) for conservation and sustainable use is introduced. Field inventories and DNA-based results on *U. laevis* and *S. torminalis* revealed distinct genetic clusters in each species, guiding both *in situ* and *ex situ* FGR use and conservation strategies. In addition, our results are emphasizing the critical role of seed orchards in generating high-quality seed material due to the observed higher genetic diversity in seed orchards compared to natural populations.

Ulmus laevis

The European white elm (*U. laevis*) has a large natural distribution range in Europe, extending from the Pyrenees in the west to their eastern limit of distribution in the Urals (Fig. 11.4). In Germany, the largest populations are concentrated in the northeast of the country; in Bavaria, however, *U. laevis* is an ecologically important rare tree species. The main natural distribution of the species in Bavaria is along the main river systems (Main, Danube, and Isar). The preferred European white elm habitats are groundwater-affected soils, riparian forests, floodplain soils, and boggy soils.

In Bavaria, *U. laevis* covers only a small distribution area; its economic importance for timber production is low to medium, and its vulnerability is likewise medium. The European white elm is assessed as a tree species with high priority, with gene conservation measures rather urgently needed. However, until now, there has been a lack of knowledge on the genetic structure and diversity of European white elm populations in Bavaria. Kavaliauskas et al. (2021) designed a study to assess this genetic structure and diversity and select putative seed stands and gene

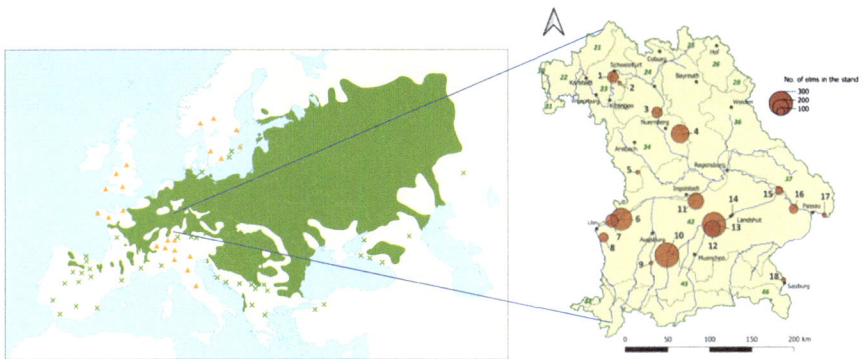

Fig. 11.4 Distribution map of *U. laevis* (European white elm) according to Caudullo et al. (2017). Legend: *filled green area*: Native range. *Filled green X*: Isolated population. *Filled yellow triangle*: Introduced and naturalized (synanthropic). Magnified (right): Distribution of sampled populations of European white elm in Bavaria. The size of each circle corresponds to the respective population size. (Kavaliauskas et al. 2021)

conservation units (GCU) for European white elm. Studies on the genetic variation of European white elm across Europe show a relatively low genetic variation (Venturas et al. 2013), and the 18 populations analyzed in Bavaria confirm this moderate-to-low genetic diversity, e.g., mean expected heterozygosity He = 0.343, the mean number of effective alleles Ne = 1.77, and rarified allelic richness Ar = 2.72 (Kavaliauskas et al. 2021).

Analysis of molecular variance (AMOVA) showed that 89% of the total genetic variation exists within populations and 11% between populations. This can be viewed as an indication of the isolation of individual populations when compared to other tree species (e.g., 4% for *Sorbus torminalis*). Bayesian cluster analysis using the STRUCTURE software revealed a possible structure of four genetic clusters (the highest delta K of 6.5 was K = 4, indicating four genetic groups) within the investigated European white elm populations in Bavaria (Fig. 11.5). One cluster follows the Main River basin (cluster K1) with the highest proportion of green. The other clusters (one main cluster and two subclusters) dominate the central part of Bavaria, following the basin of the Danube River and its tributaries. However, in all studied populations, an admixture with other genetic clusters was observed (Fig. 11.5).

In situ conservation is the favored protection measure for the European white elm. Due to the decline of suitable habitats, its natural distribution area became fragmented. The two identified main clusters should be considered for the future so that the existing gene pool can be retained. Different studies recommend using the allelic richness value (Ar) as the primary parameter for GCU selection (Petit et al. 1998; Foulley et al. 2006), and Ar has therefore become the main value for GCU selection up to now. We suggest dividing European white elm populations into three categories. Populations with a value of Ar > 3 are recommended for *in situ* conservation (populations 1, 2, 11, 14, and 18 in Fig. 11.5.). For these most precious

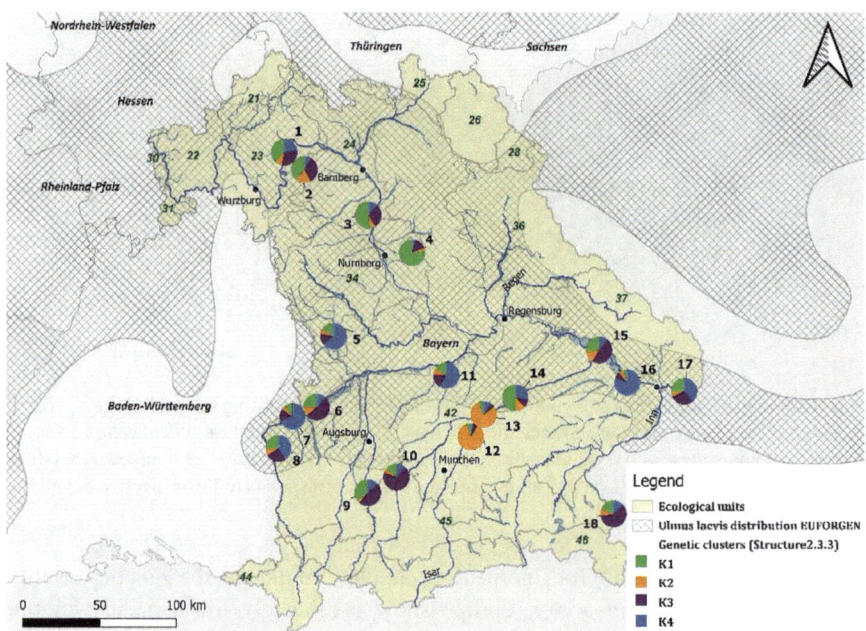

Fig. 11.5 Population genetic structure of *U. laevis* populations in Bavaria based on STRUCTURE clusters

populations, regeneration should only be carried out with their respective own reproductive material. Within the second category with Ar > 2.6, which is worthy of conservation in terms of forest genetics, the same procedure should be applied to populations 6 and 17 in Fig. 11.5. With this selection of GCUs, all relevant genetic conservation zones in Bavaria are represented. For populations, which are at the lower limit of this value, enrichment planting by introducing forest reproductive material (FRM) from suitable source populations should be considered. In addition to the genetic diversity parameters, gene flow between populations is particularly important. For this reason, the establishment of stepping-stone populations should be considered in the long term. Furthermore, additional GCUs should be registered at the European level; only two German populations of *U. laevis* are included in the European Information System on Forest Genetic Resources database (EUFGIS, http://portal.eufgis.org/). These data fill existing gaps regarding the sustainable use and conservation of rare and scattered tree species by providing insights into the genetic variation and genetic structure of these species, thereby allowing us to better plan genetic conservation measures and select GCU in Bavaria. We, therefore, suggest that all gene conservation units should go through all the steps in Fig. 11.2. As *ex situ* measures, the establishment of conservation seed orchards, which can simultaneously be used for conservation and the production of high-quality reproductive material, are planned. Therefore, at least 80–100 clones should be selected and a seed orchard established to maintain a large effective population size.

Sorbus torminalis

Ongoing changes in environmental conditions and climate are prompting forest owners to reconsider their methods of forest management and enhance the adaptation and resilience of their forest stands through the establishment of mixed forests (Dorren et al. 2004; Knoke et al. 2008; Bauhus et al. 2017). In recent years, the wild service tree (*Sorbus torminalis* (L.) Crantz) has been one of the species of increasing interest among forest owners in Europe owing to its highly valuable wood, high tolerance for drought and pathogens, and relatively fast growth. Despite its wide distribution, it is considered a rare forest tree species; it is light-demanding and has low competitive capability (Demesure-Musch and Oddou-Muratorio 2004; Bednorz 2007; Hemery et al. 2010; Welk et al. 2016). The wild service tree is a post-pioneer tree that colonizes disturbed areas and forest edges and can grow on all types of soil (Demesure et al. 2000; Welk et al. 2016). Many researchers (Spiecker 2006; Angelone et al. 2007; Paganova 2007; Pyttel et al. 2013 and references therein) claim that the formation of high forests and the transformation of coppice forests into high forests are reasons for continuing species loss. Overall, there is a lack of knowledge on the distribution and genetics of scattered broadleaves such as the wild service tree. Furthermore, there is a lack of legal guidelines for the sustainable use and protection of rare and scattered tree species such as *Sorbus torminalis* in the FRM regulations in Germany based on the German Act on Forest Reproductive Material (FoVG, 2003).

Therefore, in 2016, the AWG implemented a project on the wild service tree in southern Germany (Bavaria [BY] and Baden-Württemberg [BW]), which aimed to analyze the genetic structure and diversity of the species, delineate provenance regions, identify and propose new seed stands, and discuss GCU selection based on its distribution as well as the structure and extent of its genetic variation. The workflow for FGR selection and conservation for other rare tree species (see Fig. 11.2) was followed.

During the project, 998 trees from 34 natural wild service tree populations were sampled in BY and BW (Fig. 11.6). In addition, natural wild service tree populations were compared with a seedling seed orchard of 93 trees in Neudorf (BY) and a cohort of 56 plus trees (e.g. Fig. 11.6) selected from the 34 natural populations (46 plus trees in BY and ten in BW; Kavaliauskas et al. (2021)).

The results for the wild service tree populations in Southern Germany revealed medium differentiation among studied populations ($F_{ST} = 4\%$), probably because of the fragmented distribution across the sampled area. The level of overall genetic diversity within stands was comparatively high (He = 0.782) (Kavaliauskas et al. 2021). Other studies also found relatively high genetic diversity within wild service tree populations in the species' main distribution range in Western and Central Europe, but certain populations may be endangered due to spatial isolation and low connectivity resulting in a lack of gene flow between them (Demesure et al. 2000; Bednorz 2007; Angelone et al. 2007). Therefore, active conservation measures should be considered to protect and maintain the level of genetic diversity among natural wild service tree populations.

Fig. 11.6 Selected plus tree of *Sorbus torminalis* in Bavaria. (Photo: AWG/ Muhidin Šeho)

Half of the sampled populations (nine in BY and eight in BW) exhibit higher than average allelic richness (Ar ≥ 9.2) and were thus considered potential GCUs. We ultimately proposed approving the five populations with the highest allelic richness (Ar ≥ 10.0) as GCUs.

Further analysis of our data showed evidence of isolation by distance (IBD) based on the Mantel test, which is in line with other studies conducted in Poland (Bednorz and Kosiński 2006), Switzerland (Angelone et al. 2007), and France (Demesure et al. 2000). The Bayesian clustering method implemented in STRUCTURE 2.3.3 (Evanno et al. 2005) revealed the existence of four genetic clusters within the 34 studied wild service tree populations in Southern Germany (Fig. 11.7). We used genetic data based on genetic markers to infer population clustering and delineated four regions of provenance for the wild service tree in Southern Germany. The final decision regarding the number and distribution of provenance regions resulted from a synthesis of species distribution, genetic differentiation, and clustering and geographical proximity in Southern Germany. Thus, our results and identified provenance regions enabled us to select seed stands in a similar way as for the species regulated under the FoVG (2003), following the distribution of genetic diversity and phenotypic quality in a representative manner within a given provenance region. A total of 12 forest stands (three in BW and nine in BY) fulfilling genetic diversity and phenotypic quality selection parameters were proposed as potential seed stands. In addition, a comparison of genetic variation among natural wild service tree populations (34 populations) with the already existing Neudorf seedling seed orchard and the newly selected cohort of plus trees (56 trees) revealed a higher genetic diversity in the seed orchard and the cohort of plus trees than in natural populations (Kavaliauskas et al. 2021). In *Tilia platyphyllos,* the comparison

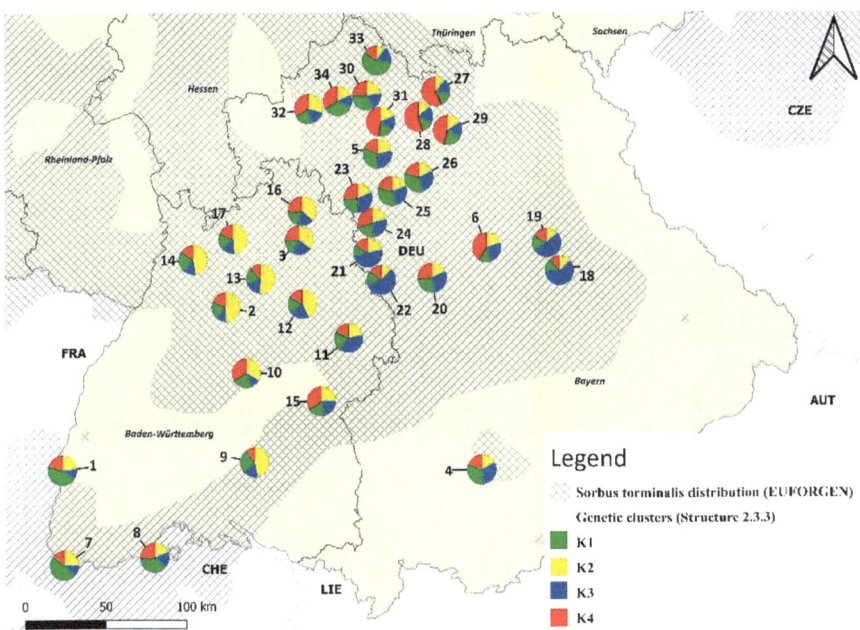

Fig. 11.7 Population genetic structure based on Bayesian clustering (STRUCTURE 2.3.3). The pie charts show the relative proportions of the four genetic clusters within the studied wild service tree populations. (Kavaliauskas et al. 2021)

of genetic variation of natural stands and seed orchards revealed that all parameters of genetic diversity were higher in seed orchards (Rau et al. 2023, Kavaliauskas et al. 2024). This confirms that seed orchards are highly important as an *ex situ* measure for ensuring high genetic diversity in wild service tree seeds. A new seed orchard including 93 selected plus trees to ensure high-quality FRM production and use in Southern Germany was established in 2022. Our study thus provides a straightforward approach for FGR selection from the delineation of provenance regions to the selection of seed stands, GCUs, and plus trees, enabling us to create a network of *in situ* and *ex situ* sites for wild service trees in different environments as a basis for further FGR use and conservation. The sustainable use of GCUs for moderate seed harvest allows the species' distribution to be increased and economic, as well as ecological goals to be reached.

Case Study of Species Genetic Conservation After Pathogen Outbreak

Fraxinus excelsior

The common ash (*Fraxinus excelsior* L.) is an important component of mixed forest ecosystems in Europe and can be found across a wide range of growing conditions from nutrient-rich floodplain forests to calcareous drier sites (Pautasso et al. 2013;

Fig. 11.8 Long-term observation plot with ash trees of varying susceptibility levels. (Photo: AWG/B. Fussi)

Kölling and Walentowski, 2002). It was therefore considered a promising tree species under climate change. Ash trees account for 1–2% of the forest area in Europe, with higher densities in some regions (Enderle et al. 2019). In Bavaria, more than 120 ash seed stands are registered, which confirms the high relevance of the species. Seeds were harvested regularly until ash dieback (caused by the pathogen *Hymenoscyphus fraxineus* introduced from Eastern Asia) appeared; disease impact nearly resulted in the loss of this economically and ecologically valuable tree species (Enderle et al. 2017). As an ecological keystone species in floodplain forests, loss of ash would mean a loss of habitat for several other species and, potentially, the extinction of species dependent on ash trees (Pautasso et al. 2013); e.g., larvae of *Euphydryas maturna* feeds exclusively on ash leaves. Ash trees are hosts to a wide range of taxa: in the UK, it was estimated that 953 species are associated with European ash and that the 69 most strongly associated species might rapidly become extinct if the ash populations were to decline (Mitchell et al. 2014).

In recent decades, ash populations in Europe have been threatened by ash dieback caused by the fungus *Hymenoscyphus fraxineus* (T. Kowalski) (Baral et al. 2014). The disease has already led to massive salvage felling and mortality and endangers the future utilization of ash in European forestry as a whole (Enderle et al. 2017). A comprehensive survey with a pan-European perspective (data from ICP Forests Level I) found a substantial increase in defoliation and mortality over time, indicating that crown defoliation has almost doubled during the past three decades; the study forecasts that the overall mean defoliation will likely reach 50% as early as 2030 (George et al. 2022). However, lower susceptibility toward ash dieback has been found to be present in a small fraction of ash individuals, which exhibit few symptoms, possibly providing a solution to sustaining the species in European forests (e.g., McKinney et al. 2014; Rigling et al. 2016; Enderle, 2019) through intensive selection and breeding efforts. Observations in a clonal and a progeny trial in Bavaria since 2014 (Fig. 11.8) have revealed that about 30 percent of all offspring showed high resilience to ash dieback (Seidel et al. 2025). The genomic basis of ash dieback tolerance in *F. excelsior* is more and more understood and can complement phenotypic selection and breeding efforts. Found SNPs with the highest predictive power located within genes related to plant defense and phenology and provided insights into a multifaceted defense response, according to which a combination of direct defense mechanisms and phenological avoidance of pathogen spread constitutes tolerance to ash dieback.

Resistance can be controlled by only a few genes, but it more often involves numerous genes, each with a small effect (White et al. 2007). Low-frequency alleles unevenly distributed across the species range are a major challenge for conservation considerations (White et al. 2007). Less susceptible genotypes need to be tested under heavy disease pressure in the field or with artificial inoculations and planted together in seed orchards or artificial populations (gene reserves). In natural populations, less susceptible genotypes should be conserved and their reproduction and natural regeneration stimulated. The overall goal is to maintain several forms of resistance as well as high genetic diversity to withstand the disease in the long term (Jump et al. 2009). Static *ex situ* conservation measures like cryopreservation of genetically diverse and less susceptible genotypes (tissue, seed, and pollen) might complement conservation efforts (Mirjani et al. 2022).

In Southern Germany, genetic variation between damaged and undamaged subpopulations was compared to see whether genetic differences between these two collectives exist (Fussi and Konnert, 2014). The authors detected higher proportions of heterozygotic individuals (observed heterozygosity) among less susceptible trees compared to susceptible ones, indicating that heterozygotic individuals might be able to withstand ash dieback better than homozygotic individuals. This means a higher variability at the individual level for heterozygotic individuals, which might therefore possess higher plasticity and responsiveness against the disease. Namkoong et al. (1998) and Tessier du Cros et al. (1999) have previously also suggested that heterozygotic individuals could be more resistant to environmental stresses.

Several studies based on genetic markers have revealed high genetic variation within stands (e.g., for Italy: Ferrazzini et al. 2007; for Germany: Höltken et al. 2003; Heuertz et al. 2004a, 2004b; Hebel et al. 2006; Fussi and Konnert, 2014; and for Austria: Heinze and Fussi, 2017), and a generally higher level of genetic variation was detected in Central and Western Europe than in Southeastern Europe (Heuertz et al. 2004a). By contrast, genetic differentiation was greater among populations in Southeastern Europe than in Central Europe (Heuertz et al. 2004a). Levels of genetic differentiation in Austria were low, with some clustering of stands along the Danube River compared to two stands in the Alps (Heinze and Fussi, 2017) and higher in Germany (Hebel et al. 2006b; Fussi and Konnert, 2014). Different gene flow intensities during postglacial recolonization likely explain these patterns (Heuertz et al. 2004a).

Using chloroplast haplotypes, distinct regions within Europe were identified, with certain haplotypes found to be typical for different regions (Heuertz et al. 2004b). Higher chloroplast variation with four haplotypes was detected in Southern Germany (Heuertz et al. 2004b), especially in populations from southeastern

Bavaria (Fussi and Konnert, 2014); this is caused by historical contact between chloroplast lineages from different refugia in Italy, the eastern Alps, and the Balkan Peninsula (Heuertz et al. 2004b).

Habitat fragmentation can reduce the genetic connectivity of plant populations. Therefore, pollen and seed dispersal patterns must be studied to understand and prevent the consequences of reduced habitat size and spatial isolation (Sork and Smouse, 2006). Ash has wind-dispersed pollen and seeds. Studies on *Fraxinus pennsylvanica* estimated mean seed dispersal distances between 47 and 60 m, with long-distance dispersal reaching up to 150 m (Schmiedel et al. 2013). Semizer-Cuming et al. (2017) examined the genetic connectivity of ash trees in an isolated forest area where only individual ash trees occurred within a radius of 5 km around the study site. It was found that most analyzed seeds and seedlings descended from local ash trees (no seed transport beyond 100 m) (Semizer-Cuming et al. 2017).

Pollen was found to be more mobile, with 50% of pollen dispersed within approximately 270 m, while 10% can be dispersed further than ca. 1400 m and 1% can reach distances of more than 4400 m (Semizer-Cuming et al. 2017). Landscape structure and wind speed influenced this pattern more strongly for ash seed than for pollen dispersal (Semizer-Cuming et al. 2017). However, the cardinal wind direction associated with the highest pollen impaction was also identical to the predominant wind direction (Eisen et al. 2022).

Pollen flow studies using pollen traps in two seed orchards revealed that 50% of pollen was found within 200 m and more than 10% at 500 m (Eisen et al. 2022). To ensure the cross-pollination of healthy ash trees, the distance between ash individuals or stands should not be too great, and barriers like large conifer stands should be avoided. In years with little flowering, more intensive gene flow from outside trees to the seed orchard was detected (Eisen et al. 2022), and it is thus recommended to conduct seed harvesting from ash trees in the orchards preferentially in full-mast years. Based on parentage analyses in a seed orchard and a floodplain forest, pollination success decreased substantially with increasing distance to the mother tree (mean distance of 76 m and 166 m, respectively) (Eisen et al. 2023). However, despite the dense tree cover in the floodplain forest, pollen was transported over long distances there as well (greater than 550 m), suggesting that nonlocal sources also play a role in pollination. This is supported by the amount of foreign pollen input identified in the seed orchard (66.5%) (Eisen et al. 2023).

Heuertz et al. (2004a) indicated that from a historical perspective, effective seed dispersal occurred mainly over short distances, whereas pollen dispersal seemed to be effective over longer distances as well. In some cases, gene flow by pollen into fragmented populations can be extensive; 46–95% were detected in each of the three remnant populations (Bacles et al. 2005) with a very localized and restricted part and another part extending over long distances, with dispersal occurring over several kilometers (Bacles et al. 2005). Effective dispersal distances (average: 328 m) were greater for fragmented populations than the values reported for contiguous populations (Bacles et al. 2005).

Self-pollination was detected only to a very small extent, with no major influence on reproduction (Eisen et al. 2023). Common ash trees with low susceptibility to ash dieback have higher reproductive fitness compared to highly susceptible trees, as determined for males (Eisen et al. 2023), and in an even more pronounced fashion for female success (Semizer-Cuming et al. 2019). Overall, this leaves healthy ash trees overrepresented as seed and pollen parents for next-generation seedlings (Semizer-Cuming et al. 2021). Long-distance pollen transport contributes to the connectivity between ash trees in the landscape. Additionally, both healthy and slightly diseased fathers/pollinators provide a greater contribution to pollination, thus potentially improving the health of the next generation of ash trees. Moreover, gene flow between stepping-stone populations is necessary to ensure a positive impact on the genetic diversity of ash populations in the future. However, the ash dieback disease itself causes major loss of ash (through natural death and active withdrawal), leading to declining populations with restricted effective gene flow between the remaining—and possibly less susceptible—trees.

European ash usually regenerates very successfully and can be very competitive on suitable sites. This should be considered for *in situ* conservation when silvicultural management options are developed with the aim of increasing resistance against ash dieback in future ash populations. Once there is enough natural ash regeneration, basically, no further management actions are necessary for the next 10–15 years. Regular thinning of older stands is still strongly recommended, along with the promotion of complementary tree species and sound-looking ash individuals (Enderle et al. 2019). Individual tree vigor, crown defoliation, and epicormic shoots as measures of individual vitality are the most important criteria in the selection of ash trees for thinning. Foresters tend to clear-cut heavily affected stands because they look for ash trees without any symptoms; however, such trees rarely exist. Instead, the focus should be on less affected individuals (Fig. 11.9), and these should be maintained as it is essential for the long-term conservation of ash to allow these trees to reproduce. Otherwise, fragmentation of ash populations and a crucial loss of genetic diversity must be expected.

Selection against the most susceptible genotypes at generation turnover represents good news for the potential recovery of ash forests, facilitating the establishment of dieback-resistant ash and—together with substantial gene flow over longer distances—generating hope for the species' future in European forests (Semizer-Cuming et al. 2019, 2021; Eisen et al. 2022, 2023). However, removing trees that appear less affected and eliminating natural regeneration will promote the ongoing fragmentation of ash populations caused by ash dieback and eventually could lead to the loss of the species. As an *ex situ* measure, healthy and less damaged plus trees are being collected in Germany to be grafted and planted as gene banks in several places (FraxGen-Project, https://www.fraxforfuture.de/). Overall, around 600 trees have been collected. Compared to the total area of ash in Germany, however, this can only be a first step, and further efforts will need to be undertaken to secure as many healthy ash trees as possible. The overall goal is to establish seed orchards of healthy ash trees to produce high-quality seeds and restore healthy ash forests. This must be considered in the light of potential new threats. For a few years, *Agrilus planipennis* (emerald ash borer, EAB), a wood-boring beetle native to East Asia, has been on its way from Russia toward Eastern Europe (Valenta et al. 2017). All ash

Fig. 11.9 Less susceptible plus tree grafted for the newly established seed orchard in Bavaria. (Photo: AWG/H. Seidel)

species native to Europe and North America are known to be susceptible to EAB attacks, which cause high tree mortality even among formerly healthy trees. Given current expansion rates, EAB is expected to reach Central Europe within 15–20 years (Valenta et al. 2017). Most likely, EAB will become a major pest in Europe; therefore, early and dedicated response is needed to reduce the level of ash mortality and cope with this new pest in the future.

Conclusion and Outlook

Both foresters and society, policymakers, conservation actors, and the scientific community need to be concerned about *in situ* and *ex situ* conservation of FGR due to environmental and climatic changes that may increase forest susceptibility to pests and stochastic events. It is crucial to recognize that forest genetic diversity and variability are vital because of their importance for adaptation and fulfilling ecological and economic forest functions now and in the future. The increasing loss of biological and genetic diversity threatens the stability of entire forest ecosystems. These concerns can promote efforts to evaluate, monitor, and conserve genetic variability (*in situ* and *ex situ*) and protect endangered species and their genetic resources. The examples provided in this chapter offer insights into the development of *in situ* and *ex situ* conservation measures for different species groups, such as rare species or those suffering from diseases.

References

Adams WT, Burczyk J (2000) Magnitude and implications of gene flow in gene conservation reserves. In: Young AG, Boshier D, Boyle TJB (eds) Forest conservation genetics: principles and practice. CABI Pub, New York, pp 215–244

Aitken SN, Yeaman S, Holliday JA, Wang T, Curtis-McLane S (2008) Adaptation, migration or extirpation: climate change outcomes for tree populations. Evol Appl 1(1):95–111

Angelone S, Hilfiker K, Holderegger R, Bergamini A, Hoebee SE (2007) Regional population dynamics define the local genetic structure in *Sorbus torminalis*. Mol Ecol 16(6):1291–1301. https://doi.org/10.1111/j.0962-1083.2006.03202.x

Aravanopoulos FA (2018) Do silviculture and forest management affect the genetic diversity and structure of long-impacted forest tree populations? Forests 9(6):355

Bacles CFE, Ennos RA (2008) Paternity analysis of pollen-mediated gene flow for Fraxinus excelsior L. in a chronically fragmented landscape. Heredity 101:368–380. https://doi.org/10.1038/hdy.2008.66

Bacles CF, Burczyk J, Lowe AJ, Ennos RA (2005) Historical and contemporary mating patterns in remnant populations of the forest treefraxinus excelsior L. Evolution 59(5):979–990

Bajc M, Aravanopoulos F, Westergren M, Fussi B, Kavaliauskas D, Alizoti P et al (2020) Manual for forest genetic monitoring. Slovenian Forestry Institute, Silva Slovenica Publishing Centre, Ljubljana. https://doi.org/10.20315/SFS.167

Baral HO, Queloz V, Hosoya T (2014) Hymenoscyphus fraxineus, the correct scientific name for the fungus causing ash dieback in Europe. IMA Fungus 5(1):79–80

Bauhus J, Forrester DI, Gardiner B, Jactel H, Vallejo R, Pretzsch H (2017) Ecological stability of mixed-species forests. In: Mixed-species forests. Springer, Berlin, Heidelberg, pp 337–382

Bednorz L (2007) Conservation of genetic resources of Sorbus torminalis in Poland. Dendrobiology 58:3–7

Bednorz L, Kosiński P (2006) Genetic variability and structure of the wild service tree (*Sorbus torminalis* (L.) Crantz) in Poland. Silvae Genet 55(1–6):197–202. https://doi.org/10.1515/sg-2006-0027

Caballero A, Rodríguez-Ramilo ST, Avila V, Fernández J (2010) Management of genetic diversity of subdivided populations in conservation programmes. Conserv Genet 11(2):409–419. https://doi.org/10.1007/s10592-009-0020-0

Danusevičius D, Kerpauskaitė V, Kavaliauskas D, Fussi B, Konnert M, Baliuckas V (2016) The effect of tending and commercial thinning on the genetic diversity of Scots pine stands. Eur J For Res 135(6):1159–1174. https://doi.org/10.1007/s10342-016-1002-7

Danusevicius D, Rajora OP, Kavaliauskas D, Baliuckas V, Augustaitis A (2023) Genetic diversity and fine-scale spatial genetic structure of unmanaged old-growth versus managed second-growth populations of Scots pine (*Pinus sylvestris* L.) in Lithuania. Eur J For Res:1–21. https://doi.org/10.1007/s10342-023-01556-x

De Dato G, Teani A, Mattioni C, Marchi M, Monteverdi MC, Ducci F (2018) Delineation of seed collection zones based on environmental and genetic characteristics for *Quercus suber* L. in Sardinia, Italy. iForest 11(5):651

Demesure B, Le Guerroué B, Lucchi G, Prat D, Petit RJ (2000) Genetic variability of a scattered temperate forest tree: *Sorbus torminalis* L. (Crantz). Ann For Sci 57(1):63–71. https://doi.org/10.1051/forest:2000101

Demesure-Musch B, Oddou-Muratorio S (2004) EUFORGEN technical guidelines for genetic conservation and use for wild service tree (*Sorbus torminalis*). Bioversity International

Doonan JM, Budde KB, Kosawang C, Lobo A, Verbylaite R, Brealey JC et al (2023) Multiple, single trait GWAS and supervised machine learning reveal the genetic architecture of Fraxinus excelsior tolerance to ash dieback in Europe. Plant Cell Environ 48(5):3793–3809

Dorren LK, Berger F, Imeson AC, Maier B, Rey F (2004) Integrity, stability and management of protection forests in the European Alps. For Ecol Manag 195(1–2):165–176. https://doi.org/10.1016/j.foreco.2004.02.057

Dubreuil M, Riba M, Gonzalez-Martinez S, Vendramin GG, Sebastiani F, Mayol M (2010) Genetic effects of chronic habitat fragmentation revisited: strong genetic structure in temperate tree, *Taxus baccata (Taxaceae)*, with great dispersal capability. Am J Bot 97(2):303–310. https://doi.org/10.3732/ajb.0900148

Ducousso A, Bordacs S (2004) EUFORGEN technical guidelines for genetic conservation and use for pedunculate and sessile oaks (*Quercus robur and Q. petraea*). International Plant Genetic Resources Institute, Rome. 6 pages

Dudley N, Parrish J, Redford K, Stolton S (2010) The revised IUCN protected area management categories: the debate and ways forward. Oryx 44(4):485–490. https://doi.org/10.1017/S0030605310000566

Eisen AK, Fussi B, Šikoparija B, Jochner-Oette S (2022) Aerobiological pollen deposition and transport of *Fraxinus excelsior* L. at a small spatial scale. Forests 13(3):424

Eisen AK, Semizer-Cuming D, Jochner-Oette S et al (2023) Pollination success of *Fraxinus excelsior* L. in the context of ash dieback. Ann For Sci 80:22. https://doi.org/10.1186/s13595-023-01189-5

Enderle R, Fussi B, Lenz HD, Langer G, Nagel R, Metzler B (2017) Ash dieback in Germany: research on disease development, resistance and management options. In: Vasaitis R, Enderle R (eds) Dieback of European Ash (*Fraxinus* spp.): consequences and guidelines for sustainable management. SLU, Uppsala, pp 89–105

Enderle R, Stenlid J, Vasaitis R (2019) An overview of ash (*Fraxinus* spp.) and the ash dieback disease in Europe. CABI Rev 2019:1–12

European Union (EU) directive 1999/105/CE of 22 December (1999) Marketing of forest reproductive material. Access to European Union law web. https://eur-lex.europa.eu/eli/dir/1999/105/oj. Accessed 26 Oct 2021

European Union (EU) communication from The Commission to the European Parliament, The European Council, The Council, The European Economic And Social Committee and The Committee of the Regions December (2019) The European Green Deal. Access to European Union law web. https://eur-lex.europa.eu/legal-content/EN/TXT/?uri. Accessed 28 Jan 2025

Evanno G, Regnaut S, Goudet J (2005) Detecting the number of clusters of individuals using the software STRUCTURE: a simulation study. Mol Ecol 14:2611–2620. https://doi.org/10.1111/j.1365-294X.2005.02553.x

Ferrazzini D, Monteleone I, Belletti P (2007) Genetic variability and divergence among Italian populations of common ash (*Fraxinus excelsior* L.). Ann For Sci 64(2):159–168

Fitzpatrick SW, Gerberich JC, Kronenberger JA, Angeloni LM, Funk WC (2015) Locally adapted traits maintained in the face of high gene flow. Ecol Lett 18(1):37–47

Foulley JL, Ollivier L (2006) Estimating allelic richness and its diversity. Livest Sci 101(1–3):150–158

FoVG (2003) Forstvermehrungsgutgesetz (FoVG) (The German Act on Forest Reproductive Material) 01.01.2003. https://www.ble.de/DE/Themen/Wald-Holz/Forstliches-Vermehrungsgut/forstliches-vermehrungsgut_node.html. Accessed 27 Oct 2021

Fussi B, Konnert M (2014) Genetic analysis of European common ash (Fraxinus excelsior L.) populations affected by ash dieback. Silvae Genetica 63(1–6):198–212

Fussi B, Westergren M, Aravanopoulos F, Baier R, Kavaliauskas D, Finzgar D, Alizoti P, Bozic G, Avramidou E, Konnert M, Kraigher H (2016) Forest genetic monitoring: an overview of concepts and definitions. Environ Monit Assess 188(8):493. https://doi.org/10.1007/s10661-016-5489-7

Gautam S, Timilsina S, Shrestha M (2021) The effects of forest management activities on genetic diversity of forest trees. Indones J Soc Environ Issues (IJSEI) 2(2):110–118

Generhaltungskonzept (2015) Konzept zur Erhaltung und zur nachhaltigen Nutzung forstlicher Genressourcen in Bayern. (Concept for the conservation and sustainable use of forest genetic resources in Bavaria). https://awg.bayern.de/mam/cms02/asp/bilder/generhaltungskonzept_bayern.pdf. Accessed 29 Jan 2024

George JP, Sanders TG, Timmermann V, Potočić N, Lang M (2022) European-wide forest monitoring substantiate the neccessity for a joint conservation strategy to rescue European ash species (*Fraxinus* spp.). Sci Rep 12(1):4764

Gougherty AV, Keller SR, Fitzpatrick MC (2021) Maladaptation, migration and extirpation fuel climate change risk in a forest tree species. Nat Clim Chang 11(2):166–171

Hebel I, Haas R, Dounavi A (2006) Genetic variation of common ash (*Fraxinus excelsior* L.) populations from provenance regions in southern Germany by using nuclear and chloroplast microsatellites. Silvae Genetica 55(1):38–43

Heinze B, Fussi B (2017) Pre-disease levels of genetic diversity and differentiation among common Ash (*Fraxinus excelsior* L.) Seedlots in Austria. Balt For 23(1):198–208

Hemery GE, Clark JR, Aldinger E, Claessens H, Malvolti ME, O'connor E et al (2010) Growing scattered broadleaved tree species in Europe in a changing climate: a review of risks and opportunities. Forestry 83(1):65–81. https://doi.org/10.1093/forestry/cpp034

Heuertz M, Hausman JF, Hardy OJ, Vendramin GG, Frascaria-Lacoste N, Vekemans X (2004a) Nuclear microsatellites reveal contrasting patterns of genetic structure between western and southeastern European populations of the common ash (*Fraxinus excelsior* L.). Evolution 58(5):976–988

Heuertz M, Fineschi S, Anzidei M, Pastorelli R, Salvini D, Paule L et al (2004b) Chloroplast DNA variation and postglacial recolonization of common ash (*Fraxinus excelsior* L.) in Europe. Mol Ecol 13(11):3437–3452

Hoban, S., Bruford, M., da Silva, J. M., Funk, W. C., Frankham, R., Gill, M. J., … , Laikre, L. (2023). Genetic diversity goals and targets have improved but remain insufficient for clear implementation of the post-2020 global biodiversity framework. Conserv Genet, 24, 181–191. https://doi.org/10.1007/s10592-022-01492-0

Höltken AM, Tahtinen J, Pappinen A (2003) Effects of discontinuous marginal habitats on the genetic structure of common ash (*Fraxinus excelsior* L.). Silvae Genetica 52(5–6):206–211

IUCN (1994) Guidelines for protected area management categories. IUCN, Gland/Cambridge

Jost L, Archer F, Flanagan S, Gaggiotti O, Hoban S, Latch E (2018) Differentiation measures for conservation genetics. Evol Appl 11(7):1139–1148

Jump AS, Marchant R, Peñuelas J (2009) Environmental change and the option value of genetic diversity. Trends Plant Sci 14(1):51–58

Kavaliauskas D, Fussi B, Westergren M, Aravanopoulos F, Finzgar D, Baier R, Alizoti P, Bozic G, Avramidou E, Konnert M, Kraigher H (2018) The interplay between forest management practices, genetic monitoring, and other long-term monitoring systems. Forests 9(3):133. https://doi.org/10.3390/f9030133

Kavaliauskas D, Šeho M, Baier R, Fussi B (2021) Genetic variability to assist in the delineation of provenance regions and selection of seed stands and gene conservation units of wild service tree (*Sorbus torminalis* (L.) Crantz) in southern Germany. Eur J For Res 140(3):551–565

Kavaliauskas D, Fussi B, Rau B, Šeho M (2022a) Assessing genetic diversity of European white elm (*Ulmus laevis* Pallas) in Bavaria as an indicator for *in-situ* conservation and sustainable use of the species genetic resources. Eur J For Res 142(1):145

Kavaliauskas D, Danusevičius D, Baliuckas V (2022b) New insight into genetic structure and diversity of Scots pine (*Pinus sylvestris* L.) populations in Lithuania based on nuclear, chloroplast and mitochondrial DNA markers. Forests 13(8):1179

Kavaliauskas D, Fussi B, Šeho M (2024) Strong genetic differentiation and diversity of highly fragmented *Tilia platyphyllos* populations in Bavaria. Eur J For Res. (In preparation)

Kelleher CT (2018) Evolution and conservation of trees – a review of salient issues. In: Roberts JA (ed) Annual plant reviews. https://doi.org/10.1002/9781119312994.apr0621

Kleinschmit JR, Kownatzki D, Gregorius HR (2004) Adaptational characteristics of autochthonous populations—consequences for provenance delineation. For Ecol Manag 197(1–3):213–224. https://doi.org/10.1016/j.foreco.2004.05.037

Klumpp R, Dhar A (2011) Genetic variation of *Taxus baccata* L. populations in the Eastern Alps and its implications for conservation management. Scand J For Res 26(4):294–304. https://doi.org/10.1080/02827581.2011.566888

Knoke T, Ammer C, Stimm B, Mosandl R (2008) Admixing broadleaved to coniferous tree species: a review on yield, ecological stability and economics. Eur J For Res 127(2):89–101. https://doi.org/10.1007/s10342-007-0186-2

Kölling C, Walentowski H (2002) Die Rolle der Esche (*Fraxinus excelsior*) in einheimischen Waldgesellschaften. Berichte aus der LWF 34:6–20

Komárková M, Novotný P, Cvrčková H, Máchová P (2022) The genetic differences and structure of selected important populations of the endangered *Taxus baccata* in The Czech Republic. Forests 13(2):137. https://doi.org/10.3390/f13020137

Lande R, Barrowclough GF (1987) Effective population size, genetic variation, and their use in population management. In: Viable populations for conservation, vol 87. Cambridge University Press, Cambridge, pp 87–124

Lefevre F, Koskela J, Hubert J, Kraigher H, Longauer R, Olrik DC, Schuler S, Bozzano M et al (2013) Dynamic conservation of forest genetic resources in 33 European countries. Conserv Biol 27(2):373–384

Lefevre F, Bojkovski D, Bou Dagher Kharrat M, Bozzano M, Charvolin-Lemaire E, Hiemstra SJ, Kraigher H, Laloe D, Restoux G, Sharrock S, Sturaro E, van Hintum T, Westergren M, Maxted N, Gen Res Bridge Expert Panel (2024) European genetic resources conservation in a rapidly changing world: three existential challenges for the crop, forest and animal domains in the 21st century. Genet Res 5(9):13–28. https://doi.org/10.46265/genresj.REJR6896

Liesebach H, Eusemann P, Höltken AM, Tröber U, Kuchma O, Karopka M et al (2024) Effective population size of adult and offspring cohorts as a genetic monitoring tool in two stand-forming and wind-pollinated tree species: *Fagus sylvatica* L. and *Picea abies* (L.) Karst. Conserv Genet:1–15

Linares JC (2013) Shifting limiting factors for population dynamics and conservation status of the endangered English yew (*Taxus baccata* L., Taxaceae). For Ecol Manag 291:119–127. https://doi.org/10.1016/j.foreco.2012.11.009

Marshall DR, Brown AHD (1975) Optimum sampling strategies in genetic conservation. In: Frankel OH, Hawkes JG (eds) Crop genetic resources for today and tomorrow, vol 2. CUP Archive, pp 53–80

McKinney LV, Nielsen LR, Collinge DB, Thomsen IM, Hansen JK, Kjaer ED (2014) The ash dieback crisis: genetic variation in resistance can prove a long-term solution. Plant Pathol 63:485–499. https://doi.org/10.1111/ppa.12196

Mirjani L, Ghamarizare A, Spahbodi K (2022) Preservation of the genetic resources of Fraxinus excelsior L. in cryopreservation. Iran J Rangelands Forests Plant Breed GenetRes 30(1):108–117. https://doi.org/10.22092/ijrfpbgr.2022.357696.1409

Mitchell RJ et al (2014) Ash dieback in the UK: a review of the ecological and conservation implications and potential management options. Biol Conserv 175:95–109

Namkoong G (1998) Forest genetics and conservation in Europe. Conservation of forest genetic resources in Europe. International Plant for Genetic Resources Institute, Rome, pp 3–10

Neel MC, Cummings MP (2003) Effectiveness of conservation targets in capturing genetic diversity. Conserv Biol 17(1):219–229. https://doi.org/10.1046/j.1523-1739.2003.01352.x

Paganová V (2007) Ecology and distribution of *Sorbus torminalis* (L.) Crantz. in Slovakia. Hortic Sci 34(4):138–151

Paul M, Hinrichs T, Janssen A, Schmitt H-P, Soppa B, Stephan BR, Doerflinger H (2000) Konzept zur Erhaltung und nachhaltigen Nutzung forstlicher Genressourcen in der Bundesrepublik Deutschland. Sächsische Landesanstalt für Forsten. 66 S

Pautasso M, Aas G, Queloz V, Holdenrieder O (2013) European ash (*Fraxinus excelsior*) dieback– a conservation biology challenge. Biol Conserv 158:37–49

Pérez-Pereira N, Wang J, Quesada H, Caballero A (2022) Prediction of the minimum effective size of a population viable in the long term. Biodivers Conserv 31(11):2763–2780

Petit RJ, Elmousadik A, Pons O (1998) Identifying populations for conservation on the basis of genetic markers. Conserv Biol 12(4):844–855. https://doi.org/10.1111/j.1523-1739.1998.96489.x

Pyttel P, Kunz J, Bauhus J (2013) Growth, regeneration and shade tolerance of the Wild Service Tree (*Sorbus torminalis* (L.) Crantz) in aged oak coppice forests. Trees 27(6):1609–1619. https://doi.org/10.1007/s00468-013-0908-7

Rajora OP, Mosseler A (2001) Challenges and opportunities for conservation of forest genetic resources. Euphytica 118(2):197–212. https://doi.org/10.1023/A:1004150525384

Rajora OP, Rahman MH, Buchert GP, Dancik BP (2000) Microsatellite DNA analysis of genetic effects of harvesting in old-growth eastern white pine (*Pinus strobus*) in Ontario, Canada. Mol Ecol 9(3):339–348. https://doi.org/10.1046/j.1365-294x.2000.00886.x

Rau B, Kavaliauskas D, Fussi B, Šeho M (2023) Bewertung von Erntebeständen der Sommerlinde in Bayern. AFZ DerWald 8(2023):18–21

Rigling D, Hilfiker S, Schöbel C, Meier F, Engesser R, Scheidegger C et al (2016) Ash dieback: biology, disease symptoms and recommendations for control. Merkblatt für die Praxis 57

Santos-del-Blanco L, Olsson S, Budde KB, Grivet D, González-Martínez SC, Alía R, Robledo-Arnuncio JJ (2022) On the feasibility of estimating contemporary effective population size (Ne) for genetic conservation and monitoring of forest trees. Biol Conserv 273:109704

Schmiedel D, Huth F, Wagner S (2013) Using data from seed-dispersal modelling to manage invasive tree species: the example of *Fraxinus pennsylvanica* Marshall in Europe. Environ Manag 52:851–860

Šeho M, Fussi B, Rau B, Kavaliauskas D (2022) Conservation and sustainable use of forest genetic resources of English yew (*Taxus baccata* L.) in Bavaria. SilvaWorld 1(1):52–68. https://doi.org/10.29329/silva.2022.462.06

Seidel H, Šeho M, Fussi B (2025) Hope for ash conservation and propagation—single individuals can be highly resistant to an invasive pathogen. J Plant Dis Prot 132(1):1–15

Seidel H, Šeho M, Fussi B (submitted) Few resilient individuals as the potential key for tree breeding towards tolerance against an invasive pathogen, submitted to Journal of Plant Diseases and Protection

Semizer-Cuming D, Kjær ED, Finkeldey R (2017) Gene flow of common ash (*Fraxinus excelsior* L.) in a fragmented landscape. PLoS One 12:e0186757. https://doi.org/10.1371/journal.pone.0186757

Semizer-Cuming D, Finkeldey R, Nielsen LR et al (2019) Negative correlation between ash dieback susceptibility and reproductive success: good news for European ash forests. Ann For Sci 76:16. https://doi.org/10.1007/s13595-019-0799-x

Semizer-Cuming D, Chybicki IJ, Finkeldey R et al (2021) Gene flow and reproductive success in ash (*Fraxinus excelsior* L.) in the face of ash dieback: restoration and conservation. Ann For Sci 78:14. https://doi.org/10.1007/s13595-020-01025-0

Sork VL, Smouse PE (2006) Genetic analysis of landscape connectivity in tree populations. Landsc Ecol 21(6):821–836

Spiecker H (2006) Minority tree species–a challenge for multi-purpose forestry. Nature based forestry in central Europe. Alternative to industrial forestry and strict preservation. Studia Forestalia Slovenica 126:47–59

Tessier du Cros E, Màtyàs C, Kriebel H (1999) Contribution of genetics to the sustained management of global forest resources—conclusions and recommendations. In: Màtyàs C (ed) Forest genetics and sustainability, forestry sciences, vol 63. Kluwer, Dordrecht, pp 281–287

Tigano A, Friesen VL (2016) Genomics of local adaptation with gene flow. Mol Ecol 25(10):2144–2164

Valenta V, Moser D, Kapeller S, Essl F (2017) A new forest pest in Europe: a review of Emerald ash borer (*Agrilus planipennis*) invasion. J Appl Entomol 141(7):507–526

Venturas M, Fuentes-Utrilla P, Ennos R, Collada C, Gil L (2013) Human-induced changes on fine-scale genetic structure in *Ulmus laevis* Pallas wetland forests at its SW distribution limit. Plant Ecol 214(2):317–327

von Wuehlisch G (2008) EUFORGEN technical guidelines for genetic conservation and use for European beech (*Fagus sylvatica*). Biodiversity International, Rome. 6 pages

Vornam B, Recarli N, Gailing O (2004) Spatial distribution of genetic variation in a natural beech stand (*Fagus sylvatica* L.) based on microsatellite markers. Conserv Genet 5:561–570

Welk E, De Rigo D, Caudullo G (2016) *Sorbus torminalis* in Europe: distribution, habitat, usage and threats. In: San-Miguel-Ayanz J, De Rigo D, Caudullo G, Durrant TH, Mauri A (eds) European atlas of forest tree species. Publ. Off. EU, Luxembourg, p e01090d+

White TL, Adams WT, Neale DB (eds) (2007) Forest genetics. Cabi

Open Access This chapter is licensed under the terms of the Creative Commons Attribution 4.0 International License (http://creativecommons.org/licenses/by/4.0/), which permits use, sharing, adaptation, distribution and reproduction in any medium or format, as long as you give appropriate credit to the original author(s) and the source, provide a link to the Creative Commons license and indicate if changes were made.

The images or other third party material in this chapter are included in the chapter's Creative Commons license, unless indicated otherwise in a credit line to the material. If material is not included in the chapter's Creative Commons license and your intended use is not permitted by statutory regulation or exceeds the permitted use, you will need to obtain permission directly from the copyright holder.

Practical Guidance for Rapid Biodiversity Assessment in Central European Forests

12

Janine Oettel, Cornelia Amon, Martin Steinkellner, Owen Bradley, Christoph Leeb, Frederik Sachser, and Katharina Lapin

Stepping stones—fieldwork in the forest (Photo: BFW/Florian Winter)

J. Oettel (✉) · C. Amon · M. Steinkellner · O. Bradley · F. Sachser · K. Lapin
Department of Forest Biodiversity & Nature Conservation, Austrian Research Centre for Forests, Vienna, Austria
e-mail: janine.oettel@bfw.gv.at; cornelia.amon@bfw.gv.at; martin.steinkellner@bfw.gv.at; owen.bradley@bfw.gv.at; frederik.sachser@bfw.gv.at; katharina.lapin@bfw.gv.at

C. Leeb
Department of Forest Biodiversity & Nature Conservation, Austrian Research Centre for Forests, Vienna, Austria

Natural History Museum, Vienna, Österreich, Austria
e-mail: Christoph.leeb@nhm.at

Abstract

Establishing biodiversity surveys is crucial for consistently monitoring ecosystems and informing conservation strategies. Rapid biodiversity assessment (RBA) approaches survey multiple species efficiently, supporting conservation planning and aiding in the evaluation of protected areas. Different assessment types include baseline inventory, species-specific, change, indicator, and resource assessments. Indicator selection is pivotal and requires reliability, representability, and replicability. Indicators linked to structural elements can provide comprehensive evaluations of forest biodiversity to assist informed decision-making. We propose an RBA protocol for Central European Forest ecosystems, with the aim of gathering relevant data to assess biological diversity. It encompasses structural elements and different species groups, incorporates insights from established monitoring systems, and prioritises species for monitoring under changing climate conditions. This modular survey approach combines standard assessments providing fundamental information on forest structure, including elements like deadwood and tree-related microhabitats (TreMs), with intensive surveys focusing on specific taxonomic groups—namely, vascular plants, fungi, birds, bats, and saproxylic beetles—with trained taxonomists involved. Enhancing our comprehension of biodiversity patterns within forest landscapes can provide valuable insights into the structural connectivity of these ecosystems. Our proposed RBA protocol serves not only as a guideline for exploring this important topic but also as a foundation for subsequent population genetic analyses, investigations of species interactions, and studies on the fundamental principles of dispersal and adaption. Ongoing research efforts aim to expand our knowledge to hitherto unrepresented species groups such as soil-dwelling organisms and pollinators, thereby enabling even more comprehensive assessments in the future.

Keywords

Forest structure · Habitat · Indicator · Monitoring · Multi-taxa · Sampling · Connectivity · Decision-making

General Principles and Definitions

Biodiversity assessments are required for various purposes; historically, they have been used to compile catalogues of species in specific geographic regions or habitat types. Covering all taxa—even for small habitats—has rarely been done due to constraints regarding expertise, time, and costs. Sampling diverse organisms demands different methods, dedicated effort, and resources, and limited resources therefore impede species-specific monitoring systems, instead favouring multispecies landscape-level approaches for efficiency (Franklin 1993; Manley et al. 2005). Certain protocols acknowledge the trade-off in precision concerning individual species by adopting an approach that evaluates multiple species simultaneously, known as rapid multispecies assessment (Nemitz and Huettmann 2015). Such systems offer increased adaptability, particularly in terms of sudden shifts in research focus (Watson and Novelly 2004).

Within forest ecosystems, rapid biodiversity assessment plays a crucial role in monitoring and promoting ecological connectivity by swiftly identifying key species and habitats, thereby supporting conservation planning (Heezik et al. 2023; Kipson et al. 2011; Sutherland 2000). By assessing biodiversity patterns, RBA is a precondition to gathering information that can be used to maintain essential ecological corridors and enable species movement across fragmented landscapes so as to support genetic diversity and overall ecosystem health and resilience. Effective RBA requires a well-defined conceptual framework and scope guiding its design and implementation. It serves as a vital tool for gathering essential information when limited data is available, supporting informed decision-making and conservation across diverse ecosystems (Sutherland 2000; Ward and Larivière 2004). A summary of key principles of rapid biodiversity assessments is provided in Fig. 12.1.

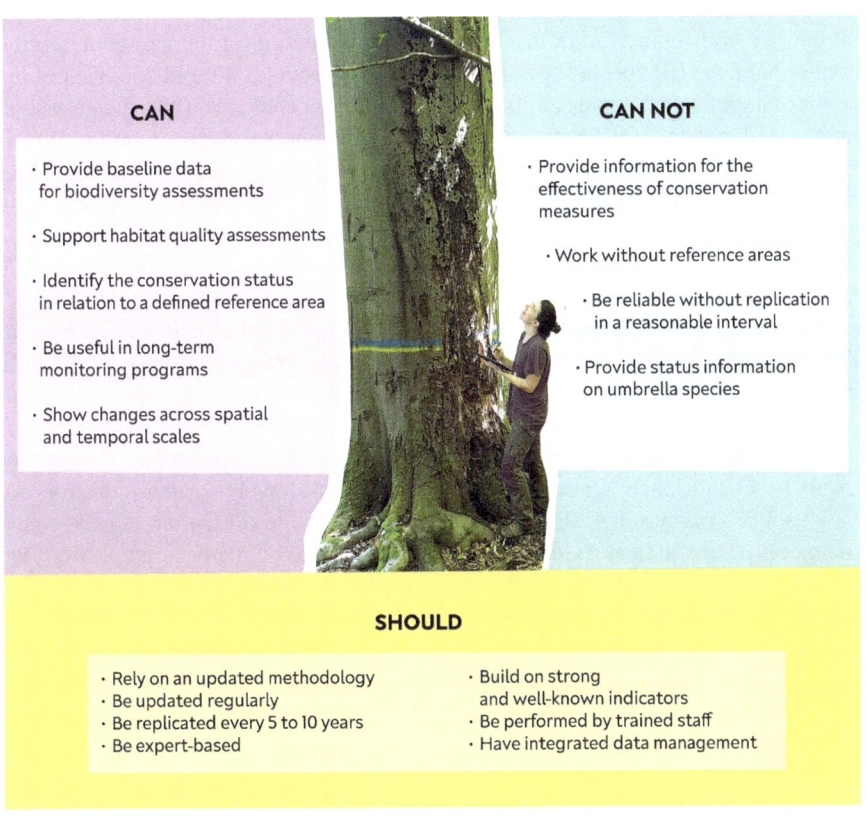

Fig. 12.1 Summary of key principles of rapid biodiversity assessments (RBA) and what it can—cannot—should cover

Experts emphasise the need for cost-effective sampling techniques yielding comparable biodiversity measurements, which are fundamental for establishing "biodiversity baselines" (Kipson et al. 2011; McNellie et al. 2020; Willis et al. 2010). The determination of baselines for biodiversity is essential for understanding the current state of ecosystems, allowing scientists and conservationists to recognise changes in biodiversity over time, identify potential threats, and design effective strategies for preservation and restoration (Fulton and Harcombe 2002). Baselines serve as vital benchmarks to monitor the health of ecosystems and track the success of conservation efforts. In the context of connectivity, biodiversity baselines offer critical insights into species and habitat distribution in different regions (Willis et al. 2005). This understanding helps to identify relevant corridors and protected areas supporting connectivity between habitats.

Designing a Rapid Assessment Concept

In general, five types of assessment can be distinguished which are applied based on the respective requirements (SCBD 2005): (i) baseline inventory—focuses on overall biological diversity rather than extensive or detailed information about specific taxa or habitats, (ii) species-specific assessment—provides a rapid appraisal of the status of a particular species or taxonomic group in a given area, (iii) change assessment—undertaken to determine the effects of human activities or natural disturbances on the ecological integrity and associated biodiversity of an area, (iv) indicator assessment—assumes that biological diversity in terms of species and community diversity can inform us about the overall health of particular ecosystems, and (v) resource assessment—aims to determine the potential for sustainable use of biological resources in a given area.

The selection of assessment type and suitable indicators for an RBA is crucial, as they form the foundation for reliable conservation strategies (Obrist and Duelli 2010). The indicators must be reliable, representable, and replicable to ensure accurate assessment of forest ecosystems (Kerr et al. 2000). One approach involves using indicators linked to organisms or structural elements representing forest biodiversity. For instance, Oettel and Lapin (2020) developed 44 such indicators for various forest ecosystems, aligning them with measurable silvicultural management practices. These indicators encompass species richness, habitat complexity, and ecosystem services and provide a comprehensive evaluation of forest biodiversity. By employing robust indicators, conservationists and policymakers can obtain valuable insights into the status of biodiversity, identify potential threats, and develop targeted interventions to ensure the ecological connectivity of forest ecosystems, fostering sustainable management practices and safeguarding biodiversity for future generations.

However, prior to undertaking an assessment, it is necessary to define the objectives and scope of the assessment (adapted from Maragos and Cook (1995)):

- Define the purpose and objectives of the rapid assessment;
- Define geographic scope based on the objectives and constraints;

- Select survey team and assign responsibilities;
- Undertake a review of existing data, literature, maps, and aerial photographs;
- Select field sites, relying on the above inputs and steps;
- Schedule and perform field work;
- Finalise and publish technical reports.

Here, we propose an internationally valid rapid biodiversity protocol for sampling in Central European Forest ecosystems that delivers comparable results in areas intended to improve the connectivity of forest ecosystems. We pursued a multi-taxa approach with the aim of collecting as much data as possible on the state of biological diversity within a given forest. A selection of different species groups ranging from the mammals and birds to saproxylic beetles is included in the RBA. In addition, different aspects of the forest structure are assessed as biodiversity proxies. The survey methodology and parameters proposed are based on several well-established monitoring systems in mountain-rich countries in Europe, e.g. from Switzerland (Düggelin et al. 2020) and Austria (BMNT 2019; Hauk et al. 2020), and further incorporate international standards where possible. The selection of species is based on Schindler et al. (2017), who proposed prioritised monitoring efforts to assess biodiversity under changing climate conditions. Accordingly, trees and vascular plants were prioritised for forest monitoring while fungi, soil organisms, bryophytes, birds, and arthropods (e.g. spiders, beetles, bees, and ants) were strongly recommended for inclusion.

As resources are often limited in biodiversity assessments, we propose a combination of standard surveys across all sites and intensive surveys on a smaller number of sites with specific research objectives (hypothesis-driven approach). During standard plot surveys, information that can be immediately assessed at the time of sampling is collected. This includes records of tree-related microhabitats covering saproxylic insect galleries, perennial fungus groups, and woodpecker breeding and feeding holes. Trained taxonomists are not required. Instead, a more general approach is adopted: Details of species groups are documented, and all observations recorded as accurately as possible, often accompanied by photos. In addition to directly assessing the presence and abundance of these taxonomic groups, other parameters such as habitat information are recorded as indirect measures of biodiversity. Additional intensive surveys involve the collection of an expanded set of parameters in greater detail, carried out by trained taxonomic experts. This includes recording of vascular plants, birds, and bats, and more comprehensive documentation of fungus species and saproxylic beetles. Selected species groups possess different dispersal abilities, which is advantageous when interpreting their presence or absence in terms of connectivity.

Defining a Sampling Design: Sample Plot Selection

To evaluate the ecological attributes of a specific area or site, different sampling techniques tailored to the respective dimensions and configurations are employed. For small areas with a size of 0.5 to 2.0 ha, one sample plot is established to provide a representation of the entire area (Fig. 12.2). When dealing with medium-sized areas (between 2 ha and 5 ha), more than one sampling plot should be established. The first sample plot is carefully chosen representatively, with subsequent plots intentionally positioned at least 100 metres away from the initial plot. For areas exceeding 5 ha in size and featuring unfavourable geometries such as elongated shapes, the same method can be applied. This approach ensures a representative and comprehensive sampling strategy. For areas measuring 5 ha or more and featuring favourable geometries, a systematic grid of sample plots evenly spaced at 100 × 100-m intervals across the entire area (see Fig. 12.2) following national grid standards is established (see, e.g., Statistik Austria (2023) for Austria). This ensures compatibility with data from other sources (e.g. on population density) for more detailed analysis. To minimise potential disturbances to the sampled area, a buffer zone with no survey activities extending 30 metres from the forest edge is considered appropriate.

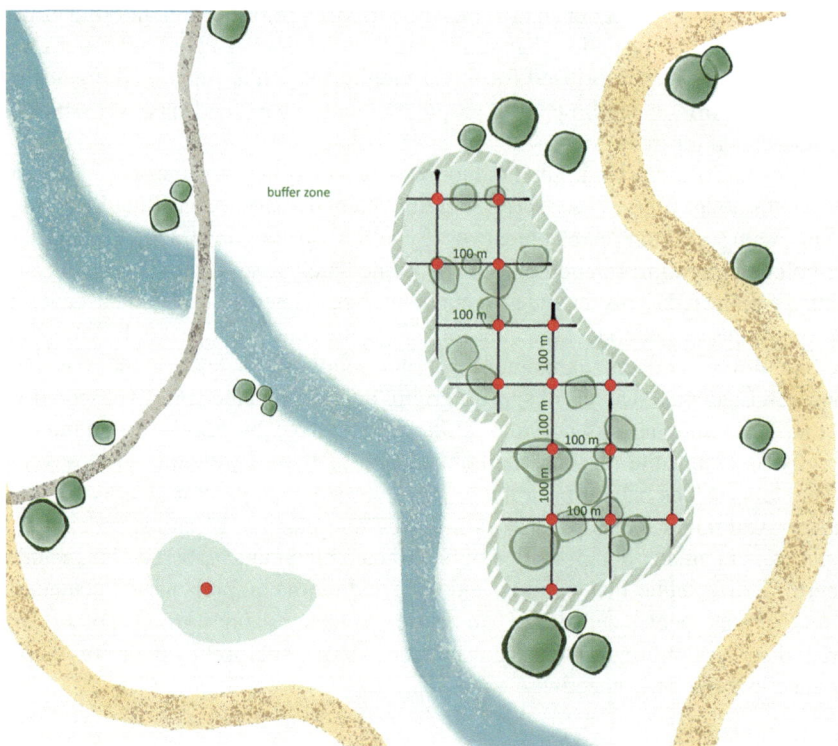

Fig. 12.2 Scheme for selecting sample plots in small (up to 2 ha) to medium (between 2 ha and 5 ha), and large (more than 5 ha) forest areas. For small and medium areas, one or more standard plots representative for the area are established. If more than one plot is established, 100 m should be maintained between plots. For large areas, a systematic grid of standard plots with a grid cell size of 100 × 100 m is established. No survey activities occur within buffer zones

The Standard Protocol: Guidance for Field Surveys in Forests

At each sample plot, a circular plot with an area of 300 m² (radius of 9.77 m) is established for evaluating site and forest stand characteristics, making observations and assessments, studying soil characteristics, and conducting tree surveys. The centre of each circular plot should be permanently marked.

Site and Forest Stand Description

The forest stand characteristics are detailed by way of various parameters including elevation in metres above sea level and slope direction (aspect) using a Suunto compass. The topographic meso-relief is categorised as either a shedding, neutral, or accumulation zone depending on the movement of material into or out of the area.

The forest type or association of the area is determined by considering socio-ecological factors. It is recommended to use existing classification schemes at the national level, e.g. as provided by national forest inventories (see Hauk et al. (2020) for Austria), or at the international level, such as the categories of European forest types (EEA 2006), for this determination.

The forest management is categorised as either intensive, extensive, or non-existent based on the stand structure. Non-existent management can be identified by the absence of stumps, extensive management by the presence of individual stumps, and intensive management by the presence of clearcutting or evenly distributed stumps. If necessary, forest owners should be contacted for clarification.

Observations and Influences

Observations and influences include factors relating to site conditions, wildlife ecology, and other factors affecting site dynamics (Table 12.1). The distance to the circular plot centre is estimated in four categories (<10 m, 10–24 m, 25–50 m, 51–100 m). Features at over 100 m are not considered. The following categories are considered:

Soil Characteristics

Using a systematic approach, a moderate yet fundamental set of soil data is collected at each circular plot. The dataset comprises organic horizon types and thicknesses, soil depth, dominant parent material, and soil moisture condition. To determine soil depth, the mean penetration depth of three "Pürckhauer" soil sounding probes inserted at representative positions within the plot is calculated. The probe is hammered into the soil until bedrock or immovable stone is encountered.

The average thickness of litter, fermentation, and humus layers is measured in millimetres. The humus type is classified into seven distinct categories (see

Table 12.1 Categories of observations and influences with details to be assessed at the circular plot level

Observations and influences	Details
Roads and trails	Forest road (gravel road)
	Hiking trail (unpaved road)
	Skidding track
Hunting facilities	Hide, feeding station, salt lick, stalking path, hunting clearing
Anthills	–
Waterbodies	Standing water (containing water at the time of survey)
	Running water (containing water at the time of survey)
	Seep, puddle (damp spots including small bodies of water)
Terrain shape	Broken terrain, e.g. ditches or rock walls
Forest outer edge	Field
	Meadow or pasture
	Paved public road
	Waterbody
Forest inner edge (change in stand type)	Regeneration (<2 m tree height)
	Young stand (≥2 m tree height, DBH < 12 cm)
	Middle-aged forest (12 ≤ DBH > 30 cm)
	Mature forest (30 cm ≤ DBH < 50 cm)
	Old forest (DBH ≥ 50 cm)
Disturbances (affecting vegetation development)	Wind throw (more than 2 trees)
	Wind or snow breakage (more than 2 trees)
	Bark beetle infestation (more than 2 trees)
	Avalanche
	Mud flow
	Fire
Other	Additional observations not covered by the abovementioned categories

Table 12.2), and soil moisture levels are assessed ranging from dry to wet conditions. The identification of parent material is contingent on the availability of digital maps, including those provided by geological institutions (e.g. Geologische Bundesanstalt (2023) for Austria).

Stand Structure

Both horizontal and vertical structural diversity are examined within the circular plot. Horizontal structural diversity is determined by assessing the degree of crown closure, which quantifies the extent to which the ground is covered by the tree canopy (Keller et al. 2013). Specifically, the top layer with a crown cover of at least 3/10 and shrubs is considered in this evaluation (see Table 12.3).

Vertical structure is evaluated by analysing the characteristics of vegetation layers and can be calculated using the standard deviation of measured tree heights following the method proposed by Mura et al. (2015). Additionally, the presence of multi-layered structure (plenter structure) due to single tree selective cutting is evaluated.

Table 12.2 Seven distinct categories of the humus type

Humus type	Description
Mull	Characterised by the periodic absence of organic matter accumulation on the surface owing to rapid decomposition process and the mixing of organic matter and mineral soil material by bioturbation. With mull humus, there is either no or a very thin fermented litter layer
Mull/moder or moder/mull	Clear indications of soil organism activity, humus and mineral substances more or less mixed together. However, there is no mixing of humus and clay-rich soil
Moder	More decomposed than raw humus but characterised by an organic matter layer on top of the mineral soil with a diffuse boundary between the organic matter layer and A horizon; difficult to separate one layer from another. This humus type develops in moderately nutrient-poor conditions, usually under a cool, moist climate
Raw humus (aeromorphic mor)	Usually thick (5–30 cm) organic matter accumulation that is largely unaltered owing to a lack of decomposers. This kind of organic matter layer develops in extremely nutrient-poor and coarse-textured soils under vegetation producing a litter layer that is difficult to decompose, over a thin A horizon
Hydromorphic humus	Greasy, wet humus form created under the influence of water; putrefied humus formation; black colour, characteristic "peat smell". Thickness of the organic layer is less than 30 cm
Turf	Practically no biological decomposition activity, periodic wetting with organic material comprising approx. 70% of the total volume
None	No humus is present

Standing and Lying Trees

Within each circular plot (300 m^2), all living and dead standing trees with a diameter at breast height (DBH) of ≥10.0 cm are recorded. This assessment encompasses several key pieces of information for each tree, including geographic location (direction and horizontal distance from the plot centre), tree species, DBH, or mid-diameter (MDM) for trees shorter than 1.3 m, tree height, and current status. The status of trees is categorised according to an adapted classification following Hunter (1990); see Table 12.4.

A line intersect method is employed to record lying deadwood (E. Van Wagner 1968; Keller et al. 2013; Roth et al. 2003). For this purpose, 4 lines are measured in the main cardinal directions starting from the centre of the circular plot, each 11.0 m long (slope distance). Any lying deadwood intersecting one of these lines at least 1 m from the centre point with a minimum diameter of 10.0 cm at the point of intersection is recorded. For each intersection of a line and tree, the species, diameter, and current status of the tree are recorded. The status is visually assessed following the classification in Hunter (1990), as presented in Table 12.4.

Whenever possible, the cause of death for any recorded (lying and standing) dead trees is selected from one of the following categories: anthropogenic causes (e.g. evident saw marks), windthrow, snow break, competition, age-related factors, bark beetle infestation, fungal attacks, unknown reason.

Table 12.3 Six categories of horizontal structure (canopy cover) with schemes and descriptions

Horizontal structure	Scheme	Description
Overlapping		The treetops overlap and compete with each other
Closed		The treetops touch but do not overlap
Loose		There is space for less than one more crown between the treetops
Light		There is space for one crown between the treetops
Spacious		There is space for several more crowns between the treetops
Clumped		Closed groups of trees unconnected to each other

Tree-Related Microhabitats

Tree-related microhabitats (TreMs) are clearly defined tree-related structures that provide crucial habitats for a wide range of organisms including animals, plants, lichens, and fungi, many of which may have specialised ecological requirements during at least part of their life cycle (Bütler et al. 2020; Kraus et al. 2016; Larrieu et al. 2014) (Fig. 12.3).

In accordance with the standardised inventory method developed by Larrieu and Bouget (2016) and adapted by Bütler et al. (2020), TreMs are assessed on all standing living and dead trees within the circular plot (300 m^2) by inspecting each tree

12 Practical Guidance for Rapid Biodiversity Assessment in Central European Forests

Table 12.4 Status/Scheme of standing and lying Tree

Status of standing tree	Scheme	Status of lying tree	Scheme
Alive, healthy		–	–
Alive, sick		–	–
Alive, dying off, bark present, fine branches (<3 cm diameter) present		–	–
Dead, bark present, fine branches (<3 cm diameter) present		Undecomposed, hard wood, bark present, fine branches (<3 cm diameter) present, trunk without ground contact	
Dead, bark loosens, most fine branches missing		Beginning to decompose, hard wood, bark present, fine branches no longer present, trunk has partial ground contact	
Dead, bark loosens, no fine branches present		Moderate decomposition, peeling bark, trunk resting on the ground, clear visual signs of decomposition	
Dead, less than half of the total height broken		Severe decomposition, soft wood, no bark, trunk moderately decomposed and partly underground	
Dead, more than half of the total height broken		Very strong decomposition, wood soft, no bark present, trunk strongly decomposed and a large part underground	

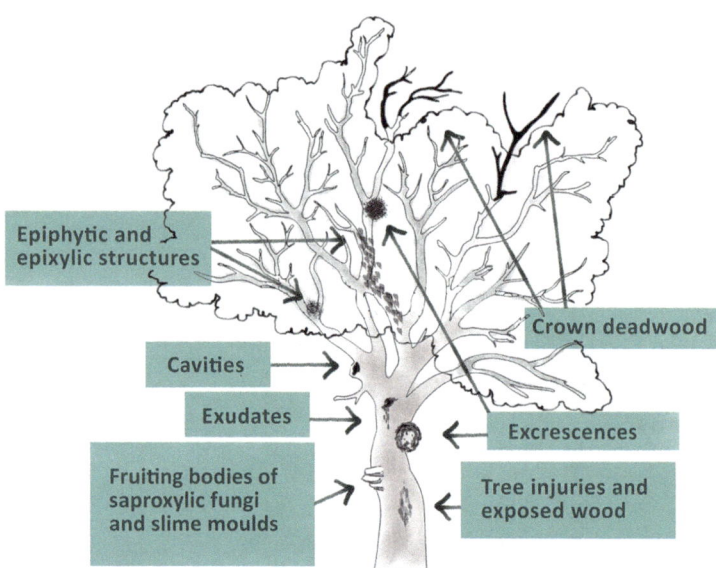

Fig. 12.3 Illustration of a tree providing an overview of categories of tree-related microhabitats on a habitat tree. (adapted version from Bütler et al (2020))

from all sides using binoculars. The survey distinguishes 42 TreMs in seven categories (see Fig. 12.3 and Table 12.5) following the typology in Bütler et al. (2020):

Signs of Vertebrates

Indirect and direct signs of the presence of vertebrate species are systematically documented within the area of each circular plot. All findings are recorded, and if possible, photographs from multiple angles and distances are taken for further clarification and specification. Firstly, the observed signs are separated into predefined categories encompassing living individuals, faeces, feathers, hair, bones, antlers, horns, carcasses, resting places, mating sites, pellets, footprints, tracks, eggs, nests, breeding sites, holes, and dens. Subsequently, the species group or functional group to which each observation can be attributed is identified. These groups include ruminant ungulates (including deer species, chamois, mouflon, and ibex), wild boar, carnivores, grouse species, woodpecker species, owls, raptors, reptiles, and amphibians.

Expanding Modules: The Intensive Survey Plots

The intensive surveys are to be carried out by taxonomic experts and include the recording of vascular plants, bird and bat vocalisations, and the detailed documentation of fungus species and saproxylic beetles.

Table 12.5 List of categories, groups, and microhabitats following the typology in Bütler et al. (2020)

Category	Group	TreMs
Cavities	Woodpecker breeding holes	Small, entrance <4 cm diameter
		Medium, entrance 4–7 cm diameter
		Large, entrance >10 cm diameter
		Multi-story holes (>3 holes below each other, entrance >3 cm diameter)
	Trunk cavities	Trunk cavity with ground contact (>10 cm diameter)
		Trunk cavity without ground contact (>10 cm diameter)
		Half-open trunk cavity (>30 cm diameter)
		Hollow trunk (chimney) with/without ground contact (>30 cm diameter)
		Hollow branch cavity (>10 cm diameter, >50 cm deep)
	Insect feeding galleries and boreholes	Insect feeding galleries and boreholes (>2 cm diameter or >300 cm^2 area)
	Dendrotelms and other holes	Dendrotelm (water-filled tree cavity) (>15 cm diameter)
		Large woodpecker feeding holes (>10 cm diameter and >10 cm deep)
		Small woodpecker feeding holes (<10 cm diameter and min 3 holes)
		Bark-lined trunk concavity (>10 cm opening and >10 cm depth)
		Trunk root concavity (>10 cm diameter opening and >45° ceiling angle)
Tree injuries and exposed wood	Wounds and exposed wood—sapwood	Loss of bark (>300 cm^2 area)
		Fire wound (>600 cm^2 area)
		Bark shelter (open at the bottom, min. 1 cm open, 10 cm wide and 10 cm long)
		Bark pocket (open at the top, min. 1 cm open, 10 cm wide and 10 cm long)
	Wounds and exposed wood—sapwood and heartwood	Broken trunk (min. diameter at the break point >20 cm)
		Large branch fracture (open heartwood)
		Crack/crevice (>30 cm long, >1 cm wide, 10 cm deep)
		Lightning scar (>30 cm long, >1 cm wide, 10 cm deep)
		Two-trunk tree crack (>30 cm long)
Crown deadwood	Crown deadwood	Dead branch (>10 cm diameter or >3 cm diameter and >10% of the crown)
		Dead top (>10 cm diameter)
		Remnants of broken branch (>20 cm diameter and >50 cm length)
Excrescences	Witch's brooms and water sprouts	Witch's broom (>50 cm diameter)
		Dense agglomeration of shoots (>5 branches)
	Burrs and cankers	Burr (>20 cm diameter, no rotten wood)
		Canker (>20 cm diameter, with rotten wood)

(continued)

Table 12.5 (continued)

Category	Group	TreMs
Fruiting bodies of saproxylic fungi and slime moulds	Fungi (groups)	Perennial polypores (>5 cm in diameter) Annual polypores (>5 cm in diameter or group of >10) Other basidiomycota [coralloid] (>5 cm diameter or group of >10) Large ascomycetes (>3 cm in diameter or area >100 cm^2) Large myxomycetes (slime moulds) (>5 cm in diameter)
Epiphytic and epixylic structures	Epiphytes (only if >10% coverage)	Bryophytes or liverworts (in 4 steps: 11–25, 26–50, 51–75, 76–100%) Crustose (crusty) lichens (in 4 steps: 11–25, 26–50, 51–75, 76–100%) Foliose (leaf) lichens (in 4 steps: 11–25, 26–50, 51–75, 76–100%) Fruticose (shrub and beard) lichens (in 4 steps: 11–25, 26–50, 51–75, 76–100%) Ivy or lianas (in 4 steps: 11–25, 26–50, 51–75, 76–100%) Ferns (>5 fronds) Mistletoe (>20 cm diameter)
	Nests	Vertebrate nest (>10 cm diameter) Invertebrate nest (>10 cm diameter)
	Micro soils	Micro soil (bark) Micro soil (crown)
Exudates	Sap or resin flow	Active sap flow (>10 cm length) Active resin flow (>10 cm length)

Vascular Plants

The vegetation assessment encompasses all vascular plants including herbs, shrubs, and trees as well as bryophyte cover. This assessment is conducted outside the circular plot within a rectangular sample area of 20 × 20 m (400 m^2). The first corner of this area is positioned at a horizontal distance of 15 m from the centre of the circular plot, selecting the direction that best represents the entire area (Fig. 12.4). Once this first corner is set, the plot is demarcated in the most suitable orientation, typically following the direction of contour lines. Ground regeneration sub-plots measuring 5 m^2 (2.24 × 2.24 m horizontal distance) and 25 m^2 (5.0 × 5.0 m horizontal distance) are situated in the corner closest to the circular plot. Tree regeneration with a height of ≤50 cm is surveyed within the 5 m^2 sub-plot, and regeneration with a height between 50 cm and 130 cm within the 25 m^2 sub-plot (see Fig. 12.4).

All vascular plant species within the rectangular plot are recorded and their coverage estimated using Braun-Blanquet (1928) values with an adaptation (including "2 m", "2a", and "2b" from Barkman et al. (1964) instead of just "2"). Whenever in situ identification is not possible and the required permissions have been granted, a voucher sample of affected species is to be taken and stored according to herbarium methods to be identified later.

Fig. 12.4 Scheme of sample plot with different survey scales including circular plot (forest structure and tree assessment) and rectangular plot (vegetation and regeneration survey)

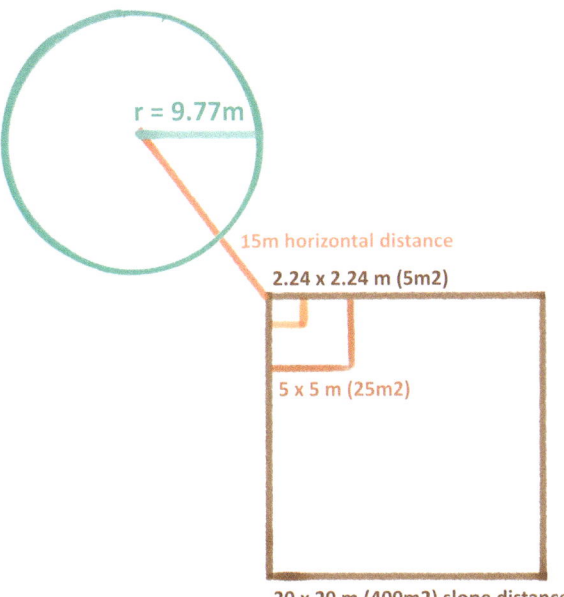

Fig. 12.5 Illustration of forest layers consisting of moss layer, herb layer, shrub layer, and tree layer

Vertical structural diversity is assessed via the development of the vegetation layers of a stand (see Fig. 12.5). For this, the percentage of coverage of each existing layer is estimated. Coverage can reach a maximum of 100% per layer, but the sum of all layer coverage levels can exceed 100%.

The layers are identified within each rectangular plot with the help of a reference point for visual estimates (especially for the critical 5 m threshold). The dominant projection belonging to a particular layer defines the layer of the individual.

Tree species regeneration and browsing is surveyed in the abovementioned subplots of the rectangular vegetation plot, with the corresponding tree species determined for all perennial regeneration. The height of perennial regeneration is categorised into 10 cm height classes up to the previous year's shoot tip. The previous year's shoot and winter buds are also examined for browsing, with a shoot considered unbrowsed if its terminal bud from the previous year was able to sprout.

Fungus Species

The survey area for the investigation of fungi matches the dimensions of the rectangular vegetation survey area (400 m^2). Every fungal fruiting body discovered within the survey area is to be sampled, including all substrate types such as soil, bark, and dead and living wood. This comprehensive assessment involves both photographing and sampling of flesh fruiting bodies. For Central Europe, this survey should be carried out in September and October when most fungal fruiting bodies can be found.

For each fungus found, one mature fruiting body is carefully extracted without causing damage by using a knife and safely stored in a paper bag labelled with details of the collection site, including the collector's name as well as the collection date, elevation, vegetation community, humidity level, and soil characteristics. Immediately upon returning from the field, the fungus samples must be dried using a fruit dryer until they become "crispy" for later identification.

In the field, photos of different growth stages of fruiting bodies are taken from the top, bottom, and side along with a picture of a cross-section cut. Photos should always include a scale for size estimation. When a fungus is picked, its aroma and colour are noted. If the mushroom releases a liquid (e.g. when gently scratched), the colour of the liquid is recorded. In addition, the tree species near the fungal growth is recorded. It is of the utmost importance to work with fresh to moderately aged fruiting bodies for photographic and descriptive purposes. Older fruiting bodies may have already undergone colour and shape changes and should therefore be excluded from consideration.

Bird and Bat Recording

The survey of birds and bats is carried out through passive acoustic monitoring. In recent years, this method has gained popularity in species assessments (Sugai et al. 2019) since it reduces required manpower in the field and provides standardised and replicable data collection as well as storable recordings enabling detailed analysis in the lab. Recording devices (e.g. AudioMoth from Open Acoustic Devices) are strategically positioned at each sample plot during two separate time periods extending over several weeks and covering the spring and late summer seasons with a

Fig. 12.6 Audio recording devices on beech and spruce trees (Photo: BFW)

focus on birds and bats, respectively (see Fig. 12.6). Recording settings should be configured to match the vocalisation frequencies and activity periods of the target species groups. This encompasses the recording of bird songs and calls in audible frequency ranges (32 kHz) as well as high-frequency calls of bats and arboreal small mammals (250 kHz). Overall recording duration, the length of actual sound recordings, and the standby (sleep mode) periods of the device need to be adjusted to the specific research question.

After the field season, recordings are analysed by artificial neural networks trained to identify species based on their calls and generate a list of detected species. However, it is essential that at least a subsample of the algorithm's predictions is validated by human experts to estimate the false identification rate.

It is important to note that the use of passive audio recorders in the public space may be subject to national data protection regulations (e.g. GDPR in EU countries) and thus require administrative approval. It may also be necessary to identify instances of human speech on records and automatically remove such data before the recordings are made accessible to humans. Artificial neural networks implemented in software for the analysis of bird and bat sounds are often capable of this task.

Saproxylic Beetles

For conducting insect surveys, particularly targeting saproxylic beetle species in forests, the utilisation of flight interception traps—also known as window traps—is a commonly used method (e.g. Larsson Ekström et al. 2021; Parisi et al. 2020;

Fig. 12.7 Flight interception trap construction and installation: Two window traps are deployed per sample plot. The study period extends over 15 weeks per year (Photo: BFW)

Sitzia et al. 2015). Here, traps consisting of two plexiglass discs measuring 40 × 60 cm, equipped with a catch funnel and 11 × 13 cm catch container were used. For dry trapping, a microbiocidal net is placed in the trap container. As this assessment aims to capture only beetles that are present within the sampling plot, no attractant is used. This decision is dependent on the respective research focus, however. The traps are securely suspended from wooden gallows and typically positioned at a height of 2 m. A visual representation of the trap design and installation is provided in Fig. 12.7.

In each survey plot, two traps are deployed for 15 weeks, with their catch containers regularly emptied every 3 weeks. To ensure precise record-keeping, every collected sample is labelled with collection date, site name, and a unique sample number. The gathered specimens subsequently undergo identification within a laboratory setting and are preserved in alcohol.

The Next Step: Enhancing Connectivity Assessments with Structural and Genetic Insights

Incorporating forest structure alongside elements such as deadwood and TreMs into our rapid biodiversity assessment protocol helps to understand biodiversity patterns and thus creates a framework for assessing structural connectivity within forested areas. Deadwood, for example, serves as a crucial habitat for a multitude of species, fostering biodiversity and supporting ecological processes like nutrient cycling.

TreMs, on the other hand, offer specialised microenvironments that provide habitat and refuge for a range of species. When taken into consideration, these structural attributes help to develop a more detailed understanding of the connectivity within forest landscapes.

In addition, intensive survey modules like the survey of saproxylic beetles by way of flight interception traps enable subsequent population and landscape genetic analyses. The actual gene flow between populations can be studied based on the identification of captured beetles, allowing conclusions regarding the functional connectivity within forest ecosystems.

While the proposed RBA currently incorporates survey protocols for some specific species groups, many other organisms likewise contribute significantly to the ecological role of forest ecosystems. Among these, soil-dwelling organisms emerge as key players with their contribution to nutrient cycling and soil health in general. Furthermore, arthropods including spiders, bees, and ants exert a profound impact on ecosystem dynamics, influencing crucial processes from pollination to pest control. Although these groups are not covered in our present study, we recognise their importance and the need for further investigation.

References

Barkman JJ, Doing H, Segal S (1964) Kritische Bemerkungen und Vorschläge zur quantitativen Vegetationsanalyse. Acta Bot Neerl:394–419
BMNT (2019) Waldstrategie Österreich 2020+. https://www.bmnt.gv.at/forst/oesterreich-wald/waldstrategie-2020/waldstrategie_paper.html.
Braun-Blanquet J (1928) Pflanzensoziologie—Grundzüge der Vegetationskunde. Springer
Bütler R, Lachat T, Krumm F, Kraus D, Larrieu L (2020) Field guide to tree-related microhabitats. Descriptions and size limits for their inventory, p 59. www.wsl.ch/fg-trems
Düggelin C, Abegg M, Bischof S, Brändli U-B, Cioldi F, Fischer C, Meile R (2020) Schweizerisches Landesforstinventar. Anleitung für die Feldaufnahmen der fünften Erhebung 2018-2026. WSL Berichte 90:288
Van Wagner CE (1968) The line intersect method in forest fuel sampling. For Sci:14
EEA (2006) European forest types. Categories and types for sustainable forest management reporting and policy. ISSN 1725-2237
Franklin JF (1993) Preserving biodiversity. Ecol Appl 3(2):202–220. https://doi.org/10.2307/1941820
Fulton MR, Harcombe PA (2002) Fine-scale predictability of forest community dynamics. Ecology 83(5):1204. https://doi.org/10.2307/3071935
Geologische Bundesanstalt. (2023). Multithematische geologische Karte von Österreich 1:1.000.000. Map
Hauk E, Niese G, Schadauer K (2020) Instruktion für die Feldarbeit der Österreichischen Waldinventur 2016–2018
Van Heezik Y, Barratt BIP, Burns BR, Clarkson BD, Cutting BT, Ewans R, Freeman C, Meurk C, Shanahan DF, Simcock R, Souter-brown G, Stanley MC, Stanley R, Thorsen MJ, Wake S, Woolley CK, Zink R, Seddon PJ (2023) A rapid assessment technique for evaluating biodiversity to support accreditation of residential properties. Landsc Urban Plan 232. https://doi.org/10.1016/j.landurbplan.2023.104682
Hunter MLJ (1990) Wildlife, forests and forestry. Principles of managing forests for biological diversity. Prentice Hall

Keller M, Kaufmann E, Meile R, Lanz A, Schwyzer A, Stierlin R, Strobel T, Ulmer U, Brändli U, Duc P (2013) Schweizerisches Landesforstinventar. Feldaufnahme-Anleitung 2013 (M. Keller (toim)). https://www.lfi.ch/publikationen/publ/anleitungen.php

Kerr JT, Sugar A, Packer L (2000) Indicator taxa, rapid biodiversity assessment, and nestedness in an endangered ecosystem. Conserv Biol 14(6):1726–1734. https://doi.org/10.1111/j.1523-1739.2000.99275.x

Kipson S, Fourt M, Teixidó N, Cebrian E, Casas E, Ballesteros E, Zabala M, Garrabou J (2011) Rapid biodiversity assessment and monitoring method for highly diverse benthic communities: a case study of mediterranean coralligenous outcrops. PLoS One 6(11). https://doi.org/10.1371/journal.pone.0027103

Kraus D, Bütler R, Krumm F, Lachat T, Larrieu L, Mergner U, Paillet Y, Rydkvist T, Schuck A, Winter S (2016) Catalogue of tree microhabitats—reference field list. Integrate+ Technical Paper. 16 p, March, 16. https://doi.org/10.13140/RG.2.1.1500.6483

Larrieu L, Bouget C (2016) Tree-related microhabitats (TreMs) as key elements for forest biodiversity. Integrate+ Conference 2016

Larrieu L, Cabanettes A, Gonin P, Lachat T, Paillet Y, Winter S, Bouget C, Deconchat M (2014) Deadwood and tree microhabitat dynamics in unharvested temperate mountain mixed forests: a life-cycle approach to biodiversity monitoring. For Ecol Manage 334:163–173. https://doi.org/10.1016/j.foreco.2014.09.007

Larsson Ekström A, Bergmark P, Hekkala AM (2021) Can multifunctional forest landscapes sustain a high diversity of saproxylic beetles? For Ecol Manage 490. https://doi.org/10.1016/j.foreco.2021.119107

Manley PN, Van Horne B, Roth JK, Zielinski WJ, McKenzie MM, Weller TJ, Weckerly FW, Vojta C, Service F (2005) Multiple species inventory and monitoring technical guide. Gen Tech Rep. GTR-WO-073:204

Maragos JE, Cook CW (1995) The 1991-1992 rapid ecological assessment of Palau's coral reefs. Coral Reefs 14(4):237–252. https://doi.org/10.1007/BF00334348

McNellie MJ, Oliver I, Dorrough J, Ferrier S, Newell G, Gibbons P (2020) Reference state and benchmark concepts for better biodiversity conservation in contemporary ecosystems. Glob Chang Biol 26(12):6702–6714. https://doi.org/10.1111/gcb.15383

Mura M, McRoberts RE, Chirici G, Marchetti M (2015) Estimating and mapping forest structural diversity using airborne laser scanning data. Remote Sens Environ 170:133–142. https://doi.org/10.1016/j.rse.2015.09.016

Nemitz D, Huettmann F (2015) GRID sampling for a global rapid biodiversity assessment: methods, applications, results, and lessons learned. F. Huettmann (Toim), Central American biodiversity—conservation, ecology and a sustainable future. Springer, pp 435–460. https://doi.org/10.1007/978-1-4939-2208-6

Obrist MK, Duelli P (2010) Rapid biodiversity assessment of arthropods for monitoring average local species richness and related ecosystem services. Biodivers Conserv 19(8):2201–2220. https://doi.org/10.1007/s10531-010-9832-y

Oettel J, Lapin K (2020) Linking forest management and biodiversity indicators to strengthen sustainable forest management in Europe. Ecol Indic 122(107275)

Parisi F, Frate L, Lombardi F, Tognetti R, Campanaro A, Biscaccianti AB, Marchetti M (2020) Diversity patterns of Coleoptera and saproxylic communities in unmanaged forests of Mediterranean mountains. Ecol Indic 110:105873. https://doi.org/10.1016/j.ecolind.2019.105873

Roth A, Kennel E, Knoke T, Matthes U (2003) Die Linien-Intersekt-Stichprobe: Ein effizientes Verfahren zur Erfassung von liegendem Totholz? Line intersect sampling: an efficient method for sampling of coarse woody debris? Forstwissenschaftliches Centralblatt 122(5):318–336. https://doi.org/10.1007/s10342-003-0001-7

SCBD (2005) Guidelines for the rapid ecological assessment of biodiversity in inland water, coastal and marine areas. Montreal, Canada, CBD technical series no. 22 and the secretariat of the Ramsar convention, gland, Switzerland, Ramsar Technical Report no. 1

Schindler S, Oberleitner I, Rabitsch W, Essl F, Stejskal-Tiefenbach M (2017) Monitoring von Klimawandelauswirkungen auf die Biodiversität. Endbericht von StartClim2016.A in StartClim2016: Weitere Bei- träge zur Umsetzung der österreichischen Anpassungsstrategie, Auftraggeber: BMLFUW, BMWF, ÖBf, Land Oberösterreich

Sitzia T, Campagnaro T, Gatti E, Sommacal M, Kotze DJ (2015) Wildlife conservation through forestry abandonment: responses of beetle communities to habitat change in the eastern Alps. Eur J For Res 134(3):511–524. https://doi.org/10.1007/s10342-015-0868-0

Sugai LSM, Silva TSF, Ribeiro JW, Llusia D (2019) Terrestrial passive acoustic monitoring: review and perspectives. Bioscience 69(1):5–11. https://doi.org/10.1093/biosci/biy147

Sutherland WJ (2000) The conservation handbook. The conservation handbook (number January 2000). https://doi.org/10.1002/9780470999356

Ward DF, Larivière MC (2004) Terrestrial invertebrate surveys and rapid biodiversity assessment in New Zealand: lessons from Australia. N Z J Ecol 28(1):151–159

Watson I, Novelly P (2004) Making the biodiversity monitoring system sustainable: design issues for large-scale monitoring systems. Austral Ecol 29(1):16–30. https://doi.org/10.1111/j.1442-9993.2004.01350.x

Willis KJ, Bailey RM, Bhagwat SA, Birks HJB (2010) Biodiversity baselines, thresholds and resilience: testing predictions and assumptions using palaeoecological data. Trends Ecol Evol 25(10):583–591. https://doi.org/10.1016/j.tree.2010.07.006

Willis KJ, Gillson L, Brncic TM, Figueroa-Rangel BL (2005) Providing baselines for biodiversity measurement. Trends Ecol Evol 20(3):107–108. https://doi.org/10.1016/j.tree.2004.12.003

Open Access This chapter is licensed under the terms of the Creative Commons Attribution 4.0 International License (http://creativecommons.org/licenses/by/4.0/), which permits use, sharing, adaptation, distribution and reproduction in any medium or format, as long as you give appropriate credit to the original author(s) and the source, provide a link to the Creative Commons license and indicate if changes were made.

The images or other third party material in this chapter are included in the chapter's Creative Commons license, unless indicated otherwise in a credit line to the material. If material is not included in the chapter's Creative Commons license and your intended use is not permitted by statutory regulation or exceeds the permitted use, you will need to obtain permission directly from the copyright holder.

Part III

Restoration, Social Dynamics, and Policy Frameworks

Restoring Forest Landscape Connectivity: Why, Where, and How?

13

Johanna A. Hoffmann, Demel Teketay, Mesele Negash, and Hafte Mebrahten Tesfay

A view of the ecoduct (a wildlife crossing bridge) on National Route 101 in Argentina in Misiones Province. (Photo: Helissa)

J. A. Hoffmann (✉)
Department of Forest Biodiversity & Nature Conservation, Austrian Research Centre for Forests, Vienna, Austria
e-mail: johanna.hoffmann@bfw.gv.at

D. Teketay
Department of Range and Forest Resources (RFR), Faculty of Natural Resources, Botswana University of Agriculture and Natural Resources (BUAN), Gaborone, Botswana
e-mail: dteketay@buan.ac.bw

© The Author(s) 2025
K. Lapin et al. (eds.), *Ecological Connectivity of Forest Ecosystems*,
https://doi.org/10.1007/978-3-031-82206-3_13

Abstract

As global challenges like forest degradation, biodiversity loss, and fragmentation intensify, conservation and restoration of forest ecosystems have become important challenges of our time. Sustainable restoration efforts extend beyond tree planting. However, they require clear objectives aligned with consideration of ecological and social resilience. Forest restoration planning benefits from the integration of spatial tools and connectivity measures, ensuring ecosystem stability, species mobility, and adaptability to climate change. Forest landscape restoration is a framework of diverse planned interventions to support biodiversity and the resilience of the forest ecosystem, taking integration into a wider landscape into account. In shifting the focus from a small-scale stand-specific to a landscape-wide perspective, the role of ecological connectivity becomes pivotal. This chapter therefore integrates landscape ecology measures for connectivity in restoration ecology and discusses four questions influencing the planning of a restoration action, including connectivity considerations: (1) What is to be restored (defining the restoration objectives)? (2) Where should the restoration activities be focused (determining spatial and connectivity measures)? (3) How can the restoration best be implemented (applying restoration methods)? (4) How successful is the restoration (determining monitoring measures)? The choice of suitable connectivity measures and restoration methods depends strongly on the landscape and the ecological, economic, and social framework conditions. Challenges for restoring connectivity span technical, financial, biophysical, and social aspects that require collaborative stakeholder engagement and adaptive management to overcome. They also encompass species-specific restoration, invasive species management, and international cooperation for restoring or conserving connected forest ecosystems.

Keywords

Ecological connectivity · Forest landscape restoration · Ecosystem resilience · Climate adaptation · Biological corridors · Habitat fragmentation · Forest regeneration

---

M. Negash
Wondo Genet College of Forestry and Natural Resources, Department of Agroforestry, Hawassa University, Hawassa, Ethiopia

European Forest Institute, Manziana, Rome, Italy
e-mail: meselen@hu.edu.et; mesele.tesemma@efi.int

H. M. Tesfay
Department of Forest Biodiversity & Nature Conservation, Austrian Research Centre for Forests, Vienna, Austria

BOKU University, Department of Ecosystem Management, Climate and Biodiversity, Institute of Forest Ecology, Vienna, Austria
e-mail: hafte-mebrahten.tesfay@bfw.gv.at; hafte.tesfay@boku.ac.at

Why Consider Connectivity in Forest Restoration Efforts?

There is high hope that forests and trees can solve or at least mitigate human-made global problems like climate change and land degradation (Garrett et al. 2022; Messier et al. 2019). With the Bonn Challenge and the UN proclamation of the Decade on Ecosystem Restoration, a global political effort to bring up to 350 million ha of land (Frietsch et al. 2023; IUCN 2015, 2020) under restoration by 2030 (1t.org 2023), policy-driven restoration efforts of forest ecosystems have reached a peak (Stanturf and Mansourian 2020). This can be an effective way to mobilize public action and join forces in the private and public sectors to engage in conservation and restoration efforts (Mansourian 2021; Sewell et al. 2020). However, simply planting trees is not enough to bring back habitats and stabilize ecosystems. The act of restoring entails much more; it must come with clear objectives for the benefit of the respective ecosystem and its society (César et al. 2021; Mansourian 2021; Marshall et al. 2023). It needs to be embedded in the sociocultural context as well as in that of the landscape (Stanturf and Mansourian 2020), the habitat, and the present species composition. The wider spatial context and spatial processes (such as habitat structure, seed distribution, species migration, fragmentation, and edge effects) have been identified as crucial for the success of sustainable restoration. Therefore, spatial elements are embedded in the term forest landscape restoration (IUCN and WRI 2014; Lemieux et al. 2022; Marshall et al. 2023; Timpane-Padgham et al. 2017).

Forest Fragmentation Threatens Biodiversity

Forests and other natural areas face numerous challenges caused mainly by anthropogenic pressure. Land-use change, agricultural systems, and the excessive consumption of natural resources threaten the stability of forest ecosystems (JRC 2019; Laurance et al. 2011) as well as their provisioning, supporting, and regulating ecosystem services that benefit human beings (Aznar-Sánchez et al. 2018; Chazdon 2008; Primack 2014). These issues are reflected in the many signs of degradation of the structure and function of forest ecosystems around the world (World Research Institute (WRI) 2015). Biodiversity loss is one major challenge, and one of the seven severest threats to forest biodiversity is fragmentation. Together with habitat loss, habitat degradation, overexploitation, invasive species, diseases, and climate change, habitat fragmentation leads to species extinction, ecosystem degradation, loss of ecosystem services, and erosion of genetic diversity (Butchart et al. 2010; Laurance et al. 2011; JRC 2019).

Besides the active destruction of forest areas (e.g., by deforestation), fragmentation possesses an additional deteriorating impact on the affected habitats and extends the negative effects to neighboring areas (Laurance et al. 2011). Driven by factors such as agriculture, urbanization, grazing, industrial activity, and linear barriers such as roads, railways, pipelines, fences, and canals, fragmentation has reconfigured more than 50% of the Earth's landscapes (Keeley et al. 2019). Highly fragmented and isolated habitats can result in population decline to the point of extinction of forest-dependent species (Lanta et al. 2020; Primack 2014) by reducing their adaptive capacity (Biere et al. 2012; Butchart et al. 2010). By increasing barriers and decreasing undisturbed contiguous habitats, fragmentation restricts movement, species reproduction (Biere et al. 2012; Butchart et al. 2010), access to food, and dispersal of seeds and genetic material (Biere et al. 2012; Thomas et al. 2014; van Diggelen et al. 2012). It can also cause ecosystem degradation and greater susceptibility to invasive species, disturbances like storms and fires, and pests (Laurance et al. 2011; Messier et al. 2019). Climate change, overexploitation, diseases, and other threats further amplify this pressure on ecosystems, exacerbating the negative impacts of biodiversity loss, habitat degradation, and destruction (Butchart et al. 2010; Primack 2014; JRC 2019; With 2019b).

Ecological Connectivity and Forest Landscape Restoration: A Key Combination

Ecological connectivity (EC) is crucial for maintaining biodiversity. EC represents the extent to which a habitat facilitates the flow of material and ecological processes such as seed dispersal, organism movement, and gene exchange (Unnithan Kumar and Cushman 2022; Thomas et al. 2014; Wang et al. 2021). EC is thus also critical for the stability, resilience, and adaptation capacity of species and the ecosystems they live in (Keane et al. 2018; Messier et al. 2019; van Diggelen et al. 2012; With 2019a). The possibility to shift ranges along climatic gradients, for example, facilitates climate change adaptation (Krosby et al. 2010; Timpane-Padgham et al. 2017; Travers et al. 2021).

For over 30 years, conservation science has emphasized the importance of enhancing connectivity conservation (Keeley et al. 2019)—yet less than a third of the world's protected areas are adequately connected (Saura et al. 2017). Along comes the recognition that isolated protected areas alone are insufficient to conserve biodiversity (Estreguil et al. 2013), and connectivity restoration is now required in light of the fragmented state of the world's forests (Tuyisingize et al. 2022; van Diggelen et al. 2012; JRC 2019). Conserving or reestablishing connectivity has become a key recommendation in restoration efforts (Belote et al. 2020; Global Land Outlook (GLO) 2022). Until recently, however, forest restoration was mainly focused on addressing site-specific problems and issues rather than considering the larger landscape context (Baldwin et al. 2012; Mansourian 2021; Tongway and Ludwig 2012). In the year 2000, nongovernmental organization (NGO) initiatives led by the World Wildlife Fund (WWF) coined the term forest landscape restoration (FLR), which refers to "a planned process that aims to regain ecological integrity and enhance human wellbeing in deforested or degraded landscapes" (WWF and IUCN 2000). Besides emphasizing the importance of including social dimensions in active restoration efforts, this definition acknowledges the role of the wider landscape context and its multifunctional character in providing various benefits (Baldwin et al. 2012; César et al. 2021; van Diggelen et al. 2012). A primary goal of

FLR is to protect and enhance biodiversity by enhancing habitat availability and habitat quality (see Chap. 10) as well as increasing landscape connectivity (Chazdon and Guariguata 2018). Bringing together the fields of restoration ecology and landscape ecology (Aronson and van Andel 2012; Mansourian 2021), connectivity analysis has developed into an important focus of conservation and restoration science (Correa Ayram et al. 2016; Keeley et al. 2019; Krosby et al. 2010; Unnithan Kumar and Cushman 2022).

Approach to Restoring Forest Connectivity

Generally, the concept of FLR is a set of diverse interventions within a wider landscape that leads to the regrowth of trees, supporting the biodiversity and stability of the ecosystem and providing livelihoods and ecosystem services for local people (Hanson et al. 2015). A great variety of potentially misleading "re-" terms are used to refer to FLR, such as *reforestation, rehabilitation, reclamation, regeneration, forest recovery*, and *ecological* or *ecosystem restoration* (Hanson et al. 2015). While restoring forests means evoking change in the landscape, this does not necessarily mean returning to a previous state; in light of climate change and intensively altered ecoscapes, it can also sometimes mean creating a "novel ecosystem" (Mansourian 2021). In contrast to small-scale site restoration approaches, FLR considers forest patches to be interconnected units within the landscape and makes room for the connectivity concept (Chazdon et al. 2016).

Enabling the transformation of a fragmented and degraded landscape to a more connected and coherent one requires precise consideration of the restoration goals and applied methods as well as evaluation using clear and comprehensive measures. The optimal restoration method depends strongly on the local requirements and objectives (IUCN 2015; IUCN and WRI 2014; Marshall et al. 2023; Messier et al. 2019) and can vary greatly when changing scale, subject (the species under focus), or local context. Figure 13.1 provides a schematic overview of the applied science of restoration ecology when prioritizing connectivity. The following four questions represent a brief summary of the complexity of practical considerations and approaches in the planning of restoration actions (adapted from Hanson et al. 2015), which will be individually elaborated in the following sections:

1. *What* is to be restored? (Objectives)
2. *Where* should the restoration efforts be implemented? (Connectivity measures)
3. *How* should the restoration be implemented? (Restoration Methods)
4. *How successful* are the implemented methods? (Monitoring)

The decisions in each section are of course interdependent, and the approach is therefore iterative (Chazdon and Guariguata 2018; Tongway and Ludwig 2012), meaning that it runs through several loops to adjust methods or measures, to optimize success (Spathelf et al. 2018). The objective, in particular, affects the selection of restoration methods, connectivity measures (see Chap. 1), and monitoring schemes (see Chap. 8). Furthermore, the choices of restoration methods and

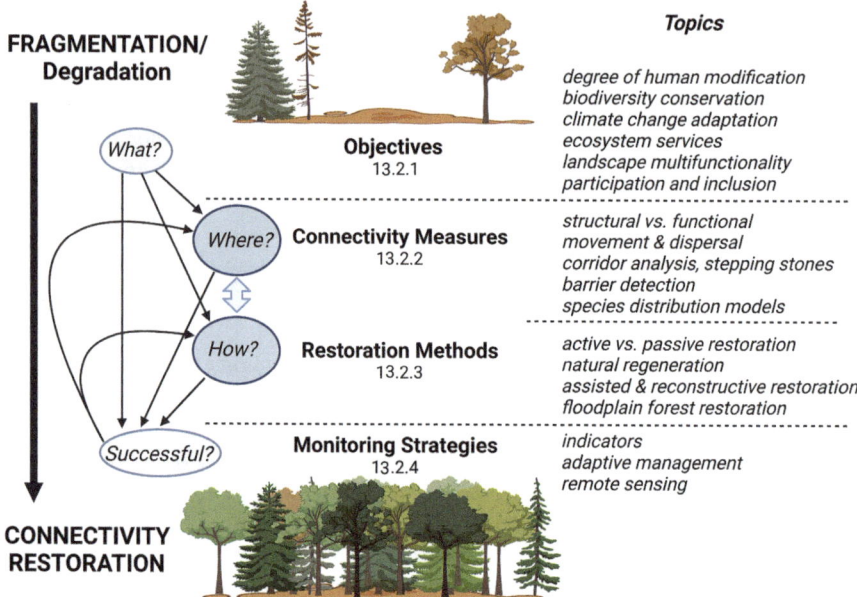

Fig. 13.1 Schematic overview of the four critical questions in the planning process of forest landscape restoration (FLR) actions when prioritizing habitat connectivity: (1) What do we want to restore and why? (Objectives); (2) Where should the restoration efforts be implemented? (Connectivity measures); (3) How should the restoration be implemented? (Methods); (4) How successful are the implemented methods? (Monitoring). Exemplary aspects when considering these questions are identified in bullet points. Each question is elaborated in a section of this chapter

connectivity measures also depend on the specific characteristics of the landscape, the level of degradation, the target species, and available resources. Finally, challenges and opportunities associated with the connectivity restoration of forest ecosystems are discussed. This also addresses the complex interactions with the social sphere.

Objectives: What Is to Be Restored?

Clearly defining objectives is considered essential for restoration and conservation planning (Lindenmayer and Hobbs 2007; Marshall et al. 2023; Messier et al. 2019; Hanson et al. 2015) since the objectives determine the restoration methods and technical (e.g., connectivity) measures as well as how the success or failure of an action is evaluated (Gann et al. 2019). Broadly, the objectives of habitat restoration and conservation align, yet their approaches diverge. Conservation primarily aims to safeguard existing areas of high-quality habitat, while restoration concentrates on revitalizing degraded land. The maintenance of landscape flows is a common goal of restoration and conservation; ideally, a coordinated strategy combining both approaches should be employed to ensure that biodiversity and ecological processes are maintained (Baldwin et al. 2012).

Before starting any activity, the broader landscape context needs to be understood and contextualized. While awareness and support for forest restoration are generally a global phenomenon, the respective measures and objectives need to be examined from the local, landscape, and regional perspectives since different drivers at these spatial levels all influence restoration success (Lohbeck et al. 2021). Obviously, restoration objectives depend strongly on the causes of disturbance or degradation (Chazdon 2008), the degree of human modification (Belote et al. 2020; Locke et al. 2019), and the configuration of the landscape (Meli et al. 2019), as well as on the local potential and need for intervention (Chazdon 2008; Chazdon et al. 2016; IUCN and WRI 2014). Degradation causes commonly include deforestation activities; increased human impact through agriculture or shifting cultivation; or forest disturbances caused by fires, storms, or landslides Hanson et al. 2015. Objectives must consider pathways of recovery from the given degradation, and restoration methods need to be selected accordingly (Gann et al. 2019).

According to Locke et al. (2019), the degree of human modification of the targeted landscape helps determine the intervention objective and restoration action. These degrees vary among (1) heavily modified ecoscapes, (2) shared ecoscapes, and (3) large wild areas. In heavily modified ecoscapes like intensive agricultural lands or urban areas, realistic objectives involve safeguarding any natural remnants, ensuring the protection of endangered species and ecosystems, considering habitat restoration where feasible, and (re)establishing connectivity in selected priority zones. In shared ecoscapes where large natural areas are interspersed with human land use—e.g., public lands with mining activities or highways—bypassing or bridging major barriers to enable connectivity between large and most small core areas is an adequate goal. In large wild areas like the Amazon rainforest or the Carpathian Mountains, protecting and maintaining biodiversity while reducing or preventing human disturbance is recommended. As shared ecoscapes represent over 55% of the world's land surface (Locke et al. 2019), landscape multifunctionality as a connectivity-related objective is considered in areas with strong human influence (César et al. 2021; Meli et al. 2019). It aims to restore multiple functions in an existing landscape—as provided by protected areas, different forest types, management structures, land-use types, or other sources—combining and connecting them with each other (Baldwin et al. 2012; César et al. 2021). In other words, the inclusion of ecological connectivity requirements in integrated land-use planning maximizes the benefits of restoration for the entire landscape. This can be seen as a continuation of the land-sharing principle that harmonizes biodiversity conservation and agricultural production by combining both objectives on the same land (Meli et al. 2019). Land sparing, on the other hand, aims to separate high-yield farming in certain locations from natural protected areas in other areas to prevent the conversion of natural habitats to agricultural lands. This can be a necessary measure when natural forest remnants are limited in regions with little land-use conflict, like abandoned agricultural regions or high slopes (Hartup et al. 2022; Meli et al. 2019).

Besides landscape connectivity and heterogeneity, principles of ecological restoration include biodiversity conservation, climate change adaptation and mitigation, promotion of ecosystem services, and conservation of native biota (César et al. 2021; Lindenmayer and Hobbs 2007; Mansourian 2021; UNEP-WCMC, FFI, and ELP 2020). As modern restoration planning often focuses on cross-level synergies of restoration benefits rather than trying to achieve only one particular goal, the integration of human and governance aspects into the realization of objectives is crucial for restoration success (César et al. 2021; van Diggelen et al. 2012). To achieve success, a clear understanding of the objectives and their definition in the context of the current state of the conservation resources, the arsenal of conservation actions possible in that context, and the potential for effective action is essential. Especially in areas with higher population densities and shared land use, objectives should be established through active participation and the inclusion of all stakeholder groups (Höhl et al. 2020; IUCN and WRI 2014). In large natural areas, the inclusion of native groups and the promotion of sustainable resource use are crucial (Locke et al. 2019). Objectives should also include short-term, medium-term, and long-term activities and prospects as well as considering best practice examples from the implementation of past restoration efforts (Gann et al. 2019).

Connectivity Measures: Where to Restore?

Surprisingly, the value of spatial tools such as geographic information systems (GIS) for forest restoration planning was long overlooked (Keeley et al. 2019) even though GIS offers powerful analytical capabilities for assessing landscape patterns, identifying ecological corridors, and understanding movement or dispersal patterns. Spatial analyses support landscape planning and restoration efforts by providing data relevant to planning, design, and implementation of actions to avoid, reduce, and reverse habitat fragmentation (Keeley et al. 2021). They can also enhance the understanding of necessities and potentials for restoration and help in monitoring restoration success (DeLuca et al. 2010; Keeley et al. 2021; Messier et al. 2019). However, they should always be part of a wider variety of considerations regarding where to restore (Höhl et al. 2020).

Countless connectivity measures ranging from simple to complex can be found in the literature (see Chap. 1), and selecting suitable measures for restoration, conservation, and monitoring can be very difficult (Keeley et al. 2021; Kindlmann and Burel 2008). In general, connectivity measures can be divided into measures addressing *structural* and *functional* connectivity depending on whether the perspective of the species or the structure of the landscape is chosen. The same species can encounter varying degrees of connectivity across diverse landscapes, and different species may experience different levels of connectivity within the same landscape (Kindlmann and Burel 2008). A combination or a well-considered selection of structural and functional measures helps pursue a meaningful measure of connectivity in different contexts (Martínez-Richart et al. 2024).

Structural connectivity is based on the entire landscape structure from a general perspective. These models are valuable because comprehensive data on distributions and habitat preferences exist only for a limited number of species (Baldwin et al. 2012). They can be easily computed using landscape analysis tools commonly supported by GIS software (Scolozzi 2009; de la Sancha et al. 2021). Landscape indices can often serve as proxies for connectivity and habitat quality, describing the overall composition of the landscape as well as edge effects and habitat sizes, proximity to disturbances, the total amount of core habitat, or the rate of fragmentation (Keeley et al. 2021).

Functional connectivity considers species-specific behavioral responses to landscape elements and the entire landscape structure (Goicolea et al. 2022; Keeley et al. 2019; Kindlmann and Burel 2008). Identifying functional connectivity networks within a region can involve visualizing linkages or corridors through direct observations of the movement patterns of specific species of interest (Kurvits 2011; WWF Tigers Alive 2020). However, gathering empirical data of this sort is resource-intensive, and available information is thus limited or nonexistent for most species. For this reason, functional connectivity is far more difficult to assess than structural connectivity. Creating connectivity models that account for the unique requirements of multiple species can lead to considerable complexity and uncertainty. As a result, connectivity analyses often make use of spatial models assessing the *potential* connectivity of a landscape from a species perspective rather than the *actual* connectivity (With 2019b); these are then coupled with human expertise (Liu et al. 2018). Species distribution modeling can serve as an insightful measure of potential habitats of a focal species and their distribution patterns (Messier et al. 2019). Here, external environmental and bioclimatic factors necessary for a species' survival, reproduction, or dispersal can be included in the analysis (Goicolea et al. 2022).

Corridor analyses can focus on the needs of a single species, typically large charismatic mammals like giant pandas (Yin et al. 2006) or tigers (Kurvits 2011), often referred to as "flagship species", or "umbrella species" when the conservation of their migratory pathways also provides habitats for many other species. Examples of umbrella species are elephants (Li et al. 2023), the gray wolf (WWF 2020), or the grizzly bear (WWF 2020). Multispecies approaches combining and overlaying corridor calculations for several species for a more comprehensive picture also exist (Fig. 13.2) (Albert et al. 2017; Liu et al. 2018).

Barrier and constraint detection represents the counterpart to corridor identification. Connectivity does not necessarily need to identify existing corridors to be maintained and protected; they might also identify factors limiting successful dispersal (McRae et al. 2012; Vasudev et al. 2015). McRae et al. (2012) propose the use of a method that includes resistance surfaces to identify barriers that strongly reduce movement between two locations in a landscape. Barrier detection can assist practitioners in prioritizing a connectivity strategy, highlighting corridors that traverse multiple barriers and allowing efforts to be channeled toward more viable pathways. For restoration actions, barrier identification can provide valuable information on the potential location of green bridges by identifying conflicts between corridors and highways or other linear structures that impede movement (Bergesen et al. 2018).

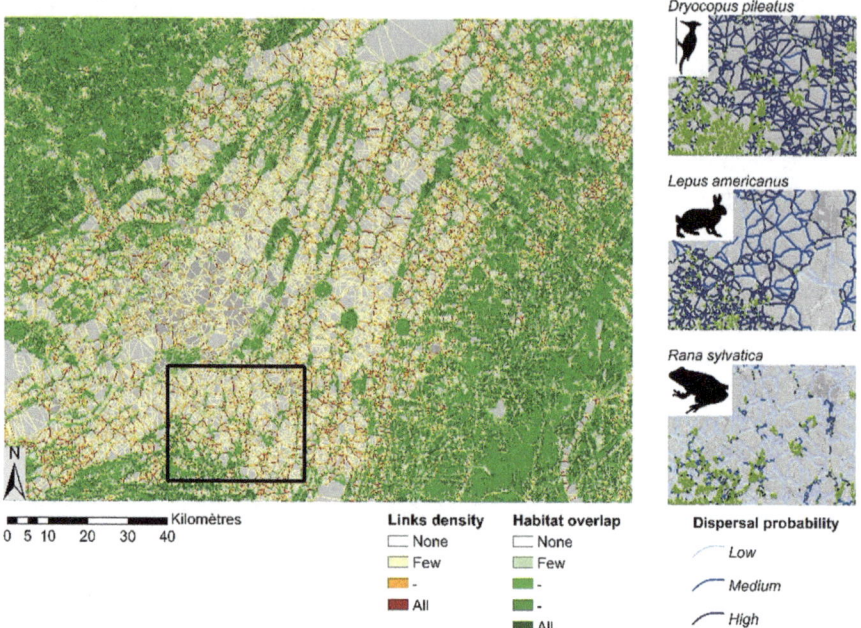

Fig. 13.2 Corridor identification and connectivity analysis with multispecies overlap using graph theory in a highly fragmented urban area around Greater Montreal, Canada. (Albert et al. 2017)

A variety of spatial algorithms have been employed to analyze connectivity or identify corridors. Various **software tools** designed to support the calculation of connectivity measures exist (Correa Ayram et al. 2016). CONEFOR is based on graph theory and quantifies the importance of habitat patches in structural landscape connectivity (Wang et al. 2021), FRAGSTAT can integrate structural and functional connectivity analyses, the tool CIRCUITSCAPE models movement patterns and simulates gene flow based on circuit theory, and LINKAGE MAPPER (Liu et al. 2018) calculates corridors based on the least-cost path principle. All these tools and approaches depend on sound input information on landscape resistance, a measure of the cost and requirements for the movement of animals through the landscape. This can be based on empirical data on gene flow, genetic distances, habitat use, and movement paths or alternatively on spatial data like land-cover classes and the location of roads and rivers in combination with expert knowledge on habitat preferences, dispersal requirements, and perceived barriers (Liu et al. 2018; Rudnick et al. 2012). Advanced habitat modeling techniques use species distribution data in combination with environmental variables to predict suitable habitats and potential corridors. These models can also provide insights into how landscapes might change over time, affecting connectivity (Miranda et al. 2021).

Selecting Connectivity Measures

According to Keeley et al. (2021), careful selection of a small handful of metrics provides "a highly defensible and comprehensive assessment of connectivity" and can be achieved by following a scheme that classifies the degree of human impact. However, any index values can only be effectively applied when compared with other values for the same index (e.g., over the course of time or different restoration stages), not between different indices (Keeley et al. 2021).

The choice of metrics for evaluating connectivity depends on the restoration objective, the focal species, and the scale of the restoration (Keeley et al. 2021; Kindlmann and Burel 2008). When dealing with specific species at a local or regional level, functional connectivity metrics are preferable, provided that data are available to parameterize models for those species (Keeley et al. 2021). On a global scale, structural connectivity metrics are more practical although they should incorporate matrix permeability in a non-species-specific manner, by accounting for human impact and linear infrastructure like roads and highways (Saura et al. 2017).

Restoration Methods: How to Restore?

When improving ecological connectivity is a restoration objective, the different processes of material exchange and gene flow are important to keep in mind. Interventions can be derived from them with the entire landscape considered. The most common differentiation of restoration methods is between active and passive methods depending on the intensity of human intervention (or conversely, the intensity of natural processes). This binary differentiation leaves some unclarity regarding practices in between, however, and the terms are used ambiguously in the literature (Atkinson and Bonser 2020). Table 13.1 lists restoration practices ranging from passive to active interventions, divided into three categories as proposed by the International Principles and Standards for the Practice of Ecological Restoration (Gann et al. 2019): (1) natural restoration with minimal human intervention, (2) assisted restoration with intermediate human intervention, and (3) regenerative restoration with maximum human intervention. Alternate terms like regeneration instead of restoration can also be found in the literature (Hanson et al. 2015).

Natural Restoration

Natural restoration (or natural spontaneous restoration) relies on the resilience of the forest ecosystem. It is best understood as a gradual recovery mechanism for the structure, functioning, and composition of the ecosystem. Natural regeneration is often the least expensive and most efficient choice in scenarios with a high likelihood of natural recovery where degradation is relatively low and populations for recolonization and recovery exist nearby (Gann et al. 2019). This can happen through resprouting, germination from soil seed banks, or seed rain, or dispersal by

Table 13.1 Restoration practices featuring increasing levels of human intervention ranging from passive to active restoration, with examples from existing research and connectivity restoration actions. (adapted from Atkinson and Bonser 2020; Gann et al. 2019; enhanced with examples from connectivity restoration)

	Restoration method	Examples of connectivity restoration interventions
Passive	**1. *Natural restoration*** *aka natural (spontaneous) regeneration* "letting the forest gradually recover on its own"	– Abandonment of unsuitable agricultural lands for regrowth of forest areas for elephant corridor habitats (Evans et al. 2017) – Protection from further degradation to support natural regeneration and secondary succession (Chazdon and Guariguata 2016)
Active	**2. *Assisted restoration*** *aka natural (assisted) regeneration* "supporting forest recovery by eliminating disturbances and improving conditions"	– Enrichment planting of indigenous trees (Mwang'ombe 2005; Mangueira et al. 2019; Palma et al. 2020) and removal of exotic trees (Mwang'ombe 2005) – Leaving of retention patches with deadwood and habitat trees that can serve as steppingstone habitats in commercial forests (Gustafsson et al. 2020; Storch et al. 2020; Lapin et al. 2024) – Reduction of competition from weedy species and recurring disturbances like fire, grazing, or wood harvesting (JRC 2019) – Combating of invasive species (Yirdaw et al. 2014) within the forest management plan (Makoni 2020)
	3. *Reconstructive restoration* *aka artificial regeneration* "extensively (re)building forest ecosystems"	– Forest establishment by **planting saplings from nurseries** in degraded clear-cut areas (Hartup et al. 2022; Chapman and Chapman 1999) – Creation of **wildlife corridors and bridges** (Jackson 2000; Smith et al. 2015) – Creation of **agroforestry systems** by adding linear elements of natural vegetation like hedgerows and tree lines to enhance landscape heterogeneity, habitat connectivity (e.g., for bats), and soil regeneration (ENoP 2023; Haggar et al. 2019)

wind or animals from nearby forest areas (Woods et al. 2020). Within mosaic landscapes, natural regeneration has the potential to occur across various patches if topsoil is retained and damage is relatively low (Chazdon and Guariguata 2016). The pace and traits of natural restoration are significantly shaped by factors such as forest biome, climate, soil conditions, recurring disturbances at the stand level, previous land utilization, surrounding vegetation, and the diversity of species in the region (Chapman and Chapman 1999; Chazdon and Guariguata 2016; Teketay 2005). Although the sequences of recovery stages — also known as succession — can vary across ecosystems, it is likely that native pioneer species have an increased ability to recuperate after natural disturbances or stressors to which they have adapted (Gann et al. 2019). If there is enough habitat connectivity and ecosystem resilience, animal species can return to an area, and plant species may rebound across a wide array of spatial scales.

Assisted Restoration

Even under favorable conditions, naturally regenerating forests are unlikely to fully reclaim the complete species composition found in the original ecosystem due to large-scale habitat loss, reduction in animal populations, and the impacts of climate change. Assisted restoration (or natural assisted regeneration) aims to accelerate ecological succession by eliminating disturbances like fire, grazing, or wood harvesting or reducing obstacles to ecological succession like soil degradation, invasive plants, or insufficient seed banks. Typically, assisted restoration is cost-effective compared to reconstructive restoration and proves effective in transforming degraded vegetation into productive forests. It is therefore recommended for intermediate degradation (Méndez-Toribio et al. 2021). Biotic, abiotic, or management interventions may be necessary to enhance commercial value, increase the presence of species with poor dispersion abilities, safeguard endangered species, reinforce genetic diversity (Chazdon and Guariguata 2016), or increase the resilience and adaptability of the forest ecosystem. An example of biotic intervention is enrichment planting, where native trees are selectively added to an existing or regenerating forest ecosystem (Lohbeck et al. 2021; Mangueira et al. 2019). Abiotic interventions involve improving substrate conditions, reshaping watercourses, and restoring environmental flows (Gann et al. 2019). Management interventions can create more favorable conditions for a resilient and diverse forest landscape.

Connectivity in restoration considerations should not only target large-scale forest ecosystems in conservation areas. Active forest management plans can include standing and lying deadwood and habitat trees in retention patches (see Chap. 10). These retention patches can conserve biodiversity and enhance habitat quality on a smaller scale, for example for saproxylic beetles (Haeler et al. 2021; Zumr et al. 2021). In addition, they can serve as refuges for vulnerable forest species or even as stepping stones between habitats within a more homogeneous matrix of production forests (Gustafsson et al. 2020). This approach is implemented in forest management plans for multiuse forest landscapes in Europe (see Chap. 22).

Assisting the regeneration of the ecosystem can also include the removal of invasive or non-native species. *Prosopis africana* invasion has been recognized as an emerging threat to plant biodiversity in the Forest Resource Strategy in Ethiopia (EBI 2014). The species extensively covers grazing lands, croplands, and areas along river courses in the northeastern and southern parts of Ethiopia (Yirdaw et al. 2014), affecting nutrient cycling, pollination, regeneration, and biodiversity of native plants. These characteristics make *P. africana* a dangerous invader, as evidenced by its rampant spread in southeastern Ethiopia and elsewhere in the tropics (Shiferaw et al. 2004). Effective management can be achieved through the utilization of the plants for fuelwood and charcoal production, which not only impedes the species' spread but also generates income for local populations (Yirdaw et al. 2014).

Reconstructive Restoration

In cases where the natural recovery capabilities of ecosystems are severely limited, maximum intervention methods like reconstructive restoration (or artificial regeneration) can become necessary. These limitations include human-induced degradation (Atkinson and Bonser 2020) with consequences like unsuitable soil substrate, absence or alteration of habitat niches, resource scarcity, herbivory, competition, lack of propagule availability, and dormancy cues for seed germination (Chalermsri et al. 2020; Teketay 2005). Time can also be a legitimate limitation. The need to reduce soil erosion and attain other important ecosystem functions as soon as possible can be reasons to actively plant rather than waiting for natural succession (Fischer and Fischer 2012). Furthermore, planting activities and global restoration efforts are aimed at reconstructing forest ecosystems and are subject to economic and political pressure to quickly achieve climate or development goals (IUCN 2020). Artificial regeneration often involves extensive intervention to mimic the original (or potential) natural vegetation, species composition, and ecological processes. The reintroduction of a significant portion of desirable biota is often essential for the interaction with abiotic components to drive further ecosystem recovery (Gann et al. 2019) (Fig. 13.3). While being the most tangible and quantifiable approach, reconstructive restoration is also quite expensive, complex, and uncertain in terms of suitable strategies and outcomes (Méndez-Toribio et al. 2021).

Fig. 13.3 Tree nurseries in the Făgăraş Mountains in Romania are part of large-scale EU forest restoration efforts combating degradation and clear-cutting. The produced Norway spruce (*Picea abies*) saplings are planted on degraded and deforested lands, representing reconstructive restoration. While Norway spruce is well-adapted to the mountainous areas of eastern Europe, a further goal is to slowly convert spruce monocultures back to mixed mountain forests. (Photo: Johanna A. Hoffmann)

Selecting Restoration Methods

Appropriate selection of restoration intervention measures depends on the degree of disturbance as well as on ecological knowledge, the objectives, the available budget, and the restoration scale (Méndez-Toribio et al. 2021). Natural restoration is highly recommended for low-disturbance regimes where there is a high potential for natural recovery (Gann et al. 2019). Natural regeneration has also been reported to yield better biodiversity recovery than artificial restoration due to the greater functional complexity in a naturally restored ecosystem (Hartup et al. 2022). Chazdon and Guariguata (2016) suggest internal and external indicators that measure the natural regeneration capacity of a site. Pertaining to the ecological memory of an ecosystem, its surroundings, and the intensity of degradation and disturbance, the most important indicators describe the proximity to mature forest or vegetation areas, levels of soil disturbance (including soil seed banks and topsoil organic matter), and the abundance and diversity of seed-dispersing animals (mammals and birds). Occasionally, natural regeneration can be feasible even in extensively damaged areas like abandoned quarries and mines, although it is likely to be a gradual and long-term process. In terms of ecological connectivity, fragmentation paired with the horizontal dispersal distances of seeds and the maximum ranges of mobile species can create limitations that may require reconstructive restoration to overcome (Woods et al. 2020).

Although natural regeneration methods are widely recommended where possible by research and practitioners, artificial restoration via tree planting takes center stage in the priorities of NGOs and marketing-based funding initiatives (Bosshard et al. 2021; Lohbeck et al. 2021). In many large-scale restoration efforts, it might be appropriate to apply a combination of all three described methods where various degrees of degradation and recovery potential exist across a site. Certain areas might benefit from a natural regeneration strategy, while others require assisted regeneration, and additional zones may require a reconstructive approach (Gann et al. 2019). River restoration is an example of the integration of different restoration approaches (see Box 13.1).

> **Box 13.1 Danube River: Connectivity Restoration**
> **Riparian zones shelter a diverse range of species and—given their linear habitat structure—in particular serve as potential corridors for wildlife movement. Due to their significance for connectivity, rivers and riverbanks often play a prominent role in restoration projects.**
>
> European rivers have been altered significantly by humans over the past decades. The restoration of the Lower Danube floodplain, for example, required reversing hybrid poplar monoculture plantations; invasion by non-native species; pollution; dykes; and degraded, eroded, and rectified riverbanks, as well as reinstating the flood protection function of the river landscape
>
> (continued)

(Mansourian et al. 2019). Consequently, restoration activities spanning two decades ranged from conservation through threat reduction to active planting. This included the protection of old natural forest remnants, felling of monoculture stands, dyke removal, soil preparation, planting of native alluvial softwood saplings, cattle reintroduction for invasive species control, and mechanical removal of invasive species. Numerous community-based initiatives concentrate on restoring riparian ecosystems, presenting unique possibilities for safeguarding connectivity. Since these interventions took place over a long period of time and across various countries, long-term learning and testing strategies with clear objectives, performance indicators, and monitoring strategies could be applied in these restorations (Baldwin et al. 2012).

Monitoring: When Is Connectivity Restoration Successful?

Monitoring means assessing the success of a restoration project by measuring the recovery of the ecosystem. Existing laws often require monitoring to ascertain whether resources have been allocated effectively and implement an adaptive management approach (César et al. 2021; DeLuca et al. 2010). For this purpose, a well-designed monitoring scheme should be put in place and solid indicators selected (Belote et al. 2020; Keeley et al. 2021; Tongway and Ludwig 2012). This demands ongoing effort and resources. A before/after comparison of relevant indicators is necessary to track changes, and to evaluate the success or failure of restoration, an ideal reference ecosystem should be determined to identify targeted values (Aronson and van Andel 2012; Gann et al. 2019).

Ideally, the indicators measure the achievement of the set objective; they can range widely from measures of *habitat quality* (e.g., tree mortality, soil organic matter, soil erosion, and water quality) and *biodiversity indicators* (e.g., species richness and vegetation composition) to concrete *connectivity measures* (e.g., seed dispersal, movement patterns, species range, and landscape metrics) or *species conservation* parameters (e.g., endangered species occurrences, species distribution, and habitat conditions). Various indicators are described in Gann et al. (2019), Keeley et al. 2021, Kindlmann and Burel (2008), and Jetz et al. (2019). There are also frameworks identifying essential steps in the selection of suitable monitoring schemes (see Chap. 8) (Block et al. 2001), including the identification of meaningful and cost-effective response variables, the establishment of thresholds for adaptive management to change the restoration strategy, the development of a sampling design, and data collection (see Chap. 12). To avoid redundancies in observation and analysis as well as high costs, the potential of already available ecological data sources should be assessed and exploited. Freely available satellite data along with existing maps of wildlife corridors, forest cover, and land-use changes can often allow a viable basic assessment (IUCN and WRI 2014).

Key Challenges in Restoring Connectivity

Technical and Financial Aspects

The availability of spatial data (and particularly its public availability) is often limited and varies immensely among countries, continents, and ecoregions (Brown et al. 2015; Chapman and Chapman 1999; Jetz et al. 2019; Theobald 2013). Insufficient data accessibility can pose a constraint when deciding on feasible local restoration methods (e.g., forest regeneration and nursery production) (Krosby et al. 2010), calculating metrics for both structural and functional connectivity (Keeley et al. 2021), and monitoring restoration success.

Even though geographical information system (GIS) data cannot replace local knowledge and research in the field, they can serve as a powerful tool supporting effective restoration planning and monitoring. The prevailing underutilization of GIS in restoration planning can be attributed to various factors including limited access to technology, a lack of awareness regarding its potential, and inadequate capacity-building efforts (FAO 2022). In some regions, historical and cultural factors may influence the adoption of spatial tools in conservation practices. Financial and regulatory constraints can also limit investments in GIS infrastructure and training, impeding its widespread use for restoration planning purposes (FAO 2022).

Similar issues are faced with regard to the amount and availability of scientific studies analyzing connectivity (Correa Ayram et al. 2016; Martínez-Richart et al. 2024). Functional connectivity metrics face greater challenges due to data limitations, but estimation of these metrics is becoming more viable with the increasing availability of movement data and advancements in deriving species-specific resistance from land cover data (Keeley et al. 2019, 2021; Martínez-Richart et al. 2024). Significant improvements in functional connectivity assessment can also be anticipated because of the decreasing costs and increasing precision of metrics linked to gene flow (Keeley et al. 2021).

Although applying adaptive management strategies is highly recommended for successful restoration (Spathelf et al. 2018), monitoring and strategy readjustment are time-consuming and costly (DeLuca et al. 2010). Financial support is usually only ensured for a relatively short period of time, and long-term monitoring is thus rarely included in project planning (IUCN and WRI 2014; Méndez-Toribio et al. 2021). Perhaps even more significant is the challenge of gathering and maintaining the essential talent pool to establish and sustain monitoring programs, which requires long-term commitment. Particularly during periods of financial limitations, voluntary or collaborative efforts become essential, and each of the participating factions will require skilled and capable personnel to formulate, supervise, and address issues occurring in the context of monitoring activities (DeLuca et al. 2010).

The lack of available data, bias in the scientific knowledge acquisition concerning regional foci and biomes is another issue. Research on forest ecosystems is predominantly focused on temperate broadleaf and mixed forests, a circumstance attributable to the fact that the majority of corresponding study locations are situated in Europe and North America (Chapman and Chapman 1999; Correa Ayram et al. 2016; Martínez-Richard et al. 2024). Across all biomes, there are still many research gaps concerning species-specific dispersal traits (Vasudev et al. 2015), distances, and multispecies connectivity approaches (Méndez-Toribio et al. 2021).

Issues with transitioning academic knowledge into public policies can also be observed (Correa Ayram et al. 2016). For example, while research widely recommends applying methods of natural regeneration, NGOs and governmental initiatives predominantly emphasize tree planting, only rarely acknowledging natural regeneration as a viable restoration approach. This discrepancy likely arises from the fact that evaluating and conveying the outcomes of tree planting is simpler, faster, and more quantifiable compared to those of natural regeneration (Lohbeck et al. 2021).

Biophysical Aspects

Besides the described technical challenges, restoration methods strongly depend on the respective geographic region and ecosystem. In terms of active planting activities, the selected planting material often lacks genetic diversity and is not properly aligned with restoration objectives. Seed availability in nurseries or specific desired traits like foliage, robust root development, or resistance to pests are frequently prioritized over the needed roles within the ecological process or ecosystem function. Planting projects aimed at enhancing primary productivity, supporting pollination services, or facilitating seed dispersal are widely lacking (Méndez-Toribio et al. 2021). Also, insufficient supply of high-quality native seedlings from nurseries as well as inadequate irrigation and tending activities are repeatedly identified as causes for low survival rates and restoration failure (Höhl et al. 2020; Méndez-Toribio et al. 2021). Species-specific needs are often ignored in reforestation projects, and planting is frequently carried out with whatever species are available regardless of their suitability requirements (Krosby et al. 2010). There is also a need to balance the fostering of populations of individual species with maximizing overall species diversity at a site (Lamb and Gilmour 2003). Stepping stones, conservation of old forests, and securing biological corridors help facilitate the movement of plants and animals in finding suitable habitats enabling natural succession and spontaneous regeneration (Höhl et al. 2020). Promoting connectivity also bears some risks, however: invasive species can outcompete native species for resources and alter an ecosystem's composition and structure. For example, some alien tree species can easily colonize open restoration sites (Lindenmayer and Hobbs 2007), and restoration efforts therefore need to include measures to prevent their establishment and spread.

Maintaining general habitat connectivity is likely to increase the resilience of an ecosystem to climate change (Krosby et al. 2010), especially with regard to potential future range shifts. Habitats at high altitudes will be particularly strongly affected by climate change (Lenoir et al. 2008), causing species to gradually move up to higher altitudes and resulting in limited geographical and altitudinal ranges and face local extermination or total extinction. In temperate mountainous landscapes, climate niche shifts should be made possible, especially for endemic species to ensure their survival (Semenchuk et al. 2021). Methods to ensure climate-smart interconnection of protected areas are a possibility for tackling the issue of "locked-in" endemic species; while such methods have yet to be researched in greater detail, they appear to hold considerable potential (Hanson et al. 2020; Hole et al. 2009; Semenchuk et al. 2021).

Social and Institutional Aspects

There are numerous challenges associated with the realization of restoration projects—and many pitfalls come down to social factors (César et al. 2021). Therefore, measures and regulations to stop cycles of destruction, deforestation, and degradation are as important as conservation and restoration efforts. Forest as a resource is often caught up in the dynamics of power, corruption, and social justice (Grgić 2020; Sewell et al. 2020). Fragmentation and disturbances need to be channeled and regulated rather than only being reversed after they have already occurred. Sustainability initiatives need to work with local communities to promote sustainable practices that reduce the intensity of human activities on forest habitats while simultaneously empowering and benefiting those communities (Chazdon 2008; Höhl et al. 2020). In restoration efforts as well, collaboration and stakeholder involvement is of key importance alongside comprehension of the respective ecological processes (Höhl et al. 2020; IUCN and WRI 2014). Participatory approaches are more likely to produce more viable outcomes than centralized decision-making (Lamb and Gilmour 2003). Ensuring transparent, equitable, and sound stakeholder engagement is crucial for all successful restoration and rehabilitation initiatives (Seyoum et al. 2015). From issue identification to benefit sharing, inclusive practices are necessary to achieve favorable restoration outcomes and extend effects beyond the lifespan of a restoration effort (Höhl et al. 2020; Moges et al. 2021). The larger the spatial scale under investigation, the more stakeholders are involved and the more signs and causes of forest degradation can be found and need to be addressed—and pursuing priority objectives without turning a blind eye to issues considered "less urgent" consequently becomes more difficult. Stakeholders may include local communities and administrations, NGOs, universities, research institutions, and others. Men, women, and the youth are equally important players in the process of restoration and rehabilitation and should be consulted during each stage of the resource planning process (Basnett et al. 2017; IUCN and WRI 2014; Méndez-Toribio et al. 2021; Seyoum et al. 2015). To reduce the likelihood of conflicts, appropriate criteria should be applied to categorize stakeholders in terms of their proximity to the restoration target forest, preexisting rights, and level of dependency on the forest (Lamb and Gilmour 2003).

Forest landscape restoration is a long-term process and requires certainty concerning land ownership or resource usage rights. The more insecure the land ownership, the more intense the utilization of the forest resource over the short term (César et al. 2021; McLain et al. 2021; Moges et al. 2021). Owing to tenure insecurity, a significant portion of forest lands in the developing world now falls within the public domain and has become a de facto open-access resource (Lamb and Gilmour 2003). Natural resource management therefore must be based on secure and agreed access and usage rights. In some cases, there is no legal transfer of land management and usage rights to the community other than the de facto understanding that the community owns the land (Yirdaw et al. 2014), resulting in ambiguous ownership of trees within restoration areas. Various forest restoration activities in developing countries are conducted on communal land (Lamb and Gilmour 2003). Under the condition that the targeted forest land is open-access, it makes sense to look for institutional arrangements such as community associations to assume responsibility for managing the restored land. This helps prevent conflicts regarding the resources and ensure the sustainability of the restored forest (Lamb and Gilmour 2003). Even if restoration potential exists (Bastin et al. 2019), agricultural land competes with the natural world, and unclear ownership threatens the long-term acceptance of the restoration (Meli et al. 2019). Land-sharing approaches may sometimes be a better fit than altering the entire composition of a landscape. The choice between intervening in a highly fragmented landscape (also in terms of ownership) or focusing on a region where intervention is still economically feasible and realistic is difficult, but it should be included in the planning and decision-making processes.

Decision support tools like Restoration Opportunity Assessment Mapping (ROAM) can be helpful in weighing options against each other and identifying opportunities following structural guidance in the implementation of an inclusive and feasible restoration project (Chazdon and Guariguata 2018; IUCN and WRI 2014). Especially in case of land conflicts with agriculture and ownership uncertainties, it is important to balance necessity and opportunity for the site of a restoration activity. Systematic approaches can help identify priority areas for FLR actions by including social and economic data in addition to connectivity and ecosystem measures (Chazdon and Guariguata 2018). These approaches do not strictly follow a top-down or bottom-up strategy; instead, they are based on optimization principles that can encompass a broad spectrum of social, political, economic, and ecological aspects. If enough spatial data relating to costs as well as other social prioritization criteria such as land tenure and ownership or the existence of local restoration engagement and active communities are available, this information should be included in a *multi-criteria spatial prioritization* framework. Applying and assessing multiple criteria in systematic approaches like ROAM increase transparency, evaluate trade-offs and identify potential land-use conflicts, and enable adaptation to divergent stakeholder needs (Chazdon and Guariguata 2018).

Last but not least, effectively connecting natural areas and habitats requires binding legal international frameworks and a great deal of cooperation across borders between governments and nongovernmental organizations (NGOs) as well as partnerships between private and public stakeholders (Keeley et al. 2019; Locke and Rissman 2012). Collaboration is crucial for sharing knowledge and coordinating conservation endeavors like removing barriers or creating bridges. Such efforts are particularly vital for endangered animal species facing imminent risk of extinction across all or most of their natural habitat due to the scarcity of surviving individuals

(Primack 2014), as well as for migratory species that rely on cross-country ecological corridors (Bergesen et al. 2018; Kurvits 2011). Similarly, floodplain restoration of large river systems requires intensive transboundary collaboration (Mansourian et al. 2019) (see Box 13.1).

Governments and international financial institutions can support ecological connectivity by providing financial incentives and grants (IUCN 2020). These incentives have been utilized in various regions: for example, both the United States Farm Bill and similar initiatives in Europe allocate substantial funds each year to incentivize agricultural and private forest practices aligned with conservation goals (Krosby et al. 2010; Locke and Rissman 2012). In addition, there are community-based initiatives already in progress aimed at implementing connectivity conservation plans (CCPs) in large-scale restoration efforts (Keeley et al. 2019). Examples of CCPs for forest ecosystems are the Yellowstone to Yukon Conservation Initiative in the United States, the Alpine Ecological Network in Europe, the Mata Atlántica in Brazil, the Gondwana Link in Australia, and the Albertine Rift Initiative in Africa.

Conclusion

Ecological connectivity and ecosystem restoration are interdependent. Properly considering forest connectivity has the potential to increase the success of long-term restoration and biodiversity conservation strategies. Preserving or reestablishing healthy, resilient, and biodiverse ecosystems and simultaneously restoring degraded, fragmented habitats, are key management tools for enhancing connectivity within a landscape. This is also reflected in the principles "connect to restore" and "restore to connect." Surprisingly, spatial tools were long overlooked in forest restoration planning even though they provide a tractable and powerful methodology. Spatial analyses support landscape planning and restoration efforts by informing the planning, design, and implementation of measures to avoid, reduce, and reverse habitat fragmentation. Geographic information systems can assist practitioners in making informed decisions based on the ecological requirements of target species, topographic features, and existing landscape characteristics. They provide powerful analytical capabilities for assessing landscape patterns, identifying ecological corridors, and determining movement and dispersal patterns. Connectivity measures can also enhance the understanding of necessities and potentials in the initial phases of restoration activities (planning and decision-making) as well as help monitor restoration success and failure.

The significance of ecological connectivity for ecosystem restoration has been increasingly acknowledged globally, internationally, and nationally in legislation and treaties like the EU Habitats Directive, the Convention on Biological Diversity (Aichi Target 11), and countless national connectivity and corridor regulations. The most recent legislation is the addition to the EU Nature Restoration Law in the year 2024. It defines forest connectivity as one of eight biodiversity indicators for forest ecosystems (Article 12) that member states need to improve by 2030 (EU Regulation 2024/1991). Additionally, scientific knowledge about the theory and practice of mitigating fragmentation by establishing links between habitats has found implementation in various connectivity conservation plans, thereby opening a multitude of funding opportunities and helping initiate publicly funded large-scale projects. In many countries where laws and policies on wildlife management and land-use planning do not specifically address connectivity conservation, NGOs and community initiatives are key to realizing large-scale restoration efforts, providing leadership and funding and ensuring that the public plays a central role.

Nevertheless, transforming ambitious objectives into tangible outcomes poses a constant challenge. Effective restoration must address the complexity of ecosystems and the changing nature of site conditions while also accommodating stakeholder needs. It requires navigating diverse and often unstable sociopolitical conditions, complicated economic and legal circumstances, ongoing deforestation and forest degradation, and constrained technical capabilities. As connectivity measures alone are only a part of any comprehensive restoration planning process, they should serve as an addition to local expert knowledge while being embedded in the local social context (see Box 13.2).

Box 13.2 Ethiopian Church Forests: Fragmented Biodiversity Hotspots

The Ethiopian Church Forests (ECFs) are impressive examples of fragmentation and habitat isolation. In the northern highlands of Ethiopia, in the South Gondar Zone, nearly all the native forests have been cleared to make way for wheat fields and grazing land. In some areas, agriculture and grasslands occupy more than 80% of the surface area, while forests are highly fragmented (Kindu et al. 2022). **Patchy remnants of rich old-growth Afromontane forests can be found around the Ethiopian Orthodox Tewahido Churches (EOTC), however.**

The forests are visible from a great distance and offer a majestic appearance, usually located on small hills overlooking the surrounding villages (Fig. 13.4). With some of them having existed for a very long time—the oldest is around 765 years old—they are considered holy places by the Tewahido faithful and represent a powerful and socially respected institution (Wassie 2007). But whereas the main purpose of most churches is to offer a place for worship, burials, meditation, and religious festivities, ECFs are also ancient sanctuaries for various organisms ranging from microbes to large animals, many of which have almost disappeared from most parts of northern Ethiopia (Kindu et al. 2022; Wassie et al. 2005; Woldemedhin and Teketay 2016). Their vegetation features numerous indigenous trees, woody plants, shrubs, and liana species recorded in various studies (Aerts et al. 2016; Wassie Eshete 2007; Woldemedhin and Teketay 2016). Distributed widely across the landscape, these forests possess high conservation value. In total, 394 ECFs have been identified in satellite images each with a size of 2 ha on average and generally separated from the nearest neighboring forest by about 2 km (Fig. 13.5) (Aerts et al. 2016). They serve as in situ conservation and

(continued)

Fig. 13.4 A church forest on a hilltop. (Image credit: Alemayehu Wassie Eshete)

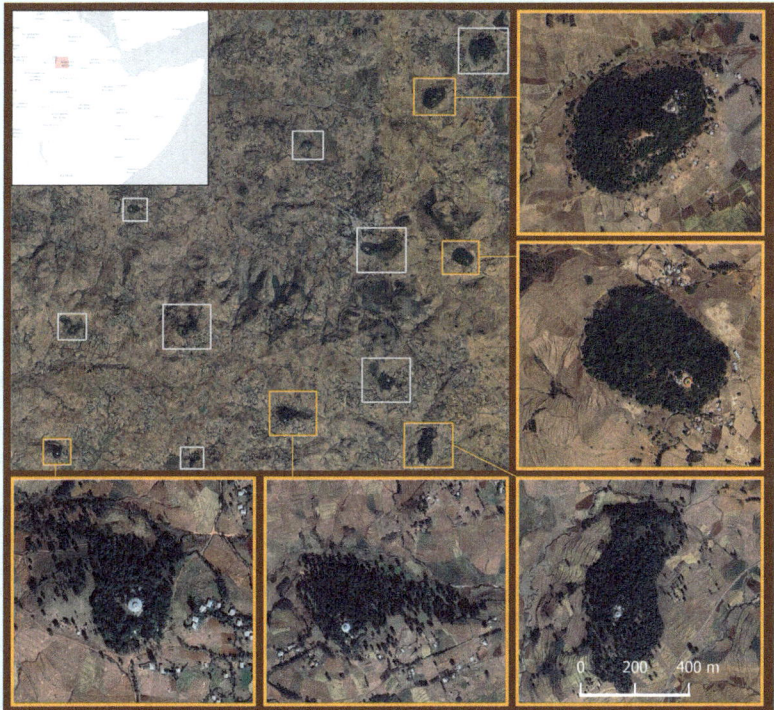

Fig. 13.5 Satellite images of five Church Forests in South Gondar, Ethiopia. These rich remnants of ancient old-growth forests are home to numerous rare native species. However, their isolation within highly fragmented agricultural areas comes with decreased adaptive capacity and poses the risk of losing biodiversity and stability in the face of climate change and higher management pressure. Possibilities of enhancing their connectivity through restoration measures are therefore being investigated. (BFW/Johanna A. Hoffmann)

(continued)

biodiversity hotspots, especially for indigenous tree and shrub species of Ethiopia, which in turn provide prestige for the religious sites.

Long-term conservation of biodiversity and the evolutionary potential of species in individual ECFs are, however, threatened by their isolation, the small size of their tree populations, and various disturbances (Kindu et al. 2022). Especially in the face of climate change, forest management interventions and integration within the landscape are necessary and should be supported by environmental education and other forms of community engagement (Aerts et al. 2016). Recently, there has been national and international engagement regarding fragmentation assessment and connectivity enhancement in the Libokemekem district (Demissie et al. 2022; Kindu et al. 2022). Possible landscape management activities may include connecting these remnant forests with vegetation corridors following natural terrain features like streams, creating buffer areas around them, excluding cattle interference, reducing the intensity of wood harvesting, and developing additional vegetation patches in the landscape. Spatial analysis estimates show that this could result in 29% of the land being covered with vegetation or forest. Soil and water conservation work has been recommended to promote participation by local communities and other stakeholders. The overall situation offers great potential for connectivity enhancement measures, which may ultimately sustain these forests and help restore the whole landscape (Wassie Eshete 2007).

References

1t.org (2023) A platform for the trillion trees community. https://www.1t.org/
Aerts R, Van Overtveld K, November E, Wassie A, Abiyu A, Demissew S, Daye DD, Giday K, Haile M, TewoldeBerhan S, Teketay D, Teklehaimanot Z, Binggeli P, Deckers J, Friis I, Gratzer G, Hermy M, Heyn M, Honnay O et al (2016) Conservation of the Ethiopian church forests: threats, opportunities and implications for their management. Sci Total Environ 551–552:404–414. https://doi.org/10.1016/j.scitotenv.2016.02.034
Albert CH, Rayfield B, Dumitru M, Gonzalez A (2017) Applying network theory to prioritize multispecies habitat networks that are robust to climate and land-use change: prioritizing a network for biodiversity. Conserv Biol 31(6):1383–1396. https://doi.org/10.1111/cobi.12943
Aronson J, van Andel J (2012) Restoration ecology and the path to sustainability. In: Restoration Ecology. Wiley, pp 293–304. https://doi.org/10.1002/9781118223130.ch22
Atkinson J, Bonser SP (2020) "Active" and "passive" ecological restoration strategies in meta-analysis. Restor Ecol 28(5):1032–1035. https://doi.org/10.1111/rec.13229
Aznar-Sánchez J, Belmonte-Ureña L, López-Serrano M, Velasco-Muñoz J (2018) Forest ecosystem services: an analysis of worldwide research. Forests 9(8):453. https://doi.org/10.3390/f9080453
Baldwin RF, Reed SE, McRae BH, Theobald DM, Sutherland RW (2012) Connectivity restoration in large landscapes: modeling landscape condition and ecological flows. Ecol Restor 30(4):274–279. https://doi.org/10.3368/er.30.4.274
Basnett BS, Elias M, Ihalainen M, Valencia AMP (2017) Gender matters in forest landscape restoration

Bastin J-F, Finegold Y, Garcia C, Mollicone D, Rezende M, Routh D, Zohner CM, Crowther TW (2019) The global tree restoration potential. Science 365(6448):76–79. https://doi.org/10.1126/science.aax0848

Belote RT, Beier P, Creech T, Wurtzebach Z, Tabor G (2020) A framework for developing connectivity targets and indicators to guide global conservation efforts. BioScience 70(2):122–125. https://doi.org/10.1093/biosci/biz148

Bergesen HO, Parmann G, Thommessen OB (2018) Convention on the conservation of migratory species of wild animals (CMS). In: Bergesen HO, Parmann G, Thommessen OB (eds) Yearbook of international cooperation on environment and development 1998–99, 1st edn. Routledge, pp 154–155. https://doi.org/10.4324/9781315066547-40

Biere A, van Andel J, van de Koppel J (2012) Populations: ecology and genetics. In: van Andel J, Aronson J (eds) Restoration ecology, 1st edn. Wiley, pp 73–86. https://doi.org/10.1002/9781118223130.ch7

Block WM, Franklin AB, Ward JP, Ganey JL, White GC (2001) Design and implementation of monitoring studies to evaluate the success of ecological restoration on wildlife. Restor Ecol 9(3):293–303. https://doi.org/10.1046/j.1526-100x.2001.009003293.x

Brown ME, Rowland J, Wood E, Tieszen LL, Lance K, Khamala E, Siwela B, Adoum A (2015) Review of remote sensing needs and applications in Africa. https://doi.org/10.13140/RG.2.1.1101.3849

Butchart SHM, Walpole M, Collen B, van Strien A, Scharlemann JPW, Almond REA, Baillie JEM, Bomhard B, Brown C, Bruno J, Carpenter KE, Carr GM, Chanson J, Chenery AM, Csirke J, Davidson NC, Dentener F, Foster M, Galli A et al (2010) Global biodiversity: indicators of recent declines. Science 328(5982):1164–1168. https://doi.org/10.1126/science.1187512

Bosshard E, Jansen M, Löfqvist S, Kettle CJ (2021) Rooting forest landscape restoration in consumer markets—a review of existing marketing-based funding initiatives. Frontiers in Forests and Global Change 3:589982. https://doi.org/10.3389/ffgc.2020.589982

César RG, Belei L, Badari CG, Viani RAG, Gutierrez V, Chazdon RL, Brancalion PHS, Morsello C (2021) Forest and landscape restoration: a review emphasizing principles, concepts, and practices. Land 10(1):28. https://doi.org/10.3390/land10010028

Chalermsri A, Ampornpan L, Purahong W (2020) Seed rain, soil seed bank, and seedling emergence indicate limited potential for self-recovery in a highly disturbed, tropical, mixed deciduous forest. Plants 9(10):1391. https://doi.org/10.3390/plants9101391

Chapman CA, Chapman LJ (1999) Forest restoration in abandoned agricultural land: a case study from East Africa. Conserv Biol 13(6):1301–1311. https://doi.org/10.1046/j.1523-1739.1999.98229.x

Chazdon RL (2008) Beyond deforestation: restoring forests and ecosystem services on degraded lands. Science 320(5882):1458–1460. https://doi.org/10.1126/science.1155365

Chazdon RL, Brancalion PHS, Laestadius L, Bennett-Curry A, Buckingham K, Kumar C, Moll-Rocek J, Vieira ICG, Wilson SJ (2016) When is a forest a forest? Forest concepts and definitions in the era of forest and landscape restoration. Ambio 45(5):538–550. https://doi.org/10.1007/s13280-016-0772-y

Chazdon RL, Guariguata MR (2016) Natural regeneration as a tool for large-scale forest restoration in the tropics: prospects and challenges. Biotropica 48(6):716–730. https://doi.org/10.1111/btp.12381

Chazdon RL, Guariguata MR (2018) Decision support tools for forest landscape restoration: current status and future outlook. Occasional Paper 183. Center for International Forestry Research (CIFOR). https://doi.org/10.17528/cifor/006792

Correa Ayram CA, Mendoza ME, Etter A, Salicrup DRP (2016) Habitat connectivity in biodiversity conservation: a review of recent studies and applications. Prog Phys Geogr Earth Environ 40(1):7–37. https://doi.org/10.1177/0309133315598713

de la Sancha NU, Boyle SA, McIntyre NE (2021) Identifying structural connectivity priorities in eastern Paraguay's fragmented Atlantic Forest. Sci Rep 11(1):16129. https://doi.org/10.1038/s41598-021-95516-3

DeLuca TH, Aplet GH, Wilmer B, Burchfield J (2010) The unknown trajectory of forest restoration: a call for ecosystem monitoring. Journal of Forestry, pp 288–295

Demissie F, Yeshitela K, Kindu M (2022) Ecological status and plan for connectivity of fragmented forests as a means of degraded land restoration in South Gonder, Ethiopia. In: Kindu M, Schneider T, Wassie A, Lemenih M, Teketay D, Knoke T (eds) State of the art in Ethiopian Church forests and restoration options. Springer, pp 245–265

EBI (2014) Ethiopia's revised national biodiversity strategy and action plan. Ethiopian Biodiversity Institute (EBI)

ENoP (2023) Reviving agroforestry landscapes in the era of climate change: for people, nature and local economy. European Network of Political Foundations (ENoP)

Estreguil C, Caudullo G, San Miguel J (2013) Connectivity of Natura 2000 forest sites. JRC executive report. European Commission. Joint URL: https://Publications.Europa.Eu/En/Publication-Detail/-/Publication/F520524b-Ee69-4fcc-A71e-360c5448a38b/Language-En

EU Regulation (2024/1991) of the European Parliament and of the Council of 24 June 2024 on Nature Restoration and Amending Regulation (EU) 2022/869 (Text with EEA Relevance). retrieved from https://eur-lex.europa.eu/eli/reg/2024/1991/oj

Evans LJ, Goossens B, Asner GP (2017) Underproductive agriculture aids connectivity in tropical forests. For Ecol Manage 401:159–165. https://doi.org/10.1016/j.foreco.2017.07.015

FAO (2022) The state of the world's forests 2022. Forest pathways for green recovery and building inclusive, resilient and sustainable economies. FAO. https://doi.org/10.4060/cb9360en

Fischer A, Fischer H (2012) Restoration of temperate forests: an European approach. In: van Andel J, Aronson J (eds) Restoration ecology, 1st edn. Wiley, pp 145–160. https://doi.org/10.1002/9781118223130.ch12

Frietsch M, Loos J, Löhr K, Sieber S, Fischer J (2023) Future-proofing ecosystem restoration through enhancing adaptive capacity. Commun Biol 6(1):377. https://doi.org/10.1038/s42003-023-04736-y

Gann, G. D., McDonald, T., Walder, B., Aronson, J., Nelson, C. R., Jonson, J., Hallett, J. G., Eisenberg, C., Guariguata, M. R., Liu, J., Hua, F., Echeverría, C., Gonzales, E., Shaw, N., Decleer, K., and Dixon, K. W. (2019). International principles and standards for the practice of ecological restoration. 2nd edn. Restoration ecology, 27(S1). https://doi.org/10.1111/rec.13035

Garrett L, Lévite H, Besacier C, Alekseeva N, Duchelle M (2022) The key role of forest and landscape restoration in climate action

Global Land Outlook, (GLO) (2022, April 19) Ecological connectivity: an essential component of ecosystem restoration. UNCCD. https://www.unccd.int/resources/global-land-outlook/ecological-connectivity-essential-component-ecosystem-restoration

Goicolea TG, Mateo R, Aroca-Fernández MJ, Gastón A, García-Viñas JI, Mateo-Sánchez MC (2022) Considering plant functional connectivity in landscape conservation and restoration management. Biodivers Conserv 31(5–6):1591–1608. https://doi.org/10.1007/s10531-022-02413-w

Grgić V (2020) Report on deforestation in the Republic of Croatia

Gustafsson L, Bauhus J, Asbeck T, Augustynczik ALD, Basile M, Frey J, Gutzat F, Hanewinkel M, Helbach J, Jonker M, Knuff A, Messier C, Penner J, Pyttel P, Reif A, Storch F, Winiger N, Winkel G, Yousefpour R, Storch I (2020) Retention as an integrated biodiversity conservation approach for continuous-cover forestry in Europe. Ambio 49(1):85–97. https://doi.org/10.1007/s13280-019-01190-1

Haeler E, Bergamini A, Blaser S, Ginzler C, Hindenlang K, Keller C, Kiebacher T, Kormann UG, Scheidegger C, Schmidt R, Stillhard J, Szallies A, Pellissier L, Lachat T (2021) Saproxylic species are linked to the amount and isolation of dead wood across spatial scales in a beech forest. Landsc Ecol 36(1):89–104. https://doi.org/10.1007/s10980-020-01115-4

Haggar J, Pons D, Saenz L, Vides M (2019) Contribution of agroforestry systems to sustaining biodiversity in fragmented forest landscapes. Agric Ecosyst Environ 283:106567. https://doi.org/10.1016/j.agee.2019.06.006

Hanson C, Buckingham K, DeWitt S, Laestadius L (2015) The restoration diagnostic: a method for developing forest landscape restoration strategies by rapidly assessing the status of key success factors. World Resources Institute. https://doi.org/10.13140/RG.2.1.4914.1846

Hanson JO, Rhodes JR, Butchart SHM, Buchanan GM, Rondinini C, Ficetola GF, Fuller RA (2020) Global conservation of species' niches. Nature 580(7802):232–234. https://doi.org/10.1038/s41586-020-2138-7

Hartup J, Ockendon N, Pettorelli N (2022) Active versus passive restoration: forests in the southern Carpathian Mountains as a case study. J Environ Manage 322:116003. https://doi.org/10.1016/j.jenvman.2022.116003

Höhl M, Ahimbisibwe V, Stanturf JA, Elsasser P, Kleine M, Bolte A (2020) Forest landscape restoration—what generates failure and success? Forests 11(9):938. https://doi.org/10.3390/f11090938

Hole DG, Willis SG, Pain DJ, Fishpool LD, Butchart SHM, Collingham YC (2009) Projected impacts of climate change on a continent-wide protected area network, pp 420–431. https://doi.org/10.1111/j.1461-0248.2009.01297.x

IUCN (2020) Bonn challenge 2020 report

IUCN (2015, March 12) Demystifying the world's forest landscape restoration opportunities. https://www.iucn.org/content/demystifying-worlds-forest-landscape-restoration-opportunities

IUCN and WRI (2014) A guide to the restoration opportunities assessment methodology (ROAM): assessing forest landscape restoration opportunities at the national or sub-national level. Working Paper (Road-test edition). IUCN, Gland, Switzerland, 125 p

Jackson SD (2000) A strategy for mitigating highway impacts on wildlife

Jetz W, McGeoch MA, Guralnick R, Ferrier S, Beck J, Costello MJ, Fernandez M, Geller GN, Keil P, Merow C, Meyer C, Muller-Karger FE, Pereira HM, Regan EC, Schmeller DS, Turak E (2019) Essential biodiversity variables for mapping and monitoring species populations. Nat Ecol Evol 3(4):539–551. https://doi.org/10.1038/s41559-019-0826-1

Joint Research Centre (JRC) (2019) FAO, state of the world's forests: Forest fragmentation, Publications Office, European Commission, https://data.europa.eu/doi/10.2760/145325

Keane RE, Loehman RA, Holsinger LM, Falk DA, Higuera P, Hood SM, Hessburg PF (2018) Use of landscape simulation modeling to quantify resilience for ecological applications. Ecosphere 9(9):e02414. https://doi.org/10.1002/ecs2.2414

Keeley ATH, Beier P, Creech T, Jones K, Jongman RH, Stonecipher G, Tabor GM (2019) Thirty years of connectivity conservation planning: an assessment of factors influencing plan implementation. Environ Res Lett 14(10):103001. https://doi.org/10.1088/1748-9326/ab3234

Keeley ATH, Beier P, Jenness JS (2021) Connectivity metrics for conservation planning and monitoring. Biol Conserv 255:109008. https://doi.org/10.1016/j.biocon.2021.109008

Kindlmann P, Burel F (2008) Connectivity measures: a review. Landsc Ecol. https://doi.org/10.1007/s10980-008-9245-4

Kindu M, Schneider T, Wassie A, Lemenih M, Teketatay D, Knoke T (2022) State of the art in Ethiopian church forests and restoration options. Springer Nature Switzerland AG

Krosby M, Tewksbury J, Haddad NM, Hoekstra J (2010) Ecological connectivity for a changing climate: ecological connectivity. Conserv Biol 24(6):1686–1689. https://doi.org/10.1111/j.1523-1739.2010.01585.x

Kurvits T (2011) Living planet, connected planet: preventing the end of the world's wildlife migrations through ecological networks: a rapid response assessment. UNEP, GRID-Arendal

Lamb D, Gilmour D (2003) Rehabilitation and restoration of degraded forests

Lanta V, Mudrák O, Liancourt P, Dvorský M, Bartoš M, Chlumská Z, Šebek P, Čížek L, Doležal J (2020) Restoring diversity of thermophilous oak forests: connectivity and proximity to existing habitats matter. Biodivers Conserv 29(11–12):3411–3427. https://doi.org/10.1007/s10531-020-02030-5

Lapin K, Hoffmann JA, Braun M, Oettel J (2024) Identification and prioritization of stepping stones for biodiversity conservation in forest ecosystems. Conserv Sci Pract 6(7):e13161. https://doi.org/10.1111/csp2.13161

Laurance WF, Camargo JLC, Luizão RCC, Laurance SG, Pimm SL, Bruna EM, Stouffer PC, Bruce Williamson G, Benítez-Malvido J, Vasconcelos HL, Van Houtan KS, Zartman CE, Boyle SA, Didham RK, Andrade A, Lovejoy TE (2011) The fate of Amazonian forest fragments: a 32-year investigation. Biol Conserv 144(1):56–67. https://doi.org/10.1016/j.biocon.2010.09.021

Lemieux CJ, Beazley KF, MacKinnon D, Wright P, Kraus D, Pither R, Crawford L, Jacob AL, Hilty J (2022) Transformational changes for achieving the Post-2020 global biodiversity framework ecological connectivity goals. FACETS 7:1008–1027. https://doi.org/10.1139/facets-2022-0003

Lenoir J, Gégout JC, Marquet PA, de Ruffray P, Brisse H (2008) A significant upward shift in plant species optimum elevation during the 20th century. Science 320(5884):1768–1771. https://doi.org/10.1126/science.1156831

Li W, Liu P, Yang N, Chen S, Guo X, Wang B, Zhang L (2023) Improving landscape connectivity through habitat restoration: application for Asian elephant conservation in Xishuangbanna Prefecture, China. Integr Zool 1749–4877:12713. https://doi.org/10.1111/1749-4877.12713

Lindenmayer DB, Hobbs RJ (2007) Managing and designing landscapes for conservation: moving from perspectives to principles

Liu C, Newell G, White M, Bennett AF (2018) Identifying wildlife corridors for the restoration of regional habitat connectivity: a multispecies approach and comparison of resistance surfaces. PLoS One 13(11):e0206071. https://doi.org/10.1371/journal.pone.0206071

Locke CM, Rissman AR (2012) Unexpected co-benefits: forest connectivity and property tax incentives. Landsc Urban Plann 104(3–4):418–425. https://doi.org/10.1016/j.landurbplan.2011.11.022

Locke H, Ellis EC, Venter O, Schuster R, Ma K, Shen X, Woodley S, Kingston N, Bhola N, Strassburg BBN, Paulsch A, Williams B, Watson JEM (2019) Three global conditions for biodiversity conservation and sustainable use: an implementation framework. Natl Sci Rev 6(6):1080–1082. https://doi.org/10.1093/nsr/nwz136

Lohbeck M, Rother DC, Jakovac CC (2021) Editorial: enhancing natural regeneration to restore landscapes. Front For Glob Change 4:735457. https://doi.org/10.3389/ffgc.2021.735457

Makoni M (2020) Africa's invasive species problem. Lancet Planetary Health 4(8):e317–e319. https://doi.org/10.1016/S2542-5196(20)30174-1

Mangueira JRSA, Holl KD, Rodrigues RR (2019) Enrichment planting to restore degraded tropical forest fragments in Brazil. Ecosyst People 15(1):3–10. https://doi.org/10.1080/21513732.2018.1529707

Mansourian S (2021) From landscape ecology to forest landscape restoration. Landsc Ecol 36(8):2443–2452. https://doi.org/10.1007/s10980-020-01175-6

Mansourian S, Doncheva N, Valchev K, Vallauri D (2019) Lessons learnt from 20 years of floodplain forest restoration: the lower Danube landscape

Marshall AR, Waite CE, Pfeifer M, Banin LF, Rakotonarivo S, Chomba S, Herbohn J, Gilmour DA, Brown M, Chazdon RL (2023) Fifteen essential science advances needed for effective restoration of the world's forest landscapes. Philos Trans R Soc B Biol Sci 378(1867):20210065. https://doi.org/10.1098/rstb.2021.0065

Martínez-Richart AI, Zolles A, Oettel J, Petermann JS, Essl F, Lapin K (2024) A review of structural and functional connectivity studies in European forests. Landsc Ecol 40(1):10. https://doi.org/10.1007/s10980-024-02028-2

McLain R, Lawry S, Guariguata MR, Reed J (2021) Toward a tenure-responsive approach to forest landscape restoration: a proposed tenure diagnostic for assessing restoration opportunities. Land Use Policy 104:103748. https://doi.org/10.1016/j.landusepol.2018.11.053

McRae BH, Hall SA, Beier P & Theobald DM (2012) Where to Restore Ecological Connectivity? Detecting Barriers and Quantifying Restoration Benefits. PLoS ONE, 7(12), e52604. https://doi.org/10.1371/journal.pone.0052604

Meli P, Rey-Benayas JM, Brancalion PHS (2019) Balancing land sharing and sparing approaches to promote forest and landscape restoration in agricultural landscapes: land approaches for forest landscape restoration. Perspect Ecol Conserv 17(4):201–205. https://doi.org/10.1016/j.pecon.2019.09.002

Méndez-Toribio M, Martínez-Garza C, Ceccon E (2021) Challenges during the execution, results, and monitoring phases of ecological restoration: learning from a country-wide assessment. PLoS One 16(4):e0249573. https://doi.org/10.1371/journal.pone.0249573

Messier C, Bauhus J, Doyon F, Maure F, Sousa-Silva R, Nolet P, Mina M, Aquilué N, Fortin M-J, Puettmann K (2019) The functional complex network approach to foster forest resilience to global changes. For Ecosyst 6(1):21. https://doi.org/10.1186/s40663-019-0166-2

Miranda L d S, Awade M, Jaffé R, Costa WF, Trevelin LC, Borges RC, Brito RM d, Tambosi LR, Giannini TC (2021) Combining connectivity and species distribution modeling to define conservation and restoration priorities for multiple species: a case study in the eastern Amazon. Biol Conserv 257:109148. https://doi.org/10.1016/j.biocon.2021.109148

Moges Y, Haile M, Livingstone J (2021) Integration of forest landscape restoration in Ethiopia's nationally determined contributions.

Mwang'ombe J (2005) Restoration and Increase of Forest Connectivity in Taita Hills: survey and suitability assessment of exotic plantations for restoration. Project Report. The East African Wild Life Society

Oliveira-Junior ND d, Heringer G, Bueno ML, Pontara V, Meira-Neto JAA (2020) Prioritizing landscape connectivity of a tropical forest biodiversity hotspot in global change scenario. For Ecol Manage 472:118247. https://doi.org/10.1016/j.foreco.2020.118247

Palma AC, Goosem M, Stevenson PR, Laurance SGW (2020) Enhancing plant diversity in secondary forests. Frontiers in Forests and Global Change 3:571352. https://doi.org/10.3389/ffgc.2020.571352

Primack RB (2014) Essentials of conservation biology (5th edn)

Rudnick D, Ryan SJ, Beier P, Cushman SA, Dieffenbach F, Epps C, Gerber LR, Hartter JN, Jenness JS, Kintsch J, Merenlender AM, Perkl RM, Perziosi DV, Trombulack SC (2012) The role of landscape connectivity in planning and implementing conservation and restoration priorities. Issues in ecology

Saura S, Bastin L, Battistela L, Madrici A, Dubois G (2017) Protected areas in the world's ecoregions: how well connected are they? Ecol Indic 76:144–158

Scolozzi, R. (2009). Habitat potential and connectivity assessment to support land-use planning: A case study in an Alpine valley floor.

Semenchuk P, Moser D, Essl F, Schindler S, Wessely J, Gattringer A, Dullinger S (2021) Future representation of species' climatic niches in protected areas: a case study with Austrian endemics. Front Ecol Evol 9:685753. https://doi.org/10.3389/fevo.2021.685753

Sewell A, van der Esch S, Löwenhardt H (2020) Goals and commitments for the restoration decade: a global overview of countries' restoration commitments under the Rio Conventions and other pledges. The Hague

Seyoum Y, Birhane E, Hagazi N, Esmael N, Mengistu T, Kassa H (2015) Enhancing the role of forestry in building climate resilient green economy in Ethiopia

Shiferaw H, Teketay D, Nemomissa S, Assefa F (2004) Some biological characteristics that foster the invasion of Prosopis juliflora (Sw.) DC. at Middle Awash Rift Valley Area, north-eastern Ethiopia. J Arid Environ 58(2):135–154. https://doi.org/10.1016/j.jaridenv.2003.08.011

Smith DJ, van der Ree R, Rosell C (2015) Wildlife crossing structures: an effective strategy to restore or maintain wildlife connectivity across roads. In: Handbook of road ecology, pp 172–183. https://doi.org/10.1002/9781118568170.ch21

Spathelf P, Stanturf J, Kleine M, Jandl R, Chiatante D, Bolte A (2018) Adaptive measures: integrating adaptive forest management and forest landscape restoration. Ann For Sci 75(2):55. https://doi.org/10.1007/s13595-018-0736-4

Stanturf JA, Mansourian S (2020) Forest landscape restoration: state of play. R Soc Open Sci 7(12):201218. https://doi.org/10.1098/rsos.201218

Storch I, Penner J, Asbeck T, Basile M, Bauhus J, Braunisch V, Dormann CF, Frey J, Gärtner S, Hanewinkel M, Koch B, Klein A, Kuss T, Pregernig M, Pyttel P, Reif A, Scherer-Lorenzen M, Segelbacher G, Schraml U et al (2020) Evaluating the effectiveness of retention forestry to enhance biodiversity in production forests of Central Europe using an interdisciplinary, multiscale approach. Ecol Evol 10(3):1489–1509. https://doi.org/10.1002/ece3.6003

Teketay D (2005) Seed and regeneration ecology in dry Afromontane forests in Ethiopia: I. Seed production to population structure. Trop Ecol 46:29–44

Theobald DM (2013) A general model to quantify ecological integrity for landscape assessments and US application. Landsc Ecol 28(10):1859–1874. https://doi.org/10.1007/s10980-013-9941-6

Thomas E, Jalonen R, Loo J, Boshier D, Gallo L, Cavers S, Bordács S, Smith P, Bozzano M (2014) Genetic considerations in ecosystem restoration using native tree species. For Ecol Manage 333:66–75. https://doi.org/10.1016/j.foreco.2014.07.015

Timpane-Padgham BL, Beechie T, Klinger T (2017) A systematic review of ecological attributes that confer resilience to climate change in environmental restoration. PLoS One 12(3):e0173812. https://doi.org/10.1371/journal.pone.0173812

Tongway D, Ludwig J (2012) Planning and implementing successful landscape-scale restoration, pp 30–42. https://doi.org/10.1002/9781118223130.ch4

Travers TJP, Alison J, Taylor SD, Crick HQP, Hodgson JA (2021) Habitat patches providing south–north connectivity are under-protected in a fragmented landscape. Proc R Soc B Biol Sci 288(1957):20211010. https://doi.org/10.1098/rspb.2021.1010

Tuyisingize D, Eckardt W, Caillaud D, Ngabikwiye M, Kaplin BA (2022) Forest landscape restoration contributes to the conservation of primates in the Gishwati-Mukura landscape, Rwanda. Int J Primatol 43(5):867–884. https://doi.org/10.1007/s10764-022-00303-0

UNEP-WCMC, FFI and ELP (2020) Funding ecosystem restoration in Europe: a summary of trends and recommendations to inform practitioners, policymakers and funders. 24 p

Unnithan Kumar S, Cushman SA (2022) Connectivity modeling in conservation science: a comparative evaluation. Sci Rep 12(1):16680. https://doi.org/10.1038/s41598-022-20370-w

van Diggelen R, Hobbs RJ, Miko L (2012) Landscape ecology. In: Restoration ecology. Wiley, pp 45–58. https://doi.org/10.1002/9781118223130.ch5

Vasudev D, Fletcher RJ, Goswami VR, Krishnadas M (2015) From dispersal constraints to landscape connectivity: lessons from species distribution modeling. Ecography 38(10):967–978. https://doi.org/10.1111/ecog.01306

Wang Z, Yang Z, Shi H, Han L (2021) Effect of forest connectivity on the dispersal of species: a case study in the Bogda World Natural Heritage Site, Xinjiang, China. Ecol Indic 125:107576. https://doi.org/10.1016/j.ecolind.2021.107576

Wassie A (2007) Ethiopian church forests: opportunities and challenges for restoration. [PhD Thesis]. Wageningen University

Wassie A, Teketay D, Powell N (2005) Church forests in North Gonder administrative zone, Northern Ethiopia. For Trees Livelihoods 15:349–373

Wassie Eshete A (2007) Ethiopian church forests: Opportunities and challenges for restoration = Kerkbossen in Ethiopië

With KA (2019a) Essentials of landscape ecology. Oxford University Press. 641 p

With KA (2019b) Landscape connectivity. In: With KA (ed) Essentials of landscape ecology. Oxford University Press. https://doi.org/10.1093/oso/9780198838388.003.0005

Woldemedhin TT, Teketay D (2016) Forest conservation tradition of the Ethiopian Orthodox Tewahdo Church: a case study in West Gojjam Zone, north-western Ethiopia, pp 57–73

Woods CL, Bitew Mekonnen A, Baez-Schon M, Thomas R, Scull P, Abraha Tsegay B, Cardelús CL (2020) Tree community composition and dispersal syndrome vary with human disturbance in sacred church forests in Ethiopia. Forests 11(10):1082. https://doi.org/10.3390/f11101082

World Research Institute (WRI) (2015) The restoration diagnostic

WWF (2020) Case studies in connectivity conservation. Global innovations in corridor identification, protection, and restoration

WWF and IUCN (2000) Forests reborn: a workshop on forest restoration. In: Proceedings of the WWF/IUCN international workshop on forest restoration, Segovia, Spain, 3–5 July 2000

WWF Tigers Alive (2020) Landscape connectivity science and practice: ways forward for large ranging species and their landscapes. Workshop Report, WWF Tigers Alive, WWF International

Yirdaw E, Tigabu M, Lemenih M, Negash M, Teketay D (2014) Rehabilitation of degraded forest and woodland ecosystems in Ethiopia for sustenance of livelihoods and ecosystem services. In: Katila P, Galloway G, de Jong W, Pacheco P, Mery G (eds) Forests under pressure—local responses to global issues, IUFRO world series, vol 32, pp 299–313

Zumr V, Remeš J, Pulkrab K (2021) How to increase biodiversity of Saproxylic Beetles in commercial stands through integrated forest management in Central Europe. Forests 12(6):814. https://doi.org/10.3390/f12060814

Open Access This chapter is licensed under the terms of the Creative Commons Attribution 4.0 International License (http://creativecommons.org/licenses/by/4.0/), which permits use, sharing, adaptation, distribution and reproduction in any medium or format, as long as you give appropriate credit to the original author(s) and the source, provide a link to the Creative Commons license and indicate if changes were made.

The images or other third party material in this chapter are included in the chapter's Creative Commons license, unless indicated otherwise in a credit line to the material. If material is not included in the chapter's Creative Commons license and your intended use is not permitted by statutory regulation or exceeds the permitted use, you will need to obtain permission directly from the copyright holder.

Assisted Migration as a Climate Change Adaptation Strategy

14

Erik Szamosvári, Debojyoti Chakraborty, Silvio Schüler, and Marcela van Loo

Tree nursery/Tulln an der Donau. (Photo: BFW)

E. Szamosvári (✉) · D. Chakraborty · S. Schüler · M. van Loo
Department of Forest Growth, Silviculture & Genetics, Austrian Research Centre for Forests, Vienna, Austria
e-mail: erik.szamosvari@bfw.gv.at; debojyoti.chakraborty@bfw.gv.at; silvio.schueler@bfw.gv.at; marcela.vanloo@bfw.gv.at

© The Author(s) 2025
K. Lapin et al. (eds.), *Ecological Connectivity of Forest Ecosystems*,
https://doi.org/10.1007/978-3-031-82206-3_14

Abstract

In the twenty-first century, the warming climate poses major threats to forest ecosystems. Assisted migration has emerged as a proactive adaptation and conservation strategy to mitigate the impacts of climate change and safeguard biodiversity. This approach comprises the human-assisted movement and dispersal of species and populations to areas predicted to be suitable under future climate conditions. Assisted migration is the subject of much debate in the scientific literature. While it offers potential benefits in terms of promoting biodiversity, sustaining forest productivity, and conserving wildlife habitats, it also raises concerns about invasion potential, hybridization, and unforeseen impacts on ecosystems. Its implementation therefore requires careful scientific assessment, risk analysis, and an ecosystem-based approach. In the following, we discuss not only the pros and cons but also the knowledge gaps and further challenges associated with assisted migration as a tool for combating the impacts of climate change and a strategy for sustainably maintaining climate-adapted and resilient ecosystems.

Keywords

Adaptation · Assisted migration · Climate change · Conservation tool · Restoration

Assisted Migration as a Climate-Driven Strategy

For thousands of years, humans have been domesticating, transporting, and relocating plant and animal species for different purposes such as agriculture, horticulture, and silviculture (Hewitt et al. 2011; Schaal 2019). Human-supported species migration is thus not new, but novel reasons for it have recently emerged: adaptation to a warming climate and altered environmental conditions, fostering biodiversity, supporting species in need, and mitigating ecological and economic damage. Ecosystems, biodiversity, and human well-being are threatened by the environmental changes we are currently facing. Adequate adaptation measures to anthropogenic climate change (CC) and the halt of habitat destruction are significant challenges of our time. According to the Intergovernmental Panel on Climate Change (IPCC), warming of around 2 °C by the end of the century is expected under the moderate climate scenario RCP 4.5, while the pessimistic RCP-8.5 scenario projects an increase of up to 5 °C (Intergovernmental Panel On Climate Change (Ipcc) 2023). In addition to the rise in temperature, changes in precipitation and an increased frequency of extreme weather events, as well as new pathogens, pest outbreaks, and the dispersal of invasive species, have been predicted (Seidl et al. 2017; Pureswaran et al. 2018; Forzieri et al. 2021). New environmental conditions affect species and habitats differently; for example, forest biodiversity and local tree species compositions in Europe will likely be altered in many regions in the future (Buras and Menzel 2019; Chakraborty et al. 2021). The effects on species can be manifold: some may persist through local adaptation or migration to new habitats, while others may disappear from given regions and/or be replaced by other native or even non-native species (Wiens 2016; Dyderski et al. 2018). Long-lived species like trees may lack the ability to adapt to the new environmental conditions or migrate to new suitable habitats due to an adaptation lag and/or limited natural migration capacity (Aitken et al. 2008). To mitigate the consequences of CC and preserve biodiversity along with its benefits for future generations, deliberate and planned human interventions and actions are required. Assisted migration is a term describing a climate adaptation strategy based on human-assisted movement and assistance in the dispersal of genotypes, populations, and species threatened by CC in their natural dispersal ranges to areas that are predicted to be suitable under future climate conditions but would not be accessible (barriers, lack of time) without anthropogenic actions (Williams and Dumroese 2013; Sáenz-Romero et al. 2020; Benomar et al. 2022). According to Hewitt et al. (2011), the concept was first introduced by Peters and Darling in 1985 (Peters and Darling 1985) as a strategy and possible option when discussing the effects of greenhouse gases and their impact on nature reserves, suggesting the relocation of species to other habitats when the current one has become unsuitable or will no longer be suitable due to changing environmental conditions. Since then, several terms and names have been used for the same or similar notions, such as assisted colonization, managed relocation, or facilitated migration (Hällfors et al. 2014). Whatever the chosen label, the goals of all these concepts are generally identical: preventing species and population extinction, minimizing economic loss, aiding the adaptation of species, sustaining ecosystem services and biodiversity, and mitigating the impacts of CC on species (Williams and Dumroese 2013; Twardek et al. 2023). There has been intense debate among scientists, conservationists, and decision-makers over the use of assisted migration from the very beginning—primarily due to the involved risks and benefits, existing knowledge gaps, and the lack of specific guidelines and frameworks (McLachlan et al. 2007; Hewitt et al. 2011; Bucharova 2017; Benomar et al. 2022), as well as social acceptance (Klenk 2015; Pelai et al. 2021). Despite this discussion, assisted migration is a main proactive climate-based translocation approach that could potentially help species survive and thrive by significantly reducing the projected negative consequences of CC in the future.

A Strategy to Mitigate the Consequences of Climate Change

Assisted migration has been one of the "hot topics" over the past few decades in the discussion on ecosystem adaptation to CC and has been proposed more often in recent years as a proactive management and adaptation strategy (Bolte et al. 2009). The taxa most discussed and researched in the existing literature is the kingdom of plants—especially tree species—with the majority of human-assisted relocations taking place in North America and Europe (Twardek et al. 2023). Tree species are of particular interest because of their economic, ecological, and cultural value, as well as due to the long history of provenance trials in forest genetics and silviculture (Hewitt et al. 2011; Mauri et al. 2023). In the field of forestry, assisted migration offers the opportunity to preserve current forest cover and timber production levels by selecting suitable resilient and potentially adapted seed sources, or even new tree species, for the future (Sousa-Silva et al. 2018). However, the potential of assisted migration as a conservation tool is likewise significant: the movement of ecosystem engineers along with foundation and keystone species, for instance, could facilitate ecological connectivity and colonization by other species in the same communities (Sáenz-Romero et al. 2020). The categorization of assisted migration can vary depending on the context and perspective of different researchers and practitioners. Some authors in the literature distinguish between three categories of assisted migration (e.g., Williams and Dumroese 2013; Benomar et al. 2022), while others only make a distinction between two (e.g., Peterson St-Laurent et al. 2019; Sáenz-Romero et al. 2020; Twardek et al. 2023). Here we use the categorization according to the latter:

Assisted species migration—intentional translocation of species or populations *outside* of their current natural distribution range

Assisted population migration—intentional translocation of species or populations *within* their current natural distribution range (also referred to as assisted gene flow) (Fig. 14.1)

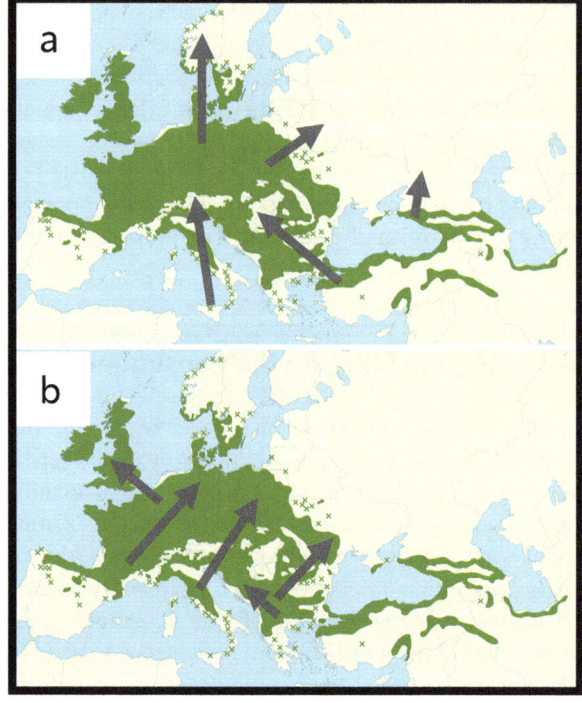

Fig. 14.1 The two categories of assisted migration are demonstrated using the distribution map of sessile oak (*Quercus petraea (Matt.) Liebl.*) (Caudullo et al. 2017) as a theoretical example. Image (**a**) shows the assisted species migration beyond the current distribution range. Image (**b**) illustrates assisted population migration within the current distribution range of the species

Risks and Benefits

Although the idea of human-supported translocation of species was introduced decades ago, the debate concerning this approach is still ongoing. Assisted migration raises ethical, ecological, economic, legal, and political questions (Schwartz et al. 2012).

One of the main benefits of assisted migration is the possibility of fostering biodiversity, supporting the economic and ecological importance of forests, and preserving species and wildlife habitats in the face of CC. Especially in the case of vulnerable species—e.g., owing to rarity, poor dispersal capacity, long generation times, or genetic isolation—with highly threatened habitats such as alpine ones and low migration capacity due to geographic barriers and/or fragmented landscapes, this strategy could be crucial (Hewitt et al. 2011; Erickson et al. 2012). Assisted migration could also support ecological connectivity, promote rapid local adaptation, preserve forest productivity and health, reduce ecosystem service loss, and increase sociopolitical and economic benefits in the future (Krosby et al. 2010; Gray et al. 2011; Williams and Dumroese 2013; Mauri et al. 2023) (Fig. 14.2). For forest trees, assisted migration may offer the chance to speed up the still ongoing postglacial tree migration, as many European trees have not yet colonized the full extent of their climatic niche across Europe (Svenning and Skov 2004). Moreover, given the frequent adaptation lags observed especially within marginal tree populations (Mátyás 1989; Fréjaville et al. 2020; Leites and Benito Garzón 2023), assisted migration will likely help counteract increasing maladaptation (Frank et al. 2017) and safeguard vulnerable populations (Schueler et al. 2014).

Despite these valuable promises of assisted migration, however, the downsides should also be considered. A limited understanding of interactions caused by species movement raises issues and increases uncertainty among scientists. The unknown risks of invasion potential, competition with native species, pest and disease outbreaks, establishment failure, hybridization between native and non-native species, and unpredicted impacts on local ecosystems and communities entail significant responsibility while implementing assisted migration (Aitken et al. 2008; Bucharova 2017). It should be also noted that these uncertainties potentially increase together with the greater migration distance (Winder et al. 2011). As the species are already well-integrated into the ecosystem, the impacts of assisted migration within a species' existing distribution range on ecosystems are likely to be significantly smaller, if not negligible, compared to assisted migration beyond the distribution range.

Nevertheless, it is reasonable that even considering the uncertainties, challenges, and potential for increased effort and costs, scientists and policymakers should value assisted migration as a strategy for CC adaptation. A "business as usual" approach or inaction eliminates risks but also potential future benefits in terms of reducing the vulnerability of current populations and species to the impacts of climate change (Butt et al. 2020; Klisz et al. 2023).

Fig. 14.2 The figure illustrates a conceptual representation of biodiversity strategies to improve connectivity and mitigate the impacts of climate change. Assisted migration involves the human-supported deliberate translocation of species to habitats, which will be more likely suitable in the future, enabling them to adapt to changing environmental conditions. This action can serve as a crucial step, especially when, in addition to the slow natural migration capacity, intermediate habitats or ecological refuges along migration routes are lacking. While assisted migration stands out as a proactive measure against climate change, complementary strategies such as stepping stones and green corridors contribute to the resilience and sustainability of ecosystems in a rapidly changing world

Implementation of Climate-Based Translocation

During the implementation of AM, several key considerations based on scientific knowledge and careful evaluation should underpin the decision-making process. These include understanding the species' biology, habitat requirements, potential shifts in distribution due to CC, and the ecological and social impacts of introducing the species into a new area (O'Neill et al. 2017). Since assisted migration can facilitate the translocation of species to more suitable habitats if their current range becomes less favorable, ecological site characteristics and climate projections need to be carefully considered to identify suitable target areas that can sustain the best-fitting and potentially preadapted populations or species in the long term (Jordan et al. 2023). Consequently, conducting a comprehensive risk assessment of both the target species and the designated area(s) is essential to mitigate potential adverse effects on the recipient ecosystems (Hällfors et al. 2016; Bucharova 2017). During the implementation of AM, it is imperative to adopt an ecosystem-based approach rather than concentrating solely on individual species. To maintain the integrity of the new ecosystem, it is essential to consider how introduced species will interact with other species and their environment (Hewitt et al. 2011). Continuous monitoring and adaptive management strategies are also critical requirements. Regular evaluation of the results of assisted migration efforts helps assess the intervention's success and make necessary adjustments to management practices (Williams and Dumroese 2013). Like other strategies, assisted migration necessitates collaboration and consultation with a diverse array of stakeholders, including scientists, conservationists, local communities, and decision-makers (Klenk 2015; Pelai et al. 2021). Understanding and incorporating various perspectives, knowledge bases, and concerns is essential for assuring successful and sustainable implementation (Butt et al. 2020). To support expert assessments and decision-making frameworks, the approach introduced by Hällfors et al. (2017) can be particularly useful. It quantifies the needs and potentials of different species for assisted migration and aids future evaluation. This method is based on the prediction of future range changes, species distribution models, and dispersal abilities, among other assumptions.

Current and Future Directions and Knowledge Gaps

As assisted migration continues to develop as a strategy and method, it introduces several important future trajectories and areas of knowledge gaps that researchers and practitioners need to address. To effectively counteract the impacts of CC, some key aspects requiring further exploration include:

1. Enhancing climate change predictions, scenarios, and species distribution models to better predict ecosystem changes, select the best-fitting populations and/or species, and identify suitable relocation destinations and deployment sites. While certainly a demanding task, we should aim to increase the effectiveness and precision of assisted migration strategies to minimize future uncertainties and risks (Leech et al. 2011).
2. Deepening the understanding of the long-term ecological and genetic impacts of assisted migration on introduced species and host ecosystems. This includes assessing potential interactions with native species, changes in community dynamics, and the consequences of increased genetic mixing (Butt et al. 2020; Twardek et al. 2023). For the genetic impact of assisted migration, also see Chaps. 20 and 21.
3. Expanding both short-termed and long-termed, small- and large-scale, multi- and individual species–level experiments and studies in this field to enhance our knowledge and lessen the uncertainties of assisted migration effects on ecosystems (Bucharova 2017; Butt et al. 2020; Jordan et al. 2023). As already stated above, implementing adaptive management strategies and precise monitoring is fundamental for assessing the outcomes of assisted migration and making necessary adjustments in light of changing circumstances or unexpected outcomes. Comparisons of both successful and unsuccessful assisted migration projects

can provide valuable insights into the factors contributing to the effectiveness of interventions and their potential risks (Hällfors et al. 2016).
4. Integrating assisted migration with complementary strategies like habitat restoration, management of protected areas, and enhancement of landscape connectivity to ensure a holistic approach (Krosby et al. 2010; Loss et al. 2011).
5. Developing legal frameworks and guidelines akin to the Expert Panel on Adaptation of Forests to Climate Change (Seppala et al. 2009) or the amendments to the Chief Forester's Standards (Ministry of Forests, Government of British Columbia 2022) for seed use. The further development of appropriate recommendation tools like the Seedlot Selection Tool (SST) (https://seedlotselectiontool.org/sst/) by the Climate Change Resource Center or forest vulnerability and transfer tools such as the Seed4Forest (https://www.seed4forest.org/) of the Austrian Research Centre for Forests (BFW) likewise seems crucial. Such tools would ensure that decision-making is adequately managed (Mauri et al. 2023). Also, collaboration with stakeholders, forest managers, conservationists, and decision-makers is highly recommended as we must begin looking ahead together to ensure that biodiverse and sustainable ecosystems are in place by the end of the century, even if CC scenarios remain uncertain.

Practical Applications of Assisted Migration

Here we showcase three examples of applying assisted migration for ecological restoration, biodiversity promotion, and species conservation across different regions of the world, namely Spain, Mexico, and China. These instances were identified through a Google Scholar search conducted in July and August of 2023, using the combination of keywords "assisted migration," "assisted translocation," "field trial," "forest," and "test," alongside geographic regions including "Europe," "America," "Asia," and "Australia."

1. Enriching Diversity and Bolstering Resilience
   The potentials of assisted migration to diversify sub-Mediterranean pinewoods with different native *Quercus* species were explored by Martín-Alcón et al. (2016) in Spain. The authors investigated early-year establishment, survival rates, responses to different environmental conditions, and the effects of plant material origin in field trials. In their study, reproductive material of four oak species (*Quercus coccifera* L., *Quercus ilex* L., *Quercus faginea* Lam., and *Quercus pubescens* Wild.) from local and warmer populations was sown and planted at low (around 1000 m), intermediate (around 1250 m), and high (around 1500 m) elevations in three pinewood forests of the Catalonian Pre-Pyrenees and monitored for the effects of translocation at three altitudinal ranges. They found that both sowing (with a 50% emergence rate) and planting (with a 76.3% survival rate) were successful at the different sites within the first 3 years. Additionally, the authors reported species- and provenance-specific responses to climate, particularly to extreme cold temperatures at higher elevations and

drought periods, which influenced emergence and survival success. They observed that biotic interactions, canopy cover, and shrub protection helped mitigate the negative impacts of translocation during harsher periods. Overall, the study emphasizes the importance of selecting appropriate species and provenances to respective environmental characteristics as a key requirement for successful translocation efforts.

2. Promoting Climate-Smart Restoration and Conservation

   In west-central Mexico, within the core area of the Monarch Butterfly Biosphere Reserve, seedlings of sacred fir (*Abies religiosa* (Kunth) Schltdl. and Cham.) from potentially future-adapted provenances were translocated to an elevation at around 3400 m (Carbajal-Navarro et al. 2019). In this study, the authors explored the use of nurse plants (e.g., shrubs providing a protective microhabitat for seedlings) in combination with assisted migration as a potential tool for ecological forest restoration and biodiversity promotion. Their study underscores the conservation importance of resilient forests of sacred firs, which serve as overwintering host plants for the monarch butterfly (*Danaus plexippus* L.) and numerous other species. At two sites—one degraded and heavily disturbed due to grazing and logging and another relatively undisturbed area with a similar climate and vegetation type—two-year-old seedlings from locations 20–35 km away at elevations between 2960 m and 3450 m were planted in the open or under the shade of nurse plants. Observations and detailed temperature measurements after one and a half years at one site and after three and a half years at the other revealed much higher survival rates (10% vs. 94% and 18% vs. 72%) of seedlings protected by nurse plants and leading to a successful early-stage establishment of populations shifted 400 meters upward. These findings underscore the significance of integrating assisted migration with other reforestation management practices.

3. Safeguarding Endangered Species

   Ren et al. (2016) published a conservation translocation study focusing on the critically endangered southern Chinese tree species *Manglietia longipedunculata* (Q.W. Zeng and Law), which exhibits low reproductive success, with only a dozen individuals known to exist in the wild. The aim of their study was to determine the effectiveness of conservation measures and contribute to the survival of the species by increasing both the population size and the species range. Grafted plants and emerged seedlings produced *ex situ* were introduced at two sites—one within the existing population and the other more than 200 km north of the current distribution, considering the potential effects of climate change. Among other parameters, the study compared the survival, growth, and ecophysiological traits of the seedlings. Five years after the translocation, the results revealed moderately good survival (64%) and growth of emerged seedlings, with even better performance (96%) of grafted seedlings, at both sites. The study demonstrates the ability of innovative conservation techniques to potentially protect or restore threatened species, or even ecosystems, from the impacts of CC.

Conclusion

The future impacts of climate change on ecological conditions can differ significantly between habitats, and the intensity of selection pressure will vary among different populations around the world. The novel and rapidly changing environment and therefore the process of selection paired with insufficient ability for local adaptation will most likely eliminate genotypes and species at certain sites, challenging the paradigm "local is the best" over the course of time. While new climatic conditions can reveal the adaptive potential of populations, such adaptation may require more time than ongoing climate change allows. Assisted migration offers an opportunity to help genotypes, populations, and species adapt to new environmental conditions, especially those that do not have the natural ability to shift their range in response to CC. Considering the often-debated risks and benefits of AM, with careful planning and collaboration, it holds promise as a proactive tool to safeguard biodiversity, support ecosystem resilience, foster ecosystem services, maintain forest productivity, and mitigate the impacts of CC. Further studies and evidence-based guidelines will contribute to ensuring safer implementation. Overall, assisted migration—also combined with other approaches—should be viewed as a complex strategy and tool within a broader spectrum of choices. Its effectiveness and appropriateness as a climate change adaptation strategy will depend on the specific circumstances, populations and species, and ecosystems involved in any given effort. Applied with care and tied in with other strategies, assisted migration can address the challenges of CC and its impacts on biodiversity. The overarching goal should always be to preserve and protect ecosystems and the species within them while simultaneously minimizing potential risks and fostering sustainable multipurpose use for future generations.

References

Aitken SN, Yeaman S, Holliday JA, Wang T, Curtis-McLane S (2008) Adaptation, migration or extirpation: climate change outcomes for tree populations. Evol Appl 1(1):95–111. https://doi.org/10.1111/j.1752-4571.2007.00013.x

Benomar L, Elferjani R, Hamilton J, O'Neill G, Echchakoui S, Bergeron Y, Lamara M (2022) Bibliometric analysis of the structure and evolution of research on assisted migration. Curr For Rep. https://doi.org/10.1007/s40725-022-00165-y

Bolte A, Ammer C, Löf M, Madsen P, Nabuurs G-J, Schall P, Spathelf P, Rock J (2009) Adaptive forest management in central Europe: climate change impacts, strategies and integrative concept. Scand J For Res 24:473–482. https://doi.org/10.1080/02827580903418224

Bucharova A (2017) Assisted migration within species range ignores biotic interactions and lacks evidence. Restor Ecol 25:14–18. https://doi.org/10.1111/rec.12457

Buras A, Menzel A (2019) Projecting tree species composition changes of European forests for 2061-2090 under RCP 4.5 and RCP 8.5 scenarios. Front Plant Sci 9. https://doi.org/10.3389/fpls.2018.01986

Butt N, Chauvenet A, Adams V, Beger M, Gallagher R, Shanahan D, Ward M, Watson J, Possingham H (2020) Importance of species translocations under rapid climate change. Conserv Biol 35. https://doi.org/10.1111/cobi.13643

Carbajal-Navarro A, Navarro-Miranda E, Blanco-García A, Cruzado-Vargas AL, Gómez-Pineda E, Zamora-Sánchez C, Pineda-García F, O'Neill G, Gómez-Romero M, Lindig-Cisneros R, Johnsen KH, Lobit P, Lopez-Toledo L, Herrerías-Diego Y, Sáenz-Romero C (2019) Ecological restoration of Abies religiosa forests using nurse plants and assisted migration in the monarch butterfly biosphere reserve, Mexico. Front Ecol Evol 7. https://www.frontiersin.org/articles. https://doi.org/10.3389/fevo.2019.00421

Caudullo G, Welk E, San-Miguel-Ayanz J (2017) Chorological maps for the main European woody species. Data Brief 12:662–666. https://doi.org/10.1016/j.dib.2017.05.007

Chakraborty D, Móricz N, Rasztovits E, Dobor L, Schueler S (2021) Provisioning forest and conservation science with high-resolution maps of potential distribution of major European tree species under climate change. Ann For Sci 78. https://doi.org/10.1007/s13595-021-01029-4

Dyderski MK, Paź S, Frelich LE, Jagodziński AM (2018) How much does climate change threaten European forest tree species distributions? Glob Chang Biol 24(3):1150–1163. https://doi.org/10.1111/gcb.13925

Erickson V, Aubry C, Berrang P et al (2012) Genetic resource management and climate change: genetic options for adapting national forests to climate change. USDA, Forest Service White Paper. http://climatechange.ecoshare.info/files/2010/11/Genetic_Options.pdf

Forzieri G, Girardello M, Ceccherini G, Spinoni J, Feyen L, Hartmann H, Beck PSA, Camps-Valls G, Chirici G, Mauri A, Cescatti A (2021) Emergent vulnerability to climate-driven disturbances in European forests. Nat Commun 12(1) Article 1. https://doi.org/10.1038/s41467-021-21399-7

Frank A, Howe GT, Sperisen C, Brang P, Clair JB, St. Schmatz DR, Heiri C (2017) Risk of genetic maladaptation due to climate change in three major European tree species. Glob Chang Biol 23(12):5358–5371. https://doi.org/10.1111/gcb.13802

Fréjaville T, Vizcaíno-Palomar N, Fady B, Kremer A, Benito Garzón M (2020) Range margin populations show high climate adaptation lags in European trees. Glob Chang Biol 26(2):484–495. https://doi.org/10.1111/gcb.14881

Gray LK, Gylander T, Mbogga MS, Chen P-Y, Hamann A (2011) Assisted migration to address climate change: recommendations for aspen reforestation in western Canada. Ecol Appl 21(5):1591–1603. https://doi.org/10.1890/10-1054.1

Hällfors MH, Aikio S, Schulman LE (2017) Quantifying the need and potential of assisted migration. Biol Conserv 205:34–41. https://doi.org/10.1016/j.biocon.2016.11.023

Hällfors MH, Aikio S, Fronzek S, Hellmann JJ, Ryttäri T, Heikkinen RK (2016) Assessing the need and potential of assisted migration using species distribution models. Biol Conserv 196:60–68. https://doi.org/10.1016/j.biocon.2016.01.031

Hällfors M, Vaara E, Hyvärinen M, Oksanen M, Schulman L, Siipi H, Lehväpirta S (2014) Coming to terms with the concept of moving species threatened by climate change—a systematic review of the terminology and definitions. PLoS One 9:e102979. https://doi.org/10.1371/journal.pone.0102979

Hewitt N, Klenk N, Smith AL, Bazely DR, Yan N, Wood S, MacLellan JI, Lipsig-Mumme C, Henriques I (2011) Taking stock of the assisted migration debate. Biol Conserv 144(11):2560–2572. https://doi.org/10.1016/j.biocon.2011.04.031

Intergovernmental Panel On Climate Change (Ipcc) (2023) Climate change 2022—impacts, adaptation and vulnerability: working group II contribution to the sixth assessment report of the intergovernmental panel on climate change, 1st edn. Cambridge University Press. https://doi.org/10.1017/9781009325844

Jordan R, Harrison P, Breed M (2023) The eco-evolutionary risks of not changing seed provenancing practices in changing environments. https://doi.org/10.22541/au.168570191.19000524/v1

Klenk NL (2015) The development of assisted migration policy in Canada: an analysis of the politics of composing future forests. Land Use Policy 44:101–109. https://doi.org/10.1016/j.landusepol.2014.12.003

Klisz M, Chakraborty D, Cvjetkovi'c B, Grabner M, Lintunen A, Mayer K, George J, Rossi S (2023) Functional traits of boreal species and adaptation to local conditions, chapter 12. In: Girona MM, Morin H, Gauthier S, Bergeron Y (eds) Boreal forests in the face of climate change: sustainable management, vol 74. Springer. https://doi.org/10.1007/978-3-031-15988-6

Krosby M, Tewksbury J, Haddad N, Hoekstra J (2010) Ecological connectivity for a changing climate. Conserv Biol 24:1686–1689. https://doi.org/10.1111/j.1523-1739.2010.01585.x

Leech S, Almuedo P, O'Neill G (2011) Assisted migration: adapting forest management to a changing climate. J Ecosyst Manag:12. https://doi.org/10.22230/jem.2011v12n3a91

Leites L, Benito Garzón M (2023) Forest tree species adaptation to climate across biomes: building on the legacy of ecological genetics to anticipate responses to climate change. Glob Chang Biol 29(17):4711–4730. https://doi.org/10.1111/gcb.16711

Loss SR, Terwilliger LA, Peterson AC (2011) Assisted colonization: integrating conservation strategies in the face of climate change. Biol Conserv 144(1):92–100. https://doi.org/10.1016/j.biocon.2010.11.016

Martín-Alcón S, Coll L, Ameztegui A (2016) Diversifying sub-Mediterranean pinewoods with oak species in a context of assisted migration: responses to local climate and light environment. Appl Veg Sci 19(2):254–266. https://doi.org/10.1111/avsc.12216

Mátyás C (1989) Adaptation lag: a general feature of natural populations. (Invited lecture)

Mauri A, Girardello M, Forzieri G, Manca F, Beck PSA, Cescatti A, Strona G (2023) Assisted tree migration can reduce but not avert the decline of forest ecosystem services in Europe. Glob Environ Chang 80:102676. https://doi.org/10.1016/j.gloenvcha.2023.102676

McLachlan JS, Hellmann JJ, Schwartz MW (2007) A framework for debate of assisted migration in an era of climate change. Conserv Biol 21(2):297–302. https://doi.org/10.1111/j.1523-1739.2007.00676.x

Ministry of Forests, Government of British Columbia (2022) Amendment to the Chief Forester's standards for seed use completes the transition to climate based seed transfer. https://www2.gov.bc.ca/assets/gov/farming-natural-resources-and-industry/forestry/tree-seed/chief-forester-s-standards-for-seed-use/memos-and-letters/270423_adm_cf_cbst_update_memo_may_2022.pdf (last visited August of 2023)

O'Neill G, TongLi W, Ukrainetz N, Charleson L, McAuley L, Yanchuk A, Zedel S (2017) A proposed climate-based seed transfer system for British Columbia. Technical report—Ministry of Forests, Lands and Natural Resource Operations, British Columbia, no.099. https://www.cabdirect.org/cabdirect/abstract/20173054868

Pelai R, Hagerman SM, Kozak R (2021) Whose expertise counts? Assisted migration and the politics of knowledge in British Columbia's public forests. Land Use Policy 103:105296. https://doi.org/10.1016/j.landusepol.2021.105296

Peters RL, Darling JDS (1985) The greenhouse effect and nature reserves: global warming would diminish biological diversity by causing extinctions among reserve species. Bioscience 35(11):707–717. https://doi.org/10.2307/1310052

Peterson St-Laurent G, Hagerman S, Findlater KM, Kozak R (2019) Public trust and knowledge in the context of emerging climate-adaptive forestry policies. J Environ Manage 242:474–486. https://doi.org/10.1016/j.jenvman.2019.04.065

Pureswaran DS, Roques A, Battisti A (2018) Forest insects and climate change. Curr For Rep 4(2):35–50. https://doi.org/10.1007/s40725-018-0075-6

Ren H, Liu H, Wang J, Yuan L, Cui X, Zhang Q, Fu L, Chen H, Zhong W, Yang K, Guo Q (2016) The use of grafted seedlings increases the success of conservation translocations of Manglietia longipedunculata (Magnoliaceae), a critically endangered tree. Oryx 50(3):437–445. https://doi.org/10.1017/S0030605315000423

Sáenz-Romero C, Mendoza-Maya E, Gómez-Pineda E, Blanco-García A, Endara-Agramont AR, Lindig-Cisneros R, López-Upton J, Trejo-Ramírez O, Wehenkel C, Cibrián-Tovar D, Flores-López C, Plascencia-González A, Vargas-Hernández JJ (2020) Recent evidence of Mexican temperate forest decline and the need for ex situ conservation, assisted migration, and translocation of species ensembles as adaptive management to face projected climatic change impacts in a megadiverse country. Can J For Res 50(9):843–854. https://doi.org/10.1139/cjfr-2019-0329

Schaal B (2019) Plants and people: our shared history and future. Plants People Planet 1(1):14–19. https://doi.org/10.1002/ppp3.12

Schueler S, Falk W, Koskela J, Lefèvre F, Bozzano M, Hubert J, Kraigher H, Longauer R, Olrik DC (2014) Vulnerability of dynamic genetic conservation units of forest trees in Europe to climate change. Glob Chang Biol 20(5):1498–1511. https://doi.org/10.1111/gcb.12476

Schwartz M, Hellmann J, McLachlan JM, Sax D, Borevitz J, Brennan J, Camacho A, Ceballos G, Clark J, Doremus H, Early R, Etterson J, Fielder D, Gill J, Gonzalez P, Green N, Hannah L, Jamieson D, Javeline D, Zellmer S (2012) Managed relocation: integrating the scientific, regulatory, and ethical challenges. Bioscience 62:732–743. https://doi.org/10.1525/bio.2012.62.8.6

Seidl R, Thom D, Kautz M, Martin-Benito D, Peltoniemi M, Vacchiano G, Wild J, Ascoli D, Petr M, Honkaniemi J, Lexer MJ, Trotsiuk V, Mairota P, Svoboda M, Fabrika M, Nagel TA, Reyer CPO (2017) Forest disturbances under climate change. Nat Clim Change 7(6) Article 6. https://doi.org/10.1038/nclimate3303

Seppala R, Buck A, Katila P (2009) Adaptation of forests and people to climate change—a global assessment report. IUFRO world series, vol 22, Helsinki, 224 p. http://www.iufro.org/science/gfep/adaptaion-panel/the-report

Sousa-Silva R, Verbist B, Lomba Â, Valent P, Suškevičs M, Picard O, Hoogstra-Klein M, Cosofret C, Bouriaud L, Ponette Q, Verheyen K, Muys B (2018) Adapting forest management to climate change in Europe: linking perceptions to adaptive responses. Forest Policy Econ:90. https://doi.org/10.1016/j.forpol.2018.01.004

Svenning J-C, Skov F (2004) Limited filling of the potential range in European tree species. Ecol Lett 7(7):565–573. https://doi.org/10.1111/j.1461-0248.2004.00614.x

Twardek WM, Taylor JJ, Rytwinski T, Aitken SN, MacDonald AL, Van Bogaert R, Cooke SJ (2023) The application of assisted migration as a climate change adaptation tactic: an evidence map and synthesis. Biol Conserv 280:109932. https://doi.org/10.1016/j.biocon.2023.109932

Wiens JJ (2016) Climate-related local extinctions are already widespread among plant and animal species. PLoS Biol 14(12):e2001104. https://doi.org/10.1371/journal.pbio.2001104

Williams MI, Dumroese RK (2013) Preparing for climate change: forestry and assisted migration. J For 111(4):287–297. https://doi.org/10.5849/jof.13-016

Winder R, Nelson E, Beardmore T (2011) Ecological implications for assisted migration in Canadian forests. For Chron 87(6):731–744. https://doi.org/10.5558/tfc2011-090

Open Access This chapter is licensed under the terms of the Creative Commons Attribution 4.0 International License (http://creativecommons.org/licenses/by/4.0/), which permits use, sharing, adaptation, distribution and reproduction in any medium or format, as long as you give appropriate credit to the original author(s) and the source, provide a link to the Creative Commons license and indicate if changes were made.

The images or other third party material in this chapter are included in the chapter's Creative Commons license, unless indicated otherwise in a credit line to the material. If material is not included in the chapter's Creative Commons license and your intended use is not permitted by statutory regulation or exceeds the permitted use, you will need to obtain permission directly from the copyright holder.

Forest Genetic Resources Under Climate Change and International Framework: Conservation Measures of Serbia and Greece

15

Branislav Trudić, Srđan Stojnić, Evangelia V. Avramidou, and Ermioni Malliarou

Mediterranean forests are especially important for soil conservation yet particularly challenged by climate change. (Photo: BFW/Heino Konrad)

B. Trudić (✉)
Faculty of Biology, University of Belgrade, Belgrade, Serbia

S. Stojnić
Institute of Lowland Forestry and Environment, University of Novi Sad, Novi Sad, Serbia
e-mail: srdjan.stojnic@uns.ac.rs

E. V. Avramidou
Laboratory of Forest Genetics and Biotechnology, Institute of Mediterranean Forest Ecosystems, ELGO-DIMITRA, Athens, Greece
e-mail: avramidou@fria.gr

Abstract

Advancements in genetic science and conservation are crucial tools for enhancing the resilience of forest ecosystems against climate change, emphasizing the need for conservation strategies that protect genetic diversity. Strategic management and policy development are imperative to utilize these technologies effectively in conservation efforts, and this is evident in the participation of Greece and Serbia in international organizations, such as the Food and Agriculture Organization (FAO) of the United Nations, the United Nations Environmental Programme (UNEP), and the European Forest Genetic Resources Programme (EUFORGEN). Greece and Serbia are actively engaged within the specified organizations, and as such, their efforts in the conservation of forest genetic resources (FGRs) amidst climate change are structured under the guidance and influence of these stakeholders. Both countries recognize collaborative endeavors between research entities and conservation initiatives and underscore the potential of innovative genetic and biotechnological approaches in FGR management. Genetic monitoring and evaluation play a vital role in pinpointing species and populations at risk and guiding conservation priorities and actions. Examples of successful conservation practices are given both from Serbia and Greece, illustrating that combining traditional methods with modern genetic insights is key to sustaining forest ecosystems. Such comprehensive approaches are essential in countering the adverse effects of climate change on forests, thereby contributing to the conservation of global biodiversity and the stability of natural environments.

Keywords

Climate change · FAO · Forest genetics · Greece · Serbia

Forest Genetic Resources and Climate Change

Environmental shifts influence genetic diversity in intricate, multifaceted, and unpredictable ways, affecting natural distribution ranges, species interactions, and timing and reducing genetic diversity and adaptive evolution potential (Aravanopoulos 2016). Forest ecosystems can mitigate climate effects only if they are well-adapted to their environments and capable of adapting to future changes (Hof et al. 2017). Effective adaptation to and mitigation of climate change require proactive forest genetic resources (FGR) management, which involves responsible monitoring and planning to support timely local adaptation and decrease tree mortality (Joyce et al. 2018; Isabel et al. 2019). Climate change presents significant

---

E. Malliarou
Forest Research Institute, ELGO-DIMITRA, Thessaloniki, Greece
e-mail: emalliarou@elgo.gr

challenges for FGR conservation. Shifting climates will impact protected areas and forest stands used for *in situ* conservation, necessitating conservation planning to consider various climate change scenarios and their implications for species suitability at specific sites, potentially requiring adjustments in protected areas. European countries have established conservation stands for "dynamic gene conservation"; however, by 2100, up to 65 percent of these stands may face conditions outside their current climatic ranges (Schueler et al. 2014). To preserve genetic diversity, it has been suggested to collect seeds from threatened populations at the edges of species' distribution ranges and conserve them *ex situ* before local extinction occurs (Gaisberger et al. 2017). Climate change could decrease genetic and species diversity within ecosystems, potentially leading to significant productivity declines (Weiskopf et al. 2020). Moreover, *in situ* and *circa situm* conservation in open conditions may be affected by climate change, impacting biodiversity and posing economic and human well-being risks. Climate change modeling, though imperfect, is crucial for FGR conservation as it offers insights into regional weather and climate shifts, enabling better protection measures. Despite its flaws, many conservationists view *ex situ* approaches as essential for preserving forest tree diversity. Many tree species exhibit significant genetic variability in key traits such as drought tolerance, cold hardiness, and the timing of flowering and fructification. This variability allows for rapid adaptation to directional and continuous environmental changes. However, the predicted magnitude and speed of climate change likely exceed the adaptive capacity of tree populations, particularly at the receding edges of their distributions. Assisted migration, a conservation strategy involving the intentional relocation of forest reproductive material (FRM) to new geographic areas where the climate is expected to become suitable, may be necessary to address these challenges. This strategy would require the movement of increased quantities of germplasm across national boundaries for both planting and research activities (Argüelles-Moyao and Galicia, 2023). Unfortunately, international transfer of tree germplasm for research has become increasingly difficult and costly in recent years (Koskela et al. 2010). It is also important to avoid indiscriminate movement of poorly adapted germplasm and consider potential problems such as the risk of introduced species becoming invasive and reducing native biodiversity (Loo et al. 2011). Climate change affects all the three abovementioned conservation approaches (*in situ, ex situ,* and *circa situm*), especially those developed and pursued in semi-open and open environments. As Verkerk et al. (2020) have discussed, climate-smart forestry is needed to (a) increase the total forest area and avoid deforestation, (b) connect mitigation with adaptation measures to enhance the resilience of global forest resources, and (c) use wood for products that store carbon and replace emission-intensive fossil and nonrenewable products and materials. This will not be possible without including FGR conservation and its implications for natural forest landscapes and planted forests, especially in evidence-based policies that consider them. Contextual strategic planning and genetic monitoring regarding climate change scenarios will be a requirement for sustainable future management of all types of forests and forest genetic resources. In other words, the management of FGRs and forests themselves will essentially be inseparable; it will not be possible to plan one without the other.

This chapter examines FGR conservation management in Greece and Serbia as case studies. It addresses how environmental shifts impact genetic diversity and the need for effective climate change adaptation. The analysis includes relevant international policies and explores climate change implications, challenges, and strategies for sustainable FGR management. These case studies aim to show how proactive FGR management can support local adaptations, reduce tree mortality, and maintain biodiversity amid changing climatic conditions.

International Institutional Framework and Stakeholders Regulating Forest Genetic Resources and Climate Change

The Food and Agriculture Organization (FAO) of the United Nations and Its Commission on Genetic Resources for Food and Agriculture (CGRFA)

The main authorities in international forest genetic resources (FGRs) policy are the Food and Agriculture Organization (FAO) of the United Nations and its Commission on Genetic Resources for Food and Agriculture (CGRFA), along with the United Nations Environmental Programme (UNEP) and its Secretariat of the Convention on Biological Diversity (CBD). Despite their importance, FGRs are often overlooked in national policies on forests, biodiversity, and climate change. To address this, the FAO advises countries to analyze the impacts of FGRs on sustainable forest management, biodiversity conservation, and climate change adaptation and develop national strategies and coordination mechanisms (FAO 2014).

The FAO's Forestry Programme, initiated in the 1950s, supports countries in enhancing FGR management and promotes regional and international cooperation. Key priorities include improving FGR information access; conserving FGRs *in situ* and *ex situ*; ensuring sustainable use and management of FGRs; and addressing policies, institutions, and capacity-building (FAO 2014).

The CGRFA facilitates intergovernmental dialogue on genetic resources through the Intergovernmental Technical Working Group on FGR (ITWG-FGR), established in 2009. The CGRFA is a forum for shaping FGR policy and exchanging knowledge, producing global reports on FGR status every 10 years. Greece and Serbia, both members of the CGRFA, contributed to the second global assessment and are engaged in FAO and EUFORGEN initiatives, aligning their national policies with FAO recommendations and demonstrating the importance of international cooperation in FGR conservation.

UNEP and CBD

The United Nations Environment Programme (UNEP) has set the global environmental agenda since 1972, promoting sustainable development and working through regional and liaison offices. It addresses environmental challenges via seven subprogrammes, including Climate Action and Nature Action, and collaborates with 193 member states and various stakeholders through the UN Environment Assembly. UNEP hosts the secretariats of critical multilateral environmental agreements (https://www.unep.org/).

Since 1993, the Convention on Biological Diversity (CBD), an autonomous UNEP secretariat, has been pivotal in biodiversity conservation, including forest genetic resources (FGRs). The CBD aims to conserve biological diversity, use it sustainably, and equitably share benefits from genetic resources. It includes the Cartagena Protocol on Biosafety and the Nagoya Protocol on Access and Benefit-sharing, both critical for managing FGRs (https://www.cbd.int/).

Greece and Serbia, as CBD signatories, are committed to sustainable biodiversity and FGR management. By ratifying the Nagoya Protocol (CBD, 2011), they ensure fair sharing of benefits from genetic resources, and the Cartagena Protocol protects ecosystems from genetically modified organisms. Both countries participate in the UN Decade on Ecosystem Restoration, promoting forest restoration to enhance biodiversity and ecosystem services, crucial for climate change mitigation and local community support.

The UN Decade on Ecosystem Restoration (2021–2030), led by UNEP and FAO, aims to prevent and reverse ecosystem degradation globally. It emphasizes the importance of FGRs and forest reproductive materials (FRMs) for successful restoration, advocating for national strategies and action plans for climate change adaptation and mitigation (https://www.decadeonrestoration.org/).

Other Relevant Stakeholders

Other international stakeholders operating at the global and regional levels include the European Forest Genetic Resources Programme (EUFORGEN), the Asia Pacific Forest Genetic Resources Programme (APFORGEN), the International Union of Forest Research Organizations (IUFRO), and Bioversity International. EUFORGEN and APFORGEN are directly involved in the implementation of FAO's global plans of action, with FGRs representing one of the key topics within their working groups and task forces. IUFRO is especially significant since it is the only forestry-related organization that directly assembles scientists and practitioners dealing with FGRs. Through its various activities, working groups, and task forces, IUFRO covers almost all FGR-related topics including climate-smart management and networking. Figure 15.1 shows the top-to-bottom policy relations among major actors in the field of FGRs. It should be kept in mind that this relation is not linear and that governments also have the capability to directly influence the work of global organizations through membership within them. One of the latest initiatives in the field of

Fig. 15.1 Outline of decision-making framework from the global (left) to the local levels (right) in the fields of forestry and FGRs

international FGR policy is the creation of Voluntary Guidelines on a National Strategy for FGR (unpublished, provisional title by lead author) coordinated by the UN Food and Agriculture Organization, aimed at supporting countries in creating and adopting FGR-related strategies to promote and integrate systematic approaches to their conservation and utilization at the national level. These guidelines are expected to be published in 2025. Countries are not expected to adopt everything devised in the guidelines but instead can use everything significant for contextual and participatory design of their own evidence-based practical policies. The document is being cocreated with some of the abovementioned international stakeholders in the field of FGRs.

Conservation Measures Under Climate Change: Examples from Serbia and Greece

Forest genetic resources conservation is nowadays most implemented in *in situ* (within the distribution range of a species, often at the original location) and *ex situ* (outside the distribution range of a species) conservation practices, putting emphasis on the geo-locational approach of individual conservation measures targeting species and populations. Since the ecological demands of specific species determine the conditions of the conservation effort, these two approaches are the most common in forest conservation management. *Circa situm* is the third type, mostly recognized as a separate conservation practice especially suited to tropical agroforestry and dryland systems. *Circa situm* conservation is the preservation of planted and/or remnant trees and wildings in farmland where natural forest or woodland containing the same trees was previously found but has been lost or modified significantly through agricultural expansion. This is the most obvious setting to consider for the role of agroforestry in conservation as practice and function coincide geographically (Dawson et al. 2013). *Circa situm* stipulates serious planning of species distribution within the targeted agroforestry system with the goal of supporting it (Boshier et al. 2004; Dawson et al. 2013). In its first State of the World's Forest Genetic

Resources report (2014), the FAO defines *circa situm* conservation as a "type of conservation that emphasizes the role of regenerating saplings in linking vegetation remnants in heavily modified or fragmented landscapes, such as those of traditional agroforestry and farming systems."

The most common approach to conserving the genetic diversity of forest trees is the establishment of *in situ* dynamic conservation units with the aim of capturing the current genetic diversity as well as supporting continued evolutionary processes within populations (de Vries et al. 2015).

In Situ and Ex Situ Conservation of FGRs: Case Studies from Serbia

EUFORGEN's pan-European strategy for the genetic conservation of forest trees envisages the identification of dynamic gene conservation units (GCUs) to protect the adaptive and neutral diversity of forest tree species across Europe (de Vries et al. 2015). So far, only nine GCUs covering six broad-leaved and four coniferous tree species have been established in Serbia. Considering the pivotal role of genetic diversity for the adaptation, survival, and evolution of forest trees in a changing climate, as well as the fact that climate models have shown that some of the tree species in Serbia may be severely endangered by climate change (Stojanović et al. 2013, 2014, 2021), additional efforts have recently been made to conserve neutral and adaptive genetic diversity. As part of earlier activities within the framework of EUFORGEN, two GCUs were established within the largest complexes of pedunculate oak (Fig. 15.2., population MO) and Norway spruce (population ZL)

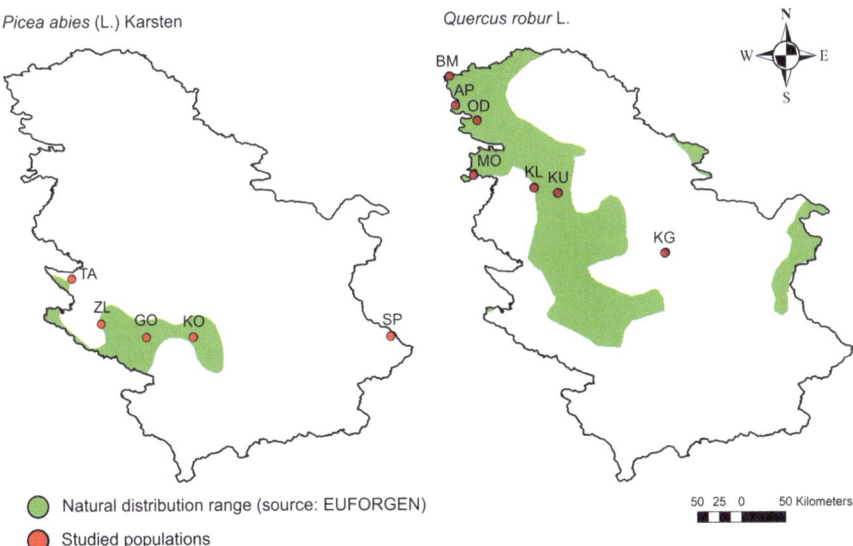

Fig. 15.2 Spatial distribution of studied Norway spruce (left) and pedunculate oak (right) populations in Serbia

forests in Serbia, respectively. Apart from the predictions of climate models, the reasons for the establishment of conservation units among these populations were twofold: (1) a decreasing growth trend and progressive mortality of pedunculate oak and Norway spruce trees across their natural distribution ranges (Kesić et al. 2016; Matović et al. 2018; Stojanović et al. 2015), and (2) high genetic diversity previously observed in these populations. Furthermore, the selected pedunculate oak population consists of varieties characterized by different phenological forms (early, intermediate, and late flushing). A recent study investigating the relative gene expression involved in different metabolic pathways (including drought tolerance) showed that the most significant differences in relative gene expression levels existed between genotypes belonging to different phenological types (early vs. late) rather than between trees with different physiological status (vital vs. senescent) (Trudić et al. 2021a).

Considering recent progress in the assessment of genetic diversity and structure of pedunculate oak and Norway spruce populations in Serbia (Stojnić et al. 2019; Trudić et al. 2021b; Kesić et al. 2021) as well as EUFORGEN recommendations, additional GCUs were identified to help preserve the neutral and adaptive genetic diversity of these species at the pan-European level. Concerning pedunculate oak, it has been proposed to set aside additional dynamic units to include populations located in a different ecological zone (Fig. 15.2, population KG) from the initially selected population MO, a population characterized by exceptional genetic diversity (e.g., the highest expected heterozygosity, allelic richness, and private allelic richness) (population KU), and populations in which a significant growth decline was observed (population BM). Finally, the selected populations are located along a north–south transect so that this approach also ensures the preservation of the species' spatial genetic diversity. In the case of Norway spruce, it is necessary to establish at least one more GCU including the easternmost population SP (Fig. 15.2), which is genetically the most distinct from other populations and the only one situated in the mesic phytogeographic region (Stojnić et al. 2019).

In addition, several molecular studies have been conducted on the species originating from natural populations in Serbia and Greece, respectively (Table 15.1). Besides the main tree species and their core populations, these efforts also targeted endemic species and their marginalized populations as well as tree species significant for agroforestry.

Ex situ conservation of forest genetic resources implies the maintenance of populations (provenances), individuals (clones), or reproductive material of different tree species in specially established sites outside their natural distribution with the aim of preserving the genetic variability of the original populations as much as possible (Mataruga et al. 2013). One of the forms of *ex situ* conservation is provenance trials, which can be defined as experiments in which plants coming from seeds collected in different parts of a species' range (i.e., provenances) are grown under the same ecological conditions (Wright 1976). According to Gömöry (2010), the main aim of provenance trials is to identify provenances characterized by vigorous growth and adaptability so as to use them as a source of seeds for future reforestation; Eriksson and Ekberg (2001) state another important goal of such research, namely, to trace

Table 15.1 Molecular studies on tree species originating from natural populations in Serbia and Greece

Serbia			Greece		
Tree species	Used markers	Reference	Tree species	Used markers	Reference
Abies alba	SSR	Popović et al. (2017), (2019)	Acer campestre	SSR	Wahlsteen et al. (2023)
Abies alba var. pyramidalis	EST-SSR	Trudić et al. (2016)	Abies borisii regis	SSR	Tourvas et al. (2017)
Picea abies	EST-SSR	Stojnić et al. (2019)			
Picea omorika	SSR	Aleksić et al. (2022)	Betula pendula	SSR	De Dato et al. (2020)
Picea omorika	cpSSR	Nasri et al. (2008)	Cupressus sempervirens	AFLP, SSR, MSAP	Avramidou et al. (2015), (2017)
Picea omorika	EST-SSR	Aleksić et al. (2009)			
Picea omorika	EST-SSR, mtDNA	Aleksić and Geburek (2014), Aleksić et al. (2017)	Fagus sylvatica	SSR	Malliarou et al. (2016)
Pinus nigra	RAPD	Lučić et al. (2010)	Pinus halepensis	SSR	Papadima (2014)
Pinus sylvestris	RAPD	Lučić et al. (2011)	Pinus nigra	AFLP, MSAP	Katsidi et al. (2023)
Pinus sylvestris	SSR	Lučić et al. (2014)	Platanus sp.	SSR	Malliarou et al. (2014), Farsakoglou et al. (2014)
Populus nigra	SSR	Čortan et al. (2016)	Populus sp.	SSR	Malliarou et al. (2014), Farsakoglou et al. (2014)
Populus sp.	AFLP, SSR	Orlović et al. (2009)	Prunus avium	SSR	Avramidou et al. (2010)
Quercus robur	SSR	Trudić et al. (2021a), Kesić et al. (2021)	Taxus baccata	SSRs, MSAP	Dalmaris et al. (2020)
Sorbus sp.	SSR	Galović et al. (2012)	Juniperus drupacea Labill.	AFLP, MSAP	Avramidou et al. (2023)

the adaptation that is taking place as well as determine the environmental factors that have the greatest influence on it. To use the full potential of this conservation approach, it is desirable to establish trials at several locations in the form of a network since testing provenances in only one locality reveals limited information on their real capacities and makes it impossible to test the "provenance × locality" interaction. By contrast, establishing trials in different environmental conditions allows provenances featuring specific adaptability, as well as those characterized by good general adaptability, to be identified (Stojnić et al. 2015b). European beech

Fig. 15.3 European beech provenance trial at Fruška Gora Mountain, Serbia, in 2010 (left) and 2022 (right)

provenance trials were initiated at two sites in Serbia in 2007 (Fig. 15.3) as part of a pan-European network with the aim of assessing the genetic variability and adaptive capacity of different provenances (von Wuehlisch 2004). Besides the value for the conservation of beech genetic resources itself, numerous studies have been carried out at these trial sites to date, most of them oriented around evaluating variability patterns of different functional traits (e.g., physiological, biochemical, morphological, and anatomical) within and among different provenances as well as assessing the provenances' stability and adaptability.

The results obtained from the beech provenance trials in Serbia revealed high variability both at the intra- and inter-provenance level (Štajner et al. 2013; Stojnić et al. 2016a; Vaštag et al. 2019), showing the presence of mainly discontinuous (ecotypic) patterns of genetic variation (Stojnić et al. 2016b). In addition, several studies demonstrated that phenotypic plasticity can be an important adaptive strategy in helping beech trees to cope with changed climatic conditions (Stojnić et al. 2013, 2015a, c). Finally, the results allow the identification of provenances exhibiting good growth in low-yield environments, as well as of provenances specifically adapted to favorable environments (Stojnić et al. 2015b).

In Greece, forest ecosystems cover a large part of land area (49.4%, of which 25.5% is high forest, Aravanopoulos (2010)) and host a great wealth of biodiversity, especially forest tree genetic resources (Aravanopoulos et al. 2019). The geographical location of the country combined with its high topographical and geological diversity explains its rich biodiversity. In total, 22% of the forest species of the Mediterranean are found in Greece, and 10% of these species are endemic to the country (Commission on Genetic Resources for Food and Agriculture 2010). Greece, with 103 biodiversity sites, is ranked third and fourth, respectively, in Europe and in the Mediterranean basin (Derneği 2010). Greece's conservation efforts for forest genetic resources (FGRs) exemplify a comprehensive approach to preserving biodiversity and ensuring ecosystem resilience amidst changing climatic conditions. The country's initiatives span both *in situ* and *ex situ* conservation strategies, showcasing a commitment to safeguarding its rich genetic heritage. This narrative explores the intricate measures Greece has adopted to protect and manage its forest genetic resources, highlighting the integration of scientific research, stakeholder collaboration, and policy development in these efforts (Alizoti et al. 2019).

In Situ Conservation Efforts in Greece

In situ conservation aims to maintain a population in the environment where it originally evolved and to which it has adapted and can be evolutionarily dynamic (Finkeldey and Gregorius 1994). This ensures the genetic diversity of species, and that species can keep pace with environmental changes and remain adapted (Koskela et al. 2013). One of the main approaches to conservation at species and ecosystem levels is the Natura 2000 network comprising 419 sites, of which 241 are Sites of Community Importance (SCI) and 202 are Special Areas of Conservation (SAC), while 24 sites are both SCI and SAC. The area covered by the Natura 2000 network is approximately 4,300,000 hectares, which corresponds to 27.2% of land and 4.5% of the European Natura 2000 network, placing Greece in tenth place among its 28 EU states (Aravanopoulos et al. 2019). Furthermore, *in situ* conservation in Greece targets the preservation of genetic diversity within natural habitats, acknowledging the intrinsic value of maintaining species in their original environments. Notably, the conservation of monumental trees, such as the Plane Tree of Hippocrates on Kos Island and the Plane Tree of Pausanias in Aigion, underscores the cultural and ecological significance of these living landmarks (https://www.monumentaltrees.com/en/grc/). Beyond individual trees, Greece's active participation in the EUFGIS network has led to the establishment of 15 gene conservation units (GCUs) for key species, including *Pinus halepensis* and *Abies cephalonica*. These units play a crucial role in conserving the genetic variation essential for species adaptation and survival (Koulelis et al. 2023).

Ex Situ Conservation Efforts in Greece

Ex situ conservation is the conservation of components of biological diversity outside their natural habitats and is applied when conservation *in situ* is not possible (Skrøppa 2005). This conservation protects populations or individuals at risk of natural disasters or genetic degradation and secures their reproductive material. Regarding *ex situ* conservation, Greece's initial venture occurred through a project funded by the United Nations Development Programme, UNSF/FAO/GRE:20/230, aimed at fortifying forest resources. This initiative launched an afforestation programme, establishing pilot provenance trial areas in 11 locations nationwide. Predominantly, *Pinus radiata*, *Pinus pinaster*, and *Pinus brutia* were the species planted. Subsequently, another UNDP-funded project, "Forest Development—Afforestation," established three pilot areas for afforestation, focusing on *Pinus pinaster*, *Pinus nigra*, and *Pinus brutia*. Additionally, the Food and Agriculture Organization (FAO) initiated a network of provenance trials in Greece for *Pinus halepensis* and *Pinus brutia* to assess existing variation in adaptive traits (Matziris

1993, 1995). *Ex situ* conservation plantations (i.e., seed orchards, provenance, and provenance-progeny trials) were established within breeding programmes carried out by the Aristotle University of Thessaloniki (Laboratory of Forest Genetics and Tree Breeding) and the Institute of Mediterranean and Forest Ecosystems (IMFE— Laboratory of Silviculture, Forest Genetics and Biotechnology) (Alizoti et al. 2019).

Breeding Programmes in Greece

As part of the breeding programmes, six first-generation clonal seeds orchards were created in the decade 1978–1987 for *Pinus nigra* (Arn.), *Pinus halepensis* (Mill.), and *Abies borisii regis* (Mattf.) due to their economic and ecological values that are particularly important for Greece. These species are ecologically essential species that grow extensively in the country and produce wood and non-wood products of high economic value, supporting the agricultural economy. A special effort was made for *Pinus nigra* to establish four clonal seed orchards in different geographical zones. The aim was to meet the needs of each specific area for improved FRM that could be used either for commercial plantations or for reforestation purposes (Alizoti et al. 2010). All clonal seed orchards serve as *ex situ* conservation plantations, harboring elite phenotypes/genotypes derived from different natural populations, some of which are marginal and carry potentially specific adaptation alleles. Four of these seed orchards have been progeny tested, more specifically *Pinus nigra*, *Pinus halepensis,* and *Abies borisii regis*. These progeny trials will be able to grow into *ex situ* stands containing much of the original genetic diversity.

Seed Production Areas in Greece

The lack of efforts to conserve FGRs is compensated to some extent by the seed production areas (SPAs) of the Hellenic Forest Service. These areas concern 170 forest stands of 12 forest species that cover a total area of approximately 11,500 hectares and are used to produce seeds for artificial reforestation purposes. These areas are subject to a special management plan aimed at protection. Therefore, specific areas and areas under different levels of biodiversity protection are areas where genetic diversity can be maintained at some level (Aravanopoulos et al. 2019). Furthermore, the Central Forest Seed Bank was established in 1992 and its objective is the collection, processing, preservation, management, and quality control of forest seeds to produce high-quality seedlings in forest nurseries. Annually, 2 tons of seeds of various forest species are collected, and after proper handling, they are stored in refrigerators and freezers, where they can be kept viable for many decades (Paitaridou 2009). Stored seeds can prove particularly important in the case of extinction of rare or endemic species in their natural environment due to stochastic events or biotic/abiotic factors (Aravanopoulos and Muller-Starck 2003).

Forest Genetic Monitoring in Greece

Forest genetic monitoring (FGM) is an important method that monitors temporal changes occurring in the genetic variation and structure of targeted tree populations. It can verify how well genetic diversity is maintained over time. It includes three markers, natural selection, genetic drift, and the gene flow coupling system, and their evaluation is based on demographic and genetic verifiers (Aravanopoulos 2011). In the framework of the LIFE13 ENV/SI/000148 "Life for European Forest Genetic Monitoring System" (LIFEGENMON) project, one stand of *Abies borisii regis* and one stand of *Fagus sylvatica* were monitored. The results obtained showed that both species are adapting well to climate change while maintaining their genetic diversity. In addition, within the framework of the project and based on the results, a handbook on forest genetic monitoring was created, which is freely available online (http://www.lifegenmon.si/lifegenmon-manual-for-forest-genetic-monitoring/).

In Greece, the use of molecular markers (Table 15.1.) has significantly advanced the conservation of FGRs, enabling precise identification and preservation of genetic diversity within forest species. This innovative approach facilitates the assessment of genetic variation and the adaptation strategies of species, informing both *in situ* and *ex situ* conservation efforts. By integrating molecular markers into their conservation strategies, Greece enhances its ability to monitor and protect the genetic integrity of its forests, ensuring the resilience and sustainability of forest ecosystems for future generations. The integration of both *in situ* and *ex situ* conservation efforts within a broader forest management framework highlights Greece's holistic approach to FGR conservation. Collaborations between research institutions, universities, forest services, and other stakeholders are vital for the success of conservation initiatives. These partnerships facilitate the sharing of knowledge, the development of conservation strategies, and the effective utilization of forest genetic resources. Greece's conservation endeavors reflect a deep recognition of the importance of forest genetic diversity for ecosystem health, species adaptation, and overall biodiversity. Through a combination of *in situ* and *ex situ* strategies, informed by scientific research and stakeholder collaboration, Greece is making significant strides in conserving its forest genetic resources for future generations. This concerted effort is crucial for ensuring the resilience of forest ecosystems and their capacity to adapt to environmental changes, ultimately contributing to global biodiversity conservation goals.

Conclusion

Forest ecosystems face significant challenges due to the rapid pace of climate change, necessitating urgent and effective conservation measures for FGRs. The adaptability and evolutionary potential of forest species heavily depend on the genetic diversity within and among tree populations. Strategic conservation practices, both *in situ* and *ex situ*, are critical for capturing and maintaining this genetic diversity, allowing forests to adapt to changing environmental conditions. International cooperation and alignment with global conservation standards and policies, such as those set by the FAO and CBD, enhance the effectiveness of FGR conservation efforts. The experiences of Serbia and Greece offer valuable insights into the practical implementation of conservation strategies that can serve as models for other regions. Proactive management of FGRs, including the establishment of GCUs and the utilization of advanced genetic analysis techniques, is fundamental to understanding and preserving the genetic basis for forest resilience. Conservation strategies must be flexible and adaptive, incorporating the latest scientific knowledge and responding to emerging threats to forest ecosystems. Public awareness and stakeholder engagement are essential for the successful implementation of FGR conservation strategies, ensuring broad support and cooperation. Future policies must integrate FGR conservation into wider forest management and climate adaptation frameworks to address the multifaceted challenges posed by climate change. Ultimately, the conservation of FGRs is a cornerstone of sustainable forestry, biodiversity preservation, and climate change mitigation, requiring a committed and collaborative global effort.

References

Aleksić JM, Schueler S, Mengl M, Geburek T (2009) EST-SSRs developed for other picea species amplify in *Picea omorika* and reveal high genetic variation in two natural populations. Belg J Bot 142:89–95

Aleksić JM, Geburek T (2014) Quaternary population dynamics of an endemic conifer, *Picea omorika*, and their conservation implications. Conserv Genet 15:87–107

Aleksić JM, Piotti A, Geburek T, Vendramin GG (2017) Exploring and conserving a "microcosm": whole-population genetic characterization within a refugial area of the endemic, relict conifer *Picea omorika*. Conserv Genet 18:777–788

Aleksić JM, Mataruga M, Daničić V, Cvjetković B, Milanović Đ, Vendramin GG, Avanzi C, Piotti A (2022) High pollen immigration but no gene flow via-seed into a genetic conservation unit of the endangered *Picea omorika* after disturbance. For Ecol Manag 510:120115

Alizoti PG, Kilimis K, Gallios P (2010) Temporal and spatial variation of flowering among Pinus nigra Arn. clones under changing climatic conditions. For Ecol Manag 259(4):786–797

Alizoti PG, Aravanopoulos FA, Ioannidis K (2019) *Ex situ* conservation of forest genetic resources in Greece. Forests of southeast Europe under a changing climate: conservation of genetic resources, pp 291–301

Aravanopoulos FA, Muller-Starck G (2003) Artificial seed transfer. In: Kremer A, Reviron MP (eds) Dynamics and conservation of genetic diversity in forest ecosystems, Forest genetics, vol 9, pp 333–340

Aravanopoulos F (2010) The importance of the biodiversity of Hellenic forest and land-based ecosystems for Europe and the Mediterranean. In: Ananiadou-Tzimopoulou M (ed) Proceedings Conference Environmental Council. Aristotle University Press, Aristotle University of Thessaloniki, Thessaloniki, pp 1–11

Aravanopoulos F (2011) Genetic monitoring in natural perennial plant populations. Botany 89:75–81

Aravanopoulos FA (2016) Conservation and monitoring of tree genetic resources in temperate forests. Curr For Rep 2:119–129

Aravanopoulos FA, Alizoti PG, Farsakoglou AM, Malliarou E, Avramidou EV, Tourvas N (2019) State of biodiversity and forest genetic resources in Greece in relation to conservation. Forests of southeast Europe under a changing climate: conservation of genetic resources, pp 73–83

Argüelles-Moyao A, Galicia L (2023) Assisted migration and plant invasion: importance of belowground ecology in conifer forest tree ecosystems. Can J For Res 54(1):110–121

Avramidou E, Ganopoulos IV, Aravanopoulos FA (2010) DNA fingerprinting of elite Greek wild cherry (*Prunus avium* L.) genotypes using microsatellite markers. Forestry 83:527–533

Avramidou EV, Doulis AG, Aravanopoulos FA (2015) Determination of epigenetic inheritance, genetic inheritance, and estimation of genome DNA methylation in a full-sib family of *Cupressus sempervirens* L. Gene 562(2):180–187

Avramidou EV, Doulis AG, Aravanopoulos FA (2017) Linkage and QTL mapping in *Cupressus sempervirens* L. provides the first detailed genetic map of the species and identifies a QTL associated with crown form. Tree Genet Genomes 13:1–14

Avramidou EV, Korakaki E, Malliarou E, Boutsios S (2023) Studying the genetic and the epigenetic diversity of the endangered species *Juniperus drupacea* Labill. towards safeguarding its conservation in Greece. Forests 14(6):1271

Boshier DH, Gordon JE, Barrance AJ (2004) Prospects for *circa situm* tree conservation in mesoamerican dry-forest agro-ecosystems. In: Frankie GW, Mata A, Vinson SB (eds) Biodiversity conservation in Costa Rica, learning the lessons in the seasonal dry forest. University of California Press

CBD (2011) Nagoya protocol on access to genetic resources and the fair and equitable sharing of benefits arising from their utilization to the convention on biological diversity. Retrieved from https://www.cbd.int/abs/doc/protocol/nagoya-protocol-en.pdf

Commission on Genetic Resources for Food and Agriculture (2010) The second report on the state of the world's plant genetic resources for food and agriculture. Food & Agriculture Organization

Čortan D, Schroeder H, Šijačić-Nikolić M, Wehenkel C, Fladung M (2016) Genetic structure of remnant black poplar (*Populus nigra* L.) populations along biggest rivers in Serbia assessed by SSR markers. Silvae Genetica 65(1):12–19

Dalmaris E, Avramidou EV, Xanthopoulou A, Aravanopoulos FA (2020) Dataset of targeted metabolite analysis for five taxanes of hellenic *Taxus baccata* l. populations. Data 5(1):22

Dawson IK, Guariguata MR, Loo J, Weber JC, Lengkeek A, Bush D et al (2013) What is the relevance of smallholders' agroforestry systems for conserving tropical tree species and genetic diversity *in circa situm*, *in situ* and *ex situ* settings? A review. Biodivers Conserv 22:301–324

De Dato GD, Teani A, Mattioni C, Aravanopoulos F, Avramidou EV, Stojnić S et al (2020) Genetic analysis by nuSSR markers of Silver Birch (*Betula pendula* Roth) populations in their southern European distribution range. Front Plant Sci 11:310

Derneği D (2010) Ecosystem profile: Mediterranean basin biodiversity hotspot final report for submission to the CEPF Donor Council July 27:2010

De Vries S, Alan M, Bozzano M, Burianek V, Collin E, Cottrell J, Ivanković M, Kelleher C, Koskela J, Rotach P (2015) Pan-European strategy for genetic conservation of forest trees and establishment of a core network of dynamic conservation units. In: European forest genetic resources programme (EUFORGEN); Bioversity International, Rome

Eriksson G, Ekberg I (2001) An introduction to forest genetics. Swedish University of Agricultural Sciences, Uppsala, p 166

FAO (2014) The state of the world's forest genetic resources, Rome

Farsakoglou AML, Avramidou EV, Doulis AG, Aravanopoulos FA (2014) Molecular genetic diversity analysis of sycamore (*Platanus orientalis* L.) and natural mutation of sycamore (*Platanus orientalis var. cretica* Dode) using microsatellite markers (SSR). Proceedings of the 15th Pan-Hellenic conference of the Hellenic scientific society for plant genetic breeding, October 15–17, Larissa, p 87

Finkeldey R, Gregorius HR (1994) Genetic resources: selection criteria and design. Conservation and manipulation of genetic resources in forestry. Proceedings

Gaisberger H, Kindt R, Loo J, Schmidt M, Bognounou F et al (2017) Spatially explicit multi-threat assessment of food tree species in Burkina Faso: a fine-scale approach. PLoS One 12(9):e0184457

Galović V, Orlović S, Trudić B, Pekeč S, Vasić S (2012) Intra-loci polymorphism of Sorbus spp.: on the territory of Republic of Serbia. Topola 189–190:87–97

Gömöry D (2010) Geographic patterns in the reaction of beech provenances to transfer. Proceedings of the Workshop and MC Meeting of the COST Action E52 "Evaluation of Beech Genetic Resources for Sustainable Forestry", Thessaloniki, May 5–7, 2009, pp 90–97

Hof AR, Dymond CC, Mladenoff DJ (2017) Climate change mitigation through adaptation: the effectiveness of forest diversification by novel tree planting regimes. Ecosphere 8(11):e01981

Isabel N, Holliday J, Aitken S (2019) Forest genomics: advancing climate adaptation, forest health, productivity, and conservation. Evol Appl 13(1):7–10

Joyce S, Shand F, Tighe J, Laurent SJ, Bryant RA, Harvey SB (2018) Road to resilience: a systematic review and meta-analysis of resilience training programmes and interventions. BMJ Open 14(6):e017858

Katsidi EC, Avramidou EV, Ganopoulos I, Barbas E, Doulis A, Triantafyllou A, Aravanopoulos FA (2023) Genetics and epigenetics of Pinus nigra populations with differential exposure to air pollution. Front Plant Sci:14

Kesić L, Cseke K, Orlovic S, Stojanović DB, Kostić S, Attila B, Attila B, Stojnic S, Avramidou EV (2021) Genetic diversity and differentiation of Pedunculate oak (*Quercus robur* L.) populations at the southern margin of its distribution range—implications for conservation. Diversity 13:371

Kesić L, Matović B, Stojnić S, Stjepanović S, Stojanović D (2016) Climate change as a factor reducing the growth of trees in the pure Norway spruce stand (*Picea abies* (L.) H. Karst.) in the National park "Kopaonik". Topola 197–198:25–34

Koskela J, Vinceti B, Dvorak W, Bush D, Dawson I, Loo J et al (2010) The use and movement of forest genetic resources for food and agriculture. In: Background study paper no. 44. The commission on genetic resources for food and agriculture, food and agriculture organization of the United Nations, Rome, Italy

Koskela J, Lefèvre F, Schueler S, Kraigher H, Olrik DC, Hubert J et al (2013) Translating conservation genetics into management: Pan-European minimum requirements for dynamic conservation units of forest tree genetic diversity. Biol Conserv 157:39–49

Koulelis PP, Proutsos N, Solomou AD, Avramidou EV, Malliarou E, Athanasiou M, Petrakis PV (2023) Effects of climate change on Greek forests: a review. Atmosphere 14(7):1155

Loo J, Fady B, Dawson I, Vinceti B, Baldinelli G (2011) Climate change and forest genetic resources—state of knowledge, risks and opportunities. [Technical Report] background study paper n.56

Lučić A, Isajev V, Rakonjac L, Ristić D, Kostadinović M, Babić V, Nikolić A (2011) Genetic divergence of scots pine (*Pinus sylvestris* L.) populations in Serbia revealed by RAPD. Arch Biol Sci 63(2):371–380

Lučić A, Mladenović-Drinić S, Stavretović N, Isajev V, Lavadinović V, Rakonjac LJ, Novaković M (2010) Genetic diversity of austrian pine (*Pinus nigra* Arnold) populations in Serbia revealed by RAPD. Arch Biol Sci 62(2):329–336

Lučić A, Popović V, Nevenić M, Ristić D, Rakonjac L, Ćirković-Mitrović T, Mladenović-Drinić S (2014) Genetic divergence of scots pine (*Pinus sylvestris* L.) populations in Serbia revealed by SSR markers. Arch Biol Sci 66(4):1485–1492

Malliarou ES, Avramidou EV, Aravanopoulos FA (2014) DNA fingerprinting *Platanus* sp and *Populus* sp. Clones with nuclear microsatellites (nSSR). In: Proceedings of the 15th conference of Hellenic scientific society for plant genetics and breeding, Larisa, pp 29–294

Malliarou E, Avramidou E, Ganopoulos I, Aravanopoulos FA (2016) Preliminary results of SSR genetic diversity in Fagus sylvatica in the frame of a genetic monitoring programme. In: Proceeding of 18th Pan-Hellenic forest science conference, Edessa, pp 951–955

Mataruga M, Isajev V, Orlović S (2013) Forest genetic resources. University book, University of Banja Luka, Faculty of Forestry, Republic of Srpska, p 399

Matović B, Stojanović D, Kesić L, Stjepanović S (2018) Uticaj klime na prirast i vitalnost smrče na Kopaoniku. Topola 201/202:99–116

Matziris D (1993) Variation in cone production in a clonal seed orchard of black pine. Silvae Genet 42:136–141

Matziris D (1995) Provenance variation of *Pinus radiata* grown in Greece. Silvae Genet 44:88–89

Nasri N, Bojović S, Vendramin GG, Fady B (2008) Population genetic structure of the relict Serbian spruce, *Picea omorika*, inferred from plastid DNA. Plant Syst Evol 271:1–7

Orlović S, Galović V, Zorić M, Kovačević B, Pilipović A, Galic Z (2009) Evaluation of interspecific DNA variability in poplars using AFLP and SSR markers. Afr J Biotechnol 8(20)

Paitaridou D (2009, October 15–16) The role of the central forest seed bank in forest restoration. In: Proceedings of the international conference on 'New approaches to the restoration of black pine forests', Sparta. LIFE 07 NAT/GR/000286 project: 'Restoration of *Pinus nigra* forests on Mount Parnonas (GR 2520006) through a structured approach'

Papadima A (2014) Genetic analysis of *Pinus halepensis* seed orchard. Phd thesis (available at: https://thesis.ekt.gr/thesisBookReader/id/38803#page/1/mode/2up)

Popović V, Lučić A, Rakonjac L, Cvjetković B, Mladenović-Drinić S, Ristić D (2017) Assessment of genetic diversity of silver fir (*Abies alba* Mill.) in Serbia using SSR markers. Genetika 49(3):979–988

Popović V, Lučić A, Rakonjac L, Milovanović J, Mladenović Drinić S, Ristić D (2019) Application of SSR markers for assessment of genetic differentiation of silver fir (*Abies alba* Mill.) originating from Javor mountain. Genetika 51(3):1103–1112

Schueler S, Falk W, Koskela J, Lefevre F, Bozzano M, Hubert J, Kraigher H, Longauer R, Olrik DC (2014) Vulnerability of dynamic genetic conservation units of forest trees in Europe to climate change. Glob Chang Biol 20:1498–1511

Skrøppa T (2005) *Ex situ* conservation methods. Conservation and management of forest genetic resources in Europe. Arbora Publishers, Zvolen, pp 567–583

Štajner D, Orlović S, Popović B, Kebert M, Stojnić S, Klašnja B (2013) Chemical parameters of oxidative stress adaptability in beech. Journal of Chemistry. Article ID 592695. https://doi.org/10.1155/2013/592695

Stojanović DB, Krzić A, Matović B, Orlović S, Duputie A, Djurdjević V, Galić Z, Stojnić S (2013) Prediction of the European beech (*Fagus sylvatica* L.) xeric limit using a regional climate model: an example from Southeast Europe. Agric For Meteorol 176:94–103

Stojanović DB, Matović B, Orlović S, Kržič A, Trudić B, Galić Z, Stojnić S, Pekeč S (2014) Future of the main important forest tree species in Serbia from the climate change perspective. South-East Eur For 5:117–124

Stojanović DB, Levanič T, Matović B, Orlović S (2015) Growth decrease and mortality of oak floodplain forests as a response to change of water regime and climate. Eur J For Res 134:555–567

Stojanović DB, Orlović S, Zlatković M, Kostić S, Vasić V, Miletić B, Kesić L, Matović B, Božanić D, Pavlović L, Milović M, Pekač S, Đurđević V (2021) Climate change within Serbian forests: current state and future perspectives. Topola 208:39–56

Stojnić S, Sass-Klaassen U, Orlović S, Matović B, Eilmann B (2013) Plastic growth response of European beech provenances to dry site conditions. IAWA J 34:475–484

Stojnić S, Orlović S, Miljković D, Galić Z, Kebert M, von Wuehlisch G (2015a) Provenance plasticity of European beech leaf traits under differing environmental conditions at two Serbian common garden sites. Eur J For Res 134:1109–1125

Stojnić S, Orlović S, Ballian D, Ivanković M, Šijačić-Nikolić M, Pilipović A, Bogdan S, Kvesić S, Mataruga M, Daničić V, Cvjetković B, Miljković D, von Wuehlisch G (2015b) Provenance by site interaction and stability analysis of European beech (*Fagus sylvatica* L.) provenances grown in common garden experiments. Silvae Genetica 64:133–147

Stojnić S, Orlović S, Trudić B, Živković U, von Wuehlisch G, Miljković D (2015c) Phenotypic plasticity of European beech (*Fagus sylvatica* L.) stomatal features under water deficit assessed in provenance trial. Dendrobiology 73:163–173

Stojnić S, Orlović S, Trudić T, Kesić L, Stanković M, Šijačić-Nikolić M (2016a) Height and root-collar diameter growth variability of European beech provenances from Southeast Europe. Topola 197–198:5–14

Stojnić S, Orlović S, Miljković D, von Wuehlisch G (2016b) Intra- and inter-provenance variation of leaf morphometric traits in European beech (*Fagus sylvatica* L.) provenances. Arch Biol Sci 68:781–788

Stojnić S, Avramidou VE, Fussi B, Westergren M, Orlović S, Matović B, Trudić B, Kraigher H, Aravanopoulos FA, Konnert M (2019) Assessment of genetic diversity and population genetic structure of Norway spruce (*Picea abies* (L.) Karsten) at its southern lineage in Europe. Implications for conservation of forest genetic resources. Forests 10:258

Tourvas N, Malliarou E, Avramidou E, Aravanopoulos FA (2017) A comparative analysis of molecular genetic diversity of fir (Abies sp.) mature tree and regeneration cohorts in the frame of a genetic monitoring programme. In: Proceeding of 18th Pan-Hellenic forest science conference, Edessa, pp 962–969

Trudić B, Kiš A, Stojnić S, Orlović S, Panjković B (2016) Spatial genetic profile of marginal population of silver fir (*Abies alba* (var. "pyramidalis")) from Serbia using SSR markers. In: VII international scientific agriculture symposium, "Agrosym 2016", 6–9 October 2016, Jahorina, Bosnia and Herzegovina. Proceedings. University of East Sarajevo, Faculty of Agriculture, pp 2910–2916

Trudić B, Draškić G, Le Provost G, Stojnić S, Pilipović A, Ivezić A (2021a) Expression profiles of 11 candidate genes involved in drought tolerance of pedunculate oak (L.). Possibilities for genetic monitoring of the species. Silvae Genetica 70(1):226–234

Trudić B, Avramidou E, Fussi B, Neophytou C, Stojnić S, Pilipović P (2021b) Conservation of *Quercus robur* L. genetic resources in its south-eastern refugium using SSR marker system–a case study from Vojvodina province, Serbia. Austrian J For Sci 138(2):117–140

Vaštag E, Kovačević B, Orlović S, Kesić L, Bojović M, Stojnić S (2019) Leaf stomatal traits variation within and among fourteen European beech (*Fagus sylvatica* L.) provenances. Genetika 51:937–959

Verkerk PJ, Costanza R, Hetemäki L, Kubiszewski I, Leskinen P, Nabuurs GJ et al (2020) Climate-smart forestry: the missing link. Forest Policy Econ 115:102164

von Wuehlisch G (2004) Series of international provenance trials of European beech. Proceedings from the 7th international beech symposium IUFRO research group 1.10.00 "Improvement and Silviculture of Beech". 10–20 May 2004, Tehran, Iran, pp 135–144

Wahlsteen E, Avramidou EV, Bozić G, Mediouni RM, Schuldt B, Sobolewska H (2023) Continental-wide population genetics and post-Pleistocene range expansion in field maple (*Acer campestre* L.), a subdominant temperate broadleaved tree species. Tree Genet Genomes 19(2):15

Weiskopf SR, Rubenstein MA, Crozier LG, Gaichas S, Griffis R, Halofsky JE, Hyde KJW, Morelli TL, Morisette JT, Muñoz RC, Pershing AJ, Peterson DL, Poudel R, Staudinger MD, Sutton-Grier AE, Thompson L, Vose J, Weltzin JF, Powys WK (2020) Climate change effects on biodiversity, ecosystems, ecosystem services, and natural resource management in the United States. Sci Total Environ 733:137782

Wright JW (1976) Introduction to forest genetics. Academic Press, Inc, San Diego, p 463

Internet Sources

Commission for Genetic Resources in Food and Agriculture. FAO. https://www.fao.org/cgrfa/en/
Convention on Biological Diversity. https://www.cbd.int/
United Nations Environmental Programme. https://www.unep.org/

Open Access This chapter is licensed under the terms of the Creative Commons Attribution 4.0 International License (http://creativecommons.org/licenses/by/4.0/), which permits use, sharing, adaptation, distribution and reproduction in any medium or format, as long as you give appropriate credit to the original author(s) and the source, provide a link to the Creative Commons license and indicate if changes were made.

The images or other third party material in this chapter are included in the chapter's Creative Commons license, unless indicated otherwise in a credit line to the material. If material is not included in the chapter's Creative Commons license and your intended use is not permitted by statutory regulation or exceeds the permitted use, you will need to obtain permission directly from the copyright holder.

Managing Forest Health in Connected Landscapes

16

Gernot Hoch, Katharina Lapin, and Maarten de Groot

Forest of dead trees. Forest dieback in the Harz National Park, Lower Saxony, Germany, Europe. Dying spruce trees, drought, and bark beetle infestation, late summer of 2020. (Photo: K I Photography)

G. Hoch (✉)
Department of Forest Protection, Austrian Research Centre for Forests, Vienna, Austria
e-mail: gernot.hoch@bfw.gv.at

K. Lapin
Department of Forest Biodiversity & Nature Conservation, Austrian Research Centre for Forests, Vienna, Austria
e-mail: katharina.lapin@bfw.gv.at

M. de Groot
Department of Forest Protection, Slovenian Forestry Institute, Ljubljana, Slovenia
e-mail: maarten.degroot@gozdis.si

© The Author(s) 2025
K. Lapin et al. (eds.), *Ecological Connectivity of Forest Ecosystems*,
https://doi.org/10.1007/978-3-031-82206-3_16

Abstract

Managing forest health is a critical challenge for forest managers and policymakers worldwide, especially in connected forests where ecological and socio-economic linkages are tightly intertwined. Conservation and sustainable management of forest ecosystems require the development and implementation of comprehensive strategies that address the complex interactions between natural and human-driven stressors affecting forest health. In this chapter, we review the main drivers of forest health degradation and provide an overview of the approaches and tools that can be used to monitor and manage forest health. We highlight the importance of integrating practical knowledge with scientific information to enhance the resilience of forest ecosystems to global environmental changes. Finally, we discuss the challenges and opportunities associated with managing forest health in connected forest landscapes, and we suggest possible strategies to improve forest health management.

Keywords

Pests · Fungal pathogens · Invasive species · Climate change · Plant protection · Tree health

Introduction

Forest health refers to the overall well-being and functionality of a forest ecosystem. This includes the physical, chemical, and biological processes that occur within the forest as well as the relationships between various components of the ecosystem such as trees, other plants, animals, fungi, soil, and water. Definitions of forest health should combine ecological and utilitarian aspects (Kolb et al. 1994). A healthy forest thus sustains the complexity of the ecosystem and provides ecosystem services for human needs. In the context of ecological connectivity, forest health is closely linked to the ability of the forest ecosystem to support biodiversity and maintain ecological processes across spatial scales (Pautasso et al. 2015). Ecological connectivity refers to the connection between different forest habitats by corridors or stepping stones and allows for the movement of species and the exchange of genetic material between them. Forest health and connectivity are linked in various ways. A decline in forest health may negatively affect connectivity; on the other hand, high connectivity may not only facilitate species movement but also the spread of pests and pathogens—particularly invasive alien ones (see also Chap. 17).

Forest health must be distinguished from tree health. Individual trees are generally attacked by numerous parasitic organisms such as herbivorous insects or pathogenic fungi. Some act as primary pests, attacking and damaging a vigorous tree without other predisposing factors. Defoliators such as leaf/needle-feeding caterpillars are typically primary pests. Others function as secondary pests that attack weakened trees. Bark beetles, long-horned beetles, wood wasps, and other insects that develop under the bark or in the wood of living trees are examples. Their degree of aggressiveness can vary; some species may switch to primary attack once they reach very high population levels, as is the case with a small number of bark beetle species (e.g. from the genera *Dendroctonus* or *Ips*). Overall, damage to individual trees will always occur in forest ecosystems and impair tree health. While this damage will remain innocuous most of the time, certain conditions may trigger the mass proliferation of a pest or pathogen species, leading to an outbreak damaging large numbers of trees. Whether such an outbreak will also impair forest health depends on the amount or degree of damage as well as the constitution and the resilience of the affected forest ecosystem. Forest stands suffering insect or disease outbreaks must be viewed in their specific landscape and societal context. There may be interests of neighbours that need to be protected, for example, by preventing spillover of a pest from an affected stand or legal requirements to control certain pests and diseases. Forest managers therefore will not always have a choice whether to implement forest protection measures or which ones to apply. These measures should ideally be proactive by controlling conditions that can trigger pest or disease outbreaks. When curative measures are necessary, they will often be limited to cutting and removal of infested trees.

A healthy forest ecosystem is one that is resilient to environmental stressors and can recover from disturbances caused by pathogen or insect infestations and abiotic stressors such as drought or windthrow (Trumbore et al. 2015). This resilience is closely tied to the connectivity of the forest as the latter facilitates the movement of species and the exchange of genetic material between different patches of habitat, thereby increasing genetic diversity and promoting ecosystem resilience (Pearson et al. 2021). Therefore, maintaining and enhancing ecological connectivity can be critical for promoting forest health and resilience.

Changing Environmental Conditions Challenging Forest Health

Two drivers of global environmental change can have severe impacts on forest health: climate change and invasion by alien pests and pathogens. Even forests adapted to certain levels of disturbance will be challenged by novel stressors caused by global environmental change (Trumbore et al. 2015). Forests exhibit varied responses to climate change and the associated abiotic challenges including wildfires, storm damage, drought events, and changes in precipitation patterns. Climate predictions for Europe show regionally diverse changes, including an escalation in high-temperature extremes, drought events, and heavy precipitation events, all of which will negatively impact forest productivity and vitality (Martinez et al. 2022; Lindner and Verkerk, 2022).

Increasing temperatures can have significant impacts on forest health. Higher temperatures can lead to increased stress on trees, making them more susceptible to disease, insect infestations, and other types of damage. High temperatures can also increase the frequency and intensity of forest fires, which can further exacerbate forest health problems. One of the primary impacts of increased temperatures on forest health is drought stress (Hayden et al. 2011). As temperatures rise, evapotranspiration rates increase, which can cause soil moisture to decrease rapidly. This can lead to water stress in trees, which can cause wilt, reduce growth rates, and increase vulnerability to pests and diseases (Netherer et al. 2021). Greater damage from secondary pests or pathogens, such as species of bark beetles or fungi infecting woody organs, following severe drought (Jactel et al. 2012a, b) is a likely consequence (Fig. 16.1).

Trees experiencing physiological stress are more susceptible to potential pests, which can lead to increased occurrence of pest outbreaks as a result of climate change. In addition, previously insignificant pests or pathogens that were not considered harmful may cause more damage due to the predisposition of trees to diseases. The spread of new harmful organisms through immigration or introduction is also contributing to an increase in tree damage. Drought and high temperatures can have a direct detrimental impact on trees. Under extreme drought conditions, the hydraulic collapse of a tree can lead to its fast death (Arend et al. 2021). Carbon starvation can be another consequence of drought damage (McDowell, 2011). Following the extreme drought in the summer of 2018 in Western and Central Europe, high mortality was recorded among beech as well as other trees (Schuldt et al. 2020). Higher temperatures can also lead to changes in the timing and severity of insect outbreaks. Insects are poikilothermic animals, meaning that within species-specific limits, warmer temperatures speed up their development, leading to larger populations and more severe outbreaks. Insect species with flexible numbers of generations per year (e.g. the European spruce bark beetle, *Ips typographus*) can further increase population growth at higher temperatures. Exacerbated by the weakening of host trees due to climatic extremes, bark beetle outbreaks have reached unprecedented levels during the past decade (Hlásny et al. 2021a). The emergence of diseases such as sooty bark disease of maple (caused by the fungus *Cryptostroma corticale*) or pine dieback (caused by *Diplodia sapinea*) is likewise driven by high temperatures and drought (Brodde et al. 2019; Muller et al. 2023).

Another impact of increased temperatures on forest health is the expansion of the range of species and the subsequent proliferation of pests that have always been present but only in low populations causing little damage. As temperatures rise, some species may be able to expand their ranges into areas previously considered unsuitable for their survival. This has been demonstrated, for example, for the pine processionary moth (Battisti et al. 2005). In addition, the establishment of alien pests or pathogens can be supported by climate change. Invasion can lead to increased competition with native species and further stress on already vulnerable forest ecosystems. Climate warming can also increase the detrimental impact of invasive harmful organisms, for example, as increasing the likelihood for expression of lethal wilt in trees infested with the pine wood nematode, *Bursaphelenchus xylophilus* (Gruffudd et al. 2016) or allowing faster population growth (Fig. 16.2). Overall, the impacts of increasing temperatures on forest health are complex and multifaceted. Addressing these impacts will require a combination of approaches including reducing greenhouse gas emissions to mitigate climate change,

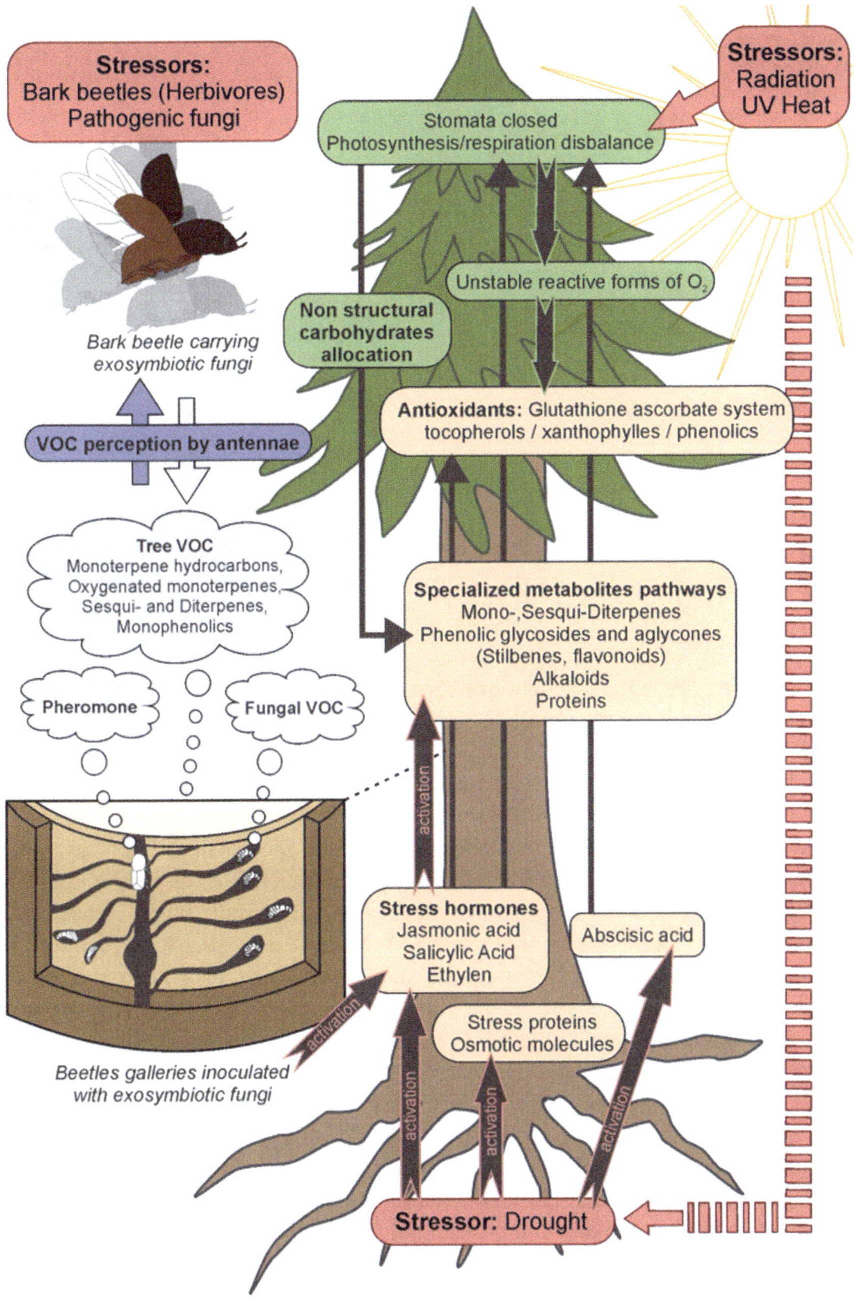

Fig. 16.1 Drought can cause a reduction in water and nutrient flow in plants, which in turn may impact their ability to defend against herbivorous insects and pathogens. In response to drought, plants may reduce the production of chemical defences, making them more susceptible to insect damage and disease. (originally published by Netherer et al. (2021). Journal of Pest Science, 94(3), 591–614; released under a Creative Commons Attribution 4.0 International License (CC BY 4.0))

Fig. 16.2 The oak lace bug, *Corythucha arcuata*, is an example of an invasive pest species that is rapidly spreading and causing damage to forests and urban trees in many parts of Europe. The spread of this insect is supported by human transport activities. Higher temperatures contribute to faster population growth since more generations can develop per year. (Photo: BFW)

implementing forest management practices that promote resilience, and developing strategies to monitor and control the spread of pests and diseases.

Considering the effects of global environmental change on forest health, the introduction and spread of invasive alien pests and pathogens play a critical role. These pests and pathogens can severely impair forest ecosystem functioning and, in some cases, even threaten the existence of entire tree species in a region. Examples of devastating diseases caused by fungi are the Dutch elm disease (caused by *Ophiostoma ulmi* and *Ophiostoma novo-ulmi*), chestnut blight (caused by *Cryphonectria parasitica*), and more recently, ash dieback (caused by *Hymenoscyphus fraxineus*). Nematodes and insects can also have threatening impacts on their host tree species: the pine wood nematode has destroyed vast numbers of pines in its invaded ranges in East Asia and on the Iberian Peninsula, and the emerald ash borer, *Agrilus planipennis*, is devastating ash stands in North America. Many other invasive pests and pathogens will not threaten a tree species with extinction but still negatively impact tree and forest health. A significant reduction of resources needed by specialised herbivores can also occur when invasive pests or pathogens do not kill trees but cause, for example, a loss of palatable foliage for leaf-feeders. This is a likely consequence of the extremely high populations of the oak lace bug, *Corythucha arcuata*, that occur every year once the species is established in an area (Paulin et al. 2020). Once established, the organism can begin to spread in the new area. This can occur via natural dispersal as well as with the aid of human activities. The connectivity of the landscape can thus be a factor regarding both establishment and spread.

Habitat Connectivity in the Context of Forest Pests and Diseases

Forest pests and disease outbreaks, on the one hand, may negatively affect connectivity that make it less useful for other native organisms, while on the other hand, connectivity may support the establishment and spread of (invasive alien) pests and pathogens. Pests and diseases can impact connectivity in a number of ways and affect functional as well as structural connectivity. If they cause tree mortality, their presence can lead to the fragmentation of forest habitats, making it difficult for species to move between different areas (Helander et al. 2007). This can result in isolated fragments of habitat that are too small to support viable populations of certain species, thus leading to local extinctions and reduced biodiversity, i.e. negatively affecting forest health. Pests and diseases can alter the structure of forest habitats by killing trees, reducing canopy cover, and changing the composition of plant communities. This can lead to changes in microclimates, soil moisture, and other environmental conditions that can impact the distribution and abundance of species. Pests and diseases can impact the availability of resources such as food, water, and shelter, which can have negative effects on the species that depend on these resources. For example, a disease that kills a particular tree species will reduce the resources for specialised herbivores. Furthermore, it may reduce the availability of habitats for other animals, such as nesting sites for birds that rely on that tree. This is of particular importance in the case of invasive pests or pathogens such as the abovementioned fungus *Hymenoscyphus fraxineus* or the pine wood nematode.

Landscape patterns can significantly affect the probability of pest and disease infestation in forest patches. When forests are arranged in fragmented patches rather than as interior habitats, they may be more vulnerable to the effects of climate change and other stressors. This can further increase their susceptibility to pests and diseases as weakened trees are more vulnerable to infestation. In addition, when forests are located close to other lands or fragmented into smaller patches, they may be more vulnerable to invasion by pests and pathogens from adjacent areas. In the European Union, the fact that 40% of woodlands are located within a distance of 100 m from other types of land means that many forests are likely to be located in areas with a high degree of human activity, such as urban or agricultural areas (Estreguil et al. 2013). This proximity to human activity can increase the likelihood of introduction and spread of invasive species, pests, and diseases that can negatively affect forest health. Typically, invasive pests and pathogens are introduced into new areas distant from previously infested areas by human activities (e.g. transport of infested plants or plant material, hitchhiking on vehicles, etc.). New populations can be established in these areas and consequently spread into adjacent forests. Examples are *Phytophthora* spp., which frequently contaminate nurseries (Jung et al. 2016) and are spread to the soil when reared plants are planted; the bacterium *Xylella fastidiosa*, an EU priority quarantine pest that can be spread with plants for planting and is then transmitted to plants in the vicinity by vectors; or the oak lace bug, which spreads easily as a hitchhiker and often establishes new populations near traffic infrastructure (Csóka et al. 2020; Hoch et al. 2023).

On the other hand, landscapes with high connectivity and a higher percentage of suitable host tree species may support the spread of an invasive pest or pathogen over a larger area following its establishment. The presence of suitable hosts can function as stepping stones for newly introduced harmful organisms, as has been shown, for example, for the pine scale *Matsucoccus feytaudi* (Rigot et al. 2014). The effect of connectivity depends on the biology and dispersal capacity of the specific invasive pest or pathogen. For example, the spread of a slowly dispersing, relatively large beetle like the Asian long-horned beetle, *Anoplophora glabripennis*, will most likely be aided by a continuous availability of suitable host trees. The oak lace bug with its notorious hitchhiking abilities, on the other hand, can easily utilise human transportation for dispersal; host tree availability will nevertheless be important for the success of the species. Root pathogens of the genus *Phytophthora* depend on water bodies for dispersal; when infested planting material is introduced near a river or a creek, a quick regional spread is likely.

Prevention and Mitigation Strategies

To address the issue of tree diseases and response to symptoms, it is important to identify the disease and determine the pest insect or pathogen responsible for the outbreak. Studying the causes of disease processes can also help prevent and manage outbreaks in the future. This includes identifying factors that contribute to the spread of pest insects and pathogens, such as environmental conditions, host susceptibility, and the presence of vectors or carriers. By understanding the underlying causes of tree diseases, we can develop more effective management strategies to mitigate the impact of outbreaks and protect forest health.

Analysing the effects of pathogenic organisms or pest insects is important to understand how they interact with both biotic and abiotic factors in their environment. This helps identify the key drivers of disease outbreaks such as environmental stressors or host susceptibility and develop effective management strategies to mitigate their impact. To better understand the threat posed by important pests and pathogens, it is essential to study their life cycles including the different stages they undergo, host preferences, and modes of transmission. This information can help identify potential pathways for pathogen introduction, transmission, and spread as well as ways to disrupt them. Corridors and stepping stones can be of particular importance for surveillance in order to detect a spreading (invasive) pest or pathogen early. Surveys should therefore consider such sites since detecting invaders there may allow for the control of the pest at an early stage or disrupt the connections, thereby contributing to stopping or at least slowing their spread.

Considering the possible harmful effects of pathogenic organisms is critical. This includes assessing the economic and ecological impacts of disease outbreaks on plant health, biodiversity, and ecosystem services. Understanding the impacts of pathogenic organisms can inform decision-making processes for the development and implementation of management strategies.

To reduce the impact of insects and pathogens on trees and on forest health, it is important to develop suitable transnational strategies and processes to prevent or manage pest insect and pathogen outbreaks (Wingfield et al. 2015). This includes the establishment of monitoring programmes to identify potential outbreaks and the use of preventative measures to reduce the spread of pathogens. Effective management strategies may involve a combination of approaches, including the use of biological control agents, cultural practices, chemical treatments, and quarantine measures to prevent the introduction and spread of pathogens. Additionally, education and outreach programmes can help increase awareness regarding tree diseases and encourage the adoption of best management practices. Well-informed professionals and interested citizens can also play a crucial role in the early detection of emerging pests or pathogens and introduced aliens. Early detection followed by rapid response is typically key to successful control. Experts can provide advice on countermeasures and management strategies for specific pathogens, including guidance on the most effective control methods and treatment options. This can include recommendations on tree species selection, cultural practices, and chemical treatments as well as advice on the potential ecological and economic impacts of different management strategies or the implementation of quarantine regulations. By taking a holistic approach to pest and pathogen management, we can help reduce the impact of diseases on trees and protect the health and biodiversity of forests.

Adaptive Forest Management and Forest Pest Management

Forest managers can take various measures to control pests and diseases and protect the connectivity of forest ecosystems. These measures include prophylactic, silvicultural approaches, such as an increase in tree diversity, use of resistant tree species, or proper sanitation measures. Measures can be curative and aim to prevent the spread of pests and diseases, like the removal of infected trees. Measures can also be taken to block the infectious process by preventing the introduction and spread of pests and diseases. This may involve implementing quarantine measures like buffer zones around infested areas (thereby reducing connectivity), monitoring the spread of pests and diseases, and restricting the movement of potentially infected plant material or the pest itself. Important pathways for this movement are living plants, wood and wood products, seeds, and hitchhiking (Liebhold et al. 2012; Meurisse et al. 2019). The most likely pathways depend on the biology of the respective species.

Many countries have clear regulations on how outbreaks of invasive quarantine pests and pathogens are to be managed (e.g. EU Plant Health Law: Regulation (EU) 2016/2031). Measures to eradicate or contain newly introduced pests or pathogens will often involve extensive removal of host trees to prevent further dispersal and spread. Deliberate interruption of connectivity can thus be an effective pest management strategy and, in some cases, necessary to prevent further negative consequences for forest health or biodiversity. It is also important to develop and expand a practical training concept for forest managers to increase their awareness of forest pests and diseases and equip them with the skills and knowledge necessary to identify, monitor, and manage outbreaks. In addition, modernisation of the forest protection monitoring system, including the use of GIS and remote-sensing technologies (Dillon et al. 2014), can help forest managers better understand complex damage events and make informed decisions about management strategies.

> **Box 16.1 Lessons Learnt from Bark Beetle Outbreaks**
>
> The European spruce bark beetle, *Ips typographus*, is one of the most important native tree–killing insects in European forests (Fig. 16.3). It can switch from the endemic to epidemic phase driven by the availability of suitable material for breeding (weakened trees, trees broken by storm, etc.), allowing the build-up of beetle populations. Large numbers of living Norway spruce trees can be attacked and killed in the epidemic phase. The population dynamics of *I. typographus* is strongly influenced by the volume and height of available Norway spruce trees, as well as by the distance to the nearest harvested area in the previous 4 years. The effects of tree diversity only become noticeable in Central Europe when the spruce percentage is below 40% (De Groot et al. 2023). Climate variables also influence the speed of infestation and population increase (e.g. Hlásny et al. 2021b, Kärvemo et al. 2023). The effects of climate change, including higher temperatures and increased storms, have become increasingly important since the 1990s; consequently, damage by *I. typographus* has been steeply rising in Europe (Hlásny et al. 2021a).
>
>
>
> **Fig. 16.3** Adults and larvae of the European spruce bark beetle, *Ips typographus*, the major pest in Europe for Norway spruce, *Picea abies*. (Photo: BFW)
>
> (continued)

Besides long-term silvicultural measures such as changing forest structure and tree species composition, the most important tool for managing the spruce bark beetle is the removal of suitable breeding sites like wind-felled trees or their treatment, for example, by debarking or bark-scratching to avoid the build-up of bark beetle populations, as well as cutting and timely removal of newly infested trees. Once an outbreak starts and the attack on standing trees begins, the loss of large numbers of mature spruce trees is inevitable, and the pest may spread to other stands.

The current extent of outbreaks of *I. typographus* is exceptional (Hlásny et al. 2021a). However, following drought conditions, outbreaks of bark beetles specialising in other tree species can cause significant tree mortality as well, e.g. fir bark beetles, *Pityokteines* spp., or pine bark beetles like *Ips acuminatus*.

Biological Control

Biological control can be one effective and environmentally sound technique for managing pests, both native as well as alien species. Biological control refers to managing pest species by using living organisms like parasitoids, pathogenic microorganisms, or herbivores. Eilenberg et al. (2001) list four strategies of biological control: inoculation biological control, inundation biological control, classical biological control, and conservation biological control. Inundation biological control is based on the mass release of living organisms to immediately suppress pest populations. Preparations of the bacterium *Bacillus thuringiensis kurstaki* have been successfully used against leaf-feeding caterpillars for many years, and preparations based on insect viruses such as the *Lymantria dispar* nucleopolyhedrosis virus (against the spongy moth, *Lymantria dispar*, an invasive pest in North America) have proven to be highly specific and effective biopesticides. But in general, the methods available for forest protection are few. Inoculation biological control works by releasing smaller numbers of the living biocontrol agent, which is supposed to proliferate in the environment and achieve control of the pest with some time delay. The most successful strategy in forest protection featuring the release of living organisms has been classical biological control. This method employs inoculative releases of natural enemies of an alien invasive pest collected in its native range (see also Chap. 17). Current examples are the release of the parasitic wasp *Torymus sinensis* against the chestnut gall wasp, *Dryocosmus kuriphilus,* in Europe and the release of several parasitic wasps from East Asia against the emerald ash borer in North America. Increasing connectivity would increase the dispersion of the biocontrol agent and therefore improve the impact on the target organism. However, a parasitoid or predator with a wide host/prey spectrum could become an invasive alien species itself. A prominent example is the Asian lady beetle, *Harmonia axyridis*. This species was introduced in greenhouses against aphids but is now also predating on native lady beetles, hoverfly larvae, and other species (Roy and Brown 2015). Therefore, thorough testing of the host specificity of the biocontrol agent prior to release is a crucial element of such classical biological control programmes to avoid negative effects on non-target organisms. Measures to increase connectivity will likely be useful to support the establishment and spread of released natural enemies just like they supported the alien pest to be controlled. The fourth method, conservation biological control, does not release natural enemies against target pests but instead attempts to protect and enhance already-present natural enemies in order to suppress the target pests. This is done by using appropriate cultural and silvicultural practices, supporting plants that serve as food, providing habitats for natural enemies, and supporting a continuous supply of hosts for parasitic insects or insect pathogens. Increasing structural and species diversity in a forest and maintaining a certain amount of deadwood can form a part of strategies to support natural enemies of forest pests. For this biocontrol strategy, an increase in connectivity for the natural enemies of concern will also be beneficial. Close-to-nature forestry, a management practice applied in Slovenia for decades with the goal of developing mixed and uneven-aged forests with high structural diversity (Diaci 2006; Diaci et al. 2017) can also be seen as a method of conservation biological control.

Implications for Managing Forest Health in Connected Forests

Effective forest management requires a multi-criteria approach that takes into account the various ecosystem services that forests provide (Jactel et al. 2012a, b), including carbon sequestration, water regulation, biodiversity conservation, and recreational and cultural benefits. In order to ensure that forest management decisions are sustainable and effective, it is necessary to conduct multi-criteria risk analyses that assess the potential impacts of various forest management scenarios on these ecosystem services. This involves considering a range of factors including the potential impact of management practices on soil health, water quality, and biodiversity as well as the potential trade-offs and synergies between different ecosystem services. Pest or disease outbreaks of native species can be considered natural processes in the forest ecosystem. It may be appropriate to rely on the resilience of the disturbed system when it is adapted to the occurring disturbance. Significant loss of mature trees can still be the consequence. Climate change impacts will likely impair this resilience. By taking a holistic approach to forest management, we can aim to maintain forest health and ensure that our forests continue to provide a wide range of ecosystem services and meet the needs of both current and future generations (Bakhtiyari et al. 2019).

In connected forests, corridors and stepping stones must be treated as integrated parts of forest health management. Therefore, we suggest the following measures to improve corridors for biodiversity management but not allowing the propagation and dispersion of invasive alien pests, emerging pests, and native pests:

A. Recognise and implement corridors as an integrated part of the forest complex and integrated forest pest management.
B. Increase tree species diversity in the corridors to reduce its vulnerability to outbreaks. Moreover, this will prevent deterioration of the corridor after a pest or disease outbreak.
C. Carry out sanitary measures to mitigate moderate disturbances, for example, by cutting and removing infested trees.
D. Consider using semiochemicals for mass trapping or repelling pests from the corridors when appropriate systems are available.
E. Manage corridors to support the establishment and dispersion of biocontrol agents (natural enemies) by providing habitats and nutritional resources to suppress pest populations in corridors and in other fragments.
F. Focus surveillance for invasive alien pests on corridors to allow early detection. Use appropriate measures to eradicate or suppress detected infestations to avoid their spread to other forest complexes or a decrease in the quality of the corridors as a dispersal structure for biodiversity.
G. Massive tree removal can be an appropriate measure to break connectivity. This may be necessary to contain or eradicate newly introduced invasive pests or pathogens in order to prevent unacceptable damage to forest health or biodiversity.

References

Arend M, Link RM, Patthey R, Hoch G, Schuldt B, Kahmen A (2021) Rapid hydraulic collapse as cause of drought-induced mortality in conifers. Proc Natl Acad Sci U S A 118(16):e2025251118. https://doi.org/10.1073/pnas.2025251118

Bakhtiyari M, Lee SY, Warnken J (2019) Seeing the forest as well as the trees: an expert opinion approach to identifying holistic condition indicators for mangrove ecosystems. Estuar Coast Shelf Sci 222:183–194

Battisti A, Stastny M, Netherer S, Robinet C, Schopf A, Roques A Larsson S (2005) Expansion of geographic range in the pine processionary moth caused by increased winter temperatures. Ecol Appl 15:2084–2096. https://doi.org/10.1890/04-1903

Brodde L, Adamson K, Camarero JJ, Castaño C, Drenkhan R, Lehtijärvi A, Luchi N, Migliorini D, Sánchez-Miranda Á, Stenlid J, Özdag S, Oliva J (2019) Diplodia tip blight on its way to the north: drivers of disease emergence in northern Europe. Front Plant Sci 9:1818. https://doi.org/10.3389/fpls.2018.01818

Csóka G, Hirka A, Mutun S, Glavendekić M, Mikó Á, Szőcs L, Paulin M, Eötvös CB, Gáspár C, Csepelényi M, Szénási Á, Franjević M, Gninenko Y, Dautbašić M, Muzejinović O, Zúbrik M, Netoiu C, Buzatu A, Bălăcenoiu F, Jurc M, Jurc D, Bernardinelli I, Streito J-C, Avtzis D, Hrašovec B (2020) Spread and potential host range of the invasive oak lace bug [*Corythucha arcuata* (Say, 1832)—Heteroptera: Tingidae] in Eurasia. Agric For Entomol 22:61–74. https://doi.org/10.1111/afe.12362

de Groot M, Ogris N, Diaci J, Castagneyrol B (2023) When tree diversity does not work: the interacting effects of tree diversity, altitude and amount of spruce on European spruce bark beetle outbreaks. For Ecol Manage 537:120952. https://doi.org/10.1016/j.foreco.2023.120952

Diaci J, Rozenbergar D, Fidej G, Nagel TA (2017) Challenges for uneven-aged silviculture in restoration of post-disturbance forests in central Europe: a synthesis. Forests 8(10):20

Diaci J (2006) Nature-based silviculture in Slovenia: origins, development and future trends. In: Nature-based forestry in Central Europe. Diaci J (ed) Alternatives to industrial forestry and strict preservation, Ljubljana, 119–132 p

Dillon WW, Haas SE, Rizzo DM, Meentemeyer RK (2014) Perspectives of spatial scale in a wildland forest epidemic. Eur J Plant Pathol 138:449–465

Eilenberg J, Hajek A, Lomer C (2001) Suggestions for unifying the terminology in biological control. BioControl 46:387–400. https://doi.org/10.1023/A:1014193329979

Estreguil C, Caudullo G, de Rigo D, San-Miguel-Ayanz J (2013) Forest landscape in Europe: pattern, fragmentation and connectivity. EUR Sci Tech Res 25717(JRC 77295) 18 p. https://doi.org/10.2788/77842

Gruffudd HR, Jenkins TAR, Evans HF (2016) Using an evapo-transpiration model (ETpN) to predict the risk and expression of symptoms of pine wilt disease (PWD) across Europe. Biol Invasions 18:2823–2840. https://doi.org/10.1007/s10530-016-1173-7

Hayden KJ, Nettel A, Dodd RS, Garbelotto M (2011) Will all the trees fall? Variable resistance to an introduced forest disease in a highly susceptible host. For Ecol Manage 261(11):1781–1791

Helander M, Ahlholm J, Sieber TN, Hinneri S, Saikkonen K (2007) Fragmented environment affects birch leaf endophytes. New Phytol 175(3):547–553

Hlásny T, König L, Krokene P et al (2021a) Bark beetle outbreaks in Europe: state of knowledge and ways forward for management. Curr For Rep 7:138–165. https://doi.org/10.1007/s40725-021-00142-x

Hlásny T, Zimová S, Merganičová K, Štěpánek P, Modlinger R, Turčáni M (2021b) Devastating outbreak of bark beetles in The Czech Republic: drivers, impacts, and management implications. For Ecol Manage 490:119075. https://doi.org/10.1016/j.foreco.2021.119075

Hoch G, Sallmannshofer M, Connell J, Hinterstoisser W, Schafellner C (2023) Rasche Ausbreitung der invasiven Eichennetzwanze (*Corythucha arcuata*) in Österreich [Rapid spread of the invasive oak lace bug, *Corythucha arcuata*, in Austria]. Forstschutz Aktuell 68:12–18

Jactel H, Branco M, Duncker P, Gardiner B, Grodzki W, Langstrom B, Moreira F, Netherer S, Nicoll B, Orazio C (2012a) A multicriteria risk analysis to evaluate impacts of forest management alternatives on forest health in Europe. Ecol Soc 17(4)

Jactel H, Petit J, Desprez-Loustau M-L, Delzon S, Piou D, Battisti A, Koricheva J (2012b) Drought effects on damage by forest insects and pathogens: a meta-analysis. Glob Chang Biol 18:267–276. https://doi.org/10.1111/j.1365-2486.2011.02512.x

Jung T, Orlikowski L, Henricot B, Abad-Campos P, Aday AG, Aguín Casal O, Bakonyi J, Cacciola SO, Cech T, Chavarriaga D, Corcobado T, Cravador A, Decourcelle T, Denton G, Diamandis S, Doğmuş-Lehtijärvi HT, Franceschini A, Ginetti B, Green S, Glavendekić M, Hantula J, Hartmann G, Herrero M, Ivic D, Horta Jung M, Lilja A, Keca N, Kramarets V, Lyubenova A, Machado H, di San M, Lio G, Mansilla Vázquez PJ, Marçais B, Matsiakh I, Milenkovic I, Moricca S, Nagy ZÁ, Nechwatal J, Olsson C, Oszako T, Pane A, Paplomatas EJ, Pintos Varela C, Prospero S, Rial Martínez C, Rigling D, Robin C, Rytkönen A, Sánchez ME, Sanz Ros AV, Scanu B, Schlenzig A, Schumacher J, Slavov S, Solla A, Sousa E, Stenlid J, Talgø V, Tomic Z, Tsopelas P, Vannini A, Vettraino AM, Wenneker M, Woodward S, Peréz-Sierra A (2016) Widespread *Phytophthora* infestations in European nurseries put forest, semi-natural and horticultural ecosystems at high risk of Phytophthora diseases. For Pathol 46:134–163. https://doi.org/10.1111/efp.12239

Kärvemo S, Huo L, Öhrn P, Lindberg E, Persson HJ (2023) Different triggers, different stories: bark-beetle infestation patterns after storm and drought-induced outbreaks. For Ecol Manage 545:121255. https://doi.org/10.1016/j.foreco.2023.121255

Kolb TE, Wagner MR, Covington WW (1994) Concepts of forest health. J For 92(7):10–15

Liebhold AM, Brockerhoff EG, Garrett LJ, Parke JL, Britton KO (2012) Live plant imports: the major pathway for forest insect and pathogen invasions of the US. Front Ecol Environ 10:135–143

Lindner M, Verkerk H (2022) How has climate change affected EU forests and what might happen next? EFI – Key Questions on Forests in Th EU https://efi.int/forestquestions/q4_en

Martinez E, Zang CS, Buras A, Hacket-pain A, Esper J, Serrano-notivoli R, Hartl C, Weigel R, Klesse S, De Dios VR (2022) Beech forests. pp 1–9

McDowell NG (2011) Mechanisms linking drought, hydraulics, carbon metabolism, and vegetation mortality. Plant Physiol 155:1051–1059

Meurisse N, Rassati D, Hurley BP et al (2019) Common pathways by which non-native forest insects move internationally and domestically. J Pestic Sci 92:13–27. https://doi.org/10.1007/s10340-018-0990-0

Muller E, Dvořák M, Marçais B, Caeiro E, Clot B, Desprez-Loustau M-L, Gedda B, Lundén K, Migliorini D, Oliver G, Ramos AP, Rigling D, Rybníček O, Santini A, Schneider S, Stenlid J, Tedeschini E, Aguayo J, Gomez-Gallego M (2023) Conditions of emergence of the sooty bark disease and aerobiology of *Cryptostroma corticale* in Europe. In: Jactel H, Orazio C, Robinet C, Douma JC, Santini A, Battisti A, Branco M, Seehausen L, Kenis M (eds) Conceptual and technical innovations to better manage invasions of alien pests and pathogens in forests, NeoBiota, vol 84, pp 319–347. https://doi.org/10.3897/neobiota.84.90549

Netherer S, Kandasamy D, Jirosová A, Kalinová B, Schebeck M, Schlyter F (2021) Interactions among Norway spruce, the bark beetle *Ips typographus* and its fungal symbionts in times of drought. J Pest Sci 94:591–614. https://doi.org/10.1007/s10340-021-01341-y

Paulin M, Hirka A, Eötvös CB, Gáspár C, Fürjes-Mikó Á, Csóka G (2020) Known and predicted impacts of the invasive oak lace bug (*Corythucha arcuata*) in European oak ecosystems—a review. Folia Oecologica 47(2):131–139. https://doi.org/10.2478/foecol-2020-0015

Pautasso M, Schlegel M, Holdenrieder O (2015) Forest health in a changing world. Microb Ecol 69(4):826–842. https://doi.org/10.1007/s00248-014-0545-8

Pearson RM, Schlacher TA, Jinks KI, Olds AD, Brown CJ, Connolly RM (2021) Disturbance type determines how connectivity shapes ecosystem resilience. Sci Rep 11(1):1–8. https://doi.org/10.1038/s41598-021-80987-1

Rigot T, van Halder I, Jactel H (2014) Landscape diversity slows the spread of an invasive forest pest species. Ecography 37:648–658. https://doi.org/10.1111/j.1600-0587.2013.00447.x

Roy HE, Brown PMJ (2015) Ten years of invasion: *Harmonia axyridis* (Pallas) (Coleoptera: Coccinellidae) in Britain. Ecol Entomol 40:336–348. https://doi.org/10.1111/een.12203

Schuldt B, Buras A, Arend M, Vitasse Y, Beierkuhnlein C et al (2020) A first assessment of the impact of the extreme 2018 summer drought on central European forests. Basic Appl Ecol 45:86–103. https://doi.org/10.1016/j.baae.2020.04.003

Trumbore S, Brando P, Hartmann H (2015) Forest health and global change. Science 349:814–818. https://doi.org/10.1126/science.aac6759

Wingfield MJ, Brockerhoff EG, Wingfield BD, Slippers B (2015) Planted forest health: the need for a global strategy. Science 349(6250):832–836

Open Access This chapter is licensed under the terms of the Creative Commons Attribution 4.0 International License (http://creativecommons.org/licenses/by/4.0/), which permits use, sharing, adaptation, distribution and reproduction in any medium or format, as long as you give appropriate credit to the original author(s) and the source, provide a link to the Creative Commons license and indicate if changes were made.

The images or other third party material in this chapter are included in the chapter's Creative Commons license, unless indicated otherwise in a credit line to the material. If material is not included in the chapter's Creative Commons license and your intended use is not permitted by statutory regulation or exceeds the permitted use, you will need to obtain permission directly from the copyright holder.

Managing Invasive Alien Species in Forest Corridors and Stepping Stones

17

Giuseppe Brundu, Maarten de Groot, Sabrina Kumschick, Jan Pergl, and Katharina Lapin

Tree of heaven. (Photo: BFW/Anna-Maria Walli)

G. Brundu (✉)
Department of Agricultural Sciences, University of Sassari, Sassari, Italy

National Biodiversity Future Center, Palermo, Italy
e-mail: gbrundu@uniss.it

M. de Groot
Department of Forest Protection, Slovenian Forestry Institute, Ljubljana, Slovenia
e-mail: maarten.degroot@gozdis.si

Abstract

Invasive alien species (IAS) pose a significant threat to forest ecosystems by disrupting ecological networks and competing with native species. Forest habitat patches and corridors designed to enhance connectivity and biodiversity can unintentionally promote the dispersal of IAS, further compromising the ecological integrity of the forest ecosystem. This chapter discusses two main aspects related to IAS and forest connectivity: (1) the spread of IAS in the landscape and their impacts on native species and (2) the consequences of IAS on forest connectivity. Effective management of IAS is crucial to improve connectivity for native species while restricting the spread opportunity for aliens and preserve biodiversity. Ideally, a site-specific risk analysis should precede conservation or restoration efforts, determining the potential impact of IAS on the respective habitat patch's structural and functional connectivity, and vice versa. Furthermore, this chapter explores management strategies to control IAS, including physical removal, biological control, and monitoring. Citizen involvement and remote sensing play vital roles in supporting management actions, IAS detection and long-term monitoring, and habitat connectivity. Including stakeholders such as forest owners and managers in such actions ensures a collaborative approach to safeguarding forest ecosystems from the threats posed by IAS.

Keywords

Biological invasions · Introduction · Invasive alien species · Non-native species · Clearing · Connectivity · Fragmentation · Landscape matrix · Management · Pathway management · Prevention · Spread

Introduction

Invasive alien species (plants, fungi, microorganisms, and animals deliberately or accidentally introduced by humans, which threaten ecosystems, habitats, or species—hereinafter IAS) are significant contributors to global biodiversity loss, and

---

S. Kumschick
Department of Botany and Zoology, Centre for Invasion Biology, Stellenbosch University, Matieland, South Africa

South African National Biodiversity Institute, Kirstenbosch Research Centre, Cape Town, South Africa

J. Pergl
Department of Invasion Ecology, Institute of Botany, Academy of Sciences of the Czech Republic, Pruhonice, Czech Republic
e-mail: Jan.Pergl@ibot.cas.cz

K. Lapin
Department of Forest Biodiversity and Nature Conservation, Austrian Research Centre for Forests, Vienna, Austria
e-mail: katharina.lapin@bfw.gv.at

the impacts of some IAS can cause degradation of unique habitats and entire ecosystems (Dukes and Mooney 1999; Vilà et al. 2011; Simberloff et al. 2013; Archer et al. 2018; IPBES 2023). According to the Convention on Biological Diversity (CBD) definition, the term "invasive alien species" refers to alien species whose introduction by humans and/or natural or human-assisted spread threatens biological diversity, as defined in Article 8(h) of the Convention and recalled in numerous other documents from COP 4 Decision IV/1 to COP 14 Decision 14/11. The original CBD definition was broadened, however, so that an IAS is now an alien species that not only has adverse impacts on biodiversity but may also cause economic or environmental harm or negatively affect human health. In fact, IAS can cause a wide variety of impacts: they can compete with native species for resources, disrupt food webs and plant-pollinator interaction networks, hybridise with native or alien congeneric species, cause problems through herbivory and predation of native species, and act as vectors or hosts for new pest and pathogens. In addition, they spread rapidly, thus necessitating timely management decisions (e.g. Wilson et al. 2016; De Groot et al. 2022; Langmaier and Lapin 2020; Lapin et al. 2021; Vaz et al. 2018).

Many forests around the world are continually subject to severe outbreaks of IAS, which can have massive environmental and sociocultural impacts (Roy et al. 2014). For example, mangrove forests dominated by halophytic plant communities and occurring predominantly along the tropical and subtropical coastlines offer important and unique ecosystem functions and services (Biswas et al. 2018). Many mangrove species are presently threatened with extinction, for example, due to deforestation, land-use changes, and IAS. Furthermore, the planting of fast-growing alien mangrove species has been used as a tool for mangrove restoration/reforestation in America, Australia, and Africa. However, the fast growth ability of these alien species can lead to them becoming invasive as they may potentially replace co-occurring native mangroves due to higher growth performance and phenotypic plasticity (Fazlioglu and Chen 2020). Therefore, effective management strategies for IAS should consider both preventive measures and the reduction of their potential for migration and dispersal, ultimately aiming to avoid or minimise their negative impacts (McNeely et al. 2001). Alarmed by the continuous loss of biodiversity, Target 7 of the Kunming–Montreal Global Biodiversity Framework (CBD/COP/DEC/15/4, 19 December 2022) invites parties and other governments to eliminate, reduce, and/or mitigate the impacts of IAS on biodiversity and ecosystem services by identifying and managing pathways of the introduction of alien species; preventing the introduction and establishment of priority IAS; reducing the rates of introduction and establishment of other known or potential IAS by at least 50 per cent by 2030; and eradicating or controlling IAS, especially in priority sites. In this context, it is essential to improve our knowledge of IAS in forest corridors and stepping stones—although the relationships between forest connectivity or fragmentation and IAS can be multifaceted (Table 17.1.) and must be assessed on a case-by-case basis. In this chapter, the term *forest corridor* is used as an equivalent to *ecological corridor* (Hilty et al. 2020), i.e. meaning a clearly defined geographical space that is governed and managed over the long term to maintain or restore effective ecological connectivity. We include in this category wildlife crossings and similar human

Table 17.1 The table summarises the results of a non-systematic literature review highlighting the multi-faced relationships among invasive alien species' (IAS) establishment and spread, native species (NS), and forest connectivity (and its complement forest fragmentation) and the two types of corridors considered in this chapter, i.e. forest corridors (ecological corridors) and artificial corridors

Relationship or interaction	Effects on IAS	Effects on NS	References
Corridors may favour the movement of IAS. For example, forest roads can facilitate the spread of IAS.	+		Simberloff et al., 1992; Mortensen et al., 2009
Small habitat patches may facilitate the presence of IAS.	+		Lawrence et al., 2018
Landscape structure may affect the spread of IAS and the invasibility of communities in several ways. It is an important predictor used in several types of models.	+/-	+/-	Chabrerie et al., 2007; With, 2002; Roy et al., 2016; Lustig et al., 2017; Fergus et al., 2023
Landscape and connectivity metrics are a spatial tool that can be used to support and guide IAS management decisions.			Todd et al., 2014; Stewart-Koster et al., 2015; Buchholtz et al., 2023;
IAS with a high probability of random long-distance dispersal can best be managed by focusing on the largest patches, while IAS with a lower probability of random long-distance dispersal can best be managed by considering landscape configuration and connectivity of the patches.			Minor & Gardner, 2011
IAS management can enhance landscape connectivity, and algorithms on its incorporation into	-	+	Glen et al., 2013

infrastructure aiming to form safe natural corridor bridges for animals to migrate between conservancies, for example, in the case of road traversed by such corridors. On the other hand, we use the term *artificial corridor* to indicate the CBD corridors pathway category (European Commission 2020), which refers to the movement of alien species into a new region following the construction of transport infrastructures in whose absence spread would not have been possible. In the framework of the present chapter, such artificial corridors include human infrastructures built in forest environments for purposes different from restoring or promoting ecological connectivity, such as canals (connecting river catchments and lakes), tunnels, bridges, roads and railways, forest roads, and land uses different from the natural forest (i.e. including plantations with non-native trees or clearcuttings for the establishment of safety corridors for electricity lines), fragmenting forests environments or mainly accidentally linking forest patches.

In general, forest habitat patches that function as ecological corridors or stepping stones may be invaded and promote the spread of IAS (Liebhold et al. 1995; Roy et al. 2014; Langmaier and Lapin 2020). This can allow IAS to endanger protected area networks—for example, outbreaks of destructive alien insect herbivores can be facilitated by connectivity among forest patches, allowing them to disrupt many ecosystem services (Kenis et al. 2009). Conversely, existing connectivity for native species within a landscape can be impeded by the presence of IAS when they limit dispersal or increase the mortality of dispersers (Glen et al. 2013). For these reasons, IAS management can enhance both landscape connectivity and nature conservation in forests. Today, almost a third of all endemic Caribbean forest-dependent bird species are threatened with extinction, largely due to habitat loss, IAS, and over-exploitation (Devenish-Nelson et al. 2019), while endemic island forest types are threatened by IAS such as the *Scalesia pedunculata* forest in the Galapagos (Riegel et al. 2023).

Yet, in specific contexts, there may also be positive effects of IAS on connectivity, as shown by the study on Madagascan lemurs in the *Mandena* littoral forest, a matrix of littoral forest, littoral swamp, and *Melaleuca* swamp habitats. Here, the alien *Melaleuca quinquenervia* has invaded the wetland ecosystem, creating a mono-dominant habitat that currently provides the only potential habitat forest corridor for lemurs between forest fragments (Eppley et al. 2015). The multifaceted relationship between IAS and connectivity has led to controversial proposals that require further careful evaluation, such as introducing *Opuntia* spp. cacti in Madagascar because hedges of these succulents may help in maintaining viable populations of several endemic species (Andriamparany et al. 2020).

This chapter aims to explore the complex interactions between IAS and forest connectivity. We will discuss how spatial patterns of IAS distribution, driven by their dispersal and influenced by the landscape structure, can directly affect the permeability of the landscape for native species. Understanding these patterns is crucial for effective management, especially when targeting individual IAS to conserve vulnerable habitats and rare native species under protection management. Moreover, we will delve into the direct consequences of IAS on the spread of native organisms and the overall connectivity of forest ecosystems. By comprehensively examining the dispersal patterns of IAS and their impacts on the spread of native species, this chapter seeks to provide valuable insights for effective IAS management in forest ecosystems. Implementing such knowledge is essential for conservation efforts, preserving biodiversity, and maintaining the ecological integrity of forested landscapes, ultimately safeguarding the essential connectivity that supports the health and resilience of forest ecosystems (see also Chap. 16).

Dispersal of IAS in the Context of Forest Connectivity and Fragmentation

The concept of landscape connectivity dates to the 1970s and 1980s and was developed under the inclusion of several key components (Fahrig et al. 2021; see also Chap. 1). Forest connectivity can also be defined as the complement to forest fragmentation (Maes et al. 2023). While forest connectivity is vital for promoting biodiversity and ecosystem functioning, it is very vulnerable when it is not robust, and maintenance of certain ecological processes may not be enabled by increases in landscape connectivity alone (Pelletier et al. 2017). The presence of IAS can impede connectivity between habitat patches for native species by discouraging dispersal through degraded stepping stones or by increasing mortality through predation (Closset-Kopp et al. 2016; Glen et al. 2013). Floodplain forests are among the most severely IAS-affected forest habitat types. In heavily affected forest areas, the influence of IAS leads to changes in species composition and structural diversity and, in extreme cases, to changes in biotic and abiotic site conditions (Dreiss et al. 2016; Langmaier and Lapin 2020).

In a recent synthesis of biological invasion hypotheses associated with the introduction–naturalisation–invasion continuum (Daly et al. 2023), all stages of the invasion process are described in detail. In a forest context, the initial stage involves the transfer of alien species from their native range to a new location. Once introduced, the alien species can establish itself in the new forest environment. This stage involves the species surviving and reproducing with a self-sustaining population in the new location and adapting to the new biotic and abiotic conditions. Once established, the alien species may spread and occupy new forest areas, becoming invasive. Typically, an IAS spreads rapidly to new areas, either through natural dispersal mechanisms such as anemochory, zoochory, or hydrochory or through human activity such as land-use changes, trade, timber transport, or harvesting operations, and artificial corridors are common pathways for spread. This often leads to negative impacts on the invaded native ecosystem. The spread of an alien species can be described as an expansion phase in which the range or invaded area occupied by the species increases over time. Besides stochasticity, the spread is dependent on several biotic and abiotic factors such as a species' reproductive success, localised dispersal of propagules, long-distance dispersal aided by humans and landscape permeability (O'Reilly-Nugent et al. 2016), and carbon dynamics (Fridley et al. 2023). Several stages of the dispersal of alien species have been identified (Blackburn et al. 2011; Wilson et al. 2016), with some peculiarities for specific groups such as alien forest pathogens (Paap et al. 2022). IAS spread differs at various scales (Pyšek et al. 2008), and different species have different dispersal abilities (Zhou et al. 2021).

As mentioned in the introduction of this chapter, forest connectivity (and its complement, forest fragmentation) can both promote or constrain the dispersal of IAS within forests. A summary from a non-systematic literature review of such double-fold effect is reported in Table 17.1. Enhancing connectivity—for example, by establishing forest habitat patches and natural forest corridors—is a common conservation or restoration strategy aimed at maintaining species dispersal and increasing the diversity of native populations (Pirnat 2000; Hilty et al. 2020). At the same time, the presence of both forest corridors and artificial (human-made) corridors can also be one of the most important factors in facilitating the dispersal of IAS (Blackburn et al. 2011, Closset-Kopp et al. 2016). For example, roads, water channels, and electricity and gas lines can easily promote the dispersal of IAS through forests (Deeley and Petrovskaya 2022; Dalu et al. 2023). This is generally the result of multiple processes and factors—for example, the openness of such human-made areas, the disturbance they cause, or accidental transport by people via hitchhiking or bringing in contaminated soil. Forest roads contribute to forest fragmentation and can increase the openness of forests, thereby creating pathways for more light-demanding IAS. For example, studying the invasion of *Ailanthus altissima* in the Fontainebleau Forest, a peri-urban forest of Paris (France), Motard et al. (2011) concluded that *A. altissima* grows best in edge habitats or disturbed sites—i.e. along the edge of roads, railways, gardens, meadows, and riverbanks. From these preferred locations, it can spread through the underwood, where light is not a limiting factor. They also concluded that the *A. altissima* plants detected within the forest stands represented individuals favoured by temporary gaps in the canopy at least 30 years before the study when the respective forest stands were gardens, a tree nursery, or a since-abandoned vegetatively restored sandpit. At that time, the trees had been able to produce a bank of root suckers and establish durably, forming a monospecific canopy that prevents the regeneration of other tree species. *A. altissima* could therefore be a threat, particularly since open habitats as well as natural or anthropogenic gaps in the canopy occur regularly (Motard et al. 2011).

At the landscape scale, invasive alien pathogens generally first colonise areas with continuous forests before eventually spreading to isolated or scattered forest stands or trees. Therefore, a landscape with diverse habitats may provide better resistance to alien pathogen invasions—though a scattered distribution of host trees will not always allow them to escape infection (Rigot et al. 2014). At the same time, natural forests with low levels of fragmentation are generally considered to be less prone to plant invasions than other modified ecosystems (e.g. agricultural systems), whereas forest fragmentation can promote plant invasions. *Pinus radiata* has been an extremely successful invader in diverse ecosystems of the Southern Hemisphere (Richardson et al. 1994), probably because of its rapid maturation, serotiny, resilience to fires, and the high ability of its seeds to disperse by wind. A study in Chile (Bustamante et al. 2003) highlighted the process of invasion in fragmented native forest close to *P. radiata* plantations, while *Acacia melanoxylon* (Gutiérrez et al. 2024) can be a successful invader in fragmented or human-disturbed riparian forests (Fig. 17.1).

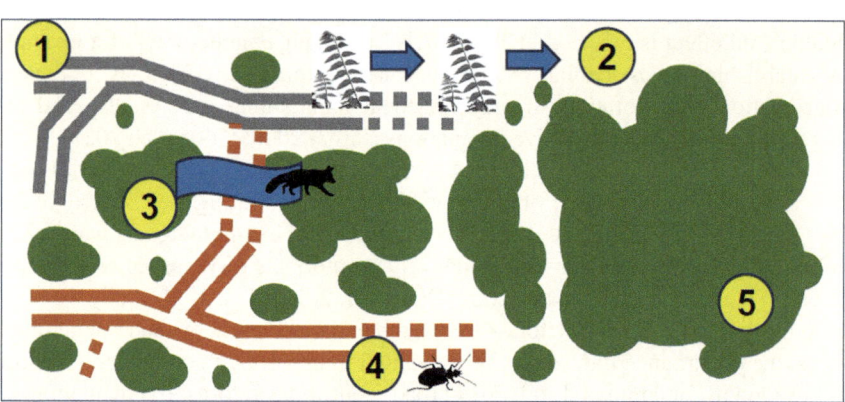

Fig. 17.1 Forest fragmentation is the splitting of large, contiguous forest areas and habitats (5) into smaller pieces of forest (green areas). Typically, these pieces are separated by several types of artificial corridors (e.g. railways (1), continuous and dotted grey lines, and roads (4), continuous and dotted brown lines in the figure), agricultural land, utility corridors, subdivisions, or other human-related land uses. Wildlife crossing (3) allows animals to cross human-made barriers safely. The fragmentation of forests is the main factor limiting their connectivity, and increasing forest connectivity is crucial for supporting biodiversity in forests. Clusters of small forest fragments can act as stepping stones. Railways can promote the spread of invasive alien trees such as *Ailanthus altissima* (2) reaching intact forest edges and roads and the spread of many insects (4) as hitchhikers

In 1999, the pinewood nematode *Bursaphelenchus xylophilus*, a causal agent of pine wilt disease that is native to North America and disperses naturally only with the aid of vector beetles of the genus *Monochamus*, was first detected in Europe—more precisely, in south-western Portugal and later at several sites in Spain and on Madeira Island. Since then, it has spread to more than 30% of Portugal, causing large-scale damage to the country's forests (De la Fuente et al. 2018). In a modelling study, De la Fuente et al. (2018) demonstrated that simulated clear-cut belts could stop the spread of the pinewood nematode only if they were wider than 30 km although thinner belts could delay the invasion. Furthermore, clear-cuts could be more effective in slowing down the invasion when combined with a reduction of the vector beetle population in the adjacent areas through mass trapping as well as with early detection and removal of infected trees. In the absence of effective containment measures, the pinewood nematode may naturally spread into Spain in about 5 years. In less than 10 years, it may reach the major forest and climatic corridors that provide a gateway for subsequent expansion towards the rest of the Iberian Peninsula and, in the longer term, towards other European countries (De la Fuente et al. 2018) (Fig. 17.2).

Fig. 17.2 (a) The colonisation of road verges by pine wildings within a large plantation of *Pinus radiata*. (b) The colonisation of a forest path by *Acacia melanoxylon* (Australian blackwood)

Prevention, Management, and Monitoring of Biological Invasions in Forest Corridors

Given that biological invasions in forests can be promoted or impeded by landscape features such as corridors (both by forest corridors and artificial corridors) and stepping stones, it is essential to incorporate this knowledge into predictions of IAS spread as well as into strategic management. The prevention of introductions and the management of IAS, whenever prevention fails, are fundamental components of any strategy aiming to conserve habitat connectivity in forest ecosystems. The impacts of IAS on connectivity can have serious negative consequences for biodiversity and ecosystem functions. Therefore, restoration efforts and effective management of IAS between habitat patches as well as within them are often necessary to improve connectivity. Importantly, IAS management can enhance landscape connectivity, and its incorporation into conservation planning may help to design optimal reserve networks. Conversely, conservation planning and connectivity modelling can optimise the targeting of IAS to achieve benefits for a wider range of taxa and ecological processes (Glen et al. 2013). The probability of IAS having an impact on forested stepping stones and corridors has been found to be influenced by various factors such as the proximity to urban areas, the age and degree of degradation of the forest stands, and climatic conditions (Basnou et al. 2016; Pino et al. 2013; Tello-García et al. 2021).

The decision to manage known, highly impactful IAS is often rather straightforward. However, prior to both the conservation and restoration of forest connectivity and forest habitat patches or forest corridors and when planning a new artificial corridor, a site-specific risk analysis should ideally be conducted. The aim of such an analysis is to assess the factual relationships between IAS and forest connectivity in the specific context, the potential impact of the present IAS on the functional and structural connectivity of the habitat patch, and the site-specific management needs and options available. Furthermore, if several IAS are present, prioritisation according to specific criteria may need to be conducted (e.g. McGeoch et al. 2016) or several species managed together to achieve the desired conservation or restoration outcomes. Importantly, several organisations, such as the Ministry of Transportation and Infrastructure, in partnership with the Invasive Species Council of British Columbia, provide invasive plant best practices from roadside maintenance operations. By applying these best practices, maintenance contractors can limit the introduction and spread of invasive plants. Besides employing the correct methods and techniques, an overall strategy based on landscape dynamics and expected spatial patterns can be fundamental to achieving success. This approach has been applied in the control of *Acacia dealbata* in a Natura 2000 site in Portugal, where Machado et al. (2022) showed that removing the patches with higher perimeter-to-area ratios (mostly small satellite patches) would be more impactful than removing larger patches or random patches with intermediate perimeter-to-area ratios first. Following this approach based on landscape dynamics, the employment of a connectivity assessment resulted in an ordered list of patches to remove sequentially (Machado et al. 2022).

If prevention fails and an alien species with the potential to become invasive has been detected in a habitat patch in a forest or close to a forest, prompt action is crucial. Outbreaks of serious or significant IAS require strategic-level plans ideally developed at a national level (contingency plans) that describe the overall aim and high-level objectives to be achieved and set out the response strategy to either eradicate or contain the IAS. Establishing a national or local action plan may be very time-consuming and should follow a standard procedure, for example, as suggested by the European and Mediterranean Plant Protection Organization (EPPO) in the standard PM 9/10 (1) (EPPO 2009). IAS eradication, control, or containment can be achieved using various methods, which can be employed and integrated into a dedicated management plan; each method has its own advantages and drawbacks depending on the biology and ecology of the IAS and the invaded landscape. For example, EPPO PM 9/29 (1) on *A. altissima* describes the control procedures aiming to monitor, contain, and eradicate *A. altissima* in the entire EPPO region (EPPO 2020). For small infestations, physical removal by manual uprooting or cutting of alien plants can be highly effective. However, for larger infested areas, the application of chemical, mechanical, and/or biological control measures is necessary to address the widespread presence of IAS. One straightforward example of the integration of methods is provided by Chabrerie et al. (2007) on *Prunus serotina*, where patch mosaic functional types (areas showing the same response to a plant invasion in a heterogeneous forest landscape) are used to predict invasion patterns in a forest landscape and produce a tailored management strategy that includes monitoring safe areas; extending cutting rotations; harvesting recently colonised stands tree by tree; promoting a multi-layered understorey vegetation; cutting down reproducing alien trees; favouring fast-growing, shade-tolerant native tree species; removing alien trees at the leading edge; and conducting soil enrichment or irrigation in heavily invaded areas.

Biological control methods (see also Chap. 17) involve the introduction of specialised natural enemies of an IAS such as insects or pathogens with the aim of reducing the population size or impeding reproduction of the IAS (Kenis et al. 2017). *Acacia longifolia* is native to Australia and was introduced into Europe between the late-nineteenth and early-twentieth centuries. Since its introduction in

Portugal, the species has become one of the most widespread IAS. It forms extensive populations within coastal ecosystems that displace native plant communities. Due to similar negative impacts recorded throughout its introduced range, the species has been the target of classical biological control using the Australian gall-forming wasp *Trichilogaster acaciaelongifoliae*. This biocontrol agent had previously been successfully employed in South Africa and was released in Portugal in 2015 (Dinis et al. 2020). Prior to releasing any biocontrol agents, a thorough assessment of risks and potential non-target effects is conducted, and regulatory approval is needed before implementation (e.g. EFSA Panel on Plant Health 2015 for *Trichilogaster*). Although biocontrol is not a full eradication of the target species, reducing its abundance may have an important effect on the native plant community and decrease the negative impact of the invader. Similarly, the use of chemical treatments, including herbicides or pesticides, should be approached with extreme caution. When chemical agents (PPPs—plant protection products) are used, they can pose risks to non-target species and the broader environment, as well as to the workers applying them. Therefore, their application should be limited, highly localised, and in line with national legislation on plant protection product (PPP) use. However, PPPs are often necessary since many woody species are able to re-sprout promptly, and thus mechanical methods alone are not effective.

Although one size does not fit all, managing heterogeneity in the landscape can have a positive effect on biodiversity. Similarly, ecological disturbances such as fire are important drivers of landscape heterogeneity that can promote diversity and create habitat structures required by certain species (Johnstone et al. 2016). International cooperation and the sharing of information and best practices are very important for the effective management of IAS in forests. The Food and Agriculture Organization (FAO) of the United Nations has helped to establish regional networks dedicated to the issue of forest pests—primarily forest IAS—and the forest sector. These networks aim to facilitate the exchange of information and mobilisation of resources; raise regional awareness; and act as links between experts, institutions, networks, and other stakeholders concerned with IAS in forests (https://www.fao.org/forestry-fao/pests/94102/en/). Furthermore, FAO provides guidelines for reducing the risk of invasive alien species in forest plantations (FAO 2006).

It should be remarked that forest patches, forest corridors, and artificial corridors, all might represent an opportunity for citizen science campaigns, i.e. for the collection and analysis of data relating to the natural world by members of the general public, typically as a part of collaborative projects with professional scientists. To fully qualify as citizen science, a project must not only rely on volunteers who participate in the detection process but also include the use of any number of tools (e.g. smartphone apps, collaborative databases, eDNA, or other technology). Using citizen science for early detection of invasive species in forests should always be an option in integrated strategies to tackle the issue. In recent times, citizen science has become possible at large scales due to the development of collaborative technology, social media and networking, and publicly accessible databases offering opportunities for anyone to participate in ecological research (Larson et al. 2020). However, not all IAS are easily surveyable by citizen science as some of them may be difficult to identify or can be found only in habitats less frequently scouted.

Furthermore, over the past decade, remote sensing has provided many important contributions to the progress of invasion science, improving our understanding of the drivers, processes, patterns, and impacts of alien species in all ecosystems (Vaz et al. 2019; Müllerová et al. 2023). For example, instruments such as light detection and ranging (LiDAR) and hyperspectral sensors are used to quantify forest characteristics at the stand-to-landscape level (Massey et al. 2023) more frequently and in combination with other methods of detecting and monitoring IAS or forest pests (Brockerhoff et al. 2023). The endemic Chilean tree *Araucaria araucana* (monkey puzzle tree) is classified as endangered in the International Union for Conservation of Nature (IUCN) Red List of Threatened Species due to its decreasing population. In addition, it is considered a natural monument under Chilean law. The spread of propagules from alien forest plantations to surrounding native forests has been documented, with competition from alien saplings threatening the regeneration of endangered *A. araucana*. In fact, using freely available medium-resolution Sentinel-2 optical satellite imagery, Martin-Gallego et al. (2020) monitored alien trees wilding from plantations within the Chilean Valdivian temperate forest, whose extent and topography limit traditional ground-based methods, achieving high levels of mapping detail and accuracy (Figs. 17.3 and 17.4).

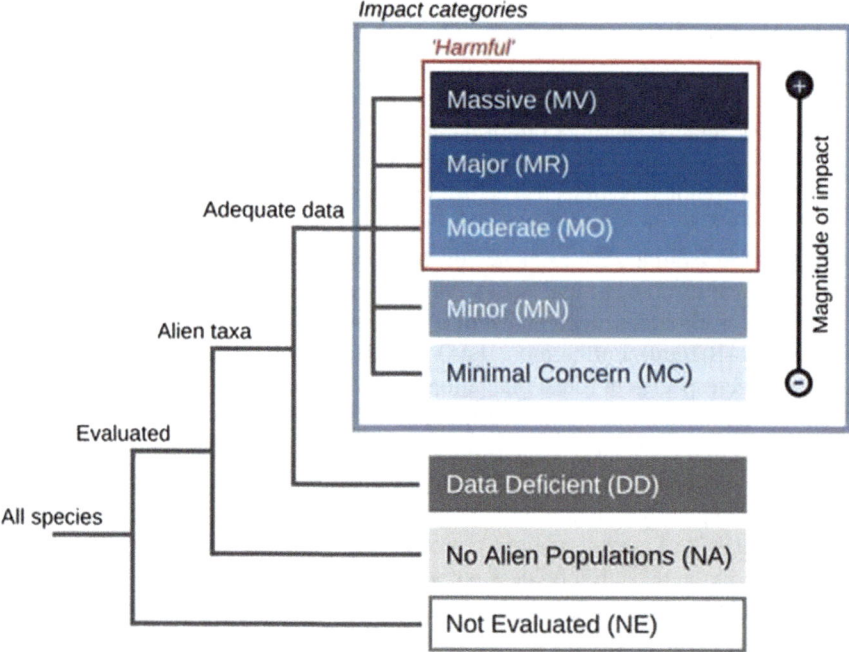

Fig. 17.3 There are eight categories for classifying taxa. The first five, known as "impact" categories, describe the increasing levels of harm caused by alien species on native biota: MC indicates negligible impact; MN suggests reduced performance; MO indicates population decline; MR denotes local extinction with potential for reversal; and MV indicates irreversible extinction and significant community structure change. DD is used when alien populations exist, but no evidence of impact is available, and the impact level is unknown; NA is used when there is no evidence of individuals outside of captivity or cultivation beyond a species' native range; and NE refers to unevaluated taxa. (IUCN 2020)

Fig. 17.4 The invasive Australian gall-forming wasp *Trichilogaster acaciaelongifoliae*. (Photo:Vuk Vojisavljevic/ iNaturalist/Flickr)

Box 17.1 Impact Assessment and Risk Analysis

Identifying the most harmful or potentially harmful alien species for the ecological connectivity of a forest is a crucial step for effective management. It is therefore highly recommended to conduct standardised impact assessments for the alien species present at a site, ideally with site-specific information on impacts.

The Environmental Impact Classification for Alien Taxa (EICAT), a standardised scoring system that classifies alien species according to the severity of their environmental impact in recipient areas, has been adopted by the International Union for Conservation of Nature (IUCN). This system considers the level of biological organisation impacted with regard to native species, as well as the potential reversibility of the impact (Hawkins et al. 2015; Kumschick et al. 2017, 2020a). EICAT offers an objective and transparent way of categorising alien species based on the degree of harm they cause to the environment in the areas they invade. Evidence of the negative effects of these species on native organisms in their introduced range is used to classify them into one of five impact categories ranging from no impact on native individuals to irreversible local population extinctions. Additionally, EICAT includes a mechanism for classifying alien species based on the specific ways in which they cause harm (IUCN 2020).

Furthermore, site-specific risk analyses including information on impacts as well as on invasion potential and management aspects are recommended to

(continued)

reach the most suitable management decisions (e.g. Kumschick et al. 2020b; Booy et al. 2017). Such analyses are key for determining which species pose the greatest threat to a forest's ecological integrity and connectivity. By conducting a thorough evaluation, managers can develop effective strategies to mitigate the negative impacts of IAS and preserve the health of the forest ecosystem (Bindewald et al. 2021).

References

Alharbi W, Petrovskii S (2019) Effect of complex landscape geometry on the invasive species spread: invasion with stepping stones. J Theor Biol 464:85–97

Andriamparany R, Lundberg J, Pyykönen M, Wurz S, Elmqvist T (2020) The effect of introduced Opuntia (Cactaceae) species on landscape connectivity and ecosystem service provision in southern Madagascar. In: Gasparatos A et al (eds) Sustainability challenges in sub-Saharan Africa II. Science for sustainable societies. Springer, Singapore. https://doi.org/10.1007/978-981-15-5358-5_6

Archer E, Dziba LE, Mulongoy KJ, Maoela MA, Walters M, Biggs R, Cormier-Salem M-C, DeClerck F, Diaw MC, Dunham AE, Failler P, Gordon C, Harhash KA, Kasisi R, Kizito F, Nyingi WD, Oguge N, Osman-Elasha B, Stringer LC, Tito de Morais L, Assogbadjo A, Egoh BN, Halmy MW, Heubach K, Mensah A, Pereira L, Sitas N (eds) (2018) Summary for policymakers of the regional assessment report on biodiversity and ecosystem Services for Africa of the intergovernmental science-policy platform on biodiversity and ecosystem services. IPBES Secretariat, Bonn

Basnou C, Vicente P, Espelta JM, Pino J (2016) Of niche differentiation, dispersal ability and historical legacies: what drives woody community assembly in recent Mediterranean forests? Oikos 125(1):107–116

Bindewald A, Brundu G, Schueler S, Starfinger U, Bauhus J, Lapin K (2021) Site-specific risk assessment enables trade-off analysis of non-native tree species in European forests. Ecol Evol 11(24):18089–18110

Biswas SR, Biswas PL, Limon SH, Yan E-R, Xu M-S, Khan MSI (2018) Plant invasion in mangrove forests worldwide. For Ecol Manage 429:480–492. https://doi.org/10.1016/j.foreco.2018.07.046

Blackburn TM, Pyšek P, Bacher S, Carlton JT, Duncan RP, Jarošík V, Wilson JRU, Richardson DM (2011) A proposed unified framework for biological invasions. Trends Ecol Evol 26(7):333–339

Booy O, Mill AC, Roy HE, Hiley A, Moore N, Robertson P, Baker S, Brazier M, Bue M, Bullock R, Campbell S, Eyre D, Foster J, Hatton-Ellis M, Long J, Macadam C, Morrison-Bell C, Mumford J, Newman J, Parrott D, Payne R, Renals T, Rodgers E, Spencer M, Stebbing P, Sutton-Croft M, Walker KJ, Ward A, Whittaker S, Wyn G (2017) Risk management to prioritise the eradication of new and emerging invasive non-native species. Biol Invasions 19:2401–2417. https://doi.org/10.1007/s10530-017-1451-z

Brockerhoff EG, Corley JC, Jactel H, Miller DR, Rabaglia RJ, Sweeney J (2023) Monitoring and surveillance of forest insects. In: Allison D, Paine TD, Slippers B, Wingfield MJ (eds) Forest entomology and pathology. Springer, Cham. https://doi.org/10.1007/978-3-031-11553-0_19

Bustamante RO, Serey IA, Pickett STA (2003) Forest fragmentation, plant regeneration and invasion processes across edges in Central Chile. In: Bradshaw GA, Marquet PA (eds) How landscapes change. Ecological studies, vol 162. Springer, Berlin. https://doi.org/10.1007/978-3-662-05238-9_9

Chabrerie O, Roulier F, Hoeblich H, Sebert-Cuvillier E, Closset-Kopp D, Leblanc I, Jaminon J, Decocq G (2007) Defining patch mosaic functional types to predict invasion patterns in a forest landscape. Ecol Appl 17:464–481. https://doi.org/10.1890/06-0614

Closset-Kopp D, Wasof S, Decocq G (2016) Using process-based indicator species to evaluate ecological corridors in fragmented landscapes. Biol Conserv 201:152–159

Dalu T, Stam EM, Ligege MO, Cuthbert RN (2023) Highways to invasion: powerline servitudes as corridors for alien plant invasions. Afr J Ecol 61:379–388. https://doi.org/10.1111/aje.13121

Daly EZ, Chabrerie O, Massol F, Facon B, Hess MCM, Tasiemski A, Grandjean F, Chauvat M, Viard F, Forey E, Folcher L, Buisson E, Boivin T, Baltora-Rosset S, Ulmer R, Gibert P, Thiébaut G, Pantel JH, Heger T, Richardson DM, Renault D (2023) A synthesis of biological invasion hypotheses associated with the introduction–naturalisation–invasion continuum. Oikos 2023:e09645. https://doi.org/10.1111/oik.09645

de Groot M, Schueler S, Sallmannshofer M, Virgillito C, Kovacs G, Cech T, Božič G, Ogris N, Hoch G, Kavčič A (2022) Forest management, site characteristics and climate change affect multiple biotic threats in riparian forests. For Ecol Manage 508:120041

de la Fuente B, Saura S, Beck PSA (2018) Predicting the spread of an invasive tree pest: the pine wood nematode in southern Europe. J Appl Ecol 55(5):2374–2385. https://doi.org/10.1111/1365-2664.13177

Deeley B, Petrovskaya N (2022) Propagation of invasive plant species in the presence of a road. J Theor Biol 548:111196. https://doi.org/10.1016/j.jtbi.2022.111196

Devenish-Nelson ES, Weidemann D, Townsend J et al (2019) Patterns in Island endemic forest-dependent bird research: the Caribbean as a case-study. Biodivers Conserv 28:1885–1904. https://doi.org/10.1007/s10531-019-01768-x

Dinis M, Vicente JR, César de Sá N, López-Núñez FA, Marchante E, Marchante H (2020) Can niche dynamics and distribution modelling predict the success of invasive species management using biocontrol? Insights from *Acacia longifolia* in Portugal. Front Ecol Evol. https://doi.org/10.3389/fevo.2020.576667

Dreiss LM, Volin JC, Closset-Kopp D, Chabrerie O, Valentin B, Delachapelle H, Decocq G, Essl F, Milasowszky N, Dirnböck T, Höfle R, Dullinger S, Essl F, Chmura D, Sierka E, Braun M, Schindler S, Essl F, Campagnaro T et al (2016) Plant invasions in temperate forests: resistance or ephemeral phenomenon? For Ecol Manage 15(1):120–130. https://doi.org/10.1016/j.jenvman.2011.01.025

Dukes JS, Mooney HA (1999) Does global change increase the success of biological invaders? Trends Ecol Evol 14(4):135–139

EFSA PLH Panel (EFSA Panel on Plant Health) (2015) Risk to plant health in the EU territory of the intentional release of the bud-galling wasp *Trichilogaster acaciaelongifoliae* for the control of the invasive alien plant *Acacia longifolia*. EFSA J 13(4):4079., 48 p. https://doi.org/10.2903/j.efsa.2015.4079

EPPO (2009) PM 9/10(1): Generic elements for contingency plans. EPPO Bulletin 39:471–474. https://doi.org/10.1111/j.1365-2338.2009.02332.x

Eppley TM, Donati G, Ramanamanjato J-B, Randriatafika F, Andriamandimbiarisoa LN, Rabehevitra D, Ravelomanantsoa R, Ganzhorn JU (2015) The use of an invasive species habitat by a small Folivorous primate: implications for Lemur conservation in Madagascar. PLoS One 10(11):e0140981. https://doi.org/10.1371/journal.pone.0140981

European Commission, Directorate-General for Environment, Harrower C, Scalera R, Pagad, Schönrogge K, Roy HE (2020) Guidance for interpretation of the CBD categories of pathways for the introduction of invasive alien species. Publications Office., https://data.europa.eu. https://doi.org/10.2779/6172

Fahrig L, Arroyo-Rodríguez V, Cazetta E, Ford A, Lancaster J, Ranius T (2021) Landscape connectivity. The Routledge handbook of landscape ecology, Fahrig, pp 67–88. https://doi.org/10.4324/9780429399480-5

FAO (2006) Responsible management of planted forests: voluntary guidelines. Planted forests and trees working paper 37/E. Rome (also available at Thwww.fao.org/forestry/site/10368/en)

Fazlioglu F, Chen L (2020) Introduced non-native mangroves express better growth performance than co-occurring native mangroves. Sci Rep 10(1):3854. https://doi.org/10.1038/s41598-020-60454-z

Fergus C, Lacher IL, Herrmann V, McShea WJ, Akre TS (2023) Predicting vulnerability of forest patches to invasion by non-native plants for landscape scale management. Ecol Appl 33(5):e2857. https://doi.org/10.1002/eap.2857

Fridley JD, Bellingham PJ, Closset-Kopp D, Daehler CC, Dechoum MS, Martin PH, Murphy HT, Rojas-Sandoval J, Tng D (2023) A general hypothesis of forest invasions by woody plants based on whole-plant carbon economics. J Ecol 111:4–22. https://doi.org/10.1111/1365-2745.14001

Glen AS, Pech RP, Byrom AE (2013) Connectivity and invasive species management: towards an integrated landscape approach. Biol Invasions 15:2127–2138

Gutiérrez J, Altamirano A, Pauchard A, Meli P (2024) Proximity to forest plantations is associated with presence and abundance of invasive plants in landscapes of south-central Chile. NeoBiota 92:129–153. https://doi.org/10.3897/neobiota.92.112164

Hawkins CL, Bacher S, Essl F, Hulme PE, Jeschke JM, Kühn I, Kumschick S, Nentwig W, Pergl J, Pyšek P (2015) Framework and guidelines for implementing the proposed IUCN environmental impact classification for Alien Taxa (EICAT). Divers Distrib 21(11):1360–1363

Hilty J, Worboys GL, Keeley A, Woodley S, Lausche B, Locke H, ... Tabor GM (2020) Guidelines for conserving connectivity through ecological networks and corridors. Best practice protected area guidelines series, vol 30, p 122

IPBES (2023) Summary for policymakers of the thematic assessment report on invasive alien species and their control of the intergovernmental science-policy platform on biodiversity and ecosystem services. Roy HE, Pauchard A, Stoett P, Renard Truong T, Bacher S, Galil BS, Hulme PE, Ikeda T, Sankaran KV, McGeoch MA, Meyerson LA, Nuñez MA, Ordonez A, Rahlao SJ, Schwindt E, Seebens H, Sheppard AW, Vandvik V (eds) IPBES secretariat, Bonn, Germany. https://doi.org/10.5281/zenodo.7430692

IUCN (2020) IUCN EICAT categories and criteria. The environmental impact classification for Alien Taxa (EICAT). IUCN Gland, Switzerland and Cambridge, UK

Johnstone JF, Allen CD, Franklin JF, Frelich LE, Harvey BJ, Higuera PE et al (2016) Changing disturbance regimes, ecological memory, and forest resilience. Front Ecol Environ 14(7):369–378

Kenis M, Auger-Rozenberg MA, Roques A et al (2009) Ecological effects of invasive alien insects. Biol Invasions 11:21–45. https://doi.org/10.1007/s10530-008-9318-y

Kenis M, Hurley BP, Hajek AE et al (2017) Classical biological control of insect pests of trees: facts and figures. Biol Invasions 19:3401–3417. https://doi.org/10.1007/s10530-017-1414-4

Kumschick S, Measey GJ, Vimercati G, De Villiers FA, Mokhatla MM, Davies SJ, Thorp CJ, Rebelo AD, Blackburn TM, Kraus F (2017) How repeatable is the environmental impact classification of Alien Taxa (EICAT)? Comparing independent global impact assessments of amphibians. Ecol Evol 7(8):2661–2670

Kumschick S, Bacher S, Bertolino S, Blackburn TM, Evans T, Roy HE, Smith K (2020a) Appropriate uses of EICAT protocol, data and classifications. NeoBiota 62:193–212. https://doi.org/10.3897/neobiota.62.51574

Kumschick S, Wilson JRU, Foxcroft LC (2020b) A framework to support alien species regulation: the risk analysis for Alien Taxa (RAAT). NeoBiota 62:213–239. https://doi.org/10.3897/neobiota.62.51031

Langmaier M, Lapin K (2020) A systematic review of the impact of invasive alien plants on forest regeneration in European temperate forests. Front Plant Sci 11:1349

Lapin K, Bacher S, Cech T, Damjanić R, Essl F, Georges F-I, Hoch G, Kavčič A, Koltay A, Kostić S, Lukić I, Marinšek A, Nagy L, Agbaba SN, Oettel J, Orlović S, Poljaković-Pajnik L, Sallmannshofer M, Steinkellner M, Stojnic S, Westergren M, Zlatkovic M, Zolles A, de Groot M (2021) Comparing environmental impacts of alien plants, insects and pathogens in protected riparian forests. NeoBiota 69:1–28. https://doi.org/10.3897/neobiota.69.71651

Larson ER, Graham BM, Achury R, Coon JJ, Daniels MK, Gambrell DK, Jonasen KL, King GD, LaRacuente N, Perrin-Stowe TI, Reed EM, Rice CJ, Ruzi SA, Thairu MW, Wilson JC, Suarez

AV (2020) From eDNA to citizen science: emerging tools for the early detection of invasive species. Front Ecol Environ 18(4):194–202. https://doi.org/10.1002/fee.2162

Lawrence A, O'Connor K, Haroutounian V, Swei A (2018) Patterns of diversity along a habitat size gradient in a biodiversity hotspot. Ecosphere 9(4):e02183. https://doi.org/10.1002/ecs2.2183

Liebhold AM, MacDonald WL, Bergdahl D, Mastro VC (1995) Invasion by exotic forest pests: a threat to forest ecosystems. For Sci 41(suppl_1):a0001–z0001. https://doi.org/10.1093/forestscience/41.s1.a0001

Machado R, Neto Duarte L, Gil A, Sousa-Neves N, Pirnat J, Santos P (2022) Supporting the spatial management of invasive alien plants through assessment of landscape dynamics and connectivity. Restor Ecol 30:e13592. https://doi.org/10.1111/rec.13592

Maes J, Bruzón AG, Barredo JI, Vallecillo S, Vogt P, Rivero IM, Santos-Martín F (2023) Accounting for forest conditions in Europe based on an international statistical standard. Nat Commun 14(1):3723. https://doi.org/10.1038/s41467-023-39434-0

Martin-Gallego P, Aplin P, Marston C, Altamirano A, Pauchard A (2020) Detecting and modelling alien tree presence using Sentinel-2 satellite imagery in Chile's temperate forests. For Ecol Manage 474:118353. https://doi.org/10.1016/j.foreco.2020.118353

Massey R, Berner LT, Foster AC, Goetz SJ, Vepakomma U (2023) Remote sensing tools for monitoring forests and tracking their dynamics. In: Girona MM, Morin H, Gauthier S, Bergeron Y (eds) Boreal forests in the face of climate change. Advances in global change research, vol 74. Springer, Cham. https://doi.org/10.1007/978-3-031-15988-6_26

McGeoch MA, Genovesi P, Bellingham PJ et al (2016) Prioritizing species, pathways, and sites to achieve conservation targets for biological invasion. Biol Invasions 18:299–314. https://doi.org/10.1007/s10530-015-1013-1

McNeely JA, Mooney HA, Neville LE, Schei PJ, Waage JK (2001) Global strategy on invasive alien species. IUCN. 50 p

Mortensen DA, Rauschert ES, Nord AN, Jones BP (2009) Forest roads facilitate the spread of invasive plants. Invas Plant Sci Manag 2(3):191–199

Motard E, Muratet A, Clair-Maczulajtys D, Machon N (2011) Does the invasive species Ailanthus altissima threaten floristic diversity of temperate peri-urban forests? C R Biol 334(12):872–879

Müllerová J, Brundu G, Große-Stoltenberg A et al (2023) Pattern to process, research to practice: remote sensing of plant invasions. Biol Invasions 25:3651–3676. https://doi.org/10.1007/s10530-023-03150-z

O'Reilly-Nugent A, Palit R, Lopez-Aldana A, Medina-Romero M, Wandrag E, Duncan RP (2016) Landscape effects on the spread of invasive species. Curr Landsc Ecol Rep 1(3):107–114. https://doi.org/10.1007/s40823-016-0012-y

Paap T, Wingfield MJ, Burgess TI et al (2022) Invasion frameworks: a forest pathogen perspective. Curr Forestry Rep 8:74–89. https://doi.org/10.1007/s40725-021-00157-4

Pelletier D, Lapointe M-EÂ, Wulder MA, White JC, Cardille JA (2017) Forest connectivity regions of Canada using circuit theory and image analysis. PLoS One 12(2):e0169428. https://doi.org/10.1371/journal.pone.0169428

Pino J, Arnan X, Rodrigo A, Retana J (2013) Post-fire invasion and subsequent extinction of *Conyza* spp. in Mediterranean forests is mostly explained by local factors. Weed Res 53(6):470–478

Pirnat J (2000) Conservation and management of forest patches and corridors in suburban landscapes. Landsc Urban Plan 52(2):135–143. https://doi.org/10.1016/S0169-2046(00)00128-6

Pyšek P, Jarošík V, Müllerová J, Pergl J, Wild J (2008) Comparing the rate of invasion by *Heracleum mantegazzianum* at continental, regional, and local scales. Divers Distrib 14:355–363

Richardson DM, Williams PA, Hobbs RJ (1994) Pine invasions in the southern hemisphere: determinants of spread and invadability. J Biogeogr 21(5):511–527. https://doi.org/10.2307/2845655

Riegl B, Walentowitz A, Sevilla C, Chango R, Jäger H (2023) Invasive blackberry outcompetes the endemic Galapagos tree daisy *Scalesia pedunculata*. Ecol Appl 33(4):e2846. https://doi.org/10.1002/eap.2846

Rigot T, van Halder I, Jactel H (2014) Landscape diversity slows the spread of an invasive forest pest species. Ecography 37:648–658. https://doi.org/10.1111/j.1600-0587.2013.00447.x

Roy BA, Alexander HM, Davidson J, Campbell FT, Burdon JJ, Sniezko R, Brasier C (2014) Increasing forest loss worldwide from invasive pests requires new trade regulations. Front Ecol Environ 12(8):457–465. https://doi.org/10.1890/130240

Simberloff D, Martin J-L, Genovesi P, Maris V, Wardle DA, Aronson J, Courchamp F, Galil B, García-Berthou E, Pascal M (2013) Impacts of biological invasions: what's what and the way forward. Trends Ecol Evol 28(1):58–66

Tello-García E, Gamboa-Badilla N, Álvarez E, Fuentes L, Basnou C, Espelta JM, Pino J (2021) Plant species surplus in recent peri-urban forests: the role of forest connectivity, species' habitat requirements and dispersal types. Biodivers Conserv 30:365–384

Vaz AS, Castro-Díez P, Godoy O, Alonso Á, Vilà M, Saldaña A, Marchante H, Bayón Á, Silva JS, Vicente JR (2018) An indicator-based approach to analyse the effects of non-native tree species on multiple cultural ecosystem services. Ecol Indic 85:48–56

Vaz AS, Alcaraz-Segura D, Vicente JR, Honrado JP (2019) The many roles of remote sensing in invasion science. Front Ecol Evol 7. https://www.frontiersin.org. https://doi.org/10.3389/fevo.2019.00370

Vilà M, Espinar JL, Hejda M, Hulme PE, Jarošík V, Maron JL, Pergl J, Schaffner U, Sun Y, Pyšek P (2011) Ecological impacts of invasive alien plants: a meta-analysis of their effects on species, communities and ecosystems. Ecol Lett 14(7):702–708

Wilson JRU, García-Díaz P, Cassey P, Richardson DM, Pyšek P, Blackburn TM (2016) Biological invasions and natural colonisations are different–the need for invasion science. NeoBiota 31:87–98

Zhou Q, Wu J, Cui X et al (2021) Geographical distribution of the dispersal ability of alien plant species in China and its socio-climatic control factors. Sci Rep 11:7187. https://doi.org/10.1038/s41598-021-85934-8

Open Access This chapter is licensed under the terms of the Creative Commons Attribution 4.0 International License (http://creativecommons.org/licenses/by/4.0/), which permits use, sharing, adaptation, distribution and reproduction in any medium or format, as long as you give appropriate credit to the original author(s) and the source, provide a link to the Creative Commons license and indicate if changes were made.

The images or other third party material in this chapter are included in the chapter's Creative Commons license, unless indicated otherwise in a credit line to the material. If material is not included in the chapter's Creative Commons license and your intended use is not permitted by statutory regulation or exceeds the permitted use, you will need to obtain permission directly from the copyright holder.

Ecological Connectivity in Urban and Semi-Urban Forests

18

Andrea Kodym, Katharina Lapin, and Debashis Sanyal

Urban forest park in British Columbia, USA (Photo: Andrew/Adobe Stock)

A. Kodym (✉) · K. Lapin
Department of Forest Biodiversity & Nature Conservation, Austrian Research Centre for Forests, Vienna, Austria
e-mail: andrea.kodym@bfw.gv.at; katharina.lapin@bfw.gv.at

D. Sanyal
National Institute of Technology, Raipur, India
e-mail: debashissanyal@rediffmail.com

© The Author(s) 2025
K. Lapin et al. (eds.), *Ecological Connectivity of Forest Ecosystems*,
https://doi.org/10.1007/978-3-031-82206-3_18

Abstract

The term 'green infrastructure' (GI) refers to a network of natural and semi-natural areas designed to provide ecosystem services. Urban green infrastructure focuses on green spaces within cities, including parks, gardens, forests, and water elements (blue infrastructure). It offers numerous benefits such as enhancing biodiversity, mitigating the urban heat island effect, acting as a carbon sink, improving air quality, aiding in stormwater management, and promoting physical and mental well-being. Urban green infrastructure also plays a crucial role in supporting regional habitat connectivity and biodiversity conservation.

Urban forests play a crucial role in urban green spaces, acting as vital connectors between rural and urban areas. They serve as stepping stones and corridors for species movement, offering shelter, nesting sites, and foraging opportunities for a diverse range of organisms. However, ensuring successful ecological connectivity requires robust community engagement. Public awareness, education, and active participation in conservation efforts are essential for implementing and maintaining connectivity measures.

Spatial urban planning encounters challenges in balancing social needs, financial expectations, and environmental sustainability. Key issues include identifying and designating ecological corridors and networks, integrating ecological principles into spatial planning frameworks, and leveraging technologies such as geographic information systems (GIS) and remote sensing for green space mapping and assessment. Restoring urban forest habitat patches is crucial for supporting species' survival and migration. Conservation efforts should consider elements like trees outside of forests and spontaneous vegetation to enhance connectivity.

Despite the benefits, managing urban green infrastructure comes with risks. These include biotic homogenisation, the introduction of non-native species, edge effects, human disturbances, lack of diversity, and pressure from increased housing density, all of which can negatively impact biodiversity. Various planning tools and strategies are available to mitigate these risks and ensure the successful implementation of ecological connectivity in urban green infrastructure.

These strategies include incorporating green infrastructure into urban master plans, establishing protected areas and wildlife corridors, promoting sustainable land-use practices, and involving local communities and stakeholders in decision-making processes. By considering different scales and site characteristics, cities can achieve effective management of ecological connectivity and create sustainable and resilient environments.

Keywords

Biodiversity conservation · Ecological connectivity · Ecosystem services · Green infrastructure · Urban forestry · Urban forest management · Urban green spaces

Introduction

Green infrastructure (GI) refers to 'a strategically planned network of natural and semi-natural areas with other environmental features, designed and manged to deliver a wider range of ecosystem services' (European Commission 2023). By this definition, it includes both green (terrestrial) and blue (aquatic) spaces. However, some authors prefer to use the term green-blue infrastructure (GBI) when referring to both types together. In the terrestrial context, GI can be found in both rural and urban settings. Specifically, urban green infrastructure (UGI) is a system designed to offer numerous benefits to both the environment and urban residents (Hansen and Pauleit 2014). It includes networks of urban forests and public parks, sports fields, linear features like street trees, and stepping stone corridors like private gardens, green roofs, and green walls. Together, these elements contribute to creating a connected and vibrant urban ecosystem (Badiu et al. 2019).

Urban land can be categorised into remnant native landscapes, managed horticultural landscapes, and abandoned ruderal landscapes. Zipperer et al. (1997) refer to these as remnant patches (which have an understory), managed patches (which have a grass cover) and emergent patches; Whitney (1985) calls them residual, managed, and ruderal. These diverse green spaces play a vital role in creating sustainable and liveable cities. They support wildlife habitats and help to enhance biodiversity, mitigate the urban heat island effect, reduce noise and pollution levels, and promote physical and mental well-being. Green spaces also aid in stormwater management, reducing the risk of flooding and enhancing water quality by filtering pollutants. In addition, these areas contribute to the improvement of air quality by capturing and filtering harmful pollutants and particulate matter (Gómez-Baggethun and Barton 2013; Integrated and Review 2020; Liu et al. 2020; Monteiro et al. 2020). Finally, they act as carbon sinks, mitigating the effects of climate change by absorbing and storing carbon dioxide (Skole et al. 2021).

Urban green infrastructure is increasingly recognised as a vital component for fostering regional habitat connectivity and supporting biodiversity. Traditionally, the planning and development of urban green infrastructure has primarily focused on improving public health by offering recreational and aesthetic value and enhancing the overall quality of urban environments for the growing population. Over the past decade, however, there has been a notable shift in the understanding and appreciation of the role that urban green infrastructure plays in promoting and safeguarding biodiversity (Filazzola et al. 2019). This evolving awareness has focused attention on the concept of UGI, which encompasses the integration of both natural and built elements to enhance habitat connectivity within urban areas (Rusche et al. 2019). By strategically incorporating green spaces along with blue elements like rivers, streams, and wetlands, cities can create networks that facilitate the movement of species, ensuring their access to essential resources and increasing their chances of survival.

In this chapter, we delve into the significance of urban green spaces with a particular focus on urban forests as key contributors to habitat connectivity in urban areas. Furthermore, we explore the manifold benefits and potential risks for biodiversity associated with urban forests. We also present a comprehensive array of planning tools and innovative strategies aimed at effectively implementing ecological connectivity in the development and maintenance of urban green infrastructure.

Connectivity of Urban and Peri-Urban Forests

Urban and peri-urban forestry (UPF) involves an integrated, interdisciplinary, participatory, and strategic approach to planning and managing tree resources in and around cities for their economic, environmental, and sociocultural benefits (FAO 2017). Recent evidence challenges the common perception that cities feature low biodiversity and highlights the importance of urban green spaces for supporting biodiversity (Aronson et al. 2017). Urban and peri-urban forests act as the foundations of green infrastructure, helping to mitigate the environmental effects of cities. Despite being highly developed and built-up, urban and semi-urban areas still maintain a connection. The presence of green spaces such as parks, gardens, and remnants of natural areas allows for some level of connectivity and interaction between urban environments and forest ecosystems. This connectivity can support the movement of species, facilitate ecological processes, and contribute to the overall biodiversity conservation efforts in urban areas. Urban forests serve as vital stepping stones and provide valuable shelter, nesting sites, and foraging opportunities for a wide range of plant and animal species. Evidence of this has been found, for example, in bird diversity, which is linked to the size and habitat quality of urban parks (Fernandez-Juricic and Jokimaki 2001).

> **Box 18.1 Definition of Urban Forests—From Large to Tiny**
> **The term 'urban forest' refers to all the trees located in urban and peri-urban areas, ranging from woodlands and large and small forests to groups of trees in parks and gardens or along streets, as well as dispersed individual trees.**
>
> The definition of urban forests can vary depending on the scale and context. Some of the common categories include the following:
>
> **Urban forest:** An urban forest typically means a large-scale forested area within an urban setting such as a city or metropolitan region. It may include public parks, nature reserves, green belts, or other extensive wooded areas. Urban forests often comprise a mix of tree species, understory plants, and wildlife habitats, providing recreational opportunities and promoting ecological balance within urban environments.

(continued)

Fig. 18.1 Newly established mini- forest at CAPE 10 in Vienna (Photo: BFW/Erik Szamosvari)

Community forest: A community forest refers to a forested area managed collectively by a community, neighbourhood, or local organisation within an urban or suburban area. These forests are often smaller in scale and may include public or privately owned land. Community forests serve as valuable green spaces by contributing to community well-being, fostering a sense of ownership and stewardship, and promoting environmental education.

Pocket forest or mini-forest: A pocket forest or mini-forest denotes a small-scale, densely forested area (typically less than 1000 m^2) commonly found within neighbourhoods or on school campuses or private properties. In Europe, this fairly recent trend began in the Netherlands in 2015. Pocket forests are designed to maximise green space in compact urban areas and provide a perfect opportunity to showcase the benefits of urban forests and engage the public (Fig. 18.1).

Urban woodlands and tree-lined streets: This category encompasses the presence of trees and vegetation along streets, boulevards, and avenues within urban areas. These green corridors enhance the visual appeal of cities and provide shade to pedestrians. Tree-lined streets are a common element of urban greening initiatives.

Regardless of the scale, urban forests play a crucial role in mitigating the adverse effects of urbanisation, improving human well-being, and promoting sustainable urban development.

Community Engagement

The management of ecological connectivity, along with the maintenance of biodiversity in urban areas, can be challenging and prone to conflicts. This is often due to a lack of effective communication regarding the proposed measures, their timelines, and their intended goals. When such information is poorly communicated, the public may perceive management actions negatively, leading to misunderstandings and conflicts. To address this, clear and transparent communication with an emphasis on the shared benefits of urban greenery for biodiversity and human well-being.

Active community involvement encompassing residents, schools, local organisations, societies, and other stakeholders is essential for promoting and achieving ecological connectivity in urban areas. Collaboration through activities and engagement in citizen science initiatives foster a sense of ownership among community members. Public awareness, education, and participation in conservation efforts are key factors to success (Connop et al. 2016).

Habitat improvement and connectivity are the major aims of the community-driven Melbourne Pollinator Corridor project in Australia. In Melbourne, more than a third of all public green space consists of nature strips (street verges) spanning across the city. Despite this extensive coverage, their diversity is low, as they typically feature grass with only a few trees or shrubs and are regularly mown. The Pollinator Corridor project was initiated to utilise this public land to create an 8-km-long wildlife corridor linking two large parks. This form of street gardening undertaken by residents involves diverse plantings aimed at increasing habitat heterogeneity and encouraging pollinators (The Heart Gardening Project 2022) (Fig. 18.2).

Challenges for Urban Spatial Planning

Challenges in incorporating ecological connectivity into spatial planning processes include striking a balance between needs driven by population growth, e.g. housing and parking spaces, distribution warehouses, meeting financial expectations, and improving the environmental sustainability of cities. One of the main difficulties is identifying and designating key ecological corridors, green spaces, and networks that effectively facilitate the movement of species across urban areas.

Integrated planning principles recognise the importance of considering a diverse range of green and open spaces, including those on private land, as essential components of a city's green network. Connectivity plays a crucial role in urban planning, operating at various spatial scales and serving different functions such as social connectivity for humans, ecological connectivity for biodiversity, and abiotic connectivity for regulating functions like water and climate (Olafsson and Pauleit 2018). Emphasising multifunctionality helps to address these diverse urban challenges and promotes synergies among urban green infrastructure principles while optimising the use of limited space (Pauleit et al. 2019).

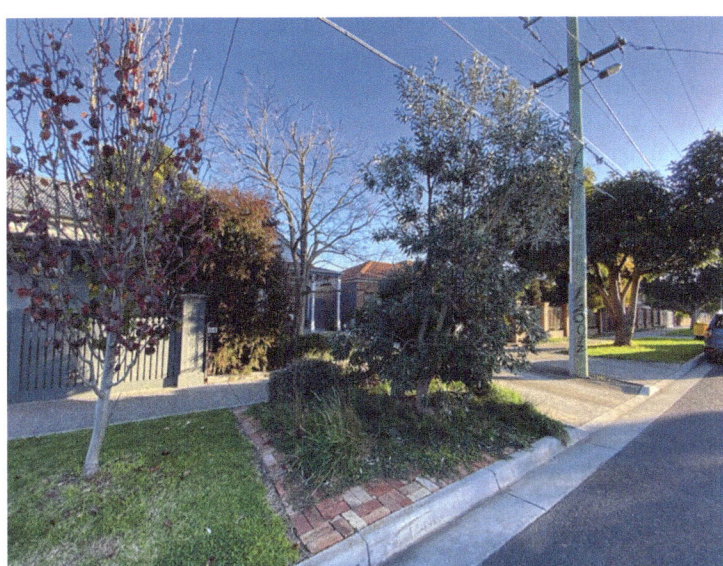

Fig. 18.2 Nature strip in Melbourne, Australia. Typical grassy area on the left, with a more diverse planting initiated by the owner visible on the right (Photo: Katherine Elizabeth Horsfall)

When attempting to improve ecological connectivity in urban areas, a major challenge lies in identifying urban and peri-urban forest (UPF) elements. Various technologies can be employed to address this challenge and aid in spatial planning processes. Geographic information systems (GIS) enable the analysis and visualisation of spatial data to allow mapping and assessment of green space distribution and connectivity (Fischer et al. 2015). Remote sensing techniques provide detailed information on land cover and vegetation, aiding in the identification and monitoring of existing green infrastructure (Li et al. 2019). Mobile mapping and crowd-sourced data collection engage the community, gathering valuable information on green spaces. Spatial modelling and simulation help to simulate species movement and evaluate planning scenarios, while data integration and open data platforms facilitate comprehensive understanding and collaboration among stakeholders (Mundher et al. 2022). By utilising these technologies, decision-makers can make informed choices, monitor progress, and create sustainable and connected urban environments.

Restoration of Urban Forest Habitat Patches

Restoring urban forest habitat patches is crucial for supporting the survival and migration of species that depend on forested environments (Alvey 2006). These habitats can be enhanced by incorporating various structural elements that provide microhabitats, such as wet and dry biotopes or piles of branches or stones (Kozák et al. 2018; Larrieu et al. 2018). By increasing structural diversity, these habitats offer valuable living spaces for a wide range of organisms (Oettel and Lapin 2020). Numerous studies have provided evidence of the importance of preserving biodiversity in urban forest patches, since they serve as havens for forest species and occasionally even endangered species (Aronson et al. 2017; Connop et al. 2016). It is thus essential to preserve existing habitat structures.

Conservation of Elements of Ecological Connectivity

Trees outside forests (TOF), regardless of whether they grow in stands, in lines, or scattered, are an important part of green infrastructure (De Foresta et al. 2013). They are commonly located in three main areas: in urban spaces, along transport corridors (roads) and in the agricultural landscape (Rouquette and Holt 2017). They play a vital role in habitat connectivity as well as in national biomass and carbon stocks, and they contribute to the livelihoods of people in many regions worldwide (Thomas et al. 2021). TOF coverage forms crucial ecological continuities that benefit various forest insects and fungi by linking forested areas (Rossi and Rousselet 2016). A promising approach to monitoring TOF is a combination of remote sensing techniques and field surveys. Recognising the significance of TOF, the United Nations Decade of Ecosystem Restoration has established a working group dedicated to monitoring trees outside forests, and the Global Restoration Observatory has formed a specialised team with the same objective (Schnell et al. 2015; UNEP 2023; Vogt et al. 2019). These initiatives highlight the importance of monitoring and conserving TOF for ecological restoration efforts.

Spontaneous Vegetation

Spontaneous vegetation, also known as spontaneous flora, refers to plant species that grow naturally and without intentional human cultivation, generally through self-seeding. In cities, they mostly colonise disturbed or abandoned areas such as vacant lots, roadsides, urban wastelands, or post-industrial sites. Urban areas exhibit elevated levels of disturbance, impervious paving, and heat retention. These elements encourage the proliferation of stress-tolerant, early successional vegetation (Del Tredici 2010).

In urban environments, spontaneous vegetation can have both positive and negative impacts. Although these plants are commonly referred to as 'weeds', a shift in attitude has also begun to occur as people become more aware of how biodiverse some of the mentioned areas are. Spontaneous vegetation can also provide aesthetic value, especially in the flowering season, and is part of the urban green infrastructure, providing ecosystem services and contributing to better ecosystem connectivity.

In most cities, these spontaneous plants are of cosmopolitan origin. Many non-native plants are well adapted to the warmer temperatures and frequent disturbances in a city, making them highly prolific under urban conditions. Some of these neophytes commonly found in European cities, like *Ailanthus altissima*, are considered invasive species since they outcompete and displace native plants. Trees like *Ailanthus* are also problematic because they can settle in tiny cracks and their roots can damage walls, concrete, and asphalt.

Observations indicate that areas with less economic prosperity often have more vacant lots and more spontaneous vegetation due to budget constraints that impede maintenance efforts like regular mowing as well as new building developments. The resulting slower turnaround rate allows plants to establish (Rink 2009). Generally, vacant lots are perceived negatively as a sign of economic decline and are associated with higher incidences of crime. However, properly managed vacant lots can also offer social benefits to the residents, for example by functioning as temporary parks with pathways for access. Considering the anticipated impacts of climate change on urban greenery, management strategies should aim to increase the ecological and social benefits of spontaneous urban vegetation (Del Tredici 2010).

Two examples from Germany demonstrate how spontaneous vegetation can be utilised to turn abandoned land into valuable green space at low cost. The coal mine *Zeche Zollverein* in the Ruhr area was closed down in 1986 and has since been renaturalised by allowing spontaneous vegetation. The 70-ha area now provides diverse habitats ranging from wet to dry and from woodland to open space. It has become home to more than 60 bird and 40 wild bee species as well as 540 fern and flowering plant species (Zollverein 2023). In Berlin, the 5.5 ha area of the former *Nordbahnhof* train station was abandoned for decades after the fall of the Berlin Wall and has developed rich spontaneous vegetation including birch woodland. In 2002, the area was revived and turned into a close-to-nature urban park integrating natural vegetation with historic traces and recreational spaces.

Unmanaged Forest Patches

Unmanaged forest patches, also known as natural forest patches, refer to areas of forest or woodland that have been left untouched or unmanaged by human intervention. These patches are characterised by the absence of deliberate forestry practices such as logging, tree planting, or active forest management. Unmanaged forest patches can occur within larger forested landscapes or in isolated areas surrounded by urban or agricultural land. They may vary in size from small patches to significant forest tracts, and often represent remnants of original or historic forests that have persisted despite surrounding land development or human activities. These forest patches play a crucial role in conserving biodiversity and serve as refuges for native fauna and flora, supporting a wide range of ecological processes such as natural succession, nutrient cycling, and species interactions. Since unmanaged forest patches are highly valuable for conservation, it is essential to consider their context within larger landscapes. Fragmentation and isolation of these patches can limit their ecological functionality and resilience.

Fig. 18.3 Tree-related microhabitat (TreM) on a city tree (Photo: Owen Bradley)

Tree-Related Microhabitats

To ensure a comprehensive understanding of tree ecosystems, it is essential to conduct a tree inventory (Alvey 2006) that includes tree-related microhabitats (TreMs)—specialised ecological niches or habitats that occur within or on various parts of trees. These microhabitats provide unique conditions and serve as habitat, food resources, nesting sites, shelter and protection from predators (Fig. 18.3).

Examples of tree-related microhabitats (TreMs) include the following (Kozák et al. 2018; Larrieu et al. 2018):

1. *Tree cavities*: Cavities are holes or hollows within tree trunks or branches. They can be formed naturally through decay or created by e.g. woodpeckers. Tree cavities serve as important nesting sites and shelters for birds, bats, squirrels, owls, and insects.
2. *Bark and bark crevices*: Bark crevices, cracks, and rough textures offer shelter, hiding places, and protection from predators. Many insects, spiders, lichens, mosses, and even small mammals utilise the bark microhabitat.
3. *Epiphytes and epiphytic gardens*: Epiphytes are plants that grow non-parasitically on the surfaces of trees. They obtain nutrients from the air, rainwater, or debris that accumulates on the tree's surface. Epiphytes such as orchids, bromeliads, ferns, and mosses create miniature gardens, providing habitats for insects, frogs, and other organisms.
4. *Lichen-encrusted surfaces*: Lichens are symbiotic organisms composed of fungi and algae or cyanobacteria, often forming colourful and intricate patterns.

Lichen-encrusted surfaces offer shelter, food, and moisture for microorganisms, insects, and small invertebrates.
5. *Leaf litter*: Leaf litter that accumulates at the base of trees provides a microhabitat for decomposers including fungi, bacteria, and invertebrates. These organisms break down the leaves, contributing to nutrient cycling and soil formation.
6. *Tree canopy*: The upper foliage and branches of trees create a complex and rich microhabitat known as the tree canopy. Canopy microhabitats support e.g. birds, arboreal mammals, insects, and epiphytic plants.

Protecting and conserving these microhabitats is important for maintaining healthy and diverse tree communities and supporting the broader ecosystem. Additionally, dead and decaying wood found within or around trees supports a wide range of organisms. It is nevertheless crucial to prioritise the safety of citizens in urban spaces: The vegetation of urban areas should not pose any risks to individuals, e.g. from falling tree branches, and should not interfere with local safety regulations.

Risks of Urban and Peri-Urban Forest Management for Ecological Connectivity

Biotic homogenisation, which refers to the replacement of localised native species with increasingly widespread species, is a significant concern in urban biodiversity conservation (Fernandez-Juricic and Jokimaki 2001). Urban areas are hotspots for non-native species, which can negatively impact ecosystems through increased competition and hybridisation. Common pathways for the introduction of non-native species include gardens and parks, where exotic plants are commonly used (Bendix 1994; Dreiss et al. 2016). Conducting site-specific risk assessments can help to mitigate the threat of non-native species by implementing targeted management strategies (Bindewald et al. 2021). Citizen science activities have also proven effective in identifying and monitoring invasive alien species in urban and peri-urban areas (Groom et al. 2021). Edge effects are another important factor to consider (Jokimäki and Huhta 2000): Depending on the overall habitat structure and species community, nesting and predation behaviour can be influenced by edge effects (Fernández-Juricic and Telleria 2000). This includes factors such as light and noise pollution, which can disrupt ecological processes and impact wildlife behaviour.

Human disturbance is a key factor affecting wildlife. Beyond fragment size, isolation, and habitat structure, human activities and recreation can have negative effects on breeding bird species (Fernández-Juricic and Telleria 2000; Wallace and Clarkson 2019), overall species richness, sensitive vegetation, and the movement of invertebrates and vertebrates. These disturbances can increase stress levels and disrupt important ecological interactions.

Residential land often provides the largest share of vegetation cover in cities. An example is Melbourne, where the share of residential green area is 46.7%,

Fig. 18.4 (a) Integrating 'messy' landscape into the Helmut-Zilk-Park, Vienna (Andrea Kodym) (b) Missing understorey layer, Kirschblütenpark, Vienna (Photo: Mark Hatfaludi)

compared to parkland (18.5%) and street trees (15.7%). However, increasing house densities and shrinking backyards contribute to a decline in tree and shrub cover on residential plots. This global trend is exacerbated by urban redevelopment and homeowners' decisions to maximise building activities. This loss of private garden areas should ideally be compensated by the establishment of more parks and public spaces, demonstrating the important interplay between public and private land (Hurley et al. 2019).

An additional risk is the lack of biodiversity in managed green spaces. Diversity can be reflected at various levels, and in terms of city trees, this includes the assortment of trees (species diversity), the diversity within individual species (genetic diversity), and the age and structure of the vegetation (ecosystem diversity). Many cities commonly face a situation where a small number of species or genera dominate the street tree population (Lohr et al. 2016; Galle et al. 2021). Nevertheless, entire avenues are frequently freshly planted with a single tree species (and maybe the same clonal variety) to this day. This lack of diversity can be attributed to traditional horticultural practices that prioritise uniformity, particularly in terms of growth habit. It also leads to a decline in genetic diversity, as varieties are typically propagated through grafting—and if seeds are used, they are often sourced from a limited number of mother trees. By contrast, tree stands that are rich in species, consist of a mix of age classes, and feature a shrub and herbaceous layer offer greater resilience and the ability to recover from disturbances. These stands also contribute to a more natural environment and habitat richness (Fig. 18.4).

A further threat is hybridisation with native populations (Afifi et al. 2023). In Central Europe, for example, wild populations of tree families such as Rosaceae (*Sorbus*, *Pyrus*, or *Malus*) are threatened by globally traded species commonly used in peri-urban areas (George et al. 2015; Konrad et al. 2020).

Urban Planning Tools and Strategies

To safeguard ecological connectivity in urban and semi-urban areas, it is essential to implement efficient management strategies tailored to site-specific biodiversity (Aronson et al. 2017). This involves enhancing the biodiversity potential by improving habitat quality in existing forest patches through coordinated and diverse management approaches across managed urban green infrastructure and other natural areas in cities.

While urban green infrastructure offers great potential for supporting habitat connectivity and biodiversity, there are also potential risks and challenges associated with its implementation. These include the loss of valuable green spaces due to urban expansion, inadequate maintenance and management of existing green infrastructure, and potential conflicts with urban development priorities. Possible countermeasures and strategies include incorporating green infrastructure into urban master plans and zoning regulations, establishing protected areas and wildlife corridors, promoting sustainable land-use practices, and involving local communities and stakeholders in the decision-making process.

Multi-Scale Management of Ecological Connectivity in Urban Areas

Achieving ecological connectivity requires a comprehensive approach that considers multiple scales. It involves addressing connectivity issues at different levels from local neighbourhoods to citywide and regional scales.

At the local scale, site-specific characteristics such as land-use patterns, green infrastructure availability, and urban design play a crucial role. Understanding these factors helps to identify key areas for implementing connectivity measures such as creating green corridors, identifying ecological hotspots, or enhancing existing green spaces. It is important to consider the socioeconomic context of the respective area—including factors like population density, community needs, and access to green spaces—to ensure that connectivity interventions are tailored to the local context and the requirements of the residents.

At the citywide and regional scales, coordination and collaboration among different stakeholders including city planners, policymakers, community groups, and environmental organisations become essential. A strategic approach is needed to identify priority areas. This requires integrating ecological connectivity into urban planning and development processes, incorporating green infrastructure requirements into land-use planning policies, and establishing partnerships to implement and manage connectivity projects across jurisdictional boundaries.

Furthermore, promoting ecological connectivity in urban areas should not only focus on physical connectivity but also consider ecological processes such as species dispersal, gene flow, and ecological functions. Protecting and restoring habitat quality within connected green spaces is crucial for supporting the survival and movement of diverse species. It is necessary to prioritise the conservation and restoration of high-quality habitats and address threats such as fragmentation, pollution, and invasive species spread that can impede or adversely affect connectivity.

Key Principles for the Planning of Ecological Connectivity in Urban and Semi-Urban Areas

1. *Assess species populations and diversity*: Conduct thorough assessments of the species present in the area to understand their population status and biodiversity, which forms the basis for connectivity planning.
2. *Identify physical and non-physical barriers*: Identify and evaluate both physical (e.g. roads, buildings, pipelines) and non-physical (e.g. social, cultural) barriers that impede ecological connectivity within the urban landscape.
3. *Acknowledge site-specific characteristics*: Consider the unique characteristics of each site, including ecological, socioeconomic, and cultural aspects, to tailor connectivity planning strategies accordingly.
4. *Increase citizen engagement*: Involve the local community, residents, and stakeholders in the planning process to increase awareness, participation, and ownership of connectivity initiatives.
5. *Establish a monitoring programme*: Develop a robust monitoring programme to assess the effectiveness of conservation actions and adapt strategies as needed to achieve desired ecological outcomes.
6. *Develop long-term solutions*: Implement connectivity plans with a long-term perspective, considering the dynamic nature of urban environments and the need for sustainable solutions that endure over time.
7. *Provide habitat and corridors*: Design and maintain green spaces that offer suitable habitats and corridors for species movement, ensuring that connectivity elements are integrated into urban design and development.
8. *Citizen science-based monitoring*: Encourage citizen participation in monitoring efforts by promoting citizen science initiatives, which involve the public in data collection and contribute to collective knowledge and understanding.
9. *Allow knowledge transfer and growth*: Share the results and experiences gained from connectivity planning and implementation to facilitate knowledge transfer, foster learning, and support continuous improvement in urban connectivity strategies.
10. *Collaborate across sectors and disciplines*: Foster collaboration and cooperation among various sectors, disciplines, and stakeholders involved in urban planning, biodiversity conservation, community development and governance. This ensures that ecological connectivity is prioritised.

References

Afifi L, Lapin K, Tremetsberger K, Konrad H (2023) A systematic review of threats, conservation, and management measures for tree species of the family Rosaceae in Europe. Flora 152244

Alvey AA (2006) Promoting and preserving biodiversity in the urban forest. Urban For Urban Green 5(4):195–201

Aronson MFJ, Lepczyk CA, Evans KL, Goddard MA, Lerman SB, MacIvor JS, Nilon CH, Vargo T (2017) Biodiversity in the city: key challenges for urban green space management. Front Ecol Environ 15(4):189–196

Badiu DL, Nita A, Iojă CI, Niță MR (2019) Disentangling the connections: a network analysis of approaches to urban green infrastructure. Urban For Urban Green 41:211–220

Bendix J (1994) Riparian landscapes. Geomorphology 11:172–173. https://doi.org/10.1016/0169-555X(94)90082-5

Bindewald A, Brundu G, Schueler S, Starfinger U, Bauhus J, Lapin K (2021) Site-specific risk assessment enables trade-off analysis of non-native tree species in European forests. Ecol Evol 11(24):18089–18110

Connop S, Vandergert P, Eisenberg B, Collier MJ, Nash C, Clough J, Newport D (2016) Renaturing cities using a regionally-focused biodiversity-led multifunctional benefits approach to urban green infrastructure. Environ Sci Policy 62:99–111

De Foresta H, Somarriba E, Temu A, Boulanger D, Feuilly H, Gauthier M (2013) Towards the assessment of trees outside forests: a thematic report prepared in the framework of the global forest resources assessment 2010; Food and Agriculture Organization of the United Nations: Rome, Italy.

Del Tredici P (2010) Nature and culture 5(3):299–315

Dreiss LM, Volin JC, Closset-Kopp D, Chabrerie O, Valentin B, Delachapelle H, Decocq G, Essl F, Milasowszky N, Dirnböck T, Höfle R, Dullinger S, Essl F, Chmura D, Sierka E, Braun M, Schindler S, Essl F, Campagnaro T et al (2016) Plant invasions in temperate forests: resistance or ephemeral phenomenon? For Ecol Manage 15(1):120–130. https://doi.org/10.1016/j.jenvman.2011.01.025

European Commission (2023) Green Infrastructure. https://environment.ec.europa.eu/topics/nature-and-biodiversity/green-infrastructure_en. Accessed on 8.8.2023

FAO (2017) Urban and Peri-urban Forestry. https://www.fao.org/forestry/urbanforestry/87025/en/

Fernandez-Juricic E, Jokimaki J (2001) A habitat Island approach to conserving birds in urban landscapes: case studies from southern and northern Europe. Biodivers Conserv 10(12):2023–2043

Fernández-Juricic E, Telleria JL (2000) Effects of human disturbance on spatial and temporal feeding patterns of Blackbird Turdus merula in urban parks in Madrid, Spain. Bird Study 47(1):13–21

Filazzola A, Shrestha N, MacIvor JS (2019) The contribution of constructed green infrastructure to urban biodiversity: a synthesis and meta-analysis. J Appl Ecol 56(9):2131–2143

Fischer LK, Kowarik I, Botzat A, Honold J, Haase D, Kabisch N (2015) Interaction of biological and cultural diversity of urban green spaces Internal Project Report on the Assessment of BCD in European City Regions Published Project Report GREEN SURGE Deliverable, p 2

Galle NJ, Halpern D, Nitoslawski S, Duarte F, Ratti C, Pilla F (2021) Mapping the diversity of street tree inventories across eight cities internationally using open data. Urban Forestry & Urban Greening 61:127099. ISSN 1618-8667. https://doi.org/10.1016/j.ufug.2021.127099

George J-P, Konrad H, Collin E, Thevenet J, Ballian D, Idzojtic M, Kamm U, Zhelev P, Geburek T (2015) High molecular diversity in the true service tree (Sorbus domestica) despite rareness: data from Europe with special reference to the Austrian occurrence. Ann Bot 115(7):1105–1115

Gómez-Baggethun E, Barton DN (2013) Classifying and valuing ecosystem services for urban planning. Ecol Econ 86:235–245. https://doi.org/10.1016/j.ecolecon.2012.08.019

Groom Q, Pernat N, Adriaens T, De Groot M, Jelaska SD, Marčiulynienė D, Martinou AF, Skuhrovec J, Tricarico E, Wit EC (2021) Species interactions: next-level citizen science. Ecography 44(12):1781–1789

Hansen R, Pauleit S (2014) From multifunctionality to multiple ecosystem services? A conceptual framework for multifunctionality in green infrastructure planning for urban areas. Ambio 43:516–529

Hurley et al (2019) Urban vegetation cover change in Melbourne. https://nespurban.edu.au/wpcontent/uploads/2019/07/urban-vegetation-cover-change.pdf. Accessed on 8.8.2023

Integrated A, Review L (2020) Green infrastructure planning principles: an integrated literature. Review:1–19

Jokimäki J, Huhta E (2000) Artificial nest predation and abundance of birds along an urban gradient. Condor 102(4):838–847

Konrad H, George J-P, Kahlert K (2020) Der Speierling in Thüringen–Populationsgenetik heute und in der Zukunft. In: Beiträge aus der angewandten Forschung zu Wald und Waldbewirtschaftung in Thüringen. Mitteilungen – Thüringen Forst 38:113–121

Kozák D, Mikoláš M, Svitok M, Bače R, Paillet Y, Larrieu L, Nagel TA, Begovič K, Čada V, Diku A (2018) Profile of tree-related microhabitats in European primary beech-dominated forests. For Ecol Manage 429:363–374

Larrieu L, Paillet Y, Winter S, Bütler R, Kraus D, Krumm F, Lachat T, Michel AK, Regnery B, Vandekerkhove K (2018) Tree related microhabitats in temperate and Mediterranean European forests: a hierarchical typology for inventory standardization. Ecol Indic 84:194–207

Li X, Chen WY, Sanesi G (2019) Remote sensing in urban forestry: recent applications and future directions. pp 1–20

Liu Z, Xiu C, Ye C (2020) Improving urban resilience through green infrastructure: an integrated approach for connectivity conservation in the Central City of Shenyang, China. Complexity 2020:1–15

Lohr VI, Kendal D, Dobbs C (2016) Urban trees worldwide have low species and genetic diversity, posing high risks of tree loss as stresses from climate change increase. Acta Hortic 1108:263–270. https://doi.org/10.17660/ActaHortic.2016.1108.34

Monteiro R, Ferreira JC, Antunes P (2020) Green infrastructure planning principles: an integrated literature review. Land 9(12):525

Mundher R, Bakar SA, Maulan S, Johari M, Yusof M (2022) Aesthetic quality assessment of landscapes as a model for urban forest areas: a systematic literature review. pp 1–22

Oettel J, Lapin K (2020) Linking forest management and biodiversity indicators to strengthen sustainable forest management in Europe. Ecol Indic

Olafsson AS, Pauleit S (2018) Green surge final project report

Pauleit S, Ambrose-Oji B, Andersson E, Anton B, Buijs A, Haase D, Elands B, Hansen R, Kowarik I, Kronenberg J (2019) Advancing urban green infrastructure in Europe: outcomes and reflections from the GREEN SURGE project. Urban For Urban Green 40:4–16

Rink (2009) Wilderness: the nature of urban shrinkage? The debate on urban restructuring and restoration in eastern Germany. Nat Cult 4(3):275–292

Rossi J-P, Rousselet J (2016) The spatial distribution of trees outside forests in a large open-field region and its potential impact on habitat connectivity for forest insects. Turkish J For 17:62–64

Rouquette JR, Holt AR (2017) The benefits to people of trees outside woods (TOWs). Report for the Woodland Trust. Natural Capital Solutions.

Rusche K, Reimer M, Stichmann R (2019) Mapping and assessing green infrastructure connectivity in European city regions. Sustain For 11(6):1819

Schnell S, Kleinn C, Ståhl G (2015) Monitoring trees outside forests: a review. Environ Monit Assess 187:1–17

Skole DL, Mbow C, Mugabowindekwe M, Brandt MS, Samek JH (2021) Trees outside of forests as natural climate solutions. Nat Clim Chang 11(12):1013–1016

The Heart Gardening Project (2022) https://theheartgardeningproject.org.au/melbourne-pollinator-corridor. Accessed on 8.8.2023

Thomas N, Baltezar P, Lagomasino D, Stovall A, Iqbal Z, Fatoyinbo L (2021) Trees outside forests are an underestimated resource in a country with low forest cover. Sci Rep 11(1):1–13

UNEP (2023) Reventing, halting and reversing loss of nature. https://www.decadeonrestoration.org/

Vogt P, Riitters K, Caudullo G, Eckhardt B (2019) FAO-state of the world's forests: forest fragmentation. Publications Office of the European Union, Luxembourg

Wallace KJ, Clarkson BD (2019) Urban forest restoration ecology: a review from Hamilton, New Zealand. J R Soc N Z 49(3):347–369

Whitney GG (1985) A quantitative analysis of the flora and plant communities of a representative midwestern U.S. town. J Urban Ecol 9(2):143–160

Zipperer WC, Susan SM, Pouyat R, Foresman TM (1997) Urban tree cover: an ecological perspective. Urban Ecosyst 1. https://doi.org/10.1023/A:1018587830636

Zollverein (2023). https://www.zollverein.de/besuch-planen/zollverein-fuer/zollverein-fuer-naturfreunde/. Accessed on 8.8.2023

Open Access This chapter is licensed under the terms of the Creative Commons Attribution 4.0 International License (http://creativecommons.org/licenses/by/4.0/), which permits use, sharing, adaptation, distribution and reproduction in any medium or format, as long as you give appropriate credit to the original author(s) and the source, provide a link to the Creative Commons license and indicate if changes were made.

The images or other third party material in this chapter are included in the chapter's Creative Commons license, unless indicated otherwise in a credit line to the material. If material is not included in the chapter's Creative Commons license and your intended use is not permitted by statutory regulation or exceeds the permitted use, you will need to obtain permission directly from the copyright holder.

Connectivity in the Social-Ecological Context and Nature's Contribution to People

19

Rosina Soler, Verónica Chillo, Paula Rodríguez, Gimena Bustamante, and Matthew Ruggirello

Cattle grazing in forest (Photo: Miheer Tevari/Unsplash)

R. Soler (✉) · P. Rodríguez · G. Bustamante · M. Ruggirello
Austral Centre for Scientific Research (CADIC – CONICET),
Ushuaia, Tierra del Fuego, Argentina
e-mail: rosina.soler@conicet.gov.ar; paula.rodriguez@conicet.gov.ar; gimenabustamante@conicet.gov.ar; matthew.ruggirello@conicet.gov.ar

V. Chillo
Forestry and Agricultural Research Institute of Bariloche (IFAB, INTA-CONICET),
El Bolsón, Río Negro, Argentina
e-mail: chillo.veronica@inta.gob.ar

© The Author(s) 2025
K. Lapin et al. (eds.), *Ecological Connectivity of Forest Ecosystems*,
https://doi.org/10.1007/978-3-031-82206-3_19

Abstract

This chapter examines the concept of forest connectivity in a globalized world, emphasizing its significance for maintaining biodiversity and ecosystem functioning as well as for human–environment relationships. From an ecological perspective, forest connectivity encompasses both structural and functional aspects and extends beyond habitat quantity and continuity. Structural connectivity focuses on the physical arrangement of landscape elements, while functional connectivity emphasizes the role of different species in promoting connectivity and biodiversity as well as enhancing resilience. But human–environment relationships also include a broader social-ecological context for multifunctional forest landscape analysis in that the impact of human activities can be as significant as ecological disturbances in shaping future forest resources. Conserving species and ecosystem services requires understanding the interactions between social and ecological systems. The Nature's Contribution to People framework recognizes the role of human societies, cultural beliefs, and practices in shaping their relationship with nature. Ecosystem services are explored along with their provision and interactions, identifying synergies and trade-offs. This chapter analyses the concept of forest connectivity in its ecological terms while simultaneously attempting to broaden its meaning to include its socio-cultural aspects as well. This integrated approach acknowledges that the well-being of people and the health of ecosystems are intricately connected, and both must be considered in decision-making processes.

Keywords

Biodiversity · Forest change · Interdisciplinarity · Multifunctional landscapes · Human–nature connections

Forest Connectivity in the Context of a Globalized World

The concept of habitat connectivity is understood as the extent to which patches are connected to one another by similar habitats or corridors capable of playing a crucial role in various processes such as survival and recolonization of deforested areas as well as the exchange of genetic material among different populations (Noss and Harris 1986; Noss 1991). This dominant assumption implies that connectivity is synonymous with the amount of habitat or habitat continuity. While this may be a desirable situation for species conservation, since habitat loss is presumably the foremost threat to it, Fahrig et al. (2021) suggest that connectivity implies a wide range of concepts differing in important ways. An example is forest connectivity—or the degree to which forest patches are linked to each other—which is frequently applied in landscape planning. Nevertheless, the movements of plant and animal species are not always restricted by forest habitats alone; they also depend on the surrounding matrix (structure, quality, etc.), stepping-stone patches, and species behaviour and requirements (Fahrig et al. 1983, 2021; Fahrig and Merriam 1985) among others.

Forest connectivity is an important concept in ecology, since it plays a crucial role in the context of maintaining biodiversity and ecosystem functioning (Bennett 1999). In the Brazilian Cerrado, Martello et al. (2023) determined that overall forest amount and patch density were the landscape predictors that explained most of the diversity of plant functional traits. However, forest connectivity is not the only factor determining biodiversity conservation, nor are these two concepts necessarily positively related. For example, forest loss in the rainforests of Mexico was directly and positively related to the abundance and richness of arboreal mammals (Cudney-Valenzuela et al. 2023). The authors maintain that this unexpected response was probably due to a crowding effect that increased local biodiversity parameters since the study had a short deforestation history. Conversely, in that work, the Shannon diversity was not affected suggesting that rare mammals were the ones most affected. Moreover, while ensuring forest connectivity is crucial, it remains insufficient if the surrounding landscape matrix is not taken into consideration. In the United States, where agriculture accounts for about half of the total land area, patches of natural vegetation like remnant forests and riparian corridors are interspersed among cropland and grazing lands thus becoming key areas for connectivity (Suraci et al. 2023). The same study also showed that ecological flow values in all types of agriculture were positively influenced by the amount of natural vegetation and negatively influenced by the amount of developed land. Overall, forest connectivity proves to be a complex term even when only its ecological implications are considered. The interactions between forest patches and other aspects such as matrix quality or landscape heterogeneity are not only a key to understanding the success of species movement and the flow of ecological processes across landscapes, but also to understanding how policy and management decisions affect human well-being.

Structural Connectivity: Influence of Forest Loss and Expansion

Structural connectivity implies a physical association or relationship between landscape elements (Crooks and Sanjayan 2006). Different structures of forest stands (young, mature, multi-age stands) increase complexity at different levels (e.g. patch, landscape levels), providing habitats for species with different requirements. For example, old-growth structures featuring mature hollow-bearing trees possess unique attributes which create essential habitats for plant and animal species that cannot easily be replaced by younger trees or artificial structures (Lindenmayer and Lawrence 2016; Le Roux et al. 2014). Meanwhile, forest gaps create suitable conditions for light-demanding plants or birds inhabiting the shrub layer. This is not only due to the occurrence of specific attributes (like nest cavities in old trees) but also a result of habitat continuity across space and time. Forests with high connectivity have a lower risk of reducing the occurrence of local species as well as genetic diversity (Gupta and Pandey 2019).

Forest structural connectivity is broadly driven by forest loss and gain. Over the past several decades, the great expansion of the agricultural frontier has been considered the main driver of deforestation, especially in the tropics and subtropics (Hoyos et al. 2013; Pendrill et al. 2022). However, global forest loss is affected by a myriad of drivers whose relative importance varies significantly by region (Curtis et al. 2018). Shifting agriculture is the main driver of forest loss in Africa, wildfire in the northern portions of Asia and North America, commodity-driven agriculture in many of the island nations of the South Pacific, and commercial forestry throughout much of the United States and Europe. It is well known that forest fragmentation modifies ecological patterns and processes by increasing the number of vegetation patches while reducing patch size and connectivity within the larger ecological network (Zhang et al. 2019), thereby causing species dispersal or migration (Bierwagen 2007). At the same time, ongoing changes in environmental drivers, land use, and disturbance regimes are forcing forests towards younger, shorter stands (McDowell et al. 2020). In recent times, forest conversion to agricultural land has decreased in some regions. There is evidence of forest cover increase over the past few decades (e.g. in Europe, the United States, and South America; see FAO and UNEP 2020) due to the abandonment of agricultural or pastoral lands and sociodemographic changes that favour forest regrowth. Spontaneous forest regrowth involves two key processes: (a) expansion of forests, where new forests emerge on previously cultivated lands, and (b) densification of forests as they mature and grow following their initial establishment (see Box 19.1). Newly growing forests constitute an opportunity in that they may facilitate the movement of forest species across landscapes, acting as corridors or stepping stones (Rautiainen et al. 2011). Additionally, new forests contribute significantly to climate change mitigation by serving as natural carbon sinks (FAO and UNEP 2020). At the same time, however, managed and unmanaged forests undergoing densification (through increasing tree density and secondary growth) and homogenization have amplified the frequency and severity of large-scale disturbances in certain forest types (Hessburg et al. 2005)—especially wildfires, insect outbreaks (Senf et al. 2017; Abatzoglou et al. 2018; Pausas and Keeley 2021), and the spread of exotic species (Lapin et al. 2019). Greater forest density allows wildfires as well as pest and pathogen outbreaks to be propagated across the forested landscape more easily; warmer and drier conditions generally also tend to exacerbate such disturbances resulting in massive forest loss (Jactel et al. 2019; Pausas and Keeley 2021).

The new global scenario of forest structure dynamics characterized by the loss and gain of forest cover presents new challenges for the concept of connectivity and its practical implementation. On one hand, there is the advantage of restoring forest habitats, thereby increasing connectivity between existing patches and expanding the physical boundaries of the forest matrix. On the other hand, there is the challenge of managing the density of new forest stands to create synergies in the landscape and avoid negative or counterproductive effects on a large scale.

Box 19.1 Fragmentation and Defragmentation in Forest Landscapes: Key Factors Influencing Habitat Connectivity

Whereas habitat fragmentation refers to the gradual segmentation of formerly connected habitats into smaller, fragmented patches of isolated habitat, new phenomena of the opposite nature—known as defragmentation—have recently been described and are occurring worldwide. Defragmentation occurs mainly as a consequence of two processes. The first of these is forest regrowth, meaning a substantial increase in both the size of the largest forest patch and the effective mesh size. This primarily stems from the growth and merging of existing forest patches, indicating a process of forest defragmentation and increasing structural connectivity. The second process is forest expansion, meaning an increase in forest cover achieved through the establishment of new forests on land previously used for other purposes (e.g. cropland, pasture, abandoned human settlements), resulting in the expansion of novel forest structures. Both forest regrowth and expansion occur as passive, non-human-assisted phenomena (e.g. the establishment of trees in peat bogs) (Payette and Delwaide 2004), but also as intentionally or unintentionally human-aided processes (e.g. the suppression of historically normal wildfire allows for previously fire-excluded tree species to establish in prairies (Briggs et al. 2002)).

While forest regrowth can clearly be interpreted as an increase in forest habitat connectivity since separated forest patches become linked in the broader matrix, the proliferation of new spontaneous forests provides stepping stones for species movement between existing forests—while at the same time enhancing fragmentation per se by increasing the number of small patches and forest edges.

It is important to point out that the transformations observed in afforested landscapes are more intricate than initially anticipated, and their causes extend beyond forest expansion alone. Factors such as landscape composition, geographic location (e.g. latitude), elevation variations, and land-use legacies across forests worldwide also contribute significantly to these changes.

Functional Connectivity

While structural connectivity deals with the physical arrangement and connectedness of landscape elements, *functional connectivity* focuses on the quality of the connections between them and on the ecological processes and interactions they facilitate. In this sense, structural connectivity does not always provide functional connectivity: Taylor et al. (2006), for example, show that in some cases, the target species or guild in a landscape does not use the connecting elements. By considering factors such as habitat suitability, resource availability, and species-specific ecological requirements, functional connectivity becomes a key tool for promoting biodiversity, supporting ecosystem functions, and enhancing resilience to environmental changes (Chapin et al. 2009; Lindenmayer and Lawrence 2016).

The consideration of functional diversity helps to bridge the gap between structure and function, providing an understanding of the relationship between taxonomic diversity and ecosystem functioning (Balvanera et al. 2006). Individuals exhibit eco-morpho-physiological characteristics related to their response to environmental drivers and disturbances (functional response traits) as well as to the role they play in different ecosystem processes (functional effect traits). In order to accurately assess functional diversity, it is therefore important to start by defining the disturbance and/or ecological process that is to be addressed (Díaz and Cabido 2001). For example, functional connectivity related to forest fires needs to consider the community composition of functional response traits related to flammability, structure, resprouting capacity, Raunkiaer life forms, dispersal capacity, and so on. The relative composition of these traits will change across a landscape primarily due to different forest types, land uses, and ecological legacies (Blackhall et al. 2017; Tiribelli et al. 2019).

Habitat connectivity implies the spatial linkage of habitat units at the landscape level, which fundamentally results in biodiversity conservation, as well as the ecological functions that directly or indirectly affect the social communities inserted in the matrix (Crooks and Sanjayan 2006; Fahrig 2013; Bastos et al. 2023). For instance, insect pollinators frequently depend on woody habitats such as forests or shrublands for nesting (Ulyshen et al. 2023). Even individual trees can benefit pollinators by providing pollen and nesting sites as well as by increasing matrix permeability, allowing the latter to subsequently move into neighbouring agricultural fields to pollinate crop species (Ricketts et al. 2008). The spatial arrangement of forests in relation to agricultural fields, as well as the mobility of pollinators within each ecosystem, are expected to impact the extent and distribution of pollination services throughout the forested landscape. Species requirements and habitat suitability thus become as important as the structurally connecting elements of the forest themselves.

At the same time, unlike bees or other highly vulnerable species, there are many organisms (plants and animals) that can safely move through impassable or partially inhabitable matrices in order to connect resources across heterogeneous landscapes. These species are referred to as *mobile link species* due to their intra-range and migratory movements providing large-scale connectivity in forested habitats (Root-Bernstein and Svenning 2017). One example of such species is the guanacos (*Lama guanicoe*, Fig. 19.1a), which inhabit forest patches with different habitat types in the Patagonian forest-steppe ecotone. In doing so, they provide multiple ecosystem functions such as nutrient input, plant dispersal by way of their natural seasonal migrations, and influence on tree growth through browsing. Other interesting examples are frugivorous bird species, as they are common in fragmented landscapes and may act as mobile links by connecting forest fragments through seed dispersal (Lundberg and Moberg 2003; Zurita and Zuleta 2009; Mueller et al. 2014). In particular, forest in the Gran Chaco ecoregion had suffered the highest deforestation

Fig. 19.1 (**a**) Mammals (*Lama guanicoe*) connect forest patches with different habitat types in the Patagonian forest-steppe ecotone (Photo: Federico González). (**b**) Birds (*Pitangus sulphuratus*) connect forest patches in a highly fragmented landscape of Gran Chaco region. (Photo: SIB, Administration of National Parks of Argentina)

and fragmentation rates in the past two decades (Hansen et al. 2013). Frugivorous birds species such as *Pitangus sulphuratus* and *Thraupis bonariensis* (Fig. 19.1b) have shown large movement rates and functional connectivity among forest patches, promoting seed dispersal of many bird-dispersed plant species in a highly fragmented landscape (Díaz Vélez et al. 2015).

Functional connectivity plays a crucial role in enhancing the resilience of ecosystems to environmental changes. At the community level, a greater number of functionally diverse species offers ecological benefits that mitigate different disturbances (Box 19.2). For instance, tree mixtures with varying root systems allow water use at different soil depths in temperate (Forrester and Bauhus 2016) and tropical forests (Montagnini and Jordan 2005). At the species level, tree species growing in mixed communities can alter their responses to stress and disturbance. Furthermore, the presence of trees of diverse ages and leaf chemical composition may help reduce pathogens as well as herbivory by individual species (Forrester and Bauhus 2016; Espelta et al. 2020) while increasing insect guild diversity (Nacif et al. 2020) thanks to niche complementarity.

Box 19.2 Anthropic Disturbance Suppression Promoting Forest Cover Gain in South America, North America, and Europe

Fire suppression and loss of indigenous land management in western North America: Before Anglo-European settlers arrived in western North America, the semi-arid forested areas away from the Pacific Ocean experienced frequent, low-intensity surface fires. Ignited by lightning or intentionally set by indigenous communities, these fires served to promote the growth of desired plant species and facilitate hunting. As a result, the landscape was dominated by fire-tolerant tree species (e.g. *Pinus ponderosa*, *Larix occidentalis*) with

(continued)

Fig. 19.2 Historical landscape conditions of the mixed-coniferous forests of inland western North America (**a**) and the deciduous forests of southern Patagonia (**c**), versus current, densified conditions across those same landscapes (**b** and **d**). Over a century after abandonment, former farms in New England, North America (**e**) have reverted to dense forest. (Gaige and Glogower 2016)

few large trees per hectare (Fig. 19.2a, b). With the relocation of indigenous groups and the arrival of new settlers, however, a policy of aggressive fire suppression was implemented, leading to increased forest continuity. This change resulted in higher tree densities, smaller tree sizes, and the replacement of fire- and drought-tolerant species with shade-tolerant, fire-sensitive ones. Over time, the formerly fire-adapted landscape became homogeneous and continuous, creating favourable conditions for devastating mega-fires and widespread pest and pathogen outbreaks in western North America.

Livestock intensity shifts in native forests of southern Patagonia: In Tierra del Fuego, the forest-grassland ecotone in southern Patagonia (Fig. 19.2c, d) experienced a rise in sheep production beginning in the early 1900s that led to fragmented forest landscapes and the conversion of forests into pasture by way of clear-cutting and fire clearance. By the end of the twentieth century, however, a shift from sheep to cattle ranching occurred, resulting in reduced livestock density as well as a socio-productive transformation. Some woodlands still exhibit significant degradation due to profound anthropic disturbances, while others have partially recovered due to the establishment of dense secondary forests. The current structure of the forests in this region is a combination of legacy structures and the vigorous resprouting of young trees, with past human disturbance and its impact on natural factors playing a more important role regarding current stand structural complexity than the extent of forest cover.

Rural abandonment in northeastern North America and southern Europe: By 1900, rural residents in northeastern North America were migrating to cities, resulting in approximately half of the agricultural land reverting to forests (Fig. 19.1e). Stone farms scattered throughout the forested landscape serve as evidence of this abandonment. This region experiences infrequent droughts and wildfires, and despite its high tree species diversity and cool climate, it faces challenges due to climate change. Similarly, since the late 1960s, rural abandonment has been occurring across Europe as young people pursue urban livelihoods. While the resulting forest expansion presents opportunities for conservation, it poses a significant threat to southern Europe, where dry, warm conditions and increasing forest connectivity have amplified the frequency, size, and impact of wildfires. Like in western North America, increased forest connectivity raises the risk of complete forest destruction and loss of associated benefits.

New Insights into Multifunctional Forest Landscapes

Scientific research conducted over the past two decades highlights the need to understand the connectivity of natural habitats not only in structural and functional terms but also by applying multifunctional approaches. Sustainable multifunctional landscapes require the integration of human activities with the ecological framework of a landscape to preserve biodiversity, critical ecosystem functions, and the uninterrupted flow of services (O'Farrell and Anderson 2010). Typically, landscape multifunctionality is considered in terms of the provision and interaction of

Fig. 19.3 Thinned forests in southern Patagonia reduce tree stand density but maintain 60% of the original structure for seed source (regeneration), microclimatic conditions, and habitat for local biodiversity. (Photo: Santiago Favoretti)

ecosystem services. On the one hand, some research areas initially focused on mapping the production and provision of ecosystem services in forest landscapes using biophysical indicators, identifying areas with the potential to provide the greatest number of ecosystem services (Egoh et al. 2008; Martínez Pastur et al. 2017). On the other hand, researchers have recognized a non-linear interaction between the provision of ecosystem services, with synergies and trade-offs occurring in multifunctional landscapes (Chillo et al. 2018; Turkelboom et al. 2018). For example, several studies have identified negative interactions between ecosystem services provided by the same biophysical structures (i.e. timber production is lower in forest gaps, while pasture production is higher). Similar trade-offs have also been identified between timber harvesting and the regulation of water quality (Little et al. 2009; Nisbet et al. 2022), carbon sequestration (Lin and Ge 2020), recreational uses (Arnberger et al. 2018), or biodiversity protection (Kangas and Ollikainen 2022). These non-linear interactions thus present a challenge in the context of land-use management (Fig. 19.3). In this sense, the consideration of "bundles" of ecosystem services represents a potential tool for managing these across landscapes, as it identifies types of services that tend to appear together. But ecosystem service bundles have been conceptualized in different ways to capture (i) the supply of ecosystem services (Raudsepp-Hearne et al. 2010), (ii) the sets of ecosystem services used by people (Hamann et al. 2015), and (iii) the sets of ecosystem services preferred by different stakeholder groups (Martín-López et al. 2012). Recent research has attempted to assess these three conceptualizations and compare their mismatches to better target management interventions (Meacham et al. 2022).

In the context of the Anthropocene and globalization, the growing challenges concerning environmental and social sustainability are deeply intertwined as a result of the interaction of numerous mutually reinforcing social and ecological processes at multiple scales (Folke et al. 2016; Biggs et al. 2021). The consideration of social-ecological systems implies a cohesive, integrated system characterized by strong interactions and feedback within and between social and ecological components that determine its overall dynamics (Folke et al. 2010). As such, social-ecological systems can be studied from a complex adaptive system approach, which considers them to be more than the sum of their social and ecological parts. This means that interactions between interdependent parts give rise to emergent system-wide patterns that cannot be predicted from the properties of the individual components. In turn, these system-wide patterns influence the behaviour of the individual, creating a feedback process that shapes the evolution of the system and allows it to adapt to changing contexts (Lansing 2003; Biggs et al. 2021). For example, different forest types (even micro-sites) are composed of different plant species with differing functional effect traits concerning litter decomposition, soil retention, primary productivity, and so on. Similarly, different social communities at the landscape level interact with these ecological components in different ways based on multiple factors. The interplay between ecosystem service production, provision, and appropriation can be considered a system-wide property of these interactions. The potential synergies and trade-offs between ecosystem services as well as governance systems at the local, regional, and international levels are evidence of a non-linear relationship between components and properties, respectively between multiple properties, of the system (Ostrom 2009).

Nature's Contribution to People (NCP) in Forest Landscapes

Human–environment relationships are embedded within a broader social-ecological context. Many economic, political, cultural, and technological processes shape forest landscapes across temporal and spatial scales. Therefore, human disturbances can be as important as ecological disturbances in shaping future forest resources (Chapin et al. 2009). Conserving species and ecosystem services depends on our understanding of social systems and their interactions with ecological systems. For this purpose, IPBES (The Intergovernmental Science-Policy Platform on Biodiversity and Ecosystem Services) has developed the Nature's Contribution to People (NCP) framework that builds on the established and well-known concept of ecosystem services (Díaz et al. 2015).

The NCP framework recognizes that human societies and their cultural beliefs and practices play a critical role in shaping their relationship with nature (Díaz et al. 2015). It thus transitions from the generalizing perspective based on a predetermined set of ecosystem services to a more context-specific perspective acknowledging unique local or cultural worldviews that can apply to specific social-ecological

Fig. 19.4 Ancient araucaria forests (*Araucaria araucana*) in Chile. This species' spiritual and religious value for local communities has promoted ethnotourism since it is perceived as being an important NCP of this forest type. (Photo: Federico González)

settings and recognizing that ecosystem values may not transfer universally (Díaz et al. 2018; Peterson et al. 2018). This shift leads to a broadening of the epistemological boundaries of the ecosystem services framework, allowing social sciences to be integrated more actively into the analysis of social-ecological systems. This means to connect social systems and relationships within and between forest-dependent communities. Forest dependency has traditionally been studied in material terms such as the provision of timber, non-timber forest products, and other goods or services (FAO 1998; Delgado et al. 2023; Derebe et al. 2023). The non-material relationships between humans and forests, however, have been addressed less frequently (Plieninger et al. 2023). These relationships are essential to understanding the complex social, cultural, and institutional processes that affect how we perceive, relate to, and ultimately impact forest ecosystems. Forest landscapes comprise a wide range of situations from urban forests within cities and small sacred forests adjacent to villages to extensive woodlands in remote wilderness areas (see also Chap. 18). The contributions of these forests to people vary across different social-ecological settings and worldviews. While urban forests may be valued for their capacity to mitigate human heat stress in residential districts (Lee et al. 2016), other forests provide preventive and therapeutic health benefits (Hansen et al. 2017) or promote low-impact tourism (e.g. ethnotourism focused on sacred forests; see Fig. 19.4) thanks to their spiritual benefits as a source of powerful forces, energy, and wisdom (Lee et al. 2017; Shakeri et al. 2021).

Forests can have differing symbolic meanings for different groups of people (e.g. relational values; Muradian and Pascual 2018), which may lead to conflicting views on how they should be managed. There is an urgent need to incorporate these non-material variables into the study of forest ecosystems, particularly given the current challenges faced by forests worldwide, which include climate change, deforestation, fragmentation, and biodiversity loss. While trying to promote or restore forest connectivity in ecological terms, some initiatives (e.g. REDD+, rewilding) can be disconnected from the realities of the areas in which they operate, resulting in social exclusion or conflicts over land use. For example, a study conducted in the Indian state of Maharashtra found that the expansion of protected areas in the state has led to the displacement of local communities and conflicts with traditional land-use practices (Gupta et al. 2022). Yasmi et al. (2013) refer to these situations as community–outsider conflicts, providing seven case studies from five countries in Southeast Asia. Lack of participation and involvement of local communities in conservation initiatives can have several implications: (i) inappropriate restoration techniques or tree species, (ii) conflict over the use of forest resources, which in turn can lead to unsustainable resource extraction, and (iii) lack of ownership and stewardship of the restored forest areas, which can result in them being neglected and vulnerable to future degradation.

In this context, the NCP framework is a tool that can provide insights for future research as it achieves culturally grounded and in-situ knowledge production by considering local traditional knowledge and relational values, and often involves long-term engagements with relevant actors (Díaz et al. 2018; Balvanera et al. 2017). By doing so, it fosters more suitable, effective, and enduring policies that ensure sustainable management and conservation of forest resources.

NCP as a Framework for Examining Ecological Connectivity

The contributions people receive from nature are inherently defined by their historical and cultural background, social status, economic situation, geographic location, and other factors (Breyne et al. 2021). Different groups of people from differing backgrounds often view the same forest as a potential source of different—and potentially conflicting—benefits (Martín-Forés et al. 2020). Valuing one contribution (e.g. recreation or wildlife) may implicitly devalue other contributions (e.g. commercial logging, fire prevention management) and pit different forest user groups against one another (Maier and Abrams 2018). Therefore, forest connectivity viewed through an NCP framework must always contextualize the affected forest user groups.

Returning to the initial notion that the structural connectivity of forests is generally driven by forest loss and gain, it seems appropriate to highlight the different human perceptions of these phenomena (see Box 19.3). The perception of increased or decreased forest cover and connectivity as either beneficial or detrimental is closely tied to where a person lives, which often significantly impacts their social-cultural values (Breyne et al. 2021). For instance, in the Pacific Northwest of the

United States, city dwellers often view the large swaths of connected forests in the Cascade Range to the east as important and desirable for the recreational opportunities they provide, as a habitat for animals, and for their role in improving water quality and security (Baur et al. 2016; Maier and Abrams 2018). However, residents living in the mountains themselves and immediately surrounded by these woodlands may perceive their connectivity as a potential risk (i.e. facilitating wildfire spread) to their homes and livelihoods (e.g. logging), thus placing the conceptions of forest connectivity of these two groups at odds (Mapes 2016; Maier and Abrams 2018). Similarly, the social perception of forest regrowth in rural Europe changes drastically depending on where people live: In rural areas, spontaneous forest regrowth is generally viewed negatively, whereas it is perceived more positively in peri-urban environments (Martín-Forés et al. 2020). Behind these differing perceptions of forest connectivity are deeply rooted historical, cultural, geographic, and economic differences. Viewing forest connectivity through contextualized NCP frameworks can help to understand where values between different user groups may overlap or contrast. This can be used to inform land management that is better aligned with the contributions diverse user groups hope to obtain from forests.

In multifunctional landscapes, it is not just structural or functional connectivity that matters but also the connection between humans and nature. In Patagonia, silvopastoral management in native temperate forests, the natural forest ecosystem, and the sociohistorical context are structuring elements that also interact with the cultural context (Peri et al. 2016; Chillo et al. 2021; see Box 19.3). This system generates different relational values of interaction with forests due to various groups of social actors having different conceptions of environmental uses that imply contrasting possibilities to implement their vision in silvopastoral systems (SPS).

Box 19.3 Conceptual Framework of the Socio-Ecosystem for the Analysis of Silvopastoral Systems in Andean Northern Patagonia

While the ecological relationships between forestry and livestock aspects have been extensively studied to improve management practices and prevent ecosystem deterioration, it is essential to incorporate the social component as well due to its role as a modulator and user of the biophysical environment. In this context, Chillo et al. (2021) have developed a heuristic model to analyse the silvopastoral systems (SPS) in Chilean and Argentine northern Patagonia as a tool for local policy development and management. In this model, SPS are understood as social-ecological systems where the sociohistorical context ties in with forest ecosystems shaping cultural aspects, relational values, and anthropogenic assets, all of which together determine management practices. This socio-ecosystem presents ecosystem services as emergent properties that are also regulated by external anthropogenic and natural drivers (Fig. 19.5). A key concept in the model is the relational values, which acknowledge that different social actors have differing perceptions of nature and approaches to

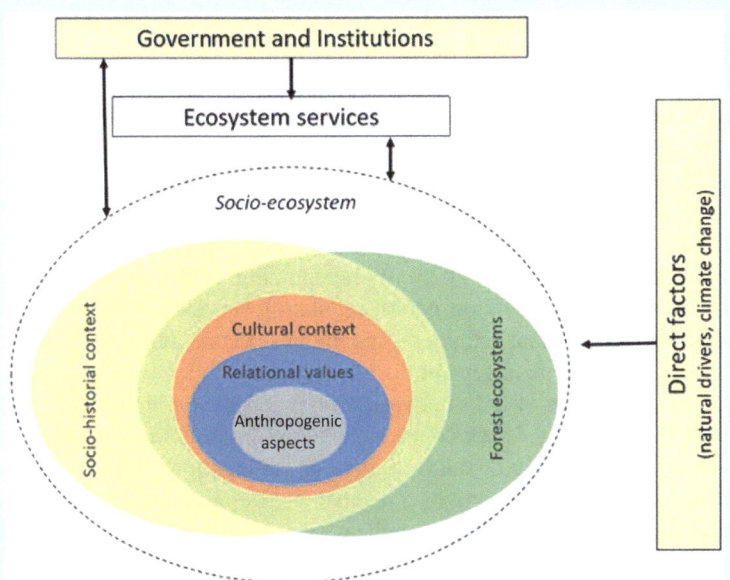

Fig. 19.5 Proposed conceptual framework of the socio-ecosystem for the analysis of silvopastoral systems in Andean northern Patagonia (extracted from Chillo et al. 2021)

environmental uses, resulting in contrasting possibilities to implement their respective vision in SPS. For example, people have developed different management strategies to achieve a higher quality of life according to their differing perceptions of "quality" and the specific resources available (funds, knowledge, services, etc). Therefore, different conditioning contexts determine different ways of valuing and using native forests—in other words, farmers' cultural views and motivations end up shaping management practices. For example, some farmers depend entirely on their agricultural income (e.g. meat or milk production) while others have a wider range of revenue sources (e.g. ecotourism, berry harvesting, mushroom collection). This can have an impact on the amount of investment into SPS and the economic risks taken by individual farmers. At the same time, some farmers are keener on innovating into new practices while others prefer to adhere to the status quo or historical legacies. This can affect their resilience to environmental degradation or external factors such as market fluctuations. The varying personal backgrounds and socio-economic resources determine specific ways of valuing and using native forests. The model developed by Chillo et al. (2021) therefore highlights the importance of considering the complex interconnections and interactions between social and ecological aspects, as well as the resulting emergent processes and key external drivers. This information is useful for aligning public policies with the needs of resource users to improve their feasibility and applicability and achieve sustainable forest management.

While the processes of globalization, including urbanization and mechanization, often contribute to the separation of society from forests, it is imperative to explore new ways to relink the ecosystem with a new kind of social system (Fischer et al. 2012). Forest–society connectivity could be understood as establishing robust connections between nature and society, as these ties are vital for effective forest conservation. Multifunctional forests present a fertile arena for various linkages between their ecosystems and users. When people directly derive tangible benefits from forest ecosystems—as emphasized in the concept of Nature's Contribution to People, they are more likely to develop a sense of stewardship concerning these environments (Plieninger et al. 2023). By recognizing the diverse contributions that forests offer—such as providing goods, regulating services, cultural significance, and supporting recreational activities—individuals become invested in their conservation and responsible management.

However, the quality of links between forests and society needs to be addressed along with their quantity. Weak connections between forests and people can result from a lack of dependency or sense of identity. As mentioned in the previous section, some forest restoration activities involve weak connections between decision-makers and forest ecosystems, causing them to end up with a limited understanding of local contexts and values. Community-based approaches recognize the unique knowledge, perspectives, and needs of local communities that have historically interacted with and depended on forest ecosystems. By empowering and involving these communities in decision-making processes, conflicts can be mitigated and forest conservation efforts can be better aligned with local values and aspirations.

Future research should delve more deeply into the social aspects of people's relationships to forests and explore the wide range of values (instrumental, relational, etc.) that shape human behaviour and decision-making concerning forests. After decades of decoupling people and forests, it is important to investigate new ways of establishing profound links between them to foster a sense of stewardship. Finally, the wide variety of social values, interests, and needs should be considered in the process of policymaking to improve feasibility and applicability and establish sustainable forest management.

References

Abatzoglou JT, Williams AP, Boschetti L, Zubkova M, Kolden CA (2018) Global patterns of interannual climate–fire relationships. Glob Chang Biol 24(11):5164–5175. https://doi.org/10.1111/gcb.14405

Arnberger A, Eder R, Allex B, Preisel H, Ebenberger M, Husslein M (2018) Trade-offs between wind energy, recreational, and bark-beetle impacts on visual preferences of national park visitors. Land Use Policy 76:166–177. https://doi.org/10.1016/j.landusepol.2018.05.007

Balvanera P, Pfisterer AB, Buchmann N, He JS, Nakashizuka T, Raffaelli D, Schmid B (2006) Quantifying the evidence for biodiversity effects on ecosystem functioning and services. Ecol Lett 9:1146–1156. https://doi.org/10.1111/j.1461-0248.2006.00963.x

Balvanera P, Calderón-Contreras R, Castro AJ, Felipe-Lucia MR, Geijzendorffer IR, Jacobs S et al (2017) Interconnected place-based social–ecological research can inform global sustainability. Curr Opin Environ Sustain 29:1–7. https://doi.org/10.1016/j.cosust.2017.09.005

Bastos JR, Capellesso ES, Vibrans AC, Marques MC (2023) Human impacts, habitat quantity and quality affect the dimensions of diversity and carbon stocks in subtropical forests: a landscape-based approach. J Nat Conserv 73:126383. https://doi.org/10.1016/j.jnc.2023.126383

Baur JW, Tynon JF, Ries P, Rosenberger RS (2016) Public attitudes about urban forest ecosystem services management: a case study in Oregon cities. Urban For Urban Green 17:42–53. https://doi.org/10.1016/j.ufug.2016.03.012

Bennett AF (1999) Linkages in the landscape: the role of corridors and connectivity in wildlife conservation (No. 1). IUCN

Bierwagen BG (2007) Connectivity in urbanizing landscapes: the importance of habitat configuration, urban area size, and dispersal. Urban Ecosyst 10(1):29–42. https://doi.org/10.1007/s11252-006-0011-6

Biggs R, Clements H, de Vos A, Folke C, Manyani A, Maciejewski K et al (2021) What are social-ecological systems and social-ecological systems research? In: Biggs R, de Vos A, Preiser R, Clements H, Maciejewski K, Schlüter M (eds) The Routledge handbook of research methods for social-ecological systems, 1st edn. Routledge, pp 3–26. https://doi.org/10.4324/9781003021339-1

Blackhall M, Raffaele E, Paritsis Tiribelli F, Morales JM, Kitzberger T, Gowda JH, Veblen TT (2017) Effects of biological legacies and herbivory on fuels and flammability traits: a long-term experimental study of alternative stable states. J Ecol 105:1309–1322. https://doi.org/10.1111/1365-2745.12796

Breyne J, Dufrêne M, Maréchal K (2021) How integrating 'socio-cultural values' into ecosystem services evaluations can give meaning to value indicators. Ecosyst Serv 49:101278. https://doi.org/10.1016/j.ecoser.2021.101278

Briggs JM, Hoch GA, Johnson LC (2002) Assessing the rate, mechanisms, and consequences of the conversion of tallgrass prairie to J*uniperus virginiana* forest. Ecosystems 5:578–586. https://doi.org/10.1007/s10021-002-0187-4

Chapin FS III, Kofinas GP, Folke C (2009) Principles of ecosystem stewardship: resilience-based natural resource management in a changing world. Springer, New York

Chillo V, Vazquez DP, Amoroso MM, Bennett E (2018) Land-use intensity indirectly affects ecosystem services mainly through plant functional identity in a temperate forest. Funct Ecol 32:1390–1399. https://doi.org/10.1111/1365-2435.13064

Chillo V, Ladio AH, Salinas Sanhueza J, Soler R, Arpigiani DF, Rezzano CA et al (2021) Silvopastoral systems in northern Argentine-Chilean andean Patagonia: ecosystem services provision in a complex territory. In: Peri PL, Martínez Pastur G, Nahuelhual L (eds) Ecosystem services in Patagonia: a multi-criteria approach for an integrated assessment. Springer, Cham, pp 115–137

Crooks KR, Sanjayan M (2006) Connectivity conservation. Cambridge University Press

Cudney-Valenzuela SJ, Arroyo-Rodríguez V, Morante-Filho JC, Toledo-Aceves T, Andresen E (2023) Tropical forest loss impoverishes arboreal mammal assemblages by increasing tree canopy openness. Ecol Appl 33(1):e2744. https://doi.org/10.1002/eap.2744

Curtis PG, Slay CM, Harris NL, Tyukavina A, Hansen MC (2018) Classifying drivers of global forest loss. Science 361(6407):1108–1111. https://doi.org/10.1126/science.aau3445

Delgado TS, McCall MK, López-Binnqüist C (2023) Non-timber forest products: small matters, big significance, and the complexity of reaching a workable definition for sustainability. Small-scale For 22(1):37–68. https://doi.org/10.1007/s11842-022-09517-9

Derebe B, Alemu A, Asfaw Z (2023) Contribution of nontimber forest products earn to livelihood in rural households and the type of use: a systematic review. Int J For Res 14. https://doi.org/10.1155/2023/9643290

Díaz S, Cabido M (2001) Vive la différence: plant functional diversity matters to ecosystem functioning. Trends Ecol Evol 16:646–655. https://doi.org/10.1016/S0169-5347(01)02283-2

Díaz S, Demissew S, Carabias J, Joly C, Lonsdale M, Ash N et al (2015) The IPBES conceptual framework—connecting nature and people. Curr Opin Environ Sustain 14:1–16. https://doi.org/10.1016/j.cosust.2014.11.002

Díaz S, Pascual U, Stenseke M, Martín-López B, Watson RT, Molnár Z et al (2018) Assessing nature's contributions to people. Science 359(6373):270–272. https://doi.org/10.1126/science.aap88

Díaz Vélez MC, Silva WR, Pizo MA, Galetto L (2015) Movement patterns of frugivorous birds promote functional connectivity among Chaco Serrano woodland fragments in Argentina. Biotropica 47:475–483. https://doi.org/10.1111/btp.12233

Egoh B, Reyers B, Rouget M, Richardson DM, Le Maitre DC, van Jaarsveld AS (2008) Mapping ecosystem services for planning and management. Agric Ecosyst Environ 127(1–2):135–140. https://doi.org/10.1016/j.agee.2008.03.013

Espelta JM, Cruz-Alonso V, Alfaro-Sánchez R, Hampe A, Messier C, Pino J (2020) Functional diversity enhances tree growth and reduces herbivory damage in secondary broadleaf forests, but does not influence resilience to drought. J Appl Ecol 57:2362–2372. https://doi.org/10.1111/1365-2664.13728

Fahrig L, Lefkovitch LP, Merriam HG (1983) Population stability in a patchy environment. In: Lauenroth WK, Skogerboe GV, Flug M (eds) Analysis of ecological systems: state-of-the-art in ecological modelling. Elsevier, New York, pp 61–67. https://doi.org/10.1016/B978-0-444-42179-1.50010-9

Fahrig L, Merriam HG (1985) Habitat patch connectivity and population survival. Ecology 66:1762–1768. https://doi.org/10.2307/2937372

Fahrig L (2013) Rethinking patch size and isolation effects: the habitat amount hypothesis. J Biogeogr 40:1649–1663. https://doi.org/10.1111/jbi.12130

Fahrig L, Arroyo-Rodríguez V, Cazetta E, Ford A, Lancaster J, Ranius T (2021) Landscape connectivity. In: Francis RA, Millington JDA, Perry GLW, Minor ES (eds) The Routledge handbook of landscape ecology. Routledge, London, pp 67–88. https://doi.org/10.4324/9780429399480

FAO (1998) Non-wood forest products and income gene-ration. Unasylva No. 198

FAO and UNEP (2020) The State of the World's Forests 2020. Forests, biodiversity and people. Rome

Fischer J, Hartel T, Kuemmerle T (2012) Conservation policy in traditional farming landscapes. Conserv Lett 5:167–175. https://doi.org/10.1111/j.1755-263X.2012.00227.x

Folke C, Carpenter SR, Walker B, Scheffer M, Chapin T, Rockström J (2010) Resilience thinking: integrating resilience, adaptability and transformability. Ecol Soc 15(4):20. https://doi.org/10.5751/ES-03610-150420

Folke C, Biggs R, Norström AV, Reyers B, Rockström J (2016) Social-ecological resilience and biosphere-based sustainability science. Ecol Soc 21(3):41. https://doi.org/10.5751/ES-08748-210341

Forrester DI, Bauhus J (2016) A review of processes behind diversity—productivity relationships in forests. Curr For Rep 2:45–61. https://doi.org/10.1007/s40725-016-0031-2

Gaige M, Glogower Y (2016) A fieldbook: great mountain forest. Yale Global Institute of Sustainable Forestry, Newhaven

Gupta SK, Pandey AC (2019) Change detection of landscape connectivity arisen by forest transformation in Hazaribagh wildlife sanctuary, Jharkhand (India). Spat Inf Res 28(4):391–404. https://doi.org/10.1007/s41324-019-00301-0

Gupta D, Sinha M, Chhatre A (2022) India's forest rights act and indigenous claims to community forest resources: a case study of Lavari, Maharashtra. World Dev Perspect 27:100449. https://doi.org/10.1016/j.wdp.2022.100449

Hamann M, Biggs R, Reyers B (2015) Mapping social–ecological systems: identifying 'green-loop' and 'red-loop' dynamics based on characteristic bundles of ecosystem service use. Glob Environ Chang 34:218–226. https://doi.org/10.1016/j.gloenvcha.2015.07.008

Hansen MC, Potapov PV, Moore R, Hancher M, Turubanova SA et al (2013) High-resolution global maps of 21st-century forest cover change. Science 342:850–853. https://doi.org/10.1126/science.1244693

Hansen MM, Jones R, Tocchini K (2017) Shinrin-yoku (forest bathing) and nature therapy: a state-of-the-art review. Int J Environ Res Public Health 14(8):851. https://doi.org/10.3390/ijerph14080851

Hessburg PF, Agee JK, Franklin JF (2005) Dry forests and wildland fires of the inland Northwest USA: contrasting the landscape ecology of the pre-settlement and modern eras. For Ecol Manage 211(1–2):117–139. https://doi.org/10.1016/j.foreco.2005.02.016

Hoyos LE, Cingolani AM, Zak MR, Vaieretti MV, Gorla DE, Cabido MR (2013) Deforestation and precipitation patterns in the arid Chaco forests of Central Argentina. Appl Veg Sci 16(2):260–271. https://doi.org/10.1111/j.1654-109X.2012.01218.x

Jactel H, Koricheva J, Castagneyrol B (2019) Responses of forest insect pests to climate change: not so simple. Curr Opin Insect Sci 35:103–108. https://doi.org/10.1016/j.cois.2019.07.010

Kangas J, Ollikainen M (2022) A PES scheme promoting forest biodiversity and carbon sequestration. Forest Policy Econ 136:102692. https://doi.org/10.1016/j.forpol.2022.102692

Lansing JS (2003) Complex adaptive systems. Ann Rev Anthropol 32(1):183–204. https://doi.org/10.1146/annurev.anthro.32.061002.093440

Lapin K, Oettel J, Steiner H, Langmaier M, Sustic D, Starlinger F et al (2019) Invasive alien plant species in unmanaged forest reserves, Austria. NeoBiota 48:71–96

Le Roux DS, Ikin K, Lindenmayer DB, Manning AD, Gibbons P (2014) The future of large old trees in urban landscapes. PLoS One 9(6):e99403. https://doi.org/10.1371/journal.pone.0099403

Lee H, Mayer H, Chen L (2016) Contribution of trees and grasslands to the mitigation of human heat stress in a residential district of Freiburg, Southwest Germany. Landsc Urban Plan 148:37–50. https://doi.org/10.1016/j.landurbplan.2015.12.004

Lee I, Choi H, Bang KS, Kim S, Song M, Lee B (2017) Effects of forest therapy on depressive symptoms among adults: a systematic review. Int J Environ Res Public Health 14(3):321. https://doi.org/10.3390/ijerph14030321

Lin B, Ge J (2020) To harvest or not to harvest? Forest management as a trade-off between bioenergy production and carbon sink. J Clean Prod 268:122219. https://doi.org/10.1016/j.jclepro.2020.122219

Lindenmayer DB, Laurance WF (2016) The ecology, distribution, conservation and management of large old trees. Biol Rev 92:1434–1458. https://doi.org/10.1111/brv.12290

Lundberg J, Moberg F (2003) Mobile link organisms and ecosystem functioning: implications for ecosystem resilience and management. Ecosystems 6:87–98. https://doi.org/10.1007/s10021-002-0150-4

Little C, Lara A, McPhee J, Urrutia R (2009) Revealing the impact of forest exotic plantations on water yield in large scale watersheds in south-Central Chile. J Hydrol 374(1–2, 162):–170. https://doi.org/10.1016/j.jhydrol.2009.06.011

Maier C, Abrams JB (2018) Navigating social forestry–a street-level perspective on National Forest management in the US Pacific northwest. Land Use Policy 70:432–441. https://doi.org/10.1016/j.landusepol.2017.11.031

Mapes LV (2016) Collateral damage: rushing to stop a fire that never came, Forest Service logged miles of big trees, critical habitat. The Seattle Times. Available at: https://projects.seattletimes.com/2016/collateral-damage/

Martello F, dos Santos JS, Silva-Neto CM, Cássia-Silva C, Siqueira KN, de Ataíde MVR et al (2023) Landscape structure shapes the diversity of plant reproductive traits in agricultural landscapes in the Brazilian Cerrado. Agric Ecosyst Environ 341:108216. https://doi.org/10.1016/j.agee.2022.108216

Martín-Forés I, Magro S, Bravo-Oviedo A, Alfaro-Sánchez R, Espelta JM, Frei T, Valdés-Correcher E, Rodríguez Fernández-Blanco C, Winkel G, Gerzabek G, González-Martínez SC (2020) Spontaneous forest regrowth in South-West Europe: consequences for nature's contributions to people. People Nat 2(4):980–994. https://doi.org/10.1002/pan3.10161

Martín-López B, Iniesta-Arandia I, García-Llorente M, Palomo I, Casado-Arzuaga I, Del Amo DG, Gómez-Baggethun E, Oteros-Rozas E, Palacios-Agundez I, Willaarts B et al (2012) Uncovering ecosystem service bundles through social preferences. PLoS One 7(6):e38970. https://doi.org/10.1371/journal.pone.0038970

Martínez Pastur G, Peri PL, Huertas Herrera A, Schindler S, Díaz-Delgado R, Lencinas MV, Soler R (2017) Linking potential biodiversity and three ecosystem services in silvopastoral managed forest landscapes of Tierra del Fuego, Argentina. Int J Biodivers Sci Ecosyst Serv Manag 13(2):1–11. https://doi.org/10.1080/21513732.2016.1260056

McDowell NG, Allen CD, Anderson-Teixeira K, Aukema BH, Bond-Lamberty B, Chini L et al (2020) Pervasive shifts in forest dynamics in a changing world. Science 368(6494):eaaz9463

Meacham M, Norström AV, Peterson GD, Andersson E, Bennett E, Biggs R, Crouzat E et al (2022) Advancing research on ecosystem service bundles for comparative assessments and synthesis. Ecosyst People 18:99–111. https://doi.org/10.1080/26395916.2022.2032356

Montagnini F, Jordan CF (2005) Tropical forest ecology: the basis for conservation and management. Springer, Berlin

Mueller T, Lenz J, Caprano T, Fiedler W, Böhning-Gaese K (2014) Large frugivorous birds facilitate functional connectivity of fragmented landscapes. J Appl Ecol 51:684–692. https://doi.org/10.1111/1365-2664.12247

Muradian R, Pascual U (2018) A typology of elementary forms of human-nature relations: a contribution to the valuation debate. Curr Opin Environ Sustain 35:8–14. https://doi.org/10.1016/j.cosust.2018.10.014

Nacif ME, Kitzberger T, Garibaldi LA (2020) Positive outcomes between herbivore diversity and tree survival: responses to management intensity in a Patagonian forest. For Ecol Manage 458:117738. https://doi.org/10.1016/j.foreco.2019.117738

Nisbet TR, Andreucci MB, De Vreese R, Högbom L, Kay S, Kelly-Quinn M et al (2022) Forest green infrastructure to protect water quality: a step-by-step guide for payment schemes. In: Nakamura F (ed) Green infrastructure and climate change adaptation: function, implementation and governance. Springer Nature, Singapore, pp 105–131. https://doi.org/10.1007/978-981-16-6791-6

Noss RF, Harris LD (1986) Nodes, networks, and MUMs: preserving diversity at all scales. Environ Manag 10:299–309. https://doi.org/10.1007/BF01867252

Noss RF (1991) Landscape connectivity: different functions at different scales. In: Hudson WE (ed) Landscape linkages and biodiversity. Island Press, Washington, pp 27–39. https://eurekamag.com/research/032/112/032112795.php

O'Farrell PJ, Anderson PM (2010) Sustainable multifunctional landscapes: a review to implementation. Curr Opin Environ Sustain 2(1–2):59–65. https://doi.org/10.1016/j.cosust.2010.02.005

Ostrom E (2009) A general framework for analyzing sustainability of social-ecological systems. Science 325:419–422. https://doi.org/10.1126/science.1172133

Pausas JG, Keeley JK (2021) Wildfires and global change. Front Ecol Environ 19(7):387–395. https://doi.org/10.1002/fee.2359

Payette S, Delwaide A (2004) Dynamics of subarctic wetland forests over the past 1500 years. Ecological monographs 74(3):373–391. https://doi.org/10.1890/03-4033

Pendrill F, Gardner TA, Meyfroidt P, Persson UM, Adams J, Azevedo T, Bastos Lima MG, Baumann M, Curtis PG, De Sy V, Garrett R (2022) Disentangling the numbers behind agriculture-driven tropical deforestation. Science 377(6611). https://doi.org/10.1126/science.abm926

Peri PL, Hansen NE, Bahamonde HA, Lencinas MV, von Müller AR, Ormaechea S et al (2016) Silvopastoral systems under native forest in Patagonia Argentina. In: Peri PL, Dube F, Varella A (eds) Silvopastoral systems in southern South America. Springer, Cham, pp 117–168. https://doi.org/10.1007/978-3-319-24109-8

Peterson GD, Harmáčková ZV, Meacham M, Queiroz C, Jiménez-Aceituno A, Kuiper JJ et al (2018) Welcoming different perspectives in IPBES. Ecol Soc 23(1). https://doi.org/10.5751/ES-10134-230139

Plieninger T, Shamohamadi S, García-Martín M, Quintas-Soriano C, Shakeri Z, Valipour A (2023) Community, pastoralism, landscape: eliciting values and human-nature connectedness of forest-related people. Landsc Urban Plan 233:104706. https://doi.org/10.1016/j.landurbplan.2023.104706

Raudsepp-Hearne C, Peterson GD, Bennett EM (2010) Ecosystem service bundles for analyzing tradeoffs in diverse landscapes. Proc Natl Acad Sci 107(11):5242–5247. https://doi.org/10.1073/pnas.0907284107

Rautiainen A, Wernick I, Waggoner PE, Ausubel JH, Kauppi PE (2011) A national and international analysis of changing forest density. PLoS One 6(5):e19577. https://doi.org/10.1371/journal.pone.0019577

Ricketts TH, Regetz J, Steffan-Dewenter I, Cunningham SA, Kremen C, Bogdanski A et al (2008) Landscape effects on crop pollination services: are there general patterns? Ecol Lett 11(5):499–515. https://doi.org/10.1111/j.1461-0248.2008.01157.x

Root-Bernstein M, Svenning JC (2017) Restoring connectivity between fragmented woodlands in Chile with a reintroduced mobile link species. Perspect Ecol Conserv 15(4):292–299. https://doi.org/10.1016/j.pecon.2017.09.001

Shakeri Z, Mohammadi-Samani K, Bergmeier E, Plieninger T (2021) Spiritual values shape taxonomic diversity, vegetation composition, and conservation status in woodlands of the northern Zagros, Iran. Ecol Soc 26:687–706. https://doi.org/10.5751/ES-12290-260130

Senf C, Seidl R, Hostert P (2017) Remote sensing of forest insect disturbances: current state and future directions. Int J Appl Earth Obs Geoinf 60:49–60. https://doi.org/10.1016/j.jag.2017.04.004

Suraci JP, Littlefield CE, Nicholson CC, Hunter MC, Sorensen A, Dickson BG (2023) Mapping connectivity and conservation opportunity on agricultural lands across the conterminous United States. Biol Conserv 278:109896. https://doi.org/10.1016/j.biocon.2022.109896

Taylor PD, Fahrig L, With KA (2006) Landscape connectivity: a return to the basics. In: Crooks KR, Sanjayan M (eds) Connectivity conservation. Cambridge University Press, Cambridge, pp 29–43. https://doi.org/10.1017/CBO9780511754821.003

Tiribelli F, Morales JM, Gowda JH, Mermoz M, Kitzberger T (2019) Non-additive effects of alternative stable states on landscape flammability in NW Patagonia: fire history and simulation modelling evidence. Int J Wildfire 28:149–159. https://doi.org/10.1071/WF18073

Turkelboom F, Leone M, Jacobs S, Kelemen E, García-Llorente M, Baró F et al (2018) When we cannot have it all: ecosystem services trade-offs in the context of spatial planning. Ecosyst Serv 29:566–578. https://doi.org/10.1016/j.ecoser.2017.10.011

Ulyshen M, Urban-Mead KR, Dorey JB, Rivers JW (2023) Forests are critically important to global pollinator diversity and enhance pollination in adjacent crops. Biol Rev Camb Philos Soc 10:111. https://doi.org/10.1111/brv.12947

Yasmi Y, Kelley LC, Enters T (2013) Community–outsider conflicts over forests: perspectives from Southeast Asia. Forest Policy Econ 33:21–27. https://doi.org/10.1016/j.forpol.2012.05.001

Zhang L, Jin G, Wei X, Xie P, Liu G, Liu Y (2019) Multiscale analysis of patch and landscape characteristics of forest fragmentation in Liaoning Province, China. Reg Environ Change 19(4):1175–1186. https://doi.org/10.1007/s10113-019-01476-w

Zurita GA, Zuleta GA (2009) Bird use of logging gaps in a subtropical mountain forest: the influence of habitat structure and resource abundance in the Yungas of Argentina. For Ecol Manage 257:271–279. https://doi.org/10.1016/j.foreco.2008.08.032

Open Access This chapter is licensed under the terms of the Creative Commons Attribution 4.0 International License (http://creativecommons.org/licenses/by/4.0/), which permits use, sharing, adaptation, distribution and reproduction in any medium or format, as long as you give appropriate credit to the original author(s) and the source, provide a link to the Creative Commons license and indicate if changes were made.

The images or other third party material in this chapter are included in the chapter's Creative Commons license, unless indicated otherwise in a credit line to the material. If material is not included in the chapter's Creative Commons license and your intended use is not permitted by statutory regulation or exceeds the permitted use, you will need to obtain permission directly from the copyright holder.

20. Conservation Initiatives to Connect the Landscape Across Indigenous and Local Communities: Perspectives from Chilean and Peruvian Biosphere Reserves

Gabriela Albarracín-Llúncor, Eduardo Jackson Filomeno, Cesar Ipenza Peralta, Andrés Moreira-Muñoz, and Andrea A. Pino Piderit

Oxapampa Valley (Photo: Carolina Perret)

G. Albarracín-Llúncor (✉)
Working Group Knowledge Systems and Innovation (AGWI), Department of Sustainable Agricultural Systems, University of Natural Resources and Life Sciences, Vienna, Austria
e-mail: gabriela.albarracin@boku.ac.at

E. J. Filomeno
Red de Iniciativas de Áreas de Conservación de Oxapampa (RIACO), Oxapampa, Peru

C. I. Peralta
Escuela de Gestión Pública, Universidad del Pacífico (UP), Lima, Peru

© The Author(s) 2025
K. Lapin et al. (eds.), *Ecological Connectivity of Forest Ecosystems*,
https://doi.org/10.1007/978-3-031-82206-3_20

Abstract

Biosphere reserves (BRs) aim to achieve biodiversity conservation, sustainable (social and economic) development, and logistical support in the form of education, research, and monitoring through complex spatial and governance arrangements in which Indigenous and local communities play an important role in their development. Landscape connectivity has been a challenge for the design of BRs, as for similar landscape-scale or regional conservation initiatives. Using Peru and Chile as examples, this paper presents conservation initiatives implemented in both countries to connect landscapes through protected areas and BRs. This is done through the concepts of Landscape Conservation (LC) mechanisms in Chile and Voluntary Conservation Initiatives (VCI) in Peru. LC expresses a shared, concerted, and consensual vision of the affected areas among the various local public and private actors, and represents a voluntary commitment. VCIs are completely voluntary, where individuals or legal entities express their willingness to protect an area of land, allowing them to become directly involved in the conservation of biodiversity by connecting large fragments of natural ecosystems. The key objective of both LC and VCI is to connect landscapes, a critical factor in landscape-scale conservation.

Keywords

Biosphere reserve · Protected area · Biodiversity conservation · Indigenous and local communities · Landscape Conservation

Introduction

Meeting the challenges of the current global biodiversity crisis requires the development of conservation strategies at the landscape scale, involving transdisciplinary programs for a common vision of the actions needed for the sustainability transition. In this sense, landscape-scale conservation can benefit from long-established global initiatives such as UNESCO's Man and the Biosphere (MAB) program and the associated World Network of Biosphere Reserves (WNBR) program.

Biosphere reserves (BR) aim to achieve (a) biodiversity conservation, (b) sustainable (social and economic) development, and (c) logistical support in the form of education, research, and monitoring (Barraclough et al. 2023; Kratzer and

---

A. Moreira-Muñoz
Instituto de Geografía, Pontificia Universidad Católica de Valparaíso, (PUCV), Valparaiso, Chile

A. A. Pino Piderit
Cape Horn International Center for Global Change Studies and Biocultural Conservation (CHIC), Puerto Williams, Chile

Transdisciplinary Center for Environmental Studies and Sustainable Human Development, Universidad Austral de Chile (UACh), Valdivia, Chile

Ammering 2019; UNESCO 1996) through complex spatial and governance arrangements (Ferreira et al. 2018; Paül et al. 2022). A critical aspect of the spatial design of BRs is their zonation scheme, which consists of a conservation core (usually a strictly protected area), a buffer zone consisting of an area that supports human activities with low environmental impacts, and a transition zone where economic activities with relatively higher impacts are allowed (UNESCO 1996).

Landscape connectivity within and beyond this zoning scheme has always been a challenge for the design of BRs, and similar landscape-scale or regional conservation initiatives. Landscape connectivity refers to "the degree to which the landscape facilitates or impedes movement between resource patches" (Taylor et al. 2006). This means that land use patterns within a landscape affect the movement, population dynamics, and structure of biotic communities, which in turn has important implications for biodiversity conservation (Zarnetske et al. 2017).

Protected areas (PAs), which typically correspond to the core areas of BRs, were established to conserve biodiversity, but they are often challenged by human land use, climate change, and invasive species, as well as social, political, and economic constraints that reduce their habitat quality at the landscape scale (Martinez-Harms et al. 2021; Mengist 2020). According to the International Union for Conservation of Nature (IUCN), PAs are "geographical areas that are recognized, dedicated and managed through legal or other effective means to achieve the long-term conservation of nature with its associated ecosystem services and cultural values" (Day et al. 2012). BRs are often classified as IUCN Category V (protected landscape), but in which humans play a major role in their development (Dudley 2008).

The Latin American and Caribbean region is one of the most biologically and culturally diverse regions in the world, and it has suffered for decades from high deforestation rates, disordered land use resulting from the extensive application of short-term productive systems, and consequently a loss of species caused by habitat destruction and land degradation (Guevara and Laborde 2008; Toledo 2008). There is an urgent need to better understand the social-ecological dynamics that can help the achievement of progressive governance transitions through recommendations for conservation policies, especially related to the biocultural values of Indigenous and local communities.

Peru and Chile have implemented conservation initiatives in recent years to connect landscapes via PA and BR. This occurs by way of the concepts of Landscape Conservation (LC) mechanism in Chile and Voluntary Conservation Initiatives (VCI) in Peru, respectively. The LC used in Chile expresses a shared, concerted, and consensual vision of affected areas among the various local public and private actors that is codified in an Agreement or Act (recommended duration: 10–15 years) constituting a voluntary commitment. LC aims to maintain or restore ecosystem composition and structure, thus enhancing biodiversity protection and supporting ecosystems on which humans depend (Travis Belote et al. 2021; Cao et al. 2023). In the case of Peru, VCIs are entirely uncompelled. Individuals or legal entities—e.g., owners of private properties—request the state to grant them a concession for a public domain area. In both cases, the individuals or legal entities express their willingness to protect a land area, allowing them to become directly involved in the conservation of biological diversity by connecting large fragments of natural ecosystems (Servicio Nacional de Areas Naturales Protegidas por el Estado 2022; Summers 2021). The key aim of both LC and VCI is to connect landscapes, a crucial factor for biodiversity conservation (Dhendup et al. 2023).

Main Characteristics of Selected Biosphere Reserves

The Oxapampa-Asháninka-Yánesha Biosphere Reserve (BIOAY)—Peru is an important conservation area due to the presence of Indigenous cultures, diverse landscape settings and PAs.

The BIOAY is the first BR to include various groups as key actors in the BR design: the Asháninka and Yánesha Indigenous communities, Austro-German descendants, and the local and regional governments (Summers 2021). These different types of actors have different ways of understanding their environment and appropriating concepts.

The BIOAY was designated in 2010 and it is one of seven BRs in the Peruvian territory. It is located in the department of Pasco Central Selva and covers 1,800,000 ha, including four PAs: Yanachaga Chemillén National Park, the Yánesha and El Sira Communal Reserves, and the San Matías-San Carlos Protected Forest (Fig. 20.1). Besides their conservation role, the main objective of the communal reserves is to guarantee the sustainable use of wild resources for the native communities of the Asháninka, Yánesha, Ashéninka, and Shipibo-Conibo ethnic groups by reducing external pressure within their territory, ensuring their participation in conservation, and improving their living conditions (Servicio Nacional de Areas Naturales Protegidas por el Estado 2022).

In Chile, there are ten BRs that protect around 100 ecosystems ranging from the arid Andes to the deciduous forests and peatlands of the southernmost islands of the American continent. Although BRs are not yet included in current legislation, they are mentioned in the draft law for the new PAs and Biodiversity Service (Republica de Chile-Senado 2014). They are also being considered in the land management plans of the current Land Management Policy (Carvajal-Mascaró et al. 2019). The southern temperate forest of Chile is recognized as a globally relevant ecoregion and part of the biodiversity hotspot of central-austral Chile (Moreira-Muñoz et al. 2020; Mittermeier et al. 2011). It is an ecoregion highly disturbed by large volcanic eruptions and almost total ice coverage during the last glaciation (Fig. 20.2).

Beyond the biogeographic history, the southern forests show an ancestral biocultural relationship with the Mapuche people, in the symbiotic relationship between man and forest, especially with several species such as the monkey puzzle tree (*Araucaria araucana*) and its fruits (piñones in Spanish) (Fig. 20.3). This temperate forest represents an area of global importance for the conservation of the landscape; it is also a center of endemism, especially for arboreal species, including monotypic genera such as *Gomortega*, *Pitavia*, *Legrandia*, *Prumnopitys*, and *Fitzroya*. Remarkable faunal elements are some of the rare marsupials of America such as the monito del monte or colocolo opossum (genus *Dromiciops*). The area is also home

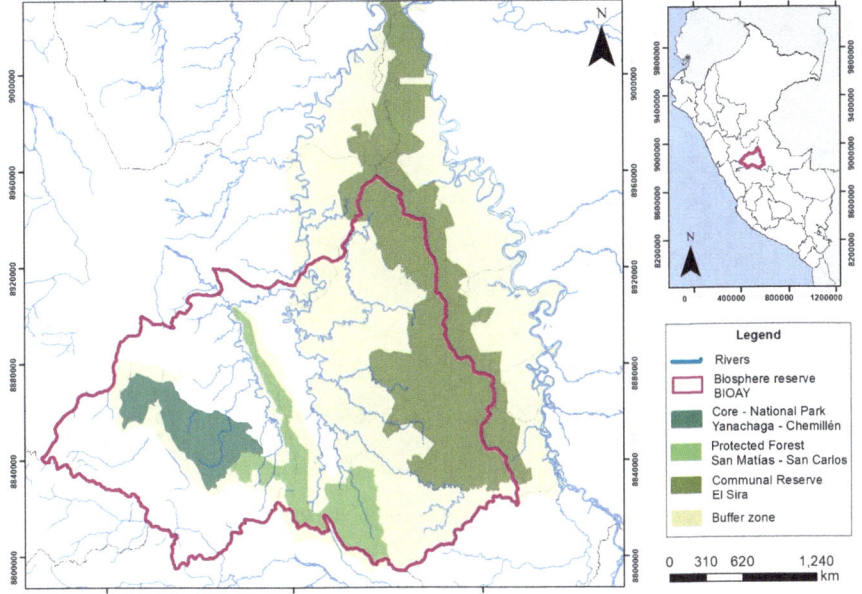

Fig. 20.1 Map showing the delimitation of the BIOAY in Pasco Central Selva, including four protected areas. Adapted by Carolina Perret

Fig. 20.2 Chilean austral temperate forest, in which a rich array of elements of geodiversity and biodiversity continuously interact. (**a**) Lago Noroeste, a glacial lake in the "Bosques Templados Lluviosos de los Andes Australes biosphere reserve" in the Futaleufú Andes (43,2°S); (**b**) Lonquimay volcano and evergreen forest in the Araucarias biosphere reserve (38,4° S). (Photo: Andrés Moreira-Muñoz)

to the Pudú or southern deer (*Pudu puda*) as well as to Darwin's Frog (*Rhinoderma darwinii*).

The Andean foothills and highs between latitude 39.5° S and 43° S belong to the "Bosques Templados Lluviosos de los Andes Australes Biosphere Reserve", declared in 2007 and encompassing 2,171,484 hectares. The BR is composed of several cores, buffers, and transition zones. The areas considered cores are 11 PAs with a total of 436,326 hectares, including Douglas Tomkins Pumalín National

Fig. 20.3 Pehuén or Araucaria (*Araucaria araucana*) in the Reserva de la Biosfera Araucarias. (Photo: Andrés Moreira-Muñoz)

Park, once considered one of the world's largest private PAs and now under Chilean public administration (Corporación Nacional Forestal 2017). The buffer and transition zones together cover 1,735,158 hectares. In all, the reserve features 19 Andean ecosystems that are included in one of the three zonation criteria.

Historically, the territory has experienced forest and human exploitation associated with the "Complejo Forestal y Maderero Panguipulli", an economic and social entrepreneurship of the 1970s that ended abruptly and violently during the Pinochet dictatorship (Pino and Cardyn 2014). Some of the major challenges faced by the "Bosques Templados Lluviosos de los Andes Australes BR" are caused by a weak management scheme related to the insufficient participatory framework. A further challenge is the landscape connectivity with the coastal ecosystems comprising a continuum from the Andes towards the Pacific Ocean. One of the iconic species connecting the conservation landscape from the Andes to the coast is the "lahuán or alerce" (*Fitzroya cupressoides*) (Fig. 20.4). It is considered to be the longest-living tree on the planet; one specimen of alerce has been estimated at 5400 years of age (Fischer 2022). One significant attempt to connect the Andean and coastal ecosystems for landscape-scale conservation is the Rio San Pedro Corridor.

Voluntary Conservation Initiatives in Peru

Peru has implemented several tools to promote VCI seeking to protect valuable biodiversity and often fragile ecosystems. VCI can relate to privately owned properties or public domain lands (Fig. 20.5). In the case of privately owned properties, the landowners—natural or legal persons—express their willingness to conserve

Fig. 20.4 Lahuán or alerce (*Fitzroya cupressoides*) in the "Alerce Costero National Park", a node connecting Andean and coastal ecosystems for landscape-scale conservation. (Photo: Andrés Moreira-Muñoz)

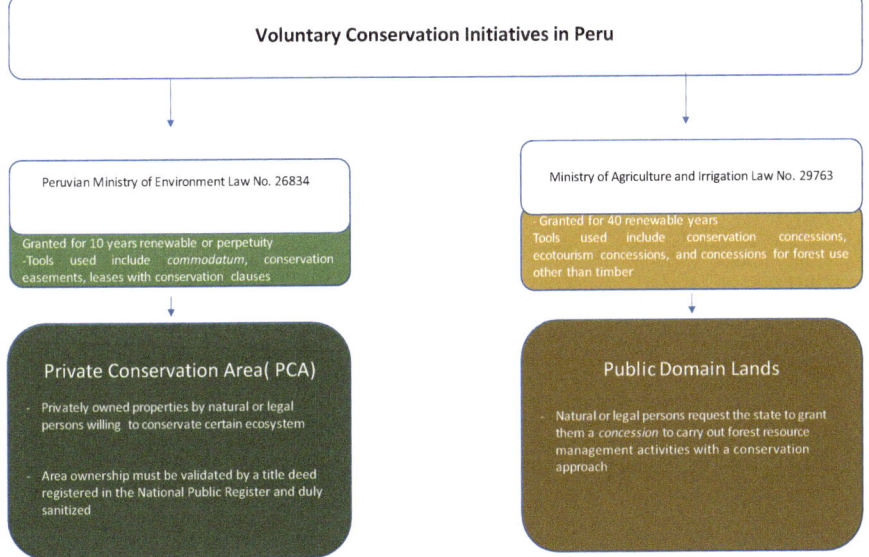

Fig. 20.5 Illustrative Definition of Voluntary Conservation Initiatives in Peru

certain ecosystems through any of several conservation tools approved for the Civil Code; they include commodatum, conservation easements, leases with conservation clauses, and others. Landowners may also seek recognition of their property as a Private Conservation Area (PCA) (Law No. 26834). Such recognition is granted by the state through the Ministry of the Environment and can be requested for a minimum of ten years (renewable) or in perpetuity (Monteferri 2019). PCAs have been employed as a tool for more than 20 years. According to the official listing, there are currently 13,944 PCAs protecting a total of 3.94 million hectars in a great diversity of ecosystems (Servicio Nacional de Areas Naturales Protegidas por el Estado 2022). The Servicio Nacional de Áreas Naturales Protegidas por el Estado—SERNANP [National Service of Natural Areas Protected by the State] defines PCAs as "properties privately owned by individuals or legal entities which contain representative samples of ecosystems belonging to landowners willing to voluntary protect that area. PCA is a tool allowing direct involvement of people, communities, non-governmental organizations and/or companies in the conservation of biological diversity by connecting large fragments of natural ecosystems" (Servicio Nacional de Areas Naturales Protegidas por el Estado 2022). For recognition as a PCA, the ownership of an area must be validated by a title deed registered in the National Public Register. In the case of public domain lands, natural or legal persons may request the state to grant them a concession to perform forest resource management activities with a conservation approach, such as conservation concessions and ecotourism concessions (Law No. 29763). Concessions are granted by the state through the Ministry of Agriculture and Irrigation for 40 years and are renewable.

Among the contributions to biodiversity conservation provided by VCIs are the following (Monteferri 2019):

- Supporting the creation of conservation corridors by connecting PAs, avoiding land fragmentation, and establishing "natural habitat islands".
- Contributing to reducing forest degradation and deforestation, avoiding land use change, and ensuring adequate management.
- Providing opportunities for research/environmental education.
- Providing ecotourism development, experiential tourism, and rural community tourism.
- Reducing the financial burden that biodiversity conservation represents for governments by helping to close the conservation financial gap through private investments.

Voluntary Conservation Initiatives in the BIOAY

The BIOAY covers the entire province of Oxapampa in the Peruvian department of Pasco, and there are several private landowners and concessionaires using various conservation tools and modalities: five PCAs, six conservation concessions, and five ecotourism concessions. There are also nine proposals for further PCAs in different development stages. Although VCIs in the BIOAY have only existed for a relatively short time, interest in creating new conservation areas in the province of Oxapampa is growing and expected to become even more dynamic in the coming years. In addition to conserving valuable ecosystems, the owners and concessionaires have been developing various activities in their respective areas so as to make them sustainable, such as research, environmental education programs, scientific tourism, ecotourism, environmental interpretation tourism, bird watching, wellness tourism, permaculture and bioconstruction, agroforestry systems, and organic food production.

In March 2017, a group of landowners and concessionaries decided to create the Red de Iniciativas de Áreas de Conservación de Oxapampa—RIACO [Oxapampa Conservation Area Initiatives Network] as a non-profit civil association with the main objectives of promoting the creation of new voluntary conservation initiatives, actively participating in the development of environmental policies, sharing experiences in conservation and sustainable economic activities, and implementing research, education, restoration and connectivity projects. RIACO currently has 14 partners that manage 15 VCIs in the provinces of Oxapampa (Dept. Pasco), Chanchamayo (Dept. Junín), and Puerto Inca (Dept. Huánuco). In the BIOAY, RIACO manages the La Suiza conservation concession (RDG N.° 028-2010-AG-DGFFS), the PCA Fundo las Neblinas (Ministerial Resolution N.° 311-2016-MINAM), the PCA Bosque de Churumazú (Ministerial Resolution N.° 330-2017-MINAM), the PCA Potsom Posho'll (Ministerial Resolution N.° 060-2021-MINAM), Ulcumano Ecolodge, El Palmeral, Osopampa, Fundo La Gorda, Fundo Los Abuelos, Fundo La Dama, Tierra de Bosques, Villa Rica I and II, and Quetzales. RIACO's associates conserve a total of 517,912 hectares of forest in the district of Chontabamba (Fig. 20.6)

Collectively, the VCIs have been contributing to the conservation of IUCN-threatened flora species such as the ulcuma (*Retrophylum rospigliosii*), walnut (*Juglans neotropica*), mountain cedar (*Cedrela montana*), and American cedar (*Cedrela odorata*), as well as threatened fauna species such as the Andean bear (*Tremarctos ornatus*), the gray woolly monkey (*Lagothrix cana*) (see Fig. 20.7), the machetero (*Dinomys branickii*), the sachavaca (*Tapirus terrestris*), the giant anteater (*Myrmecophaga tridactyla*), the giant armadillo (*Priodontes maximus*), the gray partridge (*Tinamus tao*), and frogs of the genus *Pristimantis*, among others.

Challenges and Opportunities for Voluntary Conservation Initiatives in the BIOAY

Although there has been growing interest in recent years, much more citizen participation is needed, and state and local governments still urgently need to draft appropriate measures to promote biodiversity conservation. The BIOAY's ecosystems are under constant threat, and the ecosystems services they provide to the population as well as their contribution to the region's economic activities can be only protected if inhabitants and institutions work together. The following three main pillars are fundamental for achieving this:

Fig. 20.6 Bosques de Neblina-Selva Central is located in a transition zone between the Andes and the Amazon forest. Private conservation area El Palmeral, Chontabamba—Oxapampa, Peru. (Photo: Patricia Reyna Sánchez—Aizcorbe)

Fig. 20.7 The gray woolly monkey (*L. cana*) feeds on the fruits and leaves of the palm trees (*Dictyocaryum lamarckianum*) that give their name to the El Palmeral conservation project in Chontabamba, Oxapampa. (Photo: Patricia Reyna Sánchez—Aizcorbe)

1. Strengthening of institutions necessary to protect the forest and wildlife heritage as well as the voluntary conservation initiatives against environmental threats and crimes;
2. Providing funds and technical assistance for the development of management capacities and sustainable economic activities within conservation initiatives; and
3. Implementation of effective incentives for the owners of voluntary conservation initiatives, including tax incentives, access to financing and donations, etc.

Compensation for the conservation efforts carried out by natural and legal persons on a voluntary basis with regard to valuable and fragile ecosystems should be a priority not only for the state but also for the society that benefits directly or indirectly from the work of persons and initiatives committed to safeguarding the common good.

Challenges for Voluntary Conservation Initiatives in Peru

One of the most common problems in Peru is the poor institutional framework, which leads to a lack of property deeds in PCAs, which in turn limits landowners in their ability to use their property for conservation purposes. The SERNANP does not support the process of preparing technical files, instead providing only legal recognition. Another problem is the lack of awareness among the population regarding the economic and social benefits that VCIs, and most specifically PCAs can provide. There is still a need for mechanisms promoting an increase in the number of new PCAs in biologically rich areas. While PCAs can contribute to biological conservation, it will be necessary to support the landowners and concessionaires in their area management through mechanisms of economic compensation for the following:

- Biodiversity conservation
- Carbon sequestration
- Ecosystem recovery
- Maintenance of environmental services such as water

PCAs are essential tools for conservation, and it is therefore important to support them with diverse types of compensations. An example is the Digital Compensation for Conservation in Peru (CDC) project conducted by the Catholic University Sedes Sapientiae and the Masbosques-Colombia Corporation, implemented in the departments of San Martin, Junín, and Ucayali. This project generates additional income for small farmers through compensation for biodiversity conservation on their land and the carbon capture of their agroforestry crops, forests, and reforestation areas. Compensations could prevent many PCAs from choosing not to extend their existence after completing their initial approval period (in most cases, 10 years) and help to turn them into perpetual initiatives instead. Adequate compensations and their benefits might also increase the active areas of conservation and the number of involved communities.

Besides these initiatives, there are mechanisms aimed at maintaining ecosystems in the lands of Indigenous and local communities—in the case of the BIOAY, the Yáneshas and Asháninkas in particular. These mechanisms have been linked to public initiatives such as the National Forest Conservation Program for Climate Change Mitigation (PNCBMCC) (Programa Nacional de Conservación de Bosques para la Mitigación del Cambio Climático 2023). The incentives provided by the PNCBMCC have benefited more than 15 Yánesha and Asháninka Indigenous communities in the province of Oxapampa. The department of Pasco is committed to the conservation and proper use of forests that represent a source of life and a great driver of development. The PNCBMCC is the most ambitious program of the Peruvian administration; its objective is to conserve forests by implementing mechanisms of conditional direct transfers (CDT). CDTs provide economic incentives and technical assistance to Indigenous and local communities to develop sustainable forestry activities, strengthen communal surveillance, and meet the community's basic needs by way of communal health kits, schools, promotion of sustainable economic activities, and other benefits. These measures are public-private initiatives that enable forest conservation in exchange for various types of compensation, and they should be expanded and replicated in areas of high diversity in the Amazon.

Landscape Conservation in Chile

Although there is permanent conflict over land use in the southern zone of Chile between the needs of production and conservation, it is necessary to achieve improvements and commitments towards greater sustainability of the territory and maintaining the quality of life of ancestral and rural populations associated with biocultural diversity. This is theoretically possible through what is referred to as "Landscape Conservation (LC)" or "Paisajes de Conservación" in Spanish—a mechanism for connecting forest patches and wetlands to benefit the management of ecosystem processes (Casale et al. 2014).

The establishment of LC has been promoted since 2007 by the National Commission on the Environment (CONAMA), and subsequently since 2009 by the Ministry of the Environment, as spaces for biodiversity conservation beyond PAs depending on the Servicio Nacional de Áreas Silvestres Protegidas del Estado (SNASPE) *[National Service of State Wildlife Protected Areas]*. A LC is defined as "an inhabited territory that possesses natural and cultural heritage of regional and/or national interest, delimited and managed with the purpose of implementing a consensual and effective conservation and development strategy that allows to maintain and/or improve the values of interest of the territory and contribute to the improvement of the population's quality of life" (PNUD GEF SIRAP 2013).

Landscape Conservation to Connect "Bosques Templados Lluviosos de los Andes Australes Biosphere Reserve" with Coastal Ecosystems

The initial steps date from 2004 to 2007, years prior to the establishment of the first LC in the Los Ríos region.

In the beginning, the scientific discussion focused on the need to connect the Andean Mountain range to the Coastal Range through the establishment of a biological corridor that would follow the course of the San Pedro River, whose headwaters are located in neighboring Argentina and which flows into the Pacific Ocean (Fig. 20.8). Due to a lack of information on possible but very plausible genetic differences that might exist between the populations of native species to be connected with the creation of the corridor, as well as the potential for invasive species originally present in only one of the mountain ranges to spread, the scientific consultation process eventually dismissed the idea of a full-scale corridor, preferring to prioritize a LC in the region under consideration of the different ecological attributes and conservation values.

First Landscape Conservation – "Paisajes de Conservación" in Chile

A project financed by the Global Environment Facility (GEF) and the CONAMA established the country's first LC located in the Los Ríos region between 2007 and 2013.

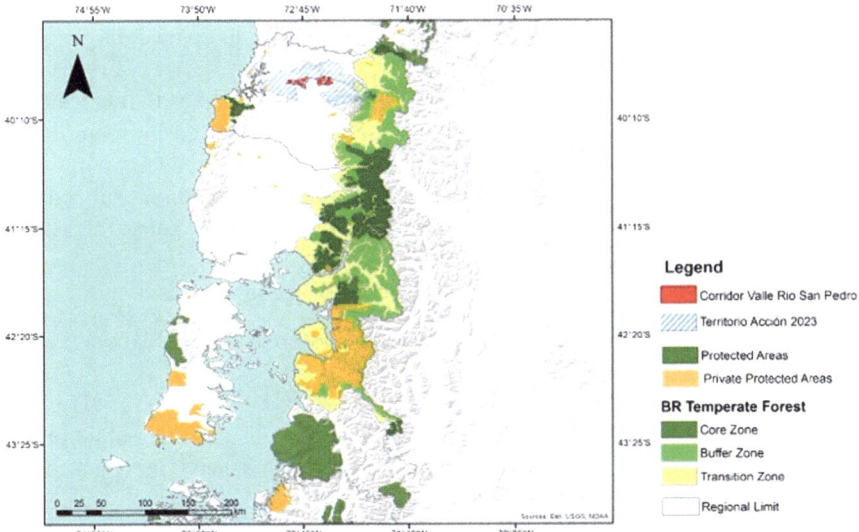

Fig. 20.8 Map of the Corridor Valle Rio San Pedro as a connection between the Chilean Andean Coastal Mountain located in the biosphere reserve "Bosques Templados Lluviosos de los Andes Australes". Adapted by Andrés Moreira-Muñoz

The process of creating this first LC considered:

1. *Ecological criteria and conservation values*

Definition and prioritization of the ecological criteria and conservation values to be used, for which scientists, conservation biologists, and ecologists were consulted. While sectors with higher conservation values than those found in the valley of the region existed after the application of the criteria matrix, the criteria serving as pillars for the formulation of the original definition were the presence of southern deciduous forest and lauriphyllous forest.

Due to the presence of patches of the northern oak (*Nothofagus obliqua*), forest fragments could be connected to each other with relatively little effort to facilitate the movement of local species.

2. *Characterization of social, cultural, and economic aspects*

The area was characterized in social, cultural, and economic terms, and a CONAMA call invited farmers to participate and become part of a network of pilot sites for biodiversity conservation. The objectives of this network were to:

- increase and improve the patches of habitat under protection
- minimize the impacts of the matrix
- promote connectivity between forest patches
- integrate biodiversity conservation into the matrix
- involve the local stakeholders so they could assist in the conservation effort.

The Transdisciplinary Center for Environmental Studies and Sustainable Human Development at the Universidad Austral de Chile led the implementation of a pilot project to promote biodiversity conservation.

Among the implemented actions were:

- The restoration of forest patches within fields in order to connect larger forest fragments;
- The installation of water troughs for livestock in pastures so as to remove those animals from the forests, thus reducing the pressure exerted by them on the regeneration of vegetation;
- Prevention of the transmission of diseases to wildlife, reduction of soil compaction and sedimentation of watercourses, and fencing off of some forest remnants; and
- Zonation of fields to define areas with different usage intensities.

Subsequently, 10 further pilot units or properties were added in 2016; at present, 22 pilot units have been created.

3. *Governance*

During the process of the installation and operation of the LC, public-private mechanisms such as governance frameworks were established. One such framework was the Regional System of Protected Areas (SIRAP) focused on the conservation of privately owned lands. Its purpose is "to implement in this space a shared territorial management model based on the needs and interests of all stakeholders living in the territory and the protection of biodiversity [...] with the objective of improving representativeness and decentralizing the management of PAs" (PNUD GEF SIRAP 2013). Regarding the governance of the LC, an initial measure was the establishment of the Municipalities Association and the Development Council. Subsequently, technical offices were installed in each of the two participating municipalities to work on reformulating the Local Development Plans (PLADECO) with the main aim of incorporating a conservation and co-management approach for the PA at the municipal level. The Municipalities Association articulates the action of both municipalities to protect and strengthen their natural heritage. The Development Council brings together the different stakeholders involved, elaborating the operating statutes and creating work plans for the management of the territory (PNUD GEF SIRAP 2013).

The conservation landscape governance structure addresses the following key issues:

- Geopolitical administration decisions often do not correspond to local conservation values.
- Fragility in the trust between municipal administration teams and Indigenous and local communities.
- Lack of connection between these new governance structures and the municipal administration structure.

Challenges and Opportunities for the Valle Rio San Pedro Corridor

Between 2007 and 2022, the local community fought against the construction of a 58 m high hydroelectric dam that would radically alter the hydrological regime of the river, affecting the biodiversity and cultural and geological heritage of the area (Fundacion Plantae 2024).

Years of demonstrations and legal appeals have so far stopped the project, but Chile's environmental legal framework is still so weak that this kind of large-scale environmental impact on a landscape scale is truly worrisome and keeps the local community and academics on their toes (Fig. 20.9).

In addition to these environmental concerns, the river and the landscape corridor have been recognized for their geoconservation aspects. In fact, several deposits of paleontological importance have been identified in the upper reaches of the San Pedro River. They include elements of Triassic (250–200 Ma), Neogene (~23 Ma), and Quaternary (~2.6 Ma) age (Abarzúa et al. 2023; Jorge et al. 2018). One of the current challenges for the Corridor is the potential progress towards an imaginary of the Southern Forest as a common heritage, and the transformation of a degraded landscape into a conservation landscape (Vergara and Carrasco 2020).

The landscape connectivity of the San Pedro Corridor, from the Andes to the coast, connecting areas under different management frameworks (coastal PAs and a BR, through forest fragments), opens the possibilities of a regional imaginary in

Fig. 20.9 Images from the Valle Rio San Pedro Corridor. (**a**), (**b**) landscape as a mosaic of uses along the San Pedro River; and (**c**) Neogene rock as an element of geoconservation at a local scale. (Photo: Andrés Moreira-Muñoz)

which the biocultural heritage forms a base of rich experience of living the temperate forest in an immersive and intensive form (Rozzi et al. 2018).

Conclusion

LC are gaining ground as mechanisms for (re)connecting landscapes where Indigenous and local communities live. Landscape connectivity is important for biodiversity conservation, as is the direct participation of Indigenous and local communities in developing appropriate conservation measures with and for them. *Landscape Conservation* (LC) in Chile and *Voluntary Conservation Initiatives* (VCI) in Peru support governance transitions related to the biocultural values of Indigenous and local communities. Also, conservation initiatives that link PAs and territories in different categories, such as in biosphere reserves, are being used as sites for various activities such as research and experimentation towards sustainability. Achieving biodiversity conservation in the context of sustainable social and economic development, and supporting education and research, are the pillars of the biosphere reserve definition.

The authors thank the Austrian Academy of Sciences (ÖAW) for financing the project "Lessons learned for the Management Policy of BIOAY – Peru" and the

National Agency for Research and Development (ANID) ANILLOS ATE230072, BECOME project—Biosphere Reserves as Effective Conservation Measures and the Biodiversa + Call—Protecting land and sea-Chile.

References

Abarzúa AM, Angélica SC, Francisca V-P, Jorge CM, Sergio CC, Esteban RS (2023) Río San Pedro Fósil: Guía paleobotánica de la selva valdiviana. Trafun Ediciones, Valdivia. https://www.researchgate.net/publication/370770812_Rio_San_Pedro_Fosil_Guia_paleobotanica_de_la_selva_valdiviana

Barraclough AD, Reed MG, Coetzer K, Price Martin F, Lisen S, Andrés M-M, Inger M (2023) Global knowledge–action networks at the frontlines of sustainability: insights from five decades of science for action in UNESCO 's world network of biosphere reserves. People Nat., Article pan3.10515. https://doi.org/10.1002/pan3.10515

Cao Y, Zhou Z, Liao Q, Shen S, Wang W, Xiao P, Liao J (2023) Effects of landscape conservation on the ecohydrological and water quality functions and services and their driving factors. Sci Total Environ 861:160695. https://doi.org/10.1016/j.scitotenv.2022.160695

Carvajal-Mascaró F, Moreira-Muñoz A, Salazar-Burrows AF, Leguia-Cruz M, Jorquera-Guajardo FI (2019) Divergencias y contradicciones en la planificación sustentable del periurbano rural metropolitano de Valparaíso. Caso Reserva de la Biosfera La Campana-Peñuelas, Chile central. RU 22(39):64–87. https://doi.org/10.22320/07183607.2019.22.39.04

Casale JF, Borsdorf A, Moreira-Muñoz A (2014) Reservas de la Biosfera de Chile: Laboratorios para la Sustentabilidad. Reservas de la Biosfera como Laboratorios para la Sustentabilidad: Paisajes de Conservación y Ordenamiento Territorial. With assistance of Moreira-Muñoz, A., Borsdorf, A. Academia de Ciencias Austriaca, Pontificia Universidad Católica de Chile, Instituto de Geografía, Santiago. (Geolibros 17, Santiago-Chile, pp 272–293

Corporación Nacional Forestal (2017) Fundación Tompkins traspasa Parque Pumalín al Estado de Chile. Ministerio de Agricultura -Gobierno de Chile

Day JC, Nigel D, Marc H, Glen H, WellsSue (2012) Guidelines for applying the IUCN protected area management categories to marine protected areas. https://www.researchgate.net/publication/266023472_Guidelines_for_Applying_the_IUCN_Protected_Area_Management_Categories_to_Marine_Protected_Areas

Dhendup T, Sharma S, Painter S, Whiteley AR, Mills LS (2023) Evidence of tiger population structure and dispersal in the montane conservation landscape of Bhutan. Global Ecol Conserv 43:e02459. https://doi.org/10.1016/j.gecco.2023.e02459

Dudley N (ed) (2008) Guidelines for applying protected area management categories. International Union for Conservation of Nature and Natural Resources (IUCN), Gland. Best Practice Protected Area Guidelines Series No. 21

Ferreira A, Zimmermann H, Santos R, von Henrik W (2018) A social–ecological systems framework as a tool for understanding the effectiveness of biosphere reserve management. Sustain For 10(10):3608. https://doi.org/10.3390/su10103608

Fischer A (2022) La historia del 'Gran Abuelo', el árbol más antiguo del mundo que data de antes de Stonehenge. National Geographic. https://www.ngenespanol.com/el-mundo/cual-es-el-arbol-mas-viejo-de-la-tierra-y-donde-se-encuentra/

Fundacion Plantae (2024) La cruzada de las comunidades de Los Lagos y Panguipulli por proteger el Río San Pedro. Edited by Ladera Sur. Chile

Guevara S, Laborde J (2008) The landscape approach. Environ Ethics 30(3):251–262. https://doi.org/10.5840/enviroethics200830331

Jorge CM, Francisca V-P, Andrea EP, Patricio CF, Abarzúa Ana M (2018) Resignificación del patrimonio paleontológico presente en el río San Pedro (Cuenca del río Valdivia, Chile). PASOS Revista de Turismo y Patrimonio Cultural 16(3):655–670. https://www.redalyc.org/journal/881/88166098007/

Kratzer A, Ammering U (2019) Rural innovations in biosphere reserves—a social network approach. J Rural Stud 71:144–155. https://doi.org/10.1016/j.jrurstud.2019.01.001

Martinez-Harms MJ, Wilson KA, Costa MDP, Possingham HP, Gelcich S, Chauvenet A et al (2021) Conservation planning for people and nature in a Chilean biodiversity hotspot. People Nat 3(3):686–699. https://doi.org/10.1002/pan3.10200

Mengist W (2020) Challenges of protected area management and conservation strategies in Ethiopia: a review paper. Adv Environ Stud 4(1) DOI: Area

Mittermeier RA, Turner WR, Larsen FW, Brooks TM, Gascon C (2011) Global biodiversity conservation: the critical role of hotspots. In: Zachos FE, Habel JC (eds) Biodiversity hotspots: distribution and protection of conservation priority areas. Scholars Portal, Berlin, pp 3–22

Monteferri B (ed) (2019) Áreas de conservación privada en el Perú: avances y propuestas a 20 años de su creación. Sociedad Peruana de Derecho Ambiental

Moreira-Muñoz A, Scherson RA, Luebert F, Román MJ, Monge M, Diazgranados M, Silva H (2020) Biogeography, phylogenetic relationships and morphological analyses of the south American genus Mutisia L.f. (Asteraceae) shows early connections of two disjunct biodiversity hotspots. Org Divers Evol 20(4):639–656. https://doi.org/10.1007/s13127-020-00454-z

Paül V, Vila-Lage R, Trillo-Santamaría J-M (2022) "The n°1 country"? A critical investigation of the booming designation of biosphere reserves in Spain. Landsc Urban Plan 222:104375. https://doi.org/10.1016/j.landurbplan.2022.104375

Pino A, Cardyn P (2014) La Reserva de la Biosfera de los Bosques Templados Lluviosos de los Andes Australes y las singularidades territoriales de la comuna de Panguipulli. In: Moreira-Muñoz A, Borsdorf A (eds) Reservas de la biosfera de Chile. Laboratorios para la sustentabilidad. Primera edición. Instituto de Geografía UC, Santiago. (Serie GEOlibros, no.17)

PNUD GEF SIRAP (2013) Guía de procedimientos básicos para el diseño ecológico territorial de un paisaje de conservación. Un ejemplo aplicado a los bosques templados fragmentados del sur de Chile. Documento de trabajo Proyecto. GEF SIRAP

Programa Nacional de Conservación de Bosques para la Mitigación del Cambio Climático (2023) Base de Datos. Ministerio del Ambiente. https://www.gob.pe/bosques

Republica de Chile-Senado (2014) Servicio de Biodiversidad y Áreas Protegidas. Comisión Mixta despachó proyecto a Sala

Rozzi R, May RH, Chapin FS, Massardo F, Gavin MC, Klaver IJ et al (2018) From biocultural homogenization to biocultural conservation: a conceptual framework to reorient society toward sustainability of life. In: From biocultural homogenization to biocultural conservation, vol 3. Springer, Cham, pp 1–17. https://doi.org/10.1007/978-3-319-99513-7_1

Servicio Nacional de Areas Naturales Protegidas por el Estado (2022) Áreas de Conservación Privada (ACP). https://www.datosabiertos.gob.pe

Summers P (2021) Construyendo una reserva de biosfera para todos. In: Albarracín-Llúncor G, Vogl Christian R (eds) Diez años de la reserva de biosfera Oxapampa—Asháninka—Yánesha (BIOAY): testimonios, reflexiones y logros. Second. 700 volumes. Austrian Academy of Sciences, Vienna

Taylor PD, Fahrig L, With KA (2006) Landscape connectivity: a return to the basics. In: Crooks K, Sanjayan M (eds) Connectivity conservation. Cambridge University Press, Cambridge, pp 29–43

Toledo VM (2008) Metabolismos rurales : hacia una teoría económico-ecológica de la apropiación de la naturaleza. Revista Iberoamericana de Economía Ecológica 7:1–26. https://raco.cat/index.php/Revibec/article/view/87196

Travis Belote R, Aplet GH, Carlson AA, Dietz MS, May A, McKinley PS et al (2021) Beyond priority pixels: delineating and evaluating landscapes for conservation in the contiguous United States. Landsc Urban Plan 209:104059. https://doi.org/10.1016/j.landurbplan.2021.104059

UNESCO (1996) Biosphere reserves: the Seville strategy and the statutory framework of the World Network

Vergara F, Carrasco N (2020) De un paisaje de degradación a un paisaje de conservación: el tránsito hacia un imaginario del bosque austral como entramado patrimonial (provincia de Valdivia, Chile). PapRevSoc 105(4):511. https://doi.org/10.5565/rev/papers.2723

Zarnetske PL, Baiser B, Strecker A, Record S, Belmaker J, Tuanmu M-N (2017) The interplay between landscape structure and biotic interactions. Curr Landsc Ecol Rep 2(1):12–29. https://doi.org/10.1007/s40823-017-0021-5

Open Access This chapter is licensed under the terms of the Creative Commons Attribution 4.0 International License (http://creativecommons.org/licenses/by/4.0/), which permits use, sharing, adaptation, distribution and reproduction in any medium or format, as long as you give appropriate credit to the original author(s) and the source, provide a link to the Creative Commons license and indicate if changes were made.

The images or other third party material in this chapter are included in the chapter's Creative Commons license, unless indicated otherwise in a credit line to the material. If material is not included in the chapter's Creative Commons license and your intended use is not permitted by statutory regulation or exceeds the permitted use, you will need to obtain permission directly from the copyright holder.

Ecological Connectivity Perspectives for Policy and Practice

21

Katharina Lapin, Janine Oettel, and Magda Bou Dagher Kharrat

Using QField for the demarcation of stepping stones in Austrian forests (Photo: BFW/Florian Winter)

K. Lapin (✉) · J. Oettel
Department of Forest Biodiversity & Nature Conservation, Austrian Research Centre for Forests, Vienna, Austria
e-mail: katharina.lapin@bfw.gv.at; janine.oettel@bfw.gv.at

M. B. D. Kharrat
European Forest Institute, Sant Pau Art Nouveau Site, Sant, Barcelona, Spain
e-mail: magda.boudagher@efi.int

Abstract

Ecological connectivity within forest ecosystems is a cornerstone of preserving biodiversity and enhancing ecosystem resilience. The interplay between ecological conditions, historical influences, and socioeconomic factors shapes the connectivity of forest landscapes, and the decisions made by policymakers and stakeholders entail consequences for ecosystem health and sustainability. The following comprehensive exploration delves into the integration of ecological connectivity concepts in national and international policies. International policies highlight the importance of connectivity for biodiversity protection. The Convention on Biological Diversity (CBD) and the Kunming-Montreal Global Biodiversity Framework emphasise the role of connectivity in halting biodiversity loss and ensuring the success of restoration efforts. Other conventions such as Ramsar and World Heritage contribute to the preservation of crucial habitats. National and transnational strategies underline the growing emphasis on ecological connectivity as a tool for linking biodiversity-rich areas. Policies across the globe reflect the global recognition of the importance of connectivity. Restoring connectivity involves a range of strategies from revitalising urban forests to preserving rural landscapes and beyond, all while considering the unique needs of diverse ecosystems. Conservation efforts to enhance connectivity include expanding protected regions, creating wildlife corridors, providing incentives for forest management, and the use of various indicators. Local strategies are key in this regard, with policymakers considering regional biodiversity data and updating connectivity models to identify potential corridors and forest patches for species movement. Recommendations include policy integration, transnational cooperation, interdisciplinary collaboration, and technological integration. Climate change mitigation policies highlight the importance of connectivity in landscapes altered by human activities, suggesting a multifaceted approach to ecosystem resilience in the face of climate change. In order to pave the way to a sustainable and resilient future for forests, it is essential to align global, national, and local initiatives and adopt a holistic approach to connectivity conservation.

Keywords

Connectivity conservation · Policy integration · Transnational cooperation · Climate resilience

Introduction

The connectivity of forest ecosystems in an ecological context is shaped by a variety of factors. One crucial factor is the environmental conditions of a specific site, as they determine the potential for connectivity between different areas. Simultaneously, human activities—and especially historical land-use developments—have played a significant role in shaping the connectivity patterns of landscapes. Moreover, socioeconomic developments are pivotal in decision-making processes, as they govern the available choices for preserving ecosystem connectivity. Land ownership arrangements have a direct impact on land and environmental management, thus affecting ecosystem connectivity. The ownership and management of land can influence decisions concerning land use and development, ultimately affecting whether ecosystems remain continuous or become fragmented. These land tenure systems can either support or impede conservation efforts aimed at preserving or restoring ecosystem connectivity. For instance, communal land tenure systems may promote collective conservation initiatives, whereas individual private ownership may prioritise economic interests over ecological considerations. The legal framework pertaining to land tenure often determines the extent to which conservation measures such as establishing wildlife corridors or protected areas can be put into practice to enhance ecosystem connectivity.

The focus of this chapter is to offer valuable insights into the integration of concepts of ecological connectivity into the decision-making processes for international and national policies. We recognise that decisions made by policymakers, land managers, and other stakeholders have far-reaching consequences on the health and resilience of our forest ecosystems. Therefore, it becomes imperative to equip decision-makers with the necessary knowledge and tools to make informed choices that will effectively preserve and enhance ecological connectivity in various forest management contexts. Through a comprehensive exploration of the challenges, strategies, and potential actions, we present existing policies and their potential for improvement, as well as highlight examples where effective connectivity strategies have been implemented.

International Policies and Strategies for Ecological Connectivity

In recent decades, connectivity has emerged as a critical aspect of nature conservation, playing a pivotal role in the efficacy of terrestrial and marine protected area systems and networks. As the threats posed by land-use development and global climate change escalate, the importance of connectivity conservation has been underscored within conservation biology and policy spheres. Its principal objective is to mitigate biodiversity loss and enhance resilience to climate change by safeguarding the integrity of protected areas and fostering ecological connectivity.

The universally agreed definition of ecological connectivity is paramount in recognising its international significance. It denotes the unhindered movement of species and the natural processes essential for sustaining life on earth. This definition is accompanied by supporting points that elucidate the crucial aspects of ecological connectivity and its relevance across various international agreements. These points emphasise the necessity of ecological interconnections and ecosystem services, while acknowledging social and cultural ties to nature and honouring the rights of Indigenous peoples and local communities IUCN (2021), UNESCO (2021), United Nations (1992), United Nations Convention to Combat Desertification (2022).

Table 21.1 International policy and ecosystem connectivity

International policy	About ecosystem connectivity
Ramsar Convention on Wetlands |1971	The convention focuses on integrating wetlands and forests to promote the importance of *connectivity*, impacting the livelihoods of millions of people worldwide.
Convention Concerning the Protection of the World Cultural and Natural Heritage | 1972	The convention adopted by the general conference of UNESCO aims to ensure the maintenance of *ecological connectivity* between each property's component parts by strengthening and improving measures for consistency and greater functional linkages between component sites of a property and its surroundings.
Convention on the Conservation of Migratory Species of Wild Animals (*CMS*) | 1983	This convention safeguards migration by preserving intact ecosystems across various habitats, particularly for migratory birds and mammals.
United Nations Convention on Biological Diversity (*CBD*) | 1992	This convention highlights the integral role of *connectivity* in preserving ecosystems and supporting biodiversity.
United Nations Framework Convention on Climate Change (*UNFCCC*) |1992	This convention states that *ecological connectivity* encompasses an internationally coordinated nature-based solution, serving as a comprehensive and indispensable element of broader global initiatives for climate change mitigation, resilience, and adaptation (UNFCCC 2023).
United Nations Convention to Combat Desertification (*UNCCD*)| 1994	This convention emphasises the importance of *ecological connectivity* in securing sustainable, enduring benefits from initiatives aimed at revitalising depleted ecosystems.
IUCN World Commission on Protected Areas (*WCPA*) |1996	The Transboundary conservation specialist group offers pertinent guidance on transboundary conservation, which is intricately linked to *connectivity* concerns.
Intergovernmental Science-Policy Platform on Biodiversity and Ecosystem Services (*IPBES*) |2012	IPBES reflects on the interlinkages between biodiversity and other relevant issues as outlined in the 2030 Agenda for Sustainable Development, facilitating integrated and cross-sectoral approaches to achieving sustainability goals, including those related to connectivity conservation. The 2019 Global Assessment on Biodiversity and Ecosystem Services by IPBES revealed that only 9.3–11.7% of protected areas are *suitably connected* (Watson et al. 2019).
UN Decade on Ecosystem Restoration |2019	This agenda acknowledges the role of connectivity in accomplishing the UN Sustainable Development Goals.
Post-2020 Global Biodiversity Framework |2021	IUCN World Conservation Congress Resolution calls to include *ecological connectivity* in the post-2020 global biodiversity framework.
Kunming-Montreal Global Biodiversity Framework (*GBF*) |2022	This framework adopted by the 15th conference of parties (COP15) to the *CBD* aims to conserve and manage at least 30% of ecosystems effectively, ensuring they are ecologically representative, *well-connected*, and governed equitably through protected areas and other conservation measures.

Table 21.1 showcases the primary international policies and their correlation with the concept of connectivity.

International policies aimed at connectivity conservation are increasingly pivotal in combating biodiversity loss and climate change (Carver 2011). By fostering cooperation and coordination, nations can synchronise their endeavours to preserve ecological connectivity, thereby securing a sustainable future for terrestrial and marine environments.

National and Transnational Strategies

In transnational policies, ecological connectivity currently serves as one element among various targets for preserving biodiversity and re-establishing the human–nature relationship. Existing national laws and policies offer opportunities to protect vulnerable areas for connectivity. However, the scope of connectivity extends beyond protected zones, leading to challenges due to segmented sectoral structures within public institutions. An illustration of a transnational initiative is the Natura 2000 non-treaty conservation network established by the EU. It aligns with the Habitats Directive, aiming to achieve robust ecological coherence among natural sites. Article 10 of the Habitats Directive and Article 3 of the Birds Directive (09/147/EC) specifically include the establishment of the necessary functional connections inside and outside the designated sites (Estreguil et al. 2013). In order to enhance the ecological continuity of protected area networks, the emphasis lies on establishing connectivity. Presently, over a third of the EU territory comprises natural and semi-natural elements linking Natura 2000 sites, primarily forests and woodlands (EEA 2020). This network's implementation has significantly enhanced the design of protected area systems for connectivity between 2010 and 2012, fostering broader transboundary protected area priorities across most EU countries with regard to the spatial patterns and connectivity of vital habitats and species in the EU. In recent times, the EU's overarching commitments for nature protection until 2030 encompass three key goals: (1) to safeguard a minimum of 30% of EU land and sea areas; (2) to incorporate ecological corridors into a comprehensive Trans-European Nature Network; and (3) to preserve at least one-third of EU protected areas, including remaining primary and old-growth forests. To achieve these goals, the EU Member States must tailor the 10% EU target (strict protection) to regional scales to ensure habitat connectivity (EC 2023).

The review of national policies reveals an increasing focus on ecological connectivity in forest ecosystems, often positioned as a contemporary tool for linking biodiversity-rich areas. Leveraging data from the ECOLEX (The Gateway to Environmental Law) database relating to transnational environmental agreements, treaties, and global environmental law, assessments were conducted for both national and transnational strategies pertaining to ecological connectivity (ECOLEX 2023). Between 2018 and 2023, we identified 27 countries or jurisdictions employing diverse policies and strategies for environmental and biodiversity conservation. The inclusion of connectivity for forest ecosystems is evident across various regions in Europe, North America, Africa, Oceania, the Caribbean, and Asia.

With regard to strategies, urban forests often exhibit indices for active habitat restoration to boost connectivity, while rural forest management policies tend to prioritise preserving areas of high connectivity importance (see Fig. 21.1). The

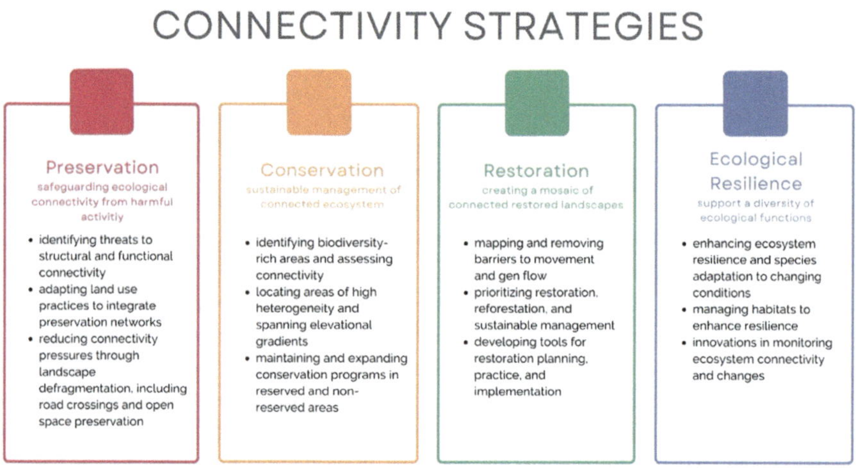

CONNECTIVITY STRATEGIES

Preservation
safeguarding ecological connectivity from harmful activitiy
- identifying threats to structural and functional connectivity
- adapting land use practices to integrate preservation networks
- reducing connectivity pressures through landscape defragmentation, including road crossings and open space preservation

Conservation
sustainable management of connected ecosystem
- identifying biodiversity-rich areas and assessing connectivity
- locating areas of high heterogeneity and spanning elevational gradients
- maintaining and expanding conservation programs in reserved and non-reserved areas

Restoration
creating a mosaic of connected restored landscapes
- mapping and removing barriers to movement and gen flow
- prioritizing restoration, reforestation, and sustainable management
- developing tools for restoration planning, practice, and implementation

Ecological Resilience
support a diversity of ecological functions
- enhancing ecosystem resilience and species adaptation to changing conditions
- managing habitats to enhance resilience
- innovations in monitoring ecosystem connectivity and changes

Fig. 21.1 Illustration of the key conservation strategies of preservation, conservation, restoration, and ecological resilience, along with corresponding actions aimed at guiding decision-makers towards effective implementation

range of intensity varies from broad goals (e.g., Japan or Vietnam) to moderate (e.g., Jordan) and detailed, measurable targets (e.g., Tanzania, North Macedonia).

Of the reviewed policies, over half (55.7%) include preservation measures like extending protected areas or establishing corridors and stepping stones. Other actions include monitoring, evaluation, urban planning, fostering habitat connectivity, and biodiversity refuges. Instruments within these policies include conservation policy identification, regulatory measures, incentives, penalties, and support for protected areas and restoration.

Connectivity for forest ecosystems is frequently central to policies targeting individual species—often large mammals—under species protection programs. For example:

- Bhutan's Elephant Conservation Action Plan maps and manages elephant migratory routes;
- Zimbabwe's National Elephant Management Plan maintains habitats and restores connectivity;
- Nepal's Red Panda Conservation Action Plan emphasises connectivity;
- Latvia's Action Plan for Baltic Lynx aims for gene flow and viable meta-populations;
- Nepal's Pangolin Conservation Action Plan prioritises connectivity;
- Action plans like that for conservation of the brown bear (*Ursus arctos*) in Europe focus on maintaining connectivity.

Connectivity Conservation Strategies and Actions

Connectivity conservation involves the collaborative efforts of individuals, communities, institutions, and businesses to safeguard, enhance, and restore ecological flows, species movement, and dynamic processes across both, intact and fragmented environments. This innovative approach unites a global movement, providing a coordinated response to protect vital natural interconnections, fortify biodiversity, and enhance resilience against climate change. The overarching strategic goal lies in enhancing landscape heterogeneity and bolstering connectivity among forest patches, which is critical for conserving diversity. Protection and restoration of natural ecosystems, combined with intergovernmental partnerships, form the foundation of a global framework to ensure ecological resilience and preserve biodiversity for future generations. Localised application of connectivity conservation strategies necessitates the integration of local forest management systems, regional markets, national biodiversity targets, and international agreements to address the impacts of climate change (Lindenmayer et al. 2000; Nuñez et al. 2013). Prioritising territories for connectivity management based on current biodiversity data and regularly updated connectivity models is crucial, as it helps to identify potential corridors and patches for species movement under changing climatic conditions (Brennan et al. 2022) Monitoring and evaluation are essential components to ensure the documentation and tracking of ecological corridors both at the national and international scale (Hilty et al. 2020). Diversification of strategies as well as their financial models and funding schemes to encompass various spatial scales, species groups, and taxa spreads the risk and increases the effectiveness of connectivity conservation interventions (Hilty et al. 2020; Locke and Rissman 2012). By weaving connectivity into multifaceted strategies, we can forge a more resilient future for biodiversity and ecosystems.

Indicators for Evaluating Connectivity

Indicators can play a pivotal role in evaluating historical management or policy endeavours, shaping policy directions, and establishing precise conservation benchmarks (Locke and Rissman 2012; Oettel and Lapin 2020). Unfortunately, the potential of biodiversity indicators remains largely untapped within critical decision-making spheres: Despite the growing scholarly engagement with connectivity, the integration of indicators into policy contexts has been relatively limited. The use of indicators holds the promise of assessing the efficacy of policy objectives in terms of biodiversity conservation, particularly on a global scale. Notably, international efforts have resulted in the formulation of two connectivity-oriented global indicators: the Protected Connected Land Indicator (Herrera et al. 2017; Koffi et al. 2019; Saura et al. 2017) and the Protected Area Representativeness & Connectedness Indicator. These indicators offer a means of evaluating the terrestrial connectivity within networks of protected areas (PAs) and other effective area-based conservation measures (OECMs).

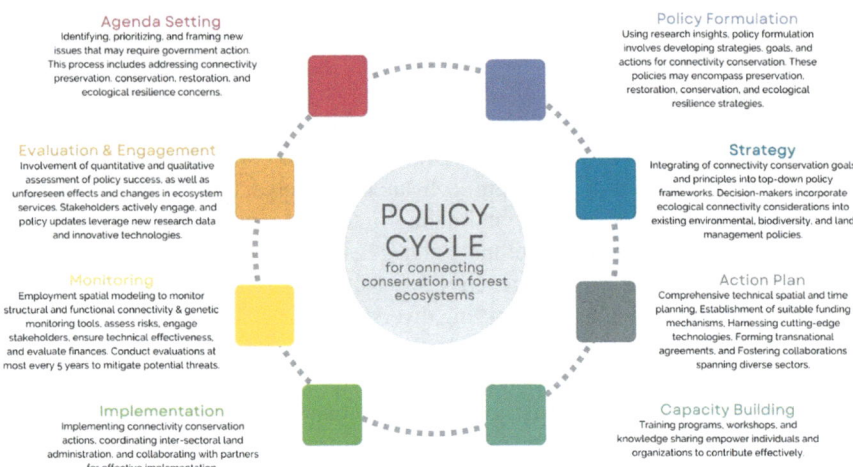

Fig. 21.2 Policy cycle for connectivity conservation in forest ecosystems

For instance, the CBD's Protected Area Representativeness & Connectedness Indices (PARC Indices) were designed to monitor progress towards CBD Aichi Target 11, which seeks to enhance the ecological representativeness and connectivity of protected areas. The PARC Indices employ biologically scaled environmental mapping and global modelling to ascertain whether terrestrial protected areas are representative and well connected. Several other examples underscore the use of indicators for ecological connectivity within national policies. Noteworthy is Sri Lanka's National Environmental Action Plan 2022–2030, which employs the indicator "Percentage of declared land extent out of identified climate-sensitive corridors". Similarly, the National Climate Change Response Strategy 2021–2026 issued by the United Republic of Tanzania utilises the indicator "Number of key habitats freely connected". However, a critical aspect that remains largely unaddressed is the thorough evaluation of such indicators' effectiveness in accurately representing the underlying biodiversity trends of interest. In this regard, Nicholson et al. (2012) emphasise that indicators should be integrated into a comprehensive indicator-policy cycle. This cycle involves embedding biodiversity monitoring indicators within a decision analysis framework employing principles of adaptive management, management strategy evaluation, and optimal monitoring. The approach creates a symbiotic relationship between monitoring activities and actionable insights, fostering both informed decision-making and continuous learning (see Fig. 21.2).

Climate Change Policies Urging for Ecological Connectivity

Climate change policies have underscored the importance of ecological connectivity within landscapes significantly influenced by human activities. This alteration of the landscape matrix has led to a lack of connectivity between individual protected areas and natural habitats. The resulting barriers to species migration and expansion have elevated the risk of biodiversity loss (Janishevski et al. 2015). Moreover, climate change constrains the migration of species along with the extent and speed of their range expansion.

The International Panel on Climate Change's (IPCC) Sixth Assessment Report in 2023 highlights adaptation measures that enhance biodiversity and ecosystem service resilience to climate change. These include minimising additional stressors, reducing fragmentation, expanding natural habitat extent, fostering connectivity and heterogeneity, and safeguarding microclimate refugia where species can persist.

Embracing connectivity in biodiversity-inclusive urban planning and dedicating space for nature within built-up environments emerge as alternatives to bolster citizens' well-being, decrease urban and infrastructural environmental impacts, and enhance overall sustainability. The pivotal role of connectivity within infrastructure is underscored in urban development policies, as evidenced by initiatives like the Australian Capital Territory's (ACT) Urban Forest Strategy 2021–2045 and the ACT Native Woodland Conservation Strategy 2019. Moreover, the European Union has provided guidance by way of a strategic framework to further support the implementation of EU-level green and blue infrastructure (EC DG Env 2023).

Conclusion and Future Directions

Recognising the significance of ecological connectivity for combating biodiversity loss and habitat fragmentation, both international and national policies emphasise the enhancement of ecosystem resilience. To fortify the effectiveness of connectivity conservation, we recommend the following actions:

1. *Policy Integration*: Embed connectivity recognition in top-down policies while nurturing bottom-up implementation.
2. *Transnational Cooperation*: Harmonise connectivity goals through transnational agreements and decision-making processes.
3. *Interdisciplinary Collaboration*: Foster cooperative approaches across sectors, particularly for contexts like riparian ecosystems and forest–agriculture interfaces.
4. *Technological Integration*: Embrace advanced earth observation and biodiversity monitoring technologies to consistently reassess policy indicators.

In summary, ecological connectivity is a fundamental strategy for mitigating biodiversity loss and enhancing ecosystem resilience. Its implementation demands a multifaceted approach involving policy recognition, private landowners, transnational cooperation, interdisciplinary collaboration, technological integration, and the re-evaluation of indicators. Since many projects aimed at restoring connectivity will be implemented on private properties, it is crucial to offer suitable incentives to encourage community-based conservation. Doing so will help to gain approval and backing from local people, which is vital for such projects to succeed. By working collectively at all levels, we can create a comprehensive framework that not only conserves biodiversity but also strengthens the planet's resilience to environmental challenges.

References

Brennan A, Naidoo R, Greenstreet L, Mehrabi Z, Ramankutty N, Kremen C (2022) Functional connectivity of the world's protected areas. Science 376(6597):1101–1104. https://doi.org/10.1126/science.abl8974

Carver S (2011) Connectivity conservation management: a global guide. Mt Res Dev 31(1):73–74

EC (2023) Biodiversity strategy for 2030. https://environment.ec.europa.eu/strategy/biodiversity-strategy-2030_en

EC DG Env (2023) Green infrastructure Promoting the use of green infrastructure in all EU policies, to help restore nature and boost biodiversity. https://environment.ec.europa.eu/topics/nature-and-biodiversity/green-infrastructure_en

ECOLEX (2023) No Title. https://www.ecolex.org/p/about/

EEA (2020) State of nature in the EU - Results from reporting under the nature directives 2013–2018. In EEA Report No 10/2020 (Issue 10)

Estreguil C, Caudullo G, San Miguel J (2013) Connectivity of Natura 2000 forest sites. JRC Executive Report. European Commission. https://Publications.Europa.Eu/En/Publication-Detail/-/Publication/F520524b-Ee69-4fcc-A71e-360c5448a38b/Language-En

Herrera LP, Sabatino MC, Jaimes FR, Saura S (2017) Landscape connectivity and the role of small habitat patches as stepping stones: an assessment of the grassland biome in South America. Biodivers Conserv 26(14):3465–3479

Hilty J, Worboys GL, Keeley A, Woodley S, Lausche B, Locke H, Carr M, Pulsford I, Pittock J, White JW (2020) Guidelines for conserving connectivity through ecological networks and corridors. Best Practice Protected Area Guidelines Series, 30. IUCN, p 122

IUCN (2021) *IUCN World Conservation Congress Adopts Resolution calling for inclusion of Ecological Connectivity in the Post-2020 Global Biodiversity Framework*. https://www.cms.int/en/news/iucn-world-conservation-congress-resolution-calls-include-ecological-connectivity-post-2020

Janishevski L, Santamaria C, Gidda SB, Cooper HD, Brancalion PHS (2015) Ecosystem restoration, protected areas and biodiversity conservation. Unasylva 245(3):19–28

Koffi B, Wilson J, Delli G, Dubois G, Mandrici A, Ehrlich D (2019) Risk and resilience indicators and indexes in Arctic ecoregions, protected areas and urban centres

Lindenmayer DB, Margules CR, Botkin DB (2000) Indicators of biodiversity for ecologically sustainable forest management. Conserv Biol 14(4):941–950

Locke CM, Rissman AR (2012) Unexpected co-benefits: forest connectivity and property tax incentives. Landsc Urban Plan 104(3–4):418–425

Nicholson E, Collen B, Barausse A, Blanchard JL, Costelloe BT, Sullivan KME, Underwood FM, Burn RW, Fritz S, Jones JPG (2012) Making robust policy decisions using global biodiversity indicators. PLoS One 7(7):e41128

Nuñez TA, Lawler JJ, McRae BH, Pierce DJ, Krosby MB, Kavanagh DM, Singleton PH, Tewksbury JJ (2013) Connectivity planning to address climate change. Conserv Biol 27(2):407–416

Oettel J, Lapin K (2020) Linking forest management and biodiversity indicators to strengthen sustainable forest management in Europe. Ecol Indic

Saura S, Bastin L, Battistella L, Mandrici A, Dubois G (2017) Protected areas in the world's ecoregions: how well connected are they? Ecol Indic 76:144–158

UNESCO (2021) Case Law - Ecological connectivity. https://whc.unesco.org/en/compendium/361

UNFCCC (2023) Climate action. https://unfccc.int/

United Nations (1992) Convention on biological diversity. Diversity 30. https://doi.org/10.1146/annurev.ento.48.091801.112645

United Nations Convention to Combat Desertification (2022) The global land outlook, 2nd edn. UNCCD, Bonn. https://www.unccd.int/resources/global-land-outlook/glo2

Watson R, Baste I, Larigauderie A, Leadley P, Pascual U, Baptiste B, Demissew S, Dziba L, Erpul G, Fazel A (2019) Summary for policymakers of the global assessment report on biodiversity and ecosystem services of the intergovernmental science-policy platform on biodiversity and ecosystem services. IPBES Secretariat, Bonn

Open Access This chapter is licensed under the terms of the Creative Commons Attribution 4.0 International License (http://creativecommons.org/licenses/by/4.0/), which permits use, sharing, adaptation, distribution and reproduction in any medium or format, as long as you give appropriate credit to the original author(s) and the source, provide a link to the Creative Commons license and indicate if changes were made.

The images or other third party material in this chapter are included in the chapter's Creative Commons license, unless indicated otherwise in a credit line to the material. If material is not included in the chapter's Creative Commons license and your intended use is not permitted by statutory regulation or exceeds the permitted use, you will need to obtain permission directly from the copyright holder.

Part IV

Case Studies in Ecological Connectivity

Austria: The Austrian Stepping-Stone Program—A Bottom-Up Approach

22

Janine Oettel and Katharina Lapin

View on the area of a stepping stone in the Austrian alps (Photo: BFW/Johann Ferenc Püspök)

Abstract

Austria is a landlocked country in Central Europe, with forests spanning diverse elevations and climates. It currently harbors around 68,000 species, but climate change and fragmentation pose threats to this rich biodiversity. Improving habitat connectivity and consequently allowing species to migrate to new suitable habitats is essential for combating both of these issues. Therefore, a national stepping-stone program with the aim of enhancing forest connectivity and

J. Oettel (✉) · K. Lapin
Department of Forest Biodiversity & Nature Conservation, Austrian Research Centre for Forests, Vienna, Austria
e-mail: janine.oettel@bfw.gv.at; katharina.lapin@bfw.gv.at

© The Author(s) 2025
K. Lapin et al. (eds.), *Ecological Connectivity of Forest Ecosystems*,
https://doi.org/10.1007/978-3-031-82206-3_22

conserving forest biodiversity was initiated. Around 750 stepping stones ranging from 0.5 to 25 hectares in size are to be excluded from regular forest management. Areas of priority include those with a presence of large amounts of deadwood, habitat trees, rare species, and special sites. Identification of these priority areas requires close collaboration with forest owners and employs a GIS-based approach featuring four indicators: Protect Value, Connect Value, Species Value, and Habitat Value. Funding schemes support the program with up to €5040 per hectare over a period of 20 years. The evaluation follows a systematic method, gathering insights and identifying strengths and weaknesses. A steering group oversees the program, and a scientific board advises on research design, data collection, and analysis. Data collection includes both habitat and biodiversity surveys. Stakeholders from the fields of forestry, nature conservation, research, and government are actively involved, supporting the program's implementation and ensuring transparency. Through collaboration, systematic evaluation, and stakeholder involvement, the program aims to safeguard biodiversity-rich forest ecosystems and ensure long-term success.

Keywords

Contractual agreement · Demarcation · Evaluation · Prioritization · Stakeholder engagement

The Study Area: Austrian Forests

Austria is a landlocked country in Central Europe with an area of about 8.34 million ha. Along with Slovenia, it is one of the most heavily forested countries in Central Europe, with a forest cover of almost 47.9% (4.02 million ha), of which 84% (3.36 million ha) are under management (BFW 2022). The forests of Austria are particularly valuable since they span a wide range of elevations (120 m to 2100 m a.s.l.) and climates (continental Pannonian, Alpine, and transitional central European climates), resulting in a considerable diversity of forest types (and thus tree species compositions) ranging from temperate lowland forests to subalpine forests (Russ 2019) as well as a great diversity of highly specialized plant and animal species. Around 68,000 species, including 2900 plant species and 54,000 animal species, have been recorded in Austria to date. Insects are the largest taxonomic group, comprising over 40,000 species (Geiser 2018). In terms of landscape fragmentation, Austria has an absolute area of 2.90 million ha with high and very high fragmentation (34%) (EEA 2021). With 1299 meshes per 1000 km^2, the country is positioned in the upper middle range among European countries. Highly fragmented forest areas amount to 0.78 million ha, constituting about 20% of the total forested area. Fragmentation from infrastructure and land use harms rare species and populations, and climate change further threatens forest ecosystems and species, resulting in significant challenges and uncertainties for sustainable forest management.

Theoretical Framework of Ecological Connectivity

The negative impacts of climate change and landscape fragmentation on biodiversity highlight the need to prioritize the conservation, restoration, and enhancement of ecologically valuable forest areas and their connectivity. Maintaining connectivity is a crucial factor in promoting biodiversity, as protected areas cannot achieve their conservation objectives if they are not functionally connected to one another (Andersson and Bodin 2009; Kadoya 2009; Travers et al. 2021). These connections support important ecological processes such as gene flow and migration, enabling the recolonization of areas with populations of endangered animal, plant, and fungus species and offering individuals and populations the chance to locate new habitats that become suitable under changing climatic conditions (Corrales and Höglund 2012; Klinga et al. 2019; Wang et al. 2008).

Suitable habitats for species affected by climate change and fragmentation include not only existing protected areas but also small stepping stones and corridors that serve as refuges for many species including specialist beetles, mosses, and lichens while also allowing for the networking of otherwise isolated patches (Gjerde et al. 2015; Perhans et al. 2007, 2009; Sverdrup-Thygeson et al. 2017). They improve the dispersal possibilities of species with limited dispersal ability (Beger et al. 2022; Drag et al. 2011; Pedley and Dolman 2020). However, effective networking requires consideration of two important aspects: (a) structural and (b) functional connectivity (Tischendorf and Fahrig 2000). Spatial networking, which refers to parameters such as distances and habitat sizes, is crucial for maintaining structural connectivity, while appropriate habitat quality ensures functional connectivity. Both aspects are necessary for improving dispersal possibilities and promoting successful species migration in the face of climate change.

Stepping stones are part of most nature conservation strategies in forest ecosystems (Gustafsson et al. 2020a, b; Lindenmayer and Franklin 2002; Wintle et al. 2019). The positive effect of this conservation measure has been primarily confirmed in terms of supplying habitats for saproxylic insect species (Gustafsson et al. 2020b; Sverdrup-Thygeson et al. 2014) but also for woodland birds, bryophytes, fungi, and lichen (Kropik et al. 2020; Larrieu et al. 2014; Sverdrup-Thygeson et al. 2014; Wiktander et al. 2001; Fig. 22.1).

Aims and Expected Program Outcomes

The Austrian national stepping-stone program is being established to enhance connectivity and species dispersal, thus making a significant contribution to the overall conservation and enrichment of biodiversity. The program aims to designate specific areas that will be excluded from regular forest management. In this context, non-utilization means refraining from logging and silvicultural measures, with hinting forming an exception. A total of approximately 750 stepping stones each covering an area of 0.5 to 25 ha and adding up to around 1500 ha are to be set aside from management for a period of 10–20 years (depending on the size of the area) through

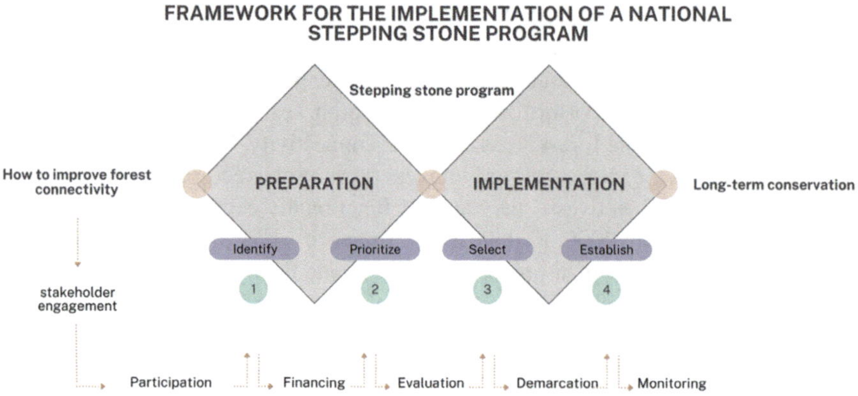

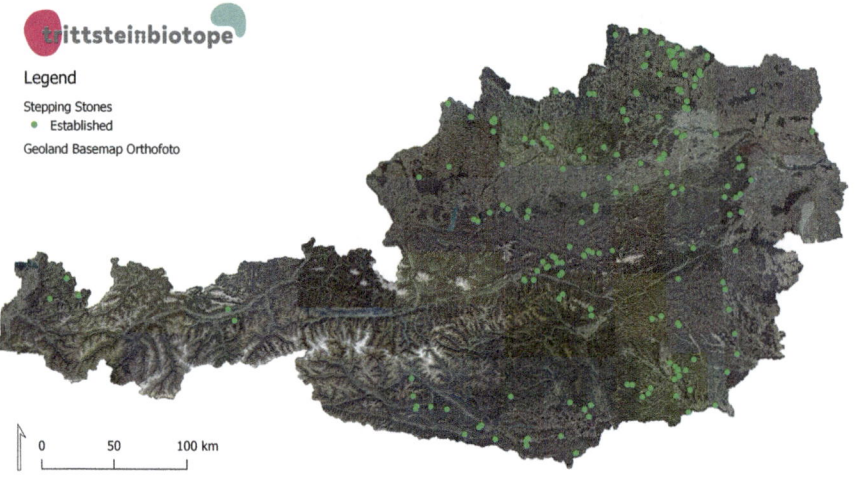

Fig. 22.1 Framework illustrating the implementation of a national stepping-stone program to improve connectivity in forests. It consits of two main phases: Preparation and Implementation. The process aims for long-term conservation

Fig. 22.2 A map showing the distribution of established stepping-stones, marked by green dots, across the country of Austria (status 01/2024). The map uses a Geoland Basemap Orthofoto for topographical representaion

contractual agreements, with per-hectare financial compensation. The current state of the program's implementation is presented in Fig. 22.2.

Four types of priority features are considered, namely areas with (1) large amounts of deadwood (minimum 20 m^3ha^{-1}), (2) habitat trees (minimum five trees per hectare), (3) an occurrence of rare and endangered species according to the national Red List, (4) special sites such as wetlands or drylands (see Fig. 22.3). The program aims for a balanced distribution of these priority areas among the federal states.

Fig. 22.3 Examples of stepping stones from one of the four priority features: (**a**) site with habitat tree, (**b**) site with rare species occurrence (e.g. *Lobaria pulmonaria*), (**c**) special site (peat with adjacent forest), and (**d**) site with large amount of deadwood

Implementation of the National Program

The following five steps are critical for implementing a nationwide conservation program aimed at improving forest connectivity through the establishment of stepping stones in forests (Fig. 22.4):

I. *Identification* of areas within forests based on priority features (see Fig. 22.3).
II. *Prioritization* of identified areas using GIS-based analysis to enhance connectivity.
III. *Selection* of prioritized areas on site according to legal boundaries.

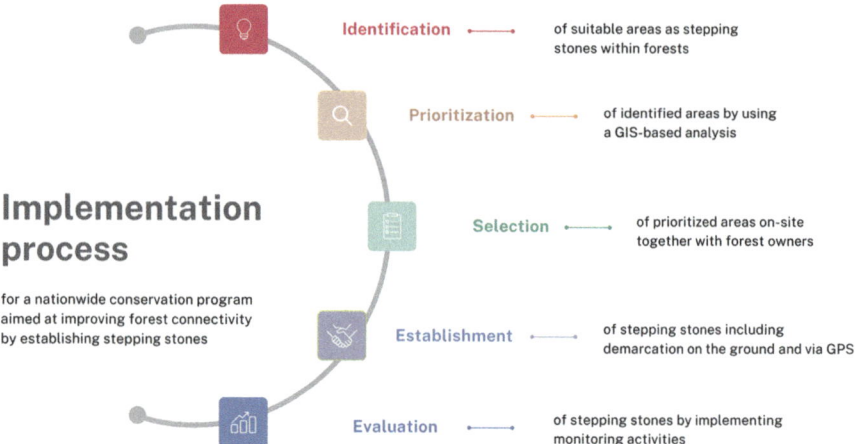

Fig. 22.4 Implementation process from identification to evaluation of stepping stones for a nationwide conservation program aiming at improving forest connectivity

IV. *Establishment* of selected areas, including on-ground demarcation and GPS-based delineation in collaboration with forest owners.
V. *Evaluation* of connectivity through monitoring activities, including standard and intensive surveys.

Forest owners voluntarily report areas for consideration via an online platform. The prioritization process involves a GIS-based analysis considering land cover types, landscape barriers, existing protected areas, and species-specific information. The aim is to establish stepping stones in regions with high connectivity potential or a buffering effect for protected and valuable habitats. The selection process considers legal regulations regarding path and forest road maintenance as well as forest phytosanitary provisions in accordance with the Austrian Forest Act. Selected areas will be clearly marked on the ground and permanently demarcated using GPS. Additionally, long-term monitoring plans will be developed in collaboration with forest owners to assess ecological impacts.

Identifying and Prioritizing Areas for Improving Forest Connectivity

The identification and prioritization of stepping stones involves utilizing national and regional biodiversity data for Austrian forest ecosystems. A combination of indicator values for structural and functional connectivity metrics is recommended to ensure future habitat connectivity. Following Lapin et al. (2024), the prioritization process for Austria includes four indicators: (1) the *Protect Value* considering patch distances to protected areas, (2) the *Connect Value*, which integrates modeled connectivity areas in Austria, (3) the *Species Value* identifying species-rich areas, and iv) the *Habitat Value* that determines biodiversity-rich areas and sites of favorable protection status. These indicators can then be combined into a prioritization value using min-max-standardization (ranging between 0 and 1). Here we assess the entire forested area of Austria (39,587 km^2) and identifies high prioritization value areas in 25% (8336 km^2) of the assessed forest area.

Selecting and Establishing Areas for Improving Forest Connectivity

When selecting the stepping stones for non-utilization, ensuring support and participation from forest owners is crucial. In Austria, where 82% of forests are privately owned—comprising small (less than 200 hectares, 54%), medium (200 to 1000 hectares, 10%), and large-scale (more than 1000 hectares, 18%) private properties—a bottom-up approach is essential to ensure the voluntary stepping-stone program's success.

Forest owners are encouraged to participate by reporting potential forest areas suitable as stepping stones. Financial compensation based on standard costs is offered for setting aside these areas from management for a defined period of up to 20 years. National funding schemes like the "Rural Development Fund (LE)" (BML 2023b) and "Waldfonds" (BML 2023a) provide up to €5040 per hectare. Potential areas are assessed based on biodiversity data, and the final selection is made during on-site visits using predefined criteria. Useful guidelines for identifying suitable areas as stepping stones for nature conservation have been provided by Mergner (2021). These guidelines include unmanaged forest areas, old-growth trees in young forest stands, deciduous trees in coniferous forest stands, areas around trees with woodpecker breeding holes, areas that are difficult to manage (such as steep slopes or riparian areas), areas with specific site conditions such as rocky outcrops, tree islands surrounded by agricultural land, and forest edges with a frequent occurrence of deadwood. After selection, experts coordinate with the forest owners to demarcate the areas, and standardized surveys of forest structure and vegetation are conducted to calculate the financial compensation. The allocation of €4.6 million to the establishment of stepping stones showcases the commitment to preserving and enhancing forest connectivity for biodiversity conservation.

Evaluation

Evaluation of the stepping-stone program in Austrian forests involves a systematic assessment to determine its effectiveness and impact in terms of improving forest connectivity for biodiversity. It aims to gather valuable insights, identify strengths and weaknesses, facilitate informed decision-making, and drive continuous improvement. A steering group has been established to oversee the program, allowing for formative evaluation with regular feedback to refine the program's design during its implementation. A summative evaluation of the program will also assess its overall effectiveness and impact, resulting in a practical handbook to ensure transparency and provide practical guidance.

The first step in the evaluation process is to clearly define the aims of the national stepping-stone program for forest connectivity. These include improving forest connectivity and conserving biodiversity. Data collection methods must be outlined to ensure the evaluation aligns with the program's aims. These methods include habitat assessments by way of standard surveys in all stepping stones as well as intensive biodiversity surveys in a selected number of areas (see also Chap. 12). The latter encompasses several functional groups and ecological assessments. A scientific advisory board has been established to support and guide data collection, analysis, and interpretation of results in the context of the program's aims.

To ensure a comprehensive evaluation, the inclusion of all relevant stakeholders from forest owners to nature conservationists, researchers, and decision-makers is essential. Transparent reporting ensures that the evaluation outcomes are accessible and understood by all stakeholders. By pursuing a comprehensive approach involving rigorous data collection, robust data analysis, and stakeholder engagement, the evaluation aims to provide valuable insights for informed decision-making.

Stakeholder Engagement

In a participatory national project such as the stepping-stone program, stakeholder involvement is of paramount importance from the very beginning. Stakeholders are directly affected by the stepping-stone program and can potentially influence or be influenced by its outcomes. The key identified stakeholders are forest owners and managers, nature conservationists, researchers, and political decision-makers (government). All of these groups are directly or indirectly impacted by the actions and development of the program (see Fig. 22.5). In addition, we see the general public indirectly linked to the program as well. Effective stakeholder management involves identifying and engaging with these stakeholders to understand their needs, concerns, and expectations as well as considering their perspectives in decision-making processes to ensure the long-term sustainability of the program.

Engaging with political decision-makers, government bodies, and funding agencies (government) during the preparation phase of the stepping-stone program is crucial for effective budget allocation and resource mobilization. It is also necessary for contractual design and in order to meet legal requirements. This proactive approach ensures the necessary legal and financial support to implement the program. To secure support and participation from forest owners, trustful communication for obtaining land contributions for stepping stones is vital. Furthermore, efforts should emphasize how biodiversity conservation and ecological connectivity will not only protect the environment but also enhance the value and sustainability of owners' lands. Providing forest owners with accessible information about their habitats and biodiversity can create a sense of responsibility, ensuring long-term interest and commitment.

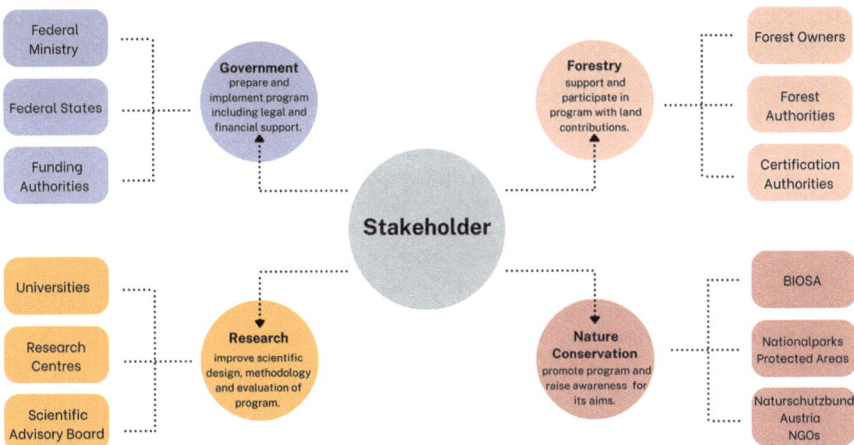

Fig. 22.5 Diagram of stakeholders engaging to ensure the long-term sustainability of the national stepping stone program in Austria. The four main groups include Government, Forestry, Research and Nature conservation. Each group is linked to specific entities contributing to the program's promotion and success

Nature conservationists are valuable allies in supporting and promoting the program. Their expertise and advocacy can help raise awareness regarding the project's conservation goals and garner further public support. In addition, data sharing between conservationists and program managers is essential for prioritizing conservation efforts and focusing on areas with the highest ecological value. Researchers play a significant role in shaping the program's methodology and evaluating its effectiveness. Collaborating with researchers provides scientific expertise to design robust monitoring and evaluation frameworks. Publishing of findings in scientific journals enables the program to share its successes and lessons learned, contributing to broader knowledge in the field of connectivity and conservation.

Overall, the program ensures that the interested general public remains well-informed through regular updates on social media platforms. It shares progress reports, significant project outcomes, key findings, and evaluation results, fostering transparency and engagement. This approach increases awareness about the importance of biodiversity conservation and ecological connectivity.

References

Andersson E, Bodin Ö (2009) Practical tool for landscape planning? An empirical investigation of network based models of habitat fragmentation. Ecography 32(1):123–132

Beger M et al (2022) Demystifying ecological connectivity for actionable spatial conservation planning. Trends Ecol Evolution 37(12):1079–1091. https://doi.org/10.1016/j.tree.2022.09.002

BFW (Bundesforschungszentrum für Wald) (2022) 'Österreichische Waldinventur'. *Waldinformationen aus erster Hand. Umfassend. Kompetent. Aktuell*

BML (2023a) Der Waldfonds - Das Zukunftspaket Für Unsere Wälder. https://www.waldfonds.at

BML (2023b) Ländliche Entwicklung 14-20. *Österreichisches Programm für Ländliche Entwicklung.* https://info.bml.gv.at/themen/landwirtschaft/eu-agrarpolitik-foerderungen/laendl_entwicklung.html

Corrales C, Höglund J (2012) Maintenance of gene flow by female-biased dispersal of black grouse Tetrao Tetrix in northern Sweden. J Ornithol 153(4):1127–1139

Drag L et al (2011) Demography and dispersal ability of a threatened Saproxylic beetle: a mark-recapture study of the Rosalia longicorn (Rosalia Alpina). PLoS One 6(6)

EEA (2021) European Environmental Agency (2021): Landscape Fragmentation 2018 in Europe. Prod-ID: DAS-283-En. Published 08 Dec 2021

Geiser E (2018) How many animal species are there in Austria? Update after 20 years. Acta ZooBot Austria 155(2):1–18. https://www.abol.ac.at/

Gjerde I, Blom HH, Heegaard E, Sætersdal M (2015) Lichen colonization patterns show minor effects of dispersal distance at landscape scale. Ecography 38(9):939–948

Gustafsson L, Hannerz M et al (2020a) Research on retention forestry in northern Europe. Ecol Process 9(1):1–13

Gustafsson L, Bauhus J et al (2020b) Retention as an integrated biodiversity conservation approach for continuous-cover forestry in Europe. Ambio 49(1):85–97

Kadoya T (2009) Assessing functional connectivity using empirical data. Popul Ecol 51(1):5–15

Klinga P et al (2019) Considering landscape connectivity and gene flow in the Anthropocene using complementary landscape genetics and habitat Modelling approaches. Landsc Ecol 34(3):521–536

Kropik M, Zechmeister HG, Moser D (2020) Climate variables outstrip deadwood amount: desiccation as the Main trigger for Buxbaumia Viridis occurrence. Plan Theory 10(1):61

Lapin K, Hoffmann JA, Braun M, Oettel J (2024) Identification and prioritization of stepping stones for biodiversity conservation in forest ecosystems. Conserv Sci Pract 6(7):1–18

Larrieu L et al (2014) Tree microhabitats at the stand scale in montane beech–fir forests: practical information for taxa conservation in forestry. Eur J For Res 133(2):355–367

Lindenmayer DB, Franklin JF (2002) Conserving Forest biodiversity: a comprehensive multi-scaled approach. Island Press, Washington, D.C.

Mergner U (2021) Das Trittsteinkonzept. Euerbergverlage, Rauhenebrach

Pedley SM, Dolman PM (2020) Arthropod traits and assemblages differ between Core patches, transient stepping-stones and landscape corridors. Landsc Ecol 35(4):937–952. https://doi.org/10.1007/s10980-020-00991-0

Perhans K et al (2007) Bryophytes and lichens in different types of Forest set-asides in boreal Sweden. For Ecol Manag 242(2–3):374–390

Perhans K et al (2009) Retention patches as potential Refugia for bryophytes and lichens in managed Forest landscapes. Biol Conserv 142(5):1125–1133

Russ W (2019) Mehr Als 4 Millionen Hektar Wald in Österreich. In: Praxisinformation: Zwischenauswertung Der Waldinventur. Federal Research Centre for Forests, Vienna, pp 3–7

Sverdrup-Thygeson A, Bendiksen E, Birkemoe T, Larsson KH (2014) Do conservation measures in Forest work? A comparison of three area-based conservation tools for wood-living species in boreal forests. For Ecol Manag 330:8–16

Sverdrup-Thygeson A et al (2017) Habitat connectivity affects specialist species richness more than generalists in veteran trees. For Ecol Manag 403(1432):96–102. https://doi.org/10.1016/j.foreco.2017.08.003

Tischendorf L, Fahrig L (2000) On the usage and measurement of landscape connectivity. Oikos 90(1):7–19

Travers TJP et al (2021) Habitat patches providing south-north connectivity are under-protected in a fragmented landscape. Proc R Soc B Biol Sci 288(1957)

Wang YH, Yang KC, Bridgman CL, Lin LK (2008) Habitat suitability Modelling to correlate gene flow with landscape connectivity. Landsc Ecol 23(8):989–1000

Wiktander U, Olsson O, Nilsson SG (2001) Seasonal variation in home-range size, and habitat area requirement of the lesser spotted woodpecker (Dendrocopos Minor) in southern Sweden. Biol Conserv 100(3):387–395

Wintle BA et al (2019) Global synthesis of conservation studies reveals the importance of small habitat patches for biodiversity. Proc Natl Acad Sci 116(3):909–914

Open Access This chapter is licensed under the terms of the Creative Commons Attribution 4.0 International License (http://creativecommons.org/licenses/by/4.0/), which permits use, sharing, adaptation, distribution and reproduction in any medium or format, as long as you give appropriate credit to the original author(s) and the source, provide a link to the Creative Commons license and indicate if changes were made.

The images or other third party material in this chapter are included in the chapter's Creative Commons license, unless indicated otherwise in a credit line to the material. If material is not included in the chapter's Creative Commons license and your intended use is not permitted by statutory regulation or exceeds the permitted use, you will need to obtain permission directly from the copyright holder.

23. Argentina: Balancing Connectivity and Production in Forest Reserves

Rosina Soler, Dardo Paredes, Martin Parodi, Sebastián Farina, and Carolina Hernández

Landscape in Bombilla Forest Reserve on the island Tierra del Fuego, Argentina, South America (Photo: Jonathan Mammani)

R. Soler (✉)
Austral Center for Scientific Research (CADIC - CONICET),
Ushuaia, Tierra del Fuego, Argentina
e-mail: rosina.soler@conicet.gov.ar

D. Paredes · M. Parodi · S. Farina
Dirección General de Desarrollo Forestal, Secretaría de Desarrollo Productivo y PyME, Gobierno de la Provincia de Tierra del Fuego AeIAS, Ushuaia, Argentina
e-mail: dparedes@tierradelfuego.gob.ar; mparodi@tierradelfuego.gob.ar; sfarina@tierradelfuego.gob.ar

C. Hernández
Secretaría de Desarrollo Productivo y PyME, Ministerio de Producción y Ambiente de la provincia de Tierra del Fuego AeIAS, Ushuaia, Argentina
e-mail: carolinahernandezcr@gmail.com

© The Author(s) 2025
K. Lapin et al. (eds.), *Ecological Connectivity of Forest Ecosystems*,
https://doi.org/10.1007/978-3-031-82206-3_23

Abstract

In various parts of the world, forest reserves are areas that support forests or systems of high forest value, meeting multiple objectives concerning both the production of goods through sustainable management and conservation of the ecosystem services those areas provide. This chapter presents the case of the Production Forest Reserves (PFR) in Tierra del Fuego, which are aimed at forest harvesting as well as the restoration of environments degraded by fires and valuation of the goods and services offered by the Fuegian forest landscapes. The implementation and monitoring of silvicultural practices, restoration, and vegetation surveys carried out by the General Directorate of Forestry (GDF) in cooperation with the scientific sector generate scientific and technical information to improve the management and sustainable use of the forest landscape. The PFRs also offer ideal sites for raising environmental awareness through educational tours, recreation, and low-impact tourism for visitors as well as other activities with the local communities. The enabling of different uses within the PFRs aims to promote the continuity of productive forest landscapes and their biodiversity in space and time.

Keywords

Biodiversity · Ecosystem services · Forest connectivity · Nature conservation · NWFP · Restoration

Introduction

Areas declared as forest reserves serve to support forests or systems of high forest value due to their development, seasonal suitability, and aesthetic or scenic beauty. The native forest area of Tierra del Fuego, Antarctica, and the Atlantic Islands (Argentina) encompasses 793,909 ha, of which 325,134 ha is covered by *Nothofagus pumilio* (41%). These forests represent 23% of the area of pure *N. pumilio* forests in Argentina. Historically and into the present, these pure forests provide the raw material for the primary timber industry, and management is carried out on private and public lands. Around 55,000 ha of *N. pumilio* forests are currently being actively harvested in the province of Tierra del Fuego, mostly in the Cordilleran forest region (Fig. 23.1, Collado and Bava 2020).

The Production Forest Reserves (PFRs) are administrative units created by Provincial Decree No. 2502 in the year 2002 within the framework of Provincial Law No. 145 of 1994 (and Regulatory Decree No. 852 of 1995). The purpose of their creation was to establish areas of public forested land within the provincial territory—in particular, production *N. pumilio* (lenga) and *N. betuloides* (guindo) forests—to ensure predictability for the local forestry industry (Forestry Industry Sectorial File—Ministry of Production and Environment 2021–2022). The province of Tierra del Fuego currently maintains nine PFRs (see Fig. 23.1, Table 23.1): Fagnano West, Bombilla, Escondido Lake, Milna River, Lainez River, Valdéz River,

23 Argentina: Balancing Connectivity and Production in Forest Reserves

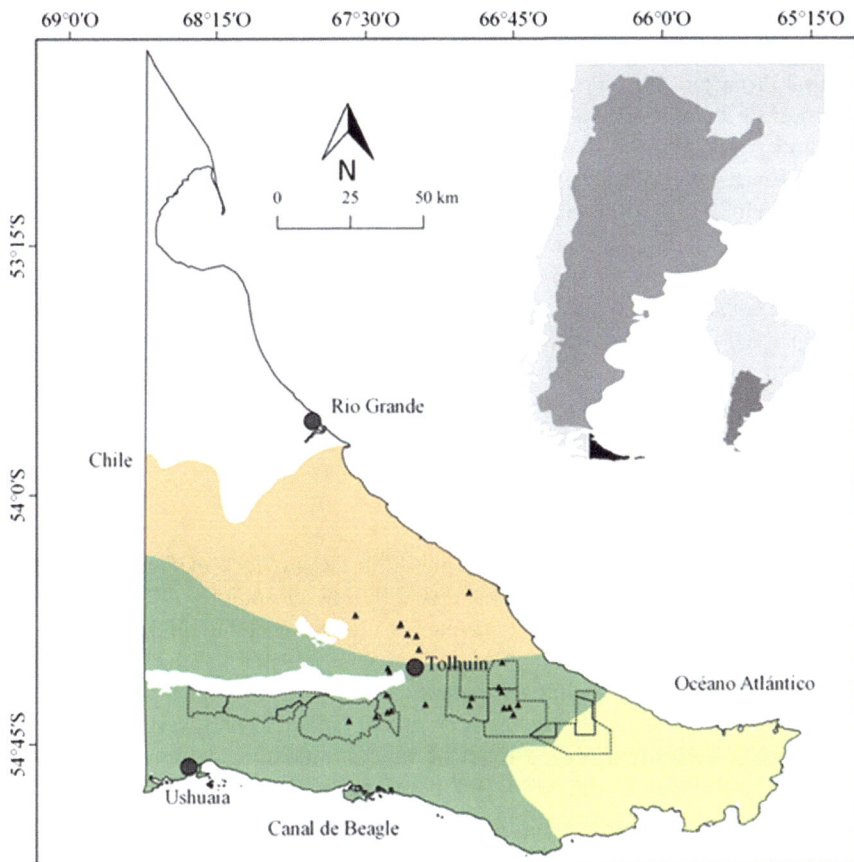

Fig. 23.1 Map showing the vegetation types, the locations of Production Forest Reserves (black contour polygons), and monitoring plots (black triangles) of the General Directorate of Forestry (GDF) in Tierra del Fuego (Argentina, South America). Orange = steppe-forest ecotone subregion, green = mountain range forest subregion, yellow = forest subregion alternating with peatland

Table 23.1 List of productive forest reserves in Tierra del Fuego indicating the respective total area in hectares and proportion (%) of the area occupied by forest

Name	Area (ha)	% Forest cover
Fagnano West	9019.80	>50%
Bombilla	18,860.89	58%
Escondido Lake	2858.91	>50%
Valdéz River	31,636.73	58%
Milna River	6373.59	68%
Lote 93	9026.33	65%
Lainez	15,337.61	63%
Irigoyen River	30,117.47	62%
Malenguena River	9620.28	>50%

Lote 93, Irigoyen River, and Malenguena River. Two of these—Irigoyen River and Malenguena River—have recently been incorporated into the new Peninsula Mitre Natural Protected Area (Provincial Law No. 1461), pursuing the same objective as the other PFRs (still managed) but with a joint administration still in planning.

Timber production is concentrated in the Bombilla, Milna River, Valdéz River, Lainez River, Lote 93, and Irigoyen River reserves, where the most timber-producing stands are located. Within the PFRs, timber extraction permits are granted to small and medium-sized producers. There is a network of roads to facilitate access for producers and visitors, rural police, forestry, and other officials. Given the management and conservation objectives of PFRs, other activities besides timber harvesting by registered forest producers, such as firewood collection and peat extraction, are permitted as well. In certain areas, low-impact tourism and recreation are allowed, and environmental awareness and scientific research activities are promoted.

Habitat and Biological Diversity

Four main types of vegetation are recognised in the province of Tierra del Fuego: (1) Patagonian steppe (north), characterised by low elevation (<200 m.a.s.l.) and gentle undulations interrupted by ravines, with the most humid areas featuring abundant herbaceous vegetation (Tuhkanen et al. 1989–1990). The climax community is the coiron steppe (*Festuca gracillima*) accompanied by other grasses (Frangi et al. 2004). (2) Magellanic peatlands or tundra (southern and western end of the archipelago) featuring a rugged relief of valleys alternating with hills exceeding 300 m.a.s.l. (Tuhkanen et al. 1989–1990; Frangi et al. 2004). (3) High Andean vegetation consisting of plate and cushion plants on rock surfaces, with few higher plants and a predominance of lichens above the treeline (Moore 1983; Frangi et al. 2004). (4) Sub-Antarctic forest, extending from sea level to 600–700 m.a.s.l. (Barrera et al. 2009; Cuevas 2002), and situated south of parallel 54°S.

The sub-Antarctic forest is dominated by deciduous formations of *N. pumilio*, *N. antarctica*, and evergreen *N. betuloides* and can be differentiated into three main regions (Fig. 23.1): steppe-forest ecotone, mountain range, and wet forests with alternating peatlands (Cabrera 1976; Allué et al. 2010; Morello et al. 2012). The steppe-forest ecotone region (44% of the area) is composed of small patches of open *N. antarctica* forest alternating with steppe at higher elevations and meadow communities (humid floodplains) in the depressions, with minor inclusions of *N. pumilio* on higher grounds. The Cordillera region extends south from the ecotone zone to the Beagle Channel and from the Chilean border in the west to approximately 66°W, where the topography is determined by the alternation of large mountain ranges (up to 1400 m.a.s.l.) and valleys; here, the dominant vegetation is deciduous *N. pumilio* forest (up to ~600 m.a.s.l.). The region of humid forests with alternating peatlands extends to the south and southeast of the Andes, where *N. betuloides* forests occupy the coldest and most humid areas, eventually accompanying *N. pumilio* forests (Magellanic mixed forests).

The PFRs are located in the Cordillera region, and the native forests are their main biophysical component. However, the reserves also include a variety of associated environments such as *N. antarctica* forests, scrublands, peatlands, small grasslands, and other disturbed habitats such as old fire sites or beaver forests. This diversity of environments enables the establishment and occurrence of a greater variety of species, increasing biodiversity at the landscape level. For example, peatlands offer a greater abundance of *N. antarctica* saplings, since it is the only woody species adapted to growing in flooded areas (Donoso 2006). On the other hand, the scrublands as areas with no forest cover and greater solar irradiation feature a greater abundance of native fruiting shrubs. The continuous forests (albeit in different stages of development) maintain the typical species of the forest interior (herbs, ferns, fungi).

The distribution of herbaceous species is also strongly linked to the type of habitat or environment. On the one hand, areas where the forest is in a state of regeneration with a closed or semi-closed canopy cover only allow the establishment of shade-tolerant species (forest interior species) such as *O. depauperata* (wild parsley) and a large variety of fungi. On the other hand, forests with a more open canopy produced by forest fires and/or beaver disturbance favour the establishment of species such as *Rubus geoides* (Patagonian wild raspberry), *Berberis microphylla* (Magellan barberry), and *Empetrum rubrum* (red crowberry).

Management of Secondary Forest and Intermediate Methods

During the last 30 years, shelterwood cutting has been the most widely applied method for harvesting and regenerating *N. pumilio* forests not only in the PFRs but throughout the island. The main drawback in the implementation of protective logging in Tierra del Fuego from the end of the last century to the present is that it has been carried out without regard for landscape type (steppe-forest ecotone, mountain range forests, or humid forests) and without taking productivity gradients (e.g., site quality), potential environmental restrictions like availability of soil water, or mass disturbance factors like wind exposure into account (Paredes et al. 2020). For this reason, the GDF has been monitoring forest structure and natural regeneration at the stand and landscape scales for several years.

Short-Term Management (5–10 Years Post-Harvest)

The development of natural regeneration in harvested forests is a significant variable for decision-making regarding the best time to initiate intermediate treatments. By monitoring harvested forests at a territorial scale, it was possible to characterise the forest structure after harvesting and collect information on the state of natural regeneration in different productive forest sites. The GDF established permanent monitoring plots in three large harvesting areas in the north, east, and south of the province to analyse the state of the harvested forests at a landscape scale and consider the connectivity of the PFRs.

The permanent plots allow different states of the short-term post-harvest forest structure to be characterised. These include the original or primary structure (EO) of the original forest before logging, the harvested structure (EC) corresponding to the trees to be felled, cleaned, and extracted from the stands, the remaining structure (ER) composed of the trees left standing immediately after harvesting, the damaged structure (ED) or remaining trees that died standing or were affected by windfalls, and the current structure (EA) comprised of the living trees standing until the present day (Fig. 23.2).

Under structural complexity, regeneration initiates its natural dynamics. It should be noted that harvested forests regenerate naturally without assistance through seeding or planting. The monitored variables include not only the number of plants per surface area and their height, but also biotic (herbivory) and abiotic (apical desiccation) damage, growth rate of the last three periods, and optimum quality of the plants considering the continuity of timber use (Table 23.2).

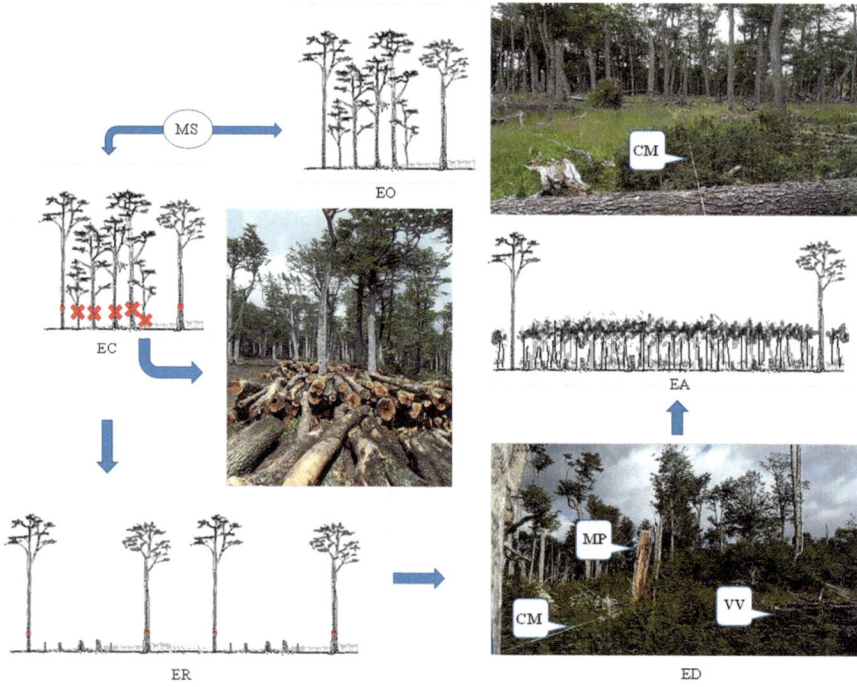

Fig. 23.2 Theoretical model of forest structure levels in forests harvested under protective logging. *EO* original structure, *MS* silvicultural marking, *EC* harvested structure: trees with red "X" indicate trees to be removed, *ER* remaining structure, *ED* damaged structure, *MP* dead standing tree, *CM* tape measure indicating rectangular plot centre, *VV* windthrown tree, *EA* current structure (Paredes 2023)

Table 23.2 Records of natural regeneration at the territorial scale for the period corresponding to 5 years (Post+5) and 10 years post-harvest (Post+10)

Period	Zone	N	A	DA	DB	TC
Post+5	East	489.6	0.3	6.7	15.6	6.7
	North	121.3	0.4	2.0	15.2	7.1
	South	219.7	0.5	0.3	10.3	9.9
Post+10	East	78.7	0.8	9.8	25.0	16.6
	North	102.6	0.4	17.5	67.9	5.5
	South	184.6	1.5	4.4	0.6	16.0

N density (plants/(ha * 1000)), *A* average height (m), *DA* abiotic damage (%), *DB* biotic damage (%), *TC* growth rate (cm/year)

The continuous monitoring of the plots allows evaluation of the post-harvest dynamics of natural regeneration in the understorey layers within the different geographic areas. During the first 5 years post-harvest, most of the natural regeneration (40–60% of seedlings) represents the lowest stratum (0.2–0.5 m). After 10 years post-harvest, the development of the surviving seedlings favours the transition to the highest stratum (>0.5 m in height) in 30–50% of the total regeneration.

Recorded plant densities (77,000 to 794,000 plants per hectare for Post+5, and between 150,000 and 301,000 plants per hectare for Post+10) fall within thresholds observed in other studies in southern Patagonia for *N. pumilio* forests post-shelterwood cutting (Caldentey et al. 1999; Silva Aguad et al. 2008; Rosenfeld et al. 2006; Cellini 2010; Martínez Pastur et al. 2011; Martínez Pastur et al. 2017). Decrease in density and increase in height post-harvest align with the natural dynamics of this species. However, biotic and abiotic damage to natural regeneration occurred in both post-harvest periods. Many of these forests are the natural habitats for *Lama guanicoe* that connect open habitats (grasslands) with open forests. Although browsing damage does not impact normal plant development, it influences height growth and causes delays in seedling establishment and growth (Martínez Pastur et al. 2016). Biotic damage is not uniform at the landscape level (e.g., greatest in the northern zone, with 67.9% of damage) due to differential habitat use by mammals and differences in habitat connectivity. North- or East-exposed sites are usually better refuges for wildlife than South-exposed sites.

In the short term, the forest structure and the vigorous natural regeneration of the secondary forest create a dynamic requiring the implementation of silvicultural measures to manage the density of the forest structure, using the extracted biomass (firewood) and maintaining the continuity of the logs until the regeneration reaches the optimal development (optimal range of 10–15 cm) to initiate intermediate treatment practices.

Medium-Term Management (10–30 Years)

The Milna River and Bombilla PFRs have been the harvesting areas in the southern part of the Fuegian territory with forest use since the 1970s, for the most part employing harvesting practices known as selection cutting. They thus correspond to an important extension of heterogeneous secondary productive forests in various advanced stages of development and with structural characteristics differing from the regular primary forests in the eastern and northern areas of the province. To sustain the production of quality wood and sawing yield, it is necessary to manage the regular regeneration stands through intermediate treatments. Within the framework of National Law No. 26331 and Provincial Law No. 869, the GDF actively participates in the planning, elaboration, and evaluation of the secondary forest management plans in the Milna River and Bombilla PFRs via the General Silvicultural Management Plan, Resolution S.D.S.yA. No. 398/15. Among the objectives of this plan is the implementation of different intermediate silvicultural treatments (e.g., thinning) in secondary forests subject to harvesting in the past as well as in other remaining, usually low-quality primary forests left out of previous management plans to increase their tree mass to production forest levels. As a result, by-products such as poles and firewood are extracted. In addition, monitoring plots are established for follow-up and to determine future interventions within the framework of sustainable forest management. Although the local productive activity presently values the forest resource for timber purposes, there is a vast area of forest that can provide woody biomass for uses other than the primary processing industry. With this in mind, the GDF develops and applies silvicultural treatments providing experience, education, and training of foresters in thinning practices.

Silvicultural treatments are implemented depending on the stage of development of these forests, imitating the natural processes of the respective forest and taking the site characteristics of each stand into account (Bava 1999). The recommended treatments for these forests are selection thinning (elimination of poorly shaped and unhealthy plants within a sapling stand) and geometrical thinning (see Figs. 23.3 and 23.4).

Fig. 23.3 Stand of *N. pumilio* (PFR Valdéz River) after the application of selective thinning, where individuals of lower diameter classes and poorly shaped dominants were eliminated. This homogenised the forest stand into a well-structured and generally co-dominant state

Fig. 23.4 Stands of *N. pumilio* (PFR Valdéz River) after the application of selective thinning (left) and geometrical thinning using north-south strips (right)

Within the PFRs, there are also demonstration areas for silvicultural interventions that are applicable to sustainable native forest management. In the Valdéz River reserve, for example, intensive harvests have historically been applied, albeit without records of specific volumes or structures harvested. The practices recently applied by way of demonstration—elimination of the upper stratum, isolated stands of grasses and shrubs, and overmature individuals—are intended to generate a homogeneous forest with a single stratum and a single age class to accelerate growth and favour dynamics. In terms of biodiversity, this (1) enhances connectivity by facilitating easier movement and dispersal of species within and between habitats, (2) increases species richness, particularly those that thrive in early successional habitats or have specific habitat requirements, and (3) by enhancing management activities, such as monitoring, restoration, and invasive species control. The new generation of individuals is freed from the competition of the unmanaged upper stratum. It should be noted that the GDF implements these silvicultural practices together with small forest producers, generating a new productive opportunity in already managed forests, demonstrating silvicultural concepts, and applying silvicultural methodologies adaptive to the present stock.

Restoration of Burned Forests for Connectivity

The PFRs are also exposed to massive disturbances such as the impact of invasive species such as the beaver, informal cattle ranching, and anthropogenic fires. In December 2008, a large forest fire occurred in the central zone of the province within the PFR Lote 93, affecting approximately 3500 ha of native *N. pumilio* and

N. antarctica forests. In reaction to this event, the project "Restoration of *N. pumilio* Forests Affected by Forest Fires in Tierra del Fuego" was initiated in 2010.

Since then, the GDF has been developing concrete restoration and environmental awareness actions to recover the lost native forest, including planting *N. pumilio* trees on burned sites (Fig. 23.5). The restoration initiative aims to increase forest connectivity and promote the recovery process of ecosystem goods and functions (e.g., timber provision, recovery of the organic soil layer, watershed protection, and habitat biodiversity) of forests degraded by fires. Restoration actions are planned each year along corridors along main roads to facilitate accessibility and identification of critically degraded areas (heightened exposure, steep slopes, soil erosion, and loss of forest structure) without passive forest recovery. These priority areas are defined based on a comprehensive matrix that includes (1) phytogeography and ecological considerations (satellite imagery), (2) climatic, (3) topography with general considerations of local geomorphology (topographic maps), and (4) connectivity aspects (proximity to areas planted in previous years).

Between 15,000 and 20,000 12-cm-high saplings with lignified stems are collected from seedling banks in undisturbed forests for each planting campaign during spring. Seedling banks in areas close to the restoration zone are chosen (Fig. 23.6). The saplings are placed in growing trays and transferred to the Tolhuin Forest

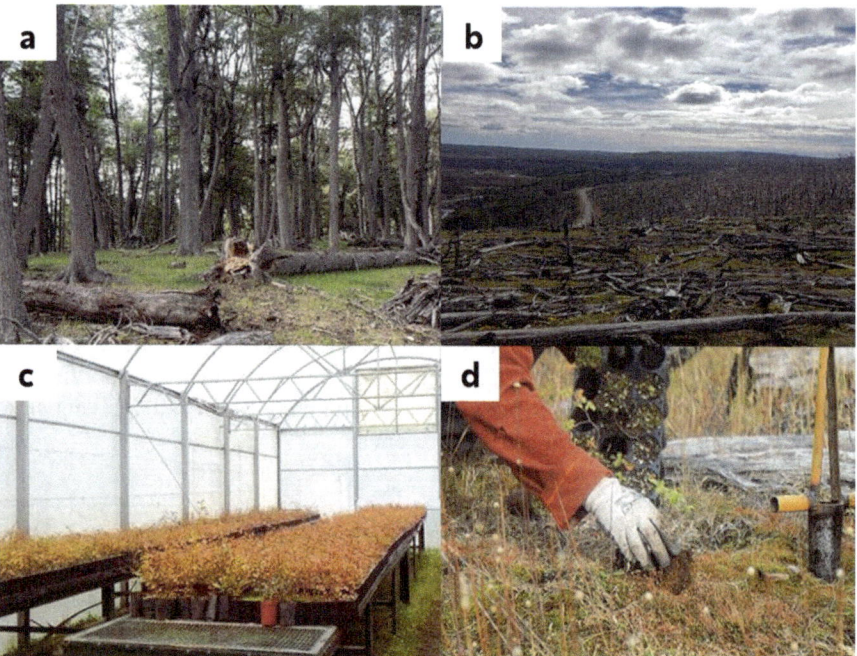

Fig. 23.5 (a) *N. pumilio* primary forest; (b) burned forest in Lote 93 during 2008; (c) production of *N. pumilio* seedlings in greenhouse; (d) restoration of burned area through assisted planting of seedlings

Fig. 23.6 (**a**) *N. pumilio* seedling bank; (**b**) technician collecting seedlings; (**c**) seedlings in the growing tray at the Tolhuin Forest Nursery; (**d**) seedling raised in the greenhouse during a growing season and exhibiting strong root development

Nursery under the supervision of the Ministry of Production and Environment where remain during the vegetative growth period (summer) in a mixed substrate (peat, perlite, and slow-release fertiliser). Their acclimatisation and development are thus promoted under optimal conditions before they are taken to their final planting site after 5 months. The growth rates of saplings in the nursery vary between 12 and 25 cm. Although this rate indicates good aerial development of seedlings, the greenhouse is primarily also intended to ensure proper development of the root system to allow the plants to survive under extreme environmental conditions—without canopy cover and exposed to direct sunlight, wind, drought, and the like—at their final destination site.

Before planting, the target sites called "regeneration nuclei" are defined through field surveys and considering site characteristics such as soil, exposure to sunlight and wind, and natural protection (e.g., large logs). At each regeneration nucleus, ten holes are made using planting tools specifically designed for the purpose (hole size = seedling container size); the seedlings are placed in them and the soil is tamped down. The development and successful application of nucleation techniques in terms of cost reduction and sustainability of restored ecosystems are widely documented (Martins 2017).

To date, more than 110 hectares have been restored by planting more than 60,000 *N. pumilio* seedlings (Fig. 23.7). The GDF continually monitors the restoration progress of the planted areas, and the results so far are very encouraging with regard to seedling adaptation and evidence of recovery of degraded sites. The restored areas play a vital role in enhancing the structural continuity of the *N. pumilio* forest, aiming to reinstate key elements that facilitate the eventual return of native species (e.g., forest interior birds, insects) but also enhance ecosystem services (e.g., organic matter decomposition, soil formation, seed dispersal).

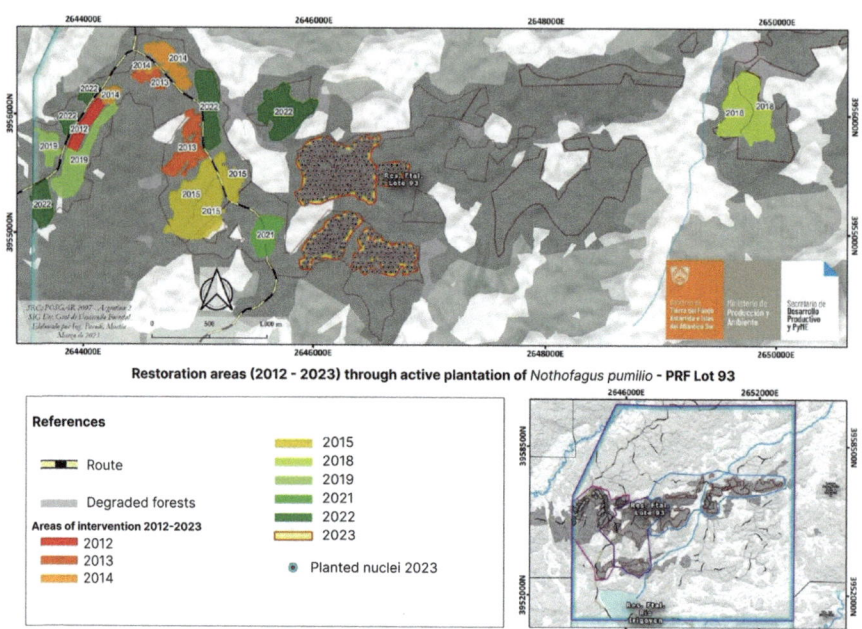

Fig. 23.7 Areas planted with *N. pumilio* in PFR Lote 93, 2012–2023 (https://prodyambiente.tierradelfuego.gob.ar/gestion-forestal-sostenible)

As part of this monitoring of restored areas began in 2018, a database of plant development metrics was generated to evaluate the performance of planting and the continuity of future canopy layers. The results regarding seedling survival (>80%), as well as sapling growth (15–65 cm height growth) and vigour, can be considered favourable for all planting techniques implemented in the territory. This experience encourages carrying on the work to recover areas affected by forest fires. However, coping with the substantial interannual fluctuations in growth rates and the detrimental effects of biotic damage, particularly herbivory, poses a significant challenge, especially in light of projected climate change scenarios. During dry years or when multiple stressors converge (such as drought coupled with herbivory), there is a heightened risk of extensive seedling mortality, which could jeopardise the connectivity of restored areas.

Along with active restoration tasks, environmental awareness and education activities are carried out with schools, clubs, volunteers, and the community in general. The objective of these activities is to bring the community closer to the forest as a medium and resource. Awareness is generated concerning the importance and consequences of exogenous disturbances causing degradation of the native forest in Tierra del Fuego—especially relating to forest fires—as well as technical measures to mitigate their impact.

Non-Timber Forest Products

Native as well as naturalised plant species have been used for nutritional and medicinal purposes (e.g., fruits, flowers, and leaves of aromatic plants) from prehistoric times to the present. The collection of non-timber forest products (NTFP) in Tierra del Fuego began to be recognised as a rural productive activity a few years ago, although it is still carried out under informal conditions. However, small producers and collectors throughout Patagonia recognise that their gathering of NTFPs means intense daily work while preserving ancestral and specific knowledge and establishing exchange networks for that knowledge among the inhabitants of the area.

The PFRs of Tierra del Fuego offer a great diversity of NTFPs that can be used by small local producers and entrepreneurs (Soler et al. 2024). The use of these resources adds value to the PFRs and promotes habitat preservation and sustainable management of native plants. At the same time, it offers a productive alternative to strengthen and diversify the productive matrix. Therefore, the project "Non-Timber Forest Products in the Production Forest Reserves of Tierra del Fuego: Technical Sheets and Specific Sustainable Harvesting Protocols" was initiated in 2022 in order to (1) identify the most relevant Fuegian NTFPs with productive and social potential, (2) determine the distribution of NTFPs within the PFRs, and (3) make general recommendations concerning sustainable harvesting.

Through interviews with small producers and local entrepreneurs, 18 species of plants, fungi, and lichens whose fruits, leaves, flowers, bark, seeds, and other parts are used as NTFPs in Tierra del Fuego were identified. Most of these species are represented within the PFRs, although their distribution varies greatly depending on the frequency and coverage of natural (mature forest, secondary forest, scrub, etc.) and disturbed habitats (fire, beaver plantations, roads, etc.) within each reserve.

One of the most interesting results is that forests of high timber value or in the regeneration phase with a closed or semi-closed canopy cover are low in NTFP frequency and abundance. This is because only shade-tolerant species established in this type of habitat are found there, whereas habitats with more open canopy cover such as scrub, forest fire sites, streams, or beaver-impacted riparian forests offer optimal habitat conditions for a greater variety of fruiting, flowering, and herbaceous plants used as NTFPs. Non-forested or disturbed areas thus increase the variety of habitats for numerous species of both forest and associated environments increasing potential connectivity for biodiversity. Furthermore, in the context of social-ecological systems (cohesive and integrated systems between the ecological component and society), NTFPs enhance landscape diversity within PFRs, foster local knowledge about natural resources, and offer an opportunity to promote sustainable harvesting.

The GDF is beginning to take steps towards compiling knowledge on these resources and analysing their potential for productive use, for instance by incorporating certain NTFPs in the Argentine Food Code (Farina et al. 2022). However, there are currently no regulations applying to the control, permits, or registration of collection except for the collection of biological material for scientific purposes. It will be necessary for sites identified as having the highest occurrence of NTFPs to be protected as well as regulations issued to ensure future sustainability by avoiding overuse of some products and possible environmental damage. In other parts of Patagonia (e.g., harvesting of seeds of *Araucaria araucana* in Neuquén), local authorities monitor NTFP harvesting in designated areas, with permits issued to groups of formal gatherers who have been trained in sustainable harvesting practices in wild habitats.

Other Activities Within the PFRs

A further productive activity permitted within some PFRs (Milna River, Lote 93, Lainez River) is the extraction of peat for commercial purposes (substrate for horticulture) by small producers within the framework of existing regulations. Tierra del Fuego sustains the southernmost concentration of peatlands on the planet, and these ecosystems play an important role in the provision of ecosystem services as well as in climate change mitigation. Therefore, scientists are carrying out studies to provide accurate information to the provincial authorities that will contribute to the regulation and sustainable management of peatlands. An ongoing project ("Quantification and characterization of peatlands with productive potential in Tierra del Fuego Island") will provide the first comprehensive set of data on the types and volumes of peat present in the areas whose zoning allows their productive use according to provincial regulations.

There are also other indirect uses of the PFRs that enhance the value of the southern forest landscape and its biodiversity. For example, local tourism companies offer recreational and scenic tours in designated public use areas (PFR Bombilla). These areas feature instructive signboards with information on forest species, natural and cultural heritage, fire prevention, and environmental care in general. Several PFRs also include camping areas and hiking trails, which play a key role in recreation, well-being, cultural identity, and appreciation of the forest landscape among the local community. Primary and secondary school visits to the PFRs are encouraged, as is scientific research by local institutions with the aim of generating information applicable to the formulation of public policies for improving the management and sustainable use of the forest landscape and its biodiversity.

References

Allué C, Arranz JA, Bava JO, Beneitez JM, Collado L, García-López JM (2010) Phytoclimatic characterization and cartography of subantarctic native forests in Isla Grande de Tierra del Fuego (Patagonia, Argentina). Forest Syst 19(2):189–207. https://doi.org/10.5424/fs/2010192-01314

Barrera M, Frangi J, Richter L, Perdomo M, Pinedo L (2009) Structural and functional changes in *Nothofagus pumilio* forests along an altitudinal gradient in Tierra del Fuego, Argentina. J Veg Sci 11:179–188. https://doi.org/10.2307/3236797

Bava JO (1999) Aportes Ecológicos y Silviculturales a la transformación de bosques vírgenes de lenga en bosques manejados en el sector argentino de Tierra del Fuego. Tehcnical Report N°29, CIEFAP. http://www.ciefap.org.ar/digitalizado/Publ_tec_29_Aportes_ecologicos.pdf

Cabrera AL (1976) Regiones Fitogeográficas Argentinas. 2nd Ed Buenos Aires, Argentina
Caldentey J, Schmidt H, Ibarra M (1999) Modificaciones ambientales debidas al manejo del bosque de Lenga en Magallanes. Technical Report, FONDECYT 1960936. http://documentacion.ideam.gov.co/openbiblio/bvirtual/005039/tema3/CaldenteyJSchmidtH.pdf
Cellini JM (2010) *Estructura y regeneración bajo distintas propuestas de manejo de bosques de Nothofagus pumilio (Poepp et. Endl) Krasser en Tierra del Fuego, Argentina*. Doctoral Thesis. National University of La Plata, Buenos Aires
Collado L, Bava J (2020) Plan regional de conservación de bosues de N. pumilio en el Ecotono de Tierra del Fuego, en el contexto de cambio climático. SADSyCC-MPyA, Argentina
Cuevas J (2002) Episodic regeneration at the *Nothofagus pumilio* alpine timberline in Tierra del Fuego, Chile. Ecology 90:52–60. https://doi.org/10.1046/j.0022-0477.2001.00636.x
Donoso C (2006) Las especies arbóreas de los bosques templados de Chile y Argentina: autoecología, 1st edn. Marisa Cuneo, Valdivia
Farina S, Mattenet F, Soler RM, Hernandez C, Peri PL (2022) Hojas de Nothofagus antarctica (Ñire). Technical Report, DGDF, SDPyPyME, Argentina
Frangi JL, Barrera MD, Puigdefábregas J, Yapura PF, Arambarri AM, Richter LL (2004) Ecología de los bosques de Tierra del Fuego. In: Frangi AM, Goya J (eds) Ecología y Manejo de los Bosques de Argentina. National University of La Plata
Martínez Pastur G, Peri PL, Cellini JM, Lencinas MV, Barrera MD, Ivancich H (2011) Canopy structure analysis for estimating forest regeneration dynamics and growth in *Nothofagus pumilio* forests. Ann For Sci 68:587–594. https://doi.org/10.1007/s13595-011-0059-1
Martínez Pastur G, Soler RM, Ivancich H, Lencinas MV, Bahamonde H, Peri PL (2016) Effectiveness of fencing and hunting to control *Lama guanicoe* browsing damage: implications for *Nothofagus pumilio* regeneration in harvested forests. J Environ Manag 168:165–174. https://doi.org/10.1016/j.jenvman.2015.11.051
Martínez Pastur G, Cellini JM, Barrera MD, Lencinas MV, Soler R, Peri PL (2017) Influence of biotic and abiotic factors on the growth of pre- and post-harvest regeneration in *Nothofagus pumilio* forests. Bosque 38(2):247–257. https://doi.org/10.4067/S0717-92002017000200003
Martins SV (2017) Alternative Forest Restoration Techniques. InTech, Rijeka. https://doi.org/10.5772/intechopen.72908
Moore DM (1983) Flora of Tierra del Fuego. Anthony Nelson, England. Missouri Botanical Garden, USA
Morello J, Matteucci S, Rodriguez A, Silva M (2012) Ecorregiones y complejos ecosistémicos Argentinos, 1st edn. Orientación Gráfica Editora, Buenos Aires
Paredes D (2023) Structural characteristics affecting natural regeneration of harvested *Nothofagus pumilio* forests along environmental gradients in Tierra del Fuego (Argentina): Adaptation of silvicultural practices to sustainable management. Doctoral Thesis, National University of La Plata, Argentina
Paredes D, Cellini JM, Lencinas MV, Parodi M, Quiroz D, Ojeda J, Farina S, Rosas YM, Martínez Pastur G (2020) Landscape influence in the shelterwood cuts of *Nothofagus pumilio* forests in Tierra del Fuego, Argentina: changes in forest structure and regeneration performance. Bosque 41:55–64. https://doi.org/10.4067/S0717-92002020000100055
Rosenfeld JM, Navarro Cerrillo RM, Guzmán Alvarez JR (2006) Regeneration of *Nothofagus pumilio* (Poepp. et Endl.) Krasser forests after five years of seed tree cutting. J Environ Manag 78:44–51. https://doi.org/10.1016/j.jenvman.2005.03.009
Silva Aguad CP, Schmidt Van Marle H, Schmidt A (2008) Desarrollo de los bosques de Lenga (*Nothofagus pumilio*) después de la corta de regeneración. Ciencia e Investigación Forestal-Instituto Forestal 14:65–76. https://doi.org/10.52904/0718-4646.2008.92
Soler R, Pal N, Bustamante G, Rodríguez P, Miller J, Mattenet F (2024) Productos Forestales No Madereros de Tierra del Fuego, Antártida e Islas del Atlántico Sur. MPyA, Argentina
Tuhkanen, S., Kuokka, I., Hyvnen, J., Stenroos, S., Niemelä, J. (1989–1990). Tierra del Fuego as a target for biogeographical research in the past and present. Anales del Instituto de la Patagonia, 19 (2), 1–107. http://www.bibliotecadigital.umag.cl/handle/20.500.11893/1433

Open Access This chapter is licensed under the terms of the Creative Commons Attribution 4.0 International License (http://creativecommons.org/licenses/by/4.0/), which permits use, sharing, adaptation, distribution and reproduction in any medium or format, as long as you give appropriate credit to the original author(s) and the source, provide a link to the Creative Commons license and indicate if changes were made.

The images or other third party material in this chapter are included in the chapter's Creative Commons license, unless indicated otherwise in a credit line to the material. If material is not included in the chapter's Creative Commons license and your intended use is not permitted by statutory regulation or exceeds the permitted use, you will need to obtain permission directly from the copyright holder.

24. Botswana: Stand Structure and Hampered Regeneration of Woody Species in Kazuma Forest Reserve, the Busiest Elephant Corridor in Northern Botswana

Demel Teketay, Witness Mojeremane, Lawrence Akanyang, Kamogelo Makgobota, Rampart Melusi, Ronnie Mmolotsi, David Monekwe, Ismael Kopong, Gosiame Neo-Mahupeleng, Topoyame Makoi, Kakambi Obusitswe, and Ednah Kgosiesele

Roaming Elephants in Kazuma Forest Reserve, Botswana (Karl Wassermann)

D. Teketay (✉) · W. Mojeremane · L. Akanyang · K. Makgobota · R. Melusi · R. Mmolotsi · D. Monekwe · I. Kopong · G. Neo-Mahupeleng · T. Makoi · K. Obusitswe · E. Kgosiesele
Botswana University of Agriculture and Natural Resources, Department of Forest and Range Resources, Gaborone, Botswana
e-mail: dteketay@buan.ac.bw; wmojerem@buan.ac.bw; lakanyang@buan.ac.bw; kmakgobota@buan.ac.bw; rmelusi@buan.ac.bw; rmmolotsi@buan.ac.bw; dmonekwe@buan.ac.bw; ikopong@buan.ac.bw; tmakoi@buan.ac.bw; ekgosiesele@buan.ac.bw

© The Author(s) 2025
K. Lapin et al. (eds.), *Ecological Connectivity of Forest Ecosystems*, https://doi.org/10.1007/978-3-031-82206-3_24

Abstract

The Chobe District of northeastern Botswana is a semi-arid dryland supporting a diverse assemblage of wildlife and habitats of global conservation significance, including the largest elephant population in Africa. Among others, the district houses six forest reserves (FRs) representing approximately 4207 km^2 (3.83% of the country's total protected land). These FRs—including Kazuma Forest Reserve—provide many goods and ecosystem services that are essential for improving and maintaining human livelihoods in addition to performing many ecological functions. Results from field studies and consultations with the relevant stakeholders have indicated that the six FRs have declined in forest cover over the last 20 years. This development has been attributed to forest degradation caused mainly by the annual human-induced fires as well as the increasing numbers of elephants. A study was carried out to: (1) assess the species richness, diversity, and evenness; (2) determine the densities, frequencies, dominance, and importance value index; and (3) assess the population structure and regeneration status of woody species in Kazuma FR. This has resulted in changes in species composition, a decline in diversity, density, frequency, and basal area or dominance, and unstable population structures and impeded regeneration of the woody species. The results reinforce the argument that relevant authorities need to develop countermeasures in management to address the issue of deforestation and degradation of Kazuma FR and other forest reserves.

Keywords

Frequency · Density · Important value index · Population structure · Species richness · Diversity · Evenness

Introduction

The Chobe District of northeastern Botswana is a semi-arid dryland supporting a diverse assemblage of wildlife and habitats of global conservation significance, including the largest elephant population in Africa (Blanc 2017; Department of Wildlife and National Parks (DWNP) 2014; Fox et al. 2017). The district houses six forest reserves (FRs) representing approximately 4207 km^2 (3.83% of the country's total protected land), namely Chobe (154,500 ha), Sibuyu (116,100 ha), Kasane Extension (64,111 ha), Maikaelelo (54,300 ha), Kazuma (16,800 ha), and Kasane (14,931 ha). The FRs are set aside for conservation purposes, and under the 1968 Forest Act, all forms of subsistence use, or commercial activities are prohibited with only rare exemptions.

FRs in Botswana are declining over time. Knowledge about the species richness, diversity, survival rates and growth of seedlings, population structure, and natural regeneration mechanisms in these forests is very scant. No information is available on the composition and density of woody species. Knowledge about these attributes would contribute to proper management and sustainable utilisation and conservation of the remaining FRs.

Regeneration is a central component of tropical forest ecosystem dynamics and the restoration of degraded forest lands. Sustainable forest utilisation is only possible if adequate information on the regeneration dynamics and factors affecting the regeneration of important canopy tree species are available (Tesfaye et al. 2010). Studies on density and population structures of major canopy tree species can help to understand the status of species regeneration., and, thus, the management history and ecology of the respective forest or woodland (Harper 1977; Saxena et al. 1984; Hubbel and Foster 1986; Lykke 1993; Alvarez-Buyalla et al. 1996; Tesfaye et al. 2010). Plant population structure shows if the population has a stable distribution that allows continuous regeneration to take place (Pinero et al. 1984; Enright and Watson 1991; Rao et al. 1990; Tesfaye et al. 2010). If regeneration takes place continuously, the distribution of species cohorts shows a reversed J-shape curve, which is an indicator of healthy regeneration (Harper 1977; Teketay 1997).

Studies carried out in different parts of the world indicate that humans, along with uncontrolled population growth of wild animals (e.g., elephants in FRs in Botswana), have not just destroyed forests, but also the natural mechanisms of forest regeneration. The barriers to forest regeneration are caused by humans. Therefore, forest restoration is merely an attempt to remove or overcome these 'unnatural' barriers to forest regeneration. Even under the most favourable circumstances, natural forest regeneration occurs slowly, generally requiring centuries. Unprecedented rates of biodiversity loss and climate change necessitate urgent actions. Waiting centuries for forests to regenerate is, therefore, no longer an option if species that are on the verge of local extermination are to be saved, or if carbon storage by forests is to have any impact on climate change. Human-induced problems require human-made solutions, and forest restoration is one of them (Elliott et al. 2013).

The seed dispersal of trees is affected by both wind and animals, and the success of germination of dispersed seeds and recruitment of the subsequent seedlings depends on the distance of dispersal from the mother trees, the ability to overcome seed dormancy and avoidance of herbivory of seeds and seedlings as well as damage caused by fire and other factors. Frequent fire events, such as the annual human-induced and natural fires in the FRs of Botswana reduce both the density and species richness of tree seedlings and sapling communities. Burning reduces the seed rain by killing seeds and seed-producing trees, hampering the accumulation of viable seeds in the soil seed bank. Fire favours the establishment of wind-borne and light-demanding pioneer tree species at the expense of shade-tolerant climax species. Fire events also destroy soil organic matter, leading to a reduction in the moisture-holding capacity and nutrient content of the soil. The drier the soil, the less favourable it is for the germination of tree seeds. Fire also kills beneficial soil microorganisms—especially mycorrhizal fungi, nitrogen-fixing bacteria, and microbes that break down dead organic matter and recycle nutrients. Studies comparing frequently burnt areas with those protected from fire show that preventing fire accelerates forest regeneration (Elliott et al. 2013). In addition, deforestation and degradation of FRs contribute directly to the emission of greenhouse gases, including carbon dioxide, which are responsible for the prevalent climate change and global warming.

A project focusing on one of the six Botswanan FRs, namely Kazuma Forest Reserve (KFR), was implemented. To develop successful strategies for restoring KFR, it is crucial to understand the different natural and artificial pathways or routes of forest regeneration. The project discussed here was, therefore, designed to fill major scientific gaps by generating the required empirical data. It is acknowledged that the decline in the forest and woodland cover has resulted from a combination of unsustainable utilisation and frequent fires. The decrease in vegetation cover directly leads to soil erosion and loss of soil productivity, which ultimately results in degradation and desertification in areas previously covered by productive forest and rangeland (Anonymous 2011). It is, therefore, apparent that Botswana needs management plans (MPs) for its FRs that will facilitate their sustainable management. As stipulated by the Forest Policy (Anonymous 2011), the overall goal of these MPs is to optimise the contribution of the forest and range resources to the long-term socio-economic development of Botswana by ensuring an equitable and sustainable flow of benefits to all segments of the population.

This study is part of a larger study aimed at investigating the restoration pathways in KFR, with component surveys addressing topics, including vegetation, germination, seed longevity, the impact of wild animals, spatial and temporal land use and land cover changes, and socio-economic importance. Disturbances of natural and human origin affect forest and woodland dynamics, diversity, regeneration, and dominance of woody species (Lawes et al. 2007; Sapkota et al. 2009; Neelo et al. 2015a, b). Numerous studies have explained the relationship between disturbance and species richness (Vetaas 1997; Sheil 1999; Venkateswaran and Parthasarathy 2003; Lawes et al. 2007), but studies elucidating how disturbances influence stand structure, species composition, and regeneration of tree species are quite limited, and this is also the case in Botswana (Neelo et al. 2015a, b). This contribution reports the results of a study on the diversity, stand structure, and regeneration status of woody species in KFR. The specific objectives of the study were to: (1) assess the diversity (species richness, diversity, and evenness); (2) determine the stand structure (densities, frequencies, dominance, and importance value index); and (3) assess the population structure and regeneration status of woody species in KFR.

Materials and Methods

Study Area

The study was carried out in KFR, Chobe District, northern Botswana, located at 18.38713° E, 25.48297° S (Fig. 24.1). Chobe District, which covers around 22,500 km^2, is the northernmost district of Botswana (van der Sluis et al. 2017) and consists of flat woodlands on deep Kalahari sands. Soils in the district can be divided into two large complexes: There are low-lying lacustrine deposits and alluvial soils

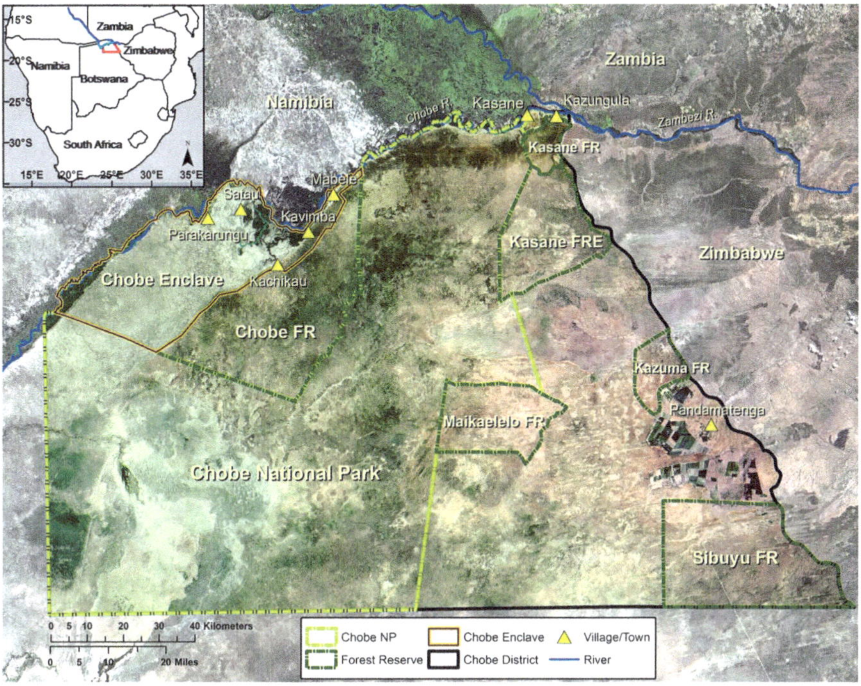

Fig. 24.1 Map showing the six forest reserves (FR, FRE) in northern Botswana (*source*: Fox et al. 2017)

along the Chobe-Linyanti drainage system, while the larger part of the territory features deep, sandy soils developed from Kalahari sands (arenosols). In geologically drier times, longitudinal dune systems developed in some areas. These soils have very low fertility, and their water retention capacity is severely limited. Their suitability is, therefore, limited to conservation for rangeland purposes (van der Sluis et al. 2017). Rainfall in Chobe District is characterised by a well-defined seasonality, with a monomodal rainy season, occurring between October and March, coinciding with the period of the highest seasonal temperatures (maximum daily temperatures reaching 34–36 °C; minimum daily temperatures of 20–22 °C). The period from June to July features the lowest temperatures (minimum daily temperature falling as low as 6 °C, and maximum daily temperature of 20–22 °C) and is, essentially, devoid of rainfall events (van der Sluis et al. 2017). The area is extremely flat, varying from a maximum altitude of 1085 m in the KFR to as low as 920 m in the Mababe depression.

KFR includes two of the global ecoregion types, namely Zambezian *Baikiaea*, and Zambezian and mopane woodlands (EcoSurv, DFRR and FCB 2018). The Zambezian *Baikiaea* woodlands comprise a mosaic of forest, thicket, and secondary grassland. The forests are dominated by the ecologically and economically important tree species *Baikiaea plurijuga* Harms, a resource that contributed strongly to the designation of the forest reserves. Much of the area's dry, and deciduous natural vegetation has remained intact until recently since the hot, semi-arid climate and nutrient-poor soils make the region unsuitable for farming. Globally, this ecoregion is listed as Vulnerable (WWF 2016). Diversity of the landscape, with hotspots in a matrix of nutrient-poor woodlands and grasslands, offering connectivity and mobility, must be preserved if the present high levels of biodiversity are to be maintained. It is the hotspot of habitats, primarily due to differences in soil moisture and nutrient content, that leads to a high level of biodiversity (Timberlake and Childes 2004). In 2015, KFR was made up of 79% savannah, 14% open woodland, 7% non-intact forest, and 1% intact forest (EcoSurv, DFRR and FCB 2018).

Fire is a recurrent annual phenomenon affecting FRs and other woodlands in different parts of Botswana. The key stakeholders felt that almost all fire in Chobe District was anthropomorphic in origin (EcoSurv, DFRR and FCB 2018). This is supported in the literature and by the timing of most fires (between July and October before the onset of the rains). Detailed mapping and analyses of fire frequency information indicated that most fires begin in the east of the district and are driven westwards by the prevailing winds. The Forest Strategy states that approximately 37% of the area of FRs is burnt annually (EcoSurv, DFRR and FCB 2018). KFR exhibits high to very high fire frequencies, with up to 70% of the FR affected. Between 2001 and 2012, for example, the percentages of KFR affected by low to medium (about once in 5 years), medium (about 3 times in 10 years), high (about once every 2 years), high to very high (6–7 times in 10 years), and very high (8–9 times in 10 years) fire frequencies were 11.6%, 18.8%, 41.4%, 17.5%, and 0.1%, respectively (EcoSurv, DFRR and FCB 2018).

KFR boasts the world's busiest elephant corridor connecting two of the largest national parks on the planet: Chobe National Park in Botswana and Mwange National Park in Zimbabwe. As a result, it is considered ideal for camping and walking safaris. However, this also means that the damage caused by elephants—especially to tree and other woody species—is much higher than in the other FRs in Botswana. Also, some portions of KFR have been degazetted to pave the way for the establishment of the Pandamatenga Commercial Farms. Its original area was 16,800 ha, of which 1200 ha (7.1%) were degazetted in 1999, leaving an area of 15,600 ha. This is the land area on which the Pandamatenga farmers erected their homesteads (EcoSurv, DFRR and FCB 2018). The Department of Forestry and Range Resources (DFRR) in the Ministry of Environment and Tourism is mandated with ensuring the ecological integrity of the forest reserves. However, the overall low level of management and policing in the FRs due to the limited personnel and budget of the DFRR has led to unacceptable levels of unregulated resource use and illegal tourism over the last decades.

Data Collection

For data collection, a total of 96 quadrats measuring 20 × 20 m were laid out systematically in KFR with at least 100 m distance between any two quadrats. Care was taken while laying down the quadrats to capture the greatest possible variation of vegetation in KFR. To determine species richness, diversity, and evenness, the number of live individuals of each woody species identified in each quadrat was recorded. The woody species were identified directly in the field using available literature (Timberlake 1980; Ellery and Ellery 1997; van Wyk and van Wyk 1997, 2007; Heath and Heath 2009; Roodt 1993, 1998; Setshogo 2002, 2005; Setshogo and Venter 2003) and with the help of locals familiar with the flora. Plant nomenclature in this article follows that of Setshogo and Venter (2003) and Setshogo (2005).

To determine the population structure and regeneration status of woody species, the data collected on the numbers and diameter at breast height (DBH) of individuals with DBH ≥ 2 cm, respectively, the numbers of individuals with DBH < 2 cm as well as seedlings and coppices were used. To determine the densities, frequencies, dominances, and importance value indices (IVIs) of the selected tree species in KFR, the number of live individuals was counted and the diameter at breast height (DBH) of individuals with DBH ≥ 2 cm was measured in each of the 96 quadrats. In the case of individuals of woody species with DBH < 2 cm, including seedlings and coppices (height < 1.5 m), their numbers were counted and recorded. A calliper and graduated measuring stick were used to measure DBH and height, respectively.

Data Analyses

Species richness (S) is the total number of different woody species recorded in the study areas. The diversity of woody species was analysed using the Shannon Diversity Index (H′) (Krebs 1989; Magurran 2004). Evenness or equitability, a measure of similarity of the abundances of the different woody species in the sampled project site, was analysed using Shannon's Evenness or Equitability Index (E) (Krebs 1989; Magurran 2004). Equitability assumes a value between 0 and 1, with 1 being complete evenness.

The mean density of each woody species per hectare was determined by converting the total number of individuals of each woody species encountered in all the quadrats to an equivalent number per hectare. The total mean density of all woody species in KFR was, then, calculated by summing the densities of all woody species recorded in all quadrats. The frequency was calculated as the proportion (%) of all quadrats in which each woody species was recorded.

The dominance of the woody species with DBH ≥ 2 cm was determined from the space occupied by a species, usually its basal area (BA). This was computed by converting the total basal area of all individuals of each woody species to the equivalent basal area per hectare (Kent and Coker 1992). The importance value index (IVI) indicates the relative ecological importance of the woody species in KFR (Kent and Coker 1992). It is determined through summation of the relative values of density, frequency, and dominance of each woody species. Relative density was calculated as the percentage of density of each species divided by the total stem number of all woody species per hectare. Relative frequency was computed as the ratio of the frequency of a species to the total frequency of all woody species.

Relative dominance was calculated as the percentage of the total basal area of a woody species out of the total basal areas of all woody species.

The population structure of each woody species in the study area was assessed through histograms constructed by using the density of individuals of each species (Y-axis) categorised into 15 diameter classes (X-axis in Fig. 24.1) (Peter 1996). Based on the profile depicted in the population structures, the regeneration status of each woody species was determined.

Results

Species, Family, and Genera Richness of Woody Species

A total of 92 woody species, representing 24 families and 57 genera were recorded (Table 39.1). The six families with the most species were Fabaceae (30.4%), Combretaceae (15.2%), Anacardiaceae (5.4%), Burseraceae (5.4%), Loganiaceae (4.3%) and Tiliaceae (4.3%). The five genera with the most species present were *Combretum* (16%), *Commiphora* (9%), *Grewia* (7%), *Strychnos* (7%), *Terminalia* (7%) and *Rhus* (5%). All other genera were represented at most with two species.

Diversity and Evenness, Mean Density, Frequency, and Dominance

The diversity (H′) and evenness (E) of the woody species were 2.8 and 0.62, respectively. The total mean density of woody species was $23{,}159 \pm 264$ individuals ha^{-1} (Table 39.1). The six species with the highest densities were *Colophospermum mopane* (6732 individuals ha^{-1}), *Diplorhynchus condylocarpon* (2599), *Rhigozum zambesiacum* (2527), *Bauhinia petersiana* (2480), *Baphia massaiensis* (1290), and *B. plurijuga* (1227). Mean densities ranged between one (several species) and 6732 (*C. mopane*) individuals ha^{-1} (Table 24.1).

The frequencies of woody species ranged between 1% (several species) and 68% (*D. condylocarpon*). The five most frequent woody species were *D. condylocarpon* (68%), *B. petersiana* (59%), *Combretum zeyheri* (53%), *Grewia monticola* (53%), and *Ochna pulchra* (50%) (Table 24.1).

The dominance of woody species ranged between 0.01 (several species) and 13 (*B. plurijuga*) m^2 ha^{-1}. *B. plurijuga*, *C. mopane*, *Amblygonocarpus andongensis*, and *Schinziophyton rautanenii* were the four most dominant species (Table 24.1).

Importance Value Index (IVI), Population Structure, and Regeneration Status

The IVI of woody species ranged between 0.12% (several species) and 49% (*C. mopane*), with *C. mopane*, *B. plurijuga*, *D. condylocarpon*, *B. petersiana*, and

Table 24.1 List of woody species recorded in KFR with their families, densities

Species	Family	DE	FR	DO	RDE	RFR	RDO	IVI
1. Colophospermum mopane	Fabaceae	6732	33	6	28	3	18	49
2. Diplorhynchus condylocarpon	Apocynaceae	2599	68	0.1	11	6	0.2	17.02
3. Rhigozum zambesiacum	Bignonaceae	2527	24	0.01	10	2	0.01	12.01
4. Bauhinia petersiana	Fabaceae	2481	59	0.01	10	5	0.03	15.03
5. Baphia massaiensis	Fabaceae	1290	48	0.01	5	4	0.01	9.01
6. Baikiaea plurijuga	Fabaceae	1227	47	13	5	4	38	47
7. Dalbergia melanoxylon	Fabaceae	808	15	0.01	3	1	0.01	4.01
8. Commiphora africanum	Burseraceae	616	27	0.01	3	2	0.01	5.01
9. Grewia monticola	Tiliaceae	552	53	0.01	2	4	0.01	6.01
10. Gymnosporia senegalensis	Celastraceae	485	20	0.2	2	2	0.5	4.5
11. Combretum zeyheri	Combretaceae	475	53	0.1	2	4	0.3	6.3
12. Brachystegia boehmii	Combretaceae	394	30	0.5	2	2	2	6
13. Terminalia sericea	Combretaceae	340	24	1	1	2	4	7
14. Ochna pulchra	Fabaceae	305	50	0.01	1	4	0.01	5.01
15. Combretum collinum	Combretaceae	274	31	0.03	1	3	0.1	4.1
16. Euclea undulata	Ebenaceae	262	20	0.02	1	2	0.1	3.1
17. Burkea africana	Fabaceae	236	35	0.4	1	3	1	5
18. Terminalia brachystemma	Combretaceae	181	25	0.4	1	2	1	4
19. Combretum hereroense	Combretaceae	155	28	1	1	2	2	5
20. Erythroxylum zambesiacum	Erythroxylaceae	151	29	0.1	1	2	0.4	3.4
21. Combretum apiculatum	Combretaceae	146	29	0.2	1	2	0.5	3.5
22. Erythrophleum africanum	Fabaceae	128	26	1	1	2	2	5
23. Rhus sp.	Anacardiaceae	120	10	0.01	1	1	0.01	2.01
24. Vitex mombassae	Lamiaceae	113	15	0.01	1	1	0.01	2.01
25. Bolusanthus speciosus	Fabaceae	104	9	0.1	0.4	1	0.3	1.7
26. Amblygonocarpus andongensis	Fabaceae	100	27	4	0.4	2	10	12.4
27. Dialium englerianum	Fabaceae	94	10	1	0.4	1	2	3.4
28. Pseudolachnostylis maprounifollia	Phyllantaceae	89	25	0.04	0.4	2	0.1	2.5
29. Vanguirea infausta	Rubiaceae	87	14	0.01	0.4	1	0.01	1.41
30. Pterocarpus rotundifolius	Fabaceae	86	10	0.01	0.4	1	0.01	1.41
31. Dichrostachys cineria	Fabaceae	62	23	0.01	0.3	2	0.01	2.41
32. Diospyros batocana	Ebenaceae	60	13	0.01	0.3	1	0.01	1.31
33. Senegalia nigrescens	Fabaceae	55	5	0.01	0.2	0.4	0.01	0.61
34. Philenoptera nelsii	Fabaceae	53	5	0.01	0.2	0.4	0.01	0.61
35. Friesodielsia obovata	Annonaceae	52	12	0.01	0.2	1	0.01	1.21
36. Ximenia caffra	Olacaceae	51	8	0.01	0.2	1	0.01	1.21
37. Rhus pyroides	Anacardiaceae	51	25	0.01	0.2	2	0.01	2.21
38. Combretum imberbe	Combretaceae	48	7	0.3	0.2	1	1	2.2
39. Ximenia americana	Olacaceae	46	17	0.01	0.2	1.4	0.01	1.61
40. Piliostigma thonningii	Fabaceae	46	6	0.2	0.2	1	1	2.2

(continued)

Table 24.1 (continued)

Species	Family	DE	FR	DO	RDE	RFR	RDO	IVI
41. *Commiphora pyramidalis*	Burseraceae	42	12	0.01	0.2	1	0.01	1.21
42. *Parinari curatellifolia*	Chrysobalanaceae	40	2	1	0.2	0.2	4	4.4
43. *Crossopteryx febrifuga*	Rubiaceae	39	4	0.01	0.2	0.3	0.02	0.52
44. *Grewia flavascens*	Tiliaceae	39	6	0.01	0.2	1	0.01	1.21
45. *Bridelia cathartica*	Phyllanthaceae	34	6	0.01	0.1	1	0.01	1.11
46. *Pterocarpus angolensis*	Fabaceae	33	17	0.2	0.1	1.2	1	2.3
47. *Croton gratissimus*	Euphorbiaceae	32	3	0.01	0.1	0.3	0.01	0.41
48. *Grewia bicolor*	Tiliaceae	28	9	0.01	0.1	0.8	0.01	0.91
49. *Combretum adenogonium*	Combretaceae	23	8	0.1	0.1	0.7	0.2	0.82
50. *Terminalia randii*	Combretaceae	23	2	0.01	0.1	0.2	0.01	0.31
51. *Lannea discolor*	Anacardiaceae	22	13	0.01	0.1	1	0.01	1.11
52. *Philenoptera violacea*	Fabaceae	19	6	0.01	0.1	1	0.01	1.11
53. Unidentified sp. 3	–	17	4	0.01	0.1	0.3	0.01	0.41
54. *Strychnos pungens*	Loganiaceae	16	10	0.01	0.1	1	0.01	1.11
55. *Cassia abbreviata*	Fabaceae	15	10	0.01	0.1	1	0.01	1.11
56. *Guibourtia coleosperma*	Fabaceae	14	10	0.01	0.1	1	0.01	1.11
57. *Albizia amara*	Fabaceae	14	3	0.01	0.1	0.3	0.01	0.41
58. *Diospyros mespiliformis*	Ebenaceae	13	6	0.5	0.1	1	1	2.1
59. *Commiphora glandulosa*	Burseraceae	12	4	0.01	0.1	0.3	0.01	0.41
60. *Ziziphus mucronata*	Rhamnaceae	11	7	0.01	0.04	1	0.02	1.06
61. *Strychnos spinosa*	Loganiaceae	11	6	0.01	0.04	1	0.01	1.05
62. *Securinega virosa*	Euphorbiaceae	10	5	0.01	0.04	0.4	0.01	0.45
63. *Julbernardia globiflora*	Fabaceae	10	3	0.04	0.04	0.3	0.1	0.4
64. *Sclerocarya caffra*	Anacardiaceae	9	4	0.02	0.04	0.3	0.1	0.35
65. *Combretum mollis*	Combretaceae	9	2	0.01	0.04	0.2	0.02	0.26
66. Unidentified sp. 2	–	8	3	0.01	0.03	0.3	0.01	0.34
67. *Azanza garckeana*	Malvaceae	8	4	0.01	0.03	0.3	0.01	0.34
68. *Vachellia* sp.	Fabaceae	8	2	0.01	0.03	0.2	0.01	0.24
69. *Grewia retinervis*	Tiliaceae	7	2	0.01	0.03	0.2	0.01	0.24
70. *Afzelia quanzensis*	Fabaceae	7	5	0.01	0.03	0.4	0.01	0.44
71. *Peltophorum africanum*	Fabaceae	6	3	0.01	0.03	0.3	0.01	0.34
72. *Schinziophyton rautanenii*	Euphorbiaceae	6	3	3	0.02	0.3	8	8.32
73. *Kirkia acuminata*	Kirkiaceae	5	3	0.01	0.02	0.3	0.01	0.33
74. *Terminalia mollis*	Combretaceae	5	2	0.01	0.02	0.2	0.02	0.24
75. *Combretum psidioides*	Combretaceae	4	3	0.01	0.02	0.3	0.02	0.34
76. *Securidaca longepedunculata*	Polygalaceae	4	4	0.01	0.02	0.3	0.01	0.33
77. *Vachellia erioloba*	Fabaceae	4	1	0.01	0.01	0.1	0.01	0.12
78. *Strychnos* sp.	Loganiaceae	4	1	0.01	0.01	0.1	0.01	0.12
79. *Albizia* sp.	Fabaceae	3	1	0.01	0.01	0.1	0.01	0.12
80. *Combretum elaeagnoides*	Combretaceae	3	1	0.01	0.01	0.1	0.01	0.12
81. *Gymnosporia heterophylla*	Celastraceae	3	1	0.01	0.01	0.1	0.01	0.12
82. *Carrisa* sp.	Apocynaceae	3	2	0.01	0.01	0.2	0.01	0.22
83. *Rhus lancea*	Anacardiaceae	3	2	0.01	0.01	0.2	0.01	0.22
84. *Ehretia rigida*	Boraginaceae	2	1	0.01	0.01	0.1	0.01	0.12

(continued)

Table 24.1 (continued)

Species	Family	DE	FR	DO	RDE	RFR	RDO	IVI
85. Unidentified sp. 1	–	2	2	0.01	0.01	0.2	0.01	0.22
86. *Commiphora mossambicensis*	Burseraceae	2	2	0.01	0.01	0.2	0.01	0.22
87. *Hyphaene petersiana*	Arecaceae	2	1	1	0.01	0.1	2	2.11
88. *Senegalia tortilis*	Fabaceae	1	1	0.01	0.01	0.1	0.01	0.12
89. *Stunduwanga*	–	1	1	0.01	0.01	0.1	0.01	0.12
90. *Markhamia obtusifolia*	Bignonaceae	1	1	0.01	0.01	0.1	0.01	0.12
91. *Strychnos cocculoides*	Loganiaceae	1	1	0.01	0.01	0.1	0.01	0.12
92. *Commiphora edulis*	Burseraceae	1	1	0.01	0.01	0.1	0.01	0.12
Total		23,159		36.2				

DE individuals ha^{-1}), *FR* frequencies, *DO* dominance, *RDE* relative densities, *RFE* relative frequencies, *RDO* relative dominance, *IVI* importance value indices

R. zambesiacum, representing the five species with the highest IVI values (Table 24.2).

The population structure of woody species exhibited seven patterns (Fig. 24.2).

1. Pattern I: Species exhibiting greater numbers of individuals in the lowest diameter class, with numbers progressively declining with increasing diameter classes (Fig. 24.2). This group is exemplified by *B. plurijuga*.
2. Pattern II: Species with a similar diameter class distribution pattern as the first group, except that individuals are missing at higher diameter classes (Fig. 24.2). This group is exemplified by *Erythrophleum africanum*.
3. Pattern III: Species with individuals only in the lower diameter classes (Fig. 24.2). This group is exemplified by *D. condylocarpon*.
4. Pattern IV: Species exhibiting both hampered seedling/coppice recruitment and missing individuals at higher diameter classes (Fig. 24.2). This group is exemplified by *Brachystegia boehmii*.
5. Pattern V: Species with individuals only in the higher diameter classes (Fig. 24.2). This group is exemplified by *Combretum apiculatum*.
6. Pattern VI: Species with missing individuals in one or more of the diameter classes (Fig. 24.2). This group is exemplified by *Combretum hereroense*.
7. Pattern VII: Species with individuals in only one diameter class (Fig. 24.2). This group is exemplified by *Guibourtia coleosperma* (only seedlings).

The seven patterns can be conveniently collapsed into two profiles: profile I (6% of the woody species), which includes patterns I and II, represents stable population structures, and profile II (> 94% of the woody species), which includes patterns III to VII and represents unstable population structures, and, thus, woody species with hampered regeneration.

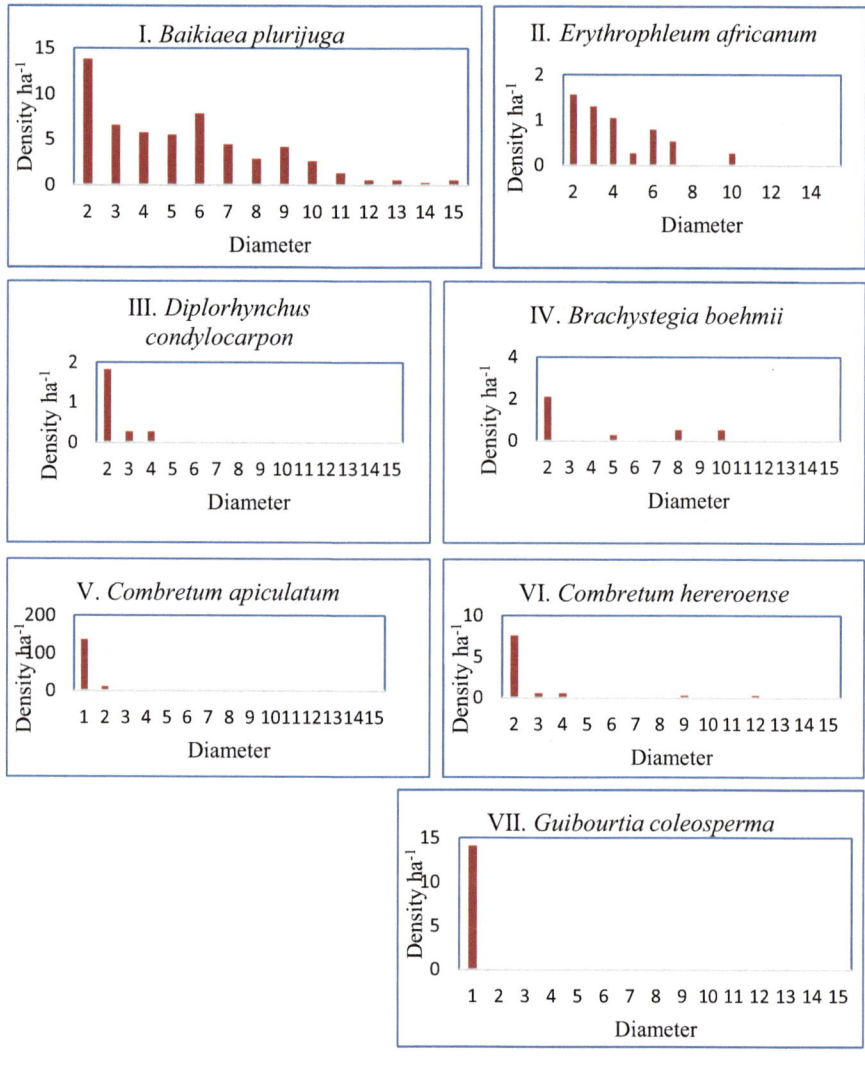

Fig. 24.2 The seven population structure patterns of woody species recorded in KFR. Diameter classes (DBH): 1 = < 5 cm; 2 = 5–9 cm; 3 = 10–14 cm; 4 = 15–19 cm; 5 = 20–24 cm; cm; 6 = 25–29 cm; 7 = 30–34 cm; 8 = 35–39 cm; 9 = 40–44 cm; 10 = 45–49 cm; 11 = 50–54 cm; 12 = 55–59 cm; 13 = 60–64 cm; 14 = 65–69 cm; 15 = ≥ 70 cm

Discussion

Our study revealed that KFR supports greater richness in terms of woody species, genera, and families (92, 24, and 37, respectively, with a diversity of 2.8 and evenness of 0.62) than Kasane FR (species/family/genus richness = 60/17/37, diversity

2.5, evenness 0.75) (BUAN Consult et al. 2020a, b), and Chobe FR (species/family/genera richness = 60, 17 and 36, diversity 2.5, evenness 0.62) (BUAN Consult et al. 2020a, b) as well as Controlled Hunting Area 1 (CH1) (species and family richness = 69, and 21) (Botswana Tourism & and BUAN Consult 2021) in northern Botswana. The mean total density of woody species (23,159 individuals ha^{-1}) in Kazuma FR was much higher than those in Kasane FR (12,767 individuals ha^{-1}) (BUAN Consult et al. 2020a, b) and Chobe FR (21,253 individuals ha^{-1}) (BUAN Consult et al. 2020a, b). The woody species with the highest densities and frequencies varied among the Kazuma, Kasane, and Chobe Forest Reserves. The diversity and evenness values recorded for Kazuma FR are also higher than those reported from open woodland areas in Shorobe (diversity = 2.18, evenness = 0.6) and Xobe (diversity = 1.5, evenness = 0.5) (Neelo et al. 2013, 2015a, b) in northern Botswana.

Like most other developing countries, Botswana is currently experiencing serious depletion of its forest resources. The expansion of agricultural activities, increased occurrence of forest fires, and high rate of urbanisation have placed considerable stress on the existing forest resources. This is evidenced by deforestation and land degradation around major population centres. Most of these environmental problems are directly linked to human activities—especially fuelwood harvesting, overgrazing, and arable agriculture (Sekgopo 2000). The greatest challenge in Botswana relating to the contribution of forests to the national economy has been the lack of information on the status of forest resources in the country. The Government of Botswana has no statistical data to help improve the management of forest resources. The use of fuelwood exerts pressure on forest resources and produces challenges for their management and conservation. This is especially the case since the rate and manner of exploitation of these resources are not known or monitored (DFRR-JICA 2017).

The Food and Agriculture Organization (FAO) (Food and Agriculture Organisation 2010; CAR (Centre for Applied Research) 2003) data based on the Botswana Country Report indicate that the land area occupied by forests has declined, with an indication that 23,670 km^2 (or 17.3%) of forest land was lost between 1990 and 2010. This loss has occurred countrywide, mainly because of forest fires, overuse of forest resources by the local communities (especially those near urban centres), and depredation due to the increase of wild animal populations. An interview with a DFRR official revealed that forest loss within protected areas was the result of dieback and lack of natural regeneration, which is aggravated by the effects of climate change and damage from growing elephant populations. The area studied for this contribution is the busiest elephant corridor located between the Mwenge (Zimbabwe) and Chobe (Botswana) National Parks and exposed to much disturbance (see Fig. 24.3i) caused by daily human and animal traffic to the river for water. As noted by Dutta and Devi (2013), overexploitation of forest resources, encroachment, and domestic animal grazing are disturbances that can affect the natural regeneration of woody species, and thus the entire forest ecosystem.

Regarding the observed population structure of the woody species, only the species categorised in the first group exhibited a stable population structure and healthy regeneration. In particular, *B. plurijuga* exhibited an exceptionally high number of

Fig. 24.3 Photographs depicting the current status of the western part of Kazuma Forest Reserve: (**a–l**): Dead and standing large trees with fire scars and toppled trees providing evidence of continued disturbance from fire and elephants; (**a–l**): Heavily deforested and degraded parts of Kazuma FR, with signs of fire and elephant disturbance; (**d–l**): Appearance of parts of the FR, strongly indicating that it is on the verge of conversion to wooded grassland. Photo L shows the electric fence erected to protect the Pandamatenga Commercial Farms. (Demel Teketay)

seedlings and an overall good stand structure. Conversely, several important tree species such as *P. angolensis* showed unstable population structures and hampered natural regeneration. This can be attributed to the various anthropogenic impacts and damage from overgrazing by animals. These results concur with those reported for the Kasane (BUAN Consult et al. 2020a) and Chobe (BUAN Consult et al. 2020b) FRs in northern Botswana as well as for a mopane-dominated woodland (Teketay et al. 2018a) and other woodlands in northern (Neelo et al. 2013, 2015a, b) and southern (Teketay et al. 2018a, b) Botswana.

From the field observations and surveys in Kazuma FR, it is evident that most areas in the FR are heavily deforested and degraded, with signs of fire (Fig. 24.2) and elephant disturbance (Fig. 24.2): In almost all parts of the FR, and especially in the western parts, dead and standing large trees with fire scars as well as toppled trees abound (Fig. 24.2). The most worrying aspect of the conducted site surveys was that the vertical stratification and horizontal distribution of woody species have been heavily affected, as evidenced by limited or absence of natural regeneration of trees/shrubs. The overall appearance of the visited areas strongly indicates that Kazuma FR is on the verge of conversion to a wooded grassland (Fig. 24.2).

Fire poses the highest risk for FRs, for while mature plants can withstand fires, young plants cannot. It should be noted that when describing the history of fire in Kazuma FR, it is critical to include the fires in Chobe District as well, since they indicate the potential risks. DFRR data show that from 2006 to 2017, Chobe District was ranked fifth in terms of total area burnt (5,305,857 km^2), while Ghanzi District had the largest area burnt (EcoSurv, DFRR and FCB 2018). Although the burnt area gradually decreased from 2006 to 2017, the numbers prove that fire is still a significant risk for the FRs in Chobe District. A study by Fox et al. (2017) revealed that fires in the district between 2003 and 2013 were almost exclusively dry-season phenomena, peaking between August and October (about 85% of detected fires). Frequent fires in the late dry season can lead to significant reductions in the amount of woody vegetation cover compared with fires occurring in the wet season (Smit et al. 2010). It is believed that most fire ignitions were anthropogenic in origin due to their locations and timing before the arrival of convective storms (Fox et al. 2017; EcoSurv, DFRR and FCB 2018). Fox et al. (2017) suggested that in addition to identifying the primary anthropogenic causes of fire, careful implementation of controlled fuel load reduction or thinning in areas identified as significant fire hot spots may represent a useful tool for managing the forest resources in Chobe District. A better understanding of the complex spatial interactions between fire, vegetation dynamics, human and wildlife activities, and climate drivers in dryland savanna areas will be essential to developing better predictive capacity and improving adaptive fire management strategies.

The field surveys undertaken reveal that almost all present woody species (>90%) have hampered population structures associated with a lack of proper regeneration. This requires prioritising the woody species for restoration/rehabilitation according to their needs and enhancing their protection and healthy regeneration, including through investigation and use of seed rain, soil seed banks, seedling banks, coppice management, and promotion of assisted/artificial regeneration. In line with this, one of the components of our research project deals with generating protocols for the germination of seeds of woody species found in Kazuma FR to raise seedlings for use in the envisaged enrichment plantings during the proposed restoration/rehabilitation efforts. The germination characteristics of seeds of 13 woody species collected from Kazuma FR have been tested, and preliminary observations reveal that some of these species exhibited dormancy induced by hard seed coats, necessitating seed treatment prior to sowing to ensure fast, uniform, and high germination percentages can be achieved in the tree nurseries in Botswana. These results will be instrumental if and when the DFRR in the Ministry of Environment and Tourism initiates programmes to raise seedlings of the woody species to be used in the restoration/rehabilitation of Kazuma FR and the other five FRs.

Future research focusing on comparative studies on herbaceous species richness, diversity, and density, horizontal and vertical distribution of soil seed banks, species richness in soil microorganisms, plant and soil biomass, plant and carbon pools above and below ground, and soil contents and properties in Kazuma FR is urgently recommended. Our results re-enforce the argument that relevant authorities need to start addressing the issue of deforestation and degradation of Kazuma and other forest reserves in Botswana. Among other things, this requires the development and implementation of forest management plans and pathways for the restoration and rehabilitation of the forest reserves.

Conclusions

Forest reserves, which cover the northern-eastern parts of Botswana, including KFR, provide many goods and ecosystem services that are essential for improving and maintaining human livelihoods in addition to performing many ecological functions. KFR houses a total of 92 woody species, representing 24 families and 57 genera with 2.8 and 0.62 diversity (H') and evenness (E) values of woody species, respectively. However, the six forest reserves, in northern Botswana, have declined in forest cover in the last 20 years. The decline in the forest cover has been attributed to forest degradation caused, mainly, by the recurrent annual fires and the increasing number of elephants or the joint actions of elephants and fires. The decline in forest cover is more evident in Kazuma Forest Reserve (KFR), our research project site, since it is a part of the longest elephant migratory corridor or route in the world, especially during the dry season.

Despite their demonstrated importance to livelihoods and the environment, FRs in Botswana are declining in size and heavily degraded due to destruction by recurrent fire and elephants and the absence of management practices. This has resulted in changes in species composition, decline of diversity, density, frequency, and basal area or dominance as well as unstable population structure and hampered regeneration of the woody species.

Acknowledgements The authors would like to acknowledge the financial support of the Research, Technology Development and Transfer Committee (RTDTC) at Botswana University of Agriculture and Natural Resources. They would also like to thank the Department of Forestry and Range Resources at the Botswana Ministry of Environment and Tourism for granting the authors a Research Permit [Ref. ENT 8/36/4 XXXX (29)] to undertake research in KFR.

References

Alvarez-Buyalla ER, Garcia-Barrios R, Lara-Moreno C, Martinez-Ramos M (1996) Demographic and genetic models in conservation biology: application and perspectives for tropical rain forest tree species. Annu Rev Ecol Syst 27:387–421

Anonymous (2011) Forest policy. Ministry of Environment, Wildlife and Tourism, Botswana Government Printer, Gaborone

Blanc JJ (2017) African elephant status report 2007: an update from the African elephant database; no. 33. IUCN, Gland

BUAN Consult and Botswana Tourism (2021) Land use and management plan for controlled hunting area No. 1 (BTO002/1005-19-20/CH1(3) (draft)). Botswana Tourism, Gaborone

BUAN Consult (Botswana University of Agriculture and Natural Resources Consult (BUAN Consult), Department of Forestry and Range Resources (DFRR) and Forest Conservation Botswana) (2020a) Chobe Forest reserve management plan (2021–2031) technical report. BUAN Consult (Botswana University of Agriculture and Natural Resources Consult (BUAN Consult), DFRR) and Forest Conservation Botswana, Gaborone

BUAN Consult (Botswana University of Agriculture and Natural Resources Consult (BUAN Consult), Department of Forestry and Range Resources (DFRR) and Forest Conservation Botswana) (2020b) Kasane Forest reserve management plan (2021–2031). BUAN Consult, DFRR and Forest Conservation Botswana, Gaborone

CAR (Centre for Applied Research) (2003) Forest management and use in Botswana: brief situation analysis and options for the Forest conservation strategy. CAR, Gaborone

DFRR-JICA (2017) Republic of Botswana - the project for enhancing national forest monitoring system for the promotion of sustainable natural resource management project completion report. Department of Forestry and Range Resources, Gaborone

Dutta G, Devi A (2013) Plant diversity. Population structure and regeneration status in disturbed tropical forests in Assam. Northeast India. J For Res 24:715–720

Department of Wildlife and National Parks (DWNP) (2014) *Aerial census of wildlife and some domestic animals in Botswana*; Department of Wildlife and National Parks. Gaborone, Monitoring Unit, Research Division

EcoSurv, DFRR and FCB (2018) Consultancy Services for the Strategic Environmental Assessment of the Chobe Forest reserves management plans. EcoSurv, Department of Forestry and Range Resources and Forest Conservation, Botswana

Ellery K, Ellery W (1997) Plants of the Okavango Delta: a field guide. Tsaro Publisher, Durban

Elliott S, Blakesley D, Hardwick K (2013) Restoring tropical forests: a practical guide. Royal Botanic Gardens, Kew

Enright NJ, Watson AD (1991) A matrix model analysis for the tropical tree *Araucaria cunninghamii*. Aust J Ecol 16:507–520

FAO (Food and Agriculture Organisation) (2010) Global Forest resources assessment: Botswana country report. FAO, Rome

Fox TJ, Vandewalle ME, Alexander KA (2017) Land cover change in northern Botswana: the influence of climate, fire, and elephants on semi-arid savanna woodlands. Land 6:73. https://doi.org/10.3390/land6040073

Harper JL (1977) Population biology of plants. Academic press, London

Heath A, Heath R (2009) Field guide to the plants of northern Botswana: including the Okavango Delta. Kew Publishing. Royal Botanic Gardens, Kew

Hubbel P, Foster B (1986) Biology, chance and history and the structure of tropical rainforest communities. In: Diamond J, Case J (eds) Community ecology. Harper and Row, New York, pp 314–329

Kent M, Coker P (1992) Vegetation description and analysis: a practical approach. John Wiley and Sons Ltd, Chichester

Krebs CJ (1989) Ecological methodology. Harper Collins Publishers, New York

Lawes MJ, Joubert R, Griffiths ME, Boudreau S, Chapman CA (2007) The effect of the spatial scale of recruitment on tree diversity in Afromontane Forest fragments. Biol Conserv 139:447–456

Lykke AM (1993) Assessment of species composition change in savanna vegetation by means of woody plants' size class distributions and local information. Biodivers Conserv 7:1261–1275

Magurran AE (2004) Measuring biological diversity. Blackwell Publishing, Malden

Neelo J, Kashe K, Teketay D, Masamba W (2015a) Ethnobotanical survey of Woody plants in Shorobe and Xobe villages. Northwest region of Botswana. Ethnobot Res Appl 14:367–379

Neelo J, Teketay D, Kashe K, Masamba W (2015b) Stand structure. Diversity and regeneration status of woody species in open and exclosed dry woodland sites around molapo farming areas in the Okavango Delta. Northeastern Botswana. Open J Forestry 5:313–328

Neelo J, Teketay D, Masamba W, Kashe K (2013) Diversity. Population structure and regeneration status of woody species in dry woodlands adjacent to Molapo farms in northern Botswana. Open J Forestry 4:138–151

Peter CM (1996) The ecology and management of non-timber forest resources. World Bank Technical Paper 322, Washington

Pinero B, Martines-Ramos M, Sarukhan J (1984) A population model of *Astrocaryum mexicanum* and a sensitivity analysis to its finite rate of increase. J Ecol 71:977–991

Rao P, Barik SK, Pandey HN, Tripathi RS (1990) Community composition and tree population structure in a sub-tropical broad-leaved forest along a disturbance gradient. Vegetatio 88:151–162

Roodt V (1993) The Shell field guide to the common trees of the Okavango Delta and Moremi game reserve. Shell Oil (Pty) Ltd., Gaborone

Roodt V (1998) Trees and shrubs of the Okavango Delta. Shell Oil (Pty) Ltd., Gaborone

Sapkota IP, Tigabu M, Oden PC (2009) Spatial distribution. Advanced regeneration and stand structure of Nepalese Sal (*Shorea robusta*) Forest subject to disturbances of different intensities. For Ecol Manag 257:1966–1975

Saxena AK, Singh SP, Singh JS (1984) Population structure of forests of Kuman Himalaya. Implications for management. J Environ Manag 19:307–324

Sekgopo M (2000) *Forestry Outlook Study for Africa (FOSA)*. FOSA Country Report, Botswana Ministry of Agriculture, www.fao.org/3/a-x6774e.pdf, accessed on 11 Aug 2019

Setshogo MP, Venter F (2003) Trees of Botswana: names and distribution. Southern African Botanical Diversity Network (SABONET)

Setshogo MP (2002) Common names of some flowering plants of Botswana. FAO, Rome

Setshogo MP (2005) Preliminary checklist of the plants of Botswana. Southern African Botanical Diversity Network, Pretoria

Sheil D (1999) Tropical forest diversity, environmental change, and species augmentation: after the intermediate disturbance hypothesis. J Veg Sci 10:851–860

Smit IP, Asner GP, Govender N, Kennedy-Bowdoin T, Knapp DE, Jacobson J (2010) Effects of fire on woody vegetation structure in African savanna. Ecol Appl 20(7):1865–1875

Teketay D, Kashe K, Madome J, Kabelo M, Neelo J, Mmusi M, Masamba W (2018a) Enhancement of diversity, stand structure and regeneration of woody species through area exclosure: the case of a mopane woodland in northern Botswana. Ecol Process 7:5. https://doi.org/10.1186/s13717-018-0116-x

Teketay D, Madome J, Kashe K (2018b) Pre-dispersal seed predation, soil seed banks and germination of seeds of Albizia harveyi E. Fourn. From northern Botswana: implication for its conservation. J Biodivers Environ Sci 12(6):57–66

Teketay D (1997) Seedling populations and regeneration of woody species in dry Afromontane Forest of Ethiopia. For Ecol Manag 98:149–165

Tesfaye G, Teketay D, Fetene M, Beck E (2010) Regeneration of seven indigenous tree species in a dry Afromontane forest, Southern Ethiopia. Flora 205:135–143

Timberlake JR, Childes SL (2004) Four Corners TBNRM Project: Biodiversity of the four corners area - technical reviews Volume 1 Chapters 1–4. Occasional publications in biodiversity No. 15

Timberlake J (1980) Handbook of Botswana acacias. Ministry of Agriculture. Gaborone, Botswana

van der Sluis T, Cassidy L, Brooks C, Wolski P, VanderPost C, Wit P, Henkens R, van Eupen M, Mosepele K, Maruapula O, Veenendaal E (2017) *Chobe District Integrated Land Use Plan*. Wageningen Environmental Research, Report 2813, Wageningen, The Netherlands

van Wyk B, van Wyk P (1997) Field guide to trees of southern Africa. Struik Nature, Cape Town

van Wyk B, van Wyk P (2007) How to identify trees in southern Africa. Struik Nature, Cape Town

Venkateswaran R, Parthasarathy N (2003) Tropical dry evergreen forests on the Coromandel Coast of India: structure, composition, and human disturbance. Ecotropica 9:45–58

Vetaas OR (1997) Spatial and temporal vegetation changes along a moisture gradient in northeastern Sudan. Biotropica 25:164–175

WWF (2016) Zambezian and Mopane Woodlands. http://www.worldwildlife.org/ecoregions/at0725

Open Access This chapter is licensed under the terms of the Creative Commons Attribution 4.0 International License (http://creativecommons.org/licenses/by/4.0/), which permits use, sharing, adaptation, distribution and reproduction in any medium or format, as long as you give appropriate credit to the original author(s) and the source, provide a link to the Creative Commons license and indicate if changes were made.

The images or other third party material in this chapter are included in the chapter's Creative Commons license, unless indicated otherwise in a credit line to the material. If material is not included in the chapter's Creative Commons license and your intended use is not permitted by statutory regulation or exceeds the permitted use, you will need to obtain permission directly from the copyright holder.

Brazil: Applied Nucleation Through Key Microsites

25

Bruna Elisa Trentin and Katharina Lapin

Study site in Dois Vizinhos, Parana, Brazil. NUC are nucleation plots, REF are plantation plots, and PAS are natural regeneration plots

B. E. Trentin (✉)
Forest Ecology and Taxonomy Laboratory, Federal University of Technology, Parana, Dois Vizinhos, Brazil

K. Lapin
Department of Forest Biodiversity & Nature Conservation, Austrian Research Centre for Forests, Vienna, Austria
e-mail: katharina.lapin@bfw.gv.at

Abstract

The restoration of ecological connectivity plays a crucial role in tropical forest conservation efforts. This chapter explores a case study by Trentin (A nucleação como alternativa à restauração passiva e ao reflorestamento com nativas para a restauração florestal. Universidade Estadual Paulista, Botucatu, 2018) employing the Brazilian approach of applied nucleation through key microsites as a means of enhancing ecological connectivity in tropical forest restoration. It examines the implementation of an integrated restoration strategy combining the productivity of plantation methods with the natural functions associated with nucleation. By utilising key microsites, which act as foci for wood species recruitment, this approach aims to develop desired composition and diversity trajectories in Neotropical forest restoration projects. The study also highlights the importance of considering site characteristics and project goals when selecting a restoration method. It emphasises the shift towards more diverse and complex forests, since high levels of biodiversity contribute to restoring ecosystem services. The choice of species for restoration goals is a challenge given the limited field testing with regard to local native species. Furthermore, it underscores the significance of monitoring the development of restored forests with regard to structural aspects, richness, functional guilds, and floristic composition. The Brazilian approach of applied nucleation through key microsites presents a promising alternative for enhancing ecological connectivity in tropical forest restoration. By allowing natural succession processes to occur while strategically incorporating plantation methods, this method aims to promote biodiversity, restore ecosystem functions, and create connectivity networks in fragmented landscapes.

Keywords

Applied nucleation · Atlantic Forest · Subtropical forest · Recovery · Succession · Natural regeneration

Introduction

Ecological connectivity refers to the uninterrupted movement and exchange of organisms, materials, and ecological processes across landscapes and is crucial for the long-term survival and functioning of ecosystems (Taylor et al. 1993). The restoration approaches discussed in this chapter are directly relevant to promoting ecological connectivity in tropical forests. Brazil has made a significant pledge toward global restoration goals by committing to restore 12 million hectares by 2030, as part of efforts to mitigate climate change. However, the restoration of forests in the country faces several challenges—especially the lack of silvicultural knowledge in many regions. While conservation efforts have historically usually focused on protecting pristine and old-growth forests, there is increasing recognition of the need to restore degraded areas to expand the extent of land with high biodiversity value and ecosystem services.

Restoring Neotropical rainforests, which are global biodiversity hotspots, is particularly challenging. One of the main obstacles in the long recovery trajectory of these ecosystems is, that it can take several decades or even centuries for the forests to regain their structure and species composition and to re-establish rare or endemic species (Garcia et al. 2016). Understanding the immense diversity of Neotropical flora is crucial for establishing reference points for restoration projects. Compared to the Asian and African tropics, the Neotropics boast a much larger number of tree species—approximately 40,000 to 50,000 (Slik et al. 2015).

To effectively restore Neotropical forests, a combination of passive and active restoration methods is necessary. Whereas active restoration methods should be utilised to expedite the recovery process, passive approaches allow for natural succession to take place. Active restoration approaches play a crucial role in accelerating the recovery process and achieving the desired restoration trajectory within a reasonable timeframe. By employing an intermediate spectrum of management, Brazil can effectively restore its forests, contribute to global restoration efforts, maximise the effectiveness of restoration projects, and achieve its ambitious restoration goals by 2030 (United Nations 2022).

Forest Restoration Approaches

Various restoration approaches are being employed in Brazil to address the challenges of forest restoration. The following four are among the most notable:

1. Passive restoration (passive): This approach acknowledges that forest succession does not always follow predetermined pathways. Rather, the interaction between site resilience, species performance, landscape, and disturbances influences the trajectory of forest recovery. By allowing natural processes to take place, passive restoration aims to support the spontaneous regeneration of forests (Vogel et al. 2015). It consists of isolating the area from disturbances to allow for natural regeneration to occur.
2. Assisted natural regeneration (passive): This approach focuses on facilitating and accelerating the natural regeneration processes of forests. It involves interventions such as removing competing vegetation, controlling invasive species, and protecting and nurturing naturally occurring seedlings and saplings. Assisted natural regeneration capitalises on the existing seed bank and the resilience of the ecosystem to drive the restoration process.
3. Tree planting (active): Active restoration through high-diversity plantations has been a popular approach in Brazil since the 1990s. This method involves the deliberate planting of a wide variety of tree species to facilitate the recovery of tropical forest ecosystems. However, the financial costs associated with this intensive silviculture technique can be high, and the growth system and tree density employed in such plantations—often similar to industrial eucalyptus plantations for pulp and paper—can limit the scalability of this approach.

4. Applied nucleation (active): This restoration technique involves the establishment of small nuclei ranging from 1 to 12 m^2 to initiate and promote forest regeneration through facilitation. These nuclei can consist of various elements including conventional woody seedlings as well as other life forms, structures, and materials like translocated seed rain, seed banks, wood debris, and perches. Although less commonly employed on a large scale, applied nucleation has shown promise in experiments and can enhance biodiversity by creating heterogeneity and facilitating natural processes.

Each of these approaches entails unique advantages and considerations. The choice of restoration method for each individual project depends on factors such as the specific context, restoration goals, available resources, and site ecological conditions. By employing a combination of these approaches and adapting them to local circumstances, Brazil can enhance the effectiveness and scalability of its forest restoration efforts.

Study Area and Method

The study area is located in the Atlantic Forest of Brazil, specifically at the experimental farm of the Federal Technological University of Paraná (UTFPR) in Dois Vizinhos.

This region is characterised by subtropical Atlantic rainforest and situated in the ecotone of Araucaria forest and semideciduous forest. The coordinates of the study area are approximately 25°41′38″ S latitude and 53°06′11″ W longitude. Its altitude is approximately 500 m, and it receives an annual average precipitation of 2044 mm per year. The climate is subtropical, with an annual average temperature of 19.2 °C. The region also experiences frost events approximately once every 2 years. In this study, they compared two active restoration methods (treatments)—high-diversity plantation versus applied nucleation, to a control group (natural regeneration with no planting). Employing a randomised block design with four replicates, each treatment was applied to one plot per block, with each plot measuring 40 × 54 m, totalling 0.9 ha per treatment. The applied nucleation treatment comprised seven techniques systematically implemented in small nuclei every 2 m within six 3 × 40 m strips spaced 6 m apart (Figs. 25.1, 25.2, 25.3, 25.4, 25.5, 25.6, and 25.7), leaving the area between strips for natural regeneration. The high-diversity plantation treatment involved planting 70 native tree species seedlings throughout the entire plot.

Results

Data from the first 6 years of the study reveal noticeable differences between high-diversity plantations and applied nucleation treatments in a landscape characterised by small and highly fragmented forest remnants of approximately 50 hectares. A

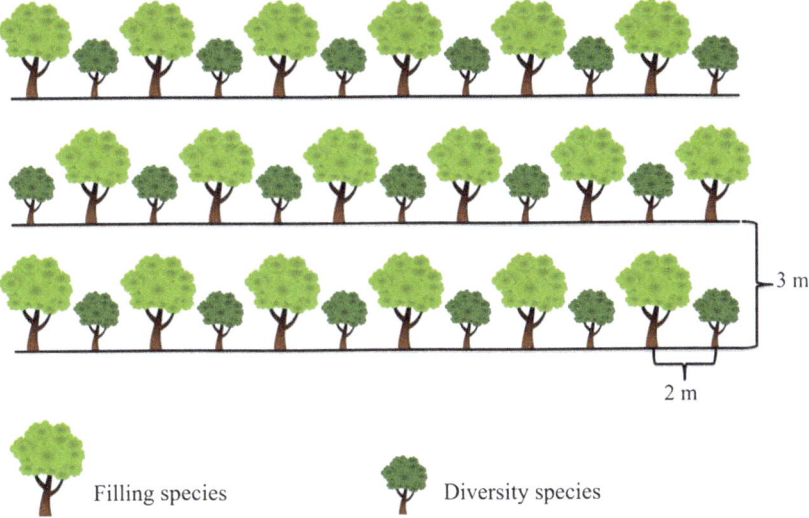

Filling species   Diversity species

Fig. 25.1 Schematic of the mixed reforestation treatment of native species in fill and diversity groups, exemplifying the distribution of species within each row and between rows. Filling species are fast growing and early shade species are used to ameliorate early site conditions while diversity species are slow-growing trees that will promote diversity over time

Fig. 25.2 (**a**) Collecting seed bank topsoil in the forest; (**b**) germinating seeds at the nursery; (**c**) transferring the seed bank to the restoration site after germination; (**d**) seed bank after a few months at the restoration site. They were taken to a nursery because it increases the success and efficiency of restoration efforts by promoting the establishment of healthy seedlings in challenging environmental conditions

Fig. 25.3 (**a**) Collecting seeds from a forest seed rain; (**b**) germinating seeds at the nursery; (**c**) transferring seed rain to the restoration site after germination; (**d**) seed rain after a few months at the restoration site

Fig. 25.4 Tree islands and bromeliad islands

Fig. 25.5 Ground coverage nuclei

total of 54,605 individuals were sampled, of which 97% were found in the understory and 3% in the overstory. The study identified 246 plant species in 66 families, of which 99% were classified to the genus and 96% to the species level. Overstory species richness was highest in plantation plots, intermediate in nucleation plots, and lowest in control plots. When considering only recruited individuals, however, nucleation plots exhibited the highest richness.

The study findings suggest that incorporating an applied nucleation approach within the restoration continuum between natural recovery and high-diversity plantations can offer a balance between recovery speed and costs. This indicates that the nucleation approach should be more widely considered when it aligns with restoration project goals. Combining passive restoration methods, restoration plantations, and applied nucleation could be essential for achieving a cost-effective restoration process that results in a diverse species pool in restored ecosystems.

With regard to cost analysis, the study revealed that in the plantation treatment, 52% of costs were attributable to maintenance, whereas this value dropped to 28% for the nucleation treatment (Bechara et al. 2021). The economic advantage of the applied nucleation system lies in the reduced number of seedlings to be planted and maintained. The cost per tree seedling maintained in the field for 3 years was US$ 4.22. Consequently, if stakeholders have a fixed limited budget for maintaining seedlings, a less dense strategy such as the nucleation approach, or nucleation with planting in larger islands (0.25 to 1 ha, 15 to 100 m apart), would allow for the restoration of a larger area. A potential strategy to make the plantation approach more cost-effective and similar to natural succession (up to 75% cheaper) might be to use

Fig. 25.6 Artificial perches and shelters

a spacing design as depicted in Fig. 25.4 but planting "filling" pioneer species between seedlings with a 6 × 4 m spacing. After 2–4 years, assessing where gaps requiring additional seedlings of pioneer or non-pioneer species for "enrichment" exist and planting those seedlings using the original 3 × 2 m spacing. A further cost-reducing approach for high-diversity plantations would be to widen the 3 × 2 m spacing commonly used in Brazil, for example by adopting a 3 × 3 m spacing, which would represent a saving of 33% (US$ 2346.02 per ha).

These findings highlight the practical implications of incorporating applied nucleation into restoration strategies, emphasising its potential for optimising both ecological and economic outcomes in the restoration of the Atlantic Forest ecosystem in Brazil.

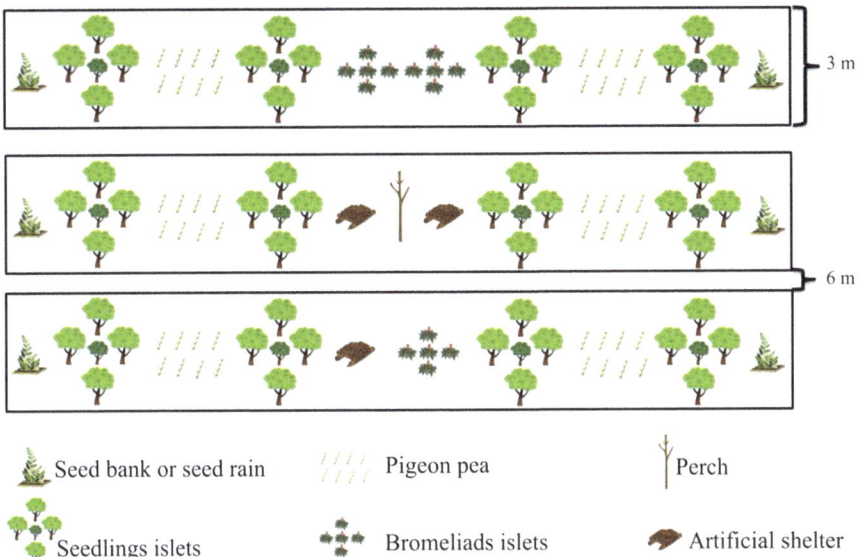

Fig. 25.7 Schematic of the nucleation treatment exemplifying the different techniques deployed and the three different banding models

The Nucleation Techniques for Restoration

Nucleation techniques offer an intermediate approach between passive and active methods in terms of effort, cost, and alignment with natural processes (Zahawi et al. 2013). These methods aim to restore habitat quality by combining various approaches including artificial shelters for fauna, perches, seed banks, ground-covering shrub/herbaceous nuclei, and the planting of native trees and bromeliads in dense nuclei. The use of nucleation techniques such as tree island planting provides important restoration options which, while they may not accelerate structural recovery as rapidly as plantations, emphasise and support natural regeneration processes, thus enabling the development of greater and more natural patterns of diversity (Vogel et al. 2016).

Brazilian nucleation techniques involve establishing restoration nuclei of different sizes within a degraded area with the aim of increasing heterogeneity and interaction among sites over time. These nuclei include topsoil seed banks and seed rain translocation from nearby natural forest remnants, artificial perches, cover crops using annual legume nodulation plants, terrestrial tank epiphyte islets, artificial shelters for animals, and tree islets consisting of pioneer trees providing shade for central non-pioneer plants.

In the context of tropical forest regeneration in abandoned pastures, nucleation models of succession start around so-called key microsites that facilitate the recruitment of wood species, spreading and merging over time. The nucleation approach originates from field observations of remnant trees acting as "nurse plants" and facilitating a sphere of regeneration. Recommended plantings in clumps or tree islands promote spatial heterogeneity, alter soil and microclimate conditions, trap wind-dispersed seeds, and serve as perches and cover for dispersing animals (Rodrigues et al. 2009).

Implementation of an Integrated Approach

Implementing an integrated approach that combines plantation productivity with natural functions such as nucleation can support the composition and diversity trajectory desired in Neotropical conservation efforts. The choice of restoration method should be based on the individual site characteristics and project goals. It is important to consider the benefits of high biodiversity in restoring ecosystem services but selecting the right species for restoration can be challenging since many local native species have not been extensively tested in the field. Conducting research on restoration regions is crucial for guiding the production of well-adapted planting material in forest nurseries.

Increasing the number of species in plantations is recommended, as there is no universally applicable number of species for tropical forest restoration. Studies have suggested planting 80–100 species in highly degraded landscapes, while in Brazil a range of 85–110 species has been recommended to achieve effective restoration (Holl et al. 2011, 2013, Carnevale and Montagnini 2002).

Monitoring the development of restored areas with a focus on structure, richness, functional guilds, and floristic composition is likewise essential (Suding 2011). It is important to note that the predictability of floristic composition may be limited. Large-scale plantations that are implemented unnecessarily or inappropriately can hinder the natural recovery processes. Restoration efforts should avoid overly engineered and mechanistic approaches, as they can result in monotonous and uniform stands that lack structural and functional diversity compared to passively restored sites.

Emphasising restoration through natural succession processes can be beneficial. Allowing natural processes to unfold rather than imposing strict controls can lead to more successful restoration outcomes. It is crucial to recognise that ecological restoration is not comparable to civil engineering and requires a nuanced understanding of ecological dynamics and the value of natural processes (Clewell and McDonald 2009).

Overall, nucleation restoration methods can play a crucial role in promoting forest connectivity by creating habitat corridors, enhancing biodiversity exchange, restoring ecological processes, and mitigating edge effects within fragmented landscapes. By integrating nucleation with broader landscape-scale conservation strategies, it is possible to enhance connectivity and resilience across fragmented forest ecosystems.

References

Bechara FC et al (2021) Performance and cost of applied nucleation versus high-diversity plantations for tropical forest restoration. For Ecol Manag 491:119088

Carnevale N, Montagnini F (2002) Facilitating regeneration of secondary forests with the use of mixed and pure plantations of indigenous tree species. For Ecol Manag 163:217–227

Clewell A, McDonald T (2009) Relevance of natural recovery to ecological restoration. Ecol Restor 27:122–124

Garcia LC, Hobbs RJ, Ribeiro DB, Tamashiro JY, Santos AM, Rodrigues RR (2016) Restoration over time: is it possible to restore trees and non-trees in high-diversity forests? Appl Veg Sci 19(4):655–666. https://doi.org/10.1111/avsc.12264

Holl KD, Zahawi RA, Cole RJ, Ostertag R, Cordell S (2011) Planting seedlings in tree islands versus plantations as a large-scale tropical forest restoration strategy. Restor Ecol 19:470–479

Holl KD, Stout VM, Reid JL, Zahawi RA (2013) Testing heterogeneity-diversity relationships in tropical forest restoration. Oecologia 173:569–578

Rodrigues RR, Lima RAF, Gandolfi S, Nave AG (2009) On the restoration of high diversity forests: 30 years of experience in the Brazilian Atlantic Forest. Biol Conserv 142:1242–1251

Slik JW, Arroyo-Rodríguez V, Aiba S, Alvarez-Loayza P, Alves LF, Ashton P, Venticinque EM (2015) An estimate of the number of tropical tree species. Proc Natl Acad Sci 112(24):7472–7477. https://doi.org/10.1073/pnas.1423147112

Suding KN (2011) Toward an era of restoration in ecology: successes, failures, and opportunities ahead. Annu Rev Ecol Evol Syst 42:465–487

Taylor PD, Fahrig L, Henein K, Merriam G (1993) Connectivity is a vital element of landscape structure. Oikos 68(3):571–573

United Nations (2022) COP15: Nations adopt four goals, 23 targets for 2030 in landmark UN biodiversity agreement. Retrieved from https://www.cbd.int/article/cop15-cbd-press-release-final-19dec2022

Vogel HF, Campos JB, Bechara FC (2015) Early bird assemblages under different subtropical forest restoration strategies in Brazil: passive, nucleation and high diversity plantation. Trop Conserv Sci 8:912–939

Vogel HF, Spotswood E, Campos JB, Bechara FC (2016) Annual changes in a bird assembly on artificial perches: implications for ecological restoration in a subtropical agroecosystem. Biota Neotrop 16:1–9

Zahawi RA, Holl KD, Cole RJ, Reid JL (2013) Testing applied nucleation as a strategy to facilitate tropical forest recovery. J Appl Ecol 50:88–96

Open Access This chapter is licensed under the terms of the Creative Commons Attribution 4.0 International License (http://creativecommons.org/licenses/by/4.0/), which permits use, sharing, adaptation, distribution and reproduction in any medium or format, as long as you give appropriate credit to the original author(s) and the source, provide a link to the Creative Commons license and indicate if changes were made.

The images or other third party material in this chapter are included in the chapter's Creative Commons license, unless indicated otherwise in a credit line to the material. If material is not included in the chapter's Creative Commons license and your intended use is not permitted by statutory regulation or exceeds the permitted use, you will need to obtain permission directly from the copyright holder.

Chile: Increasing Connectivity for Nature and People in Highly Anthropogenic Landscapes

26

Aníbal Pauchard, Eduardo Fuentes-Lillo, Darío Moreira-Arce, J. Cristóbal Pizarro, and Mónica Ortiz

Crop fields and farms at region del Maule, Chile (Photo: Jose Luis Stephens/Adobe Stock)

A. Pauchard (✉) · E. Fuentes-Lillo
Laboratorio de Invasiones Biológicas (LIB), Facultad de Ciencias Forestales, Universidad de Concepción and Institute of Ecology and Biodiversity (IEB), Concepción, Chile
e-mail: pauchard@udec.cl

D. Moreira-Arce
Departamento de Gestión Agraria, Facultad Tecnológica, Universidad de Santiago de Chile and Institute of Ecology and Biodiversity (IEB), Concepción, Chile

J. C. Pizarro
Facultad de Ciencias Forestales, Universidad de Concepción. Laboratorio de Estudios del Antropoceno, Universidad de Concepción, Concepción, Chile

Institute of Ecology and Biodiversity (IEB), Concepción, Chile
e-mail: jpizarrop@udec.cl

M. Ortiz
Institute of Ecology and Biodiversity (IEB), Concepción, Chile

Departamento de Geografía, Facultad de Arquitectura, Urbanismo y Geografía, Universidad de Concepción, Concepción, Chile

© The Author(s) 2025
K. Lapin et al. (eds.), *Ecological Connectivity of Forest Ecosystems*,
https://doi.org/10.1007/978-3-031-82206-3_26

Abstract

The central zone of Chile is currently exposed to various threats derived from historical land use, the increase in the frequency of wildfires, and the extension of non-native forest plantations. These factors have had a direct impact on the fragmentation of the native forest and the decrease in biodiversity in the region. These impacts are aggravated by the fact that only 10% of this area is protected under some category. The main challenge for the conservation of these ecosystems lies in increasing connectivity between forest fragments that have become isolated. Currently, strategies are being implemented to create natural corridors to connect these remaining forest fragments, thus preventing the isolation affecting ecosystem and species, including endangered native mammals such as *Lycalopex fulvipes* and *Leopardus guigna*. Participation of local and Indigenous communities is essential to advance public policies that promote management strategies capable of reducing biodiversity threats and promote landscape connectivity.

Keywords

Anthropogenic disturbances · Fragmentation · Biological invasions · Land use · Biological corridors

A Biodiversity Hotspot in Grave Peril

The ecosystems of south-central Chile, the transition zone from Mediterranean shrublands and forests to temperate forests and wetlands, are a major biodiversity hotspot and in great danger due to a combination of land-use change, climate change, and invasive species expansion (Myers et al. 2000; Echeverría et al. 2006; Miranda et al. 2017; Heilmayr et al. 2020; see Fig. 26.1). South-central Chile encompasses the ecosystems from the Maule region (35° S) to the Araucanía region (40° S), with less than 10% of the total area conserved within designated conservation areas (Pliscoff 2022).

Land-use changes caused by agriculture and forestry supplanting native forest areas have reduced biodiversity in the region (Fig. 26.1; Miranda et al. 2017; Heilmayr et al. 2020). Current spatial and temporal models of land change and land use indicate that Chile could experience a 90% increase in the area used for forestry and a nearly 140% increase in urbanization by the year 2080 (Benavidez-Silva et al. 2021). The areas transformed by these developments are concentrated in the central and southern regions of the country, restricting the connectivity of the remaining areas harboring great biodiversity to isolated protected areas that are distributed unevenly across Chile (Pliscoff 2022). Declining biodiversity and ecosystem functions place many of nature's contributions to humans at risk (Brauman et al. 2020; Heilmayr et al. 2020), and the loss of biodiversity and ecosystem connectivity in the central Chile region has severe consequences for the affected people.

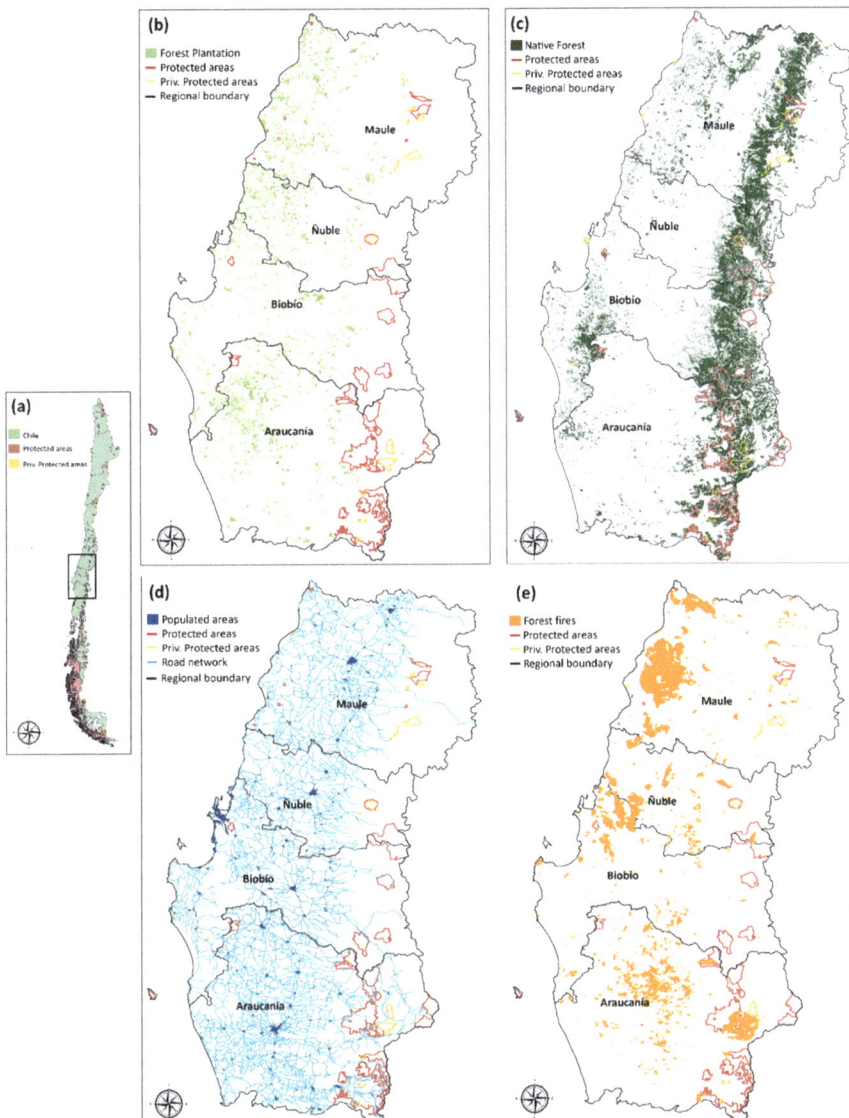

Fig. 26.1 Protected areas and native forests, and their major threats in south-central Chile: (**a**) SNASPE (National System of State-Protected Areas) (red) and private protected areas (yellow) in continental Chile. (**b**) Forest plantations (e.g., *Pinus* sp. and *Eucalyptus* sp.). (**c**) Native forests in south-central Chile. (**d**) Populated areas and the road network. (**e**) Wildfires during the past decade

The increase of anthropogenic activities and urban sprawl—including the growth of road networks, subdivision for second homes, and tourism infrastructure—significantly increases the spread of non-native invasive species as well as wildfire frequency and intensity (McWethy et al. 2018). As a result, the central zone of Chile characterized by the highest population density and per capita growth also exhibits the highest concentration of non-native species and has been extensively damaged by extreme wildfire events in recent years (Figs. 26.1 and 26.2; Fuentes et al. 2015; Fuentes-Lillo et al. 2021).

In addition to the effect of land-use change and invasive species, climate change is a key factor in this equation that contributes to biodiversity loss in Chile (Marquet et al. 2019). Current climate change models predict that by the year 2080, the richness of native plants (a reduction of 25%) and the phylogenetic diversity of native

Fig. 26.2 Characteristic elements of south-central Chilean landscapes (from top right, clockwise): (**a**) Coastal range *Araucaria araucana* forests, (**b**) wildfires affecting vegetation dominated by invasive pines, (**c**) coastal wetlands, (**d**) Darwin's fox (*Lycalopex fulvipes*) in *Araucaria araucana* forests

plants in Chile will significantly decrease both in the central regions and in the Andes Mountain range (Fuentes-Castillo et al. 2019). In this scenario, land-use change is expected to cause the loss of approximately 75% of sclerophyllous forests, 45% of native shrubs, and 43% of grasslands (Benavidez-Silva et al. 2021; Fuentes-Castillo et al. 2019).

Increasing Connectivity in South-Central Chile

In a matrix dominated by non-native tree plantations (Fig. 26.1) and experiencing increasing human pressure, forest restoration initiatives should focus on increasing the capacity of existing "natural corridors" to connect the last remnant forests. The mosaic of anthropogenic uses and the reduction of native ecosystems has led to significant impacts on both nature and people, including the displacement of nearby communities and the Indigenous Mapuche people (Torres-Salinas et al. 2016). From 1975 to 1998 alone, 40% of native forests were replaced by pine plantations, which have significant impacts on native biodiversity and water security (Uribe et al. 2020, Braun et al. 2017; Torres-Salinas et al. 2016). Promoting the connectivity of remnants of native forests is therefore critical for the persistence of biodiversity in production-oriented landscapes dominated by tree plantations, agricultural lands, and human communities.

Connectivity is also a priority for carnivores, which have significant space requirements and may avoid entering anthropogenic habitats, perceiving productive lands as a barrier to their movement across the landscape and eventually using corridors of native vegetation as potential paths of dispersal (Smith et al. 2019). The conservation of threatened carnivores such as the Darwin's fox (*Lycalopex fulvipes*) and kodkod cat (*Leopardus guigna*) in the Nahuelbuta mountain range, an anthropogenically fragmented landscape of temperate coastal forests in south-central Chile, has required the identification of habitat elements that facilitate the species' movement, such as wildlife corridors or small patches acting as stepping stones (Smith et al. 2019; Fig. 26.2). However, human–carnivore conflict has emergent population-level consequences which increases the mortality risk for carnivores in the landscape. Consequently, a unified socio-ecological framework combining landscape structure and composition with human attitudes would provide a more comprehensive approach to incorporating the suitability that certain habitat types acquire for the movement of carnivores as anthropogenic pressure intensifies (Ghoddousi et al. 2021).

Conservation efforts in the region have shown that connectivity between these forest ecosystems is not about forests alone. In fact, wetlands are dynamic ecosystems representing natural connectors between upland and aquatic systems, with many rivers connecting the Chilean Andes with the coastal range and the Pacific Ocean. Wetland connectivity provides essential habitats for migratory birds and generates conditions for unique swamp forest types; it is crucial not only for biodiversity but also for maintaining traditional ecological knowledge and biocultural memory through these resilient systems (Molares et al. 2022). Unfortunately, coastal wetlands and their forests are highly threatened by various local and global drivers of change including urbanization, energy and water production projects, industrial cellulose mills, pollution from industrial, domestic, and agricultural sources, and finally climate change (Pauchard et al. 2006; Hidalgo-Corrotea et al. 2023).

Moving Forward

Despite the potential for increasing forest connectivity in south-central Chile, no clear policy for promoting biological corridors in the region has been implemented. Furthermore, the last remnants of native forests are threatened by increasing pressure (e.g., from wildfires, Fig. 26.1e). However, biological corridors have been proposed as a central conservation tool for the region since the early 2000s, and a broad range of stakeholders have contributed to the design of such biological corridors: government agencies, NGOs, local and Indigenous communities, and forest companies and other landowners (World Wildlife Fund 2006). Private conservation initiatives along with a new private protected areas law in Chile can play an important role in the region (Martinez-Harms et al. 2021). Unfortunately, such efforts to increase connectivity remain isolated, and the lack of policy and funding support has impeded on-the-ground implementation. Over the past 10 years, conflicts regarding property ownership and rights—such as those between the forestry industry and local and Indigenous communities—have further increased the complexity of implementing landscape-scale conservation actions.

To make headway in terms of biological connectivity in south-central Chile, a broader view addressing the landscapes as socio-ecological systems is needed (e.g., Ostrom 2009). Sustainable and multifunctional landscapes that incorporate ecological and social connectivity are pivotal (Kremen and Merenlender 2018; Fischer et al. 2017). Within this approach, the participation of local communities in biodiversity protection will play a key role in maintaining the ecological processes and ecosystem services of the remnant forest habitats in south-central Chile. In conjunction, more research is needed to include social accessibility, human attitudes, and perceptions toward biodiversity in landscape planning for connectivity. Combining these attitudes and perceptions with spatially explicit information will allow conservation practitioners to identify "anthropogenic factors" as an essential component of a landscape connectivity planning framework. Therefore, creating opportunities for dialogue between private and public actors, communities, and Indigenous peoples is urgently required to forge a common vision for restoring and protecting native ecosystems and enhancing their connectivity.

References

Benavidez-Silva C, Jensen M, Pliscoff P (2021) Future scenarios for land use in Chile: identifying drivers of change and impacts over protected area system. Land 10(4):408

Brauman KA, Garibaldi LA, Polasky S, Aumeeruddy-Thomas Y, Brancalion PH, DeClerck F, Verma M (2020) Global trends in nature's contributions to people. Proc Natl Acad Sci 117(51):32799–32805

Braun, A. C., Troeger, D., Garcia, R., Aguayo, M., Barra, R., & Vogt, J. (2017). Assessing the impact of plantation forestry on plant biodiversity: A comparison of sites in Central Chile and Chilean Patagonia. Global Ecology and Conservation, 10, 159–172

Echeverría C, Coomes D, Salas J, Rey-Benayas JM, Lara A, Newton A (2006) Rapid deforestation and fragmentation of Chilean temperate forests. Biol Conserv 130(4):481–494

Fischer J, Meacham M, Queiroz C (2017) A plea for multifunctional landscapes. Front Ecol Environ 15(2):59–59

Fuentes N, Saldaña A, Kühn I, Klotz S (2015) Climatic and socio-economic factors determine the level of invasion by alien plants in Chile. Plant Ecol Divers 8(3):371–377

Fuentes-Castillo T, Scherson RA, Marquet PA, Fajardo J, Corcoran D, Román MJ, Pliscoff P (2019) Modelling the current and future biodiversity distribution in the Chilean Mediterranean hotspot. The role of protected areas network in a warmer future. Divers Distrib 25(12):1897–1909

Fuentes-Lillo E, Lembrechts JJ, Cavieres LA, Jiménez A, Haider S, Barros A, Pauchard A (2021) Anthropogenic factors overrule local abiotic variables in determining non-native plant invasions in mountains. Biol Invasions 23:3671–3686

Ghoddousi A, Buchholtz EK, Dietsch AM, Williamson MA, Sharma S, Balkenhol N, Dutta T (2021) Anthropogenic resistance: accounting for human behavior in wildlife connectivity planning. One Earth 4(1):39–48

Heilmayr R, Echeverría C, Lambin EF (2020) Impacts of Chilean forest subsidies on forest cover, carbon and biodiversity. Nat Sustain 3(9):701–709

Hidalgo-Corrotea C, Alaniz AJ, Vergara PM, Moreira-Arce D, Carvajal MA, Pacheco-Cancino P, Espinosa A (2023) High vulnerability of coastal wetlands in Chile at multiple scales derived from climate change, urbanization, and exotic forest plantations. Sci Total Environ 903:166130

Kremen C, Merenlender AM (2018) Landscapes that work for biodiversity and people. Science 362(6412):eaau6020

Marquet PA, Altamirano A, Arroyo MTK, Fernández M, Gelcich S, Górski K, Habit E, Lara A, Maass A, Pauchard A, Pliscoff P, Samaniego H, Smith-Ramírez C (eds) (2019) Biodiversidad y cambio climático en Chile: Evidencia científica para la toma de decisiones. Informe de la mesa de Biodiversidad. Comité Científico COP25; Ministerio de Ciencia, Tecnología, Conocimiento e Innovación, Santiago

Martinez-Harms MJ, Wilson KA, Costa MD, Possingham HP, Gelcich S, Chauvenet A, Bryan BA (2021) Conservation planning for people and nature in a Chilean biodiversity hotspot. People Nat 3(3):686–699

McWethy DB, Pauchard A, García RA, Holz A, González ME, Veblen TT, Stahl J, Currey B (2018) Landscape drivers of recent fire activity (2001-2017) in south-Central Chile. PLoS One 13(8):e0201195. https://doi.org/10.1371/journal.pone.0201195

Miranda A, Altamirano A, Cayuela L, Lara A, González M (2017) Native forest loss in the Chilean biodiversity hotspot: revealing the evidence. Reg Environ Chang 17:285–297

Molares S, Morales D, Aigo J, Skewes JC (2022) Cultural limnology in Patagonia: knowledge and water Management in Mapuche Rural Communities. In: Freshwaters and wetlands of Patagonia: ecosystems and Socioecological aspects. Springer International Publishing, Cham, pp 469–488

Myers N, Mittermeier RA, Mittermeier CG, Da Fonseca GA, Kent J (2000) Biodiversity hotspots for conservation priorities. Nature 403(6772):853–858

Ostrom E (2009) A general framework for analyzing sustainability of social-ecological systems. Science 325(5939):419–422

Pauchard A, Aguayo M, Peña E, Urrutia R (2006) Multiple effects of urbanization on the biodiversity of developing countries: the case of a fast-growing metropolitan area (Concepción, Chile). Biol Conserv 127(3):272–281

Pliscoff P (2022) Actualización de las áreas protegidas de Chile: análisis de representatividad y riesgo climático. Accessed from: https://www.cepchile.cl/investigacion/actualizacion-de-las-areas-protegidas-de-chile-analisis-de/

Smith JA, Duane TP, Wilmers CC (2019) Moving through the matrix: promoting permeability for large carnivores in a human-dominated landscape. Landsc Urban Plan 183:50–58

Torres-Salinas R, García GA, Henríquez NC, Zambrano-Bigiarini M, Costa T, Bolin B (2016) Forestry development, water scarcity, and the Mapuche protest for environmental justice in Chile. Ambiente & Sociedade 19:121–144

Uribe SV, Estades CF, Radeloff VC (2020) Pine plantations and five decades of land use change in Central Chile. PLoS One 15(3):e0230193

World Wildlife Fund (2006) WildFinder: online database of species distributions, version 01.06. http://www.worldwildlife.org/wildfinder

Open Access This chapter is licensed under the terms of the Creative Commons Attribution 4.0 International License (http://creativecommons.org/licenses/by/4.0/), which permits use, sharing, adaptation, distribution and reproduction in any medium or format, as long as you give appropriate credit to the original author(s) and the source, provide a link to the Creative Commons license and indicate if changes were made.

The images or other third party material in this chapter are included in the chapter's Creative Commons license, unless indicated otherwise in a credit line to the material. If material is not included in the chapter's Creative Commons license and your intended use is not permitted by statutory regulation or exceeds the permitted use, you will need to obtain permission directly from the copyright holder.

China: Ecological Restoration Projects for Connected Landscapes

27

Qiuxiao Duan and Guangzhe Liu

Saihanba forest area: part of the Three-North Shelter Forest Programme (TNSFP) area (Photo: National Forestry and Grassland Administration, 2021)

Abstract

Since the late 1970s, China has launched six national ecological restoration projects across the country to protect the environment and restore degraded ecosystems. The implementation of these national ecological restoration projects has improved ecosystem services such as soil erosion control, water retention, flood mitigation, and biodiversity conservation. As China's first world super ecological engineering project, the Three-North Shelter Forest Programme has achieved

Q. Duan (✉) · G. Liu
College of Forestry, Northwest A&F University,
Yangling, Shaanxi Province, People's Republic of China
e-mail: qiuxiao.duan@nwafu.edu.cn; gzl66106@nwafu.edu.cn

excellent results regarding maintaining national ecological security as well as promoting economic and social development. In addition, the Grain for Green Programme (GfG) is the ecological engineering project with the largest investment, the strongest policy, the widest coverage, and the highest degree of public participation in the world. The implementation of GfG has greatly increased the average forest coverage rate in the project area and significantly improved the soil erosion situation in major river basins as well as around important lakes and reservoirs.

Keywords

Ecological restoration projects · China · Three-North Shelter Forest Programme · Grain for Green Programme · Artificial forests

Six Large-Scale National Ecological Restoration Projects in China

A half-century of forest exploitation and monoculture in China has led to disastrous consequences including the degradation of forests and landscapes, loss of biodiversity, unacceptable levels of soil erosion, and catastrophic flooding events during the twentieth century (Zhang et al. 2000). To protect the environment and restore degraded ecosystems, six major national ecological restoration projects have thus been launched across the country since the late 1970s (Lu et al. 2018). The first project was the Three-North Shelter Forest Programme (TNSFP) initiated in 1978. The project is also known as the "Great Green Wall" because its massive area spans half of northern China. The second project was the Yangtze River and Zhujiang River Shelter Forest Project (abbreviated as "River Shelter Forest"), launched in 1989 across the southern parts of China with the aim of combating floods and reducing soil erosion. The Natural Forest Protection Project (abbreviated as "Forest Protection") was subsequently initiated in 1998 and has delivered numerous benefits including biodiversity conservation, reduction of soil erosion and flood risk, and prevention of other natural disasters associated with deforestation. The Grain for Green Programme (GfG, also known as the Sloping Land Conversion Programme or the Conversion of Cropland to Forest Programme) was launched in 2000 and has advanced the conversion of croplands in hilly areas to forests. GfG is regarded to be the world's largest ecological restoration programme in terms of scale and investment, and it is a typical example of ecological compensation (Zhang et al. 2000). The Beijing–Tianjin Sandstorm Source Control Project (abbreviated as "Sandstorm Control") was initiated in 2001 to promote environmental conservation by preventing sandstorm-induced soil erosion and other damage near the capital of Beijing. The last initiative, known as the Returning Grazing Land to Grassland Project (abbreviated as "Grazing Land to Grassland"), was launched in 2003 to reduce the impacts of overgrazing and promote grassland productivity. These projects together cover 44.8% of China's forests (Liu et al. 2008; Ministry of Forestry 2010) and 23.2% of China's grasslands (Ministry of Forestry 2014).

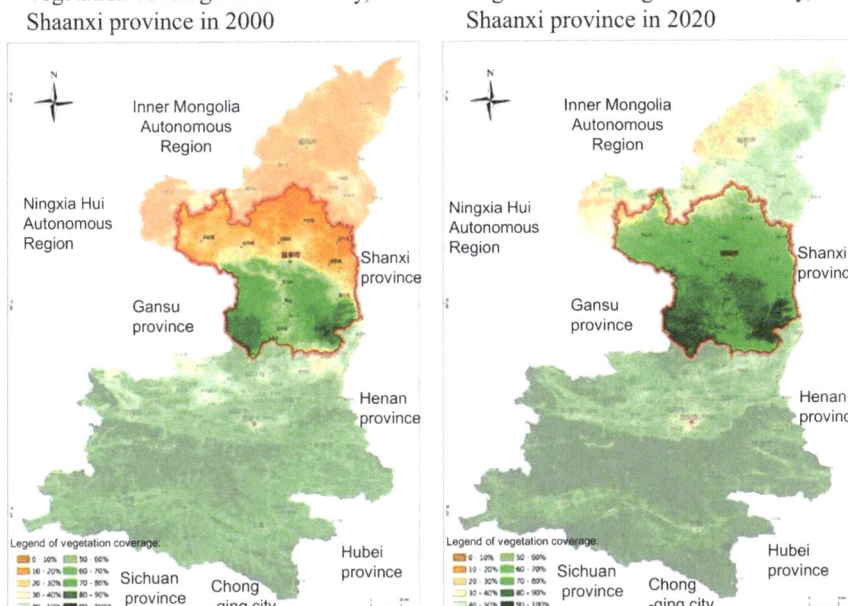

Fig. 27.1 Case study – China: Vegetation coverage of Yan'an city, Shaanxi province (before and after national ecological restoration projects. China Daily, 2021)

Recent studies have indicated that the implementation of the national ecological restoration projects has improved ecosystem services such as soil erosion control (Wang et al. 2016; Ouyang et al. 2014), water retention and flood mitigation (Ouyang et al. 2016), and biodiversity conservation (Ouyang et al. 2014; Wang et al. 2016). In addition, a series of management practices employed under the framework of these restoration projects—including afforestation, reforestation (shelter forests, forest protection, and sand control), forest tending (forest protection and sand control), transforming cropland into forest (GfG), reducing timber harvesting (forest protection), and fencing in grasslands (grassland conservation)—can increase forest area, prevent carbon loss from vegetation and soil, and subsequently increase carbon stocks and sinks (Xu et al. 2017; Bonan 2008). In this chapter, two of China's six large-scale national restoration projects will be described in detail: the Three-North Shelter Forest Programme (TNSFP) and the Grain for Green Programme (GfG) (Fig. 27.1).

The Three-North Shelter Forest Programme (TNSFP)

TNSFP is China's first world super ecological engineering project: a large-scale shelter forest system construction project approved by the State Council in 1978 in response to the severe situation of wind and sand damage along with soil erosion in the Northwestern, Northern, and Northeastern regions of the country.

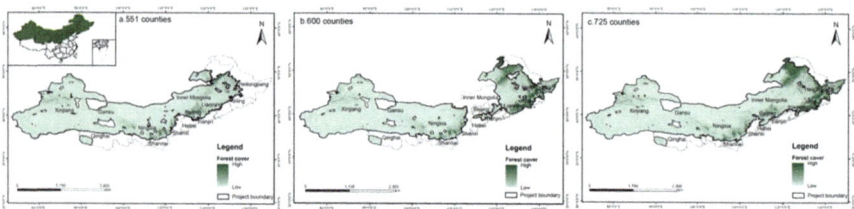

Fig. 27.2 Study area maps of TNSFP showing forest coverage and planning scope of phases III to V of the programme. Phase III (1996–2000) planning area—551 counties (**a**); Phase IV (2001–2010) area—600 counties (**b**); Phase V (2011–2020) area—725 counties (**c**). (Note: Phase I covered 406 counties, and Phase II covered 514 counties). Tree cover data derived from the annual Vegetation Continuous Fields (VCF) tree cover product by NASA's MODIS (Zhai et al. 2023)

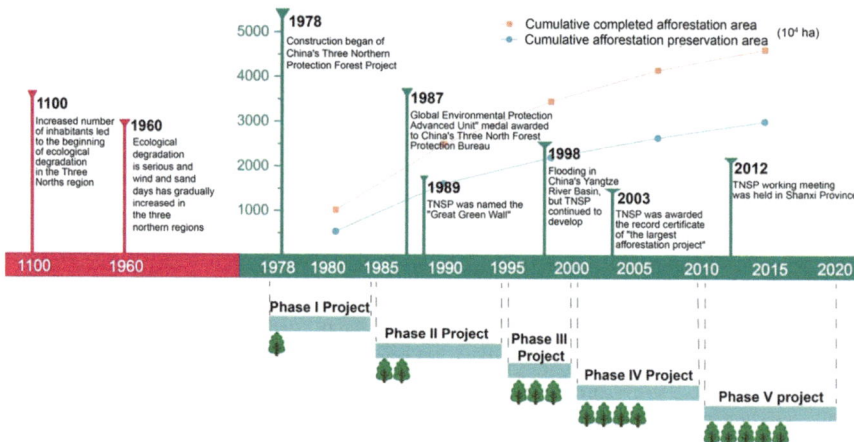

Fig. 27.3 Timeline showing ecological conditions, historical events, and achievements of TNSFP in the Three-North region of China from 1100 to 2020. Also included are the trajectories of the cumulative completed afforestation area (yellow) and cumulative afforestation preservation area (green) for Phase I to Phase V of the programme according to its 40-year report compiled in 2018 (Administration SFAG 2019; Zhai et al. 2023). TNSFP has significantly increased the total forest resources

The TNSFP system spans 4480 kilometers from east to west, 560–1460 kilometers from north to south, and covers a total area of 4.07 million square kilometers, accounting for 42.4% of the project region's land area (Fig. 27.2; Li et al. 2012). The project is scheduled to last 73 years from 1978 to 2050, divided into three stages and eight phases (Figs. 27.2 and 27.3). The total planned afforestation area is 35.08 million hectares, and the forest coverage rate will increase from 5% before the project's initiation to 15% upon conclusion (Li et al. 2012).

Background

For a long time, wind and sand damage and soil erosion were severe in Northwestern, Northern, and Northeastern China, resulting in a shortage of wood, fuel, fertilizer, and feed. Agricultural production was low and unstable. The three northern regions feature China's eight major deserts, four major sandy areas, and the vast Gobi with a total area of 1.48 million square kilometers, accounting for about 85% of the country's land area affected by wind desertification. They form the 10,000-mile Wind-Sand-Line from Heilongjiang in the east to Xinjiang in the west. The area is characterized by severe wind erosion and sand encroachment as well as frequent sandstorms.

In May 1978, an expert panel of the State Forestry Administration of China proposed the "Opinions on Creating 10 000 Miles of Shelter Forests and Transforming Nature" to the State Council. The State Forestry Administration then formulated the "Plan for the Construction of Large Protective Forests in Key Areas of Wind and Sand Damage as well as Soil Erosion in Northwestern, Northern, and Northeastern China" based on in-depth research and repeated discussions. In November 1978, the Three-North Shelter Forest Programme was officially launched and created a precedent for large-scale ecological construction in China.

Key Measures

TNSFP implements the principle of "the country, the collective, and individuals working together." The State Council established a monitoring group for the programme, tasked with studying and determining the major issues during its implementation. The State Forestry Administration oversees TNSFP, and its construction bureau is specifically responsible for project planning, supervision, and inspection. Forestry departments at all levels have established special management organizations to take charge of local project construction, forming a project management system that closely ties together central and local authorities in decision-making and implementation. Management methods and regulations have been successively formulated for engineering construction technology, planning, funding, and so on, ensuring that the entire process of the TNSFP from seed preparation to inspection and acceptance follows clear rules.

The management mechanism for engineering construction within the project has been continuously innovated. A bidding system following a process of open bidding, expert evaluation, democratic decision-making, and support based on merit has been actively promoted, especially for key construction projects. To ensure proper implementation of tasks, funds, and responsibilities, a contract system is employed in which construction job contracts are signed layer by layer. A supervision system is implemented, and quality supervision is enforced. The responsible organizations manage funding strictly, implement dedicated funds and accounts, manage them separately, and conduct regular as well as irregular special inspections. The programme has established a quality inspection and acceptance system to track and manage the entire process of afforestation quality and regularly perform multi-dimensional quality evaluations.

Appropriate measures suited to local conditions were implemented, and the engineering construction was promoted by technological progress. First, technological innovation was vigorously carried out. Based on the reality of drought and limited rainfall in the three northern regions, research groups have achieved some breakthroughs which mainly focused on drought-resistant afforestation, resulting in a 23% increase in afforestation survival rate. Aerial seeding afforestation technology has broken through the "forbidden zone" where it is not suitable for rainfall below 200 millimeters per year. Second, technological promotion/application was widely carried out. Single technology promotion was transformed into comprehensive supporting technology, including assembling supporting afforestation, forest management, and other comprehensive technical measures. Science and technology experimental demonstration zones have been established and promoted more than 1200 advanced applicable technologies and over three million hectares of forests. The afforestation preservation rate has been increased from 60% to over 85%. Third, innovation in afforestation models was vigorously carried out. In accordance with the requirements of sand prevention and control, soil and water conservation, as well as farmland protection forest construction, more than 100 afforestation models have been promoted. According to functional layout, ecological protection, ecological economy, and ecological landscape models have been promoted (Li and Feng 2021).

Implementation Effectiveness

Considerable achievements have been made over the past 40 years since the initiation of TNSFP, and the ecological situation in the project area has been noticeably improved. The forest ecosystem plays an important role in maintaining national ecological security and promoting economic and social development. First, significant ecological benefits have been achieved through effective control of wind and sand damage as well as soil erosion in key monitoring areas (Fig. 27.3). A total of 7.88 million hectares of windbreaks and sand fixation forests have reportedly been created, 336,000 square kilometers of desertification area has been brought under control, and over ten million hectares of severely decertified and salinized grasslands as well as pastures have been protected and restored (Li and Feng 2021). Moreover, 1.66 million hectares of forests for the purpose of farmland protection have been created, effectively sheltering 30.19 million hectares of farmland (Li and Feng 2021).

Second, obvious economic benefits have been achieved as well. Regional economic development has been promoted and local farmers' income and wealth have increased. TNSFP has always adhered to the goal of building an ecological and economic protective forest system. Under the premise of prioritizing ecology, it has established a number of timber forests, economic forests (similar to non-timber-oriented forests), firewood forests, and fodder forest bases, promoting the adjustment of rural industrial structure and rural economic development, effectively increasing farmers' incomes, and achieving a "win-win" situation between ecological restoration and economic development. The programme area has also established a new pattern of ecotourism development, with a forest park network as the backbone along with wetland parks, desert parks, and other supplementary areas.

Third, it has enhanced the ecological awareness of the entire society and enhanced China's international position in the field of ecological engineering. The implementation of TNSFP reflects China's will to improve the ecological landscape of its territory and stimulates the enthusiasm of the cadres and masses in the construction area to devote themselves to building a green country and has given rise to many heroic and exemplary figures as well as the cultivation of advanced models. It has forged a "Three-North spirit of hard work, tenacious struggle, unity and cooperation, perseverance, truth-seeking, pioneering and innovation, people-oriented and benefiting mankind," which has had a significant impact at home and abroad, and it is internationally known as "the largest ecological engineering in the world."

Fourth, it has strengthened international cooperation and established a model of international ecological governance. TNSFP has received widespread attention since its initiation. In 2003, TNSFP set a Guinness World Record, becoming the "largest afforestation project" in the world (Fig. 27.4).

The Grain for Green Programme

China's Grain for Green Program (GfG) is the ecological engineering project with the largest investment, the strongest policy stimulation, the widest coverage, and the highest degree of public participation in the world. GfG comprises planned and step-by-step cessation of cultivation followed by afforestation and grass planting in accordance with local conditions along with restoration of vegetation in areas with severe soil erosion, desertification, salinization, and rocky desertification, as well as in areas with low and unstable grain yields, beginning with the protection and improvement of ecological conditions.

Background

China is a traditional agricultural country. For a long time, the pressure of rapid population growth and relatively extensive modes of production led to the conversion of large areas of forest steppe and wetlands into farmland. Widespread deforestation and land reclamation caused increased soil erosion as well as aggravating water and soil loss, serious land degradation, recurring drought and flood disasters, and the general deterioration of the ecological environment. The upper and middle courses of the Yangtze and Yellow Rivers became some of the most severely eroded areas in the world due to deforestation, grassland destruction, and slope cultivation. Catastrophic floods occurred in the Yangtze River and Songhua River basins in 1998. About 223 million people were affected by the floods and 4,970,000 houses

Fig. 27.4 Certificate of Guinness World Records. In 2003, TNSFP set a Guinness record and became the "world's largest afforestation project" (the State Forestry Administration of China, 2003)

were flattened. During the flood, over 1320 people were killed and the direct economic losses amounted to 166,600 million Yuans (Zong and Chen 2000). Then the nation was aware that accelerating the establishment of forests and grasslands and improving the ecological situation had become an urgent strategic necessity and were essential for the survival and development of the entire country. Starting in 1999, the leaders of the State Council proposed policy measures such as returning farmland to forests and grasslands, blocking mountains for greening, providing food as relief, and individual contracting. Converting agricultural land back into forests and grasslands had become an inescapable path for the Chinese people to adapt to history and nature. The State Council reviewed the situation and decided to take the lead in pilot projects in the Sichuan, Shaanxi, and Gansu provinces in 1999, marking the beginning of China's Grain for Green Programme.

Key Measures

Many laws and regulations in China set out explicit provisions for the conversion of farmland to forests and grasslands. The State Council has issued the "Regulations on the Conversion of Farmland to Forests" as well as other documents. Relevant departments have formulated a series of measures to establish a comprehensive system of regulations and policies for this type of land conversion. Subsidy policies were established and gradually improved. A certain amount of raw grain and cash were subsidized annually to local farmers for areas converted from farmland to forests. Special funds were established to consolidate the achievements of land conversion and support the construction of basic grain ration fields for local farmers, rural energy construction, ecological migration, and supplementary replanting and reconstruction.

The so-called "Four to the Province" policy (goals, tasks, funds, and responsibilities) was implemented, with the provincial governments put in charge of general oversight. In addition, city and county governments would assume certain target responsibilities. The State Forestry Administration signs project construction responsibility letters with each provincial government every year; the competent department at the provincial level prepares the annual implementation plan, whereupon the county-level agencies prepare the operational design and the governments at the county or township level sign the contracts with land contractors to return farmland to forest and grassland. Ultimately, individual farmers converting agricultural land to forests form the basic unit of this national ecological engineering programme. Engineering supervision was strengthened effectively. Special management measures such as selecting the best quality of seedlings, cultivating them on site, and replenishing seedlings nearby were introduced.

A supervision system was implemented to oversee the project's progress, quality, duration, and fund utilization. A series of standards were formulated and issued, scientific and technological experimental and demonstration sites were established, and more than 100 scientific and technological support projects providing technical assistance were initiated. Consistent reviewing and monitoring were carried out and a three-tiered inspection and acceptance system consisting of county-level self-inspection, provincial-level review, and verification at the national level was implemented. According to the annual inspection and acceptance results, subsidy funds would be allocated level by level, strictly following the principle of acceptance, disclosure, and redemption.

Implementation Effectiveness

The implementation of GfG has changed the traditional farming habits of farmers who have been cultivating crops for generations, achieving a historic transformation from deforestation to converting farmland back into forests and generating significant and comprehensive benefits. The process of national greening has been accelerated and a major contribution has been made to improving China's overall ecological conditions. As an important driver of large-scale land greening, GfG has significantly increased forest and grassland vegetation since its inception (Fig. 27.5). The average forest coverage rate in the project area has risen by more than 4% (Li 2021). The soil erosion situation in major river basins such as that of the Yangtze River or the upper and middle courses of the Yellow River, as well as around important lakes and reservoirs, has been significantly improved. Land desertification in the northern areas of the country as well as rocky desertification in the southwest has been effectively curbed, habitats for wild animals have been restored, and biodiversity has been protected and improved.

Economic benefits have gradually emerged as well in the shape of contributions to alleviating poverty in the affected areas. In combination with GfG, relevant regions have established numerous ecological industry bases for the processing of woody grain, oil, and dry and fresh fruits, as well as engaging in forest understory economy, cultivating and strengthening ecological industries. The most recent funding period of GfG in particular has greatly promoted the development of economic forests. Between 2014 and 2019, 2.38 million hectares of economic forests were planted (Li 2021). The GfG Programme is primarily aimed at impoverished and remote areas, making it a powerful tool for targeted poverty alleviation. The channels for increasing agricultural revenues have been continuously expanded, making farmers' incomes more stable and diverse.

The social benefits are likewise extensive and far-reaching, with significant contributions to improving ecological production modes and rural life. Successful implementation of GfG has optimized production methods, promoted the transfer and concentration of agricultural production factors, increased grain yields and multiple cropping indices, accelerated the transition from traditional to modern agriculture, and caused many regions to break out of the vicious cycle of "becoming poorer and more cultivated, more cultivated and poorer." The younger generation has been liberated from the constraints of arable land and empowered to pursue higher dreams and goals. The concept of ecological priority and green development has been deeply enrooted in people's minds, and appreciating and protecting the natural environment has become a common trend.

The GfG Programme provides a model for global ecological governance. The large-scale conversion of farmland to forests and grasslands has contributed significantly to increasing forest coverage as well as to participation in global ecological governance. During its more than 20-year existence, GfG has helped to ensure that China maintains "both growth" of forest area and stock for many consecutive years despite the continuous decline in global forest area and stock and that the country's preserved area of artificial forests has long been ranked first in the world (Treacy et al. 2018).

Fig. 27.5 Comparison of an area before (top, in 2000) and after (bottom, in 2008) GfG in Dafang County, Guizhou Province (Photos: National Forestry and Grassland Administration 2021)

References

Administration SFAG (2019) Development report for the three-north shelterbelt system in the past 40 years: 1978–2018. China Forestry Press (in Chinese), Beijing

Bonan GB (2008) Forests and climate change: forcings, feedbacks, and the climate benefits of forests. Science (New York, NY) 320:1444–1449. https://doi.org/10.1126/science.1155121

Fei, Lu Huifeng, Hu Wenjuan, Sun Jiaojun, Zhu Guobin, Liu Wangming, Zhou Quanfa, Zhang Peili, Shi Xiuping, Liu Xing, Wu Lu, Zhang Xiaohua, Wei Limin, Dai Kerong, Zhang Yirong, Sun Sha, Xuemage AltAlternative text)Wanjun, Zhang Dingpeng, Xiong Lei, Deng Bojie, Liu Li, Zhou Chao, Zhang Xiao, Zheng Jiansheng, Cao Yao, Huang Nianpeng, He Guoyi, Zhou Yongfei, Bai Zongqiang, Xie Zhiyao, Tang Bingfang, Wu Jingyun, Fang Guohua, Liu Guirui, Yu (2018) Effects of national ecological restoration projects on carbon sequestration in China from 2001 to 2010 Significance Proceedings of the National Academy of Sciences 115(16) 4039-4044 10.1073/pnas.1700294115

Li S (2021) The grain for Green Program in China. China Green Times. http://lyj.guizhou.gov.cn/ztzl/tghl/202106/t20210624_68798698.html

Li S, Feng D (2021) The Three-North Shelterbelt Program System Construction Project in China. China Green Times. http://www.isenlin.cn/sf_B6E45E48C6D043CFAFDB8703AEA3D8F1_209_1335E561541.html

Li M, Liu A, Zou C, Xu W, Shimizu H, Wang K (2012) An overview of the "three-north" shelterbelt project in China. Forestry Study China 14:70–79. https://doi.org/10.1007/s11632-012-0108-3

Liu J, Li S, Ouyang Z, Tam C, Chen X (2008) Ecological and socioeconomic effects of China's policies for ecosystem services. Proc Natl Acad Sci USA 105:9477–9482. https://doi.org/10.1073/pnas.0706436105

Ministry of Forestry (2010) Forest resource statistics of China (2004–2008). Ministry of Forestry, Beijing

Ministry of Forestry (2014) Forest resource report of China—the 8th national forest resources inventory. Chinese Forestry Publishing House, Beijing

Ouyang Z, Wang Q, Zheng H, Zhang F, Hou P (2014) National ecosystem survey and assessment of China (2000–2010). Bulletin of Chinese Academy of Sciences. Ministry of Environment projection Chinese Academy of Sciences (2014) national ecosystem survey and assessment of China (2000–2010). China Science Publishing House, Beijing

Ouyang Z, Zheng H, Xiao Y, Polasky S, Liu J, Xu W, Wang Q, Zhang L, Xiao Y, Rao E, Jiang L, Lu F, Wang X, Yang G, Gong S, Wu B, Zeng Y, Yang W, Daily GC (2016) Improvements in ecosystem services from investments in natural capital. Science (New York, NY) 352:1455–1459. https://doi.org/10.1126/science.aaf2295

Treacy P, Jagger P, Song C, Zhang Q, Bilsborrow RE (2018) Impacts of China's grain for green program on migration and household income. Environ Manag 62:489–499. https://doi.org/10.1007/s00267-018-1047-0

Wang S, Fu B, Piao S, Lü Y, Ciais P, Feng X, Wang Y (2016) Reduced sediment transport in the Yellow River due to anthropogenic changes. Nat Geosci 9:38–41. https://doi.org/10.1038/NGEO2602

Xu W, Xiao Y, Zhang J, Yang W, Zhang L, Hull V, Wang Z, Zheng H, Liu J, Polasky S, Jiang L, Xiao Y, Shi X, Rao E, Lu F, Wang X, Daily GC, Ouyang Z (2017) Strengthening protected areas for biodiversity and ecosystem services in China. Proc Natl Acad Sci USA 114:1601–1606. https://doi.org/10.1073/pnas.1620503114

Zhai J, Wang L, Liu Y, Wang C, Mao X (2023) Assessing the effects of China's three-north shelter Forest program over 40 years. Sci Total Environ 857:159354. https://doi.org/10.1016/j.scitotenv.2022.159354

Zhang P, Shao G, Zhao G, Le Master DC, Parker GR, Dunning JB Jr, Li Q (2000) China's forest policy for the 21st century. Science (New York, NY) 288:2135–2136. https://doi.org/10.1126/science.288.5474.2135

Zong Y, Chen X (2000) The 1998 flood on the Yangtze, China. Nat Hazards 22:165–184. https://doi.org/10.1023/A:1008119805106

Open Access This chapter is licensed under the terms of the Creative Commons Attribution 4.0 International License (http://creativecommons.org/licenses/by/4.0/), which permits use, sharing, adaptation, distribution and reproduction in any medium or format, as long as you give appropriate credit to the original author(s) and the source, provide a link to the Creative Commons license and indicate if changes were made.

The images or other third party material in this chapter are included in the chapter's Creative Commons license, unless indicated otherwise in a credit line to the material. If material is not included in the chapter's Creative Commons license and your intended use is not permitted by statutory regulation or exceeds the permitted use, you will need to obtain permission directly from the copyright holder.

28

Ethiopia: Enhancing Landscape Connectivity Through Agroforests

Hafte Mebrahten Tesfay and Mesele Negash

Coffee-Fruit Tree (Photo: Adobe Stock)

H. M. Tesfay (✉)
Department of Forest Biodiversity and Nature Conservation, Federal Research and Training Centre for Forests, Natural Hazards and Landscape, Vienna, Austria

BOKU University, Department of Ecosystem Management, Climate and Biodiversity, Institute of Forest Ecology, Vienna, Austria
e-mail: hafte-mebrahten.tesfay@bfw.gv.at; hafte.tesfay@boku.ac.at

M. Negash
Hawassa University, Wondo Genet College of Forestry and Natural Resources, Hawassa, Ethiopia

European Forest Institute, Manziana, Rome, Italy
e-mail: meselen@hu.edu.et; mesele.tesemma@efi.int

Abstract

The Gedeo agroforestry cultural landscape in Ethiopia is recognized as a leading example of sustainable agroforestry practices. It mimics and resembles natural forests, characterized by vertically and horizontally diverse composition, structure, and functionality that contribute to on-farm conservation, environmental well-being, and livelihood support systems. Numerous studies have highlighted the positive impacts of agroforests in the landscape on circa situm biodiversity conservation, implying biodiversity conservation through utilization. Agroforests provide additional habitats for species that are sensitive to disturbance, conserve the gene pools of native tree species, and enhance biodiversity. They also act as buffers against forest degradation and deforestation in the surrounding natural habitats as well as connecting fragmented habitats for animal and plant species through the creation of corridors and stepping stones. Agroforests in the Gedeo landscape serve as important havens for preserving high levels of diversity. The smallholder agroforests in the region play a crucial role in conserving tropical woody plant species as circa situm reservoirs of biodiversity in agricultural environments. Previous studies have shown that the diverse species composition within agroforests contributes significantly to biomass and carbon storage, which in turn helps to mitigate climate change. For the Gedeo people, agroforestry is not just a supplementary livelihood activity but rather a mainstay. In summary, the Gedeo agroforests play a significant role in enhancing landscape connectivity and contribute to the integrity and sustainability of the agricultural production system. However, the Gedeo agroforests also face challenges including land degradation and fragmentation, the emergence of lucrative monoculture cash crops, climate change, limited market access, lack of financial resources, and insufficient technical knowledge and training.

Keywords

Agroforest · Landscape connectivity · Circa situm conservation · Gedeo landscape

The Features of the Gedeo Agroforestry Landscape

Agroforestry (AF) is an integrated strategy that takes advantage of potential synergies between agricultural and forestry technology to develop more varied, productive, financially successful, healthful, and environmentally sustainable land-use systems (Tewabech and Efrem 2014; Daizy et al. 2008). There are multiple definitions of agroforestry by different authors, including Nair (1993), Leakey (1996), and later ICRAF (2000). Although these definitions share some similarities in terms of their basic concept, the most comprehensive and explicit definition of the AF system was provided by the International Centre for Research in AF (ICRAF): "AF is a dynamic, ecologically based, natural resource management system that, through the integration of trees on farms and in the agricultural landscape, diversifies and sustains smallholder production for increased social, economic and environmental benefits" (ICRAF 2000; Fig. 28.1).

This chapter explains one of the centuries-old indigenous agroforestry systems practiced in the Gedeo Zone of Ethiopia. Although the Gedeo AF system is often cited as a model for land use, its contribution to enhancing landscape connectivity has not been described in detail. The Gedeo agroforestry landscape is located in the southeastern Rift Valley escarpment in the Gedeo Zone of the Southern Nations, Nationalities, and Peoples' Region (SNNPR) of Ethiopia. Its elevation ranges between 1300 and 3064 m, the rainfall levels between 800 and 1800 mm per year, and the mean annual temperature between 13 and 25 °C. The traditional agroclimatic classification shows that the region comprises a mid-altitude climate (Dega) and a sub-tropical climate (Weynadega) accounting for 37% and 62% of its area, respectively, with the remainder featuring a hot tropical climate (Kolla) (Mebrate 2007). The soil is primarily developed from volcanic rocks (Negash et al. 2012; Mebrate 2007; Fig. 28.2).

Historically, the Gedeo area was originally dominated by mid-altitude forest made up of species such as *Syzygium guineense*, *Podocarpus falcatus*, *Millettia ferruginea*, *Cordia africana*, *Croton macrostachyus*, *Aningeria adolfi-friederici*, and *Erythrina* spp. Farmers selectively harvested trees and practiced single-crop agriculture (Negash and Achalu 2008). Owing to high population pressure and the steepness of the land, they gradually shifted their production to cash and food-dominated perennial systems. This promoted the introduction of *Ensete ventricosum* (enset or false banana) and *Coffea arabica* into the farming system.

The Gedeo agroforestry landscape has distinct characteristics that make it highly sustainable, and it is considered an example of the best AF practice in the country. According to Habtamu and Zemede (2011), the agroforests feature a high degree of compositional, structural, and functional variation crucial for on-farm conservation, ensuring environmental well-being and enhancing livelihood support systems. The upper story of the agroforests is primarily made up of native tree species such as *Ficus* spp., *Cordia africana*, *Croton macrostachyus*, and *Millettia ferruginea* along with fruit trees such as mango (*Mangifera indica*) and avocado (*Persea americana*).

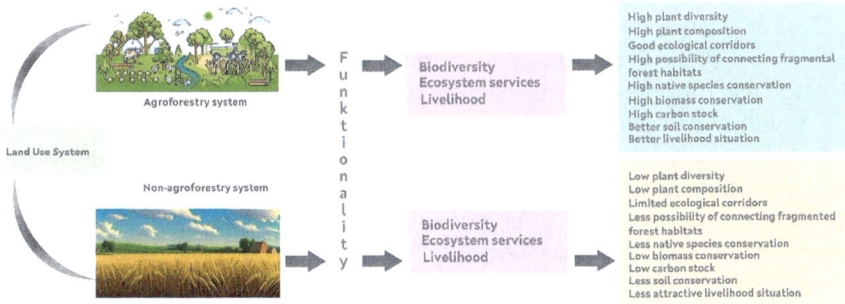

Fig. 28.1 Visual abstract of the functionality of agroforestry vs. non-agroforestry systems for ecological connectivity

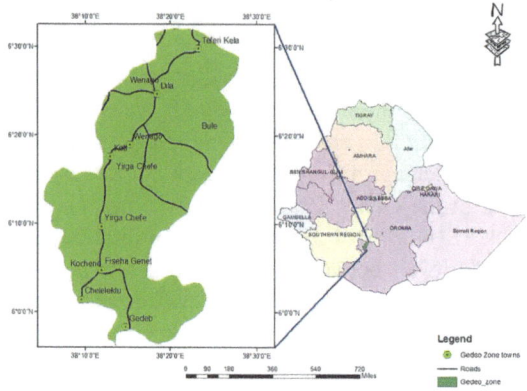

Fig. 28.2 Map of Gedeo Zone, Ethiopia (Degefa 2016)

The middle story is composed of the dominant species including coffee (*Coffea arabica L.*), an evergreen shrub, and enset (*Ensete ventricosum*). Enset is a banana-like perennial herbaceous monocarpic plant with a single thick stem; its fruits serve as a staple food in central, southern, and southwestern Ethiopia. Vegetables, spices, and herbs frequently occupy the lower story. While enset is present at all elevations, the share of coffee declines with elevation (Gebrehiwot and Maryo 2015). This system's architectural layout makes efficient use of available space, enhancing positive interactions and reducing negative ones.

The Gedeo people not only grow fruits and vegetables in their home gardens, as is the case in other home garden AF systems. Instead, all types of crops are grown together, including cash crops like coffee and fruits, staple foods such as enset and maize, and vegetables. Trees are another crucial resource for the farmers' livelihoods, providing shade for the coffee plants, fuelwood, fodder, cash income, medicinal value, and honey production as well as contributing to soil fertility and cultural values (Negash 2007). As a result, trees are integrated on the same plots of land as crops. A further component of the Gedeo AF system is livestock. Ruminants in the Gedeo Zone feed leaves of banana, enset, and *Millettia ferruginea* (Birhanu et al. 2013). In contrast to other AF systems, the Gedeo agroforests are the primary source of food production for the community (Ayele et al. 2014). Provision of diverse products to sustain the livelihoods of the farming households in the densely populated Gedeo Zone is likewise among their key features.

A further characteristic of the Gedeo agroforests is their multifaceted contribution to biodiversity conservation. As reported by several studies, agroforests have the potential to conserve biodiversity within landscapes by arranging and providing additional supportive habitats for species that do not tolerate high levels of disturbance such as epiphytic plants, rare and endemic species, and endangered species (Jose 2009); by maintaining the gene pools of native tree species in fragmented landscapes (Das and Das 2005; Harvey and Gonzalez-Villalobos 2007; Jose 2012); by playing an important role in enhancing microbial, avian, and faunal diversity

(Gillespie et al. 1995); by retaining soil and allowing water to accumulate, thereby preventing habitat degradation and loss; by protecting against the pressure of forest degradation and deforestation in the surrounding natural habitats; and by providing corridors and stepping stones to reducing habitat fragmentation (McNeely and Schroth 2006; Bhagwat et al. 2008; Haggar et al. 2019). Agroforestry plays a great role in enhancing landscape connectivity. As Haggar et al. (2019) reported, agroforestry increases land area with high biodiversity connectivity by 60–80%. In addition, the inclusion of the biodiversity-hosting capacity of the agroforestry systems substantially increases the connectivity and edge forest area by 70–100% (Haggar et al. 2019).

Circa Situm Conservation in the Gedeo Agroforests

Agroforestry as a traditional system in many regions has shown great value in terms of preserving a high level of diversity at the genetic, species, ecosystem, and landscape levels. It provides a haven for numerous trees and other plants in areas with high rates of deforestation, where they face the risk of extinction from the wild in Ethiopia. For instance, a study in the Gedeo agroforestry region identified 22 native woody species of interest for conservation according to IUCN (The International Union for Conservation of Nature). Red List and local criteria. Among them, *Pygeum africanum* and *Rhus glutinosa* were categorized as vulnerable in the wild and in need of conservation priority (Negash et al. 2012). The Gedeo landscape's smallholder agroforests may be useful for preserving tropical woody plant species through different mechanisms. The first and most effective of these is that where wild stands previously existed, farmers may have kept and/or planted trees as circa situm reservoirs of biodiversity in agricultural environments (Dawson et al. 2013). Circa situm conservation is defined by Dawson et al. (2013) as "the preservation of planted and/or remnant trees and wildings in farmland where natural forest or woodland containing the same trees was once found, but where natural vegetation has been lost or modified significantly through agricultural expansion." Interestingly, circa situm biodiversity conservation emphasizes the preservation of species while utilizing them, which means that species with utilitarian value will have a higher chance of being maintained in farmlands. Trees in farmlands may support conservation in situ by offering an alternate source of product to reduce extraction from forests. In addition, they can act as corridors or stepping stones connecting fragmented wild stands (Dawson et al. 2013), thereby benefiting wild tree stands and natural populations of other flora and fauna by creating favorable conditions for seed dispersers, pollinators, and other migrating animals between the natural forest fragments (Bhagwat et al. 2008; Doerr et al. 2010; Gilbert-Norton et al. 2010). Thirdly, the increased value offered by planting trees may lead to increased interest in including them in field gene banks, field trials, and seed collections that assist ex situ conservation. Therefore, it could be summarized that the circa situm conservation of floral and faunal diversity by agroforestry plays a significant role in enhancing landscape connectivity. Agroforestry could act as a bridge between the cropland and the forest or between the grassland and the forest.

Multiple case studies on the Gedeo agroforests have reported that dozens or even hundreds of tree and shrub species can be found in different AF systems of the Gedeo landscape (Table 28.1). Forest inventories conducted in the Gedeo Zone indicate that the agroforests frequently include numerous natural forest remnants in addition to planted trees and stands. Farmers occasionally keep these trees to ensure an ongoing (fruit, medicine) or occasional (large-trunk timber) supply of certain goods, as well as for services like soil fertility restoration or shade, for example, in shade-grown coffee production systems (Mengitu and Fitamo 2015; Mebrate et al. 2021; Tesfay et al. 2024). In addition, the Gedeo people frequently retain specific trees or stands for cultural, religious (e.g., belief in the sanctity of the natural world), and/or aesthetic reasons (Kanshie 2002).

The multi-strata AF systems in the Gedeo landscape conserve a higher proportion of woody tree and shrub species compared to other AF systems in different parts of Ethiopia and elsewhere in the tropics (Tefera et al. 2016; Negash et al. 2012). These multi-strata agroforests support agroecosystems by offering ideal habitats for woody species, thus perhaps providing an example of circa situm biodiversity conservation focusing on both production and conservation. In a relatively small space, agroforests allow a greater variety of species while also supporting a high human population density (1300 people/km^2) (Negash et al. 2012). Comparatively high tree species richness in these agroforests was also reported by Tefera et al. (2016) and Negash et al. (2012). Kindt (2002), Kirschenmann (2007), and Steffan-Dewenter et al. (2007) revealed that the favorable effect that tree species variety occasionally plays in boosting farm yields and fostering resilience supports opportunities for conservation.

Overall, analysis of the level of circa situm conservation in the Gedeo agroforests shows that they play an important role in maintaining high tree/shrub species diversity and richness owing to their usage value. This in turn helps to enhance the connectivity of the landscape by reducing habitat fragmentation and species exposure to disturbance. The contribution of the different Gedeo agroforest types to circa situm conservation is summarized in Table 28.1.

Biomass and Carbon Reserve in the Landscape Connectivity

Increasing the size of the global terrestrial carbon sink is one of the best ways to reduce the amount of CO_2 in the atmosphere. Switching to tree-based land-use systems such as forest conservation, sustainable forest management, afforestation/reforestation, and agroforestry could result in a significant increase in biomass and, as a result, more carbon storage compared to land-use systems with lower biomass stocks (like grasslands or agricultural fallows). This is due to the longer biomass lifespan of tree-based systems compared to grass-based systems or others with short lifespans (Tesfay et al. 2022a). Biomass and carbon reserves in agroforestry systems are essential for landscape connectivity as they contribute to multiple ecosystem

Table 28.1 Woody tree/shrub species richness reported in the Gedeo agroforestry landscape and photographic presentation of various Agroforestry systems

Photo references	AF Type	Study location, coverage	Description of results	References
6, 3, 1	Coffee-Fruit Tree - crop AF system; Enset - Coffee -Tree AF system; Enset -Tree AF system	Dilla Zuria district, Gedeo Zone, A total of 15 villages (5 each employing one of three AF practices) were surveyed.	A total of 39 woody plant species belonging to 35 genera and 25 families were identified in the study area. Of the identified species 33 (86%) were trees and 6 (14%) were shrubs.	(Tefera et al., 2016)
1, 3, 2	Enset AF system; Enset - coffee AF system; Fruit Tree - Coffee AF system	South-eastern escarpment of the Rift Valley, Gedeo zone. A total of 60 smallholdings with sizes around 100 m2 representing three AF practices were surveyed.	A total of 58 woody species, belonging to 49 genera and 30 families were identified. Of the woody species identified, 86% were native. The highest proportion of native woody species was recorded in Enset - AF plots (92%), followed by Enset - Coffee AF (89%) and Fruit - Coffee AF (82%).	(Negash et al., 2012)
3	Coffee -Enset - Tree AF system	Wenago district, Gedeo zone. A total of 20 sample plots in agroforests within four Peasant Associations (lowest administrative unit) were surveyed.	A total of 24 tree species belonging to 24 genera and 19 families were identified. Only tree/shrub species with a DBH ≥15 cm were considered for the inventory.	(Seta & Demissew, 2014)
4	Enset - Coffee based home garden AF system	Dilla Zuria district of Gedeo Zone, Southern Ethiopia. A total of 120 smallholdings between 250 m2 to 2000 m2 were surveyed.	In total, 75 different plant species including trees (44%) and shrubs (14,7%) were identified. The identified species were used as medicinal purposes, building material, fuel, household articles and ornamental subjects, as well as for shading and as a living fence.	(Mengitu & Fitamo, 2015)
5, 4, 1	Coffee - Fruit Tree - Enset based AF system; Coffee - Enset based AF system; Enset based AF system	Dilla Zuria district of Gedeo Zone. A total of 60 smallholdings with sizes around 100 m2 representing three AF practices were surveyed.	A total of 51 perennial woody plant species belonging to 30 families were identified, of which 31 were found in the surveyed 100 m2 smallholdings. Of these identified species, 32 (63%) were native, which two species registered as endemic. According to IUCN Red List and local criteria, 13 species were recorded as being of interest for conservation in all AF systems.	(Tesfay et al., 2024)
6, 3, 1	Coffee - Fruit Tree - Crop AF system; Enset - Coffee - Tree AF system; Enset - Tree AF system	Gedeo Zone. The survey was conducted among 9 Peasant Associations in 4 districts within the Gedeo Zone. Species inventories were conducted in a total of 108 smallholdings	A total of 234 plant species belonging to 82 families were identified. Of these, 41% were woody species; 69,2% were native species, the remaining 30,8% were non-native species.	(Mebrate et al., 2021)
7	Coffee based AF system	Yirgacheffe district, Gedeo Zone. A species inventory was conducted in 400 m2 quadrats of 38 plots	A total of 32 woody plant species representing 23 families were identified. Of the total woody species recorded, 39%, 11%, and 50% were trees, seedlings, and saplings and shrubs (including coffee plants), respectively.	(Tesfay et al., 2022b)

services, including carbon sequestration, soil health, biodiversity conservation, and resilience to climate change.

Carbon sequestration in biomass plays a significant role in mitigating rising CO_2 levels in the atmosphere when carbon is deposited in longer-lived biomass such as the woody components of agroforests (Albrecht and Kandji 2003). A further benefit of these perennial systems is that the wood component can continue to lock carbon after it is harvested (Roy 1999). This is because the trunks, stems, and branches retain their carbon content when they are turned into any kind of long-lasting product. Previous studies conducted on the Gedeo agroforests have shown that the contribution of the system to carbon sequestration in the biomass and soil is remarkable (Seta and Demissew 2014; Negash and Starr 2015; Tesfay et al. 2022a). A study conducted in the Dilla Zuria district of the Gedeo Zone showed that there is a significant biomass reserve and biomass carbon storage contribution from three indigenous AF systems (enset-based, coffee-enset-based, and coffee-fruit tree-enset-based AF) (Tesfay et al. 2022a). Mean aboveground biomass (AGB) ranging from 81.1 tons per hectare (t ha^{-1}) in enset-based AF to 255.9 t ha^{-1} in coffee-fruit tree-enset-based AF and mean belowground biomass (BGB) ranging from 26.9 t ha^{-1} in enset-based AF to 72.2 t ha^{-1} in coffee-fruit tree-enset-based AF was reported (Tesfay et al. 2022a). The total calculated biomass values in the three studied AF systems ranged from 328.1 t ha^{-1} to 560 t ha^{-1}. Similarly, the corresponding biomass carbon stock of these systems was calculated at an average value of 70 t ha^{-1} (Tesfay et al. 2022a). Other studies such as Negash (2013) and Negash and Starr (2015) have reported the average total biomass (above- and belowground, including herbaceous plants and litter) and biomass carbon stock of three AF systems in Gedeo (enset AF, enset-coffee AF, and fruit-coffee AF) as 143.7 t ha^{-1} and 67.1 t ha^{-1}, respectively. A study on the comparison of SOC stock between coffee-fruit tree-enset AF and its adjacent monocrop land showed that the former has higher SOC stock (125.5 t ha^{-1}) than the latter (95.5 t ha^{-1}) (Tesfay et al. 2022a). Due to the multilayer structure and high composition of the Gedeo agroforestry systems, the biomass, biomass carbon stock, and soil organic carbon stock are relatively higher.

Other research has calculated the total biomass carbon stock and aboveground biomass carbon stock for specific elevation and diameter at breast height (DBH) ranges. A study by Tesfay et al. (2022b) showed that the total carbon stock in coffee-based AF systems in the Yirgacheffe district of the Gedeo Zone ranged from 123.6 t ha^{-1} at 2240 meters above sea level to 97.75 t ha^{-1} at 2140 meters above sea level. Of this total amount, 20.93% of the carbon stock was contributed by biomass. From the above values, it is understood that the effect of elevation on total carbon stock is high. Seta and Demissew (2014) reported that the aboveground biomass values of AF trees in the Wenago district of the Gedeo Zone are affected by the DBH. For instance, trees that have a larger diameter class stored a large stock of AGB, whereas a small amount of AGB has been stocked by a small diameter class.

The total biomass values reported by Tesfay et al. (2022a) are greater than the global average values reported for forests and some other tropical AF practices at an average of 149 t ha^{-1}. The higher biomass values in Gedeo agroforestry might be due to good silvicultural management practice that has been employed by the farmers. The biomass carbon values reported by Tesfay et al. (2022a) and Negash and Starr (2015) in Gedeo agroforests are substantially higher than those reported for AF systems in sub-Saharan Africa (Unruh et al. 1993), in humid tropical Africa (Albrecht and Kandji 2003; Henry et al. 2009), and elsewhere in the tropics (Pandey 2002; Mutuo et al. 2005; Verchot et al. 2007; Soto-Pinto et al. 2010; Brakas and Aune 2011; Häger 2012; Schmitt-Harsh et al. 2012).

In summary, storing substantial amounts of biomass and biomass carbon stock in the Gedeo agroforests could be taken as one indicator of their contribution to enhancing the landscape connectivity. It further might contribute to climate change mitigation by sequestering large amounts of carbon in both their biomass and soil.

Local Livelihood Support Through Agroforestry

Agroforests in the Gedeo area and other well-managed AF practices provide multiple benefits including improvement of socioeconomic conditions, support for rural livelihoods, and strengthening of ecosystems within land-use systems (Kalaba et al. 2010). The diverse products provided by coffee-fruit tree-enset-based, coffee-enset-based, enset-based, coffee-fruit tree-crop, and coffee-based AF systems play a significant role in sustaining and improving the livelihoods of local communities in the Gedeo Zone. Engaging in agroforestry practices contributes to smallholder home income in the form of cash as well as extra food (Adane et al. 2019).

The AF practice in the Gedeo Zone is not simply a supplement to livelihood but in fact the mainstay of many communities. As reported by Negash (2007), Mebrate (2007), and Bishaw et al. (2013), the Gedeo Zone is one of Ethiopia's most densely populated with an average population density of 627 people per km^2 and a range of 122 to 1300 people per km^2. Although the population size is high, there is relatively less economic challenge. This is primarily because the agroforests in the area offer high productivity that helps the community maintain food security. The AF system is economically more viable than other land-use systems because high-value cash crops and staple crops are integrated and utilized efficiently (Abebe 2005). A study conducted in the Yigachefe district of the Gedeo Zone documented the high economic performance of the coffee-enset-based Gedeo AF system as compared to parkland AF (Ayele et al. 2014). This is because coffee-enset-based agroforests provide diverse products and services including construction materials, food for humans and animals, fuel, fiber, and shade. However, parkland AF, due to its limited composition of plant species and less stratified structure, provides less diverse products.

Kufa et al. (2011) revealed that AF systems practiced by smallholding farmers like those in the Gedeo Zone contribute more than 95% of the entire coffee bean production in Ethiopia. Similarly, Kanshie (2002) found that AF systems offer households a variety of benefits such as providing a source of income, access to traditional treatments for illnesses affecting both people and cattle, and employment within the Gedeo Zone. Furthermore, AF practices serve as a safety net and means of survival when natural disasters occur (Asfaw 2006).

Challenges to the Agroforestry Landscape's Connectivity

Pressure from the social, economic, technological, and demographic spheres has recently made home garden agroforestry more difficult (Habtamu and Zemede 2011). The Gedeo AF practices have been undermined by increasing population pressure in particular (Bishaw et al. 2013). The chapter shows that the Gedeo agroforests face several profound challenges to their sustainability. Among the most serious of these are described in Fig. 28.3.

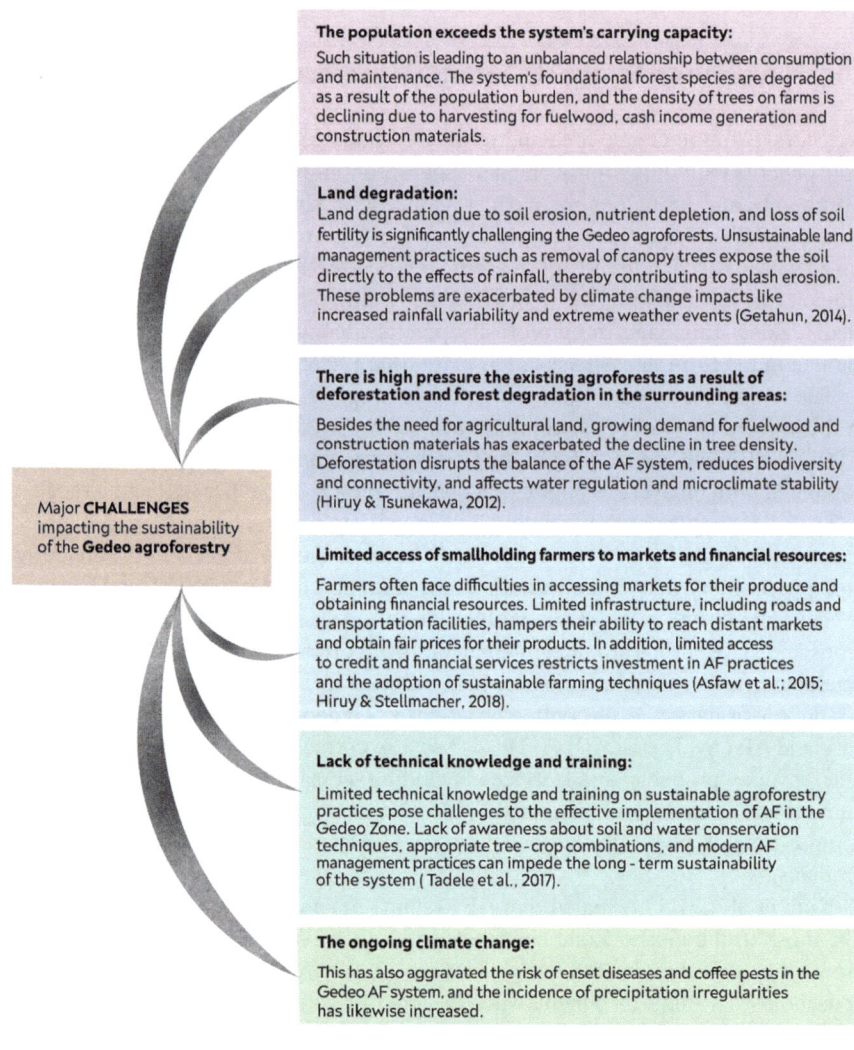

Fig. 28.3 Major challenges impacting the sustainability of the Gedeo agroforestry

References

Abebe T (2005) Diversity in Homegarden AF Systems of Southern Ethiopia. PhD thesis. Wageningen University, Wageningen. ISBN 90-8504-163-5; 143 pp

Adane F, Legesse A, Weldeamanuel T, Belay T (2019) The contribution of a fruit tree-based AF system for household income to smallholder farmers in Dale District, Sidama zone, Southern Ethiopia. Adv Plants Agric Res 9(1):78–84. https://doi.org/10.15406/apar.2019.09.00415

Albrecht A, Kandji ST (2003) Carbon sequestration in tropical AF systems. Agric Ecosyst Environ 99:15–27

Asfaw B (2006) Woody species composition and socio-economic roles of traditional AF practices across different agro-ecological zones in southeastern Langano, Oromiya. M.Sc. Thesis. Hawassa University, Wondo Genet. 39 p

Asfaw Z, Mulata Y, Assefa B, Abebe T, Duna SM, Tesfay HM, Kassa H (2015) Enhancing the role of forestry in building climate resilient green economy in Ethiopia: strategy for scalling up effective forest management practices in southern nations, nationalities and peoples regional state with particular an emphasis on AF. Center for International Forestry Research (CIFOR), Bogor. 66 p

Ayele Y, Ewunetu Z, Asfaw Z (2014) Economic evaluation of coffee-enset-based AF practice in Yirgachefe Woreda, Ethiopia: comparative analysis with parkland AF practice. J Econ Sustain Dev 5(27):2014

Bhagwat SA, Willis KJ, Birks HJB, Whittaker RJ (2008) Agroforestry: a refuge for tropical biodiversity? Trends Ecol Evol 23:261–267

Birhanu A, Getachew A, Adugna T (2013) *Millettia ferruginea*: an endemic legume tree as forage for ruminants in southern & northwestern Ethiopia. Livest Res Rural Dev 25(3)

Bishaw B, Neufeldt H, Mowo J, Abdelkadir A, Muriuki J, Dalle G, Luedeling E (2013) Farmers' strategies for adapting to and mitigating climate variability and change through AF in Ethiopia and Kenya. In: Davis CM, Bernart B, Dmitriev A (eds) Forestry communications group. Oregon State University, Corvallis

Brakas SG, Aune JB (2011) Biomass and carbon accumulation in land use Systems of Claveria, The Philippines. In: Kumar BM, Nair PKR (eds) Carbon sequestration potential of AF systems: *opportunities and challenges*, Advances in AF, vol 8. Springer, Dordrecht, pp 163–175

Daizy RB, Ravinder KK, Shibu J, Harminder PS (2008) Ecological basis of agroforestry. CRC Press Taylor & Francis Group, New York, p 383

Das T, Das AK (2005) Inventorying plant biodiversity in homegardens: a case study in Barak Valley, Assam, Northeast India. Curr Sci 89(1):155–163

Dawson IK, Guariguata MR, Loo J, Weber JC, Lengkeek A, Bush D et al (2013) What is the relevance of smallholders' AF systems for conserving tropical tree species and genetic diversity in circa situm, in situ and ex situ settings? A review. Biodivers Conserv 22(1):301–324. https://doi.org/10.1007/s10531-012-0429-5

Degefa S (2016) Home garden agroforestry practices in the Gedeo zone, Ethiopia: a sustainable land management system for socio-ecological benefits. In: Socio-ecological production landscapes and seascapes (SEPLS) in Africa, vol 28. United Nations University Institute for the Advanced Study of Sustainability, Tokyo

Doerr VAJ, Doerr ED, Davies MJ (2010) Does structural connectivity facilitate dispersal of native species in Australia's fragmented terrestrial landscapes? Systematic review no. 44. Collaboration for Environmental Evidence, CSIRO, Canberra

Gebrehiwot B, Maryo M (2015) Evaluation of land use patterns across agro-ecological and slope classes using GIS and remote sensing: the case of Gedeo zone, Southern Ethiopia. Int J Adv Remote sens 4(1):1385–1399

Getahun B (2014) Land use change, soil erosion and sustainability in the Gedeo agricultural and AF system, Southern Ethiopia. Environ Manag Sustain Dev 3(1):82–91

Gilbert-Norton L, Wilson R, Stevens JR, Beard KH (2010) A meta-analytic review of corridor effectiveness. Conserv Biol 24:660–668

Gillespie AR, Miller BK, Johnson KD (1995) Effects of ground cover on tree survival and growth in filter strips of the Cornbelt region of the midwestern US. Agric Ecosyst Environ 53(3):263–270. https://doi.org/10.1016/0167-8809(94)00577-2

Habtamu H, Zemede A (2011) Home gardens and agrobiodiversity conservation in Sabata town, Oromia regional state, Ethiopia. SINET: Ethiopian J Sci 34(1):1–16

Häger A (2012) The effects of management and plant diversity on carbon storage in coffee AF systems in Costa Rica. Agrofor Syst 86:159–174. https://doi.org/10.1007/s10457-012-9545-1

Haggar J, Pons D, Saenz L, Vides M (2019) Contribution of AF systems to sustaining biodiversity in fragmented forest landscapes. Agric Ecosyst Environ 283:106567. https://doi.org/10.1016/j.agee.2019.06.006

Harvey CA, Gonzalez-Villalobos JA (2007) AF systems conserve species-rich but modified assemblages of tropical birds and bats. Biodivers 16(8):2257–2292. https://doi.org/10.1007/s10531-007-9194-2

Henry M, Tittonell P, Manlay RJ, Bernoux M, Albrecht A, Vanlauwe B (2009) Biodiversity, carbon stocks and sequestration potential in aboveground biomass in smallholder farming systems of western Kenya. Agric Ecosyst Environ 129:238–252. https://doi.org/10.1016/j.agee.2008.09.006

Hiruy N, Stellmacher T (2018) The role of cooperatives in smallholder farmers' market participation and bargaining power: evidence from Ethiopia. J Rural Stud 59:11–20

Hiruy N, Tsunekawa A (2012) Spatio-temporal analysis of forest cover changes in the Gedeo area, Ethiopia: integration of remote sensing with socio-economic data. J For Res 17(2):133–141

ICRAF (2000) Paths to prosperity through AF. ICRAF's corporate strategy, 2001–2010. International Centre for Research in AF, Nairobi

Jose S (2009) Agroforestry for ecosystem services and environmental benefits: an overview. Agrofor Syst 76(1):1–10. https://doi.org/10.1007/s10457-009-9229-7

Jose S (2012) AF for conserving and enhancing biodiversity. Agrofor Syst 85(1):1–8. https://doi.org/10.1007/s10457-012-9517-5

Kalaba KF, Chirwa P, Syampungani S, Ajayi CO (2010) Contribution of AF to biodiversity and livelihoods improvement in rural communities of southern African regions. Environ Sci Eng:461–476. https://doi.org/10.1007/978-3-642-00493-3

Kanshie TK (2002) Five thousand years of sustainability? A case study on Gedeo land use (Southern Ethiopia). PhD dissertation. Wageningen University, Wageningen. 295 p

Kindt R (2002) Methodology for tree species diversification planning in African agroecosystems. PhD thesis. University of Gent, Gent

Kirschenmann FL (2007) Potential for a new generation of biodiversity in agroecosystems of the future. Agron J 99:373–376

Kufa T, Ayano A, Yilma A, Kumela T, Tefera W (2011) The contribution of coffee research for coffee seed development in Ethiopia. J Agric Res Dev 1(1):009–016

Leakey R (1996) Definition of agroforestry revisited. AF today 8:5–5

McNeely JA, Schroth G (2006) Agroforestry and biodiversity conservation—traditional practices, present dynamics, and lessons for the future. Biodivers 15(2):549–554. https://doi.org/10.1007/s10531-005-2087-3

Mebrate BT (2007) Agroforestry practices in Gedeo zone, Ethiopia: a geographical analysis. Ph.D. dissertation. Panjab University, Chandigarh, p 188

Mebrate A, Kippie T, Zeray N (2021) Determinant factor of plant species diversity in the organic agriculture-dominated system of Gedeo zone, Southern Ethiopia. Int J Ecol 2021:1–12

Mengitu M, Fitamo D (2015) Plant species diversity and composition of the Homegardens in Dilla Zuriya Woreda, Gedeo zone, SNNPRS, Ethiopia. Plant 3(6):80–86. https://doi.org/10.11648/j.plant.20150306.14

Mutuo PK, Cadisch G, Albrecht A, Palm CA, Verchot L (2005) Potential of AF for carbon sequestration and mitigation of greenhouse gas emissions from soils in the tropics. Nutr Cycl Agroecosyst 71:43–54. https://doi.org/10.1007/s10705-004-5285-6

Nair PKR (1993) An introduction to AF. Kluwer Academic Publishers, Dordrecht

Negash M (2007) Tree management and livelihoods in Gedeo's agroforests, Ethiopia. Forest Trees Livelihoods 170(2):157–168

Negash M (2013) The indigenous AF systems of the south-eastern Rift Valley escarpment, Ethiopia: their biodiversity, carbon stocks, and litterfall. Viikki tropical resources institute (VITRI). Ph.D. dissertation. University of Helsinki, Helsinki, p 62

Negash M, Achalu N (2008) History of indigenous agroforestry in Gedeo, Southern Ethiopia, based on local community interviews: vegetation diversity and structure in the landuse systems. Ethiop J Nat Resour 10(1):31–52

Negash M, Starr M (2015) Biomass and soil carbon stocks of indigenous agroforestry systems on the south-eastern Rift Valley escarpment, Ethiopia. Plant Soil 393:95–107

Negash M, Yirdaw E, Luukkanen O (2012) Potential of indigenous multistrata agroforests for maintaining native floristic diversity in the south-eastern Rift Valley escarpment, Ethiopia. Agrofor Syst 85:9–28

Pandey DN (2002) Carbon sequestration in agroforestry systems. Clim Pol 2:367–377

Roy C (1999) Technical and socio-economical options for decreasing CO_2 emissions and developing carbon fixation. Reports-academie d'agriculture de france 85(6):311–320

Schmitt-Harsh M, Evans TP, Castellanos E, Randolph JC (2012) Carbon stocks in coffee agroforests and mixed dry tropical forests in the western highlands of Guatemala. Agrofor Syst 86:141–157. https://doi.org/10.1007/s10457-012-9549-x

Seta T, Demissew S (2014) Diversity and standing carbon stocks of native AF trees in Wenago district, Ethiopia. JETEAS 5(7):125–132

Soto-Pinto L, Anzueto M, Mendoza J, Ferrer GJ, de Jong B (2010) Carbon sequestration through AF in indigenous communities of Chiapas, Mexico. *Agrofor Syst* 78:39–51. https://doi.org/10.1007/s10457-009-9247-5

Steffan-Dewenter I, Kessler M, Barkmann J, Bos MM, Buchori D, Erasmi S, Faust H, Gerold G, Glenk K, Gradstein SR, Guhardja E, Harteveld M, Hertel D, Hohn P, Kappas M, Kohler S, Leuschner C, Maertens M, Marggraf R, Migge-Kleian S, Mogea J, Pitopang R, Schaefer M, Schwarze S, Sporn SG, Steingrebe A, Tjitrosoedirdjo SS, Tjitrosoemito S, Twele A, Weber R, Woltmann L, Zeller M, Tscharntke T (2007) Tradeoffs between income, biodiversity, and ecosystem functioning during tropical rainforest conversion and AF intensification. Proc Natl Acad Sci USA 104:4973–4978

Tadele G et al (2017) Indigenous knowledge of soil fertility management in Gedeo AF system, Southern Ethiopia. Environ Dev Sustain 19(3):1105–1124

Tefera Y, Abebe W, Teferi B (2016) Woody plants species diversity of home garden agroforestry in three agroecological zones of Dilla Zuria District, Gedeo zone, Southern Ethiopia. Int J Fauna Biol Stud 3(3):98–106

Tesfay HM, Negash M, Godbold DL, Hager H (2022a) Assessing carbon pools of three indigenous AF Systems in the Southeastern Rift-Valley Landscapes, Ethiopia. Sustain For 14:4716. https://doi.org/10.3390/su14084716

Tesfay F, Moges Y, Asfaw Z (2022b) Woody species composition, structure, and carbon stock of coffee-based AF system along an elevation gradient in the moist mid-highlands of Southern Ethiopia. Int J Forestry Res

Tesfay HM, Laphin K, Oettel J, Negash M (2024) Plant diversity and conservation role of three indigenous AF systems of south-eastern rift- valley landscapes, Ethiopia. Diversity 16:64. https://doi.org/10.3390/d16010064

Tewabech TB, Efrem A (2014) The Flora makeup and agroforestry practices in backyard in Hiwane, Hintalo Wejerat of Tigray, northern Ethiopia. Int J Agrofor Silvic 1(9):101–109

Unruh JD, Houghton RA, Lefebvre PA (1993) Carbon storage in agroforestry: an estimate for sub-Saharan Africa. Clim Res 3:39–52

Verchot LV, Noordwijk MV, Kandji S, Tomich T, Ong C, Albrecht A, Mackensen J, Bantilan C, Anupama KV, Palm C (2007) Climate change: linking adaptation and mitigation through agroforestry. Mitig Adapt Strat Glob Change 12:901–918. https://doi.org/10.1007/s11027-007-9105-6

Open Access This chapter is licensed under the terms of the Creative Commons Attribution 4.0 International License (http://creativecommons.org/licenses/by/4.0/), which permits use, sharing, adaptation, distribution and reproduction in any medium or format, as long as you give appropriate credit to the original author(s) and the source, provide a link to the Creative Commons license and indicate if changes were made.

The images or other third party material in this chapter are included in the chapter's Creative Commons license, unless indicated otherwise in a credit line to the material. If material is not included in the chapter's Creative Commons license and your intended use is not permitted by statutory regulation or exceeds the permitted use, you will need to obtain permission directly from the copyright holder.

29. Hungary and Austria: Best Practice for Habitat and Species Connectivity: European Beech and Sessile Oak

Marcela van Loo, Erik Szamosvári, Anita Bálint,
Anikó Neuvirthné Bilics, Heino Konrad, and László Nagy

Quercus petraea (Photos: Ruckszio/Adobe Stock)

M. van Loo (✉) · E. Szamosvári
Department of Forest Growth, Silviculture and Genetics, Austrian Research Centre for Forests, Vienna, Austria
e-mail: marcela.vanloo@bfw.gv.at; erik.szamosvari@bfw.gv.at

A. Bálint · A. N. Bilics
VVÖH (Vas Vármegyei Önkormányzati Hivatal – Vas County Government Office), Szombathely, Hungary
e-mail: balint.anita@vasmegye.hu; neuvirthne.aniko@vasmegye.hu

H. Konrad
Department Forest Biodiversity & Nature Conservation, Austrian Research Centre for Forests, Vienna, Austria
e-mail: heino.konrad@bfw.gv.at

L. Nagy
Forest Research Institute – University of Sopron (SOE ERTI), Sárvár, Hungary
e-mail: nagy.laszlo@uni-sopron.hu

© The Author(s) 2025
K. Lapin et al. (eds.), *Ecological Connectivity of Forest Ecosystems*,
https://doi.org/10.1007/978-3-031-82206-3_29

Abstract

In the Austrian–Hungarian border region, scientists have initiated a bilateral collaboration to address habitat and species connectivity with respect to climate change using the management and conservation approach of assisted migration. They applied this approach to European beech and sessile oak with support from local managers and governments as part of a project. The implementation process began by employing modeling studies to assess future species distribution and vulnerability, as well as to identify suitable forest reproductive material (FRM) and appropriate locations for conservation and experimental plots. The FRM selection considered both the "local" and "future climate-adapted" provenances. The implementation process culminated in reforestation efforts, where these provenances were planted at six locations evenly distributed across Austria and Hungary. Upcoming regular inspections, silvicultural measures, and phenological observations over a 15-year trial period will validate the outcomes of the applied assisted migration strategy. Monitoring efforts will evaluate damage, growth, quality characteristics, and mortality rates of different seed sources, as well as their genetic diversity. Forest managers and research institutions share the responsibilities for management and monitoring. This project aims to enhance the resilience of European beech and sessile oak forests in the Austrian–Hungarian border region against climate change impacts. Ongoing monitoring and evaluation of the established plots will provide valuable insights into the successes and challenges of the assisted migration approach.

Keywords

Climate change · Assisted migration · *Quercus petraea* · *Fagus sylvatica* · Conservation · Experimental plots

Info Box 29.1 Terminology Clarifications for This Chapter
Occurrence: Presence of a particular species within a specific environment, habitat, or geographical location.

Vulnerability: Exposure to contingencies and stressors as well as the difficulty in coping with them. In the context of this chapter, vulnerability was estimated based on the projected decrease in the probability of species occurrence.

Provenance: A specific population or group of trees that originate from a particular geographic location.

Natural Forest: A forest with natural species composition, including foundation species, matching the climatic zone, and possessing structure and dynamics of an undisturbed forest, or resembling such a forest in a given region, site, and management system, regardless of whether it is formed naturally or is artificially created and maintained.

Background and Introduction

Forest regeneration, whether natural or artificial, relies on the use of forest reproductive material (FRM) such as seeds and young plants. Natural regeneration utilizes existing materials in a specific location, while artificial regeneration often involves transferring FRM. Seeds and seedlings have been traded in Europe for centuries, and they continue to be transferred in large quantities today as forest managers aim to minimize regeneration risks and costs. Many European countries provide recommendations, guidelines, and/or tools for selecting tree species or provenances suitable for specific sites or regions that increasingly also take future climatic conditions into consideration. This means that plants should be planted in locations where they are likely to thrive in the future as well, which is particularly important for both habitats and species heavily impacted or threatened by climate change. To determine which forest species are suitable for which habitats and vice versa, decisions must be based on scientific evidence including future climate data, vulnerability maps, species distribution modeling, and local site conditions. Social, political, economic, and ecological factors should also be considered.

In 2020, scientists from Austria (AT) and Hungary (HU) initiated a bilateral collaboration within their border region to tackle habitat and species connectivity issues in the face of climate change, employing the conservation strategy of assisted migration. Assisted migration refers to the human-facilitated relocation of species, populations, or individuals affected by climate change to new, suitable habitats, either within or beyond their current range (Hällfors et al. 2014) (see also Chap. 14). This approach is often considered when species or provenances are unlikely to migrate naturally to these habitats due to human-induced barriers or insufficient time for adaptation amidst rapidly changing climate conditions.

To mitigate the negative effects of climate change on biological diversity, both the enhancement of ecological connectivity and the application of assisted migration are often proposed. Increasing connectivity can, in many instances, achieve the same goals as assisted migration, and thus both efforts should not be dichotomized. Although conservation professionals typically avoid relocating species, especially beyond their native ranges, some have advocated for the assisted migration of species with limited dispersal capabilities, such as trees (Hoegh-Guldberg et al. 2008; Krosby et al. 2010).

Assisted migration is largely implemented in experimental settings for research purposes or intentionally during reforestation and afforestation efforts (Twardek et al. 2023). It is also viewed as a potential approach and solution to maintaining existing forest cover and safeguarding forest biodiversity and economic benefits for the future (Sousa-Silva et al. 2018). Nevertheless, the impacts of climate change on the provision of forest ecosystem services vary across spatial scales. Implementation of assisted migration may enhance the provision of forest ecosystem services in certain areas such as the Alpine and Boreal regions, while in other areas like the Mediterranean, it may help to reduce their decline (Spinoni et al. 2021).

Austria and Hungary are both significantly forested countries in Central Europe, with approximately half of Austria (ÖWI 2016–2021) and one-fifth of Hungary

(Nemzeti Földügyi Központ 2022) covered by forests. In the border region they share, deciduous forests are dominated by European beech (*Fagus sylvatica* L.) and various oak species (*Quercus* spp.). For example, in Burgenland, one of Austria's federal states bordering Hungary, these tree species account for more than a quarter of the total forest area (ÖWI 2016–2021). Given their limited natural migration capacity and local adaptation to the warming climate, both European beech and sessile oak (*Quercus petraea* (Matt.) Liebl.) are generally projected to experience a significant loss of suitable habitats and reduced distribution across Central Europe in the future (Dyderski et al. 2018; Illés and Móricz 2022).

To enhance the resilience of native European beech and sessile oak forests in the Austrian–Hungarian border region against the effects of climate change, a science-driven project, named REIN-Forest (derived from the English: *reinforcement*), targeting these two tree species, was launched by scientists from both countries, with support from local managers and government offices. The climatic conditions in the region are similar, with an maximum temperatures of 14.4–15.9° C, and an minimum temperatures of 4.7–5.8° C, along with an average annual precipitation ranging between 520 and 830 mm, except in the region of Eastern Styria (where it can reach up to 1000 mm). Downscaled projections for the RCP8.5 climate change scenario indicate significant anticipated changes for the period 2071–2100, with an estimated rise in mean annual temperature of 2–5° C and an annual precipitation ranging from −7% to 18% (European Environment Agency 2020).

A modeling study determined the future distribution and vulnerability of both species in the region. Based on these findings, a science-based local strategy accompanied by recommendations for the transfer, selection, and relocation of FRM within the area was developed. Finally, suitable locations were identified for both species, and six of these locations were chosen to establish new conservation and experimental plots. These plots now serve as sites for assessment and monitoring, enabling the validation of assisted migration in practice.

Future Distribution and Vulnerability of European Beech and Sessile Oak Forests in the Austrian–Hungarian Border Region

The species distribution models and modeling approaches of Chakraborty et al. (2021) provided the basis for modeling the future distribution, probability of occurrence, and vulnerability of European beech and sessile oak in the border area across eight Austrian administration units (Northern, Central, and Southern Burgenland, Vienna, Vienna Environs–South, Lower Austria–South, Graz, and Eastern Styria) and three Hungarian counties (Győr-Moson-Sopron, Vas, and Zala). The results of this modeling guided the development of a regional seed transfer strategy, the selection of conservation and experimental sites, and the choice of suitable FRM.

Specifically, a recent European forest cover map from the Copernicus Land Monitoring Service (COPERNICUS LMS 2018) was utilized with a focus specifically on the border area. Minor tree groups, amenity plantings, and forest strips were excluded by selecting only pixels with more than 50% tree cover density. Later, only 1 km grids with more than 75% forest coverage were considered closed forest stands for the local application of Species Distribution Models (SDMs). The distribution of forest areas based on probability classes (in 10% intervals) for both species was computed for different climate scenarios and timeframes (2041–2060, 2061–2080, and 2081–2100). Vulnerability maps were developed using the occurrence maps for the RCP 8.5 scenario.

The outcomes of modeling indicated significant shifts in the species composition of natural forests as projected by the species distribution models (Fig. 29.1). Both European beech and sessile oak are expected to experience habitat losses, or at least demographic declines, across most of the modeled area by the end of the century. The vulnerability of low-elevation *European beech* occurrences is expected to be moderate in the short term (2041–2060) and severe by the end of the century (2081–2100). The vulnerability model indicates a continuous increase in exposure to adverse climatic conditions, especially on the Hungarian side and in the Austrian border areas in Burgenland and southern Styria. At higher altitudes in the Eastern Alps, however, the beech is likely to maintain its dominance in forest stands, with only minor losses projected. The *sessile oak* is expected to diminish and potentially disappear near its lower xeric limit or it may only occur as a mixture species in thermophilic formations. The model predicts a significant decrease in the probability of occurrence, suggesting noticeable compositional changes even in the core areas currently dominated by the species. Vulnerability will also be high in lowland

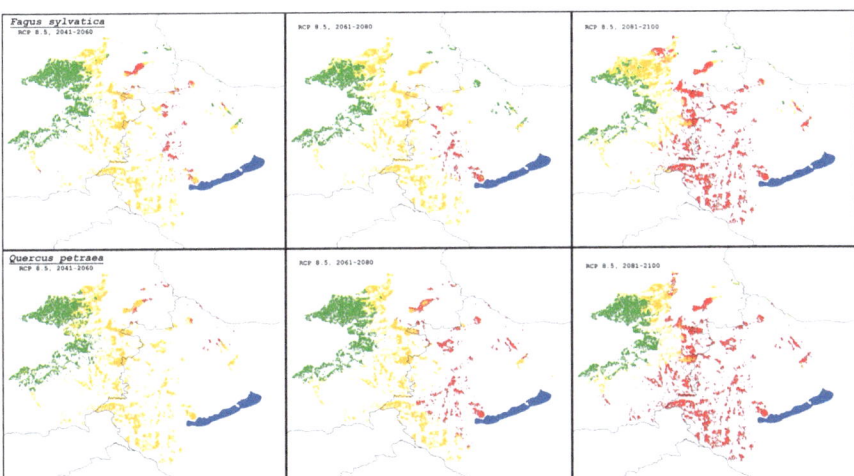

Fig. 29.1 Vulnerability of European beech (*Fagus sylvatica*, top) and sessile oak (*Quercus petraea*, bottom) in the AT–HU border region under RCP 8.5 scenario (L. Nagy & N. Móricz; SOE) for three periods (2041–2060, 2061–2080, 2081–2100). Green areas represent non-vulnerable regions with a relative decrease in the probability of occurrence of less than 15%; orange areas indicate moderately vulnerable regions with a projected decrease between 15% and 50%; red areas represent severely or highly vulnerable regions where the relative decrease in probability of occurrence exceeds 50% by the end of the twenty-first century

and colline oak stands, while there is a possibility of range extension in subalpine and alpine sites in Austria, where the sessile oak could potentially benefit from local beech habitat losses.

Establishment of Conservation and Experimental Plots

To test assisted migration in practice, a total of six conservation and experimental plots, also referred to as demonstration sites, were established. Three of these sites were designated in Austria, while the remaining three were chosen in Hungary. The selection process for these sites commenced with a bilateral open call, which was advertised in the spring of 2021, aimed at involving forest owners and managers in identifying and offering suitable areas for the demonstration sites. To aid in the selection process, various tools and resources were utilized. Model-based information, vulnerability maps, decision support systems such as App *SusSelect*, the forest vulnerability and seed transfer tool available at www.seed4forest.org, as well as climate-smart forests found at www.klimafitterwald.at/baumarten/ were employed.

These support systems helped to identify the "appropriate" provenances of both species that were best suited and adapted to future climate conditions, particularly considering the more pessimistic RCP 8.5 climate scenario. In addition to these "future climate-adapted (adapted)" provenances, the scientists from both countries also procured "local" FRM. This "local" FRM was obtained mainly from nearby certified seed sources (seed stands) considered most appropriate for present climate conditions at the selected sites.

However, challenges were encountered in acquiring the selected FRM, particularly regarding the "adapted" material: some of the desired provenances were unavailable as the respective seed stands had not been harvested during that particular year. As a result, alternative FRM from other "adapted" provenances was sought. Further seed stands were identified as potential matches based on the recommendations of the support tools, and additional FRM from the moderate climate change scenario RCP 4.5 was utilized as well.

In Austria, FRM from a total of 21 seed stands including eight "local" sources and 13 "adapted" sources was purchased, while for Hungary, FRM from two "local" and two "adapted" seed stands was obtained (Fig. 29.2). The reforestation efforts involved planting site- and species-specific combinations of "local" and "adapted" provenances on the six selected demonstration sites in the autumn of 2022. One site in each country was planted with "local" and "adapted" sessile oak provenances, another with "local" and "adapted" European beech provenances, and the third site with a mixture of "local" European beech and "adapted" sessile oak provenances (Fig. 29.3). The latter combination was chosen to account for the predicted loss of beech dominance in the area.

There were certain country-specific differences with regard to site design. These differences, largely reflected in varying plot sizes, spacing, and number of trees on site, persist due to deeply ingrained forestry traditions, experiences, and management requirements as well as expectations for long-term comparisons of

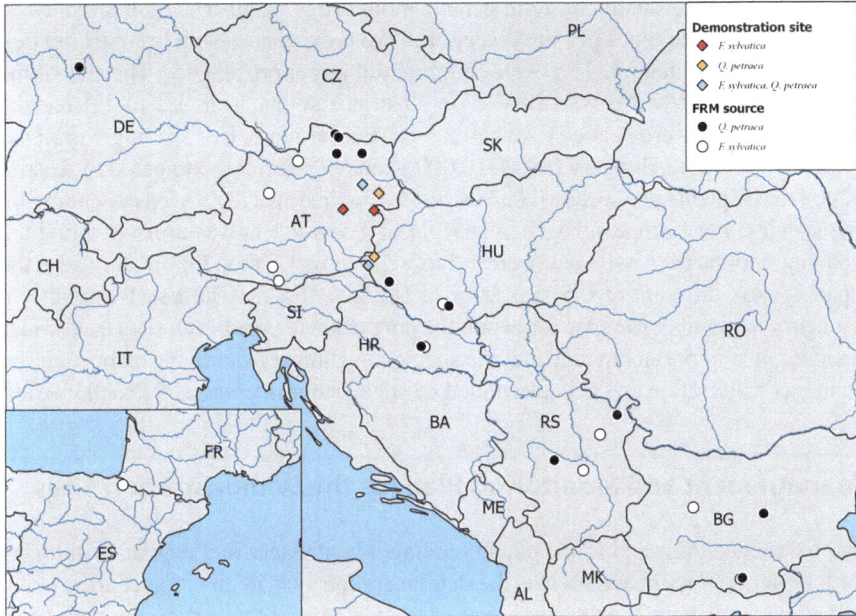

Fig. 29.2 Selected certified seed stands of European beech and sessile oak used as "local" and "future climate-adapted" FRM in the establishment of the six demonstration sites in Austria and Hungary (QGIS Version 3.4.13-Madeira)

Fig. 29.3 (**a**) European beech ("local" and "future climate-adapted" FRM) planted at Austrian demonstration site Reichenau (winter 2022). (**b**) Sessile oak ("future climate-adapted" FRM) planted at Austrian demonstration site Sparbach (spring 2023)

silvicultural aspects and long-term genetic monitoring. In Austria, each provenance was repeated three times per site (except for two provenances with low germination rates), while in Hungary, there were four repetitions per provenance. The size of the demonstration sites was also smaller in Austria, ranging from 1.2 to 1.3 ha; the Hungarian sites were between 1.2 and 2.5 ha. Furthermore, tree spacing within the sites was denser in Hungary (8000–10,000 plants per hectare) compared to Austria (2000–3500 plants per hectare). In Austria, the applied spacing varied depending on the species: For European beech, a spacing of 1.5 m × 1 m was utilized, while the spacing options for sessile oak were 1.5 m × 2.5 m or 1.5 m × 2 m. In Hungary, the spacing was different at 0.5 m × 2 m or 0.7 m × 1.5 m. The beech-oak site in Hungary featured dense spacing, while the pure oak and pure beech sites used wider spacing. It is important to note that the spacing in Hungary was at the respective site manager's discretion and not determined by specific considerations for each species.

Management and Monitoring Plan for the Demonstration Sites

In provenance trials and forest plant breeding, maintenance and regular monitoring of long-term experiments such as the demonstration sites in this project are crucial. Effective management and ongoing maintenance ensure the success of such trials and enable timely responses to unexpected changes.

For the planned 15-year trial period, various activities in keeping with local forest management practices will be implemented on all six established demonstration sites. These activities include tree and seedling care and protection, regular checking of the condition of protective fences, and implementation of necessary silvicultural measures for forest protection. These measures are outlined in contracts between the owners of the demonstration sites and the research organizations responsible for establishing them. They will be carried out at least once a year as well as after extreme weather events. Any plant losses during the first year will be replaced by supplemental planting during the initial monitoring activity. Further monitoring activities throughout the 15-year period will include observing phenology, assessing damage caused by abiotic and biotic factors, and evaluating the quality and growth characteristics of the trees at specific intervals (Table 29.1) (Liesebach et al. 2017).

The primary focus of this monitoring is to compare and validate the mortality rates and performance of different seed sources on the established plots (Fig. 29.3), thereby evaluating the FRM transfer system and the assisted migration of climate-adapted provenances based on decision support systems. Genetic monitoring may also be considered to track changes in genetic variation and assess the impact of climate change and management practices on genetic diversity (Hansen et al. 2012; Aravanopoulos et al. 2015). Responsibility for the ongoing tasks is shared, with the forest managers handling the management aspect and the involved research institutions overseeing the monitoring.

Table 29.1 The monitoring plan for the first 15-year period after establishment (2022–2037)

Assessment	1st year	2nd or 3rd year	5th year	10th year	15th year	At a later date
Survival/failure of the seedlings	X	X	X	X	X	(X)
Juvenile growth, growth rate, increment	(X)	X	X	X	X	(X)
Phenology	(X)	(X)	(X)	(X)	(X)	(X)
Vitality	X	X	X	X	X	(X)
Qualitative traits	-	X	X	X	X	(X)
Genetic monitoring	(X)	(X)	(X)	(X)	(X)	(X)

X" indicates a mandatory assessment, "(X)" denotes a voluntary assessment, and "-" signifies no assessment

Funding and Acknowledgments The research described in this study was funded as a project entitled REIN-Forest (No. ATHU150, 01.10.2020–31.12.2022) conducted within the framework of the INTERREG V-A Austria–Hungary program. The project received financial assistance from the European Regional Development Fund. The authors would like to thank the forest owners and managers who actively participated in the implementation, as well as the colleagues who contributed to the completion of various activities and outputs associated with the project, especially L. Weißenbacher and N. Móricz.

References

Aravanopoulos FA, Tollefsrud MM, Graudal L, Koskela J, Kätzel R, Soto A, Nagy L, Pilipovicˇ A, Zhelev P, Božicˇ G, Bozzano M (2015) Development of genetic monitoring methods for genetic conservation units of forest trees in Europe. European Forest genetic resources Programme (EUFORGEN). Bioversity Int. xvi+55 p

Chakraborty D, Móricz N, Rasztovits E, Dobor L, Schüler S (2021) Provisioning forest and conservation science with high-resolution maps of potential distribution of major European tree species under climate change. Ann For Sci 78(2):1–18. https://doi.org/10.1007/s13595-021-01029-4

COPERNICUS LMS (2018) Tree cover density. Copernicus. Accessed on 04.02.2021. https://land.copernicus.eu/pan-european/high-resolution-layers/forests/tree-cover-density/status-maps/tree-cover-density

Dyderski MK, Paź-Dyderska S, Frelich LE, Jagodziński AM (2018) How much does climate change threaten European forest tree species distributions? Glob Chang Biol 24:1150–1163. https://doi.org/10.1111/gcb.13925

Environment Agency (2020) Observed annual mean temperature change from 1960 to 2019 (left panel) and projected 21st century change under different emissions scenarios (right panels) in Europe. Accessed on 06.01.2023. https://www.eea.europa.eu/data-andmaps/figures/trends-in-annual-temperature-across-1

Hällfors MH, Vaara EM, Hyvarinen M, Oksanen M, Schulman LE, Siipi H, Lehvavirta S (2014) Coming to terms with the concept of moving species threatened by climate change–a systematic review of the terminology and definitions. PLoS One 9(7):e102979

Hansen MM, Olivieri I, Waller DM, Nielsen EE, GeM Working group (2012) Monitoring adaptive genetic responses to environmental change. Mol Ecol 21:1311–1329. https://doi.org/10.1111/j.1365-294X.2011.05463.x

Hoegh-Guldberg O, Hughes L, McIntyre S, Lindenmayer DB, Parmesan C, Possingham HP, Thomas CD (2008) Assisted colonization and rapid climate change. Science 321:345–346

Illés G, Móricz N (2022) Climate envelope analyses suggests significant rearrangements in the distribution ranges of central European tree species. Ann For Sci 79:35. https://doi.org/10.1186/s13595-022-01154-8

Krosby M, Tewksbury J, Haddad NM, Hoekstra J (2010) Ecological connectivity for a changing climate. Conserv Biol 24:1686–1689. https://doi.org/10.1111/j.1523-1739.2010.01585.x

Liesebach M, Ahrenhövel W, Janßen A, Karopka M, Rau HM, Rose B, Schirmer R, Schneck D, Schneck V, Steiner W, Schüler S, Wolf H (2017) Planung, Anlage und Betreuung von Versuchsflächen der Forstpflanzenzüchtung: Handbuch für die Versuchsanstellung. Thünen Report 49. Johann Heinrich von Thünen-Institut, Braunschweig, p 78. https://doi.org/10.3220/rep1496222427000

Nemzeti Földügyi Központ (2022) Erdővagyon és erdőgazdálkodás Magyarországon 2021-ben [Forestry assets and forest management in Hungary, 2021]. Budapest, NFK Erdészeti Főosztály, p 10. Accessed on 06.01.2023. https://www.oee.hu/hirek/agazati-szakmai/megjelent-az-erdovagyon-es-erdogazdalkodas-magyarszagon-2021-ben-cimu-kiadvany

Österreichische Waldinventur (2016–2021). Accessed on 06.01.2023. www.waldinventur.at

Sousa-Silva R, Verbist B, Lomba Â, Valent P, Suškevičs M, Picard O, Hoogstra-Klein MA, Cosofret V-C, Bouriaud L, Ponette Q, Verheyen K, Muys B (2018) Adapting forest management to climate change in Europe: linking perceptions to adaptive responses. Forest Policy Econ 90(2018):22–30. ISSN 1389-9341

Spinoni J, Barbosa P, Bucchignani E, Cassano J, Cavazos T, Cescatti A, Christensen JH, Christensen OB, Coppola E, Evans JP et al (2021) Global exposure of population and land-use to meteorological droughts under different warming levels and SSPs: a CORDEX-based study. Int J Climatol 41:6825–6853

Twardek WM, Taylor JJ, Rytwinski T, Aitken SN, MacDonald AL, Van Bogaert R, Cooke SJ (2023) The application of assisted migration as a climate change adaptation tactic: an evidence map and synthesis. Biol Conserv 280:109932. https://doi.org/10.1016/j.biocon.2023.109932

Open Access This chapter is licensed under the terms of the Creative Commons Attribution 4.0 International License (http://creativecommons.org/licenses/by/4.0/), which permits use, sharing, adaptation, distribution and reproduction in any medium or format, as long as you give appropriate credit to the original author(s) and the source, provide a link to the Creative Commons license and indicate if changes were made.

The images or other third party material in this chapter are included in the chapter's Creative Commons license, unless indicated otherwise in a credit line to the material. If material is not included in the chapter's Creative Commons license and your intended use is not permitted by statutory regulation or exceeds the permitted use, you will need to obtain permission directly from the copyright holder.

India: Hotspot of Connectivity Research and Conservation in Central India

Trishna Dutta and Sandeep Sharma

Indian Tiger (Photo: Bhavya Joshi/Adobe Stock)

Abstract

The persistence of wildlife in human-dominated landscapes is a challenge. Fragmentation of natural ecosystems is one of the major threats to biodiversity conservation, especially in densely populated and developing regions. Connectivity among remaining natural habitats can mitigate the negative

---

T. Dutta (✉)
European Forest Institute, Bonn, Germany
e-mail: trishna.dutta@efi.int

S. Sharma
German Centre for Integrative Biodiversity Research (iDiv) Halle-Jena-Leipzig, Leipzig, Germany

© The Author(s) 2025
K. Lapin et al. (eds.), *Ecological Connectivity of Forest Ecosystems*,
https://doi.org/10.1007/978-3-031-82206-3_30

ecological and evolutionary effects of fragmentation to a great extent. Central India has been a hotspot of landscape connectivity research in South Asia. We reviewed published research on landscape connectivity in Central India to identify the prominent trends and patterns in the literature. The overarching pattern we found is that most connectivity research in the region has been conducted on terrestrial ecosystems, especially forests, and on single species—usually the tiger—by independent research groups applying singular methods. However, recent research is tending more toward multi-species approaches within collaborative research frameworks. There is a need to integrate multiple methods, social acceptance towards wildlife, and collective action to ensure the persistence of biodiversity in the Central Indian landscape.

Keywords

Landscape · Connectivity · Corridors · Central India · Forests

Introduction

Biodiversity conservation in developing economies is challenging due to the acute juxtaposition of people's aspirations for better living conditions and the requirements of disturbance-free natural habitats for conservation. This challenge is amplified in regions with high human population density who depend on natural resources in regions that also host species of high ecological and conservation value. With growing demands on the land, these natural habitats are increasingly becoming isolated from each other, making it difficult for wide-ranging species to disperse to new habitats, a necessity for their long-term persistence.

India exemplifies these challenges. It is the world's most populous country rapidly transforming due to economic growth and increased purchasing power. With only 2.4% of the world's land area, India harbors 7–8% of all recorded species, including nearly 49,000 species of plants and over 100,000 species of animals. The country's biological diversity includes a mosaic of natural and cultural habitats, and its economy and the livelihoods of millions of people are dependent on the conservation and sustainable use of these biological resources.

The "backbone" of conservation in India rests on the protected area (PA) network, which covers merely 5.28% of the total geographic area of the country (ENVIS 2022). The number of PAs in India has increased by 18.4% between 2020 (n = 574, total area = 146,665.6 km^2) and 2022 (n = 998, total area = 173,629.52 km^2). These protected areas are a proactive measure to secure remaining habitats for biodiversity, often under the umbrella of charismatic species representative of their habitats, such as tigers (*Panthera tigris*) and elephants (*Elephas maximus*) in peninsular India, snow leopards (*Panthera uncia*) in the Himalayas, and crocodilians and freshwater dolphins in rivers. Despite this, the protected area coverage alone is quite inadequate since many of the PAs are small and isolated from each other, rendering them insufficient for the long-term persistence of the wide-ranging species they are meant to conserve.

Establishing and maintaining connectivity between PAs can increase the resilience and persistence of species in numerous ways. Connected landscapes help to create functional landscapes and enabling effective metapopulation dynamics by facilitating or increasing the probability of colonization of new habitat patches as well as recolonization of habitat patched with locally extirpated populations. Gene flow resulting from successful movement and breeding between populations helps maintain genetic diversity and prevents the deleterious evolutionary consequences of genetic drift and bottlenecks. Finally, metapopulation dynamics and genetic connectivity help to maintain the adaptive potential of species, creating opportunities for range shifts induced by ongoing and future climate change.

Central India is a region rich in biological as well as human cultural diversity. It features human-dominated landscapes with multiple uses ranging from agriculture to livestock rearing, hosts several rural and urban settlements, and it is a focus of major infrastructural development deemed necessary for the economic well-being of the entire country. It has also been a hotspot of connectivity research and conservation in South Asia (Thatte et al. 2021). In the following section, we aim to chronicle the development of connectivity research in the region, assess the current state of the art, and make pertinent suggestions for the future of connectivity research and conservation as a reflection of published information and our own experience from working in this context over the last two decades. In doing so, we highlight the challenges, conflicts, competing interests, and success stories while addressing the future trajectory of landscape conservation in the region.

Description of the Case Study Area

Central India is located at the junction of the Semi-Arid and Deccan Peninsula biogeographic zones of India (Rodgers and Panwar 1988). At a finer scale, it is composed of the Central Indian highlands biotic province along with parts of the Eastern Highlands, Central Plateau, and Chhota Nagpur biotic provinces. Due to a lack of prominent natural features, the Central Indian region has been defined and customized according to the specific questions addressed by individual researchers. Often based on a buffer around biogeographic zones or agroecological zones, conservation priorities, and political boundaries, there is much variation in the total territory delineated as Central India, ranging from 45,000 km^2 (Dutta et al. 2013a) to 729,000 km^2 (Nayak et al. 2020).

Central India represents a unique crossroad where the paths of humans and wildlife have intersected for thousands of years. Recent paleoanthropological research has revealed that *Homo sapiens* have existed in the area for the last 80,000 years (Clarkson et al. 2020). One of the most outstanding examples of this is the Bhimbetka rock shelter of Central India, a UNESCO World Heritage site that was continuously occupied by humans from around 100,000 BCE to 1000 CE. The life experiences of its inhabitants are visible in intricate paintings on the cave walls depicting interactions between people, landscapes, and wildlife since prehistoric times.

The mountain ranges of Central India, specifically the Satpura range, has been postulated as a conduit for migrations of Malayan faunal elements from Northeast India to the Western Ghats, underlining the historical importance of this region in structuring the zoogeography of South Asia. Today, the region features a rich assemblage of floral and faunal species. The major forest types found in Central India are dry teak forest, moist peninsular Sal (*Shorea robusta*) forest, *Anogeissus pendula* and *Boswellia* forest, southern dry mixed deciduous forest, northern dry mixed deciduous forest, and southern moist mixed deciduous forest (Champion and Seth 1968). Major tree species with high timber value in the region are teak (*Tectona grandis*) and Sal, along with Bija (*Pterocarpus marsupium*), Saja (*Terminalia tomentosa*), Surya (*Xylia xylocarpa*), Dhaora (*Anogeissus latifolia*), and Garari (*Cleistanthus collinus*). In addition to forests, dry deciduous scrub and grasslands form an important habitat type in the region. Historically, the entire region was occupied by large ungulate species such as the gaur (*Bos gaurus*) and Asian elephant as well as large carnivore species such as the Asiatic cheetah (*Acinonyx jubatus venaticus*) and Asiatic lion (*Panthera leo persica*) until a few centuries ago. Today, the gaur and elephant populations are confined to the eastern part of the landscape, whereas the Asiatic cheetah is locally extinct, and Asiatic lions are restricted to their westernmost range in Gujarat. The iconic species of the region is the tiger. The Central Indian forests are home to one-third of India's tiger population and represent the largest proportion of tiger-occupied forest in the country (Qureshi et al. 2022). Central India also forms all or part of the catchment area of five major rivers (Ganga, Narmada, Godavari, Mahanadi, and Tapti), thus contributing to water security for a significant proportion of the country's human population.

Ethnically, Central India is very diverse. Several of India's tribal communities such as the Gond, Baiga, Bhil, and Korku live in the region. About 70 percent of the population is rural with high dependency on forests for fuelwood, fodder, and income from forest products (DeFries et al. 2016). The livelihood of local communities is dependent on several non-timber forest products such as Tendu (*Dyospyros melanoxylon*) leaves and Mahua (*Madhuca longifolia*) flowers, which are collected during a period roughly from March to June (Mahapatra and Shackleton 2012) to produce *bidis* (a type of local cigarettes) and alcohol or flour, respectively. Disturbance regimes in the region are primarily driven by weather patterns of droughts and rainfall along with an increasing incidence rate and severity of forest fires. Some of these fires are of anthropogenic origin, as clearing the forest floor helps in the collection of Mahua flowers and helps the growth of the tender Tendu leaves, which follows the Mahua collection season.

Methods

To understand the history of connectivity research in Central India, we updated the systematic review presented by Thatte et al. (2021) with the following modifications: (a) we used the same search terms in Web of Science since we did not have access to Scopus, (b) we replaced the geography of their search with search terms for Central India, and (c) we restricted our search to 2021 and onward to find new papers since their publication. We searched All Fields on WOS from 2021 with the Boolean search string: [(connectivity OR corridor) AND (wildlife OR animal OR species OR habitat OR landscape) AND (India OR Indian OR "Central India" OR "Central Indian")]. We then screened all studies that were relevant, i.e., landscape connectivity research conducted in Central India, and extracted the following information from each of them: the species studied, the ecosystem (broadly categorized as forest, grassland and freshwater), the methods (remote sensing, species occurrences, telemetry, camera trapping, genetic, or a combination), and the research group and author composition (based on our knowledge about the research groups). Our effort is not intended to be a thorough systematic review, but rather an exercise to identify the major themes, methods, and gaps in the current body of research.

Results and Discussion

Our search resulted in 347 papers, of which 11 were relevant and included in this review. In addition, 22 papers from the Thatte et al. (2021) paper specifically referred to the Central Indian landscape, resulting in a total of 33 papers (Table 30.1).

The overarching patterns we found during the review process are that most connectivity research in the landscape has been conducted on terrestrial ecosystems (especially forests), on individual species (usually the tiger), using a singular method, and by independent research groups (Fig. 30.1). Recent research, however, leans toward multi-species approaches within collaborative research frameworks. We explain these findings below.

A Focus on Terrestrial Connectivity

Although there is a great diversity of ecosystems ranging from dry and mixed deciduous forests through naturally occurring open ecosystems such as scrub savannahs and grasslands to freshwater and riverine ecosystems in the Central Indian landscape, most connectivity research (97%) is focused on terrestrial forest ecosystems. Aside from forests (areas of closed tree canopies), open and treeless areas such as agro-pastoral lands are important habitats in the region that support a unique assemblage of plant and animal life. These habitats host a wide range of biodiversity from critically endangered grassland birds such as lesser florican (*Sypheotides indicus*) to mammals such as wolves (*Canis lupus pallipes*) and blackbuck (*Antilope cervicapra*). Another largely overlooked habitat category is aquatic connectivity and riverine or riparian connectivity. The only paper on aquatic ecosystems we found was about structural connectivity (Kantharajan et al. 2022). Rana et al. (2022) used fishing cats to design a conservation plan to safeguard wetlands in a countrywide analysis. However, fishing cats do not occupy a wide range in Central India. Other taxonomic groups such as fish, gharials and freshwater crocodiles, otters, or a

Table 30.1 List of connectivity papers on Central India along with some of their key features

New S. No.	Title	Author	Year	Total number of species	Species	Ecosystem
1	A graph theoretic approach for modeling tiger corridor network in Central India-Eastern Ghats landscape complex, India	Shanu et al.	2019	1	Tiger	Terrestrial
2	Carnivores in corridors: estimating tiger occupancy in Kanha–Pench corridor, Madhya Pradesh, India	Borah et al.	2016	1	Tiger	Terrestrial
3	Connecting the dots: mapping habitat connectivity for tigers in Central India	Dutta et al.	2016	1	Tiger	Terrestrial
4	Connectivity of tiger (*Panthera tigris*) populations in the human-influenced forest mosaic of Central India	Joshi et al.	2013	1	Tiger	Terrestrial
5	Conservation priorities for endangered Indian tigers through a genomic lens	Natesh et al.	2017	1	Tiger	Terrestrial
6	Demographic loss, genetic structure and the conservation implications for Indian tigers	Mondol et al.	2013	1	Tiger	Terrestrial
7	Fine-scale population genetic structure in a wide-ranging carnivore, the leopard (*Panthera pardus fusca*) in Central India	Dutta et al.	2013	1	Leopard	Terrestrial
8	Forest corridors maintain historical gene flow in a tiger metapopulation in the highlands of Central India	Sharma et al.	2013	1	Tiger	Terrestrial
9	Gene flow and demographic history of leopards (*Panthera pardus*) in the Central Indian highlands	Dutta et al.	2013	1	Leopard	Terrestrial

(continued)

Table 30.1 (continued)

New S. No.	Title	Author	Year	Total number of species	Species	Ecosystem
10	Genetic evidence of tiger population structure and migration within an isolated and fragmented landscape in Northwest India	Reddy et al.	2012	1	Tiger	Terrestrial
11	Genetic variation, structure, and gene flow in a sloth bear (*Melursus ursinus*) metapopulation in the Satpura-Maikal landscape of Central India	Dutta et al.	2015	1	Sloth bear	Terrestrial
12	Geospatial modeling to assess elephant habitat suitability and corridors in northern Chhattisgarh, India	Areendran et al.	2011	1	Asian elephant	Terrestrial
13	Maintaining tiger connectivity and minimizing extinction into the next century: Insights from landscape genetics and spatially explicit simulations	Thatte et al.	2018	1	Tiger	Terrestrial
14	Identifying suitable habitat and corridors for Indian Grey wolf (*Canis lupus pallipes*) in Chotta Nagpur plateau and lower Gangetic planes: A species with differential management needs	Sharma et al.	2019	1	Indian grey wolf	Terrestrial
15	Opportunities of habitat connectivity for tiger (*Panthera tigris*) between Kanha and Pench national parks in Madhya Pradesh, India	Rathore et al.	2012	1	Tiger	Terrestrial
16	Multi-scale prediction of landscape resistance for tiger dispersal in Central India	Krishnamurthy et al.	2016	1	Tiger	Terrestrial

(continued)

Table 30.1 (continued)

New S. No.	Title	Author	Year	Total number of species	Species	Ecosystem
17	Prioritizing tiger conservation through landscape genetics and habitat linkages	Yumnam et al.	2014	1	Tiger	Terrestrial
18	Spatial genetic analysis reveals high connectivity of tiger (*Panthera tigris*) populations in the Satpura-Maikal landscape of Central India	Sharma et al.	2013	1	Tiger	Terrestrial
19	Spatial variation in the response of tiger gene flow to landscape features and limiting factors	Reddy et al.	2019	1	Tiger	Terrestrial
20	Targeting restoration sites to improve connectivity in a tiger-conservation landscape in India	Dutta et al.	2018	1	Tiger	Terrestrial
21	Tiger abundance and gene flow in Central India are driven by disparate combinations of topography and land cover	Reddy et al.	2017	1	Tiger	Terrestrial
22	Human footprint differentially impacts genetic connectivity of four wide-ranging mammals in a fragmented landscape	Thatte et al.	2020	4	Tiger, leopard, sloth bear, jungle cat	Terrestrial
23	Links in a sink: Interplay between habitat structure, ecological constraints and interactions with humans can influence connectivity conservation for tigers in forest corridors	Puri et al.	2022	1	Tiger	Terrestrial
24	Genome-wide single-nucleotide polymorphism (SNP) markers from fecal samples reveal anthropogenic impacts on connectivity: Case of a small carnivore in the Central Indian landscape	Tyagi et al.	2022	1	Jungle cat	Terrestrial

(continued)

Table 30.1 (continued)

New S. No.	Title	Author	Year	Total number of species	Species	Ecosystem
25	Synthesizing habitat connectivity analyses of a globally important human-dominated tiger-conservation landscape	Schoen et al.	2022	1	Tiger	Terrestrial
26	Habitat connectivity for the conservation of small ungulates in a human-dominated landscape	Niyogi et al.	2021	4	Blackbuck, chinkara, blue bull, four-horned antelope	Terrestrial
27	Modeling landscape permeability for dispersal and colonization of tigers (*Panthera tigris*) in the greater Panna landscape, Central India	Makwana et al.	2023	1	Tiger	Terrestrial
28	Habitat connectivity for conserving cervids in a multifunctional landscape	Niyogi et al.	2023	2	Chital, sambar	Terrestrial
29	Conservation prioritization in a tiger landscape: Is umbrella species enough?	Vasudeva et al.	2022	1	Tiger	Terrestrial
30	Dog in the matrix: Envisioning countrywide connectivity conservation for an endangered carnivore	Rodrigues et al.	2022	1	Dhole	Terrestrial
31	Safe passage or hunting ground? A test of the prey-trap hypothesis at wildlife crossing structures on NH 44, Pench Tiger Reserve, Maharashtra, India	Saxena and Habib	2022	—	No specific target species	Terrestrial
32	Long-distance dispersal by a male sub-adult tiger in a human-dominated landscape	Hussain et al.	2022	1	Tiger	Terrestrial
33	Applications of Sentinel-2 satellite data for spatiotemporal mapping of deep pools for monitoring the riverine connectivity and assessment of ecological dynamics: a case from Godavari, a tropical river in India (2016–2021)	Kantharajan et al.	2022	—	No specific species	Aquatic

Papers 1–22 are from Thatte et al. (2021); the rest (23–33) were added in this review

Fig. 30.1 Location and characteristics of Central India. The map of Central India, modified from Dutta et al. (2016), shows major land uses. The photo on the top right is from the Bhimbhetka cave paintings, along with the flagship species of the landscape, the tiger (© Sandeep Sharma)

riparian tree species such as Arjun (*Terminalia arjuna*) may also be relevant candidates for freshwater conservation in the Central Indian landscape.

A Focus on Connectivity for a Single Species: The Tiger

Connectivity research in this landscape has been tiger-centric (Dutta et al. 2016). The tiger is an important umbrella and flagship species for India whose population has undergone a remarkable recovery over the last two decades under the ambit of the National Tiger Conservation Authority (earlier known as "Project Tiger"). While the tiger generates a significant amount of conservation funding and drives legal actions to set aside protected areas, the effectiveness of tigers regarding connectivity conservation for other species is unknown. A majority of the past research has been focused on identifying the corridors used by the tiger (e.g., Sharma et al. 2013a), but an important and unanswered question is to what extent tiger corridors are also used and effective for the connectivity of co-occurring species. Even when other species have been used to assess connectivity, such studies have mostly been conducted for individual species, with an emphasis on large carnivores such as leopards (Dutta et al. 2013a, b) and sloth bears (Dutta et al. 2015). More recent research has challenged this focus on large carnivores by studying small carnivores (Tyagi et al. 2022) and multi-species connectivity for carnivores (Mukherjee et al. 2021; Thatte et al. 2019) as well as for ungulates (Niyogi et al. 2021, 2022).

Multitude of Research Methods

A plethora of analytical methods have been used to perform connectivity analyses in the Central Indian landscape. Studies based entirely on remote sensing data and structural connectivity (Rathore et al. 2012) have been substantiated by research measuring functional connectivity (e.g., Sharma et al. 2013a). Methods including occupancy, camera trapping, movement, and genetics have been used to study landscape connectivity in Central India. This trend of applying diverse methodology has largely reflected the development of connectivity research (Dutta et al. 2022), which has shifted from pure remote sensing methods focused on assessing structural connectivity at a coarse scale to approaches that address ecological and evolutionary processes affecting functional connectivity at various spatio-temporal scales. However, there is yet to be a comparison made of the outcomes of the different methods in the results and interpretation, as well as an integration of different data types.

Independent Research with Room for Collaborative Work

Most of the research in the landscape has been conducted by different research groups and institutions independently. A surprising number of research papers were published within a relatively small window of time between 2013 and 2018 on tigers in the same landscape using the same data source and techniques—i.e., non-invasive genetics—albeit with different analytical approaches (Dutta et al. 2013a; Joshi et al. 2013; Reddy et al. 2017; Sharma et al. 2013a, b; Thatte et al. 2018; Yumnam et al. 2014). This body of research was largely conducted independently, with no information exchange or formal collaboration. For a country and region where challenges to conservation are abundant, one could argue that this duplication of research was a waste of valuable resources and talent. On the contrary, this deluge of scientific information indisputably highlighted the value of maintaining connectivity, primarily because all mentioned studies agreed on the central conclusion that these corridors are effective in maintaining genetic and therefore functional connectivity. Around a decade after this first wave of research performed almost in parallel, a collaborative project that synthesized five independent studies on tiger connectivity in Central India to quantify agreement on landscape resistance for potential tiger movement resulted in the identification of Consensus Connectivity Areas (CCAs) for the landscape (Schoen et al. 2022). Continued collaboration and synthesis of data and knowledge will help to increase our understanding and determine shared targets and greater goals for this landscape.

Recommendations for Future Research in Central India

The Central Indian landscape has been at the forefront of connectivity research in India, and indeed South Asia, with several independent studies pioneering and advancing our understanding of various aspects of structural and functional connectivity. To move beyond the current state of the art and make progress both on the scientific and the conservation front, we propose the following recommendations for future research in the landscape.

Connectivity research on non-forested terrestrial and aquatic ecosystems is essential for ensuring that species dependent exclusively on these habitats are also considered in long-term conservation plans for the landscape. Structural and functional connectivity can both be addressed within these ecosystems. We also encourage the integration of terrestrial and aquatic connectivity, for example through riparian forests, to integrate overall connectivity.

Some studies have already gone beyond the single-species approach, and we believe this is a good start. We encourage future research to (a) objectively evaluate several and select a set of species that represent the connectivity needs for several other co-occuring species (e.g., Dutta et al. 2023; Meurant et al. 2018), (b) use available data scattered and siloed across different research institutions and conservation NGOs to test whether tigers are indeed a good proxy for the connectivity of other species, and (c) move beyond large mammals and apply connectivity to other taxonomic groups such as birds, plants, and insects.

Integration and comparison of research approaches regarding the Central Indian landscape remain scarce. Several research methods were used to study connectivity, largely depending on the questions being asked, but it is not clear when specific approaches may be better suited than others, especially when certain types of data are difficult to obtain. For example, genetic data—often obtained noninvasively—is a gold standard for measuring long-term functional connectivity. Similarly, movement data from collared individuals are an excellent method for understanding species' contemporary space usage. However, these data may be difficult and expensive to obtain for many different species. There are several camera trapping studies in the landscape that yield by-catch data—i.e., photos of non-target species that could be used to address certain aspects of connectivity—but to effectively use them, a comparison between the various data sources with regard to a similar question would be essential to understand the limitations and potential for integrating multiple methods. This understanding could then be applied to other species for which certain types of data may be more readily available than others.

The momentum for collaborative research has recently been established (e.g., Schoen et al. 2022). Sustained cooperation in research, facilitated by cultivating mutual trust, transparency, and inclusiveness, will likely help shape a collective and shared vision for landscape conservation in Central India. Fostering a research environment that encourages healthy competition with minimal duplication and provides a true learning experience for young researchers would be a significant advantage in the future.

A stark aspect we noticed in our review is that most of the conducted research is purely ecologically, wildlife, or conservation-based. In a highly populated country like India, shared spaces are bound to be important for the long-term conservation of species and habitats. Understanding people's willingness to share space and enable the dispersal of animals through human-dominated landscapes despite anthropogenic resistance (Ghoddousi et al. 2021) will be essential for conservation

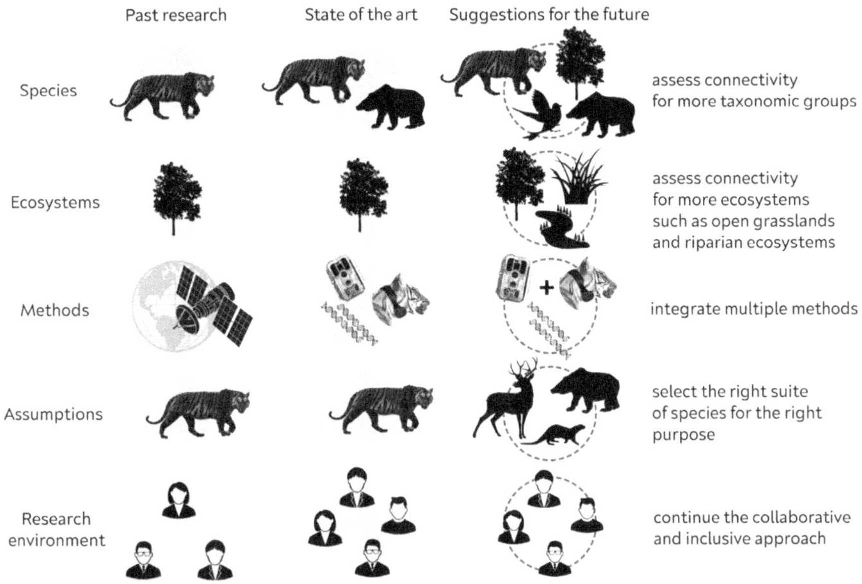

Fig. 30.2 The trajectory of connectivity research with our suggestions for the future research agenda in Central India

in India, especially concerning large terrestrial species. Carnivores are known to kill livestock and even humans, whereas ungulates cause damage to crops; both of which can lead to economic loss and social dissatisfaction with conservation efforts. Some research has recently been conducted in this direction (Puri et al. 2022), but much work remains to be done on this front. Unless these issues are integrated with ecological factors, on-the-ground conservation may be an unrealistic goal. Conservation groups and research networks such as the Network for Conserving Central India (NCCI),[1] the Satpuda Landscape Tiger Partnership (SLTP),[2] and the Coalition for Wildlife Corridors (CWC)[3] are playing a vital role in addressing holistic landscape conservation in the region.

Central India has undeniably been critical to the connectivity conservation movement in India. The country's first dedicated wildlife underpass was built on North-South National Highway 44 (earlier known as NH7) following a directive by the Mumbai High Court to the National Highway Authority of India as a mitigation measure in the course of an infrastructure project to widen the highway that intersects the Kanha–Pench corridor. A total of nine underpasses were constructed along NH44. This first legal success in securing wildlife mobility has led to similar mitigation structures being required for other infrastructure projects that pass through

---

[1] https://www.conservingcentralindia.org/.

[2] https://savingindiastigers.org/.

[3] http://corridorcoalition.org/.

critical ecological regions (Press Information Bureau 2022). A major catalyst for this may have been the abundance of studies conducted on the Central Indian landscape. We have no information on the extent to which published scientific knowledge was used to develop and site the mitigation structures, but at least the policymakers did indeed take note of the volume of scientific information that resoundingly agreed on the functionality of the existing corridors. Much remains to be done in line with the ideal mitigation hierarchy, which states that avoidance should be preferable to any other approach (Kiesecker et al. 2010).

In summary, the Central Indian landscape is rich in terms of the science it has provided to expand our understanding of connectivity. We believe that this synthesis of the history of connectivity research in the landscape can help to identify future actions and pathways to building a resilient landscape (Fig. 30.2).

References

Champion SH, Seth SK (1968) A revised survey of the forest types of India. Government of India Press, New Delhi

Clarkson C, Harris C, Li B, Neudorf CM, Roberts RG, Lane C, Norman K, Pal J, Jones S, Shipton C, Koshy J, Gupta MC, Mishra DP, Dubey AK, Boivin N, Petraglia M (2020) Human occupation of northern India spans the Toba super-eruption ~74,000 years ago. Nat Commun 11(1):Article 1. https://doi.org/10.1038/s41467-020-14668-4

DeFries R, Sharma S, Dutta T (2016) A landscape approach to conservation and development in the central Indian highlands. Reg Environ Chang 16(S1):1–3. https://doi.org/10.1007/s10113-016-1014-3

Dutta T, Sharma S, Maldonado JE, Wood TC, Panwar HS, Seidensticker J (2013a) Fine-scale population genetic structure in a wide-ranging carnivore, the leopard (Panthera pardus fusca) in Central India. Divers Distrib 19(7):760–771. https://doi.org/10.1111/ddi.12024

Dutta T, Sharma S, Maldonado JE, Wood TC, Panwar HS, Seidensticker J (2013b) Gene flow and demographic history of leopards (Panthera pardus) in the central Indian highlands. Evol Appl 6(6):949–959. https://doi.org/10.1111/eva.12078/full

Dutta T, Sharma S, Maldonado JE, Panwar HS, Seidensticker J (2015) Genetic variation, structure, and gene flow in a sloth bear (Melursus ursinus) meta-population in the Satpura-Maikal landscape of Central India. PLoS One 10(5):e0123384. https://doi.org/10.1371/journal.pone.0123384

Dutta T, Sharma S, McRae BH, Roy PS, DeFries R (2016) Connecting the dots: mapping habitat connectivity for tigers in Central India. Reg Environ Chang 16(1):53–67. https://doi.org/10.1007/s10113-015-0877-z

Dutta T, Sharma S, Meyer NFV, Larroque J, and Balkenhol N (2022) An overview of computational tools for preparing, constructing and using resistance surfaces in connectivity research. Landsc Ecol 1–30. https://doi.org/10.1007/s10980-022-01469-x

Dutta T, De Barba M, Selva N, Fedorca AC, Maiorano L, Thuiller W, Zedrosser A, Signer J, Pflüger F, Frank S, Lucas PM, Balkenhol N (2023) An objective approach to select surrogate species for connectivity conservation. Front Ecol Evol 11. https://www.frontiersin.org/articles/10.3389/fevo.2023.1078649

ENVIS (2022) Protected areas-subject area: Wildlife Institute of India, Ministry of Environment & Forests. https://wiienvis.nic.in/database/protected_area_854.aspx

Ghoddousi A, Buchholtz EK, Dietsch AM, Williamson MA, Sharma S, Balkenhol N, Kuemmerle T, Dutta T (2021) Anthropogenic resistance: accounting for human behavior in wildlife connectivity planning. One Earth 4(1):39–48. https://doi.org/10.1016/j.oneear.2020.12.003

Joshi A, Vaidyanathan S, Mondol S, Edgaonkar A, Ramakrishnan U (2013) Connectivity of Tiger (Panthera tigris) populations in the human-influenced Forest mosaic of Central India. PLoS One 8(11):e77980. https://doi.org/10.1371/journal.pone.0077980

Kantharajan G, Anand A, Krishnan P, Singh RK, Kumar K, Kumar Yadav A, Mohindra V, Shukla SP, Lal KK (2022) Applications of Sentinel-2 satellite data for spatio-temporal mapping of deep pools for monitoring the riverine connectivity and assessment of ecological dynamics: a case from Godavari, a tropical river in India (2016–2021). Environ Monit Assess 194(8):558. https://doi.org/10.1007/s10661-022-10089-6

Kiesecker JM, Copeland H, Pocewicz A, McKenney B (2010) Development by design: blending landscape-level planning with the mitigation hierarchy. Front Ecol Environ 8(5):261–266. https://doi.org/10.1890/090005

Mahapatra AK, Shackleton CM (2012) Exploring the relationships between trade in natural products, cash income and livelihoods in tropical forest regions of eastern India. Int For Rev 14(1):62–73. https://doi.org/10.1505/146554812799973217

Meurant M, Gonzalez A, Doxa A, Albert CH (2018) Selecting surrogate species for connectivity conservation. Biol Conserv 227:326–334. https://doi.org/10.1016/j.biocon.2018.09.028

Mukherjee T, Chongder I, Ghosh S, Dutta A, Singh A, Dutta R, Joshi BD, Thakur M, Sharma LK, Venkatraman C, Ray D, Chandra K (2021) Indian Grey wolf and striped Hyaena sharing from the same bowl: high niche overlap between top predators in a human-dominated landscape. Global Ecol Conserv 28:e01682. https://doi.org/10.1016/j.gecco.2021.e01682

Nayak R, Karanth KK, Dutta T, Defries R, Karanth KU, Vaidyanathan S (2020) Bits and pieces: Forest fragmentation by linear intrusions in India. Land Use Policy 99:104619. https://doi.org/10.1016/j.landusepol.2020.104619

Niyogi R, Sarkar MS, Hazra P, Rahman M, Banerjee S, John R (2021) Habitat connectivity for the conservation of small ungulates in a human-dominated landscape. ISPRS Int J Geo Inf 10(3):Article 3. https://doi.org/10.3390/ijgi10030180

Niyogi R, Shekhar Sarkar M, Shekhar Niyogi V, Hazra P, John R (2022) Habitat connectivity for conserving cervids in a multifunctional landscape. J Nat Conserv 68:126212. https://doi.org/10.1016/j.jnc.2022.126212

Press Information Bureau (9 February 2022) NHS passing through forests/wildlife sanctuaries. https://pib.gov.in/pib.gov.in/Pressreleaseshare.aspx?PRID=1796798

Puri M, Srivathsa A, Karanth KK, Patel I, Kumar NS (2022) Links in a sink: interplay between habitat structure, ecological constraints and interactions with humans can influence connectivity conservation for tigers in forest corridors. Sci Total Environ 809:151106. https://doi.org/10.1016/j.scitotenv.2021.151106

Qureshi Q, Jhala YV, Yadav SP, Mallick A (2022) Status of tigers, co-predators, and prey in India. National Tiger Conservation Authority, Government of India/Wildlife Institute of India, New Delhi/Dehradun

Rana D, Samad I, Rastogi S (2022) To a charismatic rescue: designing a blueprint to steer fishing cat conservation for safeguarding Indian wetlands. J Nat Conserv 68:126225. https://doi.org/10.1016/j.jnc.2022.126225

Rathore CS, Dubey Y, Shrivastava A, Pathak P, Patil V (2012) Opportunities of habitat connectivity for Tiger (Panthera tigris) between Kanha and Pench National Parks in Madhya Pradesh, India. PLoS One 7(7):e39996. https://doi.org/10.1371/journal.pone.0039996

Reddy PA, Cushman SA, Srivastava A, Sarkar MS, Shivaji S (2017) Tiger abundance and gene flow in Central India are driven by disparate combinations of topography and land cover. Divers Distrib 23(8):863–874. https://doi.org/10.1111/ddi.12580

Rodgers W, Panwar H (1988) Planning a wildlife protected area network in India

Schoen JM, Neelakantan A, Cushman SA, Dutta T, Habib B, Jhala YV, Mondal I, Ramakrishnan U, Reddy PA, Saini S, Sharma S, Thatte P, Yumnam B, DeFries R (2022) Synthesizing habitat connectivity analyses of a globally important human-dominated tiger-conservation landscape. Conserv Biol 36(4):e13909. https://doi.org/10.1111/cobi.13909

Sharma S, Dutta T, Maldonado JE, Wood TC, Panwar HS, Seidensticker J (2013a) Forest corridors maintain historical gene flow in a tiger metapopulation in the highlands of Central India. Proc R Soc B Biol Sci 280(1767):20131506. https://doi.org/10.1098/rspb.2013.1506

Sharma S, Dutta T, Maldonado JE, Wood TC, Panwar HS, Seidensticker J (2013b) Spatial genetic analysis reveals high connectivity of tiger (Panthera tigris) populations in the Satpura–Maikal landscape of Central India. Ecol Evol 3(1):48–60. https://doi.org/10.1002/ece3.432

Thatte P, Joshi A, Vaidyanathan S, Landguth E, Ramakrishnan U (2018) Maintaining tiger connectivity and minimizing extinction into the next century: insights from landscape genetics and spatially-explicit simulations. Biol Conserv 218:181–191. https://doi.org/10.1016/j.biocon.2017.12.022

Thatte P, Chandramouli A, Tyagi A, Patel K, Baro P, Chhattani H, Ramakrishnan U (2019) Human footprint differentially impacts genetic connectivity of four wide-ranging mammals in a fragmented landscape. Ecology. https://doi.org/10.1101/717777

Thatte P, Tyagi A, Neelakantan A, Natesh M, Sen M, Thekaekara T (2021) Trends in wildlife connectivity science from the biodiverse and human-dominated South Asia. J Indian Institute Sci 101(2):177–193. https://doi.org/10.1007/s41745-021-00240-6

Tyagi A, Khan A, Thatte P, Ramakrishnan U (2022) Genome-wide single nucleotide polymorphism (SNP) markers from fecal samples reveal anthropogenic impacts on connectivity: case of a small carnivore in the central Indian landscape. Anim Conserv 25(5):648–659. https://doi.org/10.1111/acv.12770

Yumnam B, Jhala YV, Qureshi Q, Maldonado JE, Gopal R, Saini S, Srinivas Y, Fleischer RC (2014) Prioritizing tiger conservation through landscape genetics and habitat linkages. PLoS One 9(11):e111207. https://doi.org/10.1371/journal.pone.0111207

Open Access This chapter is licensed under the terms of the Creative Commons Attribution 4.0 International License (http://creativecommons.org/licenses/by/4.0/), which permits use, sharing, adaptation, distribution and reproduction in any medium or format, as long as you give appropriate credit to the original author(s) and the source, provide a link to the Creative Commons license and indicate if changes were made.

The images or other third party material in this chapter are included in the chapter's Creative Commons license, unless indicated otherwise in a credit line to the material. If material is not included in the chapter's Creative Commons license and your intended use is not permitted by statutory regulation or exceeds the permitted use, you will need to obtain permission directly from the copyright holder.

Republic of Korea: Predicting Shifts in Forest Biodiversity

31

Yuyoung Choi, Chul-Hee Lim, Hye In Chung, Yoonji Kim, Hyo Jin Cho, Jinhoo Hwang, Florian Kraxner, Gregory S. Biging, Woo-Kyun Lee, Jin Hyung Chon, and Seong Woo Jeon

Scenic view of Mt.Seoraksan against sky (Photo: Sangoh)

Y. Choi (✉)
Division of Public Infrastructure Assessment, Environmental Assessment Group, Korea Environment Institute, Sejong, Republic of Korea
e-mail: yychoi@kei.re.kr

C.-H. Lim
Department of Forestry, Environment, and Systems, Kookmin University, Seoul, Republic of Korea
e-mail: clim@kookmin.ac.kr

H. In Chung · Y. Kim · J. Hwang
OJEong Resilience Institute (OJERI), Korea University, Seoul, Republic of Korea
e-mail: bproud0514@korea.ac.kr; yoonjik605@korea.ac.kr; i0255278@korea.ac.kr

H. J. Cho · W.-K. Lee · J. H. Chon · S. W. Jeon
Department of Environmental Science and Ecological Engineering, Korea University, Seoul, Republic of Korea
e-mail: hyojin@tomorrowuse.com; leewk@korea.ac.kr; jchon@korea.ac.kr; eepps_korea@korea.ac.kr

© The Author(s) 2025
K. Lapin et al. (eds.), *Ecological Connectivity of Forest Ecosystems*, https://doi.org/10.1007/978-3-031-82206-3_31

Abstract

This chapter addresses the critical issue of diminishing biodiversity resulting from climate change and habitat loss in Korea and is largely based on a study by Choi et al. (Journal of Environmental Management 288, 2021). International endeavors to safeguard biodiversity face the challenge of accurately quantifying and predicting its shifts. Focusing on the Republic of Korea (ROK), a region renowned for its rapid reforestation, the study seeks to evaluate the enduring biodiversity of plant species from the 1960s to the 2050s, with a specific emphasis on the consequences of reforestation efforts. Employing a fusion of global-scale methodologies and localized data, the study simulates transformations in climate change, land use, and habitat condition, culminating in an analysis of their collective influence on biodiversity. The specific methodology details can be found in Choi et al. (Journal of Environmental Management 288, 2021), with key findings presented in this chapter. It is simulated that biodiversity is deteriorating due to habitat fragmentation and reduced connectivity caused by urbanization, alongside the impacts of climate change. However, a key conclusion drawn is that consistent efforts in forest conservation can mitigate these adverse effects. This research augments our understanding of biodiversity preservation amid the intricate interplay of complex factors, exemplified by the case of the ROK.

Keywords

Biodiversity persistence · Climate change · Laud-use change · Forest management · Republic of Korea

Land Use and Forest Habitat Condition Changes from the Past to the Future (1960s–2050s)

Having undergone rapid economic development and successful reforestation in a short timeframe, the Republic of Korea (ROK) offers a unique context for exploring the integrated impact of climate change, land-use/cover change, and qualitative forest management. Following the Korean War, extensive tree planting occurred under the National Reforestation Program (1962–1987), resulting in the current forest coverage of 63% in the ROK (Kim et al. 2008; Bae et al. 2014). Despite urban

---

F. Kraxner
Agriculture, Forestry, and Ecosystem Services Research Group (AFE), Biodiversity and Natural Resources Program (BNR), International Institute for Applied Systems Analysis (IIASA), Laxenburg, Austria
e-mail: kraxner@iiasa.ac.at

G. S. Biging
Department of Environmental Science, Policy and Management, University of California, Berkeley, CA, USA
e-mail: biging@berkeley.edu

expansion and climate change causing a decline in forest cover, ongoing forest management efforts have enhanced forest conditions in the country (Lee et al. 2015; Cui et al. 2016; Kim et al. 2017).

To spatially simulate such changes, the national land cover datasets for different decades (1980s, 1990s, 2000s, 2010s) were utilized in combination with the machine learning process of the multilayer perceptron (MLP) neural network to simulate land cover changes from the past to the future. This simulation was further integrated with forest condition grades. Figure 31.1 provides a comprehensive view of land use and forest condition evolution from the 1960s to the 2050s. A key highlight is the urban expansion around strategic city hubs, shaped by government policies. In particular, urbanization is poised to escalate around metropolitan areas and the southeastern coast, mirroring the growth of new urban zones along with an extensive road network. Urban areas burgeoned from 1.13% of the total territory in the 1960s to 5.45% in the 2010s, chiefly reclaiming croplands (51% of expansion) and forests (30.5% of expansion). This urban expansion has led to fragmentation of

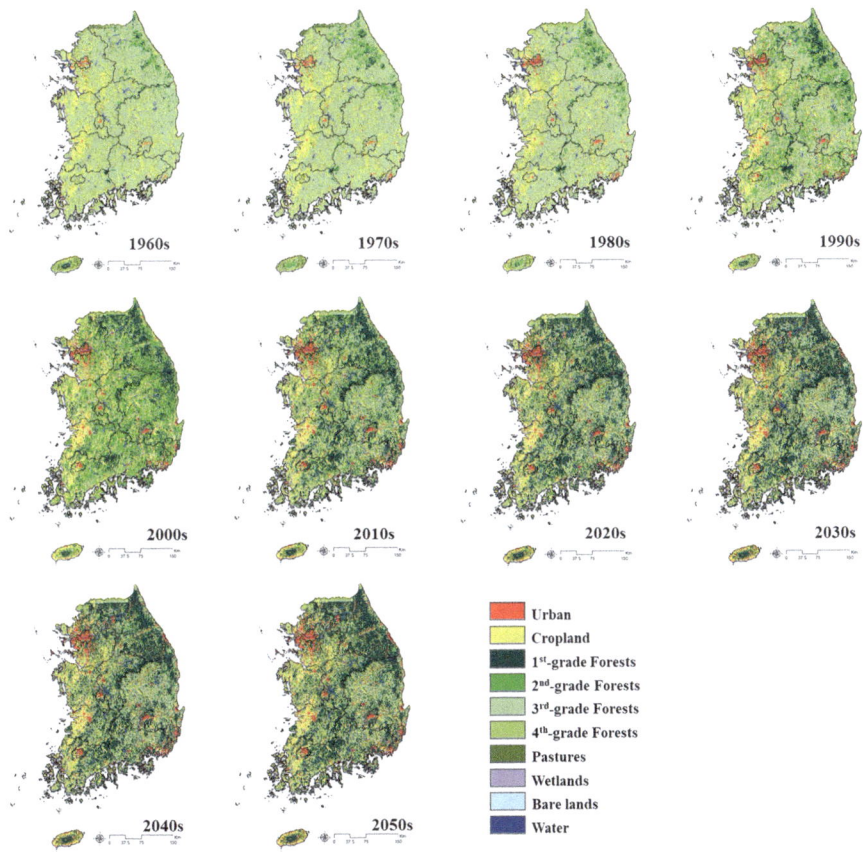

Fig. 31.1 Land-use changes with forest condition grade from the past to the future (1960–2050)

Table 31.1 Land-use and forest condition classes with descriptions

Land use and forest condition class	Explanation
1st-grade forest	Natural forests of more than 50 years old and planted forests of more than 60 years old
2nd-grade forest	Natural forests of more than 40 years old and planted forests of more than 50 years old
3rd-grade forest	Natural forests of less than 30 years old and planted forests of less than 40 years old
4th-grade forest	Forests without origin or age information
Semi-natural pasture	Land covered with herbaceous plants in use as farm, golf courses, cemeteries, etc.
Farmland	Drylands for growing grains, fruit trees, vegetables, etc.
Natural pasture	Lands naturally covered with herbaceous plants
Artificial bare land	Mining area, playgrounds, etc.
Rice paddy	Submerged farmland for growing rice
Urban	Urbanized areas including residential, industrial areas, etc.
Other croplands	A house plantation, orchard, and other cultivation areas
Natural bare land	A beach, riverbed, and rock

forests, resulting in decreased connectivity between forest habitats. However, projected into the 2050s, urban areas are poised to grow by a further 2.99%, while croplands and forests are set to diminish by 1.56% and 1.43%, respectively. Urban expansion will occur sporadically across the nation under the Shared Socio-Economic Pathways 3 (SSP3) scenario of regional rivalry, which assumes limited regulation and ongoing deforestation, for the 2050s.

In this study, the condition of forests was classified into four grades based on forest origin (artificial and natural forests) and age class (Table 31.1). Those classified as natural and with older age classes were considered to be closely resembling the expected natural state, consequently being assigned higher grades of habitat condition. By adhering to the national guidelines for the final cutting age in forests, which stipulate a final cutting age of 60 years for national forests and 40 years for public and private forests, future changes were simulated by incorporating the progression of tree age over time. As a result, forest condition improved even with a steady forest area. Enhanced conservation and reforestation efforts during the 1970s led to an increase in first-grade and second-grade forests, which rose from 27% and 41% in the 1960s to 29% and 45% in the 2000s. In particular, qualitative improvements were observed within conservation areas encompassing the major mountain ranges of the eastern part of the ROK. Following these scenarios, timber production forests, which constitute 33.9% of the total forest area, will be harvested while the aging of forests in other protected areas will lead to a consistent expansion of first-grade forests. Second-grade forests initially increase in area but then transition to first-grade status, thus maintaining a lower proportion.

An Analysis of Historical Biodiversity Persistence Changes

To assess the impact of forest management on biodiversity, this study used the concept of biodiversity persistence (BP), which reflects the portion of species anticipated to endure over an extended period (Allnutt et al. 2008; Di Marco et al. 2019; Hoskins et al. 2019). The study employed an approach inspired by Allnutt et al. (2008) that is grounded in the species–area relationship, effectively transforming proportional habitat loss into predicted species decline. The simulation of BP involved the use of generalized dissimilarity modeling incorporating habitat condition estimates. To enhance comprehension, BP (p) is translated into the extinction rate $(1 - p)$, indicating the proportion of species projected to go extinct over the long term due to climate and land-use change (Di Marco et al. 2019). We also provide an estimate of the number of potentially extinct species based on the total number of native plant species in Korea, which is 7833 (National Biodiversity Center 2019). It is crucial to note, however, that this figure is not an absolute prediction. The primary focus of this study is to compare the relative impacts of forest management, land use, and climate change rather than precisely forecasting the number of endangered species in each scenario.

Figure 31.2 illustrates changes in BP from the 1960s to the 2010s attributable to shifts in land use and forest habitat condition while maintaining the current climate conditions. Overall, the eastern mountains exhibit high BP, contrasting with lower BP in the western plains and island areas. Notably, despite urban expansion, an upward trend in BP was observed over time, owing to enhancements in forest grades. The 1960s saw mostly unstocked forests in mountainous regions, with an average BP of 89.95%. Subsequent improvements raised this figure to 91.43% in the 2010s, translating into an increase of 115 sustainable species. Furthermore, as forest grades improved, the range of BP expanded, accentuating differences between lowland and mountainous regions. This result demonstrates that the successful reforestation project in Korea prompted a significant enhancement in biodiversity even during a phase of rapid development and urbanization.

A Prediction of Future Biodiversity Persistence Changes

The prediction of BP under climate change scenarios, without considering land use and forest grade changes, is depicted in Fig. 31.3a. Notably, the 2050s scenario highlights a decline in BP in mountainous regions aligning with previous research indicating that significant alterations in bioclimatic environments in mountainous areas due to climate change have adverse effects on biotic habitats (Choi et al. 2019). Conversely, coastal areas, the southern flatlands, and islands experience positive effects from climate change, as favorable bioclimatic conditions expand. This shifts the spatial distribution trend of BP compared to the present, resulting in amplified differences between maximum and minimum values. This trend intensifies in the RCP 8.5 scenario, which assumes emissions continue to rise throughout the century without significant mitigation efforts, compared to the RCP 4.5

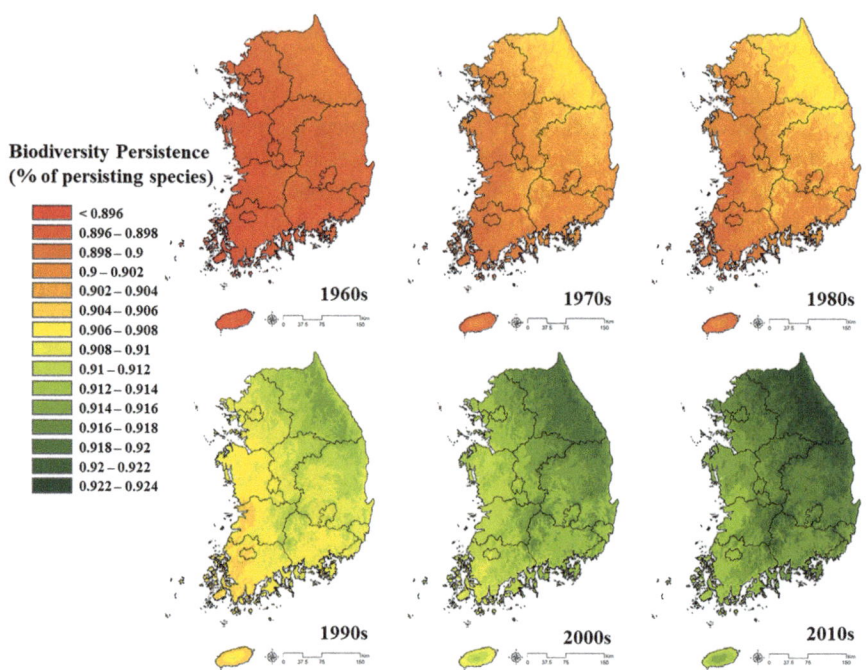

Fig. 31.2 Historical biodiversity persistence changes from the 1960s to the 2010s due to changes in land use and forest habitat condition under the current climate. Maps indicate the proportion of species expected to persist over the long term. The color legend is the same for all maps

scenario, which relatively restricts greenhouse gas emissions and aims to mitigate the impacts of climate change through sustainable energy and climate policies (Meinshausen et al. 2011). Moreover, when factoring in future land use and forest management scenarios, BP increases overall compared to considering climate change alone (Fig. 31.3b). Essentially, improved forest habitat condition positively impacts BP despite urban expansion as seen in prior findings. This implies that proper forest management can mitigate adverse climate change effects—though it may not fully offset them, necessitating additional measures like afforestation.

Time Series Analysis with Various Alternative Scenarios

Examining the time series of BP from the 1960s to the 2050s the average BP demonstrates a steady increase attributed to continuous forest management efforts. However, future projections suggest a decline in BP attributed to climate change, albeit with variations across scenarios. If we focus solely on land-use changes like urban expansion and forest reduction without enhancing forest habitat condition, BP is projected to decrease by 0.47% in the RCP 4.5 scenario and by 0.6% in the

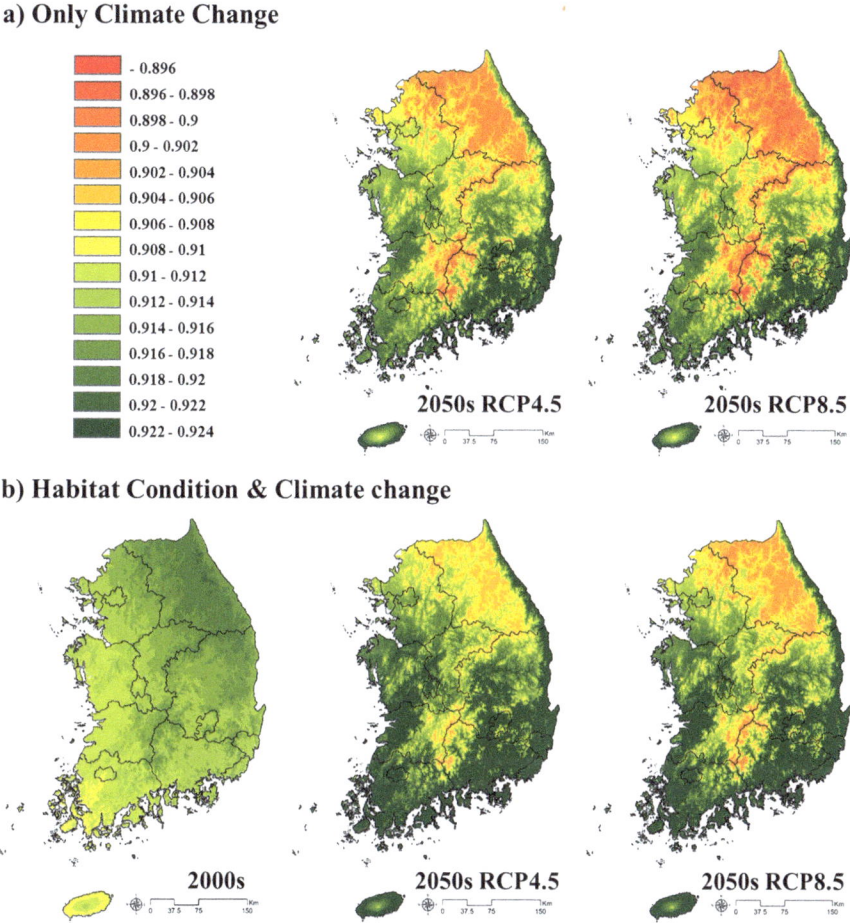

Fig. 31.3 Future biodiversity persistence changes under two different scenarios projected until 2050: (**a**) only climate change under the current habitat conditions and (**b**) changes in both habitat condition and climate. Maps indicate the proportion of species expected to persist over the long term. The color legend is the same for all maps

RCP 8.5 scenario, which assumes continued emission trends. This translates to an estimated additional 37 and 47 vascular plant species facing potential extinction under RCP 4.5 and RCP 8.5, respectively. This is due to a complex interaction: While BP in southern regions thrives with the expansion of the sub-tropical climate zone, it sharply declines in mountainous regions due to diminished cool-temperate climate zones. The crux lies in the fact that these mountainous areas house numerous rare species. If each species is assessed according to its conservation value, the adverse impact is amplified. However, applying forest management scenarios focused on enhancing habitat conditions, achievable through preserving natural forests and maintaining extended final cutting ages, could mitigate the effects of

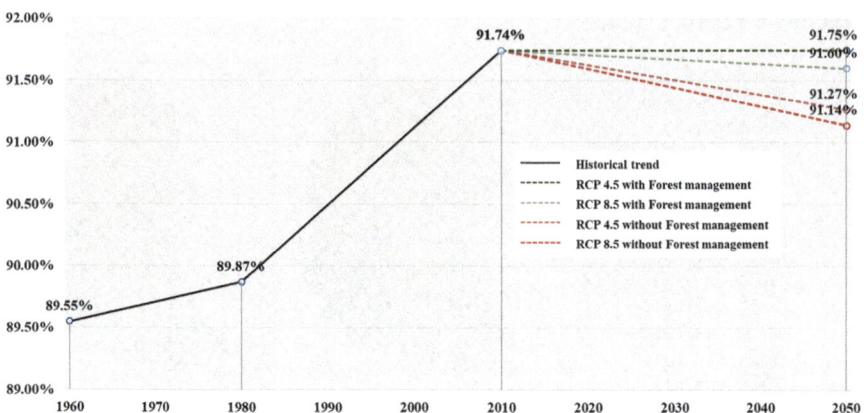

Fig. 31.4 Trends in average biodiversity persistence from the past to the future (1960–2050). The y-axis shows the percentage of species expected to persist. Future projection represents four alternative scenarios according to the combination of climate change (RCP 4.5 and RCP 8.5) with/without forest management scenario

climate change. Under RCP 4.5, the average BP remains at a level similar to that of the 2010s, while under RCP 8.5, there is a decrease of 0.14% compared to the 2010s, which is 0.46% less than the projection that considers only climate change.

Evaluation of the Effect of Forest Restoration

What would have happened to the BP if the forest that suffered damage in the 1970s was not restored through the National Reforestation Program? To gauge the net impact of forest restoration, we simulated BP under the current land cover using the forest-grade conditions of the 1960s (Fig. 31.5c). In other words, the current land cover includes urban areas accounting for 4.06%, indicating a 2.93% increase from the 1960s, and cropland areas diminished by 3.53% to constitute 21.13%. However, the forest grades from the 1960s remained entirely unchanged, distributed into 94% third grade, 4% second grade, and 2% first grade (referred to as the "only land cover change scenario").

Under this specific scenario, BP experienced a decline of 0.18% compared to the 1960s, largely attributable to urbanization. Yet, when considering the combined effects of land cover changes and enhanced forest grades in the 2000s, BP rises by 1.48% relative to the 1960s. This upward trend in BP is due to improvements in forest habitat condition effectively counterbalancing the adverse impact of urbanization, thereby bolstering BP. Consequently, the net effect of forest management was computed at approximately 1.66% (Fig. 31.6), equivalent to safeguarding roughly 130 plant species. This conclusion aligns with prior studies highlighting how the ROK's reforestation initiatives, which involve planting various tree species such as *Alnus hirsute, Pinus densiflora,* and *Quercas acutissima*, have enhanced the

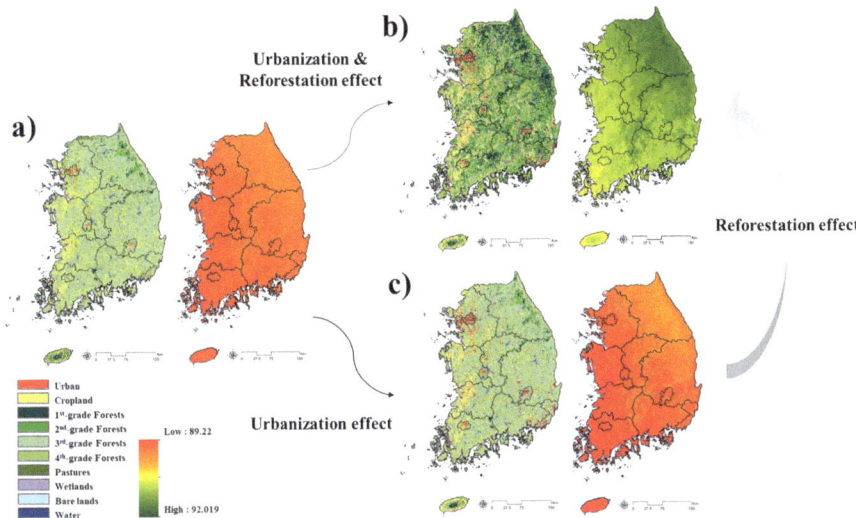

Fig. 31.5 Scenario comparison diagram to assess the effect of forest restoration on biodiversity persistence: (**a**) land cover map for the 1960s with forest grades of the 1960s, (**b**) land cover in the 2000s with forest grades of the 2000s, and (**c**) land cover in the 2000s with forest grades of the 1960s. The map on the right depicts the biodiversity persistence (BP) map corresponding to each scenario

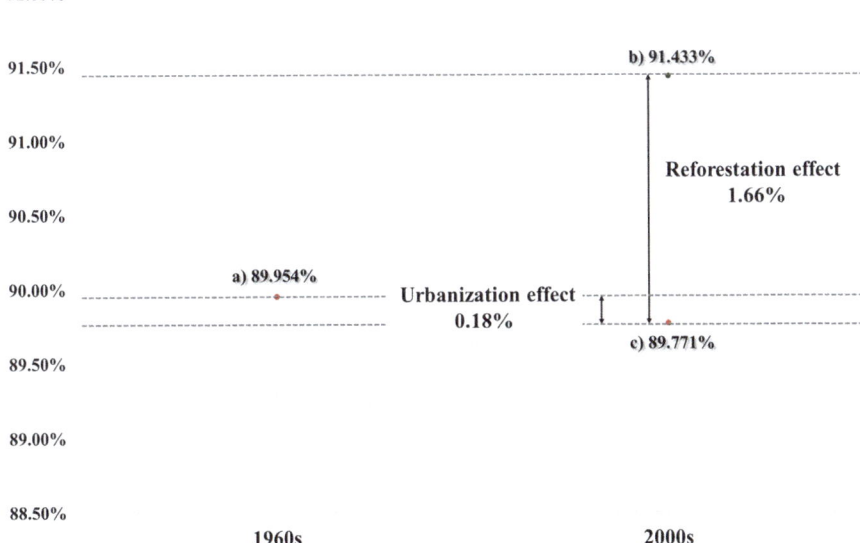

Fig. 31.6 Quantitative evaluation of restoration effect through comparison of three scenarios. The graph represents the average BP value of the scenarios in Fig. 31.5 to quantify urbanization and reforestation effects. (**a**–**c**) are the same as in Fig. 31.5

diversity of forest-dwelling organisms, e.g., mammals (Korean hare, Korean water deer, and wild boar), as well as insects and birds (Bae et al. 2014; Lee et al. 2015).

Significantly, the average BP resulting from the "only land cover change scenario" was lower than historical averages and even dipped below the projected average BP for the 2050s in the RCP 8.5 climate scenario. This underscores the potential magnification of the adverse effects of climate change by maintaining forest conditions as witnessed in the 1960s. The success of the ROK's reforestation endeavors underscores their efficacy. Furthermore, these findings emphasize the importance of proactive forest management in response to climate change, illustrating the prospective repercussions of current forest damage or depletion.

Implications and Limitations

In conclusion, this study highlights how qualitative forest management can counteract the negative effects of urbanization, accompanying habitat fragmentation and reduced connectivity, and mitigate the impacts of climate change on biodiversity. The successful reforestation in South Korea demonstrates the potential to boost biodiversity even amid development through effective strategies such as controlled development and reforestation. Nevertheless, the long-term assessment has specific limitations: The forest grade values employed in this study were categorized primarily by origin and age class, ignoring factors like tree diversity and forest structure. Additionally, this study focused on broad-scale forest management, omitting site-level practices like thinning or pruning. Moreover, due to the reliance on model-driven predictions, further validation using independent data representing actual biological variation in the ROK is essential.

References

Allnutt TF, Ferrier S, Manion G, Powell GV, Ricketts TH, Fisher BL, Lees DC (2008) A method for quantifying biodiversity loss and its application to a 50-year record of deforestation across Madagascar. Conserv Lett 1(4):173–181

Bae JS, Lee KH, Lee YG, Youn HJ, Park CR, Choi HT, Kim TG (2014) Lessons learned from the Republic of Korea's National Reforestation Programme. Korea Forest Service, Daejeon. https://www.cbd.int/ecorestoration/doc/Korean-Study_Final-Version-20150106.pdf

Choi Y, Lim CH, Chung HI, Ryu J, Jeon SW (2019) Novel index for bioclimatic zone-based biodiversity conservation strategies under climate change in Northeast Asia. Environ Res Lett 14(12):124048

Choi Y, Lim CH, Chung HI, Kim Y, Cho HJ, Hwang J, Jeon SW (2021) Forest management can mitigate negative impacts of climate and land-use change on plant biodiversity: insights from the Republic of Korea. J Environ Manag 288:112400

Cui G, Kwak H, Choi S, Kim M, Lim CH, Lee WK, Chae Y (2016) Assessing vulnerability of forests to climate change in South Korea. J For Res 27(3):489–503

Di Marco M, Harwood TD, Hoskins AJ, Ware C, Hill SL, Ferrier S (2019) Projecting impacts of global climate and land-use scenarios on plant biodiversity using compositional-turnover modelling. Glob Chang Biol 25(8):2763–2778

Hoskins AJ, Harwood TD, Ware C, Williams KJ, Perry JJ, Ota N, Purvis A (2019) Supporting global biodiversity assessment through high-resolution macroecological modelling: methodological underpinnings of the BILBI framework. BioRxiv 309377

Kim JS, Kwoun YM, Son Y, Lee SK (2008) The history of deforestation and forest rehabilitation in Korea. Division of Environmental Science and Ecological Engineering, Korea University, Seoul

Kim M, Lee WK, Son Y, Yoo S, Choi GM, Chung DJ (2017) Assessing the impacts of topographic and climatic factors on radial growth of major forest forming tree species of South Korea. For Ecol Manag 404:269–279

Lee DK, Park PS, Park YD (2015) Forest restoration and rehabilitation in the Republic of Korea. In: Restoration of boreal and temperate forests, 2nd edn. CRC Lewis, Boca Raton, pp 217–231

Meinshausen M, Smith SJ, Calvin K, Daniel JS, Kainuma ML, Lamarque JF, van Vuuren DP (2011) The RCP greenhouse gas concentrations and their extensions from 1765 to 2300. Clim Chang 109:213–241

National Biodiversity Center (2019) Biodiversity statistics of Korea (2018). National Institute of Biological Resources. Incheon, Designzip. 272pp

Open Access This chapter is licensed under the terms of the Creative Commons Attribution 4.0 International License (http://creativecommons.org/licenses/by/4.0/), which permits use, sharing, adaptation, distribution and reproduction in any medium or format, as long as you give appropriate credit to the original author(s) and the source, provide a link to the Creative Commons license and indicate if changes were made.

The images or other third party material in this chapter are included in the chapter's Creative Commons license, unless indicated otherwise in a credit line to the material. If material is not included in the chapter's Creative Commons license and your intended use is not permitted by statutory regulation or exceeds the permitted use, you will need to obtain permission directly from the copyright holder.

Mongolia: Connectivity Conservation Actions in the Khan Khentii Region

32

Jargalan Gerelsaikhan, Martin Braun, Tamir Mandakh, and Ochirvaani Soronzonbold

Riparian Zone at Confluence of Khongi and Sharlan into Eruu (Photo: O.Soronzonbold)

J. Gerelsaikhan (✉)
Laboratory of Natural Product Chemistry, Institute of Chemistry and Chemical Technology, Mongolian Academy of Sciences, Ulaanbaatar, Mongolia

Department of Environment and Forest Engineering, National University of Mongolia, Ulaanbaatar, Mongolia
e-mail: jargalan_g@mas.ac.mn

M. Braun
Department of Forest Biodiversity & Nature Conservation, Austrian Research Centre for Forests (BFW), Vienna, Austria
e-mail: martin.braun@bfw.gv.at

T. Mandakh
Department of Natural Resources Management, Khan Khentii Strictly Protected Area, Ulaanbaatar, Mongolia

O. Soronzonbold
Department of Storage, Natural History Museum of Mongolia, Ulaanbaatar, Mongolia
e-mail: soronzonbold@nhm.gov.mn

© The Author(s) 2025
K. Lapin et al. (eds.), *Ecological Connectivity of Forest Ecosystems*,
https://doi.org/10.1007/978-3-031-82206-3_32

Abstract

This chapter provides an overview of the state and structure of boreal forests in Mongolia and examines important pressures on the country's forest ecosystem along with related issues concerning habitat connectivity. Following an overview of conservation areas in Mongolia, we detail the role of Khonin Nuga Research Station as a best practice example for ecological research and sustainable resource management.

Keywords

Mongolia · Mountain forests · Boreal forests · Khan Khentii · Khonin Nuga

Introduction

The world faces critical environmental, social, and economic issues. Climate change is impacting human livelihoods, especially in mountainous and continental areas of the world. Mountain forests are vital biogeographical regions that are essential for biodiversity and often referred to as hydrological 'water towers' (specifically for northern Mongolia, cf. Gradel et al. 2019, ch. 3). These areas exert a dominant influence over fluvial regimes, aiding agriculture and socio-ecological systems downstream, yet they are increasingly threatened by climate change and human activities (FAO 2022; Foley et al. 2005). Glacial recession and altitudinal climatic shifts affect hydrological inputs to forest ecosystems (Beniston 2003) and cause habitat compression and biodiversity loss, which are additionally intensified by pests and pathogenic outbreaks linked to global warming (Dirnböck et al. 2003).

The intricate terrain of mountain forests entails both opportunities and challenges regarding ecological connectivity. On the one hand, these forests can serve as vital wildlife corridors, ensuring genetic flow between habitats and facilitating species migration. This aspect of connectivity is essential, especially as species are expected to shift to more favourable habitats as a result of climate change. On the other hand, the ruggedness of mountains can sometimes act as a barrier to mobility (e.g. Altanbagana and Naranbaatar 2022). Combined with increasing human interference and fragmentation, this natural barrier effect may lead to isolated 'habitat islands', which can become vulnerable pockets of biodiversity in vastly altered landscapes (Pauli et al. 2001).

Mongolia, a landlocked country in Central Asia, is situated in the same Palaearctic realm as Europe but retains more pristine habitats and biotopes than e.g. Central Europe (Chuluunbaatar et al. 2020; Mühlenberg et al. 2000). Located in the transition zone between the deserts of Central Asia and the boreal taiga of southern Siberia, Mongolia occupies a unique geographical position, with its northern forests forming the southernmost edge of the largest continuous forest system on earth, the Northern Eurasian boreal forest belt (Chuluunbaatar et al. 2020) representing 27% of global forests (FAO 2020a). The entirety of the country is part of the Mongolian plateau (Zhen et al. 2010) with an average elevation of 1580 m above sea level

(Yamkhin et al. 2022); this high altitude represents one of the factors of the strong continentality of the region's climate. Mongolia is disproportionately impacted by climate change and its effects resulting in environmental shifts (Taylor et al. 2021).

Mongolia is commonly categorised into 16 ecosystem types (Batbold et al. n.d.; Chimed-Ochir et al. 2010) based on an atlas of ecosystems developed in the joint Russian–Mongolian biological expeditions (Gunin and Saandar 2019; Vostokova and Gunin 2005). WWF Mongolia has adopted this approach as *fine grading* and aggregated it into a *medium grading* of 14 regions used for species habitat modelling as well as a *coarse grading* dividing Mongolia into four ecoregions for broader analyses (Chimed-Ochir et al. 2010): the Daurian steppe (28.2% of the country's total area of 1,566,500 km^2 (Namsrai et al. 2019)), Khangai (16.4%), the Central Asian Gobi (16.4%), and the Altai-Sayan (23.1%). These grading systems have become a de facto standard for ecosystem descriptions and research in Mongolia (Enkhtur et al. 2020). The chapter highlights the role of boreal forest ecosystems in Mongolia for nature conservation and presents an overview of research with relevance to ecological connectivity, with special emphasis on the Khonin Nuga Research Station located in the Khentii Mountains. Mongolia—and specifically the region around Khonin Nuga—can serve as a reference example for pristine forest ecosystems as compared to managed forests in similar biogeographic regions throughout the rest of the world and help to better understand the challenges and specificities of mountain forests around the globe.

Mountain Forests in Mongolia

As of 2020, Mongolia's forest fund[1] covers approximately 18.6 million ha (FRDC 2021) or about 12% of the country's total geographical area; it includes designated forest land and open forest land with low forest cover. A classification by forest land types according to forest fund definition and main tree species is presented in Table 32.1.[2]

Areas per international definition differ from the FRDC classification: Per international definition (FAO 2020a) taiga forests in northern Mongolia comprise about 14.1 million ha and are predominantly composed of coniferous species (85.9%) (MET 2021), while 2.6 million ha are other wooded lands (mainly saxaul forests and scrubs) (FAO 2020b) (i.e. 16.7 million ha by FAO definition).

The vegetation structure of forests within the Khentii region depends strongly on altitude and exposition (in line with international research, e.g. Martin et al., 2023). Despite the large-scale ecosystem mapping by the Russian–Mongolian biological expeditions (Gunin and Saandar 2019; Vostokova and Gunin 2005), there still is no

---

[1] The forest fund consists of the areas covered by forests, including all species of trees and scrub replanted forests as well as saxauls, as defined in the Forest Law of Mongolia (Dorjtseden 1998).
[2] Note: Definitions in the Mongolian forest fund differ from FAO and UNFCC definitions (Enkhtaivan et al. 2018; FAO 2020a); the respective areas thus differ as well, but overall distributions should be comparable.

Table 32.1 Share of forest fund types and main tree species (based on FRDC 2021)

Classification and species	Area (×1000 ha)	Share of forest fund (%)
Siberian larch (*Larix sibirica*)	7348	39.5
Scots pine (*Pinus sylvestris*)	505	2.7
Siberian pine (*Pinus sibirica*)	613	3.3
Fir (*Abies sibirica*)	20	0.1
Spruce (*Picea obovata*)	1	<0.1
Birch (*Betula platyphylla*)	1193	6.4
Poplars (*Populus laurifolia, Populus tremula*)	48	0.2
Willow (*Salix* spp.)	152	0.1
Elm (*Ulmus pumila*)	3	0.8
Dense forest	9884	<0.1
Open forest land (including saxaul and scrubs)	1967	10.6
Natural forest	*11,852*	*63.7*
Temporally unstocked forests	760	4.1
Planted forests	8	<0.1
Total forests	*12,619*	*67.9*
Other land under the forest fund	5976	32.1
Total forest fund	*18,596*	*100*

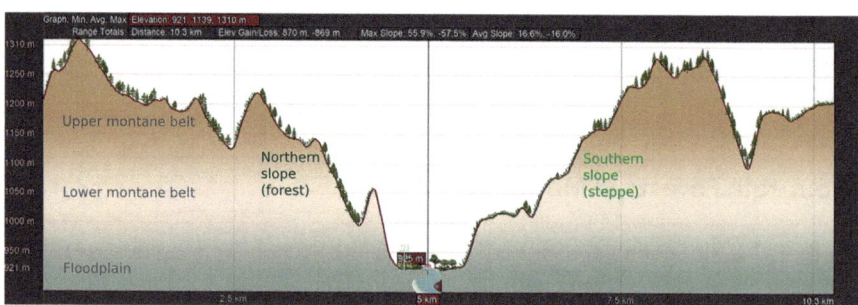

Fig. 32.1 Illustration of vegetation zones in Khan Khentii (exemplary profile of 10 km north-to-south elevation profile, with Khonin Nuga Research Station in the middle). The elevation profile was created using the Google Earth Pro application

comprehensively tested ecological classification of Mongolian forests. While one statistically founded approach was published by Kusbach et al. (2019), it has hitherto not been commonly used. Currently, taiga forests are usually classified into *dark taiga* consisting of shade-tolerant tree species and *light taiga* consisting of light-demanding species (Mühlenberg et al. 2012), with light taiga only growing where shade-tolerant species cannot exist due to climate or soil conditions, while sun-exposed southern slopes are covered by meadows and mountain steppe (Dulamsuren et al. 2005). In addition, a classification into belts is frequently used, with some authors differentiating between four belts for the region (sub-taiga, mountain taiga, pseudo-taiga, and subgolts) (Ogureeva and Bocharnikov 2014; Tsedendash 1993) while others merely differentiate between a lower and an upper montane belt (Dulamsuren et al. 2005) (cf. illustration in Fig. 32.1):

- Lower montane belt (900–1200 m a.s.l.): These forests are often referred to as sub-taiga forests and are dominated by light taiga species such as *Larix sibirica* and *Betula platyphylla*, which preferentially grow on relatively dry northern slopes.
- Upper montane belt (1200–1600 m a.s.l.): Dominated by dark taiga forests made up of *Pinus sibirica*, *Abies sibirica*, and *Picea obovata* growing at the most humid sites.

Threats to Ecological Connectivity

Over the past few decades, Mongolia's biodiversity has come under increasing pressure from economic growth, hunting, logging, land degradation, mining, and climate change (Namsrai et al. 2019). In the light taiga ecosystem, fire and pest infestation represent circumstances for natural regeneration (disaster regeneration) occurring every 80–120 years. As a result of the mentioned disturbances, higher fire frequencies, earlier regeneration, and a shorter life cycle of the light taiga forest are expected. The impacts of climate change, such as exceptional warming, rapid permafrost thawing resulting in peatland degradation (Enkhtaivan et al. 2018), and extreme weather events, as well as the resulting sudden changes in biodiversity, are similar in severity and magnitude to those in the Siberian region (Callaghan et al. 2021) and represent combined drivers of forest ecosystem degradation (Munkhjargal et al. 2020). The main threats to Mongolia's forest ecosystems, in particular, are threefold: livestock, demand for fuel and industrial wood, and forest fires (with additional and sometimes interlinked issues discussed below) (Tsogtbaatar 2004). Around 40% of Trans-Baikal forests have been more or less damaged over the last century (Schmidt-Corsitto 2017).

Forest Fires

In the Baikal catchment area, light taiga species like *Pinus* spp. and *Larix* spp. are prevalent and crucial for the region's ecological stability. The continental and semi-arid climate (high summer temperatures, high maximum temperatures, and low annual precipitation) means an elevated risk of fire. While some forest fires are natural (Schmidt-Corsitto 2017), an increase in their frequency and severity negatively affects old-growth forests. Surface fires, especially on isolated slopes, harm tree viability (Goldammer and Furyaev 1996). The early 1990s saw a rise in forest fires, which remained largely uninvestigated due to resource constraints. Between 2001 and 2021, fires caused 89% of Mongolia's total tree cover loss. However, tree cover loss from fires appears to have decreased since 2010 (Global Forest Watch 2022), even though forest fires seem to exhibit an increasing trend (based on an analysis of global CO_2 emissions; Zheng et al. 2023).

Fires are a major factor in determining the spatial and temporal dynamics of forest ecosystems (Goldammer 2002, as referenced in Kolář et al. 2020). In Mongolia,

50 to 60 forest fires occur annually on average, of which around 95% are caused by human influence, for example, by herders or antler collectors (Goldammer and Furyaev 1996; Hessl et al. 2012; Kolář et al. 2020). Fires are relatively common in north-central Mongolia due to its distinct continental climate combined with highly flammable conifers (especially pine and larch) and the mixed ground vegetation adjacent to steppe areas (Goldammer 2002, 2007, as referenced in Kolář et al. 2020). Along with overgrazing from livestock and unmanaged or illegal logging, wildfires represent a serious risk at lower elevations (Kolář et al., 2020). Lightning is the most common natural cause of forest fires in the boreal landscape.

Forest fires affect landscape diversity as well as energy flows and biogeochemical cycles, but they also have an impact on forest age, structure, species composition, and physiognomy (Grabherr 1997; Schulze et al. 2005; Wirth 2005; Goldammer and Furaev 1996 as referenced in Mühlenberg et al. 2012).

Forest Degradation, Logging, and Hunting

The integrity of forest ecosystems in northern Mongolia is threatened by various human activities. According to Tsogtbaatar (2004), the region's forests have a comparatively low carrying capacity, making them particularly vulnerable to the impacts of logging. This vulnerability is further compounded by the underreporting of timber harvest activities and the prevalence of illegal logging. Although the past decade has witnessed significant efforts to curb illegal logging practices, the repercussions of these activities continue to pose challenges to forest conservation. Wingard and Zahler (2006) also note that hunting, while an indirect factor, often contributes to forest degradation by leading to an increase in forest fires, typically due to negligent behaviour.

The hunting of mammals not only increases the pressure on Mongolia's forest ecosystems but also on the broader ecological balance. Wingard et al. (2018) and WWF (2010) highlight that hunting exacerbates the stress on these habitats, contributing to a significant decline in big game populations over the past 30 years, as detailed by Taylor et al. (2021). Furthermore, the competition between wildlife and livestock for resources—exacerbated by herding practices—has placed some large mammal species under threat, as documented by Ripple et al. (2015). This competition often results in wildlife displacement and can negatively influence the natural ecological processes within these forested areas.

Mining, Agricultural Expansion, and Herding

Since the early 1920s, Mongolia has engaged in mining, with a surge in the 1990s due to political shifts towards an open-market economy. This transition has led to increased mining activities, especially for coal and gold, accompanied by land-use changes such as deforestation and urbanisation (Endicott 2012). Influenced by government regulations and self-regulation efforts, many mining companies have adopted best practices for mining projects, including rehabilitation and post-closure management (McIntyre et al. 2016). While large-scale mines often follow these practices, smaller operations in developing regions pose significant environmental and social risks. The implementation of effective regulation is crucial for environmentally and socially responsible mining.

Since the economic transition of Mongolia from a centralised economy to a free-market economy, there has been a significant increase in the number of people engaging in nomadic livestock farming. However, some of these 'new' nomads lack adequate experience in livestock management and are unaware of the ecosystem's carrying capacity. The increase of herders has been accompanied by a massive increase in the number of livestock, leading to overgrazing and deterioration of grasslands along with a strongly increased burden on the forest steppe, as well as negative edge effects in forests and damage to forest habitats (Khishigjargal and Tsogt 2017; for birch forests, see e.g., Enkhtuya and Jaavkhlan 2022). Grazing and browsing in forested areas are destructive to plantations and tree growth, causing deformities and stunted growth. Natural regeneration can be strongly inhibited by grazing (Juřička et al. 2020a) and trampling (Tsogtbaatar 2004). In small forest patches, livestock entails major destructive impacts on remnant forest growth (Tsogtbaatar 2004). Depending on the species and plant communities, environmental change associated with forest edges—including from anthropogenic causes—can have a negative effect on biodiversity (Sukhbaatar 2018; for lichen, e.g. Boudreault et al. 2008; Enkhtuya and Jaavkhlan 2022), inducing a need for improvement in sustainable forest management (Enkhtaivan et al. 2018). Cattle and especially goats as bark strippers (Juřička et al. 2020a; Vallentine 2014) can cause significant damage, albeit on a smaller scale than e.g. climate change (Juřička et al. 2020a). Increased livestock density has led to an increase in forest fragmentation and habitat isolation. An extremified hydrological cycle is likely to further exacerbate these problems (Taylor et al. 2021).

Climate Change, Pests, and Diseases

Yu et al. (2003), as referenced in Gradel et al. (2019), identified a warming hotspot southeast of Lake Baikal affecting the forest steppe and drylands of northern China and Mongolia. Overall, significantly warmer and drier conditions, as well as more frequent droughts, are to be expected in the region (Batima et al. 2005; Dulamsuren and Hauck 2008; IWRM 2009, as referenced in Gradel et al. 2019).

Research in the Darhad Valley of North-Central Mongolia shows diverging growth responses of the Siberian larch (*Larix sibirica*) (James 2011), with growth declines observed throughout most of the Mongolian taiga since the 1950s (Dulamsuren et al. 2011; Kansaritoreh et al. 2018) and potential for further decrease due to drought stress (Dulamsuren et al. 2009). Sukhbaatar (2018) found a more resilient response and a current xeric site dominance of pine (*Pinus sylvestris* L.), with studies from the Alaskan boreal zone (Barber et al. 2000; Wilmking et al. 2004) documenting climate-change-related moisture stress in tree-ring studies and permafrost degradation leading to deterioration of forest stands and vice versa (Genxu et al. 2012), likely exacerbating a shift of forest communities away from Siberian larch to pine trees (Juřička et al. 2020b). Furthermore, Gradel et al. (2017a), as referenced in Gradel et al. (2019), documented that younger trees are more susceptible to drought than older trees, with potential adverse effects on regeneration.

In the Khan Khentii region, there has been a notable increase in pest and disease outbreaks in recent years, a phenomenon that compounds the damage from abiotic stressors such as fire and drought, which erode the resilience of tree stands (Gradel et al. 2019). Research suggests that climatic changes—especially the rise in winter temperatures and alterations in snow cover due to global warming—may enhance the survival rates of harmful insects. Such conditions are detrimental to the growth of key tree species like larch and birch (Dulamsuren et al. 2011; Khishigjargal et al. 2014; Gradel et al. 2017a, as referenced in Gradel et al. 2019).

In combination with forest fires, climatic change also facilitates the proliferation of parasites such as the Siberian gypsy moth (*Dendrolimus superans sibiricus* Tschetv.), which can lead to widespread infestations (Goldammer and Furyaev 1996). Fires of low intensity can foster fungal diseases and attract further insect infestations, thus compromising arboreal health. The most severe fires can result in profound ecological degradation, requiring up to three centuries for complete ecosystem recovery (Krasnoshekov et al. 1990; cf. Goldammer and Furyaev 1996).

Conservation Areas in Mongolia

Conservation has a long history in Mongolia. Current conservation efforts have their roots in the protected areas system established since the '*Khalkh Juram*' law of 1709, which designated 14 mountains as protected from hunting and logging (United Nations 2018), as well as in the cultivation and designation of the first government-protected area in 1778 (MNET 2017). The current centrepiece of environmental legislation is the 1995 Environmental Protection Law of Mongolia (Yembuu 2021). In addition, the 1994 law on Special Protected Areas defines four categories of protected areas: Strictly protected areas (SPA), national parks (NP), nature reserves (NR), and natural monuments (NM). The protected area system also includes local protected areas (LPA) placed under special protection either at the *aimag* (province) or the *sum* (district) level. The National Program on Protected Aeras adopted by the Mongolian Parliament in 1998 pursues the goal of establishing a system of special protected areas to cover 30% of Mongolia's territory (Chimed-Ochir et al. 2010).

Mongolia's conservation areas have been steadily increasing over the years, with 120 protected areas comprising 32.7 million ha, or 21% of the country's total area, as of 2020 (MET 2021). Of these, 21 were SPAs (13.8 million ha), 37 were national parks (13.5 million ha), 48 were nature reserves (5.3 million ha), and 14 were nature monuments (126 thousand ha; Fig. 32.2).

Furthermore, the current legal framework has designated rights and responsibilities to local authorities (at various administrative levels) to establish and manage

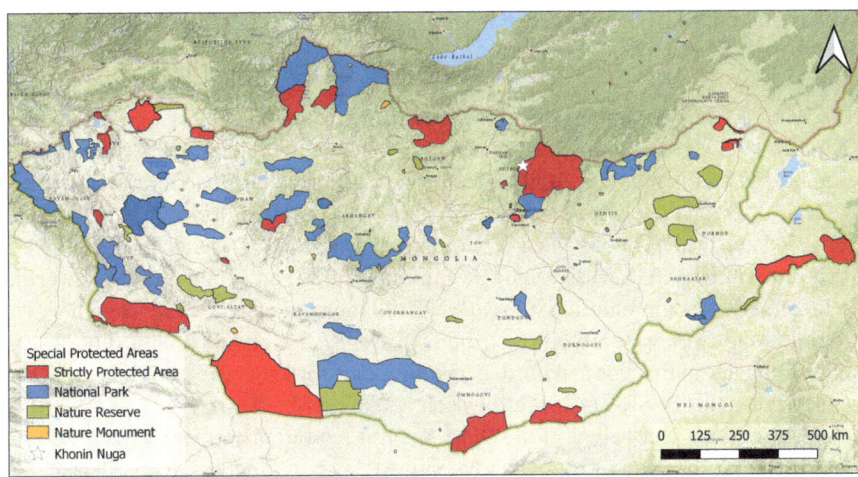

Fig. 32.2 Distribution of special protected areas within Mongolia (created using data from MET 2023; Protected Planet 2023), with Khonin Nuga Research Station depicted as a white star. Map created using QGIS 3.22.11 LTR

Table 32.2 Overview of protected areas in Mongolia (as referenced in MET 2021)

Administrative level	Category	No. of areas	Total size (million ha)	% of total territory
State level	Strictly protected area	21	13.8	8.8
	National park	37	13.5	8.6
	Nature reserve	48	5.3	3.4
	Nature monument	14	0.1	< 1
	Subtotal	*120*	*32.7*	*21*
Province/District level	Local protected areas	2745	68.7	43.9
	Total	*2865*	*101.4*	*64.9*

local protected areas (LPAs). Around 2745 areas (68.7 million ha) in 21 provinces are currently under local protection (MET 2021), although the available information about their actual protection status and role in habitat conservation is unclear since they largely aim to restrict the exploitation of mineral resources (Chimed-Ochir et al. 2010) (cf. Table 32.2).

In addition to these conservation efforts, six of the protected sites are included in the UNESCO List of World Heritage Sites, seven in the UNESCO List of World Human and Biosphere Reserves, and 11 in the Ramsar List of Wetlands of International Importance. Mongolia is currently cooperating with 14 international organisations to improve its conservation efforts (Convention on Biodiversity 2018; MET 2021).

Reference Example: Khan Khentii Region and Khonin Nuga Research Station

The Khan Khentii mountain range is part of the Trans-Baikal biogeographic region (Kim et al. 2021) as well as of the Daurian Steppe ecoregion (cf. Chimed-Ochir et al. 2010) and is recognised globally for biodiversity conservation due to its vast natural landscape. A large share of the area is covered in primary boreal forests partly located in continuous and discontinuous permafrost regions (Saruulzaya 2017).

The Khan Khentii SPA is of global importance for biodiversity conservation due to its largely pristine natural landscape (Chuluunbaatar et al. 2020). Together with the adjoining Gorkhi-Terelj NP, the Khan Khentii SPA stretches from just north-east of Ulaanbaatar to the Russian border. The SPA was set under protection in 1992 and gained the status of SPA in 1995; it has an extent of 12,270 km^2.

Khonin Nuga is a valley in the West Khentii region of northern Mongolia situated in the buffer zone of the Khan Khentii SPA where the two rivers Sharlan and Khongi meet and create the river Eruu, the upper part of the watershed of Lake Baikal (Mühlenberg et al. 2012). At the geographical coordinates 49.0871 °N, 107.2907 °E at 930 m a.s.l. (see Fig. 32.3), Khonin Nuga Research Station was established in 1997 as a cooperation project of the National University of Mongolia and the University of Göttingen, Germany (Mühlenberg et al. 2000).

The station is located in the transition zone between the southern extent of the Siberian taiga and the forest steppe. There, elements of boreal coniferous forests meet the vegetation of the Central Asian steppe (Dulamsuren 2004). The forest types around Khonin Nuga are of specific interest because they represent the southernmost extent of the dark taiga in Central Asia.

The research objectives of Khonin Nuga were to
1. Address fundamental questions of ecology and establish reference plots in an environment largely unaffected by human impact.
2. Evaluate the conservation value of the region with a special focus on the naturalness and heterogeneity of habitats as well as the occurrence of species threatened in other regions.

The results include inventories of taxa, e.g. 619 plant species found in the study area out of 2823 plant species in the entire Mongolian territory (22% of the Mongolian flora), 16 plant species from the Mongolian Red List (out of around 100 listed species) recorded in the study sites, 11 newly recorded plant species, and 63

Fig. 32.3 Site illustration: (**a**) Eruu river, (**b**) Pinus sibirica-dominated mountain forest, and (**c**) soil sampling at the lower montane belt

plant species newly recorded in the Khentii biogeographic region (34,000 km²). Extensive botanical surveys (Dulamsuren 2004; Dulamsuren et al. 2005), soil studies (Dulamsuren 2004), and habitat stratifications have been conducted at Khonin Nuga. An important contribution to the differentiation of habitats in the research area was achieved as well: Following the classification effort, 21 habitat types aggregated into eight habitat groups were identified (Gradel and Mühlenberg 2011; Mühlenberg et al. 2004).

Role of Khan Khentii Region for Connectivity

The overall ecological landscape of the Khan Khentii region is characterised by a complex mosaic of habitats ranging from dense forest ecosystems to expansive steppe grasslands, all of which contribute to a great richness of biodiversity. The region, with its varied altitudinal gradient as shown above, supports a comparatively wide range of forest communities from Siberian larch (*Larix sibirica*) and various mixed forests composed of birch (*Betula* spp.) and pine (*Pinus sylvestris*) to shrub forests and stretches of deciduous broadleaf forests dominated by Mongolian oak (*Quercus mongolica*) in the lower regions, as well as river basin floodplain meadow and meadow steppe ecosystems.

These forested areas are crucial habitats for large vertebrates such as Eurasian elk (*Alces alces*) and red deer (*Cervus elaphus*) as well as numerous avifauna species (Purevdorj et al. 2022; BirdLife International 2023), providing both foraging resources and critical reproductive habitat. The forests of the Khan Khentii mountain range also support a variety of mosses (e.g. *Ptilium crista-castrensis, Pleurozium schreberi, Hylocomium splendens*), lichens (e.g. *Buellia erubescens, Lecanora symmicta, Parmelia squarrosa*), and invertebrates (e.g. *Carabus canaliculatus* M.Adams, *Pterostichus interruptus* Dejean, *Xylotrechus hircus* Gebler) as well as amphibians (e.g. *Hyla japonica* Günther, *Salamandrella keyserlingii* Dybowski, *Rana amurensis* Boulenger), reptiles (e.g. *Gloydius halys*), and birds (e.g.) of conservation importance, including e.g. *Clanga clanga, Aquila heliaca, Falco cherruga* (Sundev et al. 2019).

The steppe ecosystem characterised by dominant grass species (e.g. *Stipa baicalensis, Phragmites communis*) transitions into the less productive Gobi Desert periphery. This gradient from forest to grassland likewise provides a variety of habitats, some of which are crucial for the needs of selected species groups (e.g. Mühlenberg et al. 2000; on the role of meadows for butterflies: Chuluunbaatar et al. 2020) as well as enabling seasonal migrations and gene flow between populations. However, the habitat continuity of some of these ecosystems—specifically habitats such as floodplain meadows—is increasingly threatened by anthropogenic climate change, which has been documented to impact phenological events and alter species distributions, potentially leading to habitat alterations. For example, changes in temperature and precipitation patterns can alter plant communities, subsequently impacting the habitats of animals like the Daurian hedgehog (*Mesechinus dauuricus*) and numerous endemic bird species (Chimed-Ochir et al. 2010).

Aquatic biodiversity in northern Mongolia is sustained by riverine systems such as the Onon River basin, which supports a range of species with limited distribution and high conservation value, including the Siberian sturgeon (*Acipenser schrenckii*) and Siberian taimen (*Hucho taimen*) (Hofmann and Battogtokh 2017). The floodplain meadows and steppe ecosystems are also important for avian species, serving as stopover sites and breeding grounds for endangered cranes (e.g. *Grus vipio* and *G. leucogeranus*) and the great bustard (*Otis tarda*). These wetlands provide essential ecosystem services such as nutrient cycling and water quality maintenance that are vital for the survival of these species and the existence of their habitats (Chimed-Ochir et al. 2010).

In view of the role of the region with regard to ecological connectivity, Khan Khentii SPA specifically is of particular interest due to its naturalness and low level of anthropogenic impact, which makes it suitable as a reference area for managed and fragmented ecosystems. Parts of northern Mongolia exhibit great heterogeneity in terms of landscape structure and (micro-)climate, creating a mosaic of different habitats (Gradel et al. 2019). Due to the relatively southern location and gradual transition to open steppe (Balandin et al., 2000, as referenced in Gradel et al., 2019), the forests around Khonin Nuga seem to be comparatively rich in species while at the same time being more sensitive to disturbances than the taiga forests further north (Mühlenberg et al. 2012). Since wildlife perceives landscapes as habitat features that facilitate or resist dispersal (Diniz et al. 2020), the responses to such disturbances are species-specific and depend on biological characteristics, localised habitat factors, and resource availability (Cushman et al. 2013; Rudnick et al. 2012). Well-connected landscapes enable high rates of migration, maintain genetic diversity and demographic stability, and increase adaptive potential (Cushman et al. 2015; Cushman and Landguth 2012; Zeller et al. 2012). The Khonin Nuga region's heterogeneity can thus be used to investigate connectivity and the dispersal of species of interest.

At present, studies on habitat connectivity are still comparatively rare. Existing studies focus on larger vertebrates such as bears (Tumendemberel et al. 2019), snow leopards (Hacker et al. 2023), wolves (Tiralla et al. 2021), and wapitis (Altanbagana and Naranbaatar 2022). These seminal studies, as well as the long-standing international cooperation at Khonin Nuga, have already laid an important foundation for further and more profound studies with regard to forest-dependent species. Publications by WWF Mongolia and the Russian–Mongolian biological expeditions have investigated the distribution of certain species, which could be used for further research.

Outlook

The Khan Khentii region highlights the role of mountain forests as connectors or barriers depending on context, underscoring the need for rigorous scientific research to guide management and conservation efforts (Rudel et al. 2005). In the face of the pressures described in the sections above that affect protected areas, habitats, and species-group-specific connectivity, the importance of management has increased (Namsrai et al. 2019). Comprehensive approaches involving local (at the *sum* and *aimag* level) and national authorities, various governmental agencies (especially the Ministry of Environment and Tourism of Mongolia, and specifically its Forest Department), and non-governmental and international organisations are continuously being implemented (for a review, cf. Namsrai et al. 2019), adapted, and expanded (for a list of environment-related legislation, cf. United Nations 2018 Annex II), and the results of the accompanying assessments are usually implemented in a timely fashion.

Likewise, the anthropogenic impact from extensive grazing, hunting, fishing, and notably from extractive industries such as mining and oil poses significant threats to habitat integrity. These activities can lead to habitat degradation and fragmentation, creating barriers that disrupt the movement patterns of large vertebrates and birds, thus impeding ecological processes like pollination and seed dispersal, leading to genetic isolation of populations (Chimed-Ochir et al. 2010).

Conservation efforts are imperative and must be focused on expanding the network of protected areas, especially in biodiversity-rich ecosystems such as the subboreal mixed forests. These forests harbour keystone species, including various ungulates, which are pivotal for ecosystem functioning. The preservation of a full range of ecosystems—from alpine tundra to high mountain steppe—is essential for maintaining habitat connectivity. A scientifically informed approach to conservation incorporating habitat protection, sustainable land-use practices, and climate adaptation strategies is necessary to mitigate anthropogenic pressures and ensure the persistence of northern Mongolia's ecological connectivity.

A communication-related problem is that there seems to be no unified approach to storing and disseminating information yet, which makes accessing data, research efforts, and insights regarding connectivity difficult. Access to funding and information exchange between various entities is a pressing problem. Internationally, publications show the need for additional funding (McCarthy et al. 2012) as well as measurable improvements with respect to conservation efforts whenever funding is available (Waldron et al. 2017) and adequate program evaluation methods are used (Ferraro and Pattanayak 2006). The implementation of sustainable forest management criteria and indicators as proposed in Erdenejav (2020) and their further expansion specifically to conservation- and biodiversity-related issues as well as species(-group)-specific research will be important contributions towards ensuring the connectivity of habitats in Mongolian mountain forests.

References

Altanbagana Y, Naranbaatar G (2022) Identifying the habitat connectivity of wapiti (Cervus canadensis) in Mongolia. Proc Institute Biology 38(1):9–28. https://doi.org/10.5564/pib.v38i1.2534

Barber VA, Juday GP, Finney BP (2000) Reduced growth of Alaskan white spruce in the twentieth century from temperature-induced drought stress. Nature 405(6787):668–673. https://doi.org/10.1038/35015049

Batbold D, Bathkuyag B, Ganzorig B, Munkhnast D, Munkhchuluun B, Onon Y, Purevdorj S, Sumiya E, Selenge G, Chimeddorj B, Enkhbayar N (n.d.) National Biodiversity Program (2015–2025). Ministry of Environment and Tourism of Mongolia; WWF; UNEP, GEF, p 59

Beniston M (2003) Climatic change in mountain regions: a review of possible impacts. Clim Chang 59(1/2):5–31. https://doi.org/10.1023/A:1024458411589

BirdLife International (2023) Important Bird Areas factsheet: Khan Khentii Strictly Protected Area. http://datazone.birdlife.org/site/factsheet/khan-khentii-strictly-protected-area-iba-mongolia/details

Boudreault C, Bergeron Y, Drapeau P, Lopez LM (2008) Edge effects on epiphytic lichens in remnant stands of managed landscapes in the eastern boreal forest of Canada. For Ecol Manag

Callaghan TV, Shaduyko O, Kirpotin SN, Gordov E (2021) Siberian environmental change: synthesis of recent studies and opportunities for networking. Ambio 50(11):2104–2127. https://doi.org/10.1007/s13280-021-01626-7

Chimed-Ochir B, Hertzman T, Batsaikhan N, Batbold D, Sanjmyatav D, Onon Y, Munkhchuluun B (2010) Filling the gaps to protect the biodiversity of Mongolia. WWF Mongolia, p 136. https://wwfeu.awsassets.panda.org/downloads/filling_the_gaps_final_report_low_resolution_2.pdf

Chuluunbaatar G, Barua KK, Mühlenberg M (2020) Habitat association and movement patterns of the violet copper (Lycaena helle) in the natural landscape of West Khentey in northern Mongolia. Int J Nematology Entomol 7(2):1–8

Convention on Biodiversity (2018) Convention on biodiversity country profiles. Mongolia—biodiversity facts. Secretariat of the Convention on Biological Diversity. https://www.cbd.int/countries/profile/?country=mn

Cushman SA, Landguth EL (2012) Multi-taxa population connectivity in the northern Rocky Mountains. Ecol Model 231:101–112. https://doi.org/10.1016/j.ecolmodel.2012.02.011

Cushman SA, McRae B, Adriaensen F, Beier P, Shirley M, Zeller K (2013) Biological corridors and connectivity. In: Macdonald DW, Willis KJ (eds) Key topics in conservation biology 2, 1st edn. Wiley, pp 384–404. https://doi.org/10.1002/9781118520178.ch21

Cushman SA, McRae H, B., & McGarigal, K. (2015) Basics of landscape ecology: an introduction to landscapes and population processes for landscape geneticists. In: Balkenhol N, Cushman SA, Storfer AT, Waits LP (eds) Landscape genetics, 1st edn. Wiley, pp 9–34. https://doi.org/10.1002/9781118525258.ch02

Diniz MF, Cushman SA, Machado RB, De Marco Júnior P (2020) Landscape connectivity modeling from the perspective of animal dispersal. Landsc Ecol 35(1):41–58. https://doi.org/10.1007/s10980-019-00935-3

Dirnböck T, Dullinger S, Grabherr G (2003) A regional impact assessment of climate and land-use change on alpine vegetation. J Biogeogr 30(3):401–417. https://doi.org/10.1046/j.1365-2699.2003.00839.x

Dorjtseden L (1998) Asia-Pacific forestry sector outlook study: country report—forestry of Mongolia (APFSOS/WP/42). FAO, Rome

Dulamsuren C (2004) Floristische Diversität, Vegetation und Standortbedingungen in der Gebirgstaiga des Westkhentej, Nordmongolei (Reihe A, Bd. 191; Berichte des Forschungszentrum Waldökosysteme). Georg-August University, Göttingen, p 292

Dulamsuren C, Hauck M, Mühlenberg M (2005) Vegetation at the taiga forest–steppe borderline in the western Khentey Mountains, northern Mongolia. Ann Bot Fenn 42(426)

Dulamsuren C, Hauck M, Leuschner C (2009) Recent drought stress leads to growth reductions in Larix sibirica in the western Khentey, Mongolia: GROWTH REDUCTIONS IN LARIX SIBIRICA. Glob Chang Biol. https://doi.org/10.1111/j.1365-2486.2009.02147.x

Dulamsuren C, Hauck M, Leuschner HH, Leuschner C (2011) Climate response of tree-ring width in Larix sibirica growing in the drought-stressed forest-steppe ecotone of northern Mongolia Annals of Forest Science 68(2) 275–282 https://doi.org/10.1007/s13595-011-0043-9

Endicott E (2012) A history of land use in Mongolia. Palgrave Macmillan, London. https://doi.org/10.1057/9781137269669

Enkhtaivan S, Nyamsuren D, Purevjav Z, Battuvshin K, Janchivdorj A, Khaltar M, Tsogt K, Chultem B-U, Byun Y, Mahmood A, Van Rijin M, Sandker M, Vickers B (2018) Mongolia's forest reference level submission to the UNFCCC. Ministry of Environment and Tourism of Mongolia; UN-REDD, p 62

Enkhtur K, Boldgiv B, Pfeiffer M (2020) Diversity and distribution patterns of geometrid moths (Geometridae, Lepidoptera) in Mongolia. Diversity 12(5):186. https://doi.org/10.3390/d12050186

Enkhtuya O, Jaavkhlan S (2022) Identification of the degradation of birch forests habitats by lichen species: (on the example of the mountain forest Ubugunt at the station Shatan Batsumber Sumona, central aimag). Mongolian J Botany 4(30):11–23. https://doi.org/10.5564/mjb.v4i30.2580

Erdenejav E (2020) Report on analytic study on availability of forest data and national C&I set for SFM in Mongolia. p. 59

FAO (2020a) Global Forest Resources Assessment 2020. FAO, Rome. https://doi.org/10.4060/ca9825en

FAO (2020b) Global Forest resources assessment 2020. Country report Mongolia. FAO, Rome, p 57. https://www.fao.org/3/cb0031en/cb0031en.pdf

FAO (2022) The state of the World's forests 2022. Forest pathways for green recovery and building inclusive, resilient and sustainable economies. FAO, Rome. https://doi.org/10.4060/cb9360en

Ferraro PJ, Pattanayak SK (2006) Money for nothing? A call for empirical evaluation of biodiversity conservation investments. PLoS Biol 4(4):e105. https://doi.org/10.1371/journal.pbio.0040105

Foley JA, DeFries R, Asner GP, Barford C, Bonan G, Carpenter SR, Chapin FS, Coe MT, Daily GC, Gibbs HK, Helkowski JH, Holloway T, Howard EA, Kucharik CJ, Monfreda C, Patz JA, Prentice IC, Ramankutty N, Snyder PK (2005) Global consequences of land use. Science 309(5734):570–574. https://doi.org/10.1126/science.1111772

FRDC (2021) Forest resources of Mongolia. Forest Research and Development Center, Ministry of Environment and Tourism, p 30. https://sudalgaa.gov.mn/vendor/pdfjs-dist/web/viewer.html?file=https://sudalgaa.gov.mn/uploads/researches/forestagency/2023/04/04/mongol-ulsyn-oyn-san-tq9.pdf

Genxu W, Guangsheng L, Chunjie L (2012) Effects of changes in alpine grassland vegetation cover on hillslope hydrological processes in a permafrost watershed. J Hydrol 444–445:22–33. https://doi.org/10.1016/j.jhydrol.2012.03.033

Global Forest Watch (2022) Mongolia deforestation rates & statistics. https://www.globalforestwatch.org/dashboards/country/MNG?category=fires

Goldammer JG, Furyaev V (1996) Fire in ecosystems of boreal Eurasia. Springer, Leiden

Gradel A, Mühlenberg M (2011) Spatial characteristics of near-natural Mongolian forests at the southern edge of the taiga. Allgemeine Forst- und Jagdzeitung 182(3/4):40–52

Gradel A, Sukhbaatar G, Karthe D, Kang H (2019) Forest management in Mongolia—a review of challenges and lessons learned with special reference to degradation and deforestation. Geography Environ Sustain 12(3):133–166. https://doi.org/10.24057/2071-9388-2019-102

Gunin PD, Saandar M (eds) (2019) Ecosystems of Mongolia. Atlas. Institute of Ecology and Evolution, Russian Academy of Sciences

Hacker C, Atzeni L, Munkhtsog B, Munkhtsog B, Galsandorj N, Zhang Y, Liu Y, Buyanaa C, Bayandonoi G, Ochirjav M, Farrington JD, Jevit M, Zhang Y, Wu L, Cong W, Li D, Gavette C, Jackson R, Janecka JE (2023) Genetic diversity and spatial structures of snow leopards (Panthera uncia) reveal proxies of connectivity across Mongolia and northwestern China. Landsc Ecol 38(4):1013–1031. https://doi.org/10.1007/s10980-022-01573-y

Hessl AE, Ariya U, Brown P, Byambasuren O, Green T, Jacoby G, Sutherland EK, Nachin B, Maxwell RS, Pederson N, De Grandpré L, Saladyga T, Tardif JC (2012) Reconstructing fire history in Central Mongolia from tree-rings. Int J Wildland Fire 21(1):86. https://doi.org/10.1071/WF10108

Hofmann J, Battogtokh D (2017) *Kharaa—Yeröö river basin atlas* (2nd edition) [Map]. Integrated Water Resources Management in Central Asia : Model Region Mongolia. https://www.igb-berlin.de/sites/default/files/media-files/download-files/MOMO_Atlas_2018_08_10.pdf

James TM (2011) Temperature sensitivity and recruitment dynamics of Siberian larch (Larix sibirica) and Siberian spruce (Picea obovata) in northern Mongolia's boreal forest. For Ecol Manag 262(4):629–636. https://doi.org/10.1016/j.foreco.2011.04.031

Juřička D, Kusbach A, Pařílková J, Houška J, Ambrožová P, Pecina V, Rosická Z, Brtnický M, Kynický J (2020a) Evaluation of natural forest regeneration as a part of land restoration in the Khentii massif, Mongolia. J For Res 31(5):1773–1786. https://doi.org/10.1007/s11676-019-00962-5

Juřička D, Novotná J, Houška J, Pařílková J, Hladký J, Pecina V, Cihlářová H, Burnog M, Elbl J, Rosická Z, Brtnický M, Kynický J (2020b) Large-scale permafrost degradation as a primary factor in Larix sibirica forest dieback in the Khentii massif, northern Mongolia. J For Res 31(1):197–208. https://doi.org/10.1007/s11676-018-0866-4

Khansaritoreh K, Schuldt B, Dulamsuren C (2018) Hydraulic traits and tree-ring width in Larix sibirica Ledeb. as affected by summer drought and forest fragmentation in the Mongolian forest steppe Annals of Forest Science 75(1) https://doi.org/10.1007/s13595-018-0701-2

Khishigjargal M, Tsogt K (2017) Forest degradation issues in forest-steppe ecotone by livestock grazing in northern Mongolia. Book of Abstracts:48–49

Kim HS, Bazarragchaa B, Lee SM, Batdelger G, Park GS, Lee J (2021) Vegetation community classification of the Sanzai area in Mongolia. J Asia-Pacific Biodivers 14(2):228–235. https://doi.org/10.1016/j.japb.2021.02.006

Kolář T, Kusbach A, Čermák P, Štěrba T, Batkhuu E, Rybníček M (2020) Climate and wildfire effects on radial growth of Pinus sylvestris in the Khan Khentii Mountains, north-Central Mongolia. J Arid Environ 182:104223. https://doi.org/10.1016/j.jaridenv.2020.104223

Kusbach A, Štěrba T, Šebesta J, Mikita T, Bazarradnaa E, Dambadarjaa S, Smola M (2019) Ecological zonation as a tool for restoration of degraded forests in northern Mongolia. Geography Environ Sustain 12(3):98–116. https://doi.org/10.24057/2071-9388-2019-31

Martin M, Sohorova E, Fenton NJ (2023) Embracing the complexity and the richness of boreal old-growth forests: a further step toward their ecosystem management. In: Girona MM, Morin H, Bergeron Y (eds) Boreal forests in the face of climate change: sustainable management. Springer International Publishing, Cham, pp 191–218

McCarthy DP, Donald PF, Scharlemann JPW, Buchanan GM, Balmford A, Green JMH, Bennun LA, Burgess ND, Fishpool LDC, Garnett ST, Leonard DL, Maloney RF, Morling P, Schaefer HM, Symes A, Wiedenfeld DA, Butchart SHM (2012) Financial costs of meeting global biodiversity conservation targets: current spending and unmet needs. Science 338(6109):946–949. https://doi.org/10.1126/science.1229803

McIntyre N, Bulovic N, Cane I, McKenna P (2016) A multi-disciplinary approach to understanding the impacts of mines on traditional uses of water in northern Mongolia. Sci Total Environ 557–558:404–414. https://doi.org/10.1016/j.scitotenv.2016.03.092

МЕТ (2021) Монгол орны байгаль орчны төлөв байдлын тайлан 2019–2020 он [Mongolian environmental status report 2019–2020]. Ministry of Environment and Tourism of Mongolia, p 201. https://met.gov.mn/public/storage/1674618662395340.pdf

MET (2023) *Geodatabase of the Forest Department of the Ministry of Environment and Tourism of Mongolia* [GIS Application]. Гео мэдээллийн сан [Geodatabase]. https://forest.gov.mn/website/page.aspx?mm_id=3096&id=17

MNET (2017) Report on the state of the environment of Mongolia. Ministry of Nature. Ministry of Nature, Environment and Tourism

Mühlenberg M, Slowik J, Samiya R, Dulamsuren C, Gantigmaa C, Woyciechowski M (2000) The conservation value of West-Khentii, North Mongolia: evaluation of plant and butterfly communities. Fragm Flor Geobot 45(1–2):63–90

Mühlenberg M, Hondong H, Dulamsuren C (2004) Large-scale biodiversity research in the southern taiga, northern Mongolia. Forset, Snow Landscape Res 78(1/2):93–118

Mühlenberg M, Appelfelder J, Hoffmann H, Ayush E, Wilson KJ (2012) Structure of the montane taiga forests of west Khentii, northern Mongolia. J For Sci 58(2):45–56. https://doi.org/10.17221/97/2010-JFS

Munkhjargal M, Yadamsuren G, Yamkhin J, Menzel L (2020) The combination of wildfire and changing climate triggers permafrost degradation in the Khentii Mountains. Northern Mongolia Atmos 11(2):155. https://doi.org/10.3390/atmos11020155

Namsrai O, Ochir A, Baast O, Van Genderen JL, Muhar A, Erdeni S, Wang J, Davaasuren D, Chonokhuu S (2019) Evaluating the management effectiveness of protected areas in Mongolia using the management effectiveness tracking tool. Environ Manag 63(2):249–259. https://doi.org/10.1007/s00267-018-1124-4

Ogureeva GN, Bocharnikov MV (2014) География разнообразия бореальных лесов у южной границы распространения и их картографирование (горы южной Сибири и Монголии) [Diversity of boreal forests n the mountains of southern Siberia and Mongolia]. Вестник Московского университета 5:53–61

Pauli H, Gottfried M, Grabherr G (2001) High summits of the Alps in a changing climate: the oldest observation series on high mountain plant diversity in Europe. In: Walther G-R, Burga CA, Edwards PJ (eds) "Fingerprints" of climate change. Springer, pp 139–149. https://doi.org/10.1007/978-1-4419-8692-4_9

Protected Planet (2023) *Shapefiles for protected areas of Mongolia*. Protected Planet. https://www.protectedplanet.net/en/country/MNG

Purevdorj Z, Munkhbayar M, Paek WK, Ganbold O, Jargalsaikhan A, Purevee E, Amartuvshin T, Genenjamba U, Nyam B, Lee JW (2022) Relationships between bird assemblages and habitat variables in a boreal Forest of the Khentii Mountain, northern Mongolia. Forests 13(7):1037. https://doi.org/10.3390/f13071037

Ripple WJ, Newsome TM, Wolf C, Dirzo R, Everatt KT, Galetti M, Hayward MW, Kerley GIH, Levi T, Lindsey PA, Macdonald DW, Malhi Y, Painter LE, Sandom CJ, Terborgh J, Van Valkenburgh B (2015) Collapse of the world's largest herbivores. Sci Adv 1(4):e1400103. https://doi.org/10.1126/sciadv.1400103

Rudel TK, Coomes OT, Moran E, Achard F, Angelsen A, Xu J, Lambin E (2005) Forest transitions: towards a global understanding of land use change. Glob Environ Chang 15(1):23–31. https://doi.org/10.1016/j.gloenvcha.2004.11.001

Rudnick D, Ryan S, Beier P, Cushman S, Dieffenbach F, Trombulak S (2012) The role of landscape connectivity in planning and implementing conservation and restoration priorities. Issues Ecol 16:1–20

Saruulzaya A (2017) *Thermokarst lake dynamics in the continuous and isolated permafrost zones, Mongolia* [dissertation]. Hokkaido University

Schmidt-Corsitto K (2017) Adaptation of key forest ecosystems to climate change. In: Bayartogtokh B, Dulamsuren C (eds) Book of abstracts. National University of Mongolia, Ulaanbaatar, pp 48–49

Sukhbaatar G (2018) Effects of scots pine (Pinus sylvestris L.) plantations on plant diversity in northern Mongolia. Mongolian J Biol Sci 18(1):59–70. https://doi.org/10.22353/mjbs.2018.16.08

Sundev G, Conaboy N, Magsar U, Khayankhirvaa T, Chuluunbaatar G (2019) Biodiversity in Mongolia. In: Pullaiah T (ed) Global biodiversity, vol 1. Apple Academic Press; CRC Press, Oakville; Boca Raton, pp 351–394

Taylor W, Hart I, Pan C, Bayarsaikhan J, Murdoch J, Caspari G, Klinge M, Pearson K, Bikhumar U, Shnaider S, Abdykanova A, Bittner P, Zahir M, Jarman N, Williams M, Pettigrew D, Petraglia M, Lee C, Dixon EJ, Boivin N (2021) High altitude hunting, climate change, and pastoral resilience in eastern Eurasia. Sci Rep 11(1):14287. https://doi.org/10.1038/s41598-021-93765-w

Tiralla N, Holzapfel M, Ansorge H (2021) Feeding ecology of the wolf (Canis lupus) in a near-natural ecosystem in Mongolia. Mamm Biol 101(1):83–89. https://doi.org/10.1007/s42991-020-00093-z

Tsedendash G (1993) *Лесорастительность Хэнтэйского нагорья [Forest vegetation of the Khentii highlands. Doctoral thesis summary in Russian]*

Tsogtbaatar J (2004) Deforestation and reforestation needs in Mongolia. For Ecol Manag 201(1):57–63. https://doi.org/10.1016/j.foreco.2004.06.011

Tumendemberel O, Zedrosser A, Proctor MF, Reynolds HV, Adams JR, Sullivan JM, Jacobs SJ, Khorloojav T, Tserenbataa T, Batmunkh M, Swenson JE, Waits LP (2019) Phylogeography, genetic diversity, and connectivity of brown bear populations in Central Asia. PLoS One 14(8):e0220746. https://doi.org/10.1371/journal.pone.0220746

United Nations (2018) Environmental performance reviews. Mongolia. United Nations, New York

Vallentine JF (2014) Grazing Management, 2nd edn. Elsevier Science

Vostokova E, Gunin P (2005) *Ecosystems of Mongolia. Atlas* (general scientific edition). Institute of Ecology and Evolution, Russian Academy of Science

Waldron A, Miller DC, Redding D, Mooers A, Kuhn TS, Nibbelink N, Roberts JT, Tobias JA, Gittleman JL (2017) Reductions in global biodiversity loss predicted from conservation spending. Nature 551(7680):364–367. https://doi.org/10.1038/nature24295

Wilmking M, Juday GP, Barber VA, Zald HSJ (2004) Recent climate warming forces contrasting growth responses of white spruce at treeline in Alaska through temperature thresholds. Glob Chang Biol 10(10):1724–1736. https://doi.org/10.1111/j.1365-2486.2004.00826.x

Wingard J, Zahler P (2006) Silent steppe: the illegal wildlife trade crisis in Mongolia. The World Bank; Wildlife Conservation Society, p 164. http://rgdoi.net/10.13140/RG.2.2.19900.77445

Wingard J, Pascual A, Bhattacharya G (2018) Silent steppe II: Mongolia's wildlife trade crisis, ten years later. The Zoological Society of London; Legal Atlas LLC, London, p 232

WWF (2010) Executive summary of environment, socio-economic baseline studies conducted in Onon river basin. WWF; ADB, Japan Fund for Poverty Reduction, p 33

Yamkhin J, Yadamsuren G, Khurelbaatar T, Gansukh T, Tsogtbaatar U, Adiya S, Yondon A, Avirmed D, Natsagdorj S (2022) Spatial distribution mapping of permafrost in Mongolia using TTOP. Permafr Periglac Process 33(4):386–405. https://doi.org/10.1002/ppp.2165

Yembuu B (ed) (2021) The physical geography of Mongolia. Springer International Publishing, Cham. https://doi.org/10.1007/978-3-030-61434-8

Zeller KA, McGarigal K, Whiteley AR (2012) Estimating landscape resistance to movement: a review. Landsc Ecol 27(6):777–797. https://doi.org/10.1007/s10980-012-9737-0

Zheng B, Ciais P Chevallier F, Yang H, Canadell JG, Chen Y, van der Velde IR, Aben I, Chuvieco E, Davis S, et al. (2023) Record-high CO_2 emissions from boreal fires in 2021 Emission emergency Science 379(6635) 912–917 https://doi.org/10.1126/science.ade0805

Zhen L, Ochirbat B, Lv Y, Wei YJ, Liu XL, Chen JQ, Yao ZJ, Li F (2010) Comparing patterns of ecosystem service consumption and perceptions of range management between ethnic herders in Inner Mongolia and Mongolia. Environ Res Lett 5(1):015001. https://doi.org/10.1088/1748-9326/5/1/015001

Open Access This chapter is licensed under the terms of the Creative Commons Attribution 4.0 International License (http://creativecommons.org/licenses/by/4.0/), which permits use, sharing, adaptation, distribution and reproduction in any medium or format, as long as you give appropriate credit to the original author(s) and the source, provide a link to the Creative Commons license and indicate if changes were made.

The images or other third party material in this chapter are included in the chapter's Creative Commons license, unless indicated otherwise in a credit line to the material. If material is not included in the chapter's Creative Commons license and your intended use is not permitted by statutory regulation or exceeds the permitted use, you will need to obtain permission directly from the copyright holder.

33. Paraguay: Toward a Landscape Restoration of the Paraguayan Atlantic Forest

Maria Laura Quevedo-Fernández, Haroldo Nicolás Silva-Imas, Lidia Florencia Pérez-de-Molas, Lila Mabel Gamarra-Ruiz-Díaz, Alba Liz González, Stella Mary Amarilla-Rodríguez, and Lucia Janet Villalba-Marin

Canopy cover in Atlantic Forest (Photo: Lila Mabel Gamarra-Ruiz-Díaz)

M. L. Quevedo-Fernández (✉) · L. F. Pérez-de-Molas · L. M. Gamarra-Ruiz-Díaz · A. L. González · S. M. Amarilla-Rodríguez · L. J. Villalba-Marin
Universidad Nacional de Asunción, Facultad de Ciencias Agrarias, San Lorenzo, Paraguay
e-mail: laura.quevedo@agr.una.py; lidia.perezmolas@agr.una.py; lila.gamarra@agr.una.py; alba.gonzalez@agr.una.py; stella.amarilla@agr.una.py; janet.villalba@agr.una.py

H. N. Silva-Imas
Itaipú Binacional, Hernandarias, Hernandarias, Paraguay
e-mail: haroldos@itaipu.gov.py

© The Author(s) 2025
K. Lapin et al. (eds.), *Ecological Connectivity of Forest Ecosystems*,
https://doi.org/10.1007/978-3-031-82206-3_33

Abstract

Forest landscape restoration (FLR) models with attractive economic potential are needed to pique the interest and to be adopted by private landowners—who own most of the land in Paraguay—to connect the fragments of the Paraguayan Atlantic Forest. This chapter contextualizes the process of deforestation, degradation, and fragmentation of Paraguay's forests while also showcasing the first forest restoration initiatives in the country and recording the incipient results for the economic approach, which considers functional groups of tree species as well as the use of forest products. To establish this economically attractive approach, an experimental plot was installed in 2021 in areas belonging to the Itaipú Binacional entity in eastern Paraguay. Eight treatments were tested in a randomized block design where strips emphasized different management goals. Monitoring was carried out for up to 18 months, and the preliminary results reveal five identified treatments delivering significant differences with superior management options with respect to the survival of the species composing the strips; in the analysis of dendrometric variables, *Eucalyptus* stands out as a facilitating exotic species. This experimental area serves as a demonstration site for adoption by interested parties.

Keywords

Forest landscape restoration · Survival · Growth · Functional groups · Diversity species · Filling species · Facilitating species · Economic approach to forest restoration · High-diversity forest restoration

Past and Present of the Landscape of the Atlantic Forest of Paraguay

Forest degradation caused by humans is as old as humankind's existence (Rodrigues 2013). Between 1990 and 2020, 420 million hectares of forests were lost worldwide to deforestation (Food and Agriculture Organization 2022). The conversion of areas into grasslands and agricultural lands is the main direct driver of deforestation in South America (Sy et al. 2015).

Paraguay is a Mediterranean country located in the center of South America, with an area of 406,752 km² and a population of 6,109,644 (Instituto Nacional de Estadística 2023); it is divided into an eastern and a western region by the Paraguay River. With 36.6% native forest cover, Paraguay is rich in forest resources (Instituto Forestal Nacional 2023), but deforestation has been a pressing problem in recent decades, mainly due to the rapid expansion of the agricultural frontier (World Bank 2020). According to Da Ponte et al. (2021), Paraguay has become one of the countries with the highest deforestation rates in the world over the last 40 years.

The Atlantic Forest shared by Brazil, Argentina, and the eastern region of Paraguay is among the most threatened tropical forests in the world (Di Bitetti et al. 2003). It has been identified as a priority for conservation on the national scale because it constitutes a center of endemism and is one of the most threatened biomes on the planet (De Egea Juvinel and Balbuena 2011).

The forest cover of the eastern region in 1945 was 8,805,000 ha (Bozzano and Weik 1992), of which only 2,693,190.01 ha remained in 2023 (Instituto Forestal Nacional 2023) as fragmented, degraded forests with deteriorated biodiversity and key ecological processes jeopardizing its balance in the medium and long term. The largest forest areas remaining today are mostly protected areas, several of which have alarming problems with illegal cultivation.

Taking into account the various connectivity and conservation initiatives to recover the ecological integrity of the Paraguayan Atlantic Forest and improve human well-being in the region, forest landscape restoration (FLR) emerges as a valid alternative. In addition to seeking to restore forest ecosystems and make them self-sustainable, it aims to provide benefits to both people and biodiversity.

Forest Restoration in the Protected Areas of Itaipú Binacional: Employed Approaches and Lessons Learned

The Itaipú Binacional hydroelectric dam is a joint venture managed by Paraguay and Brazil located on the Paraná River between the cities of Hernandarias (Paraguay) and Foz do Iguazú (Brazil). With an installed capacity of 14,000 MW through 20 generating units, it supplies approximately 10.8% of Brazil's and 88.5% of Paraguay's energy needs.

Remarkably, Itaipú oversees 120,541 hectares of land dedicated to environmental conservation, aimed at preserving biodiversity, enhancing water quality, and extending the dam's operational life. A significant portion of these areas, vital for conserving Atlantic Forests, is situated in Paraguay and includes several conservation units and a protective buffer zone around the reservoir.

Despite initial challenges like degradation due to prior agricultural use and threats from illegal activities, Itaipú has prioritized forest restoration since 1979. Employing evolving methodologies over the years—including timber species, tree diversity species, and functional groups models—it has demonstrated a steadfast commitment to environmental restoration and sustainability sustainability (Fig. 33.1).

The *forest restoration model with emphasis on native timber species* was used in the 1980s and 1990s. In this model, the most economically and culturally representative tree species were the most used. Production of these species was relatively easy, and they could also be used in tree planting work along public roads and in parks and squares. Among these species, *Handroanthus heptaphyllus, H. impetiginosus, H. pulcherrimus, H. albus, H. ochraceus,* and *Tabebuia roseo-alba* stand out. Further abundant species in this model were the legumes *Peltophorum dubium, Anadenanthera colubrina* var. *cebil, Parapiptadenia rigida, Pterogyne nitens,* and *Enterolobium contortisiliquum.* Native fruit tree species such as *Annona neosalicifolia, Psidium* spp., *Eugenia uniflora,* and *E. myrcianthes* were also used to a lesser extent. The planting density varied from 400 to 1,000 trees/ha. Maintenance

Fig. 33.1 Forest cover 2 years after establishment of forest restoration plots using the diversity model (**a**) and the functional group model (**b**)

consisted mainly of the removal of all other vegetation than the species planted during the first year, although generally this care lasted longer due to the low planting density and the aggressiveness of the invasive exotic grasses *Urochloa brizantha* and *Megathyrsus maximus*. In the sites where interventions using this model were carried out, exotic grasses were not eradicated, and forests with a simple structure were established: a single stratum composed almost exclusively of planted trees, without undergrowth and with few naturally regenerating plants. Over the years, many of these forests experienced fires and returned to their initial degraded condition.

By the beginning of the 2000s, *forest restoration work began to focus on establishing the greatest possible of native diversity species* (i.e., applying the tree diversity species emphasis model). The number of species employed depended largely on their availability in the nurseries, with species exhibiting rapid growth being prioritized. The species most commonly used in these models were, therefore, *Handroanthus heptaphyllus*, *Cordia trichotoma*, *Inga laurina*, *Parapiptadenia rigida*, *Peltophorum dubium*, and *Cordia americana*. On average, the plantations were made up of 36% pioneer species and 38% secondary species distributed randomly within the plots. The planting density used was 3 × 2.5 m, equivalent to 1,333 trees/ha. The survival of the trees planted under this design was 78% in the second year, considered within the acceptable range for plantations with native species. However, the maintenance work required to ensure the consolidation of forest cover under this model extended up to 9 years in some sites.

Only in 2017 were the first monitoring studies carried out in the areas under the forest restoration regime of Itaipú Binacional. These studies pointed to a need to reduce the formation times of closed and continuous forest cover in the maintained areas. This situation led to the establishment of the first forest restoration trial in 2018, where eight forest restoration models were evaluated considering the selection of species by functional groups for filling and diversity, the proportion of these functional groups, planting times by functional groups, and the use of green manures. The results of this trial would become decisive for interventions in new areas over the coming years.

The *forest restoration model was adopted with a functional group emphasis* in 2021—a model developed by the Forest Ecology and Restoration Laboratory of the Luiz De Queiroz Higher School of Agriculture, University of Sao Paulo, Brazil (Rodrigues 2007). This approach consists of grouping forest species into *filling* and *diversity* species, with the filling group composed of species featuring good height growth and canopy coverage to provide rapid shading of the area while maintaining maximum longevity to keep it covered or shaded for as long as possible. The species not included in the covering group due to their slower growth as well as lack of adequate ground cover and longevity are included in the so-called diversity group.

The planting density in this model is 3 × 2 m (1,667 trees/ha) with a proportion of 50% filling and 50% diversity species. Individuals within each planting line alternate between the two groups. The green manure *Cajanus cajan* is planted between the rows, contributing to coverage and the improvement of soil properties.

Itaipú Binacional's forest restoration program is one of the most important in Paraguay. The improvement of its interventions coincides with advances in scientific knowledge in ecological restoration as well as responding to findings produced by local research and lessons learned within the framework of adaptive management.

Info Box 33.1: Forest Restoration Projects in Paraguay

The first recorded forest restoration project in Paraguay was initiated by Itaipú Binacional with the installation of its forest nursery in 1979. Other initiatives have emerged since 2016, such as that of Guyra Paraguay, which has helped to restore 100 hectares with yerba mate (*Ilex paraguariensis*) and 42 hectares in the Guyra Reta Reserve. Eighty hectares corresponding to the deficit of forest-protecting watercourses have been restored since 2017 in the Naranjal district with the "Sustainable Naranjal Project," and in the following year, the Guarapi Natural Reserve began its restoration process, restoring about 19 hectares. Also, in 2018, work was carried out to identify potential biological corridors in Itapúa and propose strategies to promote forest recovery or ecological restoration of degraded areas.

In the year 2019, actions pursuing the restoration of forests protecting watercourses in Caazapá were initiated, and the designing of the technical manual on restoring forests that protect watercourses was begun.

The initiatives of 2020 were those related to the Management Plan for Invasive Exotic Species at Guasú Metropolitan Park along with a project led by the National University of Asunción that was planned for the restoration of the forest landscape of the Yaguarón district.

Since 2022, WWF Paraguay has been working on several forest restoration projects, including "Road to restoration of the Yaguarón Forest landscape," as well as supporting other initiatives such as the Trinational Atlantic Forest Restoration Network. In the same year, the Moisés Bertoni Foundation, along with other allies, began the implementation of projects to improve the recharge

(continued)

conditions of the Patiño aquifer and ensure sustainable management of the Ypacaraí Lake basin.

Likewise, in 2022, the project to build biodiversity islands in the metropolitan area of Asunción and the enrichment of municipal forest nurseries with native seeds were initiated.

There are also projects related to the restoration of productive systems through agroforestry on rural farms, as well as other initiatives in various companies.

At the end of 2022, the participatory development process of the "National Forest Restoration Plan" began, led by the National Forestry Institute with support from the Food and Agriculture Organization of the United Nations within the framework of the Paraguay + Green project. In all these initiatives, the need for collaborative and participatory work and the aim of properly documenting experiences has been evident.

Looking for New Alternatives for Forest Landscape Restoration

The next challenge is the implementation of forest restoration models by private landowners since developing and recognizing suitable models is not enough if forest restoration is not undertaken by the private landowners to whom most of the country's land belongs and who, in most cases, must bear the costs of restoration.

In this context, Itaipú Binacional, together with the Faculty of Agrarian Sciences of the National University of Asunción, is experimenting with restoration models that seek to transform forest restoration into an attractive activity from an economic point of view. These restoration models aim to offer economic returns to the producer by reducing implementation costs and generating profits through the economic valuation of ecosystem goods and services with the use of native timber, energy species, non-timber trees (*Ilex paraguariensis*—yerba mate and *Euterpe edulis*—palmetto), and *Eucalyptus* as an exotic facilitating species with a wide market in the country, combined in different treatments in strips using the functional group approach.

For experimentation with this new approach offering economic benefits, experimental plots located at reference coordinates latitude: $-25.333614°$ and longitude: $-54.638458°$ were established (Fig 33.2).

The experimental area where the study was developed belongs to the Itaipú Binacional entity; it is located within the Itaipú Biosphere Reserve in the district of Hernandarias, department of Alto Paraná, Paraguay.

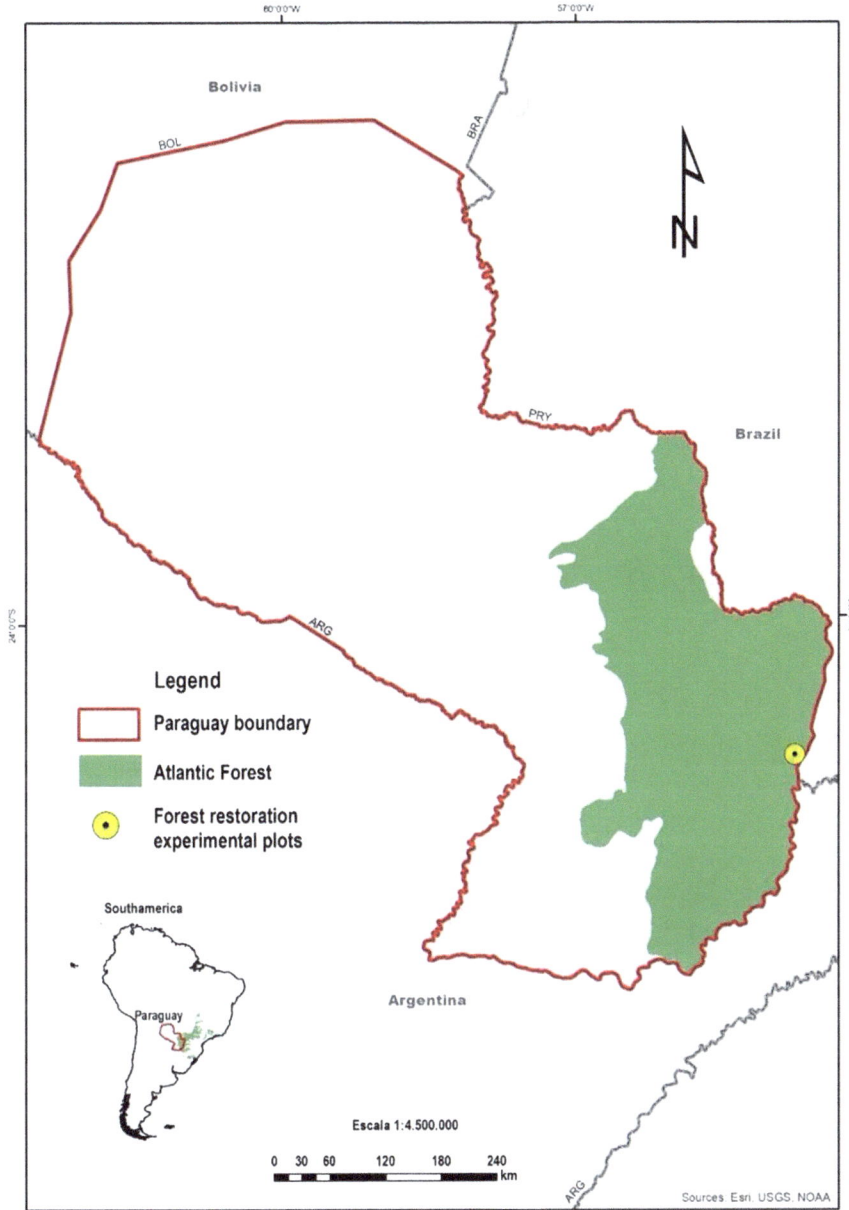

Fig. 33.2 Study area and location of the experimental plots

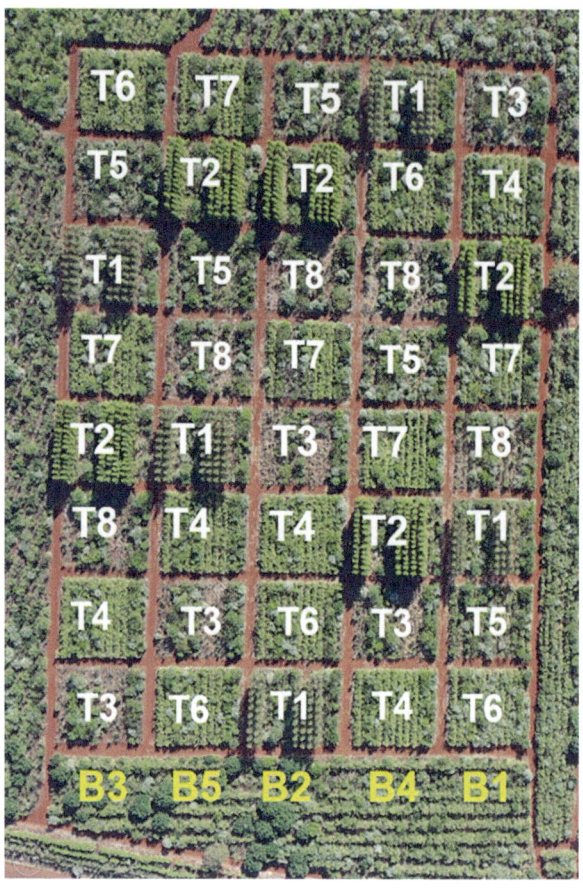

Fig. 33.3 Distribution of blocks (B1–B5) and treatments (T1–T8) in the experimental plot; drone image from June 4, 2023

Experimental Outline

The experimental plot has a randomized block design with five blocks and eight treatments, containing 6,750 trees of 30 forest species as well as the green manure *Cajanus cajan*. The total area is 5.9 hectares; each experimental unit occupies an area of 1,080 m^2 (30 × 36 m), with internal and perimeter roads four meters wide between the experimental units and along the edge (see Fig. 33.3).

Each experimental unit has 12 planting rows, and all experimental treatments are structured into strips, including harvest strips and conservation strips. The conservation strips maintain the same structure of alternating filling and diversity species, with a spacing of three meters between planting rows and two meters between plants in the same row.

In the harvest strips, which is where the treatments differ, the spacing varies between two meters (treatments 2, 5, 6, 7, and 8) and three meters (treatments 1, 3, and 4) between the plants in the planting row. The measurement area is the highlighted yellow area inside the dashed line in Fig. 33.4.

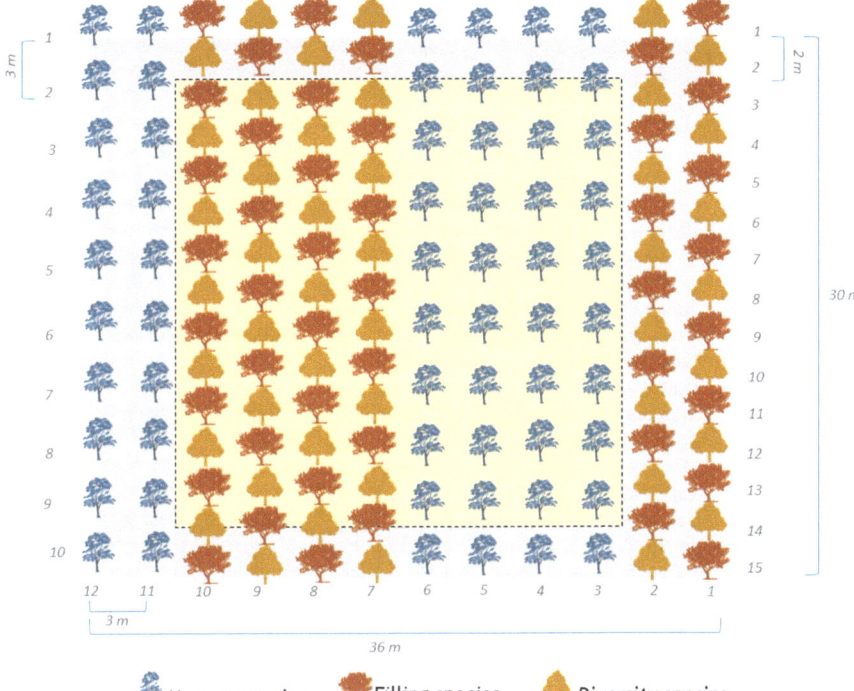

Fig. 33.4 Design of the experimental units including harvest strips and conservation strips, with diversity trees and filling trees as well as the measurement area and border area

Four months after planting, a site inspection was carried out to verify the species of each planted individual; this verification was based on macromorphological characteristics. Table 33.1 presents the family and scientific name of each planted species according to the Conosur Flora catalog, as well as the respective code used to present the results more efficiently. Each code is assembled from the first two letters of the respective genus and the first two letters of the species-specific epithet.

Table 33.1 also shows the number of individuals planted per species, the classification of species as pioneers or non-pioneers based on Barbosa et al. (2017), and the functional group (diversity and filling species, or the productive function for which the species was selected for the protection strip). Finally, each species was classified according to its seed dispersal syndrome into anemochory (by wind), autochory (self-dispersal), and zoochory (by animals).

The species in the *conservation strips* were selected from the filling and diversity functional groups according to their availability in the forest nurseries of Paraguay. The conservation strips in all experimental units have the same species distribution and are spaced at 3 × 2 m.

The *diversity species* used in the experiment are *Alchornea glandulosa* ssp. *iricurana*, *Annona neosalicifolia*, *Astronium fraxinifolium*, *Campomanesia*

Table 33.1 List of species in the experiment with family, scientific name, code used, common name in Paraguay, number of individuals planted, successional group, functional group, and seed dispersal syndrome

Family	Species	Code	Common name	No.	SG	FG	SD
Anacardiaceae	*Astronium fraxinifolium* Schott	asfr	urunde'y para	138	np	d	an
Anacardiaceae	*Lithraea molleoides* (Vell.) Engl.	limo	chichita	160	p	d	zo
Anacardiaceae	*Myracrodruon urundeuva* Allemão	myur	urunde'y mi	356	np	pne	an
Anacardiaceae	*Schinus terebinthifolia* Raddi var. *terebinthifolia*	scte	molle'i	250	p	c	zo
Anonnaceae	*Annona neosalicifolia* H. Rainer	anne	aratiku	200	np	d	zo
Aquifoliaceae	*Ilex paraguariensis* A.St.-Hil.	ilpa	yerba mate	450	np	pfnm	zo
Bignoniaceae	*Handroanthus heptaphyllus* (Vell.) Mattos	hahe	lapacho negro	146	np	pnm	an
Bignoniaceae	*Handroanthus impetiginosus* (Mart. ex DC.) Mattos	haim	lapacho rosado	124	np	pnm	an
Boraginaceae	*Cordia americana* (L.) Gottschling & J.S.Mill.	coam	guajaivi	151	np	pne	an
Boraginaceae	*Cordia trichotoma* (Vell.) Arráb. ex Steud.	cotr	peterevy	140	np	pnm	an
Caricaceae	*Jacaratia spinosa* (Aubl.) A.DC.	jasp	jakaratia	155	np	d	au
Euphorbiaceae	*Alchornea glandulosa* ssp. *iricurana* (Casar.) Secco	algl	chipa rupa	366	np	d	zo
Euphorbiaceae	*Croton urucurana* Baill.	crur	sangre de drago	157	p	c	au
Fabaceae	*Anadenanthera colubrina* var. *cebil* (Griseb.) Altschul	anco	kurupa'y kuru	360	np	pne	au
Fabaceae	*Dahlstedtia muehlbergiana* (Hassl.) M.J.Silva & A.M.G.Azevedo	damu	ka'a vusu	200	np	d	an
Fabaceae	*Enterolobium contortisiliquum* (Vell.) Morong	enco	timbo	200	np	d	zo
Fabaceae	*Inga uraguensis* Hook. & Arn.	inur	inga guasu	135	p	d	zo
Fabaceae	*Mimosa bimucronata* (DC.) Kuntze var. *bimucronata*	mibi	jukeri'i	140	p	c	an
Fabaceae	*Myrocarpus frondosus* Allemão	myfr	incienso	149	np	pnm	an
Fabaceae	*Parapiptadenia rigida* (Benth.) Brenan	pari	kurupa'y ra	221	p	pne	an
Fabaceae	*Peltophorum dubium* (Spreng.) Taub.	pedu	yvyra pyta	227	np	pnm	an
Malvaceae	*Guazuma ulmifolia* Lam. var. *ulmifolia*	guul	kamba aka	201	p	e	zo
Malvaceae	*Heliocarpus popayanensis* Kunth	hepo	apeyva	200	p	c	an
Malvaceae	*Luehea divaricata* Mart.	ludi	ka'a oveti	364	p	d	an
Myrtaceae	*Campomanesia xanthocarpa* (Mart.) O. Berg	caxa	guavira pyta	160	np	d	zo
Myrtaceae	*Eucalyptus urophylla* S.T. Blake × *Eucalyptus camaldulensis* Dehnh.	euuc	Eucalipto (VM01)	300	e	pee	au

(continued)

Table 33.1 (continued)

Family	Species	Code	Common name	No.	SG	FG	SD
Myrtaceae	Eucalyptus urophylla S.T. Blake × Eucalyptus grandis W.Hill.	euug	eucalipto (I144)	450	e	pem	au
Rutaceae	Balfourodendron riedelianum (Engl.) Engl.	bari	guatambu	140	np	pnm	an
Rutaceae	Helietta apiculata Benth.	heap	yvyra ovi	150	np	pne	an
Solanaceae	Solanum granulosoleprosum Dunal	sogr	hu'i moneha	359	p	c	zo

No. Number of individuals planted, *SG* Successional group, *p* pioneer, *np* non-pioneer, *FG* Functional group, *d* diversity species, *c* filling species, *pne* native energetic species, *pnm* native timber species, *pfnm* non-timber forest products, *SD* Seed dispersal syndrome

xanthocarpa, Dahlstedtia muehlbergiana, Enterolobium contortisiliquum, Inga uraguensis, Jacaratia spinosa, Lithraea molleoides, and *Luehea divaricata.* The filling species employed are *Croton urucurana, Heliocarpus popayanensis, Mimosa bimucronata* var. *bimucronata, Schinus terebinthifolia* var. *terebinthfolia,* and *Solanum granulosoleprosum.*

The composition of the *harvest* strip for all treatments is detailed below:

Treatment 1, composed of *Eucalyptus urophylla* x *E. grandis,* corresponds to clone I144 as a facilitating species. The individuals in the harvesting strip were pruned at 15, 18, 21, and 24 years, and nine meters of pruning are expected for the harvesting of knot-free wood. Two thinnings are planned at approximately 3 and 6 years, and a harvest is expected at approximately 10 years, providing products for lamination, sawing, and energy (firewood or chips).

Treatment 2, composed of *Eucalyptus urophylla* x *E. camaldulensis,* specifically the VM01 clone as a facilitating species, to be used for energy purposes (firewood or chips) or cellulose. No pruning or thinning activities are planned, and the harvest is expected after approximately 5 years.

Treatment 3, composed of six native timber forest species (15 individuals per species), which are: *Balfourodendron riedelianum, Cordia trichotoma, Handroanthus heptaphyllus, Handroanthus impetiginosus, Myrocarpus frondosus,* and *Peltophorum dubium.*

Treatment 4, with the same format as the previous treatment, with the exception that *Cajanus cajan* (pigeon pea or kumanda yvyra'i) was sown in the planting lines as green manure.

Treatment 5, composed of six native species for energy use (15 individuals per species), which are *Anadenanthera colubrina* var. *cebil, Cordia americana, Guazuma ulmifolia* var. *ulmifolia, Helietta apiculata, Myracrodruon urundeuva,* and *Parapiptadenia rigida.*

Treatment 6 is similar to the previous one, with the difference of green manure sown in the planting lines as in Treatment 4.

Treatment 7, composed of *Ilex paraguariensis,* with green manure likewise planted in the interrow.

Table 33.2 Planning, installation, and maintenance activities

Year	Month	Activities
2021	03	Cleaning the area, preparing the soil with fertilization, and opening roads
	05 & 06	Seedlings plantation (except yerba mate), and irrigation (B1 and B4)
	09	First replacement, yerba mate seedling plantation and green manure
	11	Manual and chemical control of weeds (in all treatments)
2022	01	Chemical weed control (in all treatments)
	03	Second replacement
	03	Chemical weed control (in T3, T4, T5, T6, T7, and T8)
	05	First yerba mate replacement
	05	Chemical weed control (in T1, T2, T3, and T7)
	08	First pruning (in T1)
	09	Chemical weed control (in T5 and T8)

Treatment 8, features nine native forest species, of which three are energetic, and six are timber species, *Guazuma ulmifolia* var. *ulmifolia*, *Anadenanthera colubrina* var. *cebil*, *Myracrodruon urundeuva*, *Balfourodendron riedelianum*, *Cordia trichotoma*, *Handroanthus heptaphyllus*, *Handroanthus impetiginosus*, *Myrocarpus frondosus*, *Peltophorum dubium*. The palmetto *Euterpe edulis* will subsequently be planted between the lines of the harvest strips.

Preparation, Implementation, Maintenance, and Monitoring

The experimental plot is in an area previously illegally occupied by an apparent agricultural use, which was recovered by Itaipú Binacional and subsequently restored. The restored area was eventually overgrown by the grass *Urochloa* sp. Table 33.2. shows a summary of the activities carried out after the planning and design of the experiment and related to the preparation, installation, and maintenance of the area.

Monitoring of the experimental plot was performed in the periods detailed in Table 33.3, and all monitoring was carried out in five days. Figure 33.5 shows images of the treatments 4 and 18 months after planting.

The data collected were related to survival, dendrometric variables (diameter at the root collar, height, crown area), soil samples for chemical, physical, and biological analysis, weed records, grass cover, necromass, and other parameters. Leaves were collected to study functional diversity, and tree phenology was recorded. The results for survival and the dendrometric variables at 18 months are briefly presented below without considering the replacement of dead individuals.

Table 33.3 Monitoring carried out on the experimental plot

Evaluated periods					
Period and months since planting	Months	Start	End	Evaluated days	Start of monitoring
1 (2 m)	Jun-Jul	1/6/2021	02/8/2021	62	02/08/2021
2 (4 m)	Aug-Sep	3/8/2021	21/9/2021	49	21/09/2021
3 (6 m)	Oct-Nov	22/9/2021	21/11/2021	60	21/11/2021
4 (8 m)	Dec-Jan	22/11/2021	06/2/2022	76	06/02/2022
5 (10 m)	Feb-mar	7/2/2022	27/3/2022	48	27/03/2022
6 (12 m)	Apr-may	28/3/2022	29/5/2022	62	29/05/2022
7 (15 m)	Jun-Aug	30/5/2022	29/8/2022	91	29/08/2022
8 (18 m)	Sep-Nov	30/8/2022	12/12/2022	104	12/12/2022

Survival and Dendrometric Variables

For the survival analysis up to 18 months after planting, an analysis of variance was carried out considering the randomized block design. For this analysis, the assumptions of ANOVA were verified, where according to the O'Neill-Mathews test with 5% significance, the variances can be considered homogeneous ($p = 0.6317$), while the Shapiro-Wilk test indicates that, with 5% significance, the residuals can be considered normal ($p = 0.4305$).

The analysis of variance revealed that there is no statistical difference between the blocks, but there is between the treatments. The Tukey test showed that treatments 1 (with 73.59% survival), 2 (80.91%), 3 (70.79%), 5 (76.59%), and 6 (71.59%) are statistically different and superior to the others in terms of the survival rates of the species making up the experiment, while treatments 8 (69.77%) and 4 (68.15%) are statistically similar and treatment 7 (33.63%) is inferior to the others (Fig. 33.6).

The low survival percentage of treatment 7 (33.63%) was mainly a result of the low survival rate of its harvest species *Ilex paraguariensis*, which had no survivors after the 2021/2022 summer period.

Ilex paraguariensis is a climax species that tolerates medium-intensity shading at any age, with greater tolerance to light during the adult phase (Carvalho 2008). It is shade-loving during the planting process, however, making it important to prevent direct afternoon sunlight from reaching the base of the seedling (Penteado and Gomes dos Reis Goulart 2019). The species also requires abundant humidity and an average temperature between 15.5 and 25.5 °C. The need for water intensifies in summer due to heat, and the demand for water during these months is even greater for young plants (de Zelada Cardozo and Gonzalez Villalba 2019).

The low survival rate of this species appears to be directly related to the extremely high temperatures and lack of precipitation during the summer: In the months of

Fig. 33.5 Photographic record created during the monitoring periods, comparing the status of the treatments 4 and 18 months after planting. The images show the treatments in block 5

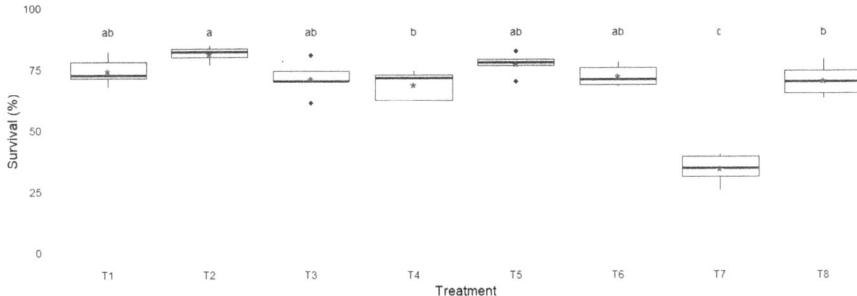

Fig. 33.6 Tukey test for comparison of survival means by treatment. Means identified with the same letters do not differ from each other (Tukey, $p < 0.05$)

November and December 2021, as well as in January and February 2022, precipitation levels were below the historical average in the study area; similarly, the average temperatures were higher than historical averages during these months, reaching temperatures of 39 °C.

In Fig. 33.7, the distribution of results obtained for all species' dendrometric variables at 18 months is shown in a boxplot: *Eucalyptus urophylla* x *E. camaldulensis* (euuc), *Eucalyptus urophylla* x *E. grandis* (euug), and *Mimosa bimucronata* var. *bimucronata* (mibi) stand out with regard to the evaluated variables.

The considerable difference in height and diameter at the root collar between the native species and *Eucalyptus urophylla* x *E. camaldulensis* (euuc, clone VM01), respectively *Eucalyptus urophylla* x *E. grandis* (euug, clone I144) is due to the genetic improvement programs to which the latter were subjected to obtain the employed hybrids.

Vilela de Resende and Silva Alves (2021) mention that genetic improvement programs are a priority in the forestry sector with the objective of obtaining superior genotypes, mainly for cloning. This is because most eucalyptus plantations are established with clones derived from hybrid plants. In addition to sheer volumetric productivity, the technological properties of wood are essential characteristics in genetic improvement programs.

However, both *Eucalyptus* varietals were easily surpassed in canopy cover by *Mimosa bimucronata* var. *bimucronata* (mibi), a hedge species commonly used for precisely this reason. All the species initially proposed as filling species until the present evaluation were correctly developed as such.

Final Considerations

Restoration of the Atlantic forest is split into three phases: the structuring phase (0–15 years), during which the first canopy is formed; a second phase of consolidation (15–30 years) corresponding to the period during which the trees of the initial canopy enter senescence and begin to die while the secondary species

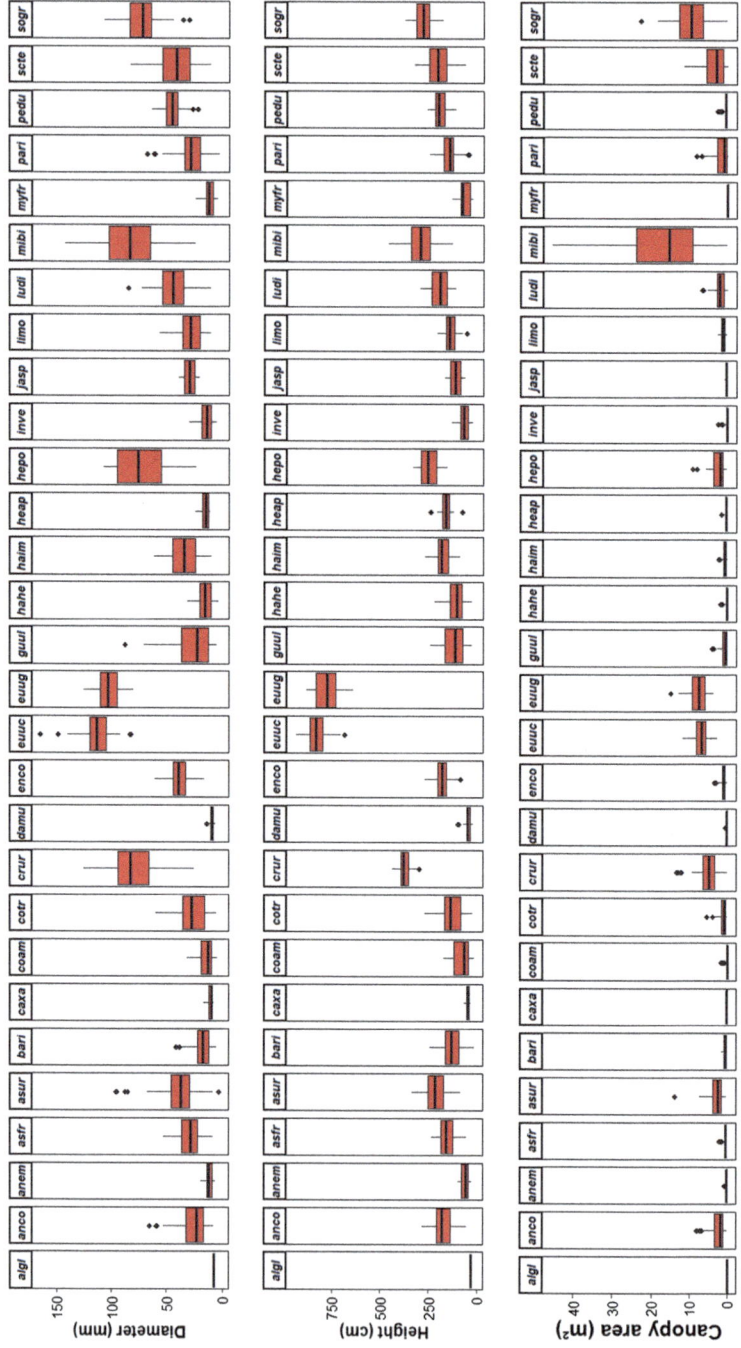

Fig. 33.7 Diameter at the root collar, height, and canopy area of all species in the study plot at 18 months after planting (for abbreviations, see Table 33.1)

simultaneously start to grow and occupy the space left by the crowns of the dying pioneer trees, slowly forming a new canopy; and the third phase of maturation (30 years and beyond) without a defined end, during which a slow accumulation of new species and forms of plant life, fauna, biomass, organic matter, nutrients, stratification, interactions, etc., occurs, along with a gradual conversion of the canopy into a mosaic of clearings, forming a dynamic mosaic dominated by climax species, although pioneer and initial secondary species will also be present (Brancalion et al. 2015).

It is important to remember that the advances and improvements in the models presented in this chapter correspond to the beginning of the first stage of forest restoration—the formation phase of the initial forest cover (first canopy). The subsequent challenge will, therefore, be to continue monitoring these models throughout the following stages, always maintaining flexibility and adaptive management for suitable decision-making in a changing world.

In addition, it is important to continue working on cost and income analysis and make the experiment described here a demonstration plot to help landowners adopt these models. It is thus essential to document our forest restoration experiences with an emphasis on aspects of monitoring and employment of adaptive management practices that will allow improvement and adjustment of the developed models or initiatives with better responses and development. This simultaneously provides economic data useful for systematizing the costs associated with restoration initiatives, a crucial aspect for their planning and projection over time.

The application of increasingly diverse and well-studied forest restoration models must be managed within a broad and equitable participatory framework that guarantees results of sustainable use and efficient conservation in terms of maintaining the provision of environmental services and other benefits for owners, local communities, and other key actors and sectors involved.

The Paraguayan initiative for the participatory formulation of the National Forest Restoration Plan is highlighted as an experience that has collected all the restoration initiatives applied in the country, synthesizes advances and limitations concerning the subject, and simultaneously emphasizes the urgency of the coordinated establishment of forest restoration initiatives adapted to the reality of the respective site and compliant with the principles governing their sustainability.

References

Barbosa LM, Shirasuna R, de Lima F, Torres Ortiz PR, Cavalheiro Barbosa K, Cavalheiro Barbosa T (2017) Lista de espécies indicadas para restauração Ecológica para diversas regiões do Estado de São Paulo

Bozzano B, Weik JH (1992) El Avance de la Deforestación y el Impacto Económico: Proyecto de Planificación del Manejo de los Recursos Naturales. Ministerio de Agricultura y Ganadería

Brancalion PH, Gandolfi S, Rodrigues R (2015) Restauração florestal. Oficina de textos

Carvalho PER (2008) Espécies arbóreas brasileiras, vol 3, 1st edn. Embrapa Informação Tecnológica

Da Ponte E, García-Calabrese M, Kriese J, Cabral N, Perez de Molas L, Alvarenga M, Caceres A, Gali A, García V, Morinigo L, Ríos M, Salinas A (2021) Understanding 34 years of Forest cover dynamics across the Paraguayan Chaco: characterizing annual changes and Forest fragmentation levels between 1987 and 2020. Forests 13(1):25. https://doi.org/10.3390/f13010025

De Egea Juvinel J, Balbuena C (2011) Adopción de los Criterios de Altos Valores de Conservación de Recursos Naturales en la Reserva San Rafael

de Zelada Cardozo NJ, Gonzalez Villalba JD (2019) Guía técnica cultivo de yerba mate. https://www.jica.go.jp/Resource/paraguay/espanol/office/others/c8h0vm0000ad5gke-att/gt_07.pdf

Di Bitetti M, Placci G, Dietz LA (2003) Una visión de biodiversidad para la ecorregión del Bosque Atlántico del Alto Paraná: Diseño de un paisaje para la conservación de la biodiversidad y prioridades para las acciones de conservación

FAO (2022) El estado de los bosques del mundo 2022. FAO, Rome. https://doi.org/10.4060/cb9360es

Instituto Forestal Nacional (2023) Reporte Nacional de Cobertura forestal y cambios de uso de la tierra 2020–2022

Instituto Nacional de Estadística (2023) TRÍPTICO DE PROYECCIONES DE LA POBLACION 2023. https://www.ine.gov.py/Publicaciones/Proyeciones%20por%20Departamento%202023/

Penteado J, Gomes dos Reis Goulart IC (2019) Erva 20 Sistema de produção de erva-mate (Embrapa Florestas)

Rodrigues RR (2007) High diversity Forest restoration in degraded areas: methods and projects in Brazil. Nova Publishers, Hauppauge

Rodrigues E (2013) Ecologia da Restauração. Editora Planta

Sy VD, Herold M, Achard F, Beuchle R, Clevers JGPW, Lindquist E, Verchot L (2015) Land use patterns and related carbon losses following deforestation in South America. Environ Res Lett 10(12):124004. https://doi.org/10.1088/1748-9326/10/12/124004

Vilela de Resende MD, Silva Alves R (2021) Genética: Estratégias de melhoramento e métodos de seleção. En *O eucalipto e a Embrapa: Quatro décadas de pesquisa e desenvolvimento*

World Bank (2020) A Forest's worth: policy options for a sustainable and inclusive Forest economy in Paraguay. World Bank, Washington, D.C. https://doi.org/10/A-Forest-s-Worth-Policy-options-for-a-sustainable-and-inclusive-forest-economy-in-Paraguay

Open Access This chapter is licensed under the terms of the Creative Commons Attribution 4.0 International License (http://creativecommons.org/licenses/by/4.0/), which permits use, sharing, adaptation, distribution and reproduction in any medium or format, as long as you give appropriate credit to the original author(s) and the source, provide a link to the Creative Commons license and indicate if changes were made.

The images or other third party material in this chapter are included in the chapter's Creative Commons license, unless indicated otherwise in a credit line to the material. If material is not included in the chapter's Creative Commons license and your intended use is not permitted by statutory regulation or exceeds the permitted use, you will need to obtain permission directly from the copyright holder.

34. Serbia: Transnational Ecological Corridor Connectivity and Invasive Plant Species (Sava River Basin)

Alen Kiš and Klara Szabados

Sava River (Photo: Adobe Stock/UroSzUp)

Abstract

Large European rivers are flowing through cultural landscapes. Their floodplains are usually narrowed, often keeping remnants of the natural vegetation in strips, with nodal extensions only at sites extremely prone to flooding.

Sava River, the largest Danube tributary by its water flow, is still well preserved considering its free streaming from Slovenia, downstream borders between Croatia and Bosnia and Herzegovina, to its confluence in Belgrade, Serbia. The river corridor is functioning as a lifeline connecting different

A. Kiš (✉) · K. Szabados
Department of Ecological Network and Biodiversity, Institute for Nature Conservation of Vojvodina Province, Novi Sad, Republic of Serbia
e-mail: alen.kis@pzzp.rs; klara.szabados@pzzp.rs

habitats. It is essential to wildlife migrations, seed transfers, and gene flow for both native and introduced species. Among the latter, many have been recognized to have invasive behavior. Invasive alien species (IAS), being recognized as the second most important cause of biodiversity loss on the global scale, are thus exploiting the river corridor, invading the natural habitats, and reducing the transnational river corridor connectivity.

A platform for joint cross-border actions against IAS was developed in the Danube Transnational Project. Differences in legislative frameworks in the economies along the Sava River are still hampering its full implementation, giving the invasive species both time and space to further develop their populations, suppressing the native species, and increasing their influence in terms of ecological, social, and economic impacts.

Keywords

River corridor · Invasive species · Cross-border challenges · Joint approach

Study Area and the Connectivity Issue

Originating in the Slovenian Alps, then forming part of the border between Croatia and Bosnia and Herzegovina while meandering along the southern boundaries of the Carpathian Basin to its confluence into the Danube at Belgrade in Serbia, the Sava River shapes the landscape, connects diverse habitats, and is affected by different policies, economies, and land use practices.

Rivers constitute the most important ecological corridors in the landscapes they flow through (Forman 1983; Gurnell et al. 1994). These lifelines for wildlife are likewise expressed in the cultural landscape of Europe since the continent's largest forest habitats have survived along its international river corridors. The riverine zones are essential habitats for preserving biological diversity (Naiman et al. 1993; Naiman and Décamps 1997), and the flooding dynamics of these habitats are key events that modify environmental conditions and determine the arrangement and succession of species. In contrast to small rivulet corridors creating narrow habitat strips suitable for few species (Gilbert-Norton et al. 2010), river plains provide large habitat matrices within near-natural landscape mosaics, functioning both as source habitats and corridors for various wildlife. Underscoring their importance for biodiversity protection, these preserved forest habitats in the landscape matrix create halo or spillover effects of plant species to neighboring areas, thereby facilitating spontaneous habitat restoration (Brudvig et al. 2009).

The Sava basin, including the river's tributaries, covers almost 100,000 km^2 (Schwarz 2016) and is one of the best-preserved and diverse river systems in Europe. According to the river basin management plan (ISRBC 2022), forests and semi-natural areas contribute 55% of the total river basin area. Along its course of 926 kilometers, the Danube's largest tributary by water discharge carves narrow gorges as well as creating extensive gravel banks and wide oxbows, thus supporting a spectrum of different riverine and floodplain habitats (Fig. 34.2).

Fig. 34.1 The morphological floodplain of the Sava River and its tributaries (Schwarz 2016)

Forest cover plays a significant role in the landscape by forming extensive connected canopy-covered areas, particularly in the upland zones of the river basin. The morphological floodplain, as the most important corridor for wildlife in the cultural landscape, is also rich in forest (Fig. 34.1). The forest matrix in the lowlands is fragmented by agricultural and urban zones, but when we narrow the focus to the active inundation area—which, however, has been trained and reduced to only 23% of its historical extent—we find a greater proportion of forest cover in the landscape (Schwarz 2016).

The water-rich river directly feeds a mosaic of forests, marshes, and meadows in the active floodplain. Landscape restoration (Fig. 34.3), in such cases, requires a holistic approach that considers the functional integrity of the habitat types (César et al. 2021). By replenishing the groundwater and moderating soil moisture at a distance behind the levees extending from Zagreb in Croatia to Belgrade in Serbia, the Sava provides additional water essential to the lowlands, promoting the growth of the famous Slavonian oak-ash-elm forests (Kozarac 1888). In recent times, the largest share of the forest tree species within the historical and active floodplain have been pedunculate oak (*Quercus robur*), narrow-leaved ash (*Fraxinus angustifolia*), and hornbeam (*Carpinus betulus*), followed by field maple (*Acer campestre*), Tatarian maple (*A. tataricum*), and other deciduous hardwood species (Schwarz 2016). Field elm (*Ulmus minor*) is not present anymore in the canopy layer due to Dutch elm disease, which caused sudden death to millions of elm trees already in the first half of the twentieth century in this region (Vajda 1952).

The hardwood forest types, with the common oak as a keystone species, cover more than 140,000 ha in the morphological floodplain, while softwood willows and poplars have significantly lower coverage, with its majority in the foreland (Fig. 34.2). In light of the heavily narrowed inundation and the above-depicted differences in the land cover types in the active and the former floodplain, the importance of maintaining ecological connectivity of the forests, grasslands and other near-natural habitats is increasing.

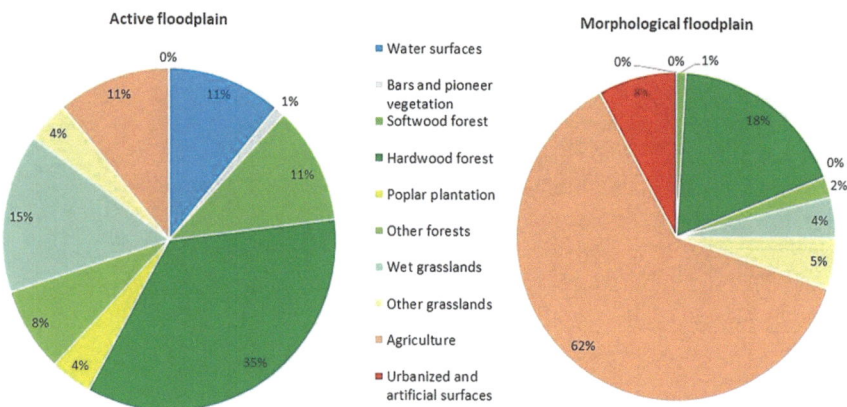

Fig. 34.2 Land cover in the morphological and active floodplain. Different forest types dominate in the active floodplain (adapted from Schwarz 2016)

Fig. 34.3 Zasavica Special Nature Reserve, Serbia: Restored landscape mosaic from open meadows and scrublands to forest in the background, provides habitats for diverse wildlife (Sava TIES, DTP2-096-2.3)

The river flow and flooding events promote the growth of hygrophilous forest trees. Seed transfer is also enhanced by the seasonally strong currents, with the water transporting both native and non-native species equally. Among the latter, species with invasive behavior cause numerous issues (United Nations Environment Programme 1992; Shine et al. 2000; IUCN 2000; CBD 2002, 2007). Invasive alien species (IAS) are plants, animals, and other organisms that are not native to an ecosystem but have been introduced there. Definitions and terminology considering invasive species have evolved and diverged from the time when fundaments in their systematic research have been set out (Elton 1958), linking their adverse impact on ecology later to social, health, and economic issues (Pyšek et al. 1995; Richardson et al. 2000, 2011; Larson 2011).

Common characteristics of IAS include rapid reproduction and growth, high dispersal ability, phenotypic plasticity, and ability to survive on various food types and in a wide range of environmental conditions. Such species are present in all taxonomic groups, including animals, plants, fungi, and microorganisms, and they can affect all types of ecosystems. IAS is considered one of the two major causes of biodiversity loss at the global scale. Some invasive species also have negative impacts on human health and can cause major economic losses. The damage to the global economy brought about by IAS was estimated at nearly 5% of the world's economy (Pimentel 2001), manifesting an upward trend in cumulative economic costs caused by those species (IPBES 2023).

With their dense canopy and vigorous trees, forests modify light, space, and other resources in the lower vegetation layers. Forest resilience against intruding plants is, therefore, strongly related to the maintained coenological complexity and developed vertical structure of the forest layers. Forest ecosystems are known for their strongly delayed response to environmental changes, i.e., lag time (Puettmann and Bauhus 2023). As invasive plant species are opportunistic, an IAS outbreak may occur after decades of "sitting and waiting" (Lapin et al. 2019) by occupying ecological niches left after the habitat perturbances. Both natural (e.g., wind damage) and anthropogenic disturbances (e.g., forest utilization, fragmentation) are often reported to promote IAS growth or spread and reduce forest resilience (Langmaier and Lapin 2020). Slow transformations of the three-dimensional forest structure associated with the impact of IAS (Joshi et al. 2015) may thus be overlooked in some forests. Structural changes in invaded forests should be monitored in the long term to determine the full impact of the present IAS, such as homogenization or shifting of species composition (Wronska-Pilarek et al. 2023).

Recent research has identified 53 invasive alien plant species with negative impacts on the regeneration of 21 native forest tree species in nearby Austria (Lapin et al. 2019). Pedunculate oak, one of the keystone species of the Sava floodplain forests, proved to be among the most affected by IAS. The study found that oak and riparian forests are frequently reported to be impacted by the structural changes caused by alien plants and are among the forest types most vulnerable to biological invasions.

In terms of the ecological corridor functionality of the Sava River, invasive alien species deteriorate the corridor's permeability for native plant species, reducing their natural transfer potentials and the related habitat restoration effects. In addition, other species whose life cycles correlated with the native plants suppressed by invaders are also affected. From a broader perspective, the process of species invasion also has negative impacts on ecosystem services since the capacity of ecosystems to provide their goods and services strongly depends on their functionality (MEA 2005). This issue also seems to be synergistically aggravated by global climate change (Masters and Norgrove 2010; Burgiel and Muir 2010). Transnational cooperation in the matter of invasive species management may become even more significant in the context of ongoing initiatives for floodplain restoration in the Sava region (Kiš et al. 2018; Glatz-Jorde et al. 2021), since reconnection of the morphological floodplain without appropriate strategy and site management may enhance IAS establishment (Resasco et al. 2014).

Problem Approach

As the Sava White Book (Schwarz 2016) argues, the river currently stands at a crossroads. This is true not only for its ecological integrity, as the river connects different economies and policies throughout its valley. The transnational ecological corridor of the Sava links EU and non-EU countries with significantly different policies and management practices when it comes to invasive alien species. As a result, a larger portion of the river's basin is not subject to systematic mapping, monitoring, reporting, and information exchange regarding the presence of invasive species in the European Alien Invasive Species Notification System (EASIN). In such an environment, the invasive plants are able to spread uncontrolled in the Sava River basin, pushing out native species, reducing soil fertility, and even causing increased flood risk (Spencer et al. 2013).

Invasive plant species degrade important rare habitat types, deteriorate corridor connectivity, and are often very difficult to eradicate. The key challenge for the effective management of invasive species in the transnational Sava basin is, therefore, the prevailing lack of cross-border cooperation. Considering the described similarities in terms of the natural environment and heterogeneities in social and political matters, the logical solution was to develop a common platform at the level of the entire river basin. This task was initiated and carried out by the SavaParks Network, consisting of 22 protected area management authorities and nature conservation organizations along the Sava River from Slovenia, Croatia, Bosnia and Herzegovina, and Serbia (https://savaparks.eu). The official SavaParks Network association was established on the event of World Wetlands Day in 2015 (2 February) in Zagreb, Croatia. The network was organized under the leadership of the EuroNature Foundation (Radolfzell, Germany), and its later work was financially supported by the Aage V. Jensen Foundation (Copenhagen, Denmark). It has assumed responsibility for conserving biodiversity in the protected areas within the Sava River basin and is committed to cross-border cooperation promoting sustainable development for the protection of natural wealth and European cultural heritage. Keeping in mind the transnational character of the river basin, the availability of funds, and the common eligibility of the involved countries, the network decided to apply to the EU Interreg Danube Transnational Programme for funding (DTP, https://www.interreg-danube.eu). Considering the described situation of high susceptibility to plant invasion within the lowland landscapes prone to flooding, improving corridor connectivity at the transnational level was the primary project objective.

A consortium with a common ground for the project's development was built considering all the differences among the parties. The SavaParks Network prevailed and successfully applied to the DTP call for proposals in 2017 with the project "Sava TIES" (Sava TIES, DTP2-096-2.3, full project title: ***Preserving Sava River Basin Habitats through Transnational Management of Invasive Alien Species***). The project was launched in June 2018 and ended in May 2021, encompassing nine project partners and 12 associated organizations.

The project consortium consisted of EuroNature Foundation (Germany), Lonjsko Polje Nature Park Public Institution (Croatia), Public Institution Green Ring (Croatia), Public Institution Ljubljansko Barje Nature Park (Slovenia), Public Company National Park Una (Bosnia and Herzegovina), Centre for Environment (Bosnia and Herzegovina), Institute for Nature Conservation of Vojvodina Province (Serbia), Public Enterprise Vojvodinašume (Serbia), and Nature Conservation Movement Sremska Mitrovica (Serbia), along with 12 associated partners from governments and management authorities.

Project Outputs

As the project's main result, the consortium formulated an IAS management toolkit referred to as "Strategic Framework for Effective Management of Invasive Species in the Sava River Basin" (hereinafter: Strategic Framework). It was tailored for conservation practice and land managers with a view to the need for cross-sectoral cooperation. A "tree" of the project outputs illustrates the roadmap from planning IAS management actions to a joint IAS database (Kiš et al. 2020). Each of the outputs and deliverables was developed to assist in a specific task within the overall process. The research began with the gathering of information about introduced plant species and the creation of a joint list of the key IAS in the Sava River basin, which includes the 32 most important invasive vascular plants in terms of their effects on the habitats and species of conservation importance in the basin.

The *review of best practices* in invasive plant species management was performed at the beginning of the project, during the phase of planning pilot activities for invasive species eradication. The project actions concerning IAS management planning began from there.

The risk assessment study provides examples of risk assessment associated with six selected invasive plant species and is based on an internationally accepted method. It describes important aspects to be considered when introducing a new plant species into the area, as well as methods for rapid response to threats.

In the *mapping and monitoring protocol for IAS Control*, several methods for mapping invasive plants and infested habitats, depending on the mapping scale and available resources, are explained. It includes a *field manual* designed for mapping invasive species at two levels: basic (layman) and expert level. The second one features additional information about habitats and species, invasion pathways, corridors, and other information useful to professionals for defining priorities in planning IAS control.

The Joint Pilot Report with Transferability Plan summarizes seven invasive species management actions implemented in four countries under varying national and local environmental and policy conditions. It provides an overview of different methods, such as mechanized grinding followed by grazing (Fig. 34.4), selective pesticide application targeting the non-palatable species, and wetland rejuvenation to achieve environmental pessimum for wooden IAS. The invasive plants selected for eradication testing are among the most challenging in the project area, as well as in many other European countries. The results are informative for both policy and physical aspects of IAS management planning.

The *Cross-Sectoral Guidelines for Effective Invasive Species Management* provides an overview of the key gaps and synergies in land use practices concerning invasive species control. It was based on the *land use study* performing basin-wide analysis of land use practices within the protected areas relevant to the introduction and/or dispersal of invasive plants. A transnational policy gap analysis was also carried out. Both gaps and synergies with regard to land governance and land use practices were revealed and subsequently addressed with a set of guidelines.

The *invasive species database* with associated *mobile applications* (for Android and Apple platforms) was developed in close cooperation and with in-kind

Fig. 34.4 Lonjsko Polje Nature Park, Croatia: Grazing is a cost-effective and nature-friendly solution to prevent IAS shrubs from resprouting after grinding (left). Dense thickets of false indigo (*Amorpha fruticosa*) before the intervention (right) (Sava TIES, DTP2-096-2.3.)

contributions from EASIN experts at the Joint Research Centre, the custodian of the European Union invasive species database. The existing application for mapping invasive species was extended with a special "Sava River basin" geographical layer and the 32 key invasive species in the basin. Additional fields to record (newly introduced) invasive species were also added. The app was translated into the relevant languages spoken in the project area. In a subsequent step, the installment of a separate river basin IAS database was planned since the consortium found that many of the Sava basin target IAS were neither listed in the EU IAS blacklist nor in the EASIN mobile app for IAS mapping.

Info-Box 34.1 Pilot Actions
Although some of the Sava River basin countries have advanced considerably in the system of mapping and reporting invasive alien species, IAS management practices still seem underdeveloped. The pilot actions began with the development of an activity planning template employing both biophysical and economic indicators. The subsequent steps included the identification of key invasive alien plant species and the affected habitats of conservation importance, determining priority sites for the IAS management pilot actions, selecting the most adequate eradication methods (Fig. 34.5), implementing the pilot actions, monitoring and reporting results, analyzing each of the methods' efficiency along with issues involving legislative limitations, and natural events such as floods.

Two years of the consequent activities, including issuing approvals for the usage of pesticides in protected areas, facing flood events, and subcontracting services, brought some quite unexpected results. Domestic goats did not browse the tree of heaven (*Ailanthus altissima*) despite positive results in

Fig. 34.5 Removing invasive knotweed (*Reynoutria* sp.) along the Una River (left). The highly valued Štrbački Buk waterfall on the Una River (right) (Sava TIES, DTP2-096-2.3.)

(continued)

some other countries. Managers of the protected areas were also surprised how bird nesting places for corncrake (*Crex crex*) in NATURA 2000 sites cannot be efficiently restored from the invasion of false indigo (*Amorpha fruticosa*) due to conservation policy restrictions. In another protected area, an already contracted service for grinding and balling several thousand hectares of false indigo thickets for a biomass power plant was foreclosed due to complexities in issuing approval for the works in another country within the EU. Even a missing word in subcontracting service for mud excavation in another case cost protected area manager additional investments, to place the mud infested by IAS in a proper place.

The full report with technical details about the seven pilot activities, the target invasive species and infested habitats, and the applied methods and their efficiency are available on the Danube Transnational Project web page (https://www.interreg-danube.eu/approved-projects/sava-ties).

Policy Gaps

At the time of its implementation, the project area included two EU member states (Slovenia and Croatia), one candidate (Serbia), and one country on the way to candidate status (Bosnia and Herzegovina). Regardless of their joint history spanning more than half a century, the four countries' policy frameworks have been diverging since the 1990s. Both natural and social differences have contributed to the relatively diverse status of invasive species in the respective national legislations. To determine the point of departure in this regard, one of the project outputs was a transnational policy gap analysis. Proportional to the period of membership in the EU, the furthest-developed mapping and monitoring system for invasive species management, including a relevant IAS database, was found in Slovenia. Croatia was on the way to developing national capacities for mapping and reporting IAS to the EU database according to EU Regulation 1143/2014 on invasive species, while the efforts in Serbia, as well as Bosnia and Herzegovina, were still limited to the local scale, with some pilot apps developed and mostly used by researchers at universities. Building on this diverse legislation and implementation status—and with various practical examples concerning conflict management and potential synergies revealed in the project outputs—the Sava TIES project developed a set of guidelines for IAS-smart land management.

Further Steps and Challenges: Impediments to Upscaling

The project partners have successfully navigated the complex and demanding project matrix. The strategic framework was successfully tested as a comprehensive toolkit for efficient IAS management across the diverse economies and national entities in the Sava basin. However, the process of its full implementation and employment seems to have come to standby after the project ends. Despite the significantly increased knowledge and experience of the involved project partners, we must concede a failure to establish the intended basin-wide IAS database and continue transnational monitoring and joint actions for invasive species management, mainly due to the lack of financial resources. The more time elapses without joint and harmonized IAS monitoring and management; the more invasive species will be allowed to spread in an uncontrolled fashion, causing massive economic and ecological losses. There is a simple basic rule when it comes to the efficiency of IAS management: The later the action, the more difficult and costly the control of the invasion.

So far, the resilience of the old, multi-layered, and species-mixed forest in the Sava River basin, supported by forest management activities focused on maintaining the highly valuable mixed oak-ash-hornbeam forest, is still keeping aggressive species from invading the forest canopy and interior forest parts. Whenever this barrier is weakened—either by natural events such as heavy storms or by anthropogenic influences such as large-scale forest regeneration cuttings—the invaders take over quickly. The cross-border forest area is prevailingly covered in old forest growth considered "mature" in forestry terms (i.e., ripe for cutting), and its rotation period is nearing its end. In addition, heavy summer storms along the Sava River in 2023 have caused serious disturbances and canopy openings in the forests, making them more susceptible to invasive plant species and their complex adverse impacts (Vilà et al. 2011).

Maintaining the cultural landscapes and these forests as social-ecological systems depends strongly on appropriate land management (Folke et al. 2011; Demeter et al. 2020). As in other European lowland oak forests, sustainable management of the Sava River basin's landscape matrix is linked with traditional land practices such as silvopasture and pannage (Molnár et al. 2021). When it comes to landscape mosaics and the control of invasive species, the traditional practice of combined mowing, silviculture, and silvopasture seems to be the most efficient method (Biró et al. 2020; Demeter et al. 2021; Kiš et al. 2018). The Sava TIES pilot actions have confirmed for various habitat types and invasive plants that building capacities in IAS management practice are the key component of transnational IAS management in the river basin.

When looking more broadly at the implementation of the European policies particularly relevant to the project area (Water Framework Directive, Flood Directive, Natura 2000 Ecological Network), the process will eventually bring all the parties sharing the Sava River basin to the same table to discuss the full implementation of the EU acquis, including the matter of invasive species. Among various other more frequently cited reasons, further advances in the management of IAS and the ecological corridor functionality of the Sava River basin depend on completing the process of accession of the Western Balkan countries to the EU. Meanwhile, the Sava River itself continues to flow through the various lands and landscapes, bringing both blessings and curses—including functioning as an ecological corridor for both native and invasive alien species.

References

Biró M, Molnár Z, Öllerer K, Lengyel A, Ulicsni V, Szabados K, Kiš A, Perić R, Demeter L, Babai D (2020) Conservation and herding co-benefit from traditional extensive wetland grazing. Agric Ecosyst Environ 300. https://doi.org/10.1016/j.agee.2020.106983

Brudvig LA, Damschen EI, Tewksbury JJ, Haddad NM, Levey DJ (2009) Landscape connectivity promotes plant biodiversity spillover into non-target habitats. Proc Natl Acad Sci USA 106(23):9328–9332. https://doi.org/10.1073/pnas.0809658106

Burgiel SW, Muir AA (2010) Invasive species, climate change and ecosystem-based adaptation: addressing multiple drivers of global change. In: Global invasive species Programme (GISP), Washington. https://doi.org/10.13140/2.1.1460.8161

CBD (2002) Alien species that threaten ecosystems, habitats or species to which is annexed Guiding principles for the prevention, introduction and mitigation of impacts of alien species that threaten ecosystems, habitats or species. Decision VI/23.Sixth Conference of the Parties. The Hague, the Netherlands

CBD (2007) Invasive alien species. Accessed at: https://www.cbd.int/invasive/introduction.shtml

César RG, Belei L, Badari CG, Viani RAG, Gutierrez V, Chazdon RL, Brancalion PHS, Morsello C (2021) Forest and landscape restoration: a review emphasizing principles, concepts, and practices. Land 2021(10):28. https://doi.org/10.3390/land10010028

Demeter L, Bede-Fazekas Á, Molnár G-C, Ortmann-Ajkai A, Varga A, Molnár A, Ferenc Horváth F (2020) The legacy of management approaches and abandonment on old-growth attributes in hardwood floodplain forests in the Pannonian Ecoregion. Eur J Forest Res 139:595–610. https://doi.org/10.1007/s10342-020-01272-w

Demeter L, Molnár ÁB, Fazekas ÁB, Öllerer K, Varga A, Szabados A, Tucakov M, Kiš A, Biró M, Marinkov J, Molnár Z (2021) Controlling invasive alien shrub species, enhancing biodiversity and mitigating flood risk: a win–win–win situation in grazed floodplain plantations. J Environ Manag 295. https://doi.org/10.1016/j.jenvman.2021.113053

Elton CS (1958) The ecology of invasions by animals and plants. T. Methuen and Co., London. https://doi.org/10.1007/978-1-4899-7214-9

Folke C, Jansson A, Rockström J, Olsson P, Carpenter SR, Chapin FS 3rd, Crépin AS, Daily G, Danell K, Ebbesson J, Elmqvist T, Galaz V, Moberg F, Nilsson M, Osterblom H, Ostrom E, Persson A, Peterson G, Polasky S, Steffen W, Walker B, Westley F (2011) Reconnecting to the biosphere. Ambio 40(7):719–738. https://doi.org/10.1007/s13280-011-0184-y

Forman RTT (1983) Corridors in a landscape: their ecological structure and function. Ekologia, Czechoslovakia 2:375–387

Gilbert-Norton L, Wilson R, Stevens J, Beard K (2010) A meta-analytic review of corridor effectiveness. Conserv Biol 24:660–668. https://doi.org/10.1111/j.1523-1739.2010.01450.x

Glatz-Jorde S, Köstenberger L, Jorde K, Grigull M, Berger V, Kirchmeir H (2021) Sava.Restore - Connecting the Floodplains for a healthy alluvial forest. Feasibility Study for Spačva—Bosut Forests Restoration. Final Report. E.C.O. Institute of Ecology, Klagenfurt. https://balkanrivers.net/uploads/files/5/Summary_Feasiblity_Study.pdf

Gurnell AM, Angold P, Gregory KJ (1994) Classification of river corridors: issues to be addressed in developing an operational methodology. Aquat Conserv 4:219–231

International Sava River Basin Commission (ISRBC) (2022). The 2nd Sava River Basin Management Plan, Zagreb.

IPBES (2023) Thematic assessment report on invasive alien species and their control of the intergovernmental science-policy platform on biodiversity and ecosystem services. In: Roy HE, Pauchard A, Stoett P, Renard Truong T (eds) IPBES secretariat, Bonn. https://www.ipbes.net/ias

IUCN (2000) IUCN Guidelines for the Prevention of Biodiversity Loss Caused by Alien Invasive Species. Prepared by the SSC Invasive Species Specialist Group. Approved by the 51st Meeting of the IUCN Council, Gland Switzerland, February 2000. https://portals.iucn.org/library/efiles/documents/Rep-2000-052.pdf

Joshi A, Mudappa D, Raman T (2015) Invasive alien species in relation to edges and forest structure in tropical rainforest fragments of the Western Ghats. J Trop Ecol 56:233–244. https://www.jstor.org/stable/30044903

Kiš A, Stojnić N, Sabadoš K et al (2018) Case study Advocating Ecosystem Services Assessment and Valuation (ESAV) in Bosut Forests area - integrating biodiversity and ecosystem services in natural resource uses and management. Institute for Nature Conservation of Vojvodina Province (INCVP), Novi Sad, Serbia in cooperation with Deutsche Gesellschaft für Internationale Zusammenarbeit (GIZ), within the Open Regional Fund for South-East Europe—Biodiversity (ORF BD) funded by the German Federal Ministry of Economic Cooperation and Development (BMZ). https://balkangreenenergynews.com/wp-content/uploads/2018/06/ESAV-case-study-Bosut-Forests-2018.pdf

Kiš A, Panjković B, Szababos K, Perić R, Molnar Z (2020) IAS mapping and monitoring in Sava River Basin—a harmonized transnational platform for successful IAS management. Poster session, P75. 11th International Conference on Biological Invasions: The Human Role in Biological Invasions - a case of Dr Jekyll and Mr Hyde? NEOBIOTA2020, 15–18 September 2020, Vodice, Croatia

Kozarac J (1888) Slavonska šuma. Vijenac, Zagreb

Langmaier M, Lapin K (2020) A systematic review of the impact of invasive alien plants on forest regeneration in European temperate forests. Front Plant Sci 11:524969. https://doi.org/10.3389/fpls.2020.524969

Lapin K, Oettel J, Steiner H, Langmaier M, Sustic D, Starlinger F, Kindermann G, Frank G (2019) Invasive alien plant species in unmanaged forest reserves. Austria:71–96. https://doi.org/10.3897/neobiota.48.34741

Larson BMH (2011) Embodied realism and invasive species. In: de Laplante K, Brown B, Peacock KA (eds) Philosophy of ecology. Elsevier, North-Holland, p 129. https://doi.org/10.1016/B978-0-444-51673-2.50005-4

Masters G, Norgrove L (2010) Climate change and invasive alien species. CABI WP 1, 30 pp

Millennium Ecosystem Assessment (MEA) (2005) ECOSYSTEMS AND HUMAN WELL-BEING: WETLANDS AND WATER synthesis. World Resources Institute (WRI), Washington, DC

Molnár Z, Szabados K, Kiš A, Marinkov J, Demeter L, Biró M, Öllerer K (2021) Preserving for the future the — once widespread but now vanishing — knowledge on traditional pig grazing in forests and marshes (Sava-Bosut floodplain, Serbia). J Ethnobiol Ethnomed 17:1–30

Naiman RJ, Décamps H (1997) The ecology of interfaces: riparian zones. Annu Rev Ecol Syst 28:621–658. https://doi.org/10.1146/annurev.ecolsys.28.1.621

Naiman RJ, Décamps H, Pollock M (1993) The role of riparian corridors in maintaining regional biodiversity. Ecol Appl 3:209–212. https://doi.org/10.2307/1941822

Pimentel D (ed) (2001) Biological invasions: economic and environmental costs of alien plant, animal and microbe species. CRC Press, Boca Raton. https://doi.org/10.1201/b10938

Puettmann KJ, Bauhus J (2023) Effects of lag time in forest restoration and management. Forest Ecosystems 10. https://doi.org/10.1016/j.fecs.2023.100131

Pyšek P, Prach K, Rejmánek M, Wade M (eds) (1995) Plant invasions - general aspects and special problems. SPB Academic Publ, Amsterdam. 263 pp

Resasco J, Haddad N, Orrock J, Shoemaker D, Brudvig L, Damschen E, Tewksbury J, Levey D (2014) Landscape corridors can increase invasion by an exotic species and reduce diversity of native species. Ecology 95:2033–2039. https://doi.org/10.1890/14-0169.1

Richardson DM, Pyšek P, Rejmánek M, Barbour MG, Panetta DF, West CJ (2000) Naturalization and invasion of alien plants: concepts and definitions. Divers Distrib 6:93–107. https://doi.org/10.1046/j.1472-4642.2000.00083.x

Richardson DM, Pyšek P, Carlton JT (2011) A compendium of essential concepts and terminology in invasion ecology. In: Richardson DM (ed) Fifty years of invasion ecology. The legacy of Charles Elton. Wiley-Blackwell, Oxford, pp 409–420. https://doi.org/10.1002/9781444329988.ch30

Schwarz U (2016) Sava white book. The river Sava: threats and restoration potential. EuroNatur/Riverwatch, Radolfzell/Wien. https://balkanrivers.net/sites/default/files/01_SavaWhite%20Book%20Study.pdf

Shine C, Williams N, Gündling L (2000) A guide to designing legal and institutional frameworks on alien invasive species, IUCN environmental policy and law paper no. 40. IUCN, Gland, Cambridge and Bonn. 138pp. https://portals.iucn.org/library/efiles/documents/EPLP-040-En.pdf

Spencer DF, Colby L, Norris GR (2013) An evaluation of flooding risks associated with giant reed (Arundo donax). J Freshw Ecol 28(3):397–409. https://doi.org/10.1080/02705060.2013.769467

United Nations Environment Programme (1992) Convention on biological diversity, June 1992. https://wedocs.unep.org/20.500.11822/8340

Vajda Z (1952) Uzroci epidemijskog ugibanja brijestova. Glasnik za šumske pokuse br. 10, Sveučilište u Zagrebu, Poljoprivredno-Šumarski faultet i Zavod za šumske pokuse. (105–197)

Vilà M, Espinar JL, Hejda M, Hulme PE, Ch Jarošík V, Maron JL, Pergl J, Schaffner U, Sun Y, Pyšek P (2011) Ecological impacts of invasive alien plants: a meta-analysis of their effects on species, communities and ecosystems. Ecol Lett 14:702–708. https://doi.org/10.1111/j.1461-0248.2011.01628.x

Wronska-Pilarek D, Rymszewicz S, Jagodziński A, Gawryś R, Dyderski M (2023) Temperate forest understory vegetation shifts after 40 years of conservation. Sci Total Environ. https://doi.org/10.1016/j.scitotenv.2023.165164

Open Access This chapter is licensed under the terms of the Creative Commons Attribution 4.0 International License (http://creativecommons.org/licenses/by/4.0/), which permits use, sharing, adaptation, distribution and reproduction in any medium or format, as long as you give appropriate credit to the original author(s) and the source, provide a link to the Creative Commons license and indicate if changes were made.

The images or other third party material in this chapter are included in the chapter's Creative Commons license, unless indicated otherwise in a credit line to the material. If material is not included in the chapter's Creative Commons license and your intended use is not permitted by statutory regulation or exceeds the permitted use, you will need to obtain permission directly from the copyright holder.

Tanzania: The Eastern Arc Mountains Forests as World Natural Heritage—Status and Future Prospects

35

Linus Munishi

Eastern Arc Mountain Forest, Tanzania (Photo: Tranquil Kilimanjaro)

Abstract

Maintaining forest connectivity for biodiversity conservation in Tanzania has become a conservation priority in response to increasing land development and road networks. Forests harbour and support diverse life forms worldwide, with the distribution and density of human populations having negative impacts on gene flow and landscape connectivity. On the following pages, the Eastern Arc Mountains (EAM) forest networks in Tanzania serve as a case study to identify and describe areas that promote connectivity by facilitating the movement of biodiversity and gene flow in a fragmented landscape. The objective is to increase

L. Munishi (✉)
School of Life Sciences and Bioengineering, The Nelson Mandela-African Institution of Science and Technology (NM-AIST), Arusha, Tanzania
e-mail: linus.munishi@nm-aist.ac.tz

© The Author(s) 2025
K. Lapin et al. (eds.), *Ecological Connectivity of Forest Ecosystems*,
https://doi.org/10.1007/978-3-031-82206-3_35

awareness and understanding regarding the drivers of connectivity in a patchwork landscape characterised by human use. We present profiles of areas now protected as Nature Forest Reserves (NFRs) whose biological and socio-economic values have been documented by scientists for over a century. The potential for biodiversity conservation and ecosystem service delivery is described while identifying the current challenges and future prospects.

Keywords

Eastern Arc Mountains · Nature Forest Reserve · Endemic species · Tourism · Forest networks

Eastern Arc Mountains Forests of Tanzania

The term 'Eastern Arc' was introduced in the mid-1980s to describe the moist forests on the eastward-facing slopes of the ancient crystalline mountains in eastern Tanzania and south-eastern Kenya, which are under the influence of the Indian Ocean climatic regime with regard to local weather and climatic conditions (Lovett 1990). These moist forests on the Eastern Arc Mountains (EAM) are remarkable for their high level of species and generic endemism (Polhill 1968). The Eastern Arc is mentioned in the WCMC review of the coverage of World Heritage Sites as an ecoregion where there is no current World Heritage property (Chape and Magin 2004). Thirteen separate mountain blocks comprise the Eastern Arc; 12 of these are found in 14 districts within five regions of Tanzania (Fig. 35.1). Around 40% of the mammal and plant species inhabiting the EAM forests are rare and/or endemic, making them one of the world's top ten biodiversity hotspots and placing them among the most important conservation areas on the planet (Burgess et al. 2007a, b; Lyakurwa et al. 2019).

For over a century, scientists have been documenting the biological value of the 12 EAM forests now protected as Nature Forest Reserves. This has led to recognition of the importance of these forests and attracted further scientists to do more research, including the amazing work on vegetation cover and classification undertaken in the 1970s (White 1983) and the pioneering work on the identification of global biodiversity hotspots by Myers (1990). Based on systematic analyses of available species data, the importance of the Eastern Arc Mountains has been recognised in the following analyses of global biological priority: Global 200 ecoregion (Olson and Dinerstein 1998), part of a global biodiversity hotspot according to Conservation International (Mittermeier et al. 1998, 2004), and part of an Endemic Bird Area (BirdLife International: ICBP 1992; Stattersfield 1998). Assemblages of biodiversity hotspots around these forests and their ecosystem connectivity play an essential role in the structure and composition of natural habitats, and their importance in supplying ecosystem goods and services is critical. The loss of some of these hotspots has begun to show signs of destabilisation of the natural systems and loss of resilience, while their recovery has the potential to restore ecological integrity.

Fig. 35.1 Map of the Eastern Arc Mountains with key connected forests and their cover (Kimaro O.D., Lyamtane, E., Kimaro D. N., Darr, D., Feger, K. H., Vancampenhout, K. 2022)

Deforestation is a major threat to the conservation of these forests and their biodiversity, especially in terms of the area's endemic plants and animals (Hall et al. 2009). Most of the remaining natural habitat on the mountains is found within nearly 150 government forest reserves, of which 107 are managed nationally for water catchment, with forest exploitation not permitted. Part of the EAM forests is also protected within the Udzungwa Mountains National Park and Amani Nature Forest Reserve in the East Usambara Mountains. Outside these reserves, most forests have been cleared except in small village burial sites, a few village forest reserves, and inaccessible areas.

In most areas of the Eastern Arc Mountains, local populations respect the reserve boundaries (where they are clear), but forest resources are used locally to provide ecosystem goods and services, including water, clean air, pollination, fuel, and building materials. Some forests—especially those adjacent to community residential areas—are heavily degraded by overutilisation and land-use change, causing deforestation and loss of natural cover. Fire is also a problem, as it can destroy these forests during the dry season. The future of biodiversity in the Eastern Arc Mountains is closely tied to management policies and the capabilities of the governmental Forest and Beekeeping Division as well as the National Parks Authority in Tanzania and the Forest Department in Kenya.

EAM Forest Status and Sustainability in the Local and Global Context

The Eastern Arc Mountains are known to biologists and conservationists as one of the world's most important areas for biodiversity. The EAM ecosystems and the populations that depend on them for their social, cultural, and economic reproduction have been identified as being particularly at risk due to increased vulnerability to deforestation (for agricultural systems), especially in semi-protected forest areas, as well as to climate change (FAO 2022). These concerns are identified in several of the UN's Sustainable Development Goals and in response to the increased recognition of the impact that climate change may have on populations that live and depend on fragile ecosystems (Butchart et al. 2010; Yannelli et al. 2022). Consequently, intervention and mitigation strategies need to be developed and integrated into forest and biodiversity conservation and management policies and implemented at national, regional, and local scales.

In the EAM region of Tanzania, deforestation and biodiversity loss impact every aspect of development and conservation. Increased human population coupled with intensive land use focused around the EAM areas poses unprecedented threats to ecosystems and livelihoods, severely challenging the adaptive capacity of these ecosystems and undermining the natural and cultural identity of the rural communities depending on them. What is more, the rate and scale of ecosystem change and the impacts it has on the services those ecosystems provide are intensifying. Management within these areas needs to respond and adapt to shifting environmental and social drivers at both the local and the global levels. While there have been many calls to address such challenges, current strategies have had limited success—as evidenced by encroachment increasing loss of biodiversity and deforestation in some of the forest networks in the region. Although improved management and investment from both the government and donor-funded projects have reduced some threats in recent years, there are challenges ranging from illegal logging, charcoal and firewood production, and pole cutting to hunting, fires, illegal wildlife trade, and the unpredictable impacts of climate change (Burgess et al. 2007b; Hall et al. 2009). Systematic analyses of available species data and economic value have proven the global and national importance of the EAM. The key facts covering the biodiversity and economic value of the EAM forests and their management are summarised below:

Biodiversity

Most forests in the EAM are large in size and mostly in a natural condition, with a proportion of them under sustainable natural resource management in which low-level, non-industrial natural resource use compatible with nature conservation is seen as one of the main aims. As one of the world's most diverse tropical montane forest ecosystems, the EAM forests chiefly comprise flowering trees and animals uniquely adapted to higher elevations in tropical climatic conditions. The structure and composition of the EAM forest ecosystem are discussed along with species diversity and distribution at different locations (Fig. 35.1). The EAM forests meet the UNESCO World Heritage Convention criteria for '*Outstanding Universal Value*'. There are 211 endemic animal species and around 550 endemic plant species found exclusively in these mountain forests. More than 70% of the species endemic to the EAM are found only in the nine interconnected forest belt reserves and nowhere else on earth; many of these unique species are also threatened with extinction (Lyakurwa et al. 2019). An analysis conducted for the UNESCO World Heritage Convention in 2005 showed that the Eastern Arc Mountains are the most important area in Africa lacking a World Heritage Site (see http://www.unep-wcmc.org/biodiversity-wh_975.html). Elevation, the major environmental gradient in mountain regions of the world, produces diverse habitats within the EAM, with some species existing only within narrow elevational ranges (Hall et al. 2009).

Carbon Storage and Sequestration in the EAM Forests

Recently, payments for environmental/ecosystem services (PES) have attracted interest from national and regional governments and are becoming one of the leading conservation policy instruments in tropical countries. However, the degree to which areas designated for PES overlap with areas that are critical for maintaining landscape connectivity for species is rarely evaluated. The Eastern Arc Mountains forests store up to ~300 tonnes of carbon per hectare (tC/ha) within the wood and roots of their trees (Platts et al. 2023). The total amount of carbon stored is 6.33 petagrams, and this is being proposed as part of the Tanzanian national contribution to reducing climate change (Platts et al. 2023). Around 35% of the carbon is stored within protected areas, and about half of this amount is in the nine sites of the proposed World Heritage Site. Estimates of habitat loss suggest that carbon storage in the Eastern Arc Mountains has declined over the last century. While forest growth within protected areas has seen them gain carbon at ~4.8 tC/ha over the past 100 years, unprotected areas have emitted a mean of ~11.9 tC/ha. This means that the proposed REDD+ mechanism could be implemented in the Eastern Arc Mountains, particularly within the proposed World Heritage Site. This could, in turn, generate revenues for the government and help to fund further conservation efforts.

Timber and Non-timber Forest Products as Well as Other Ecosystem Services Used by Adjacent Communities

There are several non-timber values in the Eastern Arc Mountains that are used by adjacent communities. These include medicinal plants, firewood, building poles, roofing thatch, food (bushmeat, wild vegetables, and fruit), controlled beekeeping, and artisanal raw materials (e.g., for baskets, mats, or dyes). However, the EAM forests and woodlands are also illegally harvested for timber. Estimates of the total economic value of both timber and non-timber forest products exceed USD 50 million per year (Platts et al. 2023). In terms of water supply, the EAM are the source of a significant proportion of Tanzania's water, providing drinking water for all the major coastal cities, irrigation water for most of eastern Tanzania, and the water used to generate more than 50% of the country's electricity.

Amani Nature Forest Reserve as a Key Area Within the EAM Forest Reserve Network

Like many other protected areas, EAM forests are essential for conserving biodiversity and storing carbon, thus representing an important buffer against climate change (Dudley 2008; Engh 2011). The creation of protected forest areas in Tanzania has a long history extending back to the German colonial period in the late 1800s. This 'forest reserve' network has expanded over time—first during the German and subsequent British colonial periods and then again after Tanzania's independence in 1961 (Burgess et al. 2007a, b). These protected forest areas act as reserves where the interaction of people and nature over time has produced a distinct character with significant ecological, biological, cultural, and scenic value; safeguarding the integrity of this interaction is therefore vital to protecting and sustaining the area and its associated nature conservation and other values. In this network of forests, conserved areas are managed with sustainable use of natural resources and the primary goal of conserving ecosystems as well as associated cultural values and traditional natural resource management systems.

Amani Nature Forest Reserve (ANFR) represents one of the largest forested blocks within the EAM and occupies the southern extremity of the East Usambara Mountains. The reserve covers an area of 8380 ha and is famous as the home of most of the species of African violet (*Saintpaulia*), named after Baron Walter von Saint Paul-Illaire, the German administrator of Tanga Province in the 1890s. ANFR is internationally renowned for its high number of endemic species and incredible diversity of plant life encompassing between 600 and 1000 different species. Among them are eight species of the section *Saintpaulia* (African violet) as well as *Leptonychia usambarensis, Cephalosphaera usambarensis*, and *Allanblackia* spp., some of which are considered threatened. In general, the floristic composition in the area is very diverse, with around 2012 vascular plant species per hectare. In addition, more than a quarter of the 30-odd species of amphibians and reptiles in the area are found nowhere else in the world. Of the 35 threatened vertebrate species in the East Usambara Mountains, 23 are found in ANFR.

With regard to plants, the submontane forests are especially rich in endemic species. However, the semi-deciduous forests dominate the lowlands of the reserve, particularly in Mnyuzi Scarp, which also has lower annual rainfall than the tall submontane and evergreen forests found in the mountains above 750 m, where rainfall is higher. Common tree species in ANFR include *Cephalosphaera usambarensis, Allanblackia stuhlmannii, Albizia gummifera, Beilschmiedia kweo, Diospyros abyssinica, Englerodendron usambarense*, and *Drypetes gerrardii*. Epiphytic lichens and bryophytes are abundant, especially on steep summits. Within this centre of biological diversity lies Amani Botanical Garden, established in 1902 and thus one of the oldest botanical gardens in Africa. It covers 340 ha and features more than 1000 species of plants from all over the world. This institution is valuable for conservation, education, and research. The fact that many of the endemic species in the EAM are found within ANFR makes it one of the most important forest blocks in Africa, and the reserve has been referred to as the African equivalent of the Galápagos Islands in terms of endemism and biodiversity.

Protection, Threats, and Management of the EAM Forest Reserves

Conversion of natural forest ecosystems for human land use leads to fragmentation, loss of habitats, and restriction of species movement (Beier et al. 2011; Bolliger and Silbernagel 2020; Hadley and Betts 2009; Lindsay et al. 2008; Liu et al. 2020; Lumsden and Bennett 2005; Rogan and Lacher Jr 2018). A decrease in habitat connectivity has adverse effects on population viability, resulting in greater extinction risk than the loss of habitat area alone (Brook et al. 2008; Caughley 1994; Lindenmayer and Fischer 2013). Better management of human-modified landscapes is central to minimising the impact of fragmentation on species movement and connectivity and ultimately ensuring the viability of populations and ecosystems. The Eastern Arc Mountains have suffered an estimated 80% loss of their historical forest area in total, with 25% lost since 1955. This forest loss has not been even across all elevations; however, the upper montane zone (> 1800 m) has lost 52% of its total palaeoecological forest area and 6% since 1955. Conversely, the submontane habitat (800–1200 m) has lost close to 93% of its palaeoecological extent, with 57% lost since 1955. A list of 123 narrowly endemic Tanzanian Eastern Arc tree taxa with defined and restricted elevational ranges was compiled and analysed with regard to mountain block locations, elevational range, and area of forest within each 100 m elevational band. Half of these taxa have lost more than 90% of their palaeoecological forest habitat within their elevational range. When the elevational range is considered, 98 (80%) of these endemic forest tree species should have their level of extinction threat within the IUCN Red List elevated (Hall et al. 2009). Although the area has been the focus of conservation efforts and botanical research for over a century, Amani Nature Forest Reserve has only been a protected area since 1997. Management aims included reserves established for production from natural forests (timber and charcoal), protection of natural forests (water catchment reserves and prevention of landslides and erosion), and plantation forestry using exotic species. After the implementation of the 'new' Forest Policy in 1998 and the Forest Act of 2002 (URT 1998, 2002), forest reserves began to be designed for the preservation of their biodiversity and habitats, and human activities were restricted. Under this new legislative framework, Nature Forest Reserves (NFRs) were recognised as forest areas of particularly high importance for globally unique biodiversity and—in most cases—put under strict management with strong protection. The first phase of declaring NFRs focused on the Eastern Arc Mountains ecoregion (Platts et al. 2011) during the period from 1997 to 2009, starting with the Amani Nature Forest Reserve in the East Usambara Mountains in 1997. International efforts to preserve the beauty of the EAM forests within its boundaries have been met with much local success. With increasing demands on natural resources from multiple stakeholders, we argue that a more detailed understanding of the interactions between different ecosystem components—and, importantly, an understanding of how different policies, land uses, and settlement histories affect these interactions and their impacts on poverty and inequality—is required. Similarly, planning for the long-term sustainable use of Africa's natural resources and ecosystems surrounding protected areas requires a longer-term historical perspective on human–ecosystem–environment interactions than is currently available. The challenges facing the forests in the EAM are the conversion of forests to farmland, fires spreading from surrounding farmlands, invasive plant species, gold mining, and the wildlife trade. In response to these challenges, the Tanzanian government has continued to work to identify and upgrade other biologically important reserves to become NFRs. These sites were initially placed under the ownership and management of the Forestry and Beekeeping Division of the Ministry of Natural Resources and Tourism before being transferred to the Tanzania Forest Services (TFS) in 2005. Following the first phase of declaring NFRs, the network was gradually expanded to cover all the different forest types in the country (including the coastal forests as well as the Northern Volcanic, Southern Highlands, Guineo-Congolian, Miombo, and Miombo-Acacia forests). The current network of NFRs in Tanzania encompasses 19 reserves, with an additional three reserves currently in the process of being established. Recently, there has been a global shift of forest management to local levels to better reconcile local livelihoods and biodiversity conservation. Within the EAM forest reserve network, there is site-based revenue generation by way of nature-based tourism, which also acts as an incentive for local people to support the conservation and management of these reserves. Tourism revenues across the 120 hotels located in the Eastern Arc Mountains were estimated at about USD 1.7 million in 2010, with about 38% of the total (~USD 650,000) coming from nature-based tourism. Most of this value comes from the reserves of the proposed Eastern Arc World Heritage Site, especially Amani Nature Forest Reserve, Udzungwa National Park, Uluguru Nature Forest Reserve, and Magamba Nature Forest Reserve. Indeed, the declaration of a World Heritage Site in the Eastern Arc Mountains could be expected to increase the number of tourists coming to the area significantly.

Conclusion

Habitat loss and fragmentation can restrict species movement and reduce connectivity, negatively impacting biodiversity and associated ecosystem services. Connectivity underpins the persistence of life and must therefore be incorporated in biodiversity conservation decisions. Yet when prioritising conservation areas and developing actions, connectivity is not regularly operationalised in spatial planning. The challenge lies in the translation of flows associated with connectivity into conservation objectives that lead to actions. The conclusions in the subsequent paragraphs are derived from this baseline study as well as the experience gained from previous studies in the EAM forests and other similar studies elsewhere.

Evidence shows that there is a disconnect between global high-level conservation goals and on-the-ground actions such as maintaining ecosystem services or persistence and the local planning of protected areas (Beger et al. 2022; Hilty et al. 2020). Conservation objectives can provide a link between high-level conservation goals and the local or regional design and implementation of functionally connected protected area networks. With current implementation gaps in terms of protected area commitments and the increasing threat of climate change, there is a tremendous opportunity to use quantifiable objectives for ecological connectivity as a vehicle for futureproofing protected area networks to help achieve local and global conservation goals. It is important to assume this perspective if we are to evaluate and know which conservation strategies have proven to be effective in preventing ecosystem degradation and should thus be encouraged in future management plans. From an evidence-based perspective, increased landscape and ecosystem connectivity in Tanzania varies across different protected areas (PAs), with specific management approaches more effective at preventing PA fragmentation than others. Areas close to national parks exhibit the least loss of forest and landscape connectivity and cover, whereas multiple-use areas have the highest rates of forest loss. However, complexity and power relations have hampered scientists' efforts to engage with the people who use and influence the use of resources at the landscape and local scales. Landscape approaches present an opportunity for science to help steer local management to address local contexts. This means that there is a need for more effective strategies for reviewing and enforcing existing policies to integrate science into management in ways which ensure that socio-economic benefits to local communities are maximised and national interests are sustained. While prioritising the designation of new PAs may be the most efficient means by which to maintain connectivity and improve biodiversity conservation, we believe that achieving such outcomes requires considering and embedding science as a key factor in the management actions, especially at the landscape level.

References

Beger M, Metaxas A, Balbar AC, McGowan JA, Daigle R, Kuempel CD, Possingham HP (2022) Demystifying ecological connectivity for actionable spatial conservation planning. Trends Ecol Evol

Beier P, Spencer W, Baldwin RF, McRAE BH (2011) Toward best practices for developing regional connectivity maps. Conserv Biol 25(5):879–892

Bolliger J, Silbernagel J (2020) Contribution of connectivity assessments to green infrastructure (GI). ISPRS Int J Geo Inf 9(4):212

Brook BW, Sodhi NS, Bradshaw CJ (2008) Synergies among extinction drivers under global change. Trends Ecol Evol 23(8):453–460

Burgess ND, Butynski TM, Cordeiro NJ, Doggart N, Fjeldså J, Howell K, Mbilinyi B (2007a) The biological importance of the eastern Arc Mountains of Tanzania and Kenya. Biol Conserv 134(2):209–231

Burgess ND, Loucks C, Stolton S, Dudley N (2007b) The potential of forest reserves for augmenting the protected area network in Africa. Oryx 41(2):151–159

Butchart SH, Walpole M, Collen B, Van Strien A, Scharlemann JP, Almond RE, Bruno J (2010) Global biodiversity: indicators of recent declines. Science 328(5982):1164–1168

Caughley G (1994) Directions in conservation biology. J Anim Ecol:215–244

Chape S, Magin C (2004) Review of the World Heritage network: biogeography, habitats and biodiversity: final draft

Dudley N (2008) Guidelines for applying protected area management categories. Iucn

Engh V (2011) Integrated conservation and development projects and efforts around Amani nature reserve, Tanzania, and their effects on livelihoods and forest conservation. Norwegian University of Life Sciences, Ås

FAO (2022) The state of the World's forests 2022. Forest pathways for green recovery and building inclusive, resilient and sustainable economies. FAO, Rome. https://doi.org/10.4060/cb9360en

Hadley AS, Betts MG (2009) Tropical deforestation alters hummingbird movement patterns. Biol Lett 5(2):207–210

Hall J, Burgess ND, Lovett J, Mbilinyi B, Gereau RE (2009) Conservation implications of deforestation across an elevational gradient in the eastern Arc Mountains, Tanzania. Biol Conserv 142(11):2510–2521

Hilty J, Worboys GL, Keeley A, Woodley S, Lausche B, Locke H, White JW (2020) Guidelines for conserving connectivity through ecological networks and corridors. Best Practice Protected Area Guidelines Series 30:122

ICBP (1992) Putting diversity on the map: priority areas for global conservation, vol 06-2-1992. International Council for Bird Preservation

Lindenmayer DB, Fischer J (2013) Habitat fragmentation and landscape change: an ecological and conservation synthesis. Island Press

Lindsay DL, Barr KR, Lance RF, Tweddale SA, Hayden TJ, Leberg PL (2008) Habitat fragmentation and genetic diversity of an endangered, migratory songbird, the golden-cheeked warbler (Dendroica chrysoparia). Mol Ecol 17(9):2122–2133

Liu W, Hughes AC, Bai Y, Li Z, Mei C, Ma Y (2020) Using landscape connectivity tools to identify conservation priorities in forested areas and potential restoration priorities in rubber plantation in Xishuangbanna, Southwest China. Landsc Ecol 35:389–402

Lovett J (1990) Classification and status of the Tanzanian forests. Mitteilungen aus dem Institut für Allgemeine Botanik in Hamburg 23:287–300

Lumsden LF, Bennett AF (2005) Scattered trees in rural landscapes: foraging habitat for insectivorous bats in South-Eastern Australia. Biol Conserv 122(2):205–222

Lyakurwa JV, Howell KM, Munishi LK, Treydte AC (2019) Uzungwa scarp nature Forest reserve: a unique hotspot for reptiles in Tanzania. Acta Herpetologica 14(1):3–14

Mittermeier RA, Myers N, Thomsen JB, Da Fonseca GA, Olivieri S (1998) Biodiversity hotspots and major tropical wilderness areas: approaches to setting conservation priorities. Conserv Biol:516–520

Mittermeier R, Robles Gil P, Hoffmann M, Pilgrim J, Brooks T, Mittermeier C, Lamoreux J, da Fonseca GAB (2004) Hotspots revisited: Earth's biologically richest and Most endangered Ecoregions. CEMEX, Mexico City

Myers N (1990) The biodiversity challenge: expanded hot-spots analysis. Environmentalist 10:243–256

Olson DM, Dinerstein E (1998) The global 200: a representation approach to conserving the Earth's most biologically valuable ecoregions. Conserv Biol 12(3):502–515

Platts PJ, Burgess ND, Gereau RE, Lovett JC, Marshall AR, McClean CJ, Marchant R (2011) Delimiting tropical mountain ecoregions for conservation. Environ Conserv 38(3):312–324

Platts PJ, Schaafsma M, Turner RK et al (2023) Inequitable Gains and Losses from Conservation in a Global Biodiversity Hotspot. Environ Resource Econ 86:381–405. https://doi.org/10.1007/s10640-023-00798-y

Polhill RM (1968) Miscellaneous notes on African species of crotalaria L.: II. Kew Bull 22(2):169–348

Rogan JE, Lacher Jr TE (2018) Impacts of habitat loss and fragmentation on terrestrial biodiversity

Stattersfield AJ (1998) Endemic bird areas of the world-priorities for biodiversity conservation. Bird Life Int

URT (1998) National forest policy. Ministry of Natural Resources and Tourism Dar es Salaam, Dodoma

URT (2002) The new Forest act, no. 7 of 7th June 2002. Ministry of Natural Resources and Tourism, forestry and beekeeping division. The United Republic of Tanzania. Government Printer Dar es Salaam, Dar Es Salaam

White F (1983) The vegetation of Africa (Vol. 20)

Yannelli FA, Bazzichetto M, Conradi T, Pattison Z, Andrade BO, Anibaba QA, Damasceno G (2022) Fifteen emerging challenges and opportunities for vegetation science: a horizon scan by early career researchers. J Veg Sci 33(1):e13119

Open Access This chapter is licensed under the terms of the Creative Commons Attribution 4.0 International License (http://creativecommons.org/licenses/by/4.0/), which permits use, sharing, adaptation, distribution and reproduction in any medium or format, as long as you give appropriate credit to the original author(s) and the source, provide a link to the Creative Commons license and indicate if changes were made.

The images or other third party material in this chapter are included in the chapter's Creative Commons license, unless indicated otherwise in a credit line to the material. If material is not included in the chapter's Creative Commons license and your intended use is not permitted by statutory regulation or exceeds the permitted use, you will need to obtain permission directly from the copyright holder.

36. Tunisia: Genetic Diversity Assessment of Cork Oak Provenance Trials in the Context of Climate Change

Boutheina Stiti, Issam Touhami, Awatef Slama, Amel Ennajah, Lamia Hamrouni, Mohamed Larbi Khouja, and Abdelhamid Khaldi

Cork oak forest of Bellif (Tunisia), 2018 (Photo: Issam Touhami)

Abstract

The cork oak (*Quercus suber* L.) is a native species in Tunisian forests; it plays a key role from an ecological and socio-economic point of view. As a result of serious biotic and abiotic problems, its ecosystem has significantly deteriorated, and its natural regeneration is random and nearly absent. Considering this situation, foresters have been assisting regeneration by creating new plantations. Within the framework of the EUFORGEN network, a project collected seedlings of

B. Stiti (✉) · I. Touhami · A. Slama · A. Ennajah · L. Hamrouni · M. L. Khouja · A. Khaldi
The National Research Institute of Rural Engineering, Water and Forestry (INRGREF).
Laboratory of Management and Valorization of Forest Resources, University of Carthage, Ariana, Tunisia

Q. suber from various origins in northern Tunisia in 1997. The aim was to identify the best-adapted material to ensure the success and sustainability of plantations. Samples of 26 populations of cork oak from the natural distribution range of the species were established in five sites with differing soil and climatic conditions. This study evaluates the behavior of these different provenances in the experimental Tunisian site in terms of survival and height growth after 7 and 14 years, respectively. The cluster analysis (UPGMA) showed two groups: one from Morocco, Spain, and Portugal and the other from Italy, Algeria, and Tunisia. Overall, the Tunisian (Fernana) and Italian (Puglia) provenances prove successful and well adapted to the edapho-climatic conditions specific to the Tunisian trial site and resistant to biotic attacks.

Keywords

Cork oak · Provenances · Genetic variability · Resilience · Nature conservation

Introduction

Tunisia's ecological transition has recently adopted a holistic vision of sustainable development, moving from warning policymakers of the risks of extinction and the need to conserve biodiversity to integrating ecosystem services into decision-making processes. New strategies consider conservation as an option and aim for relevant management based on multidisciplinarity and agroecology on the one hand and on the concept of the ecological network and landscape connectivity on the other, especially in cork oak forests (Khalfaoui et al. 2020; Bel Fekih Boussema et al. 2022).

The cork oak (*Quercus suber* L.) occupies a key position among the forest species of the Mediterranean Basin due to its high environmental, socio-economic, and landscape value. Its bark (cork) represents a valuable natural resource harnessed in several ways (APCOR 2020), and its edible fruits (acorns) are used as an alternative food source in at least 27 countries since they are rich in starch, proteins, and lipids with gluten-free flour (Vinha et al. 2016; Stiti et al. 2022). The fruits can also be used to provide up to 20% of the diet of chickens and other animals without difficulty (Stiti et al. 2021) since they are nutritionally comparable to many cereal grains (Zarroug et al. 2022). Cork oak forests also support recreational and tourism activities for both locals and tourists from abroad. They inhabit large areas in both the southern (Morocco, Algeria, Tunisia) and northern (Italy, France, Spain, and Portugal) Mediterranean region. The total area covered is around 2,123,000 hectares, 67% of which is in Europe and 33% in Africa (APCOR 2020). However, the current state of cork oak stands, especially in southern Mediterranean forests, is alarming. For several decades, the area inhabited by the species has been gradually decreasing due to a combination of biotic and abiotic factors that are constantly intensifying (Palahi 2004; Nsibi et al. 2006; Boussaidi and Rebai 2017; Touhami et al. 2020). These issues include grazing (Stiti et al. 2014), seed predation, summer droughts causing mortality rates reaching up to 100% in open areas (Natividade

1950), wildfires (Catry et al. 2012), pests and diseases (Catry et al. 2017), and lack of forest management (Mechergui et al. 2023; Touhami et al. 2023), which led to the fragmentation of landscapes and the degradation of living environment.

Natural regeneration of cork oak is difficult and almost absent (Hasnaoui 1992; Abid 2006). In Tunisia, regeneration trials by direct sowing on more than 3000 ha were undertaken in 1988–1989, but they ended in total failure (Hasnaoui 1992). Numerous direct seeding trials have suffered the same fate in other Mediterranean regions (Messaoudene 1984; Louro 1999; Löf et al. 2019). Therefore, planting was adopted as a solution to attenuate the decline of cork oak forests, especially due to the enhancement of forest nursery and seedling production techniques (Stiti et al. 2014).

In addition to genetic inheritance, the plant production stage is important to produce seedlings of high morphophysiological quality that can cope with transplant shock, establish and grow in reforestation sites, and withstand abiotic stress (Lamhamedi et al. 2000). Moreover, seeds as the result of sexual reproduction are the foremost source of the genetic variability enabling plant species to cope with unpredictable environmental conditions (Ennajeh et al. 2013). Among other things, a successful plantation thus requires a judicious choice of plant material from a genetic standpoint. A multi-site provenance trial was established in 1997 within the framework of the EUFORGEN network (Turok et al. 1997) to meet this objective. Better selection of seed sources is a guarantee for the success of artificial regeneration by planting. The goals of this project were to achieve reforestation that is well-adapted (resistance to drought and various predators) and economically more profitable (better production of high-quality cork). The trial was set up to estimate the extent of geographical variability of the cork oak as well as to ascertain certain genetic parameters—elements necessary to define a strategy of conservation and improvement to be adopted regarding the Tunisian pedoclimatic conditions (Ennajeh et al. 2013). The established common garden experiment includes 26 Mediterranean provenances sampled within the natural area of the species (Khouja et al. 2000 & Khouja et al. 2005). In this contribution, we present the results in terms of growth and vigor traits obtained from the data collected from the trial site in the northwest of Tunisia.

Sampling and Trial Site

Within the framework of a concerted action launched in 1996 and financed by the European Union, a collection of provenances was obtained from a sampling effort carried out within the entire natural range of the species. The harvested material (acorns) was raised in a nursery in Portugal and then distributed to the different participating countries. The material used in the trials set up in 1997 in northwestern Tunisia (Fig. 36.1) was made up of 26 provenances of different origins. Geographical, edaphic, and climatic characteristics of these origin sites are described in Gandour et al. 2007. The trial plot was set up in Tebaba (8° 52' E, 36° 58' N), located in the Mogod Mountains, 57 km from the Algerian border. This region corresponds to the

Fig. 36.1 Examples of provenances with the highest survival rates at the Tebaba trial site (Photo credit Issam Touhami)

Numidian flysch deposited from the Oligocene to the early Miocene and features clay and sandstone soils. The site is located at an altitude of 250 m in a humid bioclimate (Gandour et al. 2007), on a slope ranging between 12 and 15%. The average annual rainfall is 1044 mm, and the average annual temperature is 17.9 °C (min: 7.2 °C, max: 30.3 °C).

The land underwent mechanical preparation (ripping and plowing) followed by equidistant ridging of 3 m between the plantation lines.

The 26 provenances of cork oak were planted in 30 randomized blocks, each consisting of 78 plants arranged in three lines, with 26 plants per line and a spacing of 3 × 2 m. Each provenance was represented by 90 individuals randomly distributed across the blocks. A total of 2340 plants were thus planted for observation (Gandour et al. 2007; Ennajeh et al. 2013).

Genetic Diversity Among and Within Provenances by Growth Traits

Morphological Characteristics

For each surviving 6-year-old tree in each block at the trial site, seven response variables were measured: total tree height (HT), diameter at base (DB), crown width at two perpendicular directions, form (on a subjective scale based on straightness and verticality from 1 (poor quality) to 3 (good quality)), vigor (on a subjective scale from 1 (poor state) to 3 (good state)) (Alia et al. 1995), and plant survival ratio (SUR) for each provenance per block (Gandour et al. 2007).

Statistical Analyses

Prior to statistical analysis, the survival rate percentages underwent an angular (arcsin) transformation. Variation in adaptive traits among provenances was analyzed using both univariate and multivariate methods. A two-way ANOVA was conducted for all traits including population and provenance effect using the following model:

$$Yijk = \mu + Pi + Bij + Eijk \qquad (36.1)$$

where Yijk: phenotypic value of the trait (Y) in plant (k) of provenance (j) in population (i), μ: overall mean, Pi: population effect, Bij: effect of provenance (j) in population (i), and Eijk: error term specified by the effect of plant (k) of provenance (j) in population (i)

The variability of the morphological characters among provenances was measured by calculating the coefficient of variation (CV) (Gandour et al. 2007).

The levels of genetic variance within (V_G) and between populations (V_B) were evaluated using ANOVA. Estimated V_B and V_G were used to quantify the level of population differentiation (Q_{st}) for each trait (Gandour et al. 2007):

$$Q_{st} = \frac{V_B}{V_B + 2V_G} \qquad (36.2)$$

This parameter is analogous to the parameter $F_{(ST)}$ defined by Weir and Cockerham (1984) and, therefore, provides an opportunity to compare population divergence in phenotypic characters and allozyme markers. This calculation assumes that the differences between populations are genetically controlled, an assumption that is plausible when plants are grown in the same environment as in this study. Product moment and Spearman's correlations between each trait and the respective origin site characteristics were calculated following appropriate procedures (Snedecor and Cochran 1967). It is worth noting that Spearman's rank correlation is the best-known procedure for studying the degree of relationship between two variables when there is sub-normality in both pairs of variables. Geographical differentiation among provenances with regard to growth traits was explored by means of cluster analysis using an Unweighted Pair-Group Method with Arithmetic mean (UPGMA; (Sokol and Michener 1958)) and Euclidian distances as the criterion for clustering (STATISTICA for Windows version 3.10). Finally, in order to test the correlation between genetic and geographical distances among populations, a Mantel test (Mantel 1967) was performed using the program GENETIX (version 4.02). The null hypothesis refers to the absence of association between the elements of the pairs of matrices. The matrix of geographical distances was calculated by converting geographical coordinates to Cartesian coordinates (x, y, and z), and the Euclidian distance (dab) between two points (a and b) was calculated as follows (Gandour et al. 2007):

$$dab = \sqrt{(xa - xb)^2 + (ya - yb)^2 + (za - zb)^2} \qquad (36.3)$$

Results

For the seven growth traits analyzed between provenances, Gandour et al. 2007 stated that the coefficients of variation ranged from 36% for form to 59% for total tree height (Gandour et al. 2007). Despite the observed moderate variation, the two-way ANOVA revealed statistically significant differences in mean and score values among provenances for each examined trait. At the population scale, the coefficients of variation likewise ranged from 36% to 59%. With regard to mean traits across the 26 provenances, the Moroccan provenance (Oulmes: 13) was the best in terms of height (57.13 ± 31.42) and stem straightness (2.52 ± 0.6). The Spanish provenance (La Almoraima: 18) had the largest diameter and most regular crown width and form. Furthermore, total height was the most diverging variable between populations (Gandour et al., 2007). These results agree with the results already observed for these traits in other common garden experiments with cork oak (Ramírez-Valiente et al. 2009, 2014; Sampaio et al. 2021).

Furthermore, the survival rates differed significantly between provenances ($p = 0.01$), varying from 95% for the Spanish-Portuguese provenance (no. 2) to 70% for the Moroccan one (no. 10). Estimates of Q_{st} varied widely between traits. Q_{st} values, as well as the mean estimate over traits, diverged significantly from zero. They ranged from 0.137 to 0.31. The greatest Q_{st} was determined for form (0.25, $p < 0.001$) and height (0.22, $p < 0.001$). Survival and vigor (0.18, $p < 0.001$) also exhibited significant differences between populations, albeit at a reduced level. Diameter, crown width 1, and crown width 2 displayed the least differentiation (0.137, 0.148, and 0.143, respectively), though the values were still significantly greater than zero. The cluster analysis (UPGMA) showed two groups: The first group consisted of the provenances originating from Morocco, Spain, and Portugal, while the second included those from Italy, Algeria, and Tunisia. A high correlation was found between genetic distance and geographical distance ($r = 0.699$; $P = 0.039$) based on the Mantel test. The one-tail probability indicated that the null hypothesis could be rejected, suggesting a clear geographical pattern of isolation by distance in the distribution of the species' genetic variability (Gandour et al. 2007).

Actually, significant genetic variation in fitness and functional traits like survival, growth, budburst phenology, drought tolerance, or cold resistance among cork oak populations has previously been determined (Aranda et al. 2005; Gandour et al. 2007; Ramírez-Valiente et al. 2009, 2014; Sampaio et al. 2021). In addition, research using isozymes (Toumi and Lumaret 1998; Jiménez et al. 1999) and microsatellites (SSRs) (Ramírez-Valiente et al. 2009) reported high genetic variation within cork oak populations.

Genetic Variability and Adaptation to Local Conditions

Growth Traits Monitoring

The growth dynamics of the trees at the trial site were monitored in September 2004 and September 2011, 7 and 14 years after planting. Total plant height was measured along a main axis (in cm) fixed for each individual. In 2011, the circumference at the base of the trunk was measured for each individual, and provenance survival rates (%) at the ages of 7 and 14 years old were calculated (Ennajeh et al. 2013). For statistical purposes, the "provenance x block" effect was tested on our measurements using a 2-factor analysis of variance (ANOVA) at the 5% risk threshold (STATISTICA). This analysis was complemented by a multiple comparison of means using the Newman and Keuls test.

Results

From 2004 to 2011, the survival rates of the cork oak provenances decreased significantly by approximately half (Fig. 36.2a). In 2011, significant mortality was noted within blocks, and the lowest survival rates were recorded among the Santiago de Cacem provenance from Portugal (33.33%), while the provenances from Sassari and Cagliari in Italy exhibited the highest survival rates. Actually, from 2004 to 2011, certain provenances stood out from their neighbors. The Tunisian Fernana provenance showed a fairly low survival rate in 2004 compared to its neighbors, but the percentage rebounded in 2011 to be among the best (Fig. 36.2). Similarly, the Italian Puglia provenance developed from least vigorous to most vigorous provenance. By contrast, the survival rate of the Portuguese Santiago de Cacem provenance declined from 2004 to 2011 (Ennajeh et al. 2013). The Ain Johra provenance from Morocco maintained a low survival rate across both measurements, whereas the Spanish El Padro provenance consistently exhibited relatively high survival rates in 2004 and 2011 (Fig. 36.2a). The mean tree height was estimated at 114.66 ± 5.49 cm in 2011, with heights ranging from 101.88 cm to 124.64 cm for the provenances 19 (Cataluna) and 15 (Montes de Toledo), respectively. Figure 36.2b shows the tree height measurement data for the 26 provenances in 2004 and 2011.

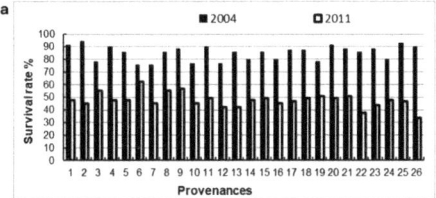
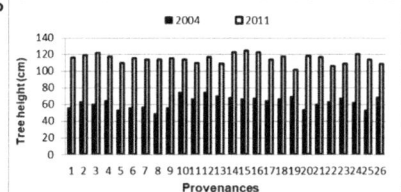

Fig. 36.2 Survival rates (**a**) and mean tree height (**b**) measured in 2004 and 2011 for each cork oak provenance at the experimental site in Tebaba, Tunisia

Given the deterioration of the climate over the past years (Ennajah 2010), certain provenances stood out. Fourteen years after planting on the Tunisian site, only the more adapted provenances were able to survive and grow without major damage. The Italian Puglia and Tunisian Fernana provenances were found to be well-adapted to the edapho-climatic conditions specific to the trial site. On the other hand, the Moroccan Oumles and Ain Johra and Portuguese Santiago de Cacem provenances were the least vigorous in this environment.

It is important to point out that in September 2006, symptoms of pathological decline of varying severity depending on the provenance were recorded at the site (Khouja et al. 2010). They were caused by a fungal attack that resulted in desiccation recorded to varying degrees across all provenances. Twenty-four percent of the trees in the system were affected, of which 87% were completely dried out. The provenance of Mekna (Tunisia) appeared to be the most sensitive, while that of Fuencaliente (Spain) was the least affected. The analyses carried out on different tree parts (shoots, branches, stem, roots) showed the presence of various pathogenic agents, including three pathogenic species: *Biscogniauxia (=Hypoxylon) mediterranea*, *Armillaria mellea* and *Ungulina annosa*. The most important range of pathogens associated with the provenances was observed on the roots, with nine species (50%). The *B. mediterranea* species dominated with a very high index (FI), ranging from 50% on branches without necrosis to 80% on branches without necrosis. Some of these seem to be involved in the etiology of the tree decline (Khouja et al. 2010). The change in behavior of certain provenances of cork oak between 2004 and 2011 could be identified as a character of adaptation to the environment following this parasitic attack. Until 2004, the Moroccan provenances (Bousafi, Ain Johra, and Oumless) had the best results in length and survival rates; likewise, the Portuguese provenance Santiago de Cacem was one of the most vigorous. Five years after the fungal attack, in 2011, these same provenances showed the lowest survival rates and average length growth. Conversely, the Italian Puglia and local Tunisian Fernana provenances showed the lowest survival rates and growth in length in 2004. However, in 2011, these provenances showed the best survival rates showed the best survival rates with fairly high growth (Ennajeh et al. 2013).

Selecting Cork Oak Provenances for Resilience and Efficiency

Besides the issue of production, the current fragile conditions of the Tunisian and Mediterranean cork oak forests, which are subject to numerous decline factors, are a cause for concern regarding the sustainability of the ecosystem. In fact, when selecting provenances for new reforestation stands or the repopulation of old-growth forests, the criterion of adaptation to environmental conditions—and in particular to parasitic attacks of all kinds—must be given priority. Indeed, the sudden onset of pathogen attacks in the common garden of cork oak provenances in Tebaba made it easier to choose the most suitable provenances in combination with other environmental constraints. The top three provenances combine high resistance to both biotic and abiotic stresses. In general, the lowest-performing provenances are also the most sensitive to frost and drought (Ennajeh et al., 2013). If the Puglia and Fernana provenances prove successful, this would justify further research into them. Indeed, certain traits contribute to the adaptive value of the individuals—in our case, resistance to fungal attack and climatic hazards (Ennajeh et al. 2013). If these traits are heritable, this criterion (the adaptation mechanism) will be transmitted to the next generations. An adaptive pattern and transmissibility study would constitute an important tool for clarifying the cork oak adaptation process.

References

Abid H (2006) La gestion durable des forêts de chêne-liège en Tunisie : cas du Projet de Gestion Intégrée des Forêts. Annales de l'INRGREF 9(1) Numéro spécial:1–20

Alia R, Gil L, Pardos JA (1995) Performance of 43 *Pinus pinaster* Ait. Provenance on 5 locations in Centrale Spain. Silvae Genet 44(1995):75–81

APCOR (2020) Cork 2020. Associação Portu-guesa de Cortiça, Santa Maria de Lamas, Portugal, pp. 68. http://www.apcor.pt/en/portfolio-posts/apcor-year-book-2020/

Aranda I, Castro L, Alia R, Pardos JA, Gil L (2005) Low temperature during winter elicits differential responses among populations of the Mediterranean evergreen cork oak (Quercus *suber* L.). Tree Physiol 25:1085–1090. https://doi.org/10.1093/treephys/25.8.1085

Bel Fekih Boussema S, Cohen M, Khebour AF (2022) Green and blue infrastructure design in a semi-arid region. Front Environ Sci 10. https://doi.org/10.3389/fenvs.2022.1061256

Boussaidi N, Rebai L (2017) Impacts of grazing on the degradation and regeneration of the kroumirie subria (north west of Tunisia). J New Sci Agric Biotechnol 44(4):2410–2429

Catry FX, Moreira F, Cardillo E, Pausas JG (2012) In: Moreira F, Arianoutsou M, Corona P, De Las HJ (eds) Post-fire management of Cork oak forests. In:"post-fire management and restoration of southern European forests". Springer, London, pp 195–222

Catry FX, Branco M, Sousa E, Caetano J, Naves P, Nobrega F (2017) Presence and dynamics of ambrosia beetles and other xylophagous insects in a Mediterranean cork oak forest following fire. For Ecol Manag 404:45–54. https://doi.org/10.1016/j.foreco.2017.08.029

Ennajah A (2010) Croissance et productivité des forêts de chêne liège (*Quercus suber* L.). Vulnérabilité aux changements climatiques. Thèse de Doctorat en Biologie. Faculté des Sciences de Tunis. 261p

Ennajeh A, Azri W, Khaldi A, Nasr Z, Selmi H, Khouja M (2013) Variabilité génétique du chêne liège (*Quercus suber* L.) en Tunisie. Bilan d'un éssai comparatif multisites de plantations de provenances diverses. Revue Internationale de Geologie, de Geographie et d'Ecologie Tropicales 37(2):191–200

Gandour M, Khouja ML, Toumi L, Triki S (2007) Morphological evaluation of cork oak (*Quercus suber* L.): Mediterranean provenance variability in Tunisia. Ann For Sci 64:549–555. https://doi.org/10.1051/forest:2007032

Hasnaoui B (1992) Chênaies du Nord de la Tunisie, écologie et régénération [oak-groves of northern Tunisia, ecology and regeneration]. These de doctorat. Universite Aix Marseille I, Marseille, p 202

Jiménez P, Agundez D, Alia R, Gil L (1999) Genetic variation in central and marginal populations of Quercus suber L. Silvae Gen 48:278–284

Khalfaoui M, Daly-Hassen H, Stiti B, Jebari S (2020) Decision-making support: valuation and mapping of new management scenarios for Tunisian Cork oak forests. Forests 11(2):197. https://doi.org/10.3390/f11020197

Khouja ML, Khaldi A, Ben Jemaa ML, Toumi L, Lumaret R (2000) Conservation and improvement of cork oak forests in Tunisia: Acquired of researches and technical applications, Congrès mondial sur le chêne liège tenu à Lisbonne (Portugal) du 19 au 21 juillet 2000, 11p

Khouja ML, Khaldi A, Sellemi H (2005) Variabilité adaptative de chêne liège en Tunisie. IOBC/WPRS Bull 28(8):231–236

Khouja ML, Ben Jamâa ML, Franceschini A, Khaldi A, Nouri M, Sellemi H, Hamrouni L (2010) Integrated protection in oak forests. IOBC/wprs Bull 57:53–59

Lamhamedi MS, Ammari Y, Fecteau B, Fortin JA, Margolis H (2000) Problématique des pépinières forestières en Afrique du Nord et stratégies de développement. Problems in forest nurseries in North Africa and guidance strategies. Cahier d'Études et de Recherches Franco-phones/Agricultures 9(5):369–380

Löf M, Castro J, Engman M, Leverkus AB, Madsen P, Reque JA, Villalobos A, Gardiner ES (2019) Tamm review: direct seeding to restore oak (Quercus spp.) forests and woodlands. For Ecol Manag 448:474–489., ISSN 0378-1127. https://doi.org/10.1016/j.foreco.2019.06.032

Louro G (1999) Avaliaçaõ da aplicação de programas de apoio à floresta na região Doalgarve. Direcção Geral das Florestas (DGF-Lisboa), Portugal 1999:6–8

Mantel NA (1967) The detection of disease clustering and a generalized regression approach. Cancer Res 27:209–220

Mechergui T, Pardos M, Boussaidi N, Jacobs DF, Catry FX (2023) Problems and solutions to cork oak (*Quercus suber* L.) regeneration: a review. iForest 16:10–22. https://doi.org/10.3832/ifor3945-015

Messoudène (1984) Résultats des essais de semis directs du chêne liège à Melata, Rapport interne. Institut National de Recherches Forestières (INRF-Algérie), p 10

Natividade JV (1950) Subericultura. DGSFA, Lisboa

Nsibi R, Souayah N, Khouja LM, Bouzid S (2006) La régénération naturelle par semis de la suberaie de Tabarka - Aïn Draham face aux facteurs écologiques et anthropiques. Geo-Eco-Trop 30(1):35–48

Palahi M (2004) New tools and methods for Mediterranean forest management and planning. Tempus IMG. Centre Tecnologic Forestal de Catalunya, Solsona

Ramírez-Valiente JA, Valladares F, Gil L, Aranda I (2009) Population differences in juvenile survival under increasing drought are mediated by seed size in cork oak (*Quercus suber* L.). For Ecol Manag 257:1676–1683. https://doi.org/10.1016/j.foreco.2009.01.024

Ramírez-Valiente JA, Valladares F, Sanchez-Gomez D, Huertas AD, Aranda I (2014) Population variation and natural selection on leaf traits in cork oak throughout its distribution range. Acta Oecol 58:49–56. https://doi.org/10.1016/j.actao.2014.04.004

Sampaio T, Gonçalves E, Faria CM, Helena Almeida MH (2021) Genetic variation among and within *Quercus suber* L. populations in survival, growth, vigor and plant architecture traits. For Ecol Manag 483:118715., ISSN 0378-1127. https://doi.org/10.1016/j.foreco.2020.118715

Snedecor GW, Cochran WG (1967) Statistical methods. The Iowa State University Press, Ames

Sokol RR, Michener CD (1958) A statistical method for evaluating systematic relationships. Univ Kansas Sci Bull 28:1409–1438

Stiti B, Piazzetta R, Khaldi A (2014) Régénération de la subéraie tunisienne: état des lieux, contraintes et avancées techniques. Forêt Méditerranéenne XXXV(2):151–160

Stiti B, Khalfaoui M, Bahri S, Khaldi A (2021) Towards optimizing acorn use in animal feeding in Tunisia: valuation and impact on natural regeneration. Bois et Forêts des Tropiques 348

Stiti B, Mezni F, Zarroug Y, Slama A, Fkiri S et al (2022) Agri-food valuation of oak fruits in human consumption: formulation of desserts. J Food Chem Nanotechnol 8(2):38–42. https://doi.org/10.17756/jfcn.2022-124

Touhami I, Chirino E, Aouinti H et al (2020) Decline and dieback of cork oak (*Quercus suber* L.) forests in the Mediterranean basin: a case study of Kroumirie, Northwest Tunisia. J For Res 31:1461–1477. https://doi.org/10.1007/s11676-019-00974-1

Touhami I, Rzigui T, Zribi L, Ennajah A, Dhahri S, Aouinti H, Elaieb MT, Fkiri S, Ghazghazi H, Khorchani A, Candelier K, Khaldi A, Khouja ML (2023) Climate change-induced ecosystem disturbance: a review on sclerophyllous and semi-deciduous forests in Tunisia. Plant Biol J 25:481–497. https://doi.org/10.1111/plb.13524

Toumi L, Lumaret R (1998) Allozyme variation in cork oak (*Quercus suber* L.): the role of phylogeography and genetic introgression by other Mediterranean oak species and human activities. Theo Appl Genet 97:647–656. https://doi.org/10.1007/s001220050941

Turok J, Varela MC, Hansen C (1997) *Quercus suber* network (EUFORGEN)– report of the third and fourth meetings of Sassari-Italy (9–12 June 1996) and Almoraima-Spain (20–22 February 1997). Publication IPGRI, Nairobi, pp 5–10

Vinha AF, Costa ASG, Barreira JCM, Pacheco R, Oliveira MBPP (2016) Chemical and antioxidant profiles of acorn tissues from *Quercus* spp.: potential as new industrial raw materials. Ind Cro Prod 94:143–151. https://doi.org/10.1016/j.indcrop.2016.08.027

Weir BS, Cockerham CC (1984) Estimating f-statistics for the analysis of population structure, Evol 38:6. https://doi.org/10.2307/2408641

Zarroug Y, Boulares M, Sfayhi D, Slimi B, Stiti B, Zaieni K, Nefissi S, Kharrat M (2022) Structural and physicochemical properties of Tunisian *Quercus suber* L. starches for custard formulation: a comparative study. Polymers 14:556. https://doi.org/10.3390/polym14030556

Open Access This chapter is licensed under the terms of the Creative Commons Attribution 4.0 International License (http://creativecommons.org/licenses/by/4.0/), which permits use, sharing, adaptation, distribution and reproduction in any medium or format, as long as you give appropriate credit to the original author(s) and the source, provide a link to the Creative Commons license and indicate if changes were made.

The images or other third party material in this chapter are included in the chapter's Creative Commons license, unless indicated otherwise in a credit line to the material. If material is not included in the chapter's Creative Commons license and your intended use is not permitted by statutory regulation or exceeds the permitted use, you will need to obtain permission directly from the copyright holder.

37. United Kingdom/Scotland: Assisted Regeneration to Restore Lost Forests

Philippa Gullett, Mark Hancock, Lucy Mason, and Andrew Weatherall

Deer Browsing in Scottish Highland (Adobe stock/Marc Scharping)

P. Gullett (✉)
Cairngorms Connect and RSPB Centre for Conservation Science, Scotland, UK
e-mail: pip.gullett@cairngormsconnect.org.uk

M. Hancock · L. Mason
RSPB Centre for Conservation Science, Etive House, Beechwood Business Park, Inverness, UK
e-mail: lucy.mason@rspb.org.uk

A. Weatherall
RSPB Policy and Advocacy Team, RSPB Scotland Headquarters, Edinburgh, UK
e-mail: andrew.weatherall@rspb.org.uk

Abstract

Scotland was once a largely forested country, but after centuries of human influence, only a tiny fraction of the original forest area remains. With one of the lowest forest covers of any European country and the majority consisting of non-native plantations, increasing the amount of native woodland is an urgent need. The natural expansion of native woodland is limited by three main factors in Scotland: browsing by wild deer, a shortage of regeneration niches, and a shortage of seed sources. In many areas, the process of natural regeneration is, therefore, very slow or entirely absent for some species.

Cairngorms Connect is a land management partnership established to restore habitats and natural processes to 60,000 ha of the Scottish Highlands. Woodland expansion is a key goal—specifically, by departing from the recent norm of planting and aiming to enable expansion by natural processes. We have devised a Before-After-Control-Impact experimental trial to test methods of accelerating the natural expansion of native broadleaves via applied nucleation. The trial compares two methods of opening up regeneration niches by cutting the dense understorey vegetation, as well as two methods of seed source establishment by sowing or planting. Through detailed monitoring and thorough documentation of management methods, this Seed Source Establishment Trial will inform restoration efforts in Scotland and further afield.

Keywords

Natural processes · Applied nucleation · Experimental trials · Woodland expansion · Deer management

A Landscape of Lost Forests

Historically, Scotland was a largely forested country, with extensive cover of deciduous and pine woodland and wood pasture (Huntley et al. 1997). Today, however, Scotland has one of the lowest proportions of woodland cover in Europe, estimated at around 19% compared to the European average of 46% (Forest Research 2022). The majority of this remaining woodland is plantation under commercial production, with non-native species accounting for around two-thirds of the total forested area; this means that native woodland (including both planted and semi-natural) covers only around 6% of the country's total land area (Forest Research 2022)—a small fraction of the roughly 50% of Scotland's territory believed to be suitable for forests (Towers et al. 2004). Given the current climate and biodiversity crises, expanding the area of native woodland is an urgent priority, as reflected in current Scottish government targets (Scottish Government 2019). Besides providing multiple benefits to people, native forests also have intrinsic value, adding a moral dimension to their restoration (Smout 2003; Hobbs 2009).

Following centuries of forest clearance and livestock grazing (Smout 2003), the natural recovery and expansion of woodland in the Scottish Highlands appears to be limited by three main factors. The first is browsing by red deer (*Cervus elaphus*) and, increasingly, roe deer (*Capreolus capreolus*) (Miller et al. 1998; Rao 2017). Deer control by lethal culling is a key part of land management throughout Scotland, with more than 100,000 individuals culled annually (Scottish Government 2020), but in many areas, deer densities are still too high to enable widespread establishment of trees (Newton et al. 2001; Scottish Government 2020). Secondly, there is a general shortage of regeneration niches across much of the Scottish Highlands (Miller et al. 1998) due to active management over the past two centuries to maintain these formerly wooded uplands as open landscapes (Bowditch et al. 2023), which has resulted in vast areas now being dominated by woody ericaceous shrubs that inhibit succession back to more natural vegetation communities, including trees of many species and growth forms (e.g. Tanentzap et al. 2013). Finally, there is a lack of seed sources in many areas following the widespread loss of woodland and active management to remove broadleaved trees from pinewood plantations. The restoration of Scotland's lost forest landscapes, therefore, requires large-scale changes in land use and collaborative efforts between multiple stakeholders.

Cairngorms Connect (cairngormsconnect.org.uk) is one such partnership of land managers committed to an ambitious 200-year vision of ecosystem restoration across a 60,000 ha area of Cairngorms National Park in the Scottish Highlands (Fig. 37.1). The partnership area spans over 1100 m of altitudinal range, with over 5000 recorded species and some of the largest surviving remnants of several of Britain's rarest habitats, including Scots pine-dominated ancient woodland and low-stature mountain woodland of willows and birches (Summers 2019). At the heart of our vision is the landscape-scale restoration of habitats, ecosystems, and natural processes. This includes the restoration and expansion of native woodlands at the landscape scale—largely by natural processes—to link together the few surviving native woodland fragments (Fig. 37.1; Gullett et al. 2023). Alongside the ambitious spatial and temporal scale of our management targets, a central tenet of Cairngorms Connect is to expand our knowledge of how to accomplish restoration most effectively by incorporating science and monitoring at the heart of what we do, aspiring to be a natural 'observatory' or 'laboratory' of restoration science.

How to Restore Forests Where They Have Long Been Lost?

In the past, efforts to increase woodland area in the Scottish Highlands have commonly relied on planting within fenced exclosures to enable seedlings to escape the browsing pressure from deer (Smout 2003). In recent years, however, arguments against the use of fencing have been mounting, primarily due to the lethal impacts of fence strikes on endangered woodland grouse species (Catt et al. 1994), which have been shown to be contributing to catastrophic population declines in the western Capercaillie *Tetrao urogallus* (Moss 2001). Fencing has also been increasingly criticised for its perceived negative visual impacts, restriction of human access, and financial cost (Warren 2002; Hobbs 2009). There has also been a gradual shift away from planting as the sole means of establishing new woodland back towards the

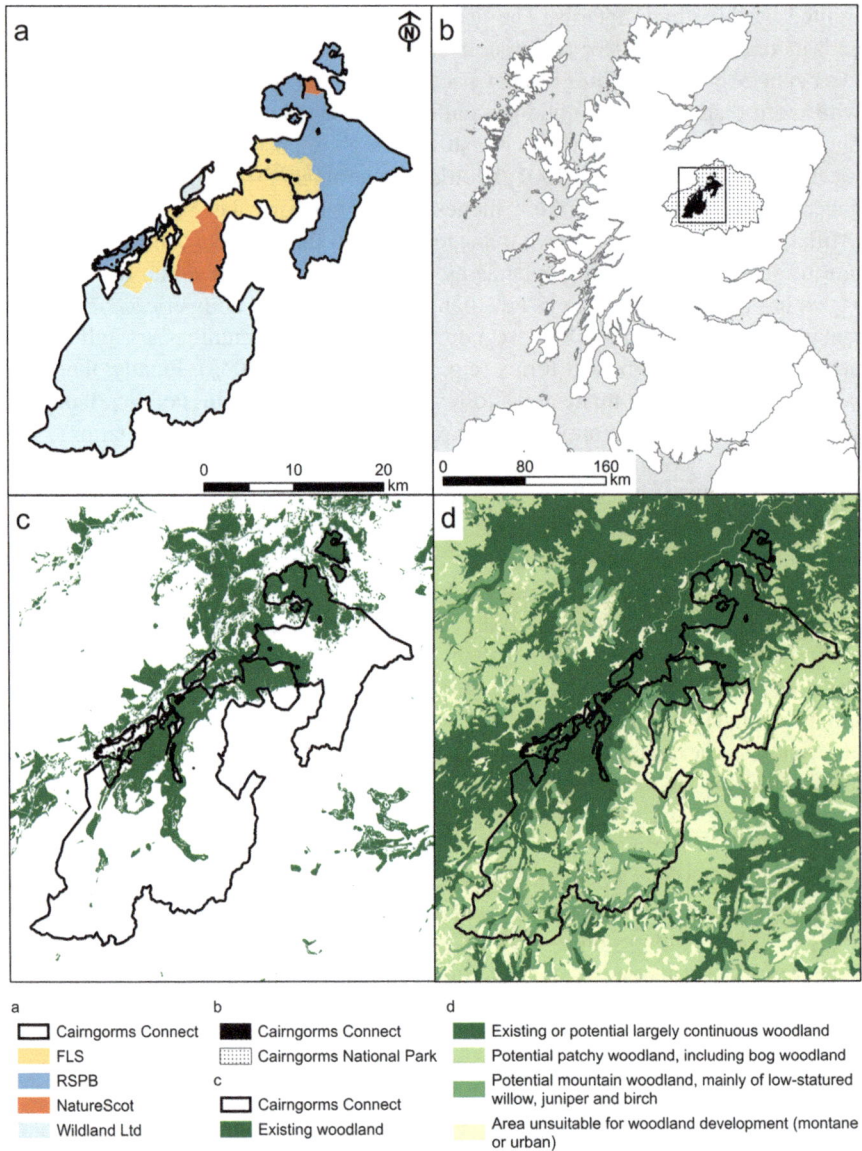

Fig. 37.1 The Cairngorms Connect (CC) partnership area and the extent of existing and potential woodland. (**a**) CC partnership area, indicating ownerships of the four partners; (**b**) location of CC within Cairngorms National Park and Scotland; (**c**) existing woodland area; (**d**) modelled area of potential future woodland

more traditional method of enabling natural regeneration that prevailed in older practices (Steven and Carlisle 1959). The potential ecological benefits of naturally regenerated vs. planted woodland, such as faster recovery of biodiversity, ecological function, and forest structure (Crouzeilles et al. 2017; Meli et al. 2017; di Sacco et al. 2021), better preservation of the local cultural and genetic woodland legacies (Summers and Cavers 2021), and greater resilience to climate change by favouring individuals adapted to current climates (Chazdon et al. 2021), are being increasingly recognised. Furthermore, restoration via natural regeneration can often be more cost-effective than more interventionist methods (Brancalion and Holl, 2020).

Across Cairngorms Connect, a widespread expansion of Scots pine has been observed over the past 30 years, believed to be largely due to the significant increase in deer culling for conservation reasons over this period (Gullett et al. 2023). However, there has been only very limited regeneration of broadleaved tree species, especially those with small windblown seeds, which is thought to be due to the lack of local seed sources and relatively poor dispersal capabilities of these species. Although such species may eventually colonise remote deforested uplands given sufficiently low browsing pressure, this process will be extremely slow—especially in colder, northern climes. This is directly at odds with the urgent need to restore ecosystems to address the current climate and biodiversity crises, reflected in time-bound government commitments to increase forest cover and improve land for biodiversity (Scottish Government 2019, 2022). Furthermore, previous work in the Cairngorms has shown that tree species highly palatable to browsers are underrepresented in patches of naturally regenerating woodland (Gullett et al. 2023). Therefore, although some people oppose planting in all circumstances (Summers and Cavers 2021), we believe there is a strong argument for increasing the level of intervention for certain tree species in areas distant from remnant woodland.

Monitoring Passive Regeneration and Trialling Active Methods

Across Cairngorms Connect, we aim to enable natural, passive regeneration where possible but intervene with more active means when passive methods are ineffective or too slow (Holl et al. 2020; di Sacco et al. 2021). In contrast to the extensive planting of trees, we are trialling a more diverse approach with large areas of natural regeneration, smaller areas of planting, and other areas under 'assisted regeneration', where we seek to speed up the natural regeneration process by artificially establishing (through planting or sowing) small pockets of trees to act as seed sources to repopulate the wider area. By including sowing as an alternative to planting, we are trialling an intervention technique that is as naturalistic as possible and may offer more of the benefits associated with natural regeneration (such as more natural spatial distribution of trees) and fewer of the drawbacks of planting (such as the unnatural absence of competition and herbivory at the early seedling stage) (Summers and Cavers 2021).

Over the past 30 years, alongside an increase in deer culling efforts, the Cairngorms Connect partners have monitored natural tree regeneration using a range of plot-based and transect-based survey methods (Gullett et al. 2023). In addition to tracking the pattern of woodland expansion across the area, an important element of this monitoring is to understand how different tree species are responding to elevated deer culling intensity in different parts of the study area. These data show a consistent, large-scale, and dramatic expansion of native woodland over the past 30 years, with an estimated annual average of approximately 164 ha of new woodland established during peak periods of expansion in core areas near mature woodland (Gullett et al. 2023). The pattern of regeneration varies between different parts of the site, however, and some species show little evidence of expansion—notably the rarer and more palatable deciduous broadleaf species such as Aspen (*Popula tremula*; Gullett et al. 2023).

There have also been several experimental trials comparing the effects of different ground preparation treatments on the establishment success of pine and broad-leaved tree seeds on open, shallow-peat moorland at RSPB Abernethy Forest, a key nature reserve managed by the Royal Society for the Protection of Birds within the Cairngorms Connect partnership area. These trials were first considered in the late 1990s when new tree seedling recruitment slowed following the development of dense stands of heather and thick bryophyte layers (which are considered hostile to tree recruitment) after deer reductions at the site (Amphlett, 2003). Trials began at a small scale, with Scots pine (*Pinus sylvestris*) seeds hand-sown into experimentally burnt moorland patches exhibiting much better establishment compared to unburnt control areas (Hancock et al. 2005). This prompted a larger-scale trial to explore pine establishment in burnt patches from natural seed rain, which again showed strong (between 10- and 30-fold) enhancement of seedling densities on burnt patches, an effect that was uninfluenced by the presence of fencing to reduce deer browsing (Hancock et al. 2009). Despite the apparent success of ground preparation through burning for pine seedling establishment, this method is not being applied in Cairngorms Connect due to issues surrounding carbon release as a result of burning, along with the potential for damage to upland peat soils and hydrological functions. As an alternative to burning, the team also investigated whether short periods of cattle grazing could similarly promote pine regeneration. Although there were some signs of improved seedling regeneration niches and pockets of new regeneration at cattle feeding areas, the low seed fall into the trial area meant that the overall results were inconclusive with regard to the usefulness of this method as a management tool to assist regeneration over large areas (Hancock et al. 2010).

Establishing Seed Sources for Assisted Regeneration

Assisted regeneration methods are increasingly being advocated in some parts of the world (e.g. García et al. 2020; Holl et al. 2020; di Sacco et al. 2021; Krishnan and Osuri 2022), yet such approaches are still novel in Britain. Furthermore, existing knowledge on the success of tree planting and sowing compared to natural regeneration under different ground preparation treatments is largely based on silvicultural requirements (e.g. Nixon and Worrell 1999), which means that knowledge about tree species less valuable for timber (like birch *Betula* spp.) or in contexts without good silvicultural potential (like high-altitude areas) is lacking (Willoughby et al. 2007). One trial in the Scottish Highlands showed that sowing of downy birch (*Betula pubescens*), alder (*Alnus glutinosa*), and rowan (*Sorbus acuparia*) seeds could successfully establish upland woodland with prior preparation of the ground surface by weed clearance, brash raking, and scarification (Willoughby et al. 2019); to our knowledge, however, there have been no similar trials comparing both sowing and planting alongside natural regeneration with and without ground preparation.

Given this lack of knowledge regarding how best to re-establish woodland in remote upland areas, an experimental trial was designed at Cairngorms Connect to test the applied nucleation model of assisted regeneration (Info Box 37.1; Corbin and Holl 2012)—a middle path between planting and natural regeneration. This trial aims to measure the relative efficacy of different methods of establishing seed source populations of broadleaf tree species in the high-altitude forest expansion zone between 450 and 650 m.a.s.l., taking into consideration financial costs and practical limitations. The native pinewoods that provide the natural seed rain for these areas are relatively depauperate in broadleaf species, so by establishing broadleaf seed sources, we hope to encourage succession to a more natural, species-rich woodland system. We are focusing on downy birch, aspen, and eared willow (*Salix aurita*), which are key species in the upland woodland habitat we seek to restore. These species have predominantly wind-dispersed seeds, most of which fall within 50 m of the parent tree (see Atkinson 1992), and are consequently poor at dispersing to remote areas; in addition, they are relatively palatable species (especially aspen and willow; Scottish Forestry, 2023) and therefore strongly suppressed by deer (Atkinson 1992; Beguin et al. 2016; Ramirez et al. 2019).

In this replicated, large-scale experiment, we are trialling two ground preparation methods (robocutting and brushcutting) and two tree introduction methods (sapling planting and seed sowing) alongside control plots with no ground preparation and only natural seed rain. For the robocutting treatment, we deployed a remote-controlled robotic mower (manufacturer: Husqvarna/Stihl, fitted with tri-blade cutting head), cutting the woody material down to the surface vegetation/moss layer while taking care not to disturb the latter directly (Fig. 37.2, top). For the brushcutting treatment, we used a handheld brushcutter fitted with a metal tri-blade to cut the woody vegetation, again making sure not to disturb the surface vegetation/moss layer directly (Fig. 37.2, bottom). After both treatments, the resulting cut plant matter was left *in situ*. For the planting treatment, we planted saplings by hand by parting the surrounding vegetation but avoiding any vegetation removal or soil preparation; the saplings were raised in the Cairngorms Connect tree nursery from local seed (birch and aspen) or cuttings (willow). Planting density was 3200 stems per ha (based on current Scottish Government guidance for establishing upland woodland; Scottish Government 2021), with a ratio of 1:1:5 for aspen/willow/birch based on their perceived relative abundances in more natural examples of the habitats we are seeking to restore. For the seed sowing treatment, we broadcast seed by hand, aiming for roughly even coverage across the entire plot, using two million seeds per hectare (the seeding rate recommended by forest research for establishing birch in Scottish upland habitats; Willoughby et al. 2019). Due to difficulties in obtaining sufficient seed quantities, only birch was sown in the seeding treatment, while aspen and willow were also planted in seeded plots to maintain consistency between treatments as far as possible; the birch seed was harvested on-site.

Fig. 37.2 Ground preparation by robocutting (top) and brushcutting (bottom), showing machines in action (left) and plots shortly after cutting (right)

Info Box 37.1 Applied nucleation

Applied nucleation is a method of assisted regeneration for woodland recovery in which small patches of trees or shrubs are planted in otherwise deforested areas and then allowed to expand outwards by natural processes. The term 'applied nucleation' was coined by Corbin and Holl (2012), who proposed applying the natural process of 'nucleation' (i.e. the pattern commonly observed in natural forest recovery whereby initial colonisers establish in distinct patches or 'nuclei', which then expand via natural successional processes; Yarranton and Morrison 1974) in a restoration context to restore diverse ecosystems to areas where forests have been degraded or destroyed.

Applied nucleation offers a middle ground between zero-intervention methods (i.e. natural regeneration) and high-intervention methods (i.e. extensive planting or sowing). It offers several major benefits including (a) development of diverse communities that more closely resemble their 'natural' counterparts compared to plantations; (b) faster establishment of woody cover and tree species diversity compared to pure reliance on natural regeneration, especially in areas more remote from seed sources; (c) financial savings compared to high-intervention methods, with increasing cost-effectiveness at greater spatial scales; (d) greater resilience to future environmental changes

(continued)

compared to plantations (Holl et al. 2020; Rojas-Botero et al. 2020; Bechara et al. 2021; Werden et al. 2021).

The schematic below shows how applied nucleation (b, d) can enable the establishment of more diverse communities than natural regeneration (a, c) in scenarios like the Cairngorms, where historic management has led to remnant forests being dominated by a single species (here: Scots pine) and depauperate in the full suite of species that should be present in natural, functioning examples of the habitat type (here: various broadleaf tree species).

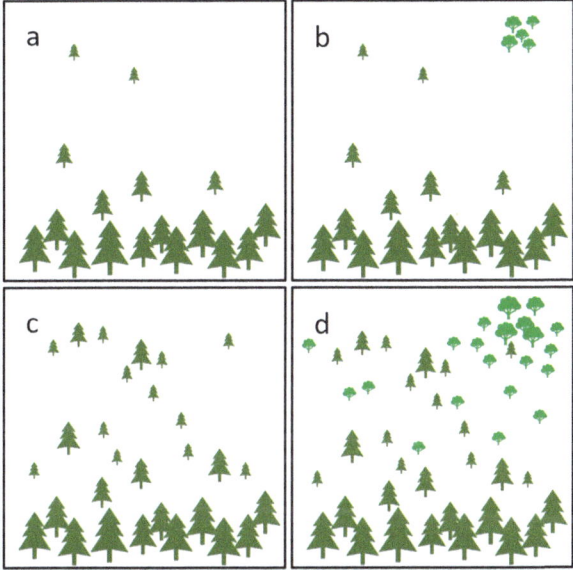

Natural regeneration vs applied nucleation in areas lacking seed sources for broadleaf species. Under natural regeneration (**a**), developing woodland is species-poor (**c**). Under applied nucleation (**b**), developing woodland is more species-diverse (**d**), more closely resembling 'natural' woodland

The trial comprises six replicate plot groups (i.e. experimental blocks), each consisting of nine study plots measuring 20 m × 20 m (0.04 ha), with each plot randomly assigned to a different treatment combination (Fig. 37.3). The six plot groups were separated into two tranches for resourcing reasons and to reduce the overall influence of individual year effects; three plot groups were treated during winter/spring 2021–2022 and the remaining three during winter/spring 2022–2023. All plot groups are located within the forest expansion zone (450–650 m altitude) in areas representative of the main type of habitat and topography in which these techniques might be used in the wider landscape. The sites consist of open heathland

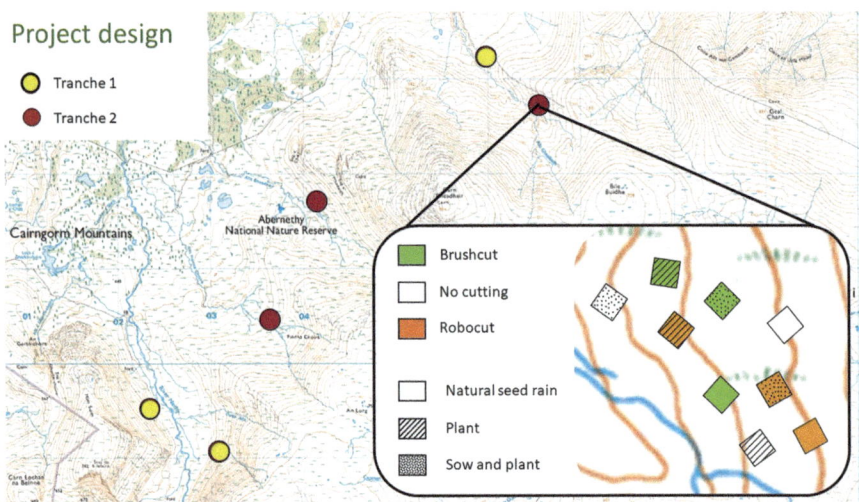

Fig. 37.3 Seed Source Establishment Trial project design showing the distribution of the six replicate plot groups across the 450–650 m altitude forest expansion zone at RSPB Abernethy. Plot groups are split into two tranches; the inset shows example treatment combinations for the nine plots (20 m × 20 m) within each plot-group. Ordnance Survey data © Crown copyright and database right 2023, RSPB licence 100030994

dominated by dwarf ericaceous shrubs, with varying proportions of *Calluna vulgaris, Erica cinerea, Vaccinium myrtillus,* and some graminoids such as *Deschampsia flexuosa*. Neighbouring plot groups are spaced at least 750 m apart (Fig. 37.3). See Fig. 37.4 for an aerial image of a typical plot-group following treatment.

Prior to treatment, we conducted detailed baseline vegetation monitoring encompassing percentage cover of key plant species, browsing intensity (i.e. percentage of shoots browsed, degree of trampling, abundance of dung), sward density and height, as well as detailed searches for existing tree seedlings. Our baseline surveys revealed just 17 tree seedlings across 1350 quadrats of 1 m² (i.e. <13 stems per hectare), with only two tree species represented: rowan (*Sorbus aucuparis*) and Scots pine. There was no evidence of any of the target tree species (downy birch, eared willow, and aspen) in any of the plots. By contrast, 6% of the quadrats contained deer dung, proving the continued presence of browsers despite the relatively high levels of active deer culling throughout the study areas. Data from the latest site-wide deer dung survey conducted in 2018 show that deer densities within the vicinity of the Seed Source Establishment Trial plots range from zero to 20 red deer per km², whilst roe deer densities are <5 per km².

Following treatment by robocutting or brushcutting (Fig. 37.4) undertaken during the winter months, we surveyed the extent and nature of cutting within the plots, including measuring the depth and density of the mulch and any remaining understorey layer that could significantly affect tree seedling establishment success. This monitoring revealed dramatic differences between the treatments (Table 37.1). The

Fig. 37.4 Aerial image of a typical plot-group approximately 9 months after cutting, comprising three brushcut plots (yellow rectangles), three robocut plots (red rectangles), and three control plots (not cut, therefore not visible)

Table 37.1 Main differences in ground characteristics before and after cutting, calculated as a mean average across 288 brushcut plots and 288 robocut plots

Variable	Brushcut	Robocut	t-test result
Vegetation height pre-cutting (cm)	45	45	n.s.
Average height of cut stems (cm)	23	13	$P < 0.0001$, $t = 11.2$, df = 499
Understorey depth pre-cutting (cm)	11	10	n.s.
Understorey depth post-cutting (cm)	11	1	$P < 0.0001$, $t = 24.4$, df = 408
Understorey cover post-cutting (%)	32	2	$P < 0.0001$, $t = 19.7$, df = 352
Brash cover post-cutting (%)	38	9	$P < 0.0001$, $t = 15.7$, df = 512
Mulch cover post-cutting (%)	4	72	$P < 0.0001$, $t = 40.8$, df = 512

On each 20 m × 20 m plot, 25 monitoring quadrats (1 × 1 m) were completed pre-cutting and 13 quadrats post-cutting. Shown here is a subset of the monitored variables comprising the average height of stems pre- and post-cutting, understorey depth pre- and post-cutting, and percentage ground cover of the three main cover types (understorey, brash, and mulch) post-cutting. Also shown are the results of two-tailed t-tests to identify significant differences between brushcut and robocut treatments for each variable

most marked difference with a likely considerable impact on tree seedling establishment success concerned the effect of the cutting method on the resulting substrate: In brushcut plots, the moss understorey remained largely alive, with the ground cover after cutting dominated by mosses (intact understorey) and brash (severed woody stems, predominantly heather); meanwhile, the robocutting treatment resulted in most of the moss understorey dying back, with ground cover after cutting dominated by mulch. Further details regarding differences in ground cover after cutting are shown in Table 37.1.

Subsequently, following tree introduction by seed sowing and sapling planting (undertaken in late winter to early spring), we will be monitoring tree seedling establishment success of both sown and planted individuals approximately 15 months after sowing/planting, including a detailed study of the evidence of browsing on young seedlings. This monitoring will be repeated approximately 7 years after sowing/planting, by which time the majority of surviving saplings should have escaped browser height, allowing us to compare the overall long-term success of our alternative treatments.

Throughout the project conception and design phase, several challenges arose relating to proposed management methods, availability of machinery and equipment, limited availability of local seed and saplings, and limitations in land available for planting due to unexpectedly large areas of deep peat (>0.3 m) across the reserve where tree planting is not permitted. Such challenges are inevitable when trialling novel techniques, and they highlight the crucial need for thoroughly documented, replicated, and controlled trials such as these (Krishnan and Osuri, 2022). Indeed, well-designed trials inform future management not only by illuminating the relative efficacy of different techniques but also by acting as a platform for refining and trialling highly novel management approaches.

This is a before-after control-impact (BACI) experimental trial, the likes of which are rarely conducted in nature conservation despite being very informative (Christie et al. 2020; Ockendon et al. 2021). Such experimental trials to test conservation interventions are crucial for enabling best practices and efficient allocation of limited resources, yet they are sorely lacking (Marshall et al., 2023; Tinsley-Marshall et al., 2022). Through thorough documentation of intervention methods and costs (Krishnan and Osuri 2022), the results of this trial will inform woodland restoration efforts elsewhere in the Scottish Highlands. More broadly speaking, the general principles of this approach (establishing pockets of trees as seed sources, remote from existing woodland, using naturalistic methods like sowing) have much wider relevance wherever woodland of natural character that includes species with windblown seeds is the management objective for extensive deforested areas. Importantly, we hope that this trial will provide evidence to inform the debate on whether planting or seed sowing is necessary for the restoration of native woodland in remote upland areas such as these, where planting continues to be criticised. This project tests an alternative method of assisted regeneration via seed sowing and cutting, comparing the results to the standard forestry practice of extensive planting. Such long-term, empirical comparisons of alternative methods of woodland restoration are sorely needed globally (García et al. 2020).

The Future for Restoring Scotland's Lost Forests

Moving forward, our vision is to see the widespread restoration of native forests and woodlands across much of upland Scotland, sustained largely through natural processes. Far from being a land devoid of people and culture, such a landscape would enable people to reconnect with forests and woodlands, providing associated health benefits (e.g. Ebenberger et al. 2014) as well as numerous other ecosystem services such as carbon sequestration (e.g. Fletcher et al. 2021), flood alleviation (e.g. Thomas and Nisbet 2006), and timber. Natural regeneration should play a key role in this vision, as it is a far more cost-effective method than planting over large spatial scales, as well as creating more structurally diverse, multifunctional habitats that provide a higher quality of ecosystem services than plantations (Chazdon et al. 2020; Fuentes-Montemayor et al. 2021). Ground preparation may be necessary in some areas to provide sufficient regeneration niches for natural regeneration to occur (Fuentes-Montemayor et al. 2021), and in areas remote from seed sources or with certain species lacking in the local species pool, we expect that applied nucleation methods will offer an important way forward. The current study will help to identify the exact shape that such methods should take.

However, widespread forest restoration will require appropriate management of grazing animals to allow tree seedlings to establish, and coordinated deer management across multiple adjoining landholdings, as well as sustainable levels of agricultural grazing are a necessity (Tanzentap et al. 2013). There is also a need for conflict management with other upland land uses such as sheep farming and intensive management of moorland for driven red grouse shooting, for which open treeless landscapes are the management aim.

Crucially, enabling this shift towards widespread restoration of native woodlands will rely on a change in government policy in Scotland. Current financial incentives prioritise non-native, near-monoculture commercial plantations, which support a forest industry that is already benefitting from a generous taxation system and high income from timber and carbon credits. Continued commercial afforestation will take place without government incentives and will continue to be profitable. We believe that financial incentives (i.e. public money) should cease to fund non-native, near-monoculture plantations and instead be used to prioritise native woodland creation; carbon credit systems should also be altered to include natural regeneration-based credits. Finally, we believe that woodland expansion should take place under an adaptive management framework using a spectrum of approaches and taking into account key site characteristics like soil type and climate (García et al. 2020; Baggio-Compagnucci et al. 2022). Under an adaptive management framework, government incentives should prioritise the preservation and expansion of ancient woodlands (i.e. continuously wooded for 250 years) given the apparent biodiversity and ecosystem service benefits that such ancient woodlands confer (Spencer and Kirby 1992; Fuentes-Montemayor et al. 2021).

There are several important knowledge gaps regarding the relative costs and benefits of the full spectrum of approaches to achieving woodland expansion, and further research is necessary to develop more comprehensive guidance for land managers and a regulatory framework that provides a fuller range of benefits for both humans and nature (see Thomas et al. 2015). Perhaps most importantly, studies are needed to understand the carbon dynamics and biodiversity value of native natural regeneration compared to plantings, as well as the impact of impoverished soil biodiversity on natural regeneration success in long-degraded landscapes. Long-term replicated trials are crucial to addressing such knowledge gaps alongside experimental mechanistic studies to more fully comprehend the processes at work. There is also a need for long-term monitoring of natural regeneration and the development of effective remote-sensing methods enabling reliable, cost-effective monitoring at a landscape scale as we are currently working towards in Cairngorms

Connect. Crucially, long-term and experimental studies (such as the Seed Source Establishment Trial described here) should be used to inform the development of process-based models of woodland expansion and enable more effective expansion planning going forward.

Acknowledgements We would like to acknowledge the crucial roles of Steve Blow, Richard Mason, and David Blair in setting up and implementing the Seed Source Establishment Trial. We also extend our thanks to Jeremy Roberts for comments on this chapter. Funding for this work came from the Endangered Landscapes Programme, RSPB Scotland, and the Scottish Government's Nature Restoration Fund, to whom we are very grateful.

References

Atkinson MD (1992) Betula pendula Roth (B. verrucosa Ehrh) and B. pubescens Ehrh. J Ecol 80(4):837–870. https://doi.org/10.2307/2260870

Baggio-Compagnucci A, Ovando P, Hewitt RJ, Canullo R, Gimona A (2022) Barking up the wrong tree? Can forest expansion help meet climate goals? Environ Sci Policy 136:237–249. https://doi.org/10.1016/j.envsci.2022.05.011

Beguin J, Tremblay J-P, Thiffault N, Pothier D, Côté SD (2016) Management of forest regeneration in boreal and temperate deer-forest systems: challenges, guidelines, and research gaps. Ecosphere 7(10):e01488. https://doi.org/10.1002/ecs2.1488

Bechara FC, Trentin BE, Engel VL, Estevan DA, Ticktin T (2021) Performance and cost of applied nucleation versus high-diversity plantations for tropical forest restoration. For Ecol Man 491. https://doi.org.uk/10.1016/j.foreco.2021.119088

Bowditch EAD, McMorran R, Smith MA (2023) Right connection, right insight engaging private estate managers on woodland expansion issues in times of uncertainty. Land Use Policy 124:106437. https://doi.org/10.1016/j.lusepol.2022.106437

Brancalion PHS, Holl KD (2020) Guidance for successful tree planting initiatives. J Appl Ecol 57(12):2349–2361. https://doi.org/10.1111/1365-2664.13725

Catt DC, Dugan D, Green RE, Moncrieff R, Moss R, Picozzi MN, Summers RW, Tyler GA (1994) Collisions against fences by woodland grouse in Scotland. Forestry 6(2):105–118. https://doi.org/10.1093/forestry/67.2.105

Chazdon RL, Lindenmayer D, Guariguata MR, Crouzeilles R, Benayas JMR, Chavero EL (2020) Fostering natural forest regeneration on former agricultural land through economic and policy interventions. Environ Res Lett 15(4):043002. https://doi.org/10.1088/1748-9326/ab79e6

Chazdon R, Falk D, Banin L, Wagner M, Wilson S, Grabowski R, Suding K (2021) The intervention continuum in restoration ecology: rethinking the active-passive dichotomy. Restor Ecol. https://doi.org/10.1111/rec.13535

Christie AP, Abecasis D, Adjeroud M et al (2020) Quantifying and addressing the prevalence and bias of study designs in the environmental and social sciences. Nat Commun 11(6377). https://doi.org/10.1038/s41467-020-20142-y

Corbin JD, Holl KD (2012) Applied nucleation as a forest restoration strategy. For Ecol Manag 265:37–46. https://doi.org/10.1016/j.foreco.2011.10.013

Crouzeilles R, Ferreira MS, Chazdon RL, Lindenmayer DB, Sansevero JBB, Monteiro L, Iribarrem A, Latawiec AE, Strassburg BBN (2017) Ecological restoration success is higher for natural regeneration than for active restoration in tropical forests. Sci Adv 3(11). https://doi.org/10.1126/sciadv.1701345

di Sacco A, Hardwick KA, Blakesley D, Brancalion PHS, Breman E, Cecilio Rebola L, Chomba S, Dixon K, Elliot S, Ruyonga G, Shaw K, Smith P, Smith RJ, Antonelli A (2021) Ten golden rules for reforestation to optimize carbon sequestration, biodiversity recovery and livelihood benefits. Glob Chang Biol 27(7):1328–1348. https://doi.org/10.1111/gcb.15498

Ebenberger M, Arnberger A, Eder M, Cervinka R, Hoeltge J, Pirgie L, Schwab M, Sudkamp J, Haluza D (2014) Green public health—benefits of woodlands on human health and wellbeing. Austrian Research Centre for Forests, Vienna

Fletcher TI, Scott CE, Hall JE, Spracklen DV (2021) The carbon sequestration potential of Scottish native woodland. Environ Res Commun 3(4):041003. https://doi.org/10.1088/2515-7620/abf467

Forest Research (2022). https://cdn.forestresearch.gov.uk/2022/09/FRFS022.pdf

Fuentes-Montemayor E, Park KJ, Cordts K, Watts K (2021) The long-term development of temperate woodland creation sites: from tree saplings to mature woodlands. Forestry Int J Forest Res 95(1):28–37. https://doi.org/10.1093/forestry/cpab027

García C, Espelta JM, Hampe A (2020) Managing forest regeneration and expansion at a time of unprecedented global change. J Appl Ecol 57(12). https://doi.org/10.1111/1365-2664.13797

Gullett PR, Leslie C, Mason R, Ratcliffe P, Sargent I, Beck A, Cowie NR, Cameron T, Hetherington D, MacDonell T, Moat T, Moore P, Teuten E, Hancock MH (2023) Woodland expansion in the presence of deer: 30 years of evidence from the Cairngorms Connect landscape restoration partnership. J Appl Ecol 60(11):2298-2308. https://doi.org/10.1111/1365-2664.14501

Hancock M, Egan S, Summers R, Cowie N, Amphlett A, Rao S, Hamilton A (2005) The effect of experimental prescribed fire on the establishment of scots pine Pinus sylvestris seedlings on heather *Calluna vulgaris* moorland. For Ecol Manag 212(1–3). https://doi.org/10.1016/j.foreco.2005.03.039

Hancock MH, Summers RW, Amphlett A, Willi J (2009) Testing prescribed fire as a tool to promote scots pine Pinus sylvestris regeneration. Eur J For Res 128(4). https://doi.org/10.1007/s10342-009-0267-5

Hancock MH, Summers RW, Amphlett A, Willi J, Servant G, Hamilton A (2010) Using cattle for conservation objectives in a Scots pine Pinus sylvestris forest: results of two trials. Eur J For Res 129(3):299–312. https://doi.org/10.1007/s10342-009-0330-2

Hobbs R (2009) Woodland restoration in Scotland: ecology, history, culture, economics, politics and change. J Environ Manag 90(9). https://doi.org/10.1016/j.jenvman.2007.10.014

Holl KD, Reid JL, Cole RJ, Oviedo-Brenes F, Rosales JA, Zahawi RA (2020) Applied nucleation facilitates tropical forest recovery: lessons learned from a 15-year study. J Appl Ecol 57(12):2316–2328. https://doi.org/10.1111/1365-2664.13684

Huntley B, Daniell JRG, Allen JRM (1997) Scottish vegetation history: the highlands. Bot J Scotl 49:163–175. https://www.tandfonline.com/doi/abs/10.1080/03746609708684864

Krishnan A, Osuri AM (2022) Beyond the passive-active dichotomy: aligning research with the intervention continuum framework of ecological restoration. Restor Ecol. https://doi.org/10.1111/rec.13828

Marshall AR, Waite CE, Pfeifer M, Banin LF, Rakotonarivo S, Chomba S, Herbohn J, Gilmour DA, Brown M, Chazdon RL (2023, 1867) Fifteen essential science advances needed for effective restoration of the world's forest landscapes. Philos Trans Royal Soc B Biol Sci 378. https://doi.org/10.1098/rstb.2021.0065

Meli P, Holl KD, Rey Benayas JM, Jones HP, Jones PC, Montoya D, Moreno Mateos D (2017) A global review of past land use, climate and active vs. passive restoration effects on forest recovery. PLoS One 12(2). https://doi.org/10.1371/journal.pone.0171368

Miller G, Cummins RP, Hester A (1998) Red deer and woodland regeneration in the Cairngorms. Scott For 52:14–20

Moss R (2001) Second extinction of capercaillie (*Tetrao urogallus*) in Scotland? Biol Conserv 101(2):255–257. https://doi.org/10.1016/S0006-3207(01)00066-0

Newton AC, Stirling M, Crowell M (2001) Current approaches to native woodland restoration in Scotland. Bot J Scotl 53(2). https://doi.org/10.1080/03746600108685021

Nixon CJ, Worrell R (1999) The potential for the natural regeneration of conifers in Britain. Forestry Commission, Bulletin 120, Edinburgh

Ockendon N, Amano T, Cadotte M et al (2021) Effectively integrating experiments into conservation practice. Ecol Solutions Evidence 2(2):e12069

Ramirez JI, Jansen PA, Poorter L (2019) Effects of wild ungulates on the regeneration, structure and functioning of temperate forests: a semi-quantitative review. For Ecol Manag 424:406–419. https://doi.org/10.1016/j.foreco.2018.05.016

Rao SJ (2017) Effect of reducing red deer Cervus elaphus density on browsing impact and growth of scots pine Pinus sylvestris seedlings in semi-natural woodland in the Cairngorms, UK. Conserv Evidence 14:22–26

Rojas-Botero S, Solorza-Bejarano J, Kollman J, Teixeira LH (2020) Nucleation increases understory species and functional diversity in early tropical forest restoration. Ecol Eng 158. https://doi.org.uk/10.1016/j.ecoleng.2020.106031

Scottish Government (2019) Scotland's forestry strategy 2019–2029. The Scottish Government, Edinburgh. https://forestry.gov.scot/component/edocman/373-scotland-s-forestry-strategy-2019-2029/download?Itemid=0

Scottish Government (2020) In: Pepper S, Barbour A, Glass J (eds) The management of wild deer in Scotland: deer working group report. Scottish Government, Edinburgh

Scottish Government (2021) Agricultural payments: Common Agricultural Policy (CAP). Scottish Rural Development Programme (SRDP). https://www.gov.scot/policies/agriculture-payments/scottish-rural-development-programme-srdp/

Scottish Government (2022) Scottish biodiversity strategy to 2045. The Scottish Government, Edinburgh. https://www.gov.scot/publications/scottish-biodiversity-strategy-2045-tackling-nature-emergency-scotland/pages/3/

Smout TC (2003) People and woods in Scotland: a history. Edinburgh University Press, Edinburgh

Spencer JW, Kirby KJ (1992) An inventory of ancient woodland for England and Wales. Biol Conserv 62(2):77–93. https://doi.org/10.1016/0006-3207(92)90929-H

Steven A, Carlisle HM (1959) The native pinewoods of Scotland. Oliver and Boyd, Edinburgh

Summers R (2019) Abernethy Forest. Inverness, UK

Summers R, Cavers S (2021) The past, present and uncertain future for Caledonian pinewoods. Scott For 75(2):19–28

Tanentzap AJ, Zou J, Coomes D (2013) Getting the biggest birch for the bang: restoring and expanding upland birchwoods in the Scottish highlands by managing red deer. Ecol Evol 3(7):1890–1901. https://doi.org.uk/10.1002/ece3.548

Thomas H, Nisbet TR (2006) An assessment of the impact of floodplain woodland on flood flows. Water Environ J 21(20):114–126. https://doi.org/10.1111/j.1747-6593.2006.00056.x

Thomas HJD, Paterson JS, Metzger MJ, Sing L (2015) Towards a research agenda for woodland expansion in Scotland. For Ecol Manag 349:149–161. https://doi.org/10.1016/j.foreco.2015.04.003

Tinsley-Marshall P, Downey H et al (2022) Funding and delivering the routine testing of management interventions to improve conservation effectiveness. J Nat Conserv 67:126184. https://doi.org/10.1016/j.jnc.2022.126184

Towers W, Hall J, Hester A, Malcolm A, Stone D (2004) The potential for native woodland in Scotland: the native woodland model. Scottish Natural Heritage, Battleby

Warren C (2002) Managing Scotland's environment. Edinburgh University Press, Edinburgh

Werden LK, Holl KD, Chaves-Fallas JM, Oviedo-Brenes F, Rosales JA, Zahawi RA (2021) Degree of intervention affects interannual and within-plot heterogeneity of seed arrival in tropical forest restoration. J Appl Ecol 58(8):1693–1704. https://doi.org/10.1111/1365-2664.13907

Willoughby I, Harrison A, Jinks R, Gosling P, Harmer R, Kerr G (2007) The potential for direct seeding of birch on restock sites. Forestry Commission, Information Note, Edinburgh

Willoughby I, Jinks RL, Forster J (2019) Direct seeding of birch, rowan and alder can be a viable technique for the restoration of upland native woodland in the UK. Forestry 92:324–338

Yarranton GA, Morrison RG (1974) Spatial dynamics of a primary succession: Nucleation. J Ecol 62:417–428. https://doi.org/10.2307/2258988

Open Access This chapter is licensed under the terms of the Creative Commons Attribution 4.0 International License (http://creativecommons.org/licenses/by/4.0/), which permits use, sharing, adaptation, distribution and reproduction in any medium or format, as long as you give appropriate credit to the original author(s) and the source, provide a link to the Creative Commons license and indicate if changes were made.

The images or other third party material in this chapter are included in the chapter's Creative Commons license, unless indicated otherwise in a credit line to the material. If material is not included in the chapter's Creative Commons license and your intended use is not permitted by statutory regulation or exceeds the permitted use, you will need to obtain permission directly from the copyright holder.

GPSR Compliance

The European Union's (EU) General Product Safety Regulation (GPSR) is a set of rules that requires consumer products to be safe and our obligations to ensure this.

If you have any concerns about our products, you can contact us on ProductSafety@springernature.com

In case Publisher is established outside the EU, the EU authorized representative is:

Springer Nature Customer Service Center GmbH
Europaplatz 3
69115 Heidelberg, Germany

Batch number: 08497527

Printed by Printforce, the Netherlands